7th Annual International Conference on Material Science and Environmental Engineering (MSEE 2019)

IOP Conference Series: Materials Science and Engineering
Volume 735

Wuhan, China
15 - 16 November 2019

ISBN: 978-1-7138-2653-8
ISSN: 1757-8981

Printed from e-media with permission by:

Curran Associates, Inc.
57 Morehouse Lane
Red Hook, NY 12571

proceedings.com

Some format issues inherent in the e-media version may also appear in this print version.

Institute of Physics
Dirac House, Temple Back
Bristol BS1 6BE UK

Phone: 44 1 17 929 7481
Fax: 44 1 17 920 0979

techtracking@iop.org

Additional copies of this publication are available from:

Curran Associates, Inc.
57 Morehouse Lane
Red Hook, NY 12571 USA
Phone: 845-758-0400
Fax: 845-758-2633
Email: curran@proceedings.com
Web: www.proceedings.com

TABLE OF CONTENTS

MATERIAL SCIENCE AND ADVANCED MATERIALS

MATERIAL ENGINEERING AND APPLICATION

ENVIRONMENT PROTECTION AND SUSTAINABLE DEVELOPMENT

ENVIRONMENTAL SCIENCE AND ENGINEERING

IOP Conf. Series: Materials Science and Engineering **735** (2020) 011001 doi:10.1088/1757-899X/735/1/011001

Preface

7th annual 2019 International Conference on Material Science and Environmental Engineering [MSEE 2019] was held on November 15-16th, 2019, Wuhan, China. MSEE2019 aims to bring researchers, engineers, and students to the areas of Material Science and Environmental Engineering. MSEE2019 features unique mixed topics of Material Science and Advanced Materials, Material Engineering and Application, Environmental Protection and Sustainable Development, Environmental Science and Engineering.

We received over 192 submissions from various parts of the world. The Technical Program Committee worked very hard to have all manuscripts reviewed before the review deadline. All the accepted papers have been submitted to strict peer-review by 2-4 referees, and selected based on originality, significance and clarity for the purpose of the conference. The conference program is extremely rich, profound and featuring high-impact presentations of selected papers and additional late-breaking contributions. We sincerely hope that the conference would not only show the participants a broad overview of the latest research results on related fields, but also provide them with a significant platform for academic connection and exchange. There is two keynote speaker and four invited sessions. The keynote speakers are internationally recognized leading experts in their research fields, who have demonstrated outstanding proficiency and have achieved distinction in their profession. The proceedings would be published by IOP Conference Series: Materials Science and Engineering (MSE).

We would like to express our sincere gratitude to all the members of Technical Program Committee and organizers for their enthusiasm, time, and expertise. Our deep thanks also go to many volunteers and staffs for the long hours and hard work they have generously given to MSEE2019. The last but not least, we would like to thank all the authors, speaker and participants for their great contributions to the success of MSEE2019.

MSEE2019 Organizing Committee

IOP Conf. Series: Materials Science and Engineering **735** (2020) 011002 doi:10.1088/1757-899X/735/1/011002

Peer review statement

All papers published in this volume of *IOP Conference Series: Materials Science and Engineering* have been peer reviewed through processes administered by the proceedings Editors. Reviews were conducted by expert referees to the professional and scientific standards expected of a proceedings journal published by IOP Publishing.

IOP Conf. Series: Materials Science and Engineering **735** (2020) 012001 doi:10.1088/1757-899X/735/1/012001

Effects of Pt Content on Room-Temperature Hydrogen Sensing Characteristics of Pt-SnO_2 Nano-Composite Ceramics

Caochuang Wang[1], Pengcheng Li[1], Yongming Hu[2], Liqun Xu[3], Zhen Zhou[3] and Wanping Chen[1]

1 School of Physics and Technology, Wuhan University, Wuhan 430072, China
2 Hubei Key Laboratory of Ferro and Piezoelectric Materials and Devices, Faculty of Physics and Electronic Technology, Hubei University, Wuhan 430062, China
3 Huaneng Wuhan Power Generation Co., Ltd., Wuhan 431415, China
Email: wpchen@whu.edu.cn

Abstract. A series of Pt-SnO_2 nano-mixtures with 0, 0.25, 1 and 2.5 wt% Pt separately were prepared using SnO_2 nanoparticles and H_2PtCl_6. Through pressing and sintering, Pt-SnO_2 nano-composite ceramics with Pt nanoparticles well dispersed have been successfully prepared from the nano-mixtures. Their resistance is found to increase monotonously and greatly with increasing Pt content, which is increased by three orders of magnitude when Pt is increased from 0 to 2.5 wt%. All those samples with a Pt content show distinct responses to hydrogen at room temperature, while their hydrogen sensitivity and their recovery time both decease dramatically with increasing Pt content. These results strongly suggest that Pt as a catalyst promotes oxygen chemisorption on SnO_2 and the reaction between hydrogen and oxygen at room temperature. Pt thus has catalytic effects on both hydrogen and oxygen, which leads to quite complicated effects of Pt content on the room-temperature hydrogen sensing behaviors of semiconducting metal oxides.

1. Introduction

As a clean and renewable energy carrier, hydrogen has been receiving more and more attention and being applied in more and more automobiles in recent years [1]. However, hydrogen is very dangerous because it is odorless, colorless, and explosive in air over a large concentration range of 4%-75%. The rapidly increasing quantity of hydrogen automobiles has caused much concern over hydrogen safety [2]. Nowadays, all commercial hydrogen sensors are not quite satisfactory for hydrogen automobiles in one way or another. For example, commercial hydrogen sensors based-on SnO_2 thick films are generally considered most suitable for hydrogen automobiles due to their low cost, high stability in harsh environment and simplicity of fabrication and use [3]. Unfortunately, they have to work at elevated temperatures (around 400 °C) [4], which brings additional security risks and increases energy consumption [5]. Extensive investigations have been conducted to develop semiconducting metal oxide (MOX) hydrogen sensors that can work at room temperature.

In the past few years, several Pt-MOX composite ceramics have been surprisingly found to possess extraordinary room-temperature hydrogen sensing capabilities [6-10]. As bulk materials, they are promising to be applied as room-temperature MOX hydrogen sensors and should be studied systematically. Generally speaking, Pt in the composites is believed to decompose hydrogen molecules into hydrogen atoms, which are chemisorbed on MOXs at room temperature and result in the observed room-temperature hydrogen sensing capabilities. According to a recent investigation, for Pd-SnO_2 composite ceramics [11], there exists an intriguing interaction between Pd and SnO_2 at high temperatures, whose product of Pd^{4+} leads to an unexpected room-temperature CO sensing capability for the composite [12]. Presently, we have conducted a study on the effects of Pt content on room-temperature hydrogen sensing characteristics of Pt-SnO_2 nano-composite ceramics, which first reveals that Pt has a catalytic effect on oxygen other than hydrogen chemisorption on SnO_2 for Pt-SnO_2 system. The results reported in this paper should be important not only for mechanism study but also for further property improvement of room-temperature MOX hydrogen sensors.

2. Experimental Methods

2.1. Materials Preparation

SnO_2 nanoparticles (70 nm) and zinc powder were dispersed in deionized water at a series of ratios and magnetically stirred. For every suspension, 0.1 M H_2PtCl_6 solution was dropped slowly into the suspension until the molar number of H_2PtCl_6 reached to one third of that of the zinc powder in the suspension. After 4h of stirring, the suspensions were centrifuged and dried in oven at 110 °C for 12 h. The dried powders were pressed into cylindrical pellets (~10 mm in diameter and ~1.5 mm thick) at 2 MPa through a hydraulic press with deionized water added as binder. The pellets were sintered in air at 1000 °C for 2h, and a pair of rectangular Au electrodes was coated on a major surface of sintered pellets through DC magnetron sputtering for gas sensing measurement.

2.2. H_2 Sensing Measurement

A commercial gas sensing measurement system (GRMS-215, Partulab Com., Wuhan, China) was used in this study. Samples were placed in a sealed quartz chamber (about 350 mL), which has four gas entrances and a gas outlet to change internal atmosphere. In the response process, specific atmospheres were introduced through mixing O_2, N_2, and 5% H_2 in N_2 at some designed ratios. The total gas flow rate was 300 mL/min. In the recovery process, air was pumped into the chamber at a rate of 1000 mL/min. Through a Keithley 2400 Source/Meter, a DC voltage of 10 V was applied between the Au electrodes of the samples and the flown electric current was recorded. The room temperature was kept at 25 °C and the relative humidity (RH) in air was measured.

2.3. Materials Characterization

X-ray diffraction (XRD) patterns were recorded on an X-ray diffractometer (BRUKER AXS D8 ADVANCE) using Cu *Kα* radiation. Microstructural analyses were performed through a scanning electron microscopy (SIRION TMP). Energy dispersive spectroscopy (EDS) analyses were recorded through OXFORD Aztec 250 instrument.

3. Results and Discussion

3.1. Phase and Microstructural Analyses

The following reactions occur when a H_2PtCl_6 solution is dropped into a suspension of SnO_2 nanoparticles and zinc powder:

$$Pt^{4+} + 2Zn \rightarrow Pt + 2Zn^{2+} \tag{1}$$

$$2H^{+} + Zn \rightarrow H_2 + Zn^{2+} \tag{2}$$

In this way, one mole H_2PtCl_6 reacts with three mole Zn and metallic Pt is deposited on SnO_2 nanoparticles. This method has been widely used to prepare nano-mixtures of noble metals (e.g. Pd, Pt) and MOXs, where the noble metals are usually a few nanometers in size [13, 14]. In this study, we have prepared a series of nano-mixtures of Pt-SnO_2, whose Pt content was 0, 0.25, 1 and 2.5 wt%, separately.

Figure 1 shows the X-ray diffraction patterns taken for the surface of four sintered pellets prepared from this series of Pt-SnO_2 nano-mixtures. For all samples, all strong peaks are from rutile SnO_2. For the samples of 1 and 2.5 wt% Pt, two small peaks from Pt can be observed. It indicates that the Sintered pellets we obtained are composites of SnO_2 and Pt, though the peaks of Pt cannot be seen for the pellet of 0.25 wt% Pt due to its low Pt content. As a matter of fact, Pt is well known for its high stability at high temperatures.

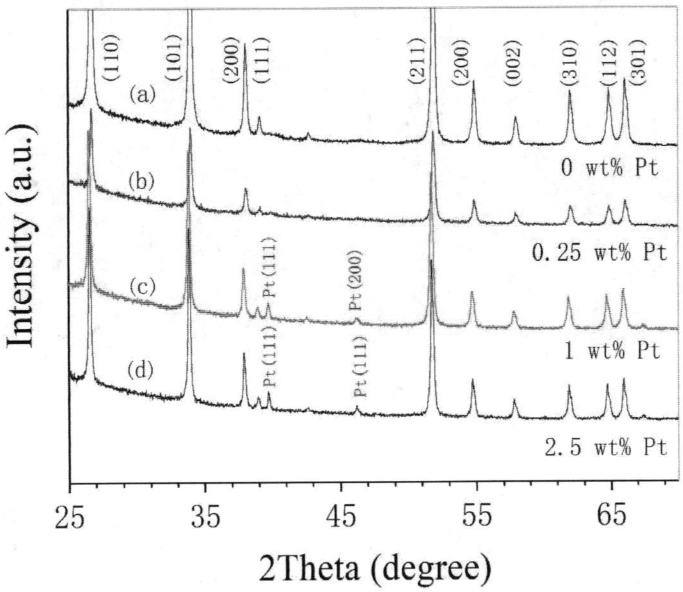

Figure 1. X-ray Diffraction Patterns Taken for the Surface of Four Sintered Pellets Prepared from Pt-SnO_2 Nano-Mixtures with 0, 0.25, 1 and 2.5 wt% Pt, Separately

As for microstructure, the four kinds of sintered pellets are very similar. Figure 2 shows an SEM micrograph with EDS analyses for a sample of 2.5 wt% Pt sintered at 1000 °C for 2 h in air. Some previous studies have clearly shown that the grain-growth of SnO_2 nanoparticles is very limited when the ceramics are sintered at temperatures as high as 1000 °C. Most grains in this micrograph are also around 100 nm in size and the microstructure appears relatively dense for a low magnification. It should be pointed out, however, that there was actually no shrinkage in diameter after the pellets were sintered at 1000 °C and numerous nanoscale pores can be observed in the microstructure for high magnifications. The EDS analyses clearly show that Pt is rather uniformly dispersed. Due to their small size, however, Pt nanoparticles cannot be directly observed in the SEM micrograph. Obviously, through a very simple and convenient process, we have obtained Pt-SnO_2 nano-ceramics with numerous nanoscale pores and with Pt rather uniformly dispersed, which should be highly attractive for gas sensing applications.

Figure 2. SEM Micrograph with EDS Analyses for a Pt-SnO_2 Nano-Ceramic Sample of 2.5 wt% Pt Sintered at 1000 °C for 2 h in Air

3.2. Room-temperature Hydrogen Sensing Characteristics

Extraordinary room-temperature hydrogen sensing capability has been reported for Pt-SnO_2 composite nano-ceramics in some previous researches [8,10]. In accordance with these researches, these three kinds of samples with 0.25, 1 and 2.5 wt% Pt separately prepared in this study all respond dramatically to hydrogen at room temperature, while the samples with no Pt exhibit no response to hydrogen at room temperature, as shown in figure 3. It has been proposed that Pt as a catalyst splits hydrogen molecules into atoms, which are chemisorbed on SnO_2 and decrease the resistance of SnO_2 at room temperature. Obviously, the interaction between hydrogen and Pt plays a vital role in the observed room-temperature hydrogen sensing capability.

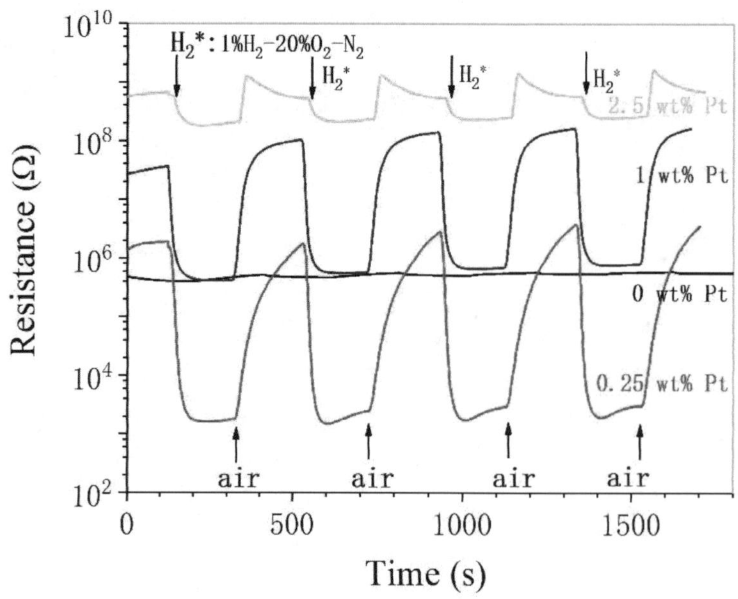

Figure 3. Room-temperature Response to 1.0% H_2-20% O_2-N_2 and Recovery in Air of 41% RH for Four Pt-SnO_2 Nano-Ceramic Samples with 0, 0.25, 1 and 2.5 wt% Pt, Separately

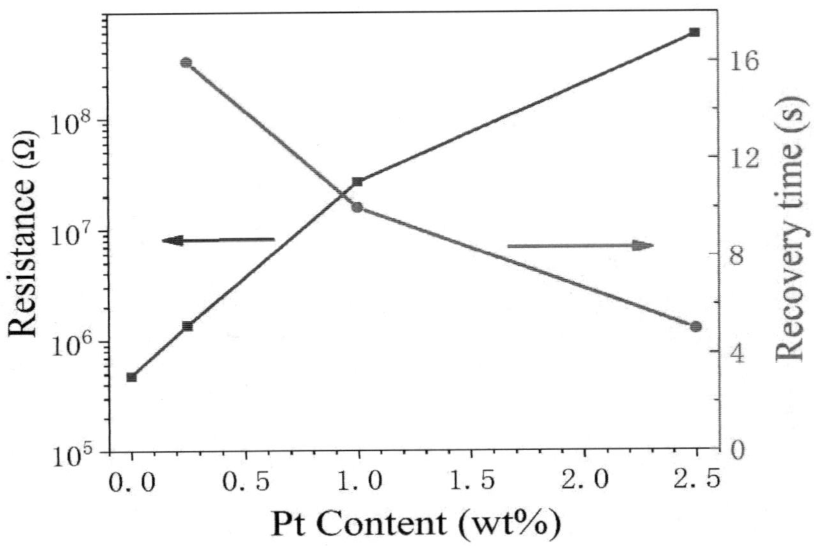

Figure 4. Resistance Dependence on Pt Content for Pt-SnO_2 Nano-Composite Ceramics with 0, 0.25, 1 and 2.5 wt% Pt Separately; Recovery Time Dependence on Pt Content for Pt-SnO_2 Nano-Composite Ceramics with 0.25, 1 and 2.5 wt% Pt Separately

It is interesting to further compare these four kinds of ceramics with different Pt contents. First of all, it can be clearly seen that their resistance in air increases steadily with increasing Pt content, as shown in figure 4. The resistance of the samples with 2.5 wt% Pt is three orders of magnitude larger than that of the samples with no Pt. Secondly, though Pt is the key to the observed room-temperature hydrogen sensing capability, the hydrogen sensitivity decreases with increasing Pt content among those three kinds of samples with 0.25, 1 and 2.5 wt% Pt separately. Thirdly, the recovery time in air decreases obviously with increasing Pt content among those three kinds of samples with a Pt content, as shown in figure 4. All these results demonstrate that Pt has more complicated effects on room-temperature hydrogen sensing behaviour of Pt-SnO_2 system than its catalytic effect on H_2.

It is well known that the resistance of SnO_2 is greatly increased by oxygen chemisorption, by which an electron-depleted layer is formed and the concentration of free electrons is decreased [15,16]. The great difference in resistance among Pt-SnO_2 composite nano-ceramics with different Pt contents strongly suggests that Pt is able to dramatically promote oxygen chemisorption on SnO_2. With increasing Pt content, more oxygen is chemisorbed on SnO_2 in a more rapid speed, so the resistance is increased and the recovery time in air is decreased. Moreover, Pt must also be able to promote the reaction between oxygen and hydrogen. So with increasing Pt content, less hydrogen is chemisorbed on SnO_2 and the hydrogen sensitivity is decreased. Both hydrogen sensitivity and recovery speed are important for hydrogen sensing, it is thus important to optimize the Pt content for Pt-SnO_2 composite nano-ceramics to achieve attractive overall room-temperature hydrogen sensing capability.

4. Conclusions

Pt-SnO_2 nano-mixtures with 0, 0.25, 1 and 2.5 wt% Pt separately were prepared using SnO_2 nanoparticles, zinc powder and H_2PtCl_6 through a replacement reaction. Pt-SnO_2 composite nano-ceramics with Pt nanoparticles well dispersed have been successfully prepared from these Pt-SnO_2 nano-mixtures. The resistance increases monotonously and greatly with increasing Pt content for all samples. And for those samples with a Pt content, obvious room-temperature hydrogen sensing behaviors are observed, which depend dramatically on the Pt content. With increasing Pt content, both their hydrogen sensitivity and their recovery time in air decease steadily. It is proposed that Pt as a catalyst promotes not only hydrogen chemisorption but also oxygen chemisorption on SnO_2 and also the reaction between hydrogen and oxygen. Pt thus has rather complicated effects on the room-temperature hydrogen sensing behaviors of MOXs and its content must be optimized to obtain overall attractive room-temperature hydrogen sensing capacity for MOXs.

5. Acknowledgments

This work was partially supported by the National Natural Science Foundation of China under Grant No. 61274073 and J1210061.

6. References

[1] Rusman N A A and Dahari M 2016 *Int. J. Hydrogen Energy* **41** 12108
[2] Moy R 2003 *Science* **301** 47.
[3] Korotcenkov G 2013 *Handbook of Gas Sensor Materials: Properties, Advantages and Shortcomings for Applications* vol 1 (New York: Springer) p 81
[4] Boon-Brett L, Bousek J, Black G, Moretto P, Castello P and Hübert T 2010 *Int. J. Hydrogen Energy* **35** 373.
[5] Hübert T, Boon-Brett L, Black G and Banach U 2011 *Sens. Actuators B* **157** 329
[6] Chen W P, Xiong Y, Li Y S, Cui P, Guo S S, Chen W, Tang Z L, Yan Z J and Zhang Z Y 2016 *Int. J. Hydrogen Energy* **41** 3307
[7] Li P C, Xiong Z H, Zhu S, Wang M Y, Hu Y M, Gu H S, Wang Y and Chen W P 2017 *Int. J. Hydrogen Energy* **42** 30186
[8] Zhu S, Li P C, Wu G T, Li Z Y, Wu P, Hu Y M, Gu H S and Chen W P 2018 *Int. J. Hydrogen Energy* **43** 21177
[9] Song C B, Wu G T, Sun B L, Xiong Y, Zhu S, Hu Y M, Gu H S, Wang Y and Chen W P 2017 *Int. J. Hydrogen Energy* **42** 6420
[10] Xiong Y, Chen W P, Li Z Y, Cui P, Guo S S, Chen W, Tang Z L, Yan Z J and Zhang Z Y 2016 *Nano Res.* **9** 3528
[11] Wang M Y, Sun B L, Jiang Z Y, Liu Y, Wang X N, Tang Z L, Wang Y and Chen W P 2018 *J. Nanosci. Nanotechnol.* **18** 4176
[12] Zhu S, Liu Y, Wu G T, Fei L F, Zhang S H, Hu Y M, Yan Z J, Wang Y, Gu H S and Chen W P 2019 *Sens. Actuators B* **285** 49
[13] Zhang H, Yang D, Ma X Y, Jiang J Y, Xu J and Que D L 2004 *Nanotechnol.* **15** 622.
[14] Lee J M, Park J, Kim S, Kim S, Lee E and Kim S J 2010 *Int. J. Hydrogen Energy* **35** 12568
[15] Hassan J, Mahdi M, Chin C, Abu-Hassan H and Hassan Z 2013 *J. Alloys Compd.* **546** 107
[16] Heiland G 1988 *Chem. Sens. Technol.* **1** 15

Experimental Study on Modified Shielding Functional Mortar

Yi Gao[1*], Yingjian Lu[1,2] and Jing Huang[1]

1 Naval Logistics Academy, Tianjin 300450, China
2 Naval University of Engineering, Wuhan 430000, China
Corresponding author e-mail: gaoyi5777@163.com
Corresponding author: Yi Gao

Abstract. The mortar board was prepared with the content of reduced iron powder, thickness and shielding fabric as variables. The results showed that appropriate content of reduced iron powder could improve the shielding efficiency of mortar, and the peak shielding efficiency of 10mm thick specimens with a dosage of 300kg/m3 reached 16.9db and the 20mm-thick specimens reached 17.1db. The 10mm-thick specimens with shielding fabric reached 73.5db, and the 20mm-thick specimens reached 59.3db

1. Introduction
Cement-based materials have a wide range of uses. The shielding effectiveness of cement-based materials prepared by adding copper-nickel powder and fiber has been studied [1]. The absorption effect of prepared cement-based materials in high frequency band was studied with graphite powder [2]. The shielding performance of copper-nickel wire mesh reinforced cement-based material is studied [3]. Mortar, as a kind of cement-based material, is often used in masonry and plastering, etc. The functional plastering mortar can reconstruct the old engineering and meet the special needs. There are few studies on cement-based materials mixed with reduced iron powder. Some scholars studied its electromagnetic wave reflectivity and transmittance in a certain frequency band [4]. As an ultrafine powder, reduced iron powder has good electromagnetic properties [5]. Adding the plastering mortar prepared by the powder can provide a choice for the reconstruction with needs of electromagnetic shielding.

2. Experiment

2.1. Raw Materials
Reduced iron powder: 200 mesh secondary; Cement: grade 42.5 ordinary Portland cement; Sand: medium river sand; Fly ash: 1250 mesh first class; Shielding fabric: copper-nickel composite fabric with shielding efficiency up to 70dB in the 1MHz-10GHz band.

2.2. Mix Proportion
According to the proper regulation, select the proportion. The specific ratio is shown in table 1.

Table 1. Mix Proportion

number	Fabric form	Thickness (mm)	Content (kg/m^3)				
			iron powder	Cement	Sand	Fly ash	Water
A1	-	10	0	450	1498	50	300
A2	-	10	100	450	1473	50	300
A3	-	10	200	450	1448	50	300
A4	-	10	300	450	1423	50	300
A5	-	10	400	450	1398	50	300
B1	surface	10	0	450	1498	50	300
B2	surface	20	0	450	1498	50	300
B3	surface	10	300	450	1423	50	300
B4	surface	20	300	450	1423	50	300
C1	-	20	0	450	1498	50	300
C2	-	20	300	450	1423	50	300

2.3. Methods of Preparation

According to the selected mix proportion. The cement and reduced iron powder were put into the mixer for 60s dry mixing at a slow speed, followed by 60s dry mixing at a fast speed. Sand and fly ash were added at a uniform velocity, and water was added at the same time. Put the mortar into the mold used in the test, remove the mold after 1d, and experiment after curing to the specified age.

2.4. Method of Experiment

According to the proper method, the shielding effectiveness at specific frequency points was tested by using the coaxial flange. The method is shown in figure 1.

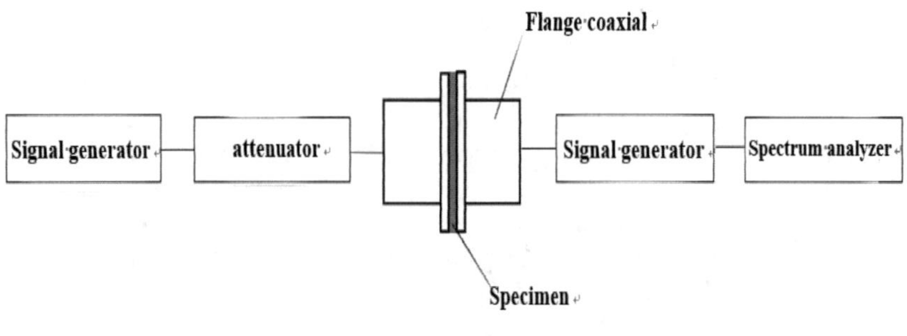

Figure 1. Measurement Method of Signal Generator/Spectrum Analyzer

3. Results of Experiment

3.1. Effect of Reduced Iron Powder on Shielding Efficiency

3.1.1. No Shielding Fabric

When no shielding fabric was added, the shielding efficiency of 10mm thick specimen first increased and then decreased. The shielding efficiency of 20mm one was the same, the effect is shown in figure 2.

IOP Conf. Series: Materials Science and Engineering **735** (2020) 012002 doi:10.1088/1757-899X/735/1/012002

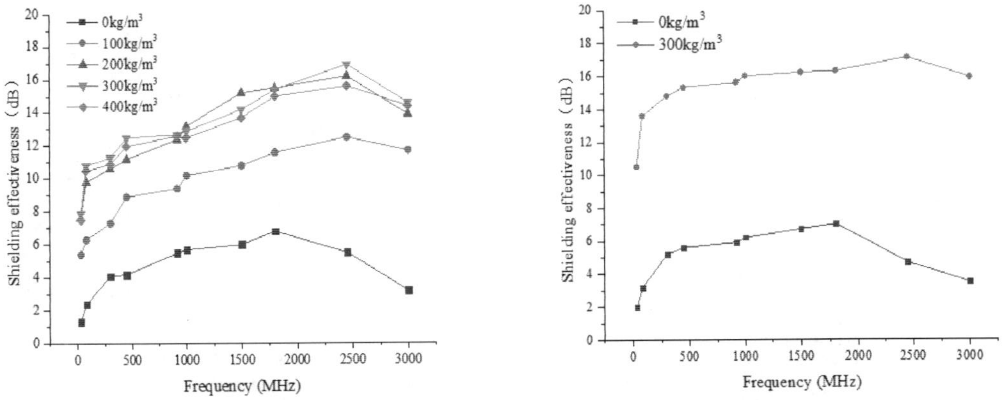

(a) 100mm×100mm×10mm (b) 100mm×100mm×20mm

Figure 2. Effect of Reduced Iron Powder on Shielding Efficiency of Functional Mortar Specimen

When the thickness is 10mm, the peak shielding efficiency of the specimens mixed with reduced iron powder moves towards high frequency. the reference mortar specimen reached the peak of 6.8db at the point of 1800MHz, and the specimen with a mixture of 300kg/m^3 reached the peak of 16.9db at 2450MHz. In the frequency band of 30MHz-915MHz, when the content is less than 300kg/m^3, the shielding efficiency gradually increases. The sample with the content of 300kg/m^3 increases most obviously, with an average of 7dB. When the content increases to 400kg/m^3, the shielding efficiency decreases. At the low-frequency point, the shielding mainly depends on the absorption. The powder belongs to the electric loss absorbing medium and has a good permeability. When the powder is added, the internal conductivity and permeability rise, so that the wave can get better absorption loss. However, as the electrical conductivity increases, the difficulty of the internal magnetic permeability material increases, which reduces the loss ability. In the frequency band of 1000MHz-3000MHz, when the content is more than 100kg/m^3, the shielding efficiency crossed. The sample with the content of 300kg/m^3 has a better shielding efficiency, and the average increase is about 10dB compared with the reference. At the high frequency point, electromagnetic shielding mainly depends on the surface reflection. The addition of powder forms an amount of conductive aggregates, so that the surface conductivity increases. When the content is 300kg/m^3, the powder has a good dispersion effect, can form a good surface conductive network. When the content increased to 400kg/m^3, the distribution of powder changed because of the density, and the conductivity decreased, leading to the decrease of shielding efficiency. When the thickness is 20mm, the reference reaches the peak of 7dB at the 1800MHz frequency point, and the specimen with a content of 300kg/m^3 reaches 17.1db at the 2450MHz frequency point. Similar to the 10mm thick ones, the powder made the peak value move towards the high frequency.

3.1.2. With Shielding Fabric

When shielding fabric is added, the shielding efficiency of specimens with different thickness decreases first and then increases. The effect is shown in figure 3.

IOP Conf. Series: Materials Science and Engineering **735** (2020) 012002 doi:10.1088/1757-899X/735/1/012002

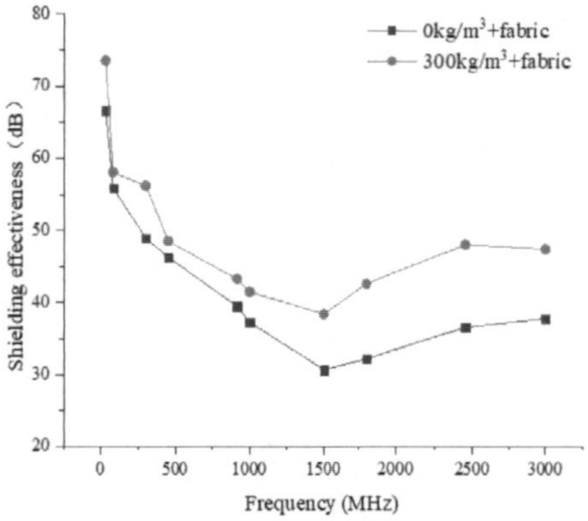

(a) 100mm×100mm×10mm

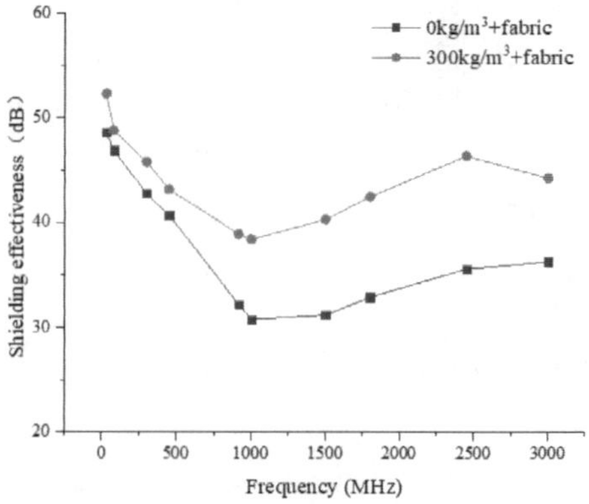

(b) 100mm×100mm×20mm

Figure 3. Effect of Reduced Iron Powder on Shielding Efficiency of Functional Mortar Specimen

When the thickness is 10mm, with the increase of frequency, the difference between the two groups' increases gradually, which is only 2.2db at the 80MHz frequency point and 11.4db at the 2450MHz. When fabric is added, the effect of powder on high frequency is better. The powder has a good conductivity. When the content is 300kg/m^3, it forms a good conductive network. When the frequency is low, the fabric can reflect most electromagnetic waves. With increasing frequency, the penetrate ability of wave is stronger, due to the impedance mismatch of wave, making electromagnetic wave gets reflection loss at the border and internal absorption. When the thickness is 20mm, the difference tends to increase gradually, which is only 2.5db at 450MHz frequency point and 10.8db at 2450MHz. Similar to the 10mm ones, when fabric is added, the powder can improve high-frequency shielding efficiency.

3.2. Effect of Thickness on Shielding Efficiency

3.2.1. No Shielding Fabric

When no shielding fabric was added, the shielding efficiency of the two groups increased first and then decreased with the increase of frequency. The effect is shown in figure 4.

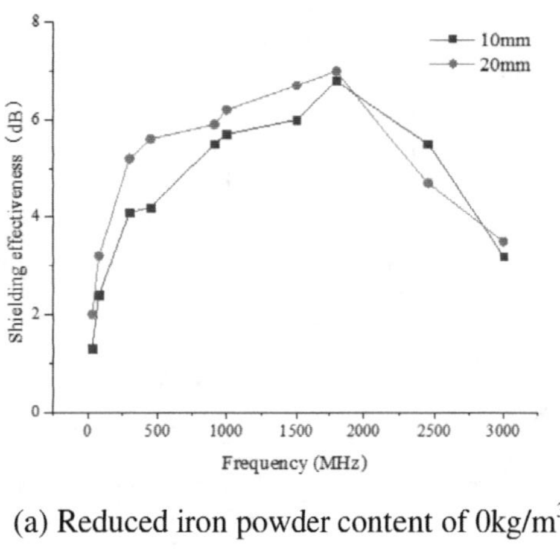

(a) Reduced iron powder content of 0kg/m^3

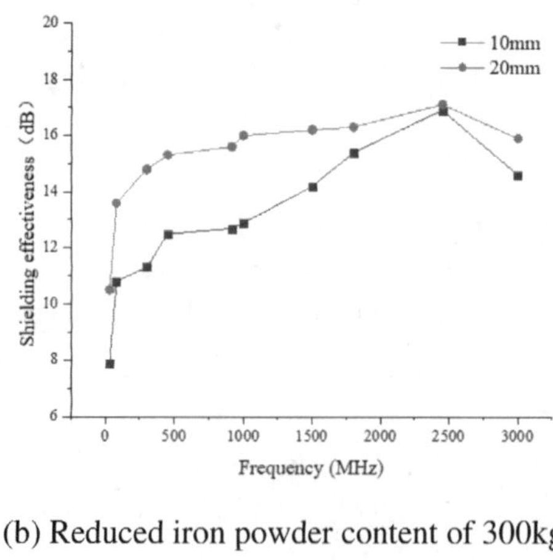

(b) Reduced iron powder content of 300kg/m^3

Figure 4. Effect of Thickness on Shielding Efficiency of Functional Mortar Specimen

Aimed at reference mortar, the shielding efficiency of 20mm thick specimen is about 1dB higher than that of 10mm one in the frequency band of 30MHz-1500MHz, with the increase of frequency, the shielding efficiency of the two specimens appears to intersect. Increasing the thickness can increase the internal reflection and absorption loss. However, there is little material with high permeability in the reference. For high frequency, thickness has little effect, and the lack of high conductivity material leads to the intersection. In addition, the larger the thickness is, the larger of the fixture is. The coaxial waveguide method requires the thickness to be below 5mm. When the thickness is larger than 5mm, the external electromagnetic wave is easy to enter the receiving cavity, causing the error. When the content is 300kg/m^3, in the frequency band of 30MHz-1000MHz, the shielding efficiency of 20mm thick one is about 3dB higher than that of 10mm. With the increase of frequency, the difference tends to decrease gradually. In the low frequency, the increase of thickness can improve the chance of wave being reflected and absorbed, thus improving the shielding efficiency. When the thickness reaches a certain critical state, increasing the thickness has a limited effect. In addition, with the error caused by the increase of the thickness, the improvement is limited. In the high frequency band, the shielding mainly depends on the surface reflection loss. The increase of thickness has little effect.

3.2.2. With Shielding Fabric

When the shielding fabric is added, the shielding efficiency of the two groups has the same trend. The efficiency first decreases and then increases. The effect is shown in figure 5.

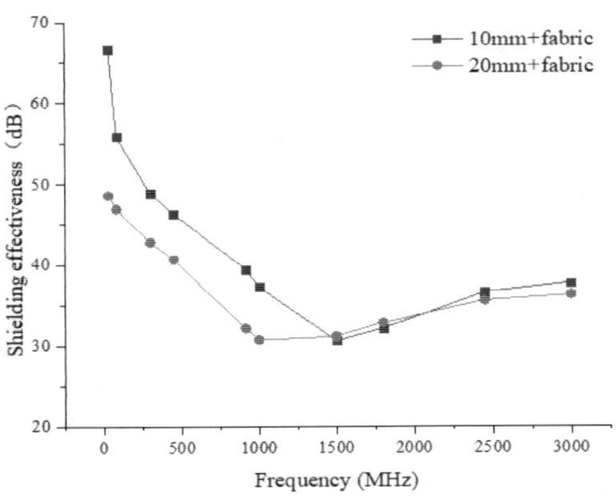

(a) Reduced iron powder content of 0kg/m^3

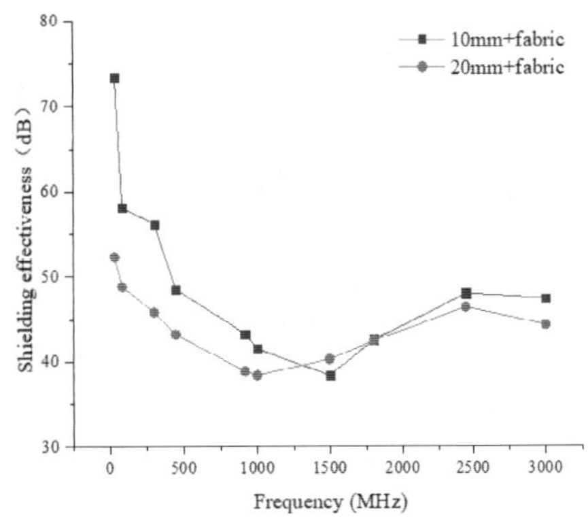

(b) Reduced iron powder content of 300kg/m^3

Figure 5. Effect of Thickness on Shielding Efficiency of Functional Mortar Specimen

The difference of shielding efficiency is more obvious in the low frequency band. When there is no powder, shielding effectiveness is mainly by fabric, thickness has little effect, when the thickness is 20 mm, the gap of fixture is too large. Also, the specimen with fabric is affected by binder, which leads to the bad roughness. Flange clamp can't completely clamped specimen, making external electromagnetic wave receiving waveguide cavity, thus to reduce the result. With increasing frequency, more electromagnetic wave through the fabric, because of the impedance mismatch differs; the wave will get lots of reflection losses, compensating for the error of the thickness. When powder is added, among the 30MHz-1GHz band, the shielding efficiency of 10mm thick specimen is higher. When the frequency is beyond 1GHz, the shielding effectiveness appears to be crossed. At the low-frequency point, increasing the thickness can increase the absorption loss, less wave penetrates the fabric, leading to a small loss in this part. At the same time, due to the effect of the excessive gap, which has a greater effect on the results of the 20mm ones. At this time, the test error brought by the thickness is larger than the absorption gain. In high point, the wave through the fabric, because the surface wave impedance mismatch, wave gets multiple reflection losses and absorption loss, compensating for the error caused by the thickness.

3.3. Effect of Shielding Fabric on Shielding Efficiency

3.3.1 Reference Mortar
Aimed at reference mortar, the shielding effectiveness of the specimens without fabric increases first and then decreases, while that of the specimens with fabric opposites. the effect is shown in figure 6.

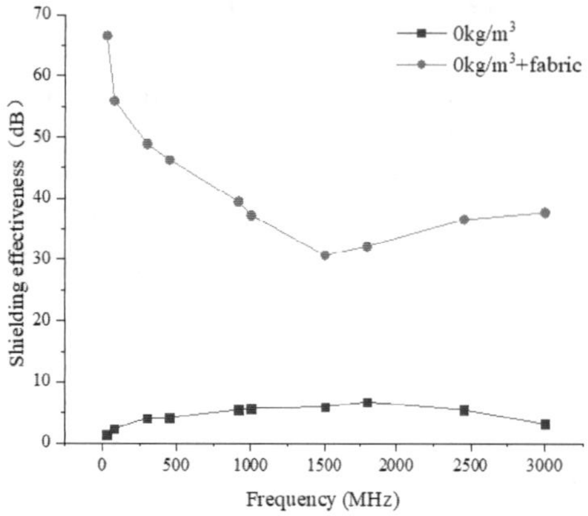

(a) 100mm×100mm×10mm

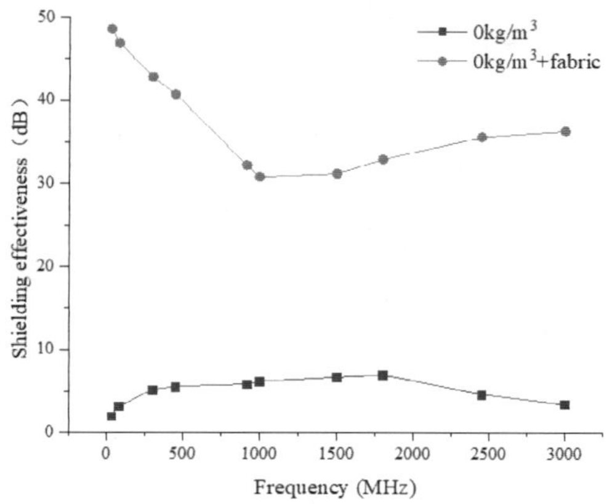

(b) 100mm×100mm×20mm

Figure 6. Effect of Shielding Fabric on Shielding Efficiency of Functional Mortar Specimen

The difference between the two groups of 10mm thick specimens decreased first and then increased. At the frequency point of 30MHz, the difference was 65.3db, at 1500MHz, the increase was 24.7db, and the average increase was 38.7db. The shielding fabric's shielding efficiency in the test frequency band can reach 70dB. It can significantly improve the shielding efficiency at the low-frequency point, but the performance at some frequency points is not ideal. With the increase of frequency, the shielding efficiency at the high-frequency point rises. The thickness of the specimen with shielding fabric is more than 10mm, which has a large test error compared with the single fabric. In addition, epoxy resin adhesive is used to bond the fabric and it will affect flatness and internal structure, so as to reduce the loss of electromagnetic wave. At the high frequency point, due to the mismatch, the electromagnetic wave gets multiple reflection losses between the specimen surface and the inner surface of the fabric. The difference between the 20mm thick specimens decreased first and then increased. At the 30MHz frequency point, the difference was up to 46.6db, at the 1000MHz frequency point, the smallest increase was only 24.6db, the average increase was 30.6db. Compared with 10mm thick ones, the increase in the shielding effect of fabric on the reference mortar specimens reduced. This is mainly because the increase in thickness causes the larger gap, which makes it easier to cause the decrease of results.

3.3.2. Mortar Mixed with Reduced Iron Powder
With the increase of frequency, the shielding efficiency first increases and then decreases, while that of the sample with fabric opposites. The effect is shown in figure 7.

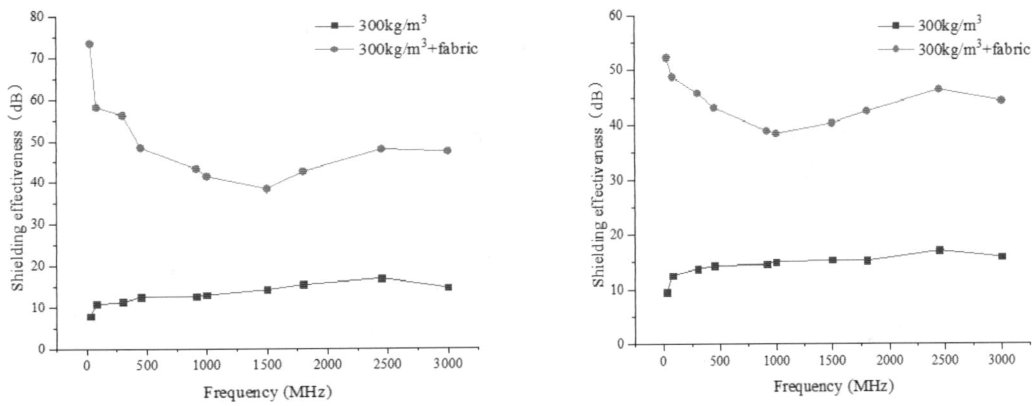

(a) 100mm×100mm×10mm (b) 100mm×100mm×20mm

Figure 7. Effect of Shielding Fabric on Shielding Efficiency of Functional Mortar Specimen

The shielding efficiency difference between the two groups of 10mm thick specimens decreased first and then increased. At the frequency point of 30MHz, the maximum increase was up to 65.6db; at 1500MHz, the minimum increase was only 24.2db; and the average increase was up to 36.3db. When shielding fabric and reduced iron powder work together, they shield electromagnetic wave in the low frequency band better. On the one hand, at the low-frequency point, the shielding fabric can reflect most wave, which can effectively improve shielding efficiency. On the other hand, the electromagnetic wave passed through the fabric got multiple reflection losses and absorbed between the fabric and the specimen surface. The combination is conducive to improve the overall shielding efficiency. The difference between the two groups of 20mm thick were the same. At the frequency point of 30MHz, the maximum increase was up to 42.8db. At the 1000MHz, the increase was only 23.4db, and the average increase was 29.3db. Compared with 10mm ones, the shielding efficiency of 20mm thick specimens is improved less by adding fabric. Because the increase of thickness will lead to greater test errors, resulting in the further reduction of shielding effectiveness at some frequency points.

4. Conclusion

(1) When no fabric is added, an appropriate content of reduced iron powder can improve the shielding efficiency, which has a more obvious effect on high frequency electromagnetic wave. The shielding efficiency of specimens with a mixture of 300kg/m^3 is better. When fabric is added, the shielding efficiency of the specimen against high frequency electromagnetic wave can be effectively improved.

(2) When no fabric is added, the increase in thickness can improve the shielding efficiency at the low-frequency point. When fabric is added, the increasing thickness will reduce the shielding efficiency.

(3) The addition of shielding fabric can significantly increase the shielding efficiency, especially in the low-frequency band. The increase is more obvious for specimens of 10mm thickness.

5. References

[1] Chen Yangru, Xiong Guoxuan, Zhang Zhibin. The application study of shielding media in cement-based materials, J. New Building Materials, 2010, 37 (10) 80-82.

[2] Han bin, Ji Zhi jiang, Zhang Zhong lun. Research on Absorbing Properties of Cement-based Coating Material J. Materials Review, 2009,23 (z1) 370-373.

[3] Dai Yinsuo, Gong Lei, Yan Fengguo, et al. Research on electromagnetic protection performance of metal wire mesh reinforced cement-based materials J. China Concrete and Cement Products ,2015(12):6-9.

[4] Zheng Guozhi, Chen Bin, Zhang Zehai, et al. Reflection and transmission performances of concrete slabs mixed with reduced iron powder J. High Power Laser and Particle Beams, 2015,27 (4) 151-156.
[5] Xu Fangxing, Zeng Guoxun, Zhang Haiyan, et al. Preparation and electromagnetic properties of flaky reduced iron powders J. Electronic components and materials, 2014, (12) 41-48.

Effect of Surface Burnishing Treatment on Microstructure and Wear Performance of ZK60 Magnesium Alloy

Haiquan Zhang, Yichen Liu and Peiquan Guo*

School of Mechanical Engineering, University of Jinan, Jinan 250022, China.
Email: oss_guopq@ujn.edu.cn

Abstract. In order to improve the surface properties of high strength magnesium alloy ZK60, the surface burnishing treatment was carried out. The effects of burnishing process parameters on surface roughness, microhardness and microstructure of ZK60 magnesium alloy were studied. And the wear performance of the untreated and the surface burnishing treated samples was compared and analyzed. A gradient microstructure layer of nanocrystalline, ultrafine, deformed and original coarse grains from the treated surface to the matrix was formed on ZK60 alloy by surface burnishing treatment. In comparison with the untreated sample, the surface roughness of the burnishing treated sample was significantly reduced and the surface microhardness was evidently improved. After surface burnishing treatment, the wear of ZK60 sample decreased by 44.58%, and the friction coefficient decreased from 0.328 to 0.293, significantly improving the wear performance.

1. Introduction

Magnesium alloy with a low density, high specific strength and specific stiffness, excellent electromagnetic shielding and machining properties is known as 'the green engineering material of the 21st century'[1-3]. However, magnesium alloy parts are inevitably subjected to high temperature, corrosion, severe wear and alternating load during service. Over time, magnesium alloy parts are prone to fatigue damage and fatigue fracture.

Because the fatigue fracture usually starts from the surface of the part, it puts forward more strict requirements on the surface performance of the part[4-5]. For this kind of failure, there are many existing surface strengthening methods, including burnishing strengthening[6], shot peening strengthening[7-8], laser impact strengthening[9-10] and surface mechanical attrition treatment[11]. Among them, surface burnishing strengthening technology is widely used. Surface burnishing strengthening treatment uses special burnishing tools to apply pressure on the surface of parts and conduct multiple burnishing, so as to change the microstructure and physical properties of the surface metal of parts[12]. Surface burnishing treatment has the advantages of no pollution, high efficiency and low cost, etc., which can be completed on the machine[13].

As a high-strength deformed magnesium alloy, the strength of ZK60 is close to the high-strength 7075 aluminum alloy. However, ZK60 alloy is not an invulgable material. It has many defects such as severe thermal cracking tendency and poor plasticity[3]. At present, there are no reports on the surface burnishing strengthening of ZK60 magnesium alloy. Therefore, the surface burnishing processing of ZK60 magnesium alloy was carried out in this paper, and the effects of burnishing process parameters on surface roughness, microhardness and microstructure of ZK60 magnesium alloy were studied. Meanwhile, the effects of surface burnishing strengthening on wear and friction coefficient of ZK60 magnesium alloy were studied.

2. Experiment Details

The test material was ZK60 magnesium alloy bar in extrusion state with a diameter of 35mm, with the chemical composition (in wt.%) of 5.679Zn, 0.613Zr, 0.095Y, 0.085Nb, 0.04Gd and balance Mg. The original microstructure of ZK60 magnesium alloy is shown in figure 1. The mean grain size is ~7μm.

Figure 1. Original microstructure of ZK60 magnesium alloy

Surface burnishing treatment of ZK60 magnesium alloy was carried out by self-developed burnishing equipment. The burnishing device was installed on CAK3665 lathe. Specific burnishing process parameters are as follows. The burnishing pressure was 600N. The burnishing speed was 6m·min^{-1}. The number of burnishing passes was 10, 20, 30, 40, 60, 80, 100, 120 and 160, respectively.

The microstructure of the cross-section of the sample with different burnishing process parameters was observed by using 4XC metallographic microscope, and the microstructure at different depths from the burnishing surface was characterized by using a JSM-7610F scanning electron microscope (SEM). Mitutoyo SJ-410 portable surface roughness tester was used to measure the surface roughness, and the probe stroke was set at 4mm. A 402MVD vickers hardness tester was used for the microhardness test. The test load was 0.25N and the retention time was 15 seconds. The friction and wear test was carried out on MFT-50 friction and wear test machine. The friction ball was Cr15 bearing steel with diameter of 6mm, applied load of 5N, running time of 10 minutes, reciprocating running distance of 4.5mm and frequency of 2. USP-Sigma white light interferometer was used to measure the three-dimensional surface morphology characteristics of the sample after friction and wear test, and the sampling area was 5.7 mm×1.08mm.

3. Results and Discussions

3.1. Effect of Different Burnishing Parameters on Microstructure of ZK60 Alloy

Figure 2 shows the effects of the number of burnishing passes on surface microstructure of ZK60 samples under the same burnishing pressure (600N) and burnishing speed (6m·min^{-1}). The results show that the thickness of grain refinement layer increases with the increase of the number of burnishing passes (from 10 to 160). However, when the number of burnishing passes reached 80, the grain refinement layer did not increase significantly. This is because the increase of the number of burnishing passes leads to the increase of twins and dislocation density in the surface layer, that is, the appearance of work hardening layer increases the resistance to further plastic deformation of the surface layer. When the resistance of plastic deformation of the surface material is in balance with the burnishing pressure applied by the test, the material cannot be further deformed. Therefore, the thickness of grain refinement layer does not increase linearly with the increase of the number of burnishing passes, but tends to a stable range (~73μm).

3.2. Microstructure of the Burnishing Treated Surface Layer

Figure 3 shows the SEM microstructure of ZK60 alloy sample after surface burnishing treatment (160 passes). A gradient microstructure layer of nanocrystalline/ultrafine, deformed and original coarse grains from the treated surface to the matrix were formed on ZK60 alloy by surface burnishing treatment. The A region in figure 3 is the nanocrystalline/ultrafine grain region of ~12μm in thickness. The B region is the deformed grain layer with a depth range of 12~75μm. In this region, the grain boundary was severely distorted and grains were irregularly distributed. The matrix was the original extruded structure, and grains were mainly isoaxial and coarse grains that were not completely broken (region C in figure 3). It can be seen from figure 3 that the grain size from the surface to the matrix gradually increases with the increase of depth, showing a gradient structure with continuous changes.

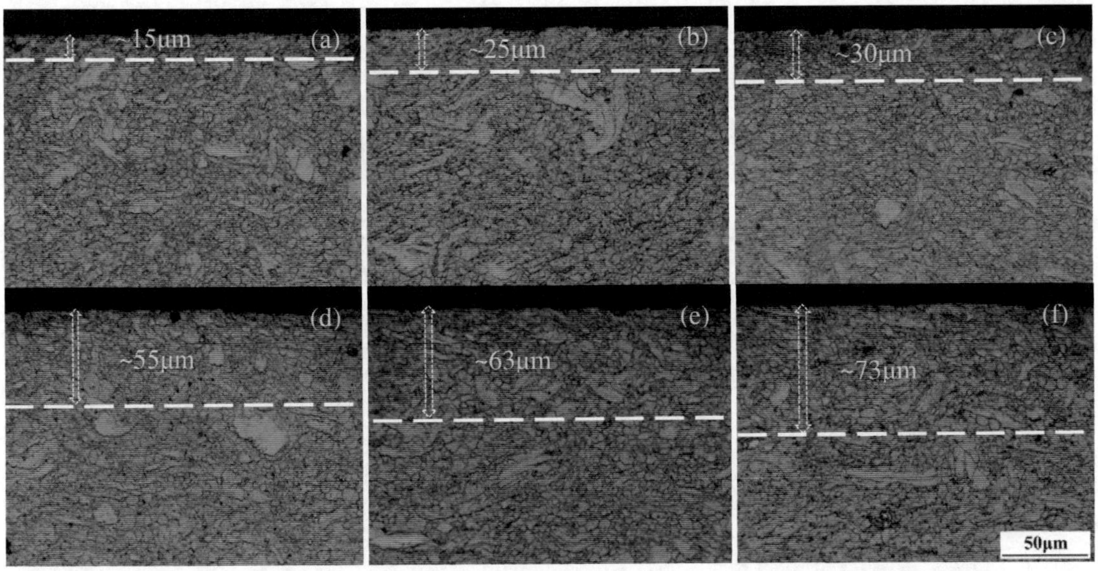

Figure 2. Cross-section morphologies of ZK60 alloy after surface burnishing treatment with different burnishing passes:(a) 10 passes; (b) 20 passes; (c) 40 passes; (d) 80 passes; (e) 120 passes; (f) 160 passes

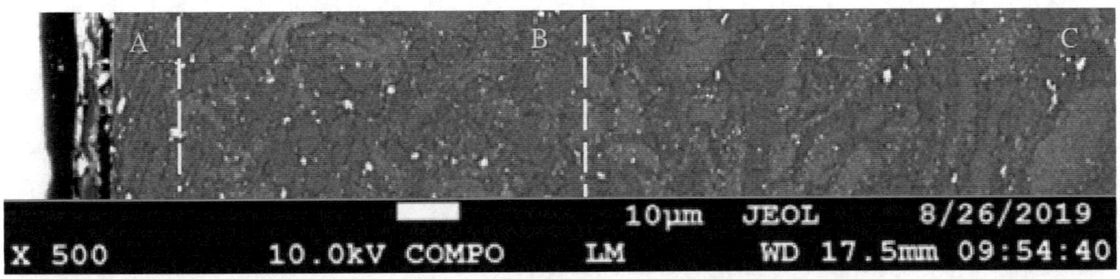

Figure 3. Cross-section SEM image of burnishing treated surface of ZK60 sample (160 passes)

3.3. Analysis of Surface Roughness and Microhardness

Contour arithmetic mean deviation Ra is a widely used basic evaluation parameter of surface roughness, which can directly reflect the peak-valley amplitude variation of the detected contour. The effect of burnishing passes on surface roughness of ZK60 magnesium alloy samples in figure 4. The Ra value of ZK60 samples was 1.258μm. After surface burnishing treatment, the Ra of the corresponding sample decreased the least when the number of burnishing passes was 10, down to 0.751μm, which was 40.3% lower than that of turning. When the number of burnishing passes was 120, the Ra of the sample decreased the most, to 0.333μm, which decreased by 73.5% compared with the untreated sample. Within the value range of burnishing passes, the surface roughness decreases first and then increases with the increase of the number of burnishing passes. The change in Ra value due to the change of the number of burnishing passes was 0.418μm.

Dependence of microhardness on depth is determined in the burnished surface layer of ZK60 sample, as shown in figure 5. Due to the low hardness of ZK60 magnesium alloy, 0.25N loading force was adopted. The diagonal length of the square indentation obtained under this load was about 23μm. When the measuring center was selected within 20μm from the surface of the workpiece, the indentation image obtained at the test point will be seriously deformed (indentation is irregular), resulting in distortion of the measurement results. Therefore, the measurement starting point was selected 20μm away from the sample surface and the interval between adjacent measuring points was 20μm.

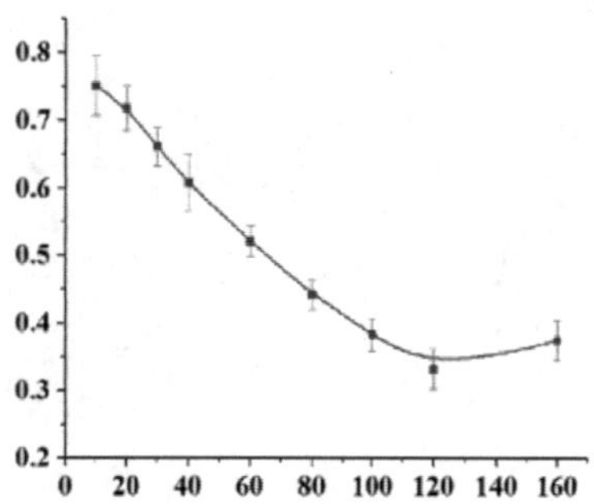

Figure 4. Effect of different burnishing passes on surface roughness of ZK60 alloy

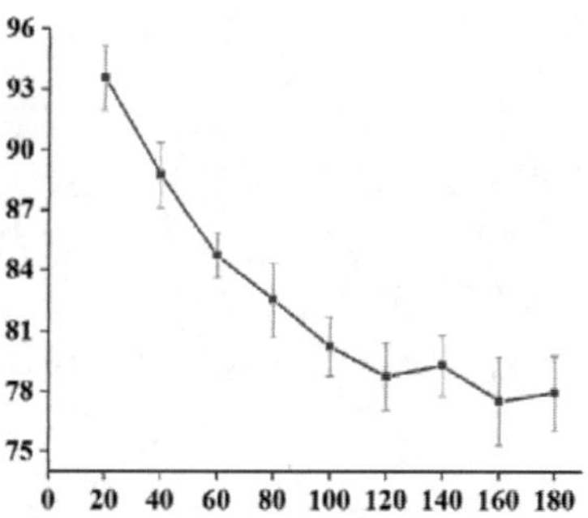

Figure 5. Variations of microhardness with distance from treated surface of the burnished sample

Turning leads to the cold hardening of the surface metal of the sample, and then the surface microhardness is improved. The microhardness of the turning sample at 20μm away from the surface was 82.07$HV_{0.25}$, which was 4.55% higher than that of the matrix material. After 120 burnishing passes, the microhardness reached 93.58$HV_{0.25}$, which increased 19.21% compared with the microhardness of the matrix material. As can be seen from figure 6, the microhardness from the surface to the matrix decreased gradually with the increase of depth.

3.4. Analysis of Wear Performance

3.4.1. Wear of ZK60 samples

Figure 6 shows the three-dimensional surface morphology of ZK60 samples after friction and wear test. The wear amount of the untreated sample was 1.66×10^{-4}g. After surface burnishing treatment (120 passes), the wear amount of the sample decreased to 9.2×10^{-5}g, reducing the wear amount by 44.58%. Therefore, surface burnishing treatment can effectively reduce the wear of ZK60 magnesium alloy and improve its wear performance.

3.4.2. Friction coefficient of ZK60 samples

The curves of the friction coefficients over time of untreated and surface burnishing treated samples is shown in Figure 7. The average friction coefficient of the untreated sample was 0.328, and it changed to 0.293 after surface burnishing treatment. After surface burnishing treatment, the friction coefficient of the sample becomes smaller. This is because a hardened layer on the surface of ZK60 samples was formed by the surface burnishing treatment which improves the surface microhardness and thus improves its wear performance.

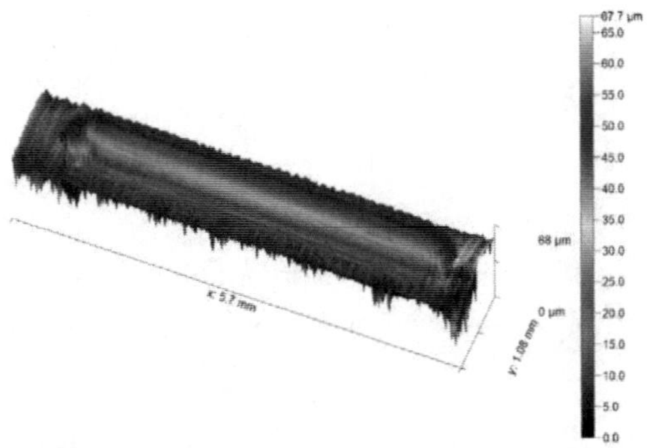

(a)Surface morphology of untreated sample

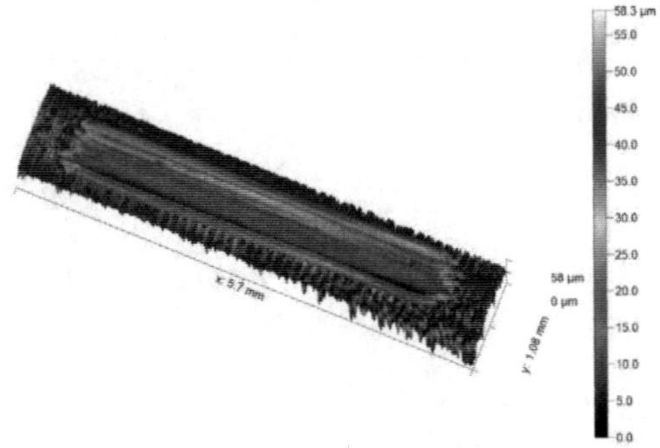

(b)Surface morphology of burnishing treated sample

Figure 6. The three-dimensional surface morphology of ZK60 sample after friction and wear test

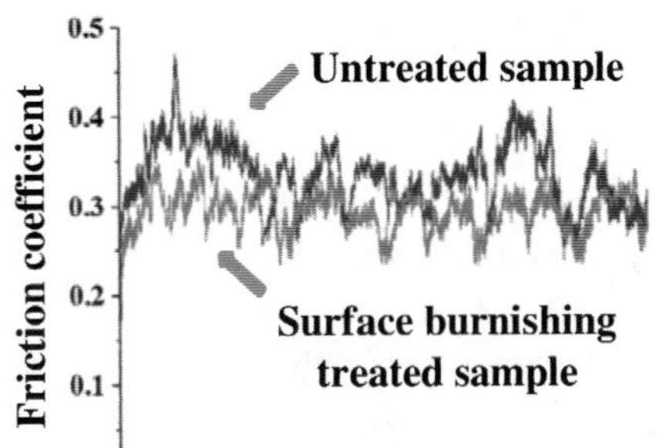

Figure 7. Friction coefficients of untreated and surface burnishing treated samples

4. Summary

A gradient microstructure layer of nanocrystalline, ultrafine, deformed and original coarse grains from the treated surface to the matrix was formed on ZK60 alloy by surface burnishing treatment. The thickness of nanocrystalline/ultrafine grain region was ~12μm. And the thickness of the deformed grain layer was ~63μm.

In comparison with the untreated sample, surface roughness of the burnishing treated sample was significantly reduced and the surface microhardness was evidently improved. When the number of burnishing passes was 120, the Ra of the sample decreased to 0.333μm, which decreased by 73.5% compared with the untreated sample. And the microhardness reached $93.58HV_{0.25}$, which increased 19.21% compared with the microhardness of the matrix material.

After surface burnishing treatment (120 passes), the wear amount of the sample decreased to 9.2×10^{-5}g, reducing the wear amount by 44.58%. Friction coefficient decreased from 0.328 to 0.293. The wear performance of ZK60 alloy was improved obviously.

5. Reference

[1] Liu Q 2010 *Acta Metallurgica Sinica* **46** 1458-72
[2] Zhang Q L, Hu Y X and Wang L L 2008 *Rare Metal Materials and Engineering* **37** 678-81
[3] Ma Y L, Pan F S and Zuo R L 2004 *R Journal of Chongqing University* **27** 80-85
[4] Xin C, Xu W, Sun Q Y, Xiao L and Sun L 2017 *Rare Metal Materials and Engineering* **46** 1954-60
[5] Song X C, Zhao D Q, Zhou J X and Yang Y S 2017 *Materials for Mechanical Engineering* **41** 1-7
[6] Zhang P and Lindemann J 2005 *Scripta Materialia* **52** 1011-15
[7] Zhang W and Tian L 2014 *Hot Working Technology* **43** 153-5
[8] Liu W C, Dong J, Zhang P, Zhai C Q and Ding W J 2009 *The Chinese Journal of Nonferrous Metals* **19** 1733-40
[9] Wang H, Huang Y H, Zhang W W, Ru H L and Wang S B 2016 *Laser and Optoelectronics Progress* **53** 1-7
[10] Li X C, Zhang Y K, Zhou J Y, Chen J F and Lu Y L 2016 *Laser Technology* **40** 5-10
[11] Sun Y, Zhu L X and Ma S W 2018 *Foundry Technology* **2** 448-51
[12] Liu Z Q, He M and Zhao J 2015 *China Mechanical Engineering* **26** 403-13
[13] Wang Y L, Zhu Y L and Yang J Q 2018 *Aeronautical Manufacturing Technology* **61** 75-83

IOP Conf. Series: Materials Science and Engineering **735** (2020) 012004 doi:10.1088/1757-899X/735/1/012004

Preparation and Characterization of Submicron TiO_2 Microspheres via Urea-formaldehyde Template

Shaosen Lin, Shiguo Du and Yanling Lu

The 3rd Department, Army Engineering University Shijiazhuang Campus, Shijiazhuang, Hebei 050003, PR China
E-mail: lin_shaosen@163.com

Abstract. In order to obtain sub-micron TiO_2 microspheres with good dispersibility and easy collection, urea-formaldehyde(UF)-TiO_2 hybrid microspheres were synthesized by polymerization-induced precipitation in acidic TiO_2 sol through the reaction of urea and formaldehyde. The hybrid microspheres and their calcined products were analyzed by scanning electron microscopy, laser particle size analyzer, thermogravimetric analyzer and X-ray Powder diffractometer. The results of morphological characterization showed that the UF-TiO_2 hybrid microspheres were spherical and well dispersed, and their particle size ranged from 0.8~5μm. The sub-micron TiO_2 microspheres with large pore volume and good dispersion were obtained by calcining and removing the organic components in the hybrid microspheres. Thermogravimetric analysis showed that the proportion of inorganic components in hybrid microspheres was about 20%. XRD analysis showed that the product of calcined hybrid microspheres was single anatase-type titanium dioxide.

1. Introduction

As an important broadband gap semiconductor material, TiO_2 has shown great application prospects in the fields of solar cells, photocatalysts, pigments, biological materials and sensor materials due to its excellent photocatalytic performance, chemical stability, environmental friendliness and low cost[1-3]. Monodisperse microspheric TiO_2 has become one of the hotspots in the field of materials science and engineering due to its good fluidity, high bulk density, non-reunification, easy recovery and reuse[4].

At present, sol-gel and hydrothermal methods are often used to prepare nano or sub micron pellets, which are difficult to achieve non agglomeration and easy collection[5]. UF microspheres have become templates for the preparation of micro-nano metal oxide powders and mesoporous materials due to their advantages of easy synthesis, more variable conditions and various reaction groups[6-8]. The surface of UF prepolymer contains a large number of polar functional groups such as amines, hydroxyl groups and carbonyl groups. These groups also play a guiding role in the nucleation and growth of TiO_2. Based on these considerations, this paper combined the characteristics of sol-gel method and polymerization-induced precipitation to prepare hybrid microsphere[9]. A hybrid microsphere was synthesized by the polymerization of urea and formaldehyde in acidic TiO_2 sol. The size, distribution and shape of UF-TiO_2 hybrid microspheres were controlled by polymerization-induced precipitation. TiO_2 microspheres were prepared by removing organic templates from hybrid microspheres at high temperature. This method will provide a new method and a new idea for the preparation of ultrafine inorganic materials with good dispersion and without agglomeration.

2. Experiments

2.1. Materials
Urea (analytical reagent), Tetrabutyl titanate ($Ti(OBu)_4$, analytical reagent), Ethanol (analytical reagent) and Acetylacetone (analytical reagent) were obtained from Tianjin Yongda Chemical Reagent Co. Ltd.; Formaldehyde solution (analytical reagent, 37.0%~40.0%) and Hydrogen nitrate (analytical reagent) were purchased from Shijiazhuang Chemical Reagent Factory; Double-distilled water was used for the preparation of all solutions in this study.

2.2. Preparation of Hybrid Microspheres
6.8 g $Ti(OBu)_4$, 0.6 g acetylacetone and 25mL ethanol were mixed in beaker. Under the condition of stirring, water was added drop by drop until the precipitation was complete. Then a certain amount of nitric acid with a concentration of 30wt% was added to the turbid solution to remove the precipitation and form a transparent TiO_2 sol. After dissolving 6 g urea in 50mL water and mixing evenly with the above-mentioned TiO_2 sol, the volume of the mixed liquid was adjusted to about 200mL with water and its pH value was adjusted to about 2.5. Under the stirring condition, 20 g formalin was slowly added into the mixed liquid. After homogeneous mixing, the stirring was stopped and the reaction vessel was sealed. After the reaction was completed, the products were separated by centrifugation, washed alternately with water and ethanol for 3 to 5 times, and dried in an air blast drying chamber at 60°C for 6 hours, resulting in the preparation of UF-TiO_2 hybrid microspheres.

2.3. Characterization
Morphology and microstructure of hybrid microspheres and TiO_2 were observed by scanning electron microscopy (SEM, Hitachi S-4800). The particle size distribution was measured using a laser particle size analyzer (Malvern, Zetasizer Nano ZS90). The X-ray diffraction (XRD) measurement of the hybrid microspheres after calcination was made using Cu Kα radiation (D8 ADVANCE, Bruker AXS). The thermal properties of hybrid microspheres were examined by a thermogravimetric analyzer (TGA, PerkinElmer Pyris Diamond, SDTQ600) at a heating rate of $10°C \cdot min^{-1}$ under N_2 atmosphere.

3. Results and Discussion

3.1. Morphology
Scanning electron microscopy analysis was performed to identify the morphology of UF-TiO_2 hybrid microspheres. Figure 1 shows SEM images of UF-TiO_2 hybrid microspheres at different magnification. These pictures together confirm the good dispersion of the hybrid microspheres. As can be seen from Figure 1 (a), UF-TiO_2 hybrid microspheres possess good spherical shape and fine particle size. Figure 1 (b) shows the morphology of single hybrid microspheres. Figure 1 (c) shows high magnification SEM image of the dotted line frame in Figure 1 (b). Two high magnification photographs show that there are nano-sized porous structure on the surface of hybrid microspheres, which lays a good foundation for the next step of calcination to remove organic templates to obtain porous TiO_2 microspheres. Figure 1 (d) is the particle size distribution of UF-TiO_2 hybrid microspheres. The result shows that the particle size of UF-TiO_2 hybrid microspheres concentrated in 0.8-5μm, and the average particle size was about 2μm. It can be predicted that ultrafine TiO_2 microspheres can be obtained by calcining this type of hybrid microspheres.

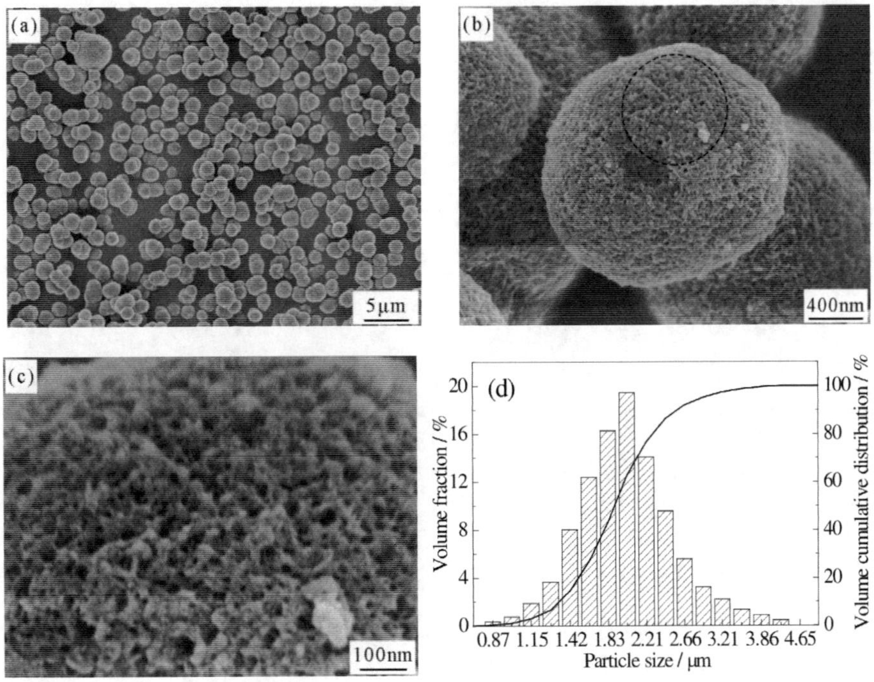

Figure 1. SEM Images and Particle Size Distribution of UF-TiO_2 Hybrid Microspheres

3.2. Thermal Analysis

The TG/DTG curves of UF microspheres (control sample, without TiO_2) and UF-TiO_2 hybrid microspheres at a heating rate of 10°C·min^{-1} were shown in Figure 2. The weight loss process of samples could be divided into three stages: slow weightlessness at room temperature to 200°C, rapid weightlessness at 200–350°C and carbonization reaction at 350–600°C.

It can be seen from Figure 2 (a) that the UF microsphere has a weight loss of about 9% before 200°C. The main reason for weight loss in this stage is the volatilization of water adsorbed in UF microspheres and other low boiling point substances. In addition, this stage also corresponds to the thermal curing process of UF, that is, the reactive groups such as unreacted methylol group and amine group in the UF will undergo condensation reaction at high temperature to remove moisture and formaldehyde. The rapid decomposition stage of UF is 200-350°C. The weight loss of the sample is about 83%. The main reasons are the breakdown of methylene ether bond, methylene bond and the removal of other small molecules. The carbonization reaction stage is 350-600°C, and the weight loss is about 4%. The main reason is the removal of nitrogen, oxygen, hydrogen and other elements. Figure 2 (b) shows that the weight loss of UF-TiO_2 hybrid microspheres is about 10% before 200°C. The main reason is the volatilization of water adsorbed in the hybrid microspheres and other low boiling point substances, including the moisture and formaldehyde removed by the thermal curing process of UF. The weight loss of samples at 200-350°C is about 62%, which is mainly caused by the thermal decomposition of organic components in hybrid microspheres and the transformation of TiO_2 sol to solid TiO_2. Among them, 252°C corresponds to the thermal decomposition of UF and 290°C corresponds to the transformation process from TiO_2 sol to solid TiO_2, The weight loss during the transformation process is caused by the loss of structural water and alcohol in the titanium dioxide sol. Carbonization occurs above 350°C, and the weight loss is about 4%. From the above analysis, it can be inferred that the content of solid TiO_2 in UF-TiO_2 hybrid microspheres is about 20%.

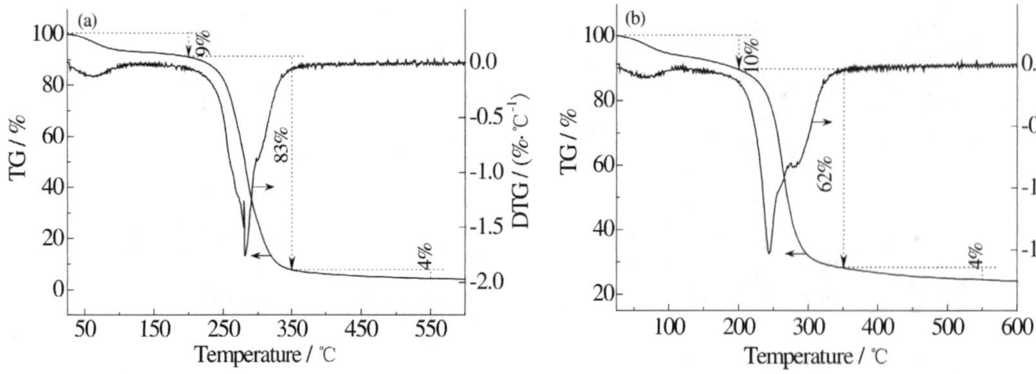

Figure 2. TG/DTG Curves of Samples: (a) UF Resin; (b) UF-TiO_2 Hybrid Microspheres

3.3. Characterization of UF-TiO_2 Hybrid Microspheres after Calcination

The as-prepared UF-TiO_2 hybrid microspheres were calcined at 600°C for 20 minutes to remove the organic components. The calcined products were collected and their structures were observed by SEM, as shown in Figure 3. From Figure 3 (a), it can be seen that the TiO_2 microspheres have regular shape and good dispersion. Figure 3 (b) is an intervention broken TiO_2 microspheres. The broken TiO_2 microspheres show hollow structure or even double-layer structure. The shell of the broken TiO_2 microspheres is composed of multi-layer or single-layer TiO_2 particles. It is inferred that the TiO_2 microspheres are produced by calcining the hybrid microspheres whose surface is composed with UF. Duc to thc limitation of UF, thc TiO_2 particlcs in the hybrid microspheres will self assemble into TiO_2 microspheres during calcination, and the TiO_2 particles tend to migrate outward to form hollow structure due to the effect of surface tension in the process of polymerization. After the organic components was removed by calcination, the TiO_2 particles with small particle size and large specific surface area formed TiO_2 microspheres.

Figure 3. Morphology and Structure of UF-TiO_2 Hybrid Microspheres after Calcination

Figure 4 is the XRD pattern of the above-mentioned TiO_2 microspheres. The positions of the diffraction peaks on each crystal plane in the pattern are consistent with those on the standard anatase-type TiO_2 card (JCPDS 21-1272). This indicates that the crystalline form of TiO_2 microspheres obtained by calcining hybrid microspheres at 600°C is anatase TiO_2.

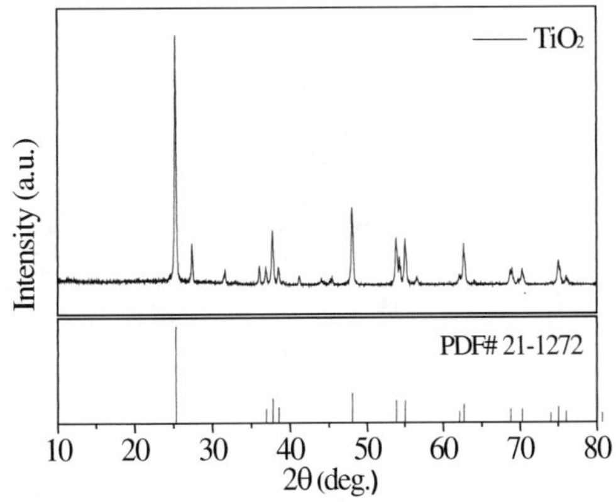

Figure 4. XRD Patterns of Hybrid Microspheres after Calcination

4. Conclusion

(1) UF-TiO_2 hybrid microspheres were successfully prepared by polymerization-induced precipitation method. The microspheres were well dispersed and no agglomeration was observed. The particle size of the microspheres ranged from 0.8-5μm. Submicron TiO_2 microspheres with good dispersion and easy collection could be obtained by removing organic template from hybrid microspheres by calcination.

(2) Thermal analysis showed that the proportion of TiO_2 in UF-TiO_2 hybrid microspheres was about 20%.

(3) XRD analysis showed that the crystalline form of TiO_2 microspheres obtained by calcining hybrid microspheres was anatase TiO_2.

5. Acknowledgements

The financial support provided by the project (51272284) of the National Natural Science Foundation of China for this work is gratefully acknowledged.

6. References

[1] Ma Y, Wang X, Jia Y, Chen X, Han H and Li C 2014 *Chem. Rev.* **114** 9987-10043
[2] Mathew S, Yella A, Gao P, Humphry-Baker R, Curchod BF, Ashari-Astani N, Tavernelli I, Rothlisberger U, Nazeeruddin MK and Grätzel M 2014 *Nat. Chem.* **6** 242-7
[3] Wang S Q, Zhu A M, Liu Q L 2010 *Chemical Industry Times* **24** 5-8
[4] Ye C Y, Quan X J, Jiang L, Zhang, L and Tan, L 2011 *Chemical Reaction Engineering and Technology* **27** 515-20
[5] Jing M X, Yan Y Z, Ping Y H and Li L K 2014 *Journal of Materials Science & Engineering* **32** 5-9
[6] Wang, Y, Yan X, Tu M, Cheng J and Zhang J 2017 *Mater. Lett.* **191** 178-81
[7] Hao Z, Guo B, Liu H, Gan L, Xu Z and Chen L 2006 *Micropor. Mesopor. Mat.* **95** 350-9
[8] Liu Z, Du Z, Song H, Wang C, Subhan F, Xing W and Yan Z 2014 *J. Colloid. Interf. Sci.* **416** 124-32
[9] Siva T, Muralidharan S, Sathiyanarayanan S, Manikandan E and Jayachandran M 2017 *J. Inorg. Organomet. P.* **27** 1-9.

IOP Conf. Series: Materials Science and Engineering **735** (2020) 012005 doi:10.1088/1757-899X/735/1/012005

Analytic Solutions of Two-photon Rabi Model Based on Bargmann Space

Zhanyuan Yan and Xuemin Yao*

Department of Mathematics & Physics, North China Electric Power University, Baoding, Hebei 071003, China

Email: xueminyao1995@163.com

Abstract. The quantum Rabi model (QRM) is identified as the simplest model to study the interaction between light and matter. In recent years, Braak proposed an alternative more efficient method to obtain the exact solvability and spectrum of the QRM in the Bargmann space, which is appropriate frame for extended squeezed coherent states. Based on this feasible method, some study for the two-photon QRM is concentrated. In this paper we diagonalized the Hamiltonian and obtain the Spectrum of the system of two-photon QRM in the special case that the level splitting of the atom was equal to 0, and compared with Travěnec's results without diagonalizing the Hamiltonian. Finally, we confirm the validity of diagonalization for the Hamiltonian in the two-photon QRM.

1. Introduction

The fundamental and concentrated study of the interaction between light and matter is a significant part of modern quantum optics [1]. It is excellent that the quantum Rabi model (QRM) is proposed as the simplest and most efficient model to solve the interaction between a two-level atom (qubit) and a single-mode boson field. It has a worldwide applications ranging from circuit quantum electrodynamics (QED) [2], cavity QED [3] and quantum information science. In order to solve this model, various attempts are practiced, such as rotating wave approximation method, coupled cluster method, Bogoliubov transformation method and so on. In 2011, Braak proposed a unprecedented method, based on Bargmann-Fock representation ,which was succeed in acquiring the exact solution to the full Hamiltonian in Rabi model [4][5]. It is often reduced to algebraic manipulation of infinite matrices and Fock basis.

More recently, many physicists are interested in the two-photon QRM [6-9], describing the interaction between a bosonic mode with energy ω and a two-level atom with level spacing Ω ($\hbar = 1$):

$$H = \omega a^{+} a + \frac{\Omega}{2}\sigma_z + g\left(a^2 + \left(a^{+}\right)^2\right)\sigma_x \tag{1}$$

Where Ω and ω are respectively frequencies of qubit and the single-mode boson field, σ_x and σ_z are the Pauli matrices, and a and a^{+} are respectively boson annihilation and creation operators.

In this paper, we will focus on the energy spectrum of the QRM under two-photon exchange. The solution of generalized two-photon Rabi model in the Bargmann space including the static bias of the qubit has been given [7]. In this process, the Hamiltonian is diagonalized. On the other hand, Travěnec gave the exact solution of the Hamiltonian (1) with Z_4 symmetry [8].We will try to diagonalize Hamiltonian to obtain the exact solution of the above Hamiltonian, and then compare the result with the Travěnec and give the relevant analysis conclusion.

The paper is organized as follows. In Sec.2, we are succeeding in obtaining the theoretical derivation of the two-photon QRM by diagonalizing the Hamiltonian. In Sec.3, the regular spectrum in the special case that the level splitting of the atom was equal to 0 and the exceptional solutions will be given. Finally, we draw some conclusions inchiding perspectives for future work in the last section.

2. Bargmann Representation Method for the Two-Photon QRM

Firstly, we make unitary transformation $s=\frac{1}{2}(\sigma_x+\sigma_z)$ to the equation (1) and turn it into the form (ω is set to be 1)

$$H=\omega a^+ a+\frac{\Omega}{2}\sigma_x+g\left(a^2+\left(a^+\right)^2\right)\sigma_z \tag{2}$$

In the Bargmann space, the boson annihilation and creation operators go over to the following expression [4]:

$$a=\frac{\partial}{\partial z},\quad a^+=z \tag{3}$$

Where z is a complex variable. The function $\psi(z)$ is required to satisfy a very crucial regulation that it should be analytic everywhere in the complex plane.

Now introduce the reflection operator T which well represents the Z_2 symmetry of the two-photon QRM and has the property $T=T^+=T^{-1}$. The form of operator T is [7]

$$T=\exp[\frac{i\pi}{2}a^+a(a^+a-1)] \tag{4}$$

In the Fock state $|N>$, it satisfies

$$T|N>=\exp\left(\frac{i\pi}{2}\left(N^2-N\right)\right)\frac{\left(a^+\right)^N}{\sqrt{N!}}|0>=i^{N^2-2N}\frac{\left(ia^+\right)^N}{\sqrt{N!}}|0> \tag{5}$$

And the operator T acting on elements $f\left(z\right)$ of the Bargmann space satisfies the condition $Tf\left(z\right)=cf\left(iz\right)$, where c related to the parity (being odd or even) of $f\left(z\right)$ is a constant which is determined by equation (5).The operator T has eigenvalues ± 1 and satisfies $T^2=1$.

In the basis $\{|\uparrow>\otimes\psi_1(z),|\downarrow>\otimes\psi_2(z)\}$, where $|\uparrow>$ and $\downarrow>$ are the eigenvectors of σ_z and $\psi_1(z)$, $\psi_2(z)$ are the wavefunctions in the Bargmann space. The Hamiltonian reads

$$H=\begin{pmatrix}\omega z\frac{\partial}{\partial z}+g\left(\frac{\partial^2}{\partial z^2}+z^2\right) & \frac{\Omega}{2}\\ \frac{\Omega}{2} & \omega z\frac{\partial}{\partial z}-g\left(\frac{\partial^2}{\partial z^2}+z^2\right)\end{pmatrix} \tag{6}$$

Next, we apply the Z_2 symmetry with a Fulton-Gouterman transformation [10]

$$U=\frac{1}{\sqrt{2}}\begin{pmatrix}1 & 1\\ \hat{T} & -\hat{T}\end{pmatrix} \tag{7}$$

And can obtain the diagonalized Hamiltonian, which reads

$$H_{diag}=\hat{U}^{\dagger}H\hat{U}=\begin{pmatrix}\omega z\frac{\partial}{\partial z}+g(z^2+\frac{\partial^2}{\partial z^2})+\Delta T & 0\\ 0 & \omega z\frac{\partial}{\partial z}+g(z^2+\frac{\partial^2}{\partial z^2})-\Delta T\end{pmatrix} \tag{8}$$

For H_{diag}, the eigenfunctions are the solutions of the equation is

$$\begin{pmatrix}\omega z\frac{\partial}{\partial z}+g(z^2+\frac{\partial^2}{\partial z^2})+\Delta T & 0\\ 0 & \omega z\frac{\partial}{\partial z}+g(z^2+\frac{\partial^2}{\partial z^2})-\Delta T\end{pmatrix}\begin{pmatrix}\psi_1(z)\\ \psi_2(z)\end{pmatrix}=E\begin{pmatrix}\psi_1(z)\\ \psi_2(z)\end{pmatrix} \tag{9}$$

Setting $\begin{cases}\hat{T}\psi_1(z)=\psi_3(z)\\ \hat{T}\psi_2(z)=\psi_4(z)\end{cases}$, we can rewrite (9) as

$$\omega z\frac{\partial}{\partial z}\psi_1(z)+g(z^2+\frac{\partial^2}{\partial z^2})\psi_1(z)+\Delta\psi_3(z)-E\psi_1(z)=0 \tag{10}$$

$$\omega z\frac{\partial}{\partial z}\psi_2(z)+g(z^2+\frac{\partial^2}{\partial z^2})\psi_2(z)-\Delta\psi_4(z)-E\psi_2(z)=0 \tag{11}$$

Applying Z_2 symmetry, we make a transformation $Tf(z)=cf(-iz)$ and obtain from (10) and (11)

$$\omega z\frac{\partial}{\partial z}\psi_3(z)-g(z^2+\frac{\partial^2}{\partial z^2})\psi_3(z)+\Delta\psi_1(z)-E\psi_3(z)=0 \tag{12}$$

$$\omega z\frac{\partial}{\partial z}\psi_4(z)-g(z^2+\frac{\partial^2}{\partial z^2})\psi_4(z)-\Delta\psi_2(z)-E\psi_4(z)=0 \tag{13}$$

We can combine (10) and (12) as one group, which contains $\psi_1(z)$ and $\psi_3(z)$ .Setting $\psi_i(z)=e^{-\kappa z^2}\varphi_i(z)(i=1,2,3,4)$ and substituting them into (10) and (12), we can get

$$g\frac{d^2}{dz^2}\varphi_1(z)+(1-4g\kappa)z\frac{d}{dz}\varphi_1(z)-(E+2g\kappa)\varphi_1(z)+\Delta\varphi_3(z)=0 \tag{14}$$

$$g\frac{d^2}{dz^2}\varphi_3(z)-(1+4g\kappa)z\frac{d}{dz}\varphi_3(z)+(4\kappa z^2-2g\kappa+E)\varphi_3(z)+\Delta\varphi_1(z)=0 \tag{15}$$

Where κ has been set to be $\kappa=\frac{1-\sqrt{1-4g^2}}{4g}$ to ensure the term $4g\kappa^2-2\omega+g$ to be removed. And we demand $g\le 0.5$ to make two-photon QRM significant. Next, we expand $\varphi_1(z)$, $\varphi_3(z)$ into power series in z

$$\varphi_1(z)=\sum_{-\infty}^{\infty} A_n z^n \tag{16}$$

$$\varphi_3(z)=\sum_{-\infty}^{\infty} C_n z^n \tag{17}$$

And then, we take (16) and (17) into ((14) and (15)) and obtain

$$gn(n-1)A_n=\left((n-2)(4g\kappa-1)+2g\kappa+E\right)A_{n\text{-}2}-\Delta C_{n-2} \tag{18}$$

$$gn(n-1)C_n=\left((n-2)(1+4g\kappa)+2g\kappa-E\right)C_{n-2}-4\kappa C_{n-4}-\Delta A_{n-2} \tag{19}$$

It is worth noting that we should make $A_n=0$ and $C_n=0$ for $n<0$ to satisfy analytic requirement in the whole complex plane .For another, we set

$$\varphi_2(z)=\sum_{-\infty}^{\infty} B_n z^n \tag{20}$$

$$\varphi_4(z)=\sum_{-\infty}^{\infty} D_n z^n \tag{21}$$

And obtain

$$gn(n-1)B_n=\left((n-2)(4g\kappa-1)+2g\kappa+E\right)B_{n\text{-}2}+\Delta D_{n-2} \tag{22}$$

$$gn(n-1)D_n=\left((n-2)(1+4g\kappa)+2g\kappa-E\right)D_{n-2}-4\kappa D_{n-4}+\Delta B_{n-2} \tag{23}$$

For equation (5), we find $c=1$ if n is even, and $c=-1$ if n is odd. The relationships $T\varphi_{3,4}(z)=c\varphi_{3,4}(iz)=\varphi_{1,2}(z)$ is important for us because it can be used to acquire the E form the above equations. It can read as

$$ce^{\kappa z^2}\sum_{n=0}^{\infty} C_n(iz)^n-e^{-\kappa z^2}\sum_{n=0}^{\infty} A_n z^n=0 \tag{24}$$

$$ce^{\kappa z^2}\sum_{n=0}^{\infty} D_n(iz)^n-e^{-\kappa z^2}\sum_{n=0}^{\infty} B_n z^n=0 \tag{25}$$

And for even parity, considering the point $z=0$, $C_0=A_0$ and $D_0=B_0$ are apparent. So the initial condition can be set as follows

$$C_0=A_0=1\,,\,D_0=B_0=1 \tag{26}$$

For odd parity, considering the first derivatives, we can set

$$C_1=A_1=1\,,\,D_1=B_1=1 \tag{27}$$

3. Numerical Analysis for the Two-Photon QRM

Now we assume that the level splitting of the atom is extremely small or equal to 0. Obviously, the two groups have the same eigenvalues. It is exactly known that the complete spectrum is given by [6]

$$E=-\frac{1}{2}+\left(N+\frac{1}{2}\right)\Xi,\quad N=0,1,2,3...... \tag{28}$$

Where another dimensionless quantity reads

$$\Xi = \sqrt{1-16g^2} \tag{29}$$

In numerical calculations, we truncate the apposite value of n in equation(24) and (25) and ensure the eigenvalues are not affected simultaneously. Now we set values as large as z=1000 or even z=10000[8]for the unaffected root. If the values of the root is convergent and the difference is within 0.01, we can accept it. It is obvious that the convergence of the eigenvalue will be poorer with the addition of the coupled coefficient g. So we have to increase the value of n.

Next we give the spectrum of the two-photon QRM when the level splitting of the atom is 0. See the Figure.

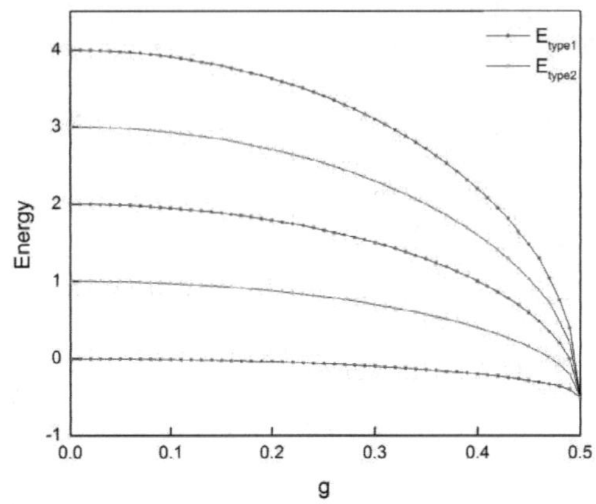

Figure 1: Spectrum of the two-photon QRM according to the Eq (24) and Eq. (25).

In conclusion, the regular spectrum by diagonalizing the Hamiltonian has common top form with the Ref [6] and Ref. [8]. According to the above discussion and the link with Jie Peng's solution [7], we confirm the validity of diagonalization for the Hamiltonian in the two-photon QRM. Next we will implant this method to apply to more systems, such as Dicke model and others, to obtain more valuable conclusions. Extensions to the QRM are in progress.

4. Acknowledgments

We acknowledge useful discussions with Jie Peng for his patient answer and feasibility advice.

5. References

[1] D. F Walls and G. J. Milburn, *Quantum optics* (Springer, 2007).
[2] J. Gambetta, A. Blais, M. Boissonneault, et al., 2008 Phys. Rev A 77, 12112.
[3] F. Brennecke, T. Donner, S. Ritter, et al., 2007 Nature 450, 268.
[4] D. Braak, 2011 Phys. Rev. Lett. 107, 100401.
[5] V. Bargmann, 1961 Comm. Pure Appl. Math.14, 187.
[6] C. Emary and R. Bishop, 2002 J. Phys. A: Math. Gen. 35, 8231.
[7] Jie Peng, Zhongzhou Ren, Guangjie Guo, Guoxing Ju and Xiaoyong Guo, 2013 Eur. Phys.J. D 67, 162.
[8] Travěnec, 2012 Phys. Rev. A 85, 043805.
[9] Q.-H. Chen, C. Wang, S. He, et al., 2012Phys. Rev. A 86, 023822.
[10] Fulton R L and Gouterman M 1961 J.Chem. Phys. 35 1059.

Providing A Long-term Protection for NCM811 Cathode Material by Al_2O_3 Coating Layer

Jing Wang[1,2,3], Qiang Yuan[1], Ran Wang[1], Guoqiang Tan[1,2], Yuefeng Su[1,2,3], Danhua Li[1] and Feng Wu[1,2,3]

1 School of Materials Science & Engineering, Beijing Key Laboratory of Environmental Science and Engineering, Beijing Institute of Technology, Beijing, 100081, China
2 National Development Center of High Technology Green Materials, Beijing, 100081, China
3 Collaborative Innovation Center of Electric Vehicles in Beijing, Beijing, 100081, China
Email: wangjingbit98@bit.edu.cn

Abstract. The degradation of Ni-rich $LiNi_{0.8}Co_{0.1}Mn_{0.1}O_2$ cathode material is successfully suppressed via a facile Al_2O_3 coating method. The coated Al_2O_3 layer on the electrode surface is inactive and amorphous, it does not react with the electrolyte and could provide a long-term protection on the electrode during cycling. The electrochemical performance of the modified NCM811 was improved dramatically both at room and elevated temperature. The capacity retention of pristine NCM811 was enhanced from 81.8% to 91.7% when the coating amount reached to 2 wt% at 25°C and much increased from 68.5% to 87.7% at 55°C. Such excellent high temperature performance clearly confirmed the effect of Al_2O_3 on the structure stability of electrode, it successfully built a favorable protection of the bulk NCM811 from direct contact with electrolyte and suppressed the undesired side reactions.

1. Introduction

Over recent years, the rechargeable lithium-ion batteries (LIBs) have widely used as the power sources for portable electronics, electric vehicles (EVs) and massive energy storage systems (ESSs). Thus, in turn, a lithium-ion battery system with higher energy density, longer cycle life, better safety and lower cost are in urgent demand [1,2] Layered Ni-rich compounds with the general formula $LiNi_xCo_yMn_{1-x-y}O_2$ (NCM, where x>0.5, x+y+z=1) are considered as most promising cathode materials for the next-generation LIBs because of their relatively high capacity (>180 mA h g^{-1}) and some of them have been applied in electric vehicles [3,4] However, the high content of Ni usually causes rapidly capacity fading and severe structure deterioration, and the major reason is the side reactions between electrode and electrolyte at the electrode surface, such as the dissolution of transition metal ions from the cathode into the electrolyte, and the solid electrolyte interface (SEI) formation on the cathode surface [5,6] Moreover, these harmful side reactions are usually accelerated at high temperature and in the deeply delithiated state at elevated voltages (>4.3 V).

To overcome the above problems, many works have been performed in recent years, and the method of surface coating is a very effective strategy to improve the cycling stability and thermal stability of the Ni-rich cathode materials. The improvements are mainly ascribed to restricting the direct contact between the electrode and electrolyte, protecting the NCM cathode materials from the attack by HF and reducing the undesired side reactions, thus a stable interface is received during long cycling. Unfortunately, many coating materials are not inactive and easily react with some components of electrolyte, leading to other extra side reactions and can't support as a persistent protect layer during extensive cycling, these materials are called as sacrificial protect layer. Al_2O_3 is chemically inert and is a desirable choice for coating the Ni-rich layered cathode material, and can provide a persistent protection during long cycling. This study therefore explores the effect of Al_2O_3 surface coating on the electrochemical and physicochemical characteristics of a NCM811 cathode.

2. Experimental

2.1. Materials Synthesis

The $Ni_{0.8}Mn_{0.1}Co_{0.1}(OH)_2$ precursor was synthesized by a conventional co-precipitation method, and then the precursor was mixed with $LiOH{\cdot}H_2O$ (5mol% excess amount), then sintered at 550 °C for 5 h preliminarily and calcined for 15 h at 750 °C in O_2 to obtain NCM811 cathode (denoted as pristine).

To produce the Al_2O_3 coating on the NCM811 cathode, a certain amount of aluminum is opropoxide was dissolved in 15 mL ethanol and then 1 g of the prepared NCM811 powder was slowly added and stirred at a certain speed. The obtained mixture was dried at 60°C and then sintered at 500°C for 5 h. The prepared materials with different coating content were denoted as A-1, A-2 and A-3, respectively.

2.2. Materials Characterization

The crystalline phases of the prepared materials were characterized by powder X ray diffraction (XRD) at the scan rate of 1° 2θ/min and ranged from 10° to 90° (Rigaku UltimaIV-185 instrument). The morphologies were observed by field emission scanning electron microscopy (FE-SEM, FEI QUANTA 250). High-resolution transmission electron microscopy (HR-TEM) was operated upon a JEOL JEM-2100 instrument.

2.3. Electrochemical Measurement

The electrochemical properties were measured in the CR2025 coin type half-cells for all samples. The working electrodes were prepared by coating slurry containing 10 wt% acetylene black, 80 wt% active materials and 10 wt% PVDF onto current collector (aluminum foil) and dried at 80°C for 24h, then punched into discs with 11mm diameter. The electrolyte was 1M $LiPF_6$ dissolved in ethylene carbonate, ethyl methyl carbonate and dimethyl carbonate (EC/EMC/DMC = 1:1:1). The cells were assembled in an Ar-filled glovebox and aged for 24h. The electrochemical test was measured between 2.8 and 4.35 V (versus Li/Li^+) on a CT2001A Land instruments. Electrochemical impedance spectroscopy (EIS) was measured by a CHI604D electrochemical workstation with a frequency range of 100 kHz to 0.01 Hz.

3. Results and Discussion

X-ray powder diffraction (XRD) was used to measure the crystal structure of different samples. As shown in figure 1a, the peak positions of all materials were indexed to the hexagonal α-$NaFeO_2$ layered structure with R-3m space group. No any other impurity peaks appeared in the all patterns and the splitting of (006)/(102) and (108)/(110) peaks, indicating that the treatment of aluminum is opropoxide has no influence on the bulk crystalline structure. However, the Al_2O_3 diffraction peaks were not appeared in the three modified samples because of their amorphous structure and low content of Al_2O_3 layer. Furthermore, the lattice parameters of pristine and A-2 are further determined by Rietveld refinement and the results are presented in figure 1b, c.

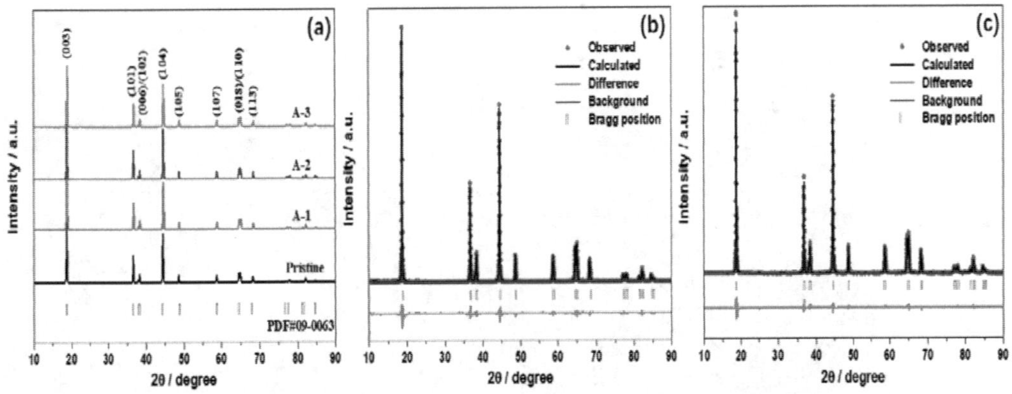

Figure 1. (a) XRD Patterns of Samples and Rietveld Refinement of (b) Pristine and (c) A-2

Table 1.The values of R_p and R_{wp} factors are lower than 15% indicating that the refinement is acceptable. The higher value of c/a than 4.94 shows that all samples preserved a well-defined layer structure after Al_2O_3 modification, moreover, the value of I_{003}/I_{104} (>1.2) reflects a low degree of Li^+/Ni^{2+} cation mixing. In a word, the XRD measurement demonstrated that the Al_2O_3 modification did not destroy the crystal structure on NCM811.

Table 1. Rietveld Refinement Results of NCM and A-2 Materials

Sample	a(Å)	c(Å)	v(Å^3)	c/a	R_{wp}(%)	R_p(%)	I003/I104
Pristine	2.87002	14.18851	101.2102858	4.9437	11	7.09	1.44
A-2	2.870836	14.19216	101.29389	4.9436	13.3	8.49	1.47

Scanning electron microscope (SEM) and high-resolution transmission electron microscopy (HR-TEM) were used to obtain the morphology of different samples and also to verify the existence and distribution of Al_2O_3 on the surface of NCM811 particles. As shown in figure 2, the NCM811 secondary particle has a spherical shape with a particle size of 6-8 μm and is composed of numerous nano-sized primary particles. Through comparing figure 2a to figure 2b-d, we found that there are no obvious changes in the overall morphology before and after modification, which indicates that the Al_2O_3 coating has no influence on the general morphology of NCM811 particles. However, with the increase of coating content, a thin and blurred Al_2O_3 layer has been clearly observed on the surface of the primary particles. Moreover, the EDS mappings of figure 2e-i further confirm the uniform distribution of Al_2O_3 on the particle surface.

Here, the HRTEM measurement was carried out to further investigate the microstructure and the TEM images of the pristine and A-2 samples as shown in figure 2j-m. Figure 2j reveals that the pristine NCM811 is rather smooth and the planar distance in figure 2k is 0.471 nm which corresponds to the (003) plane of pristine NCM811 material. In contrast, there is a clear coating layer on the primary particle surface in the A-2 sample and its thickness is about 2 nm, but any lattice fringes corresponding to Al_2O_3 are not found, indicating that the Al_2O_3 is amorphous.

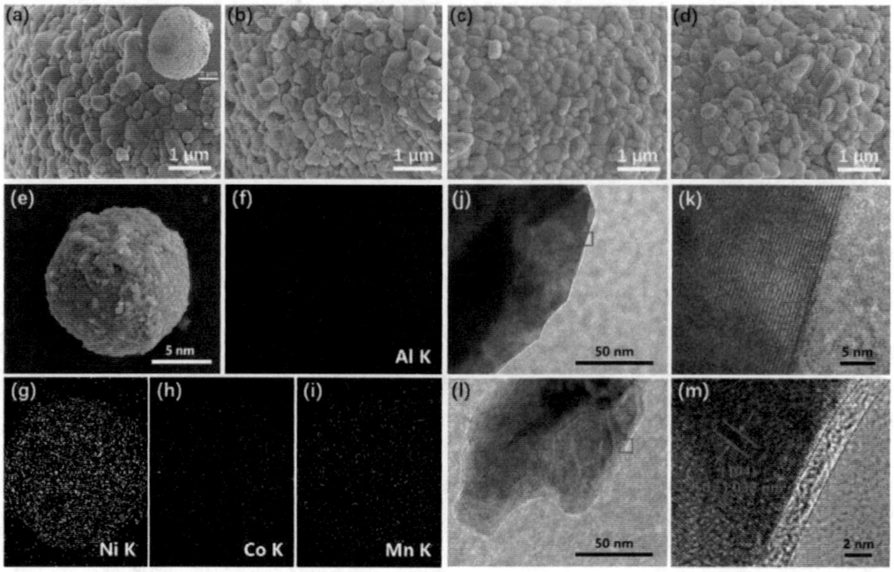

Figure 2. (a-d) SEM Images of Pristine, A-1, A-2 and A-3 Samples, Respectively. (e-i) EDS Mapping of A-2. (j) TEM and (k) HRTEM Images of Pristine $LiNi_{0.8}Co_{0.1}Mn_{0.1}O_2$. (l) TEM and (m) HRTEM Images of A-2.

The electrochemical performances of the pristine and modified samples were then tested as shown in figure 3. Figure 3a displays the initial charge-discharge profiles of all samples at 1C between the voltage range of 2.8-4.35V. Clearly, all samples show a similar electrochemical charge/discharge behavior, indicating that the Al_2O_3 coatings do not induce obvious changes to the bulk of NCM811 material. However, the initial capacities of different samples are decreased with the improvement of Al_2O_3 coating content and are 185.4, 183.6, 182.1 and 172.2 mAh/g for pristine, A-1, A-2 and A-3, respectively. The decreased initial discharge capacity is attributed to the electrochemically inert nature the Al_2O_3 coating layer. Figure 3b illustrates the long cycling performance of the four different samples at the current of 1C between 2.8 and 4.35 V. It is obvious that the capacity stability is significantly improved after Al_2O_3 modification. After 100 cycles, the pristine NCM811 delivers a discharge capacity of 151.7 mAh/g and the capacity retention is 81.8%, however, the other three samples all display a higher discharge capacity than the pristine NCM811, 158.4, 167.5 and 163.7 mAh/g for A-1, A-2 and A-3, respectively, and the corresponding capacity retentions are 89.4%, 91.7% and 92.4%, suggesting that the Al_2O_3 coating layer is beneficial for the structure stability of electrode interface.

Figure 3c, d show the cycling performance of different samples at 55°C. Note that the charge and discharge capacities are all improved at elevated temperature, but the capacity retention is decreased. The capacity retention of pristine NCM811 rapidly decreased from 81.8% of 25°C to 68.5% of 55°C after 100 cycles, implying that the undesired side reactions between electrode and electrolyte is accelerated at higher temperature. However, the effect of Al_2O_3 coating on the cycling stability is more obvious at high temperature, the capacity retention increased from 68.5% of pristine to 87.7% of A-3. But when the coating content beyond 2 wt%, the initial reversible capacity drops dramatically, so we consider that 2 wt% is the optimal coating value.

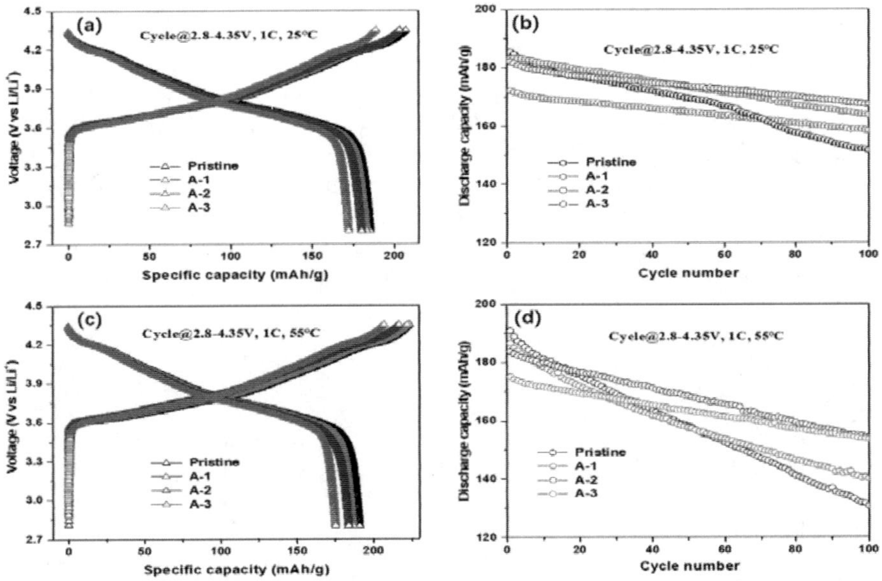

Figure 3. (a)The Initial Charge–Discharge Curves and (b) Cycling Performance at 25°C; (c) the Initial Charge–Discharge Curves and (d) Cycling Performance at 55°C

To further confirm the effect of the Al_2O_3 coating on the electrochemical properties of the NCM samples, EIS measurements were conducted for the pristine and Al_2O_3 modified electrodes during 100 cycles at room temperature, as depicted in figure 4. It can be found that every Nyquist plot contains two semicircles and a slope. The semicircle at high frequency corresponds to resistance of solid electrolyte interphase film (R_f). The second semicircle at middle high frequency refers to charge transfer resistance (R_{ct}), which reflects the transfer speed of Li^+ on the interface of electrode materials. The sloped line at low frequency represents the diffusion rate of Li^+ in bulk electrode and was defined as Warburg impedance. [7] The fitted results of Nyquist plots are shown in figure 4c.

After 1st cycle, the R_{ct} is 13.7 Ω for pristine NCM and similar with 11.6 Ω for A-2, indicating that the Al_2O_3 coating layer does not influence the Li^+ transfer through the electrode interface. However, the R_f for pristine NCM811 is 7.82 Ω, which is more than three times that of A-2, the difference is attributed to the restriction of side reactions that a thin SEI film is formed. After long cycles, both R_f and R_{ct} values for all samples are increased, but the increasing rate is different. After 100 cycles, the R_f and the R_{ct} increased to 54.8 Ω and 554.4 Ω for pristine NCM811, respectively. Such large impedance would result the severe degradation of cycling performance for pristine NCM811. However, the value of R_f and R_{ct} are just 23.1 Ω and 147.2 Ω for A-2, which indicates the long-term protection of Al_2O_3 coating layer on the electrode.

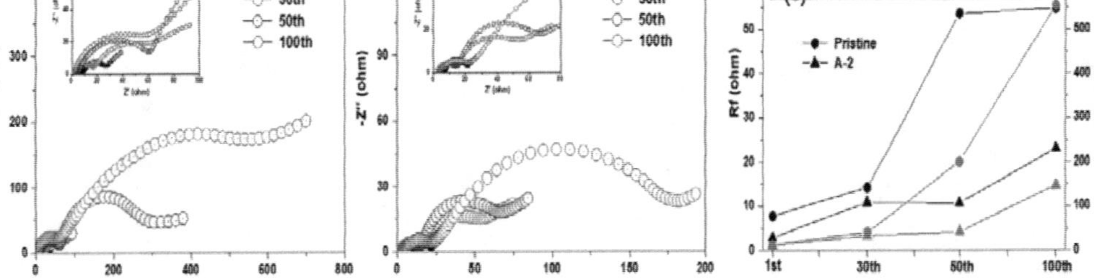

Figure 4. Nyquist Plots of (a) Pristine and (b) A-2 after Different Cycles, (c) the Fitting Results according to the Equivalent Circuit in the Inset

4. Conclusion

We have successfully improved the electrochemical properties of NCM811 via a uniform Al_2O_3 coating layer. According to the XRD, SEM and HRTEM analysis, we know that the Al_2O_3 coating layer does not influence the morphology and crystal structure of the NCM811, and exists as an amorphous phase. From the electrochemical characterizations, we confirm that the presence of Al_2O_3 at the electrode surface can suppress the undesirable side reactions and provide a long-term protection upon cycling due to its inactive nature, thus an improved cycling stability was achieved, especially at elevated temperature. Based on the above results, we believe that this modified NCM811 is very promising to be used in commercial power lithium-ion batteries.

5. Acknowledgment

This work was financially supported by the National Key R&D Program of China (2016YFB0100400) and Special fund of Beijing Co-construction Project.

6. Reference

[1] Arumugam M, Bohang S and Wangda L 2017 *Energy Storage Materials* **6** 125-139
[2] Yuefeng S et al. 2018 *ACS Appl. Mater. Interfaces* **10** 6407-14
[3] Hao W, Wujie G, Wen L, Feng W, Wenjing L, Meizhen Q and Gongchang P 2016 *ACS Appl. Mater. Interfaces* **8** 18439-49
[4] Chong S Y, Min-Jae C, Do-Wook J, Qian Z, Payam K, Kwang-Ho K and Yang-Kook S 2018 *Chem. Mater.* **30** 1808-14
[5] Bin H, Xinhai L, Zhixing W, Huajun G, Li S and Jiexi W 2018 *Journal of Power Sources* **252** 200-207
[6] Guorong H, Xianyue Q, Kaihua H, Xiangwan L, Xiang Z, Ke D, Zhongdong P and Yanbing C 2018 *Electrochimica Acta* **265** 391-399
[7] Qiyu Z, Yuefeng S, Lai C, Yun L, Liying B, Tao H, Jing W, Renjie C, Jing T and Feng W 2018 *Journal of Power Sources* **396** 734-741

Research Progress on High-temperature Oxidation of Austenitic Stainless Steel for Ultra-superCritical Boilers

Tao Song[1] and Zhiwu Wang[2]

1. Zhixing College, Hubei University, room 213, Xingsheng Road, 1th sector, Wuhan, China
2. School of Power and Mechanical Engineering, Wuhan University, room A9423, South Road of East Lake, 8th sector, Wuhan, China
Email: wzw1618 @163.com

Abstract. In this paper, four typical austenitic heat-resistant steels (TP304H steel, TP347H steel, Super304H steel and HR3C steel) are taken as examples to review the research progress of high-temperature oxidation of austenitic stainless steel for ultra-supercritical unit boilers, and the mechanism and protective measures of high-temperature oxidation of austenitic stainless steel are discussed, so as to provide references for material improvement and safe use of unit boilers.

1. Introduction

In recent years, with the rising demand for electricity, efficiency and reduce the pollution of coal-fired power plant put forward higher requirements, improve the parameters of power plant boiler is a kind of effective method to solve the problem, the development of the ultra-supercritical unit becomes the hotspot technology development at home and abroad [1-2]. The ultra-supercritical unit is a boiler whose parameters are raised to a steam temperature not less than 593°C or a steam pressure not less than 31 MPa [3-4]. The improvement of the parameters can increase the thermal efficiency of thermal power plant to 48%, and reduce the coal consumption per kilowatt-hour of power supply to 256g [5-6]. The improvement of boiler parameters of the unit also greatly improves the material performance requirements of the key parts, namely the "four tubes" of the boiler (superheater tube, reheater tube, water-cooled wall tube and economizer tube) [7-8].Austenitic stainless steel has become the preferred material for ultra-supercritical unit boiler steel due to its excellent high-temperature strength, high-temperature endurance strength and high-temperature oxidation resistance [9-10].Austenitic stainless steel has been in service for a long time in high-temperature and high-pressure environment. Oxidation of pipe wall and peeling off of oxide skin have always been the key to the service life of materials, which has attracted a large number of researchers to study the high-temperature oxidation of austenitic stainless steel [11-12].

In this paper, the main austenitic stainless steel as the object, the research status and protection mechanism of high temperature oxidation of different kinds of austenitic stainless steel were reviewed, and the existing problems of high temperature oxidation resistance of austenitic stainless steel were discussed, in order to provide reference for the improvement of materials and anti-oxidation protection in the future.

2. High Temperature Oxidation of Austenitic Stainless Steel for Ultra-supercritical Boiler

2.1. TP304H Stainless Steel

TP304H steel belongs to type 18-8 chromium-nickel austenite heat-resistant steel, which has good bending and welding properties, high durable strength, corrosion resistance and microstructure stability. The maximum service temperature of the material can reach 650°C, and the maximum oxidation resistance temperature can reach 850°C [13-14].

TP304H steel were studied at 560°C, 590°C and 620°C temperature oxidation behaviour by Ming-li Xu in Xian university of technology [15], it is found that TP304H steel oxidation kinetics equation follow Δm = KTZ parabolic law, test temperature material does not appear in the process of oxidation layer peeling situation, at the same time, the study found that the TP304H steel surface oxide growth way is: first generate granular oxide surface, and then generate crystal in oxide on the surface of objects, these crystals lateral growth form continuous oxide film layer, crystalline oxide grew up further, started the longitudinal grew up, small holes formed between crystalline oxide, with the extension of time, the crystalline oxide continue to grow up, the hole between the disappear, then on the crystalline oxide generated a lot of whisker. And with the increase of oxidation temperature, this growth process is faster.TP304H inner wall oxide is composed of two layers of oxide, the outer layer oxide is loose, the outer layer oxide is Fe_2O_3, the inner layer oxide is dense, and the inner layer is composed of (Fe, Cr) spinel oxide and NiO. The interface between the inner and outer oxides corresponds to the original metal surface.

Hongfei Feng [16] studied the double-layer structure of TP304H steel oxide film and found that the inner layer contained Cr elements and the outer layer only contained oxygen and Fe elements. Moreover, continuous holes were often formed on the interface between the inner and outer layers. The growth mechanism of the outer oxide layer is as follows: firstly, granular oxide is generated on the surface; then, it grows into continuous crystalline oxide layer; then, the grains grow vertically and pores appear; then, as the grains grow horizontally, the pores disappear and the oxide film continues to thicken. The growth process of inner layer is as follows: due to the presence of Cr in inner layer, Cr rich band will be generated in the growth process, which can effectively reduce the growth rate of oxide layer. However, due to the slow diffusion of Cr elements in the grain, it is difficult to form a continuous Cr rich belt, and the oxide layer can continue to grow beyond the Cr rich belt, and often forms a Cr rich black spot. Due to the rapid diffusion of Cr element, continuous and dense Cr-rich layer is often formed at the grain boundary, so it is difficult for the grain boundary to be oxidized, and finally, long and narrow white unoxidized gap or closed nickel-rich region is often formed.

Xingeng Li [17] and others studied shot peening in TP304H steel high temperature steam oxidation behavior of function, the results show that the surface peening deformation can effectively slow oxidation kinetics in 0.5 MPa shot peening intensity and improve the effect of antioxidant properties under the condition of 730°C, the best surface oxidation film grains are refined, Cr in membrane relative concentration increased significantly, oxide film phase composition by pure Fe and Fe, Cr mixed oxide into Cr_2O_3.It is concluded that the treatment of surface shot peening increases the concentration of dislocation, vacancy and other defects in the surface area of the alloy, improves the nucleation density in the early oxidation stage, provides a large number of short circuit diffusion channels for Cr, and reduces the critical concentration of selective oxidation of Cr. The short-circuit diffusion of Cr on the surface layer of matrix ensures the diffusion flux of the stable growth of Cr_2O_3 film, thus inhibiting the formation of Fe oxide.

2.2. TP347H Stainless Steel

TP347H steel is a niobium stable chromium-nickel austenite heat-resistant steel with high thermal strength and intergranular corrosion resistance, good oxidation resistance and good elbow and welding performance [18-19].

Dongbin Wei [20] and others using home-made lab under 800°C high temperature steam oxidation device of TP347H steel tube sample simulation test and high temperature steam oxidation, TP347H steel high temperature steam oxidation organization and composition were studied, and the results show that the TP347H steel at 800°C water vapor oxidation kinetics follows the laws of parabola, the oxide film of test samples by scanning electron microscopy (SEM) and energy spectrum analysis showed that the structure of the oxide film is divided into inside and outside two layer, in the outer layer, oxygen content gradually reduce from outside to inside, Fe content gradually reduce from the inside out; The outer oxide film only contains trace Cr and Ni elements, while Cr elements in the inner layer are enriched at the interface between matrix and inner layer, while Ni elements in the inner layer show correlation dilution.

Yanming Chen [21] and others, using analog device designed steam oxidation experiments under 700°C for 1.5, 2 and 4 h 4 oxidation experiment was carried out at different times of the early stage of TP347H steel oxidation behavior research, the results showed that under 700°C, TP347H fast oxidation, surface covered by tiny particles completely first, then give birth to many island structure, and on the island structure will grow small whisker. EDS and Raman results show that the oxide layer consists of inner and outer layers, which are mainly composed of Cr_2O_3, NiO, Fe Cr_2O_4 and Fe Cr_2O_4.

Qiang Ma et al. [22] studied the high-temperature vapor oxidation layer after 5000h operation of TP347H tube in the screen superheater of 600MW supercritical unit. The results showed that the volatilization of $CrO_2(OH)_2$ led to the formation of Fe oxide on the outer layer, and the formation of $FeCr_2O_4$ inner layer could effectively slow down the process of vapor oxidation, and the thickness of oxide layer would remain unchanged after entering the steady state oxidation stage. At the same time, the vapor oxidation resistance of TP347H was improved with the decrease of grain size.

2.3. *Super304H stainless Steel*

Super304H steel is improved and developed on the basis of TP304H steel. By adding Cu to form Cu rich phase, creep fracture strength of the material is greatly improved. By adding Nb and N elements in the composite, high-temperature strength and durability of the material are further improved [23-24]. Xuedong Li and et al. [25] studied the high-temperature steam oxidation behavior of Super304H steel, and the results showed that the oxidation rate of Super304H steel was close to a straight line at 400°C, while the oxidation rate was closer to a parabola when it was higher than 400°C. Super304H steel oxide film has a double-layer structure, and the outer oxide film particles are thick and loose, mainly composed of Fe_3O_4, while the inner oxide film particles are small and compact, mainly $FeCr_2O_4$.

Dongsheng Li and et al. [26] studied the high-temperature oxidation resistance of Super304H austenitic stainless steel by weighing method, SEM, XRD et al. It is found that the oxidation kinetics curves of the two stainless steels follow the parabolic law. Super304HS has a better oxidation resistance than Super304H, and Super304H oxidized 100 at 900°C H, the oxide film obvious loss. The method of using scanning electron microscopy (SEM), X-ray diffraction (XRD) on Super304H stainless steel on the surface of the oxide film morphology and structure were studied, the results show that at 700°C and 800°C, the two materials of oxide film, are $FeCr_2O_4$ Cr_2O_3 and small amounts of spinel structure, Super304H oxidation product of steel at 900 ℃ is mainly composed of Cr_2O_3, Fe_2O_3 and spinel $FeCr_2O_4$ Super304HS at 900°C of oxide film is mainly composed of Cr_2O_3 and spinel $FeCr_2O_4$.

Ming Le [27] studied the influence of surface modification technology on Super304H steam oxidation behavior, and the research results showed that :(1) surface modification methods, such as shot peening and aluminizing, could significantly improve the anti-steam oxidation ability of Super304H.When Super304H carried out steam oxidation in water vapor with dissolved oxygen concentration of 10ppb, its oxidation kinetics curve tended to be linear, and the oxidation film was mainly single-layer Cr_20_3.When the surface shot peening samples were oxidized in the water vapor environment with dissolved oxygen concentration of 10ppb, the oxidation kinetics curve was parabolic and the oxide film was monolayer Cr_2O_3.When the surface aluminizing sample was oxidized in the water vapor environment with dissolved oxygen concentration of 10ppb, its oxidation kinetics curve was parabolic and the oxide film was single layer $A1_2O_3$.(2) with the increase of steam temperature,Super304H oxide film gradually changed from single layer to double layer structure. The oxidation behavior of surface shot peening samples and surface aluminizing samples was not affected by temperature. (3) Dissolved oxygen concentration plays an important role in the oxidation weight gain, the morphology and composition of the oxide film of Super304H samples. When the concentration of dissolved oxygen in the steam rose to 5ppm, the oxide scale of Super304H sample developed a large-scale peeling after 500h of steam oxidation. The anti-vapor oxidation capacity of the surface shot peening samples was less affected by the concentration of dissolved oxygen. With high concentration of dissolved oxygen, the parabolic rate constant of the surface shot peening samples increased, and the oxidation activation energy increased. The vapor oxidation resistance of surface aluminized samples was not affected by the concentration of dissolved oxygen. Under the condition of high dissolved oxygen concentration, the parabolic rate constant of the surface shot peening sample increases and the oxidation activation energy increases.

Jiangbin Wang [28] studied the oxidation characteristics of Super304H stainless steel. Based on the supercritical water experiment platform, experiments were carried out in supercritical water environment with different temperatures and dissolved oxygen for Super304H austenitic steel. Through the analysis of oxidation weight gain, oxide surface morphology, oxide cross section EDS analysis figure and XRD. The effects of temperature and dissolved oxygen on the growth rate and microstructure of oxide film were investigated. It was found that the oxidation rate increased with the increase of temperature and dissolved oxygen. The oxide film is a typical three-layer structure. The outer layer is rich in iron, the inner layer is rich in chromium, and the inner layer is diffused. With the increase of temperature and dissolved oxygen, the amount of holes in the outer oxide film increased gradually. In the early stage of oxidation, nodular oxides will form, and with the increase of temperature and time, these nodular oxides will eventually completely cover the metal surface. The oxides formed in the absence of oxygen are mainly Fe_3O_4 and (Fe, Cr)$_3O_4$.Fe_2O_3 is formed in the outermost layer of the environment containing 2000ppb dissolved oxygen. In this paper, the stress factors leading to the flaking of the oxide film are analysed, and the two kinds of wedge-type flaking and buckling flaking under tensile stress are explained. It is also discussed that the temperature fluctuation and the thickness change of oxide film have significant influence on the spalling.

2.4. HR3C Stainless Steel

HR3C steel is a new type of austenitic heat-resistant steel developed by composite addition of Nb and N alloy elements on the basis of TP310, with high high-temperature strength and excellent comprehensive performance [29-30].

Zhen Yang and et al. [31] studied the high-temperature oxidation behavior of HR3C in air and water vapor at 750°C, and the results showed that continuous (Cr, Mn)$_2O_3$ films were formed on the surface of HR3C in two atmospheres. The oxidation kinetics of HR3C alloy in static air presents a piecewise parabola law. The existence of water vapor makes the (Cr, Mn)$_2O_3$ film rupture advance and greatly accelerate the oxidation of the alloy, which leads to the overall deviation of the oxidation kinetics of HR3C in the pure water vapor from the parabolic law. (Cr, Mn) the rupture of (Cr, Mn)$_2O_3$ film is the result of the combined action of the cavity in the interface between the film and the matrix and the growth stress inside the membrane, which makes the bare metal matrix react directly with the high-temperature air or water vapor to form Fe_3O_4 tumour.

Tong Lee-ying [32] huaneng yuhuan power plant are analyzed, at the end of the operation of a 5400 h and 23400 h HR3C level reheater tube steam side oxide layer of micro structure, morphology and distribution pattern of the elements of the results found that with the deepening of the oxidation, lateral HR3C steam oxide layer by layer structure developed into four layers structure, the poor Cr in the base metal near the oxide layer is the growth of the oxide layer shortcut.

3. High-temperature Oxidation Mechanism and Protection of Austenitic Stainless Steel for Ultra-supercritical Boiler

3.1. Oxidation Mechanism

Yamin Huang and et al. [33] studied the oxidation failure mechanism of austenitic stainless steel at ultra-high temperature, and the results showed that: the oxide film on the surface of stainless steel has a multi-film structure, and the oxidation environment and element distribution and diffusion at the interface between matrix and oxide layer have an important impact on the structure and performance of the oxide film. The failure behavior of the oxide film is mainly manifested by the generation of cracks in the oxide layer and the spalling of the non-dense oxide layer. Hong Xu and et al. [34] studied the regrowth and spalling behavior of oxide on the surface of austenitic heat-resistant steel. The results showed that iron ions were diffused outwards through complex fe-cr oxide to form the oxide outer layer dominated by iron oxide. When the outer oxide layer grows to a certain extent, local spalling occurs due to the size difference between the inner and outer oxide film oxide particles.

3.2. Protective Measures

The steam oxidation resistance and oxide stripping performance of steel should be considered in the design. Internal shot peening can produce crushed austenite grains and a large number of slip bands near the inner surface of the steel tube. In addition, due to the surface defects of shot peening, the nucleation density and growth rate of chromium oxide are increased, which is conducive to the formation of a single Cr_2O_3 film. During shot peening, we should pay attention to the strength of shot peening. Too low strength will lead to the diffusion and enrichment of chromium and no significant change in the nucleation rate of oxide. If the strength is too high, it will promote the diffusion of chromium as well as accelerate the diffusion of iron and oxygen ions. Both of them will cause the failure to form a single Cr_2O_3 film on the inner wall of the steel tube [35].

The grain size of the material can be controlled by controlling the amount of rolling deformation and the final solid solution treatment in the manufacturing process of the raw material. The larger the degree of cold deformation of the material, the finer the grain size will be, and the better its oxidation resistance will be. Therefore, the grain size should be refined as far as possible in the manufacturing process under the condition that other properties meet the requirements [36].

4. Summarize

High temperature oxidation of austenitic stainless steel has always been a hot topic in the industry. However, with the increasing of temperature, the oxidation of austenitic stainless steel becomes an inevitable problem, and there are many factors affecting the oxidation. Although there are a lot of researches on oxidation of austenitic stainless steel, there are always difficulties in this problem. The in-depth study of materials is of great significance to the development of materials science and engineering practice, and is also the focus of attention in the future.

5. References

[1] Aderhold J, Davydov V Yu, Fedler F, Klausing H, Mistele D, Rotter T, Semchinova O, Stemmer J and Graul J 2001 *J. Cryst. Growth* **222** 701
[2] Zhesheng J, Wei-guo D and Guoguang M 2007 *Electric Power Construction* **28** 6
[3] Baotian Z and Rongcan Z 2009 *Proceeding of the CSEE* **S1** 95
[4] Weizhong F 2008 *East China Electric Power* **36** 6
[5] Baotian Z, Naijin M, Zhaotuan L and et al. 2005 *Thermal Power Generation* **34** 1
[6] Hancai X and Zhen Z 2004 *Power System Engineering* **20** 8
[7] Yaning Y2004 *Power System Engineering* **20** 8
[8] Jianxiong M 2013 *Electric Power Construction* **34** 69
[9] Changxin F, Hongjun Z, Rongcan Z and et al 2006 *Electrical Equipment* **7** 11
[10] Zhongping Z, Mingfang Y 2000 *Power Engineering* **20** 640
[11] Xueping M, Gang W, Zhiyong M 2005 *Modern Electric Power* **22** 69
[12] Cheng J, Zou C,Qianjin L,Shibo L,Yunpeng W,Yang L 2019 *Combustion and Flame* **209** 56
[13] Zhongping Z 2000 *Materials for Mechanical Engineering* **24** 1
[14] Wang Z, Li M, Ni J and et al. 2015 *Heat Treatment of Metals* **40** 121
[15] Ming L 2016 Xi 'an technology university
[16] Mingli X 2010 Xi 'an technology university
[17] Hongfei F 2011 Xi 'an technology university
[18] Xingeng L, Min F, Jiawen H 2004 Materials Science &Technology **12** 253
[19] Lingguang Z, Gang X, Xueshan X and et al. 2008 *Journal Of The Chinese Rare Earth Society* **26** 744
[20] Hongsheng C, Rui T, Chongsheng L and Guan L 2019 *Corrosion Science* **15** 134

[21] Dongbin W, Xiaoqiu L, Kelong H and et al. 2014 *Engineering Journal of Wuhan University* **47** 552
[22] Yanming C, Qiang W, Chang L and et al. 2015 *Journal of South-Central University for Nationalities(Nat.ci.Edition)* **3** 79
[23] Qiang M, Ping L, Shou-en Y and et al. *Transactions Of Materials And Heat Treatment* 2009 **30** 172
[24] Yan Y, Shichang C, Gang Y 2002 *Materials for Mechanical Engineering* **26** 23
[25] Xinmei L, Yong Z, Zhongwen Z and et al 2009 *Transactions Of Materials And Heat Treatment* **30** 51
[26] Xuedong L, Shuzhong W, Wen L 2013 *Journal Of Engineering For Thermal Energy And Power* **28** 386
[27] Dongsheng L, Qixun D, Guojian W and et al. 2012 *Journal of Jiangsu University(Natural Science Edition)* **33** 581
[28] Ming L, Hongzhe W, Zaisong Y and et al. 2016 *Thermal Power Generation* **45** 42
[29] Jinabin W 2016 Harbin Institute of Technology
[30] Yuanyuan F, Jie Z, Xiaona L 2010 *ACTA Metallurgica Sinica* **46** 844
[31] Junjian L, Guohong C, Jiaqing W and et al. 2011 *Journal Of Hefei University Of Technology* **34** 47
[32] Zhen Y, Jintao L, Xinbao Z and et al. 2015 *Journal of Chinese Society of Power Engineering* **35** 859
[33] Liying T, Jianmin J, Hui C and et al. 2011 *Corrosion and Protection* **32** 775
[34] Ya-min H, Youming W, Chunxu P 2009 *Journal of Zhengzhou University (Engineering Science)* **30** 53
[35] Hong X, Zhiyuan L, Hetai S and et al. 2018 *Journal Of Engineering For Thermal Energy And Power* **33** 93
[36] Zhaohe G, Justyna K M, Etienne B, Xun Z, Ying C, Han L, Peter K, Ping X 2019 *Thin Solid Films* 688.
[37] Limin W, Chao L, Yi C and et al. 2017 *PTCA (Part A: PHYS. TEST.)* **53** 731

China Network Public Opinion Guidance Research Hotspots and Development Situation Analysis-Quantitative Analysis Based on CiteSpace Knowledge Map

Yanfang Liu and Pengyao Wang

School of Management, Harbin Normal University, 309 Social Science Building, Harbin Normal University, No. 1 Shida Road, Limin Development Zone, Harbin
Email: c_lyf@163.com, wangpengyao1229@163.com

Abstract. Through the quantitative analysis of related literatures, it reveals the research hotspots of china network public opinion to guide this research field, and provides theoretical research for the further development of academic research in this field and the improvement of China's network public opinion guidance work level. Using CiteSpace software to use 2009-2018 as the specified time period, the Chinese literature on the theme of "Network Public Opinion" in China Knowledge Network (CNKI) is used as the source data, based on CiteSpace's visualization technology, the key Word clustering maps, research organization timing maps, and institutional and author cooperation networks are analyzed. It is concluded that the focus of China's network public opinion guidance research mainly focuses on the four aspects of the main body, object, carrier and content of the network public opinion.

1. Introduction

With the development of information technology, network public opinion, which is based on the network for information exchange and dissemination, has begun to enter the public's attention. Network public opinion is a collection of public opinions and opinions on public events expressed through network platform [1], network public opinion is diverse and coexists with accumulation and elimination, Positive and positive network public opinion helps to build a harmonious network environment, negative network public opinion is very easy to cause the formation of network crisis. Therefore, for the healthy operation and development of the network as an information exchange platform, while fully guaranteeing the public's enthusiasm for participation in social affairs, do a good job of guiding online public opinion.

The guidance of network public opinion is a dynamic process that makes the attitude of network public change to the correct direction of network public opinion. In 2016, General Secretary Xi Jinping emphasized the need to improve the guiding power of news and public opinion at the Party's Symposium on news and public opinion work [2], in the latest report of the Nineteenth Congress, General Secretary Xi Jinping stressed once again the importance of strengthening content construction, establishing a comprehensive Network governance system [3]. Thus it can be seen, the guidance of network public opinion is an indispensable link in the development of network public opinion. it has its inevitability to become the focus of academic discussion and research.

2. Data Sources and Research Methods

2.1. Data Sources

In order to have a more comprehensive understanding of the research results on the guidance of network public opinion in China, this paper collects CNKI journals with the theme of "Network Public Opinion Guidance" from 2009 to 2018 as initial data. Search the topic "Network Public Opinion Guidance "in the column of journals, the designated period is from 2009 to 2018, a total of 2808 valid literatures were obtained through further screening. The selected literature covers a wide range and has a large number of representations, which can be used in this study.

2.2. Research Methods

Compared with other visual analysis software, CiteSpace can intuitively display the basic general situation and information of the research and development guided by network public opinion; it is more suitable for observation and analysis of a research topic [4]. Previous reviews in this field mostly analyzed and commented on the relevance and difference of related topics from the individual judgment of scholars, this way is easily influenced by the subjectivity of scholars themselves [5].

This study uses CiteSpace software to set Time Slicing as from 2009 to 2018. In node type, keywords, institutions and authors were selected and the top 50 variables were screened to generate associated maps. It can clearly show the research hotspots in this field, and understand the degree of influence and cooperation of network theory of institutions in recent years by observing charts.

3. Basic Survey of Research on Network Public Opinion Guidance

In this paper, Citespace is used to make a detailed econometric analysis from the annual volume of papers, disciplines, classification and influence of research institutions, cooperation between institutions and authors, keyword clustering and compactness.

3.1. Annual Volume of Papers Analysis

The number of publications is the primary indicator of research development in a field, the change of the number of publications with the years can clearly describe the hot issues and different stages of development in this field.Combining with Figure 1, we can divide the research into three stages: 2009-2012 is the high-speed development stage of our country's network public opinion guidance research, at this stage, the focus of network public opinion is more, for example, the events of Wukan and Guo Meimei in 2011 and the controversy between Fang Zhouzi and Han Han in 2012, these hotspots have aroused the interest of many scholars, which makes the volume of publications in this stage show a rapid growth trend. The second stage is from 2013 to 2014, during this period, the annual number of papers was maintained at about 300, which shows that the upsurge of research on the guidance of network public opinion has not yet receded. The third stage is from 2015 to 2018, at this stage, the volume of publications continues to grow but at a slower pace, but it also shows that the guidance of network public opinion in this field is still valuable and we need further attention and research. The research on the guidance of network public opinion in China is very fruitful, the attention received is also high, from the content of published articles, on the basis of the existing theory of network public opinion, experts and scholars began to pay more attention to the impact of network public opinion on people's social life, and began to emphasize that the government should adopt the necessary guiding means and effective guiding measures to help network public opinion develop healthily.

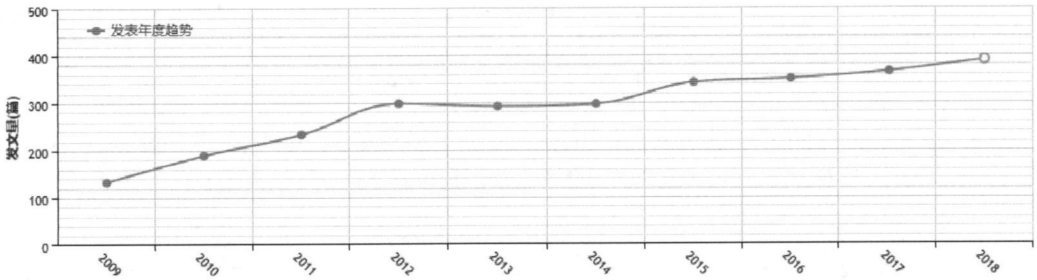

Figure 1. The Trend Map of the Number of Research Documents on Network Public Opinion Guidance in China from 2009 to 2018

3.2. Subject Analysis

Based on 2808 Chinese documents retrieved, this paper uses co-occurrence analysis to calculate the top ten subject names and the number of publications of the key word " Network Public Opinion Guidance", as shown in Table 1.Network public opinion guidance is a cross-disciplinary research area. Combined with Table 1, we can easily see that the subject of Journalism and communication occupies a considerable advantage in this research area. The number of publications of literature accounts for 72.90% of the total number of publications, which is far higher than that of other disciplines, followed by education and public management, which accounts for 16.42% and 9.47% respectively. It can be seen that at present, the research of network public opinion orientation in China mainly focuses on the fields of news, education and public administration (accounting for 98.79% of the total number of publications), while other disciplines do not pay enough attention to the research of network public opinion guidance, and the proportion of publications in the total number of publications is less than 2%.This kind of research situation makes our view on the guidance of network public opinion incomplete. To this end, scholars from other disciplines should be encouraged to enter the research field of guidance of network public opinion and put forward some views and suggestions from their own disciplinary point of view.

Table 1. Major Disciplines of Network Public Opinion Guidance Research in China

serial number	subject	number of documents	percentage /%
1	journalism and communication	2047	72.90%
2	education	461	16.42%
3	public management	266	9.47%
4	political science	154	5.17%
5	sociology	129	4.68%
6	public security	67	2.10%
7	law	59	1.78%
8	Marxism	48	1.78%
9	military science	41	1.45%
10	economics	26	1.13%

Experts and scholars in China conduct research on network public opinion guidance based on the theory of different disciplines. Through reading and sorting out a large number of documents, the research in this field can be divided into the following categories according to different disciplines:

Firstly, the relevant research based on the theory of news communication. Some scholars have explored the research of network public opinion guidance from the perspective of interrelated multi-disciplines [6]. It emphasizes that the government should correctly guide the role of network public opinion leaders in network public opinion, and plays a positive role in creating a healthy public opinion environment for society [7]. They also suggest that the network media should guide the network public opinion by establishing democratic interest expression [8].

Secondly, on the basis of educational theory research. In recent years, there have been many negative public opinion incidents in universities. It is very important for the harmonious development of campus environment to grasp the guidance of network public opinion in universities. Based on the public opinion environment of universities to analyze the current situation of guidance and explore guidance strategies, explore the practical guidance of network public opinion in Colleges and universities is the current research hotspot in this field [9]. At the same time, guiding the harmonious and healthy development of network public opinion is also a new problem faced by colleges and universities in psychological and ideological education for students [10].

Thirdly, relevant research based on public management theory. Improving network governance and strengthening the guidance of network public opinion are the important subjects of strengthening social management at present [11], each scholar and Research Institute firmly grasps the existing hot spot network public opinion event, proceeds from the realistic case, follows the intrinsic law of the system, carries on the research to the hot spot network public opinion dissemination characteristic and its guiding mechanism [12].

3.3. Classification of Research Institutions and Analysis of Their Impact

In order to find out the core research institutes of network public opinion guidance research in China, the number of papers published by various research institutes in this field between 2009 and 2018 is counted, among which the top ten are shown in Figure 2. These research institutes can be roughly divided into four categories: The first category is media colleges and universities, which mainly study how to guide network public opinion under different media. The second category is normal colleges and universities, which mainly study the management and guidance of campus network public opinion. The third category is comprehensive institutions of higher learning, which study the guidance of network public opinion from various angles such as sociology, economics and Library and information science. The fourth category is political and legal institutions of higher learning, which mainly studies the security issues in the dissemination of network public opinion and the legal norms of the guidance of network public opinion. Among them, China Media University takes the first place in the amount of publications with its great advantages, which fully demonstrates that this institution has a strong strength in the research of network public opinion guidance, and indirectly reflects its influence in the research of the direction of guiding news communication.

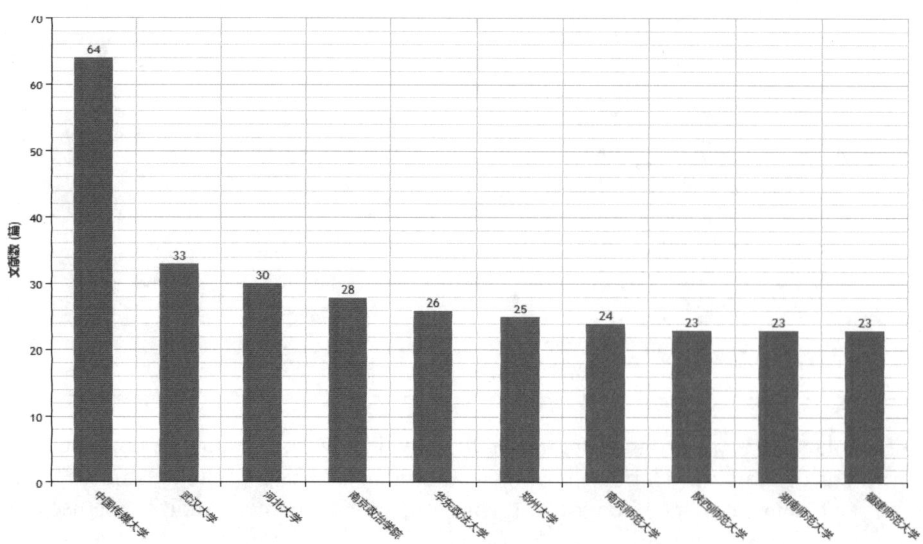

Figure 2. Top 10 Research Institutes in the Number of Papers Published between 2009 and 2018

To statistics the cooperation among research institutions in this field, the cooperation diagram of research institutions shown in Figure 3 is generated. The size of the circle is positively related to the number of articles sent. The larger the circle is, the more articles sent by the organization in that year.

It can be seen from the figure that the research institute of journalism and communication of Hebei University ranks first in the volume of publications in this field. The results in the upper left corner of the graph show that there are 178 related research institutions in the cooperative network, but the number of links indicating these inter-agency cooperative relationships is only 21, and the overall density of the network is only 0.0013. This shows that the research team of network public opinion guidance in our country is relatively scattered, and there is little cooperation research between the team and the team. In the future, research institutes in this field should strengthen their cooperation, further expand their disciplines, and conduct cooperative research of interdisciplinary and joint discussion,and on the basis of cross-disciplinary research, we should strengthen the integration of multi-disciplinary research ideas and research methods, and create more and better research results through multi-party collaboration for the frontier issues in the field of research.

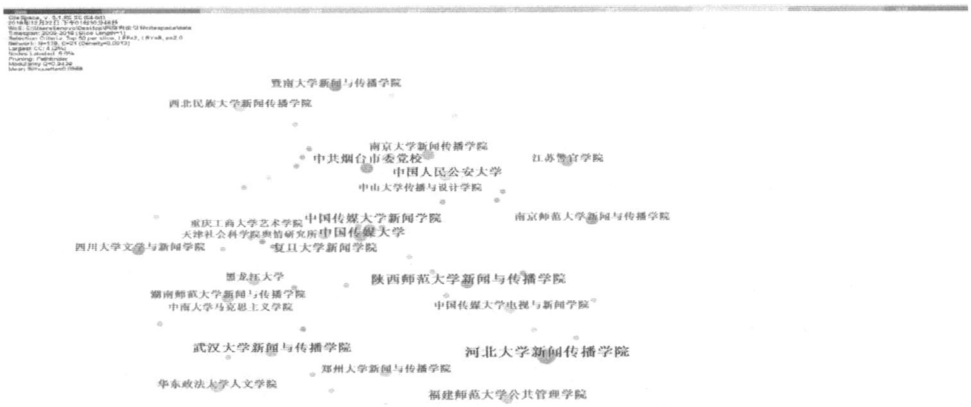

Figure 3. Research Institution Cooperation Diagram

3.4. Formatting Author Names

Authors who have published more than three times are regarded as leading authors in the field of research. From 2009 to 2018, The top three authors in this field are: Fan Jinshan (The CPC henan provincial party committee party school), Jiang Shenghong (Institute of Public Opinion of the Tianjin Academy of Social Sciences), Lin Ling (East China University of Political Science and Law). These authors are the main research force of their organizations in this field, and they have made a lot of contributions to the research of network public opinion guidance. Looking at the data in the author's cooperative map in Figure 4, we can find that the author's cooperative network contains only 40 connections, and the overall density of the network is only 0.0029, which is at a very low level of cooperation. This shows that most of the research in the field of network public opinion guidance in China is done by the author alone, and few scholars conduct cooperative research, which makes the research in this field more decentralized, which largely limits the development of research in this field. In the context of the prosperity of a large number of researchers, it must be clearly recognized that the lack of scientific research cooperation "follow the trend" research is difficult to produce a leading figure with sufficient academic influence. In this regard, the state, universities and academic groups should build a platform for learning and communication for scholars of different institutions and disciplines, and provide some support for research funds, so that these scholars can break the barriers between institutions and disciplines, establish scientific cooperation and create more innovative research results.

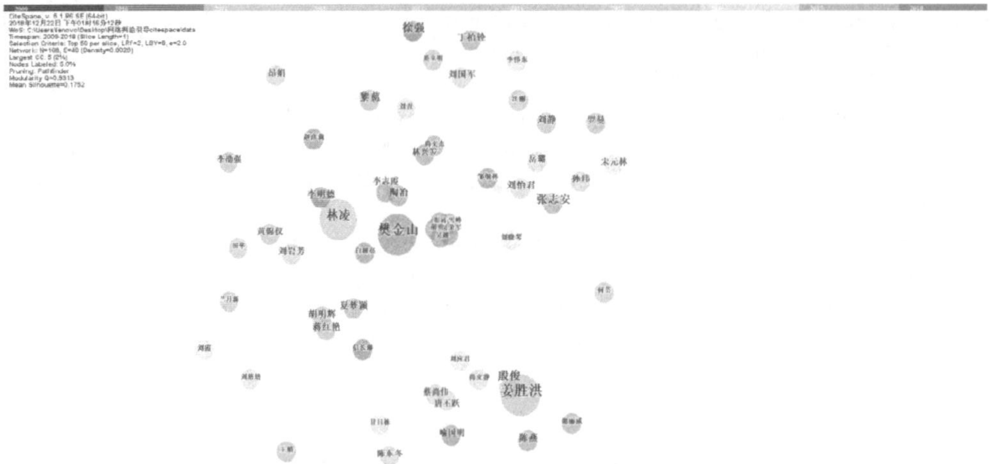

Figure 4. Author's Cooperative Diagram

3.5. Keyword Co-occurrence Clustering Analysis

In order to observe the location and objective law of keywords in social network, analyze the research hotspots in the field of network public opinion guidance, and conduct co-occurrence cluster analysis of keywords through CiteSpace. After filtering out the smaller clusters, only the main ones are displayed, there are 14 clusters as shown in Figure 5, the number of clusters is between 0 and 13. The smaller the cluster number, the larger the cluster size. The large-scale clustering mainly includes clustering 0 "emergencies", clustering 1 "network media", clustering 2 "network public opinion", clustering 3 "public opinion guidance", clustering 4 "emerging media" and clustering 5 "micro-era". These clustering results show that in the past ten years, the research focus of network public opinion guidance mainly concentrates on the following aspects: analysis of network public opinion on emergencies, the role of public opinion guidance in network media, guidance of colleges and universities to college students' network public opinion, how opinion leaders play their role in network public opinion guidance, how to make good use of the micro-era in the guidance of public opinion.

Figure 5 The calculation results in the upper left corner show that the number of network nodes in the keyword coexisting network is 227, the number of connections is 225, the overall density of the network is 0.0107, the modularity of the network is 0.8085, and the homogeneity of the network is 0.7369(The reliability will be reduced when the internal members of the cluster are smaller). Overall, the network density of keyword co-occurrence is not high, and the overall structure is still relatively scattered, which makes it difficult to form a systematic research system. Therefore, researchers need to calm down, take full account of the frontier research hotspots in this field, and concentrate more on the research topics.

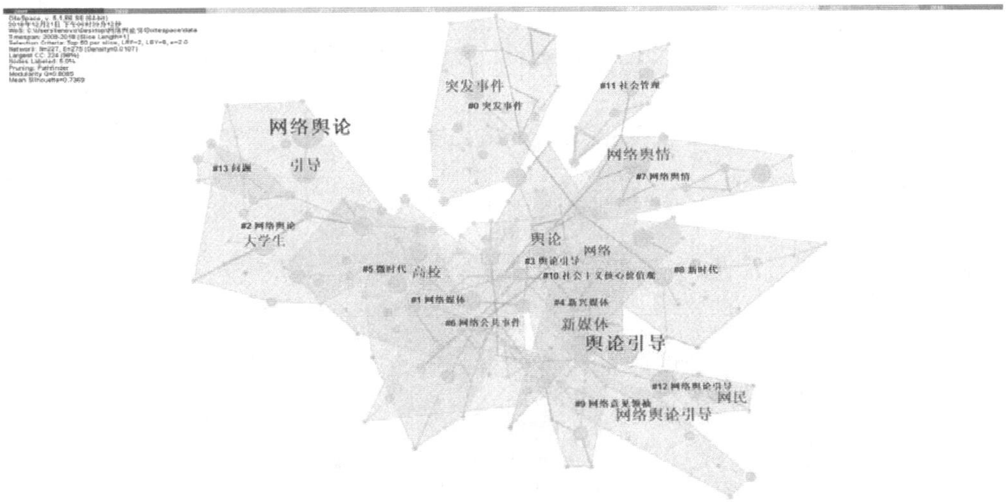

Figure 5. Keyword Co-occurrence Clustering Diagram

3.6. Closeness Analysis of Keyword Clustering

Through clusters operation and Export operation in CiteSpace, we get the data of keyword clustering compactness as shown in Table 2 and statistics of keyword clustering frequency over 100 times in 2009-2018 as shown in Table 3. From the data in these two tables, we can find that there is no positive or negative linear relationship between the size of clustering scale and its compactness (centrality). That is to say, the frequency of keywords with strong clustering compactness is not necessarily high, and the degree of clustering compactness of keywords with high clustering frequency is not necessarily strong. The compactness of clustering mainly depends on the correlation of content. The deeper the color of the clustering blocks, the higher the compactness. From the overall characteristics of research hotspots, the research topics of network public opinion guidance are rich, but the clustering characteristics between different topics are not obvious enough. Although there is a certain correlation between research hotspots, there is no representative clustering structure, and its research is still not concentrated and in-depth.

Table 2. Keyword Clustering Compactness Table

cluster number	clustering compactness	cluster number	clustering compactness
0	0.849	7	0.891
1	0.818	8	0.905
2	0.923	9	0.917
3	0.903	10	0.928
4	0.885	11	0.973
5	0.785	12	0.940
6	0.844	13	0.968

Table 3. Key Words that Appeared more than 100 Times in 2009-2018

frequency of occurrence	key word	subordinate cluster number
848	network public opinion	2
759	public opinion guidance	3
238	network public opinion guidance	12
228	guide	2
224	network public opinion	7
192	new media	3
169	public opinion	3
135	netizen	12
134	emergencies	0
117	universities	5
107	network	3
102	university student	2

4. Research Prospect of Network public Opinion Guidance

From the following aspects, this paper discusses and analyses the future research trends in the field of network public opinion guidance in China.

4.1. Research on the Subject of Network Public Opinion Guidance

The hot labels of this research direction mainly include: government, local government, colleges and universities, leading cadres of Party and government and public prosecution organs, etc. The characteristics of network anonymity, information fragmentation and the increasing number of netizens make the environment of network public opinion more complex and disorderly. Cyber rumors and violence, netizens' emotional speech and other phenomena occur from time to time. As an authoritative subject oriented by network public opinion, the government encounters many difficulties in the administrative control of network public opinion [13]. Therefore, the research on the challenges and Countermeasures of the guidance of government network public opinion has become a hotspot of current research on the guidance of network public opinion. The widespread popularization and application of the network in university campus obviously makes the campus network a magnifying glass of students' ideology and culture in public opinion, and highlights the function of public opinion [14]. Universities are the main position to receive and disseminate network public opinion. As the backbone of social development, students' value orientation and recognition of society are the important basis for future social evaluation. Therefore, how to effectively guide college students with immature social cognition and network standardization ability is the focus of college students ' work [15]. Most of the scholars who study the network public opinion guidance for college students are university teachers or administrators. Their research can be operated and operated in schools, and the results have more practical guiding significance.

4.2. Research on the Object of Network t Public Opinion Guidance

The main hot tags in this research direction are: information publishers, network opinion leaders and college students. The information publishers mentioned here are netizens. Netizens are a bounded rational group with complex elements and huge numbers. As of June 30, 2018, the number of netizens in China was 802 million, and the network penetration rate was as high as 57.7% .In the network public opinion, the netizen is both the subject and the object. As the subject, the netizen is the producer, publisher and judge of information, while as the object, the netizen is the guided object [16]. From the visual analysis of the hot keywords, we can see that the two most important groups in the research of online public opinion guiding object are the online opinion leaders and college students. Individuals or organizations with certain influence in the network platform or a certain network community are called network opinion leaders. These network public opinion leaders express their subjective opinions on hot social events through the network platform. In a short time, these comments can affect the emotional views of ordinary netizens in a wide range, and then guide the overall trend of network public opinion. However, we should clearly recognize that the role of these "leaders" is not always positive, sometimes too radical and emotional views will make the network unreasonable expression and Conformity Phenomenon and other negative effects are infinitely magnified. At present, the research on online opinion leaders mainly focuses on exploring more positive roles of online opinion leaders and consciously cultivating online opinion leaders with social responsibilities. In addition, due to the unstable values and social cognition of College students, there is often a strong personal subjectivity in the network expression, so the guidance of network public opinion of college students has become a hot topic for researchers.

4.3. Research on the Carrier of Network Public Opinion Guidance

The hot tags of this research direction are: mainstream media, traditional media, network media, new media, self-media, university new media, government official website, micro-era, micro-blog, BBS, micro-letter,etc. The network public opinion, which takes the emerging network media as the main carrier, breaks the monopoly of social information by the government and the traditional mainstream media, and makes everyone in the network become the participants of social events. The wide and rapid dissemination of new media has brought many difficulties to the government's guidance of network public opinion. Therefore, using the advantages of new media to solve the new problems arising from government guidance under the new situation is not only the focus of government work, but also the research hotspot reflecting the needs of the times [17]. Weibo and Weixin have now become indispensable tools for people to communicate and obtain information, the emergence of government microblogging provides a new way for government management. Strengthening the guidance of network public opinion has become the necessary work of government management.

4.4. Research on the Content of Network Public Opinion Guidance

The hot tags of this research direction are: sudden public events, unconventional events, network public events, network group events, natural disasters, network rumors, network public opinion, network violence and so on. Most of the incidents that need the guidance of network public opinion are unhealthy ones that have negative effects or can cause heated discussion among netizens. The abrupt and unconventional events are the beginning of the network public opinion, and the network public opinion is the driving force of the expansion of these abrupt events, which affects the trend of these events. Because of network anonymity and fragmentation of information, it may cause the generation and dissemination of negative network public opinion, and it may also lead to anomie phenomena such as network rumors and network violence. Therefore, in order to create a healthy network environment, we must adhere to the guidance of network public opinion, explain the truth in time, and strangle the bad network public opinion in its infancy. It can be seen that the guidance of network public opinion generated by sudden social events and the guidance of network anomie phenomenon have become a common concern of scholars.

In the future, the development trend of the research field of network public opinion guidance in China will be a comprehensive study of the characteristics and ways of network public opinion guidance in various environments from the specific reality. The ever-developing new media is gradually influencing the formation and development of network public opinion in its unique way, and the research on the guidance of network public opinion is becoming more and more important.

This work was financially supported by the national social science foundation of China (15BGL157), Research Projects of Teaching Reform of Higher Education in Heilongjiang Province (SJGY20180269).Thanks Pengyao Wang as the corresponding author.

5. References

[1] Deng Xinmin. Guidance of network Public Opinion and network Public Opinion [J]. Exploration, 2003 (05): 78-80.

[2] Xi Jinping emphasized at the Party's Symposium on news and public opinion work that we should adhere to the correct direction and innovate methods and means to "improve the guiding power of news and public opinion dissemination" [N]. Beijing News, 2016-02-29 (A01).
[3] Xi Jinping. To build a well-off society in an all-round way and win the great victory of socialism with Chinese characteristics in the new era [N]. People's Daily, 2017-10-28 (01).
[4] Chen Yue, Chen Chaomei, Ho Zhigang, Wang Zhiwen, etc. Citation spatial analysis principle and application: CiteSpace practical guide [M]. Beijing: Science Press, 2014.
[5] Wang Qi. Key Topics and Discipline Span of Confucius Institute Research from 2005 to 2017: Analysis of Visualization Map Based on Catespace [J]. Language Strategy Research, 2018, 3 (06): 42-55.
[6] Ding Boquan. Differentiation and Analysis of Two Problems of Public Opinion Guidance under the New Situation [J]. Nanjing Social Science, 2009 (04): 43-49.
[7] Dong Yuzhi. The Public Opinion Effect and Guidance of the Opinion Leader of Weibo in the Media Age [J]. Zhongzhou Academic Journal, 2014 (04): 173-176.
[8] Li Xiaoyan. Guidance Strategy of network Public Opinion in New Media Environment [J]. News Front, 2015 (20): 131-132.
[9] Lu Hongxun, Xie Jihua, Wang Xiaoxu. Research on the Construction of Network Public Opinion Environment in Colleges and Universities Based on the Guidance of Network Public Opinion [M]. Beijing: University of Electronic Science and Technology Press, 2013.
[10] Ren Hai. Formation, Characteristics and Guidance Strategy of Network Public Opinion on College Campus [J]. Journal of Yanshan University (Philosophy and Social Sciences Edition), 2009, 10 (02)134-136.
[11] Xiao Wentao, Xu Xiaomei. Current situation of network public opinion and guiding Countermeasures in the new media era [J]. Administrative Forum, 2012, 19 (06): 79-84.
[12] Zhang Fang. Viewing the characteristics of network public opinion dissemination in emergencies from micro-blog -- "Chongqing falling bus" incident as an example [J]. Journalism Research Guide, 2018, 9 (20): 87-88.
[13] Xie Jinlin. The government should effectively guide network public opinion [J]. Theoretical exploration, 2010 (02): 118-120.
[14] Lu Lan, Du Yulong, Zhu Yuan. Enhancing the Effectiveness of network Public Opinion Guidance in Colleges and Universities: Taking Taiyuan University of Technology as an example [J]. Journal of Social Sciences of Shanxi University, 2018, 30 (11): 67-71.
[15] Hu Minghui, Jiang Hongyan. Importance and strategy of strengthening public opinion guidance on campus network [J]. Journal of Shenyang Agricultural University (Social Science Edition), 2016, 18 (04): 450-455.
[16] Liu Yanfang, Xu Jianzhong, Ding Nan. Research on hot topic discovery guided by network public opinion in China based on Co-word Analysis and scientific knowledge map [J]. Information Science, 2017, 35 (05): 57-62+68.
[17] Lai Yiru, Zhang Zhongtao. Government Network Guidance in New Media Crisis [J]. Journal of Tianshui Administrative College, 2018, 19 (04): 35-39

Effects of Magnesium-Calcium Alloys with Different Calcium Content on Their Mechanical Properties

Yanan Wang, Guodong Liu, Yukun Wei and Yang Qiao*

School of Mechanical Engineering, University of Jinan, Jinan 250022, China.
*Corresponding author. Email: me_qiaoy@ujn.edu.cn

Abstract. Compared with traditional medical metal materials, magnesium alloys have the characteristics of small density, small elastic modulus and high specific stiffness. In order to study the effect of the addition of calcium on the mechanical properties of magnesium alloys, the tensile test and hardness test of magnesium-calcium alloy were carried out. The stress changes during the tensile test of magnesium alloy were discussed, and the microstructure of its section was further observed, so as to compare the effects of different calcium content on the mechanical properties of magnesium alloy. The results show that the addition of calcium content increases the hardness and tensile strength of the magnesium alloy, and it tends to rise first and then decrease.

1. Introduction

As we all know, magnesium is a very active metal, which has the advantages of high specific strength and low density, so it is widely used in transportation and aerospace fields[1-3]. At the same time, magnesium and magnesium alloys have excellent biodegradable properties, can be completely degraded in the human body, the elastic modulus and tissue for the human body is extremely close to, can effectively avoid stress shelter effect, promote bone tissue affected function reconstruction, so magnesium alloy as a new type of medical implant material has good application potential, and has been widely research [4-8] .However, magnesium, as a medical degradable metal material that can be implanted into the human body, is not only uncontrollable in corrosion rate, but also the key factor in the success or failure of surgery[9-10]. In this paper, vickers hardness test, quasi-static tensile test and mechanical analysis were carried out on Mg-Ca alloy with different calcium content, and then the tensile fracture was observed by SEM (Scanning Electron Microscope) to explore the influence of calcium content on the mechanical properties of Mg - Ca alloy.

2. Test Materials and Program

The selected materials are pure magnesium and magnesium calcium 15 as the intermediate alloy, and the as-cast alloy is formed after smelting in the electric resistance furnace according to the ratio. The extruded magnesium calcium alloy is formed after extrusion strengthening, and the alloy content obtained through smelting is MgCa0.8, MgCa1.2, MgCa1.6 and MgCa2.0. Its main components are shown in table 1.

Table 1. Smelting raw material ingredients(The mass fraction, %)

The raw materials	Mg	Ca	Fe	Cu	Mn	Ni	Si	Al	Zn
Pure magnesium	99.93	—	<0.01	<0.01	0.02	0.01	0.02	0.01	<0.01
Calcium magnesium 15	88.60	11.36	0.01	<0.01	0.01	<0.01	0.01	—	—

Four kinds of magnesium alloys with different compositions were prepared by melting, heating, mixing and cooling. Due to its active chemical properties, magnesium is prone to oxidation and combustion in the smelting process. Therefore, protective gas must be introduced in the smelting process. CO_2+SF_6 is selected as the protective gas in the experiment.

2.1. Vickers Hardness Measurement

In order to study the effect of different calcium content on the mechanical properties of Mg-Ca alloy, four kinds of Mg-Ca alloy with different content were tested by quasi-static tensile test. Samples were taken from different alloys and tested on a digital microhardness tester after mosaic, grinding and polishing. The pressure applied during the test was 50N. Each sample was tested at 5 points under the same condition, and the average value was taken as the hardness of the sample.

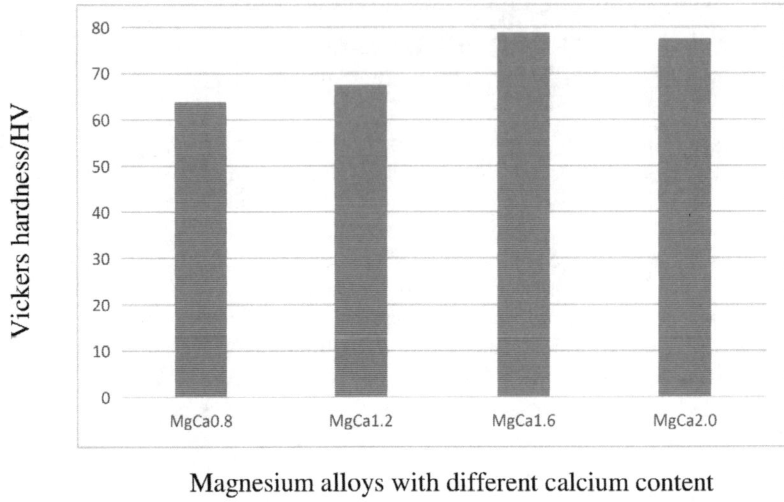

Figure 1. Effect of different calcium content on hardness of magnesium alloy

Figure 1. shows the effect of different calcium content on the hardness of magnesium alloys. The hardness of the alloy increases first and then decreases with the addition of calcium content. When the calcium content increases from 0.8% to 1.6%, the hardness of the alloy increases steadily. When the calcium content reaches 1.8%, the hardness of the alloy begins to decrease.

2.2. Quasi-Static Tensile Test

The bar material after extrusion is processed and the sample is obtained by wire cutting. Samples preparation and tensile test parameters were selected according to Chinese national standards (GB/ t228.1-2010). Compression test equipment is universal material testing machine (XYB305C), as shown in figure 1. In the compression test, the effect of the content of calcium on the mechanical properties of mg-ca alloy was mainly studied, and the tensile speed was 5 mm/min [11-12]. The sample size is shown in figure 2.

Figure 2. Universal material testing machine(XYB305C)

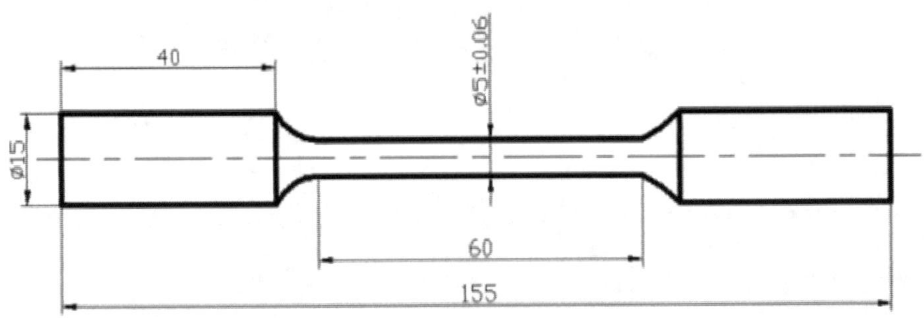

Figure 3. The sample size

The universal material testing machine was used to stretch the specimens to fracture successively, record the engineering stress and corresponding engineering strain, and draw the engineering stress-strain curve according to the original data, as shown in Figure 4.

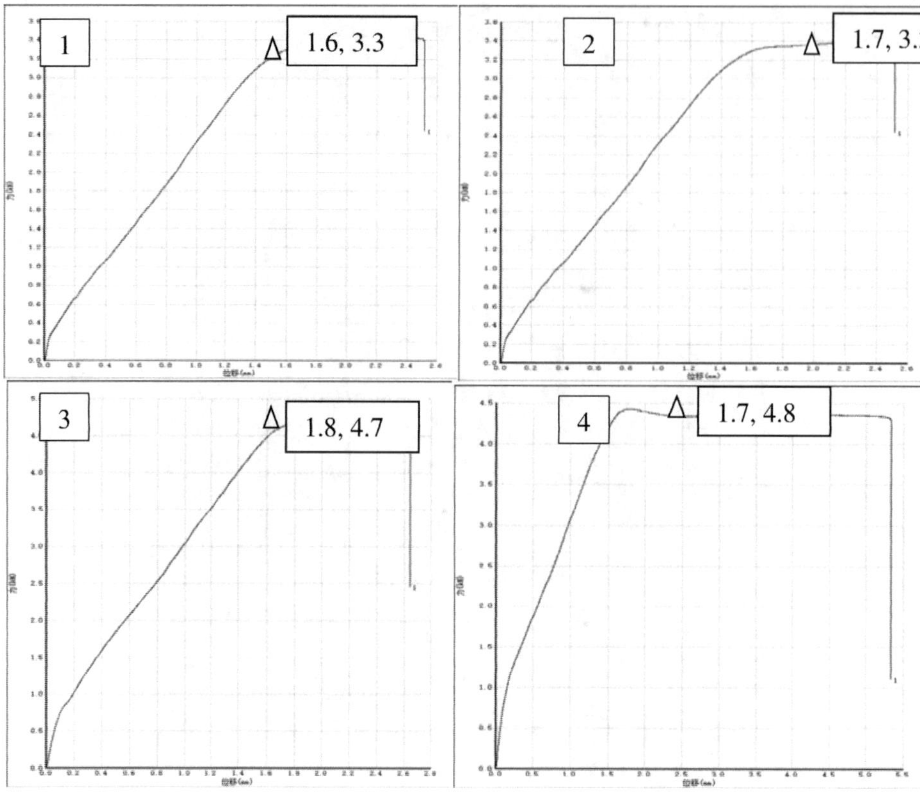

Figure 4. Stress - strain curve of magnesium - calcium alloy
(1. MgCa0.8 2. MgCa1.2. 3. MgCa1.6 4. MgCa2.0)

Quasi-static tensile test was carried out on magnesium-calcium alloy with different calcium content to obtain its stress-strain curve. During the tensile test, the tensile load on the sample gradually increased, and local plastic deformation occurred in the material, leading to the whole fracture of parallel parts [13]. It can be seen that as the calcium content increases, the force required to enter the yield state becomes greater and greater. This indicates that the addition of calcium content increases the tensile strength and mechanical properties of the alloy.

2.3. Fracture Morphology Observation

The fracture morphology of the sample was subjected to SEM analysis, the magnification is 2000, as shown in Figure 5.

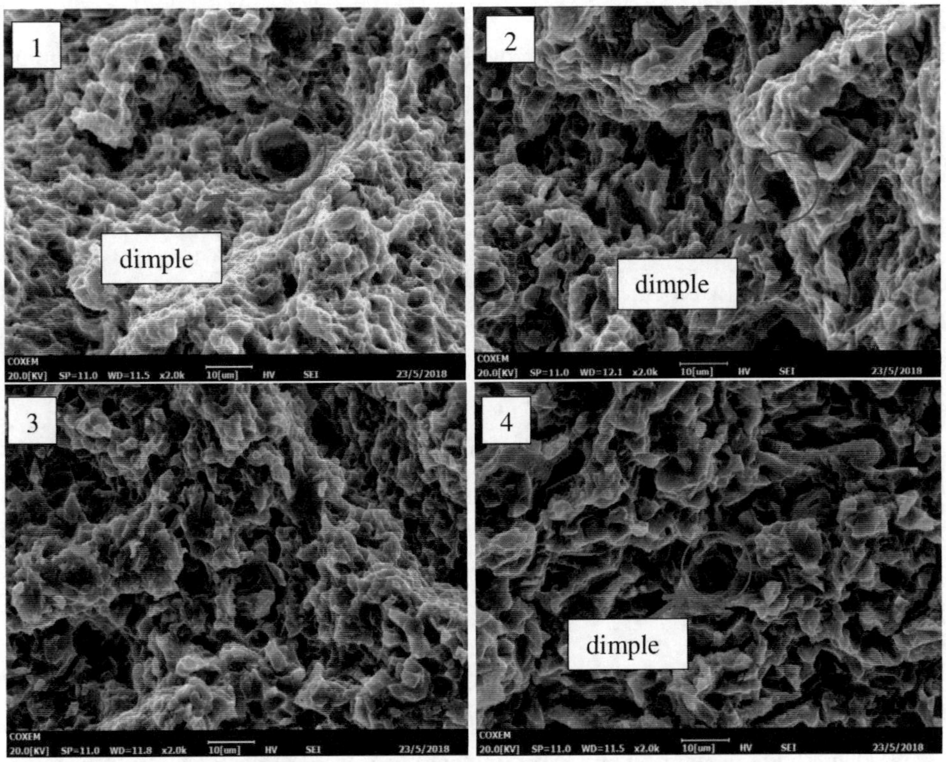

Figure 5. Magnesium-calcium alloy tensile fracture morphology
(1.MgCa0.8 2. MgCa1.2. 3. MgCa1.6 4. MgCa2.0)

As shown in figure 5 different calcium content of calcium magnesium alloys presented under SEM fracture morphology, it can be seen that the four different calcium magnesium alloys are different degrees of the toughening, belongs to ductile fracture. This is due to the dynamic recrystallization grain size and precipitation phase also experienced a greater degree of elaboration[14], so with the increase of calcium content, section presents the micro ball more and more dense, namely toughening nest is becoming more and more obvious. When calcium content reached 2%, the spread in the process of the precipitated phase within the grain of the snap is overall, there are a large number of globular carbide, alloy has the highest tensile strength.

3. Conclusions

(1)It can be seen from the stress-strain curve that the increase of calcium content in mg-ca alloy can enhance its tensile strength. And the SEM analysis was carried out on the cross section can be seen that the increase of calcium content makes crystal refining is apparent in the matrix, the existing studies have shown that grain refinement can effectively improve the corrosion resistance of the material and its tensile ability, thus to some extent, the calcium added to improve the mechanical performance of magnesium alloy has a positive effect.

(2)From the above experiments, calcium content in the range of 0.8% -- 2.0%, tensile test and hardness test results are consistent, are MgCa1.6 better results. However, whether MgCa1.6 is the magnesium calcium alloy with the best mechanical properties still needs to be further studied.

4. Acknowledgments

This study was supported by Shandong Provincial Natural Science Foundation (No. ZR2019QEE032) and Shandong key research and development program (No. 2017GGX30143)

5. References

[1] Guangyin Yuan, Jialin Niu. Research progress of biodegradable medical magnesium alloy in bone repair[J]. *Journal of Metals.* p 1168

[2] Xianhua Chen, Yuxiao Geng, Fusheng Pan. Research Progress in Magnesium Alloys as Functional Materials[J]. *Rare Metal Materials and Engineering.* pp 2269-2274.

[3] Guangyin Yuan, Xiaobo Zhang, Jialin Niu. Research progress of new biodegradable biomedical magnesium alloy JDBM[J]. *Chinese Journal of Nonferrous Metals.* p *2476.*

[4] Jie Zheng. Study on Mechanical Properties and Corrosion Behavior of Magnesium-Calcium Alloy[D]. Xi'an: Department of Materials Physics and Chemistry, Chang'an University, 2015.

[5] Rongchang Zeng, Lanyue Cui. Medical magnesium alloys: composition, organization and corrosion[J]. *Journal of Metals.* pp 1216-1235.

[6] Ruizhi Wu, Huaijing Zhang. Advanced magnesium alloy manufacturing and processing technology[M]. Beijing: *Science Press*, 2012.

[7] Wenyu Zhang. Revolutionary medical material - magnesium alloy[J]. *Chinese Journal of Nonferrous Metals.* pp *1-2.*

[8] Haiying Jiang, Hongjun Ai. Medical characteristics of biodegradable magnesium alloy implant materials[J]. *Chinese Tissue Engineering Research and Clinical Rehabilitation.* pp 6411-6415.

[9] Yanying Zhu, Guangming Wu, Qing Zhao. Research progress in magnesium-based biomedical materials[J]. *Chinese Journal of Biomedical Engineering.* pp 932-938.

[10] Lei Yi, Qian Zhou, Xiaoying Jiang. Development history and research status of magnesium alloy degradable vascular stent[J]. *Medicine and Philosophy (B).* pp 60-62.

[11] Meiqin Jiang. Analysis of Factors Affecting Tensile Test of Metallic Materials at Room Temperature[J]. Jiangsu Metallurgy. pp 27-31.

[12] Hongde Ye. Some thoughts on the standard of tensile test method for metal materials[J]. *Physical and Chemical Testing - Physical Division.* pp 394-395.

[13] Jianhong Miu, Jinkun Ding. Fracture analysis method for metal tensile specimens [J]. *Physical Test.* pp 35-40.

[14] Chuming Liu, Zijuan Liu. Research progress on dynamic recrystallization of magnesium and magnesium alloys [J]. *Chinese Journal of Nonferrous Metals.* pp 1-10.

Research on Preparation Process of Graphene by Modified Ball Milling and Its Thermal Properties

Quanmao Yu[1,*], Qiurong Liu[1], Xiaoxia Li[1], Zhong Chen[2], Tao Ren[1], Zhenyu Chen[1] and Zhengyi Huang[1]

1 Institute of Functional Materials, School of Software and Internet of Things Engineering, Jiangxi University of Finance and Economics, Nanchang 330013, PR China

2 School of Materials and Mechanical and Electrical Engineering, Jiangxi Science and Technology Normal University, Nanchang 330013, PR China

* Corresponding author. Email: bsyqm@163.com

Abstract. Graphene is a focus for its excellent properties. The high efficient low cost preparations of high pure Graphene are bottlenecks in Graphene application field. This paper reports a modified mechanical milling high pure Graphene preparation method. The method has the characteristics of less cost and can scale-production. The XRD, TEM and TG/DTA are used to character the properties, especially the thermal properties of Graphene prepared by the method.

1. Introduction

Carbon materials are well-known for their superior properties. A single layer of graphite (also called Graphene) is one of the focus materials of this type of material [1]. Graphene is the basic unit for forming two-dimensional carbon materials. With a single layer of graphite, we can get graphite, 0-dimensional fullerenes, 1-dimensional carbon nanotubes and so on. In recent years, two-dimensional materials have attracted more and more attentions, mainly because of their potential applications [2]. Graphite has many unique characteristics and distinctive properties because of its unique and stable crystal structure [3,4]. These rich and unique phenomenon arouse the world research and development community wide interests [5]. Graphene not only has excellent electrical properties, outstanding thermal conductivity (up to 5000 $wm^{-1}k^{-1}$), extraordinary specific surface area (2630 m^2/g), ultra-high carrier mobility at room temperature (over 15000 cm^2/Vs , the theoretical value is more than 1 million cm^2/Vs) [6], its young's modulus and fracture strength can also be equivalent to carbon nanotubes. Hence, Graphene has a broad application prospect in such aspects as flexible materials, photoelectric materials, semiconductor materials, electronic information materials, drug carriers, etc [7,8].

In 2004, Geim et al. prepared stable monolayers Graphene using a method called mechanicstripping at the University of Manchester [9]. Since 2004, the researches on Graphene become hot topic. The researches mainly focus on the Graphene preparation and application performance [10]. At present, there are many methods to prepare Graphene, and we can divide them into physical and chemical methods. Among them, the physical methods mainly include mechanical exfoliation method, liquid phase stripping method [11], ball-milling method, and so on. The chemical methods mainly include epitaxial growth method [12,13], oxidation-reduction method [14], chemical vapor deposition method [15,16], electric arc synthesis, in situ self-generated template method, etc. In comparison, the preparation process of Graphene by mechanical ball mill is simple, with low cost. However, there is no ideal method for preparing high-quality Graphene at low cost [17]. Therefore, the low cost produce method of high quality Graphene becomes a bottleneck in Graphene researches and application fields [18,19]. In this paper, we report a modified low cost mechanical milling method for preparing high pure Graphene and character the XRD, TEM and thermal properties of Graphene prepared by the method.

2. Experimental

2.1. Materials
The raw materials were natural graphite powder (analytical-grade, purchased from market), deionized water and some certain agents. All agents used in this work were analytical-grade reagents (A.R.) and purchased from Sinopharm Chemical Reagent Co., Ltd.

2.2. Synthesis
The process of preparing Graphene by our modified ball milling method is shown in Figure1. Firstly, the raw materials are pretreated, and then the processed raw materials and functional agents are mixed and added to a mechanical ball mill. After been grinded in a mechanical mill for the definitely time at the setting speed, samples are taken out and purified to high pure Graphene.

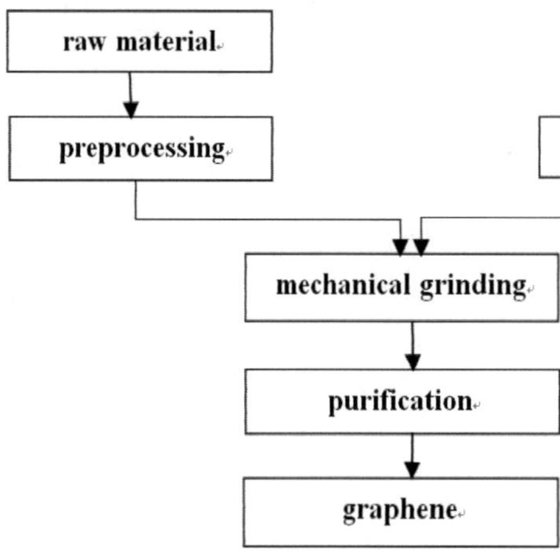

Figure 1. Schematic Diagram of Experimental Process

2.3. Characterizations
X-ray diffraction data were obtained by X-ray diffractometer (Shimazu XRD-6100 Lab, Japan, CuKα1) with the scanning speed 2°/min and step 0.02° in the range of (2θ=20°-70°) at room temperature. Transmission electron microscopy images were obtained using TEM (Tecnai G2 20 S-Twin). The thermal performance was performed by the EXSTAR6000 TG/DTA synchronous thermal analyzer (Japan) with the heating speed 5°/min and step 0.02° in the range from 50°C to 1000°C.

3. Results and Discussion
Figure 2 shows the Graphene samples (the content about 2.0% (Wt.)) produced by our modified ball milling method. The samples are visible light perspective. In order to enhance the stability and prevent the secondary lamination, the Graphene pretreated by functional agents are dispersed in liquid (in this research, we use water). The content of Graphene produced by our method can rich to 5.0% (Wt.).

Figure 3 shows the XRD patterns of raw materials (untreated graphite) and our Graphene samples. For the original graphite, there is a strong sharp diffraction peak at 26.5°, and the corresponding layer spacing is 0.34 nm. Because the samples using in XRD characterization are accumulations of Graphene flakes (as shown in figure 4), the XRD strength of our Graphene samples is very slight, the diffraction peak is wide peak around 23.5° (as shown in the figure), and the corresponding layer

IOP Conf. Series: Materials Science and Engineering **735** (2020) 012011 doi:10.1088/1757-899X/735/1/012011

spacing is about 0.38 nm. The XRD patterns and the increase of layer spacing indicate that most of the graphite is mechanical stripping into a few layered Graphene flakes by ball milling.

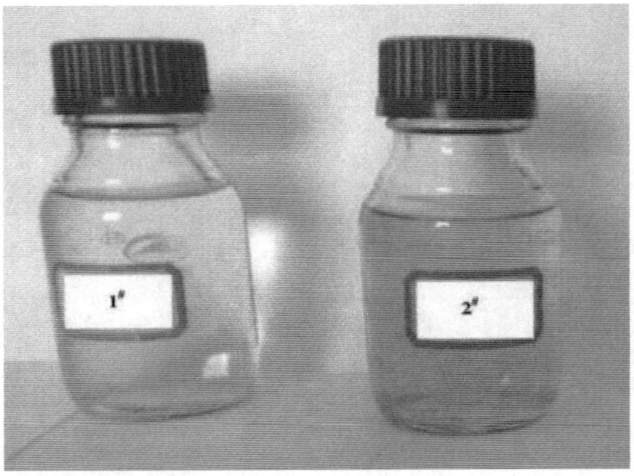

Figure 2. Graphene Produced by Modified Ball Milling Method (the Content about 2.0% (Wt.))

Figure 4 is TEM photographs of Graphene prepared by our modified ball milling method. The Graphene are two-dimensional flakes accumulated together. We can clearly see the number of layers of Graphene flakes, and most of them are below 5 layers.

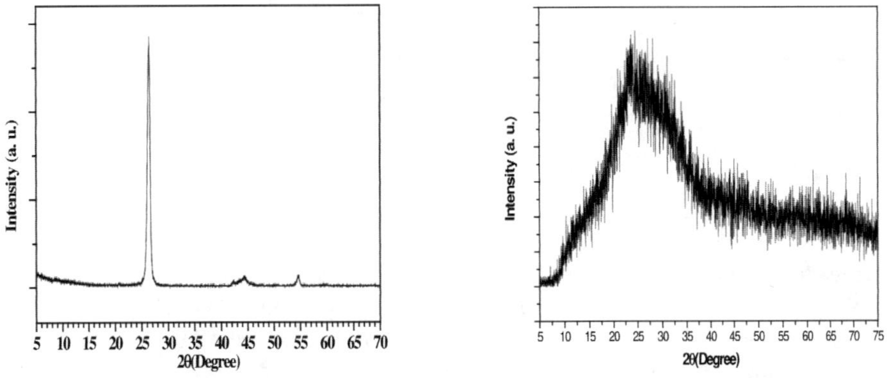

(a) (b)

Figure 3. XRD Patterns of Untreated Raw Materials (a) and Graphene (b)

Figure 5 shows the DTG characterization results of Graphene and pretreated raw materials (graphite). The line (A) represents the pretreated raw material and the line (B) is Graphene. It can be seen from the figure that the pretreated graphite raw material has a good heat resistance before 500°C (though the untreated graphite has thermal stability to over 1000°C, as all known), and when the temperature is higher than 500°C, the pretreated graphite raw material begins to decompose. The Graphene sample has a peak at a temperature below 90°C, where the weight loss is the evaporation of the adsorbed water in the sample. There is a peak between 250°C and 400°C, and the peak is 350°C, which is the decomposition of Graphene. After 400°C, the Graphene is broken down, so there is no weightlessness. This means the Graphene produced by our modified ball milling method should be applied under 250°C.

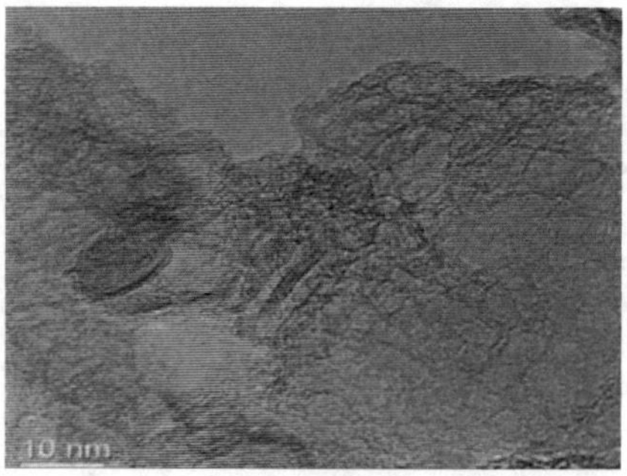

Figure 4. TEM Images of Graphene

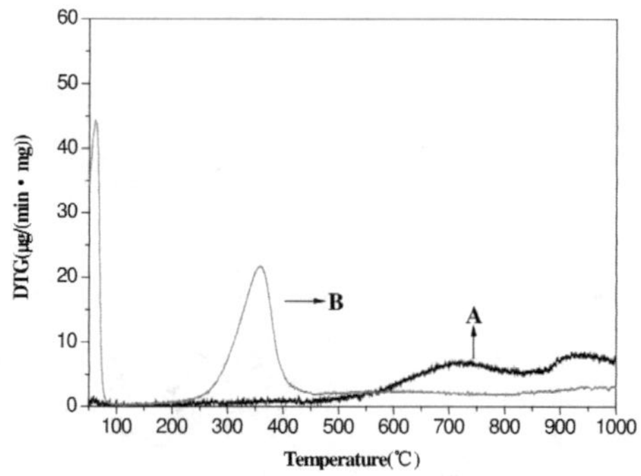

Figure 5. The DTG Curves of Graphene and Pre-treated Raw Materials

4. Conclusions

In this paper, few-layer Graphene were prepared by a simple low cost modified ball milling method. The Grapheme is less than 5 layers. The Graphene has stable thermal properties under 250°C. The Graphene and its preparation method should have great potential application.

5. Acknowledgements

This work was supported by the Key Science Project of Jiangxi Provincial Department of Education (No. GJJ180246).

6. References

[1] Geim A K and Novoselov K S 2007 *J. Nat. Mater.* **6**(3):183-91

[2] Sakamoto J, Van H J, Lukin O and Schlüter A D. Two-dimensional polymers: just a dream of synthetic chemists? 2010 *J. Angew Chem Int Ed Engl.* **48**(6):1030-69

[3] Geim A K. Graphene: status and prospects 2009 *J. Science.* **324**(5934):1530-34

[4] Wu J, Pisula W and Müllen K. Graphenes as potential material for electronics 2007 *J. Cheminform.* **38**(26):718

[5] Lu N, He G, Liu J, Liu G, Li J. Combustion Synthesis of Graphene for Water Treatment *2017 J. Ceramics International,* **44** (2017) 2463-2469

[6] Balandin A A, Ghosh S, Bao W, Calizo I, Teweldebrhan D and Miao F. Superior thermal conductivity of single-layer graphene 2008 *J. Nano Letters.* **8**(3):902

[7] Kuila T , Bose S, Mishra A K, Khanra P , Kim N H and Lee J H. Chemical functionalization of graphene and its applications 2012 *J. Progress in Materials Science.* **57**(7):1061-105

[8] Chen K, Li Y. Progress in preparation methods of graphene 2018 *J. Scientific and technological innovation Information,* (20):1-2

[9] Novoselov K S, Geim A K, Morozov S V, Jiang D, Zhang Y, Dubonos S V, Grigorieva V and Firsov A A 2004 *J. Science.* **306**: 666-9

[10] Lerf A, He H and Forster M. Structure of graphite oxide revisited‖ 1998 *J. Journal of Physical Chemistry B.* **102**(23):4477-82

[11] Bepete G, Anglaret E, Ortolani L, Morandi V, Huang K, Pénicaud A and Drummond C, Surfactant-free single-layer graphene in water 2017 *J. Nature Chemistry.* **9**:347-52

[12] Yang Z, Xu S, Zhao L, Zhang J, Wang Z, and Chen X. A new direct growth method of graphene on Si-face of 6H-SiC by synergy of the inner and external carbon sources 2018 *J , Applied Surface Science.* **436**:511-8

[13] Kim K S, Park G H, Fukidome H, Takashi S, Takushi I, Fumio K, Iwao M and Suemitsu M. A table-top formation of bilayer quasi-free-standing epitaxial-graphene on SiC(0001) by microwave annealing in air 2018 *J. Carbon.* **130**:792-98

[14] Voiry D, YangnJ, Kupferberg J, Fullon R, Jeong H Y and Chhowalla M. High-quality graphene via microwave reduction of solution-exfoliated graphene oxide 2016 *J. Science.* **353** :1413-16

[15] Tang Y, Peng P, Wang S, Liu Z, Zu X and Yu Q. Continuous production of graphite nanosheets by bubbling chemical vapor deposition using molten copper 2017 *J. Chemistry of Materials.* **29**(19):8404-11

[16] Komissarov I V, Kovalchuk Nikolai G, Labunov Vladimir A, Girel Ksenia V, Korolik Olga V, and Tivanov Mikhail S. Nitrogen-doped twisted graphene grown on copper by atmospheric pressure CVD from a decane precursor 2017 *J. Beilstein Journal of Nanotechnology,* **2017**(8):145-58

[17] Du M, Zhang G. Progress in preparation and application of graphene 2019 *J. Inorganic Chemicals Industry*, **51**(3):12-5

[18] Ren S, Rong P, Yu Q. Preparations, properties and applications of graphene in functional devices: A concise review *2018 J. Ceramics International*, **44**(11): 11940-55

[19] Gu X, Zhao Y, Sun K, et al. Method of ultrasound-assisted liquid-phase exfoliation to prepare graphene 2019 *J. Ultrasonics Sonochemistry*, **58**:104630

IOP Conf. Series: Materials Science and Engineering **735** (2020) 012012 doi:10.1088/1757-899X/735/1/012012

Compatibility of Agents for Polymer-bearing Produced Water from Oilfield

Limei Sun[1*], Jiajun Xu[1], Wenxian Xie[1],Ting Yao[1] and Haigang Zhou[2]

1. Shengli Oil Production Plant of Shengli Oilfield Branch Company, SINOPEC Dongying Shandong China 257051
2. Technical Testing Center of Shengli Oilfield Branch Company, SINOPEC Dongying Shandong China 257000

[*]Email: slmwky@163.com

Abstract. The compatibility of polymer-bearing produced water and corrosion inhibitor, bactericide has been explored from the angle of suspended solids content (ssc) of produced water from water treatment process of a station in Shengli Oilfield. The results show that cationic corrosion inhibitor and bactericide can interact with negatively charged suspended solids and polymers in produced water, resulting in the increase of ssc of produced water. Therefore, the use of cationic agents should be avoided for polymer-bearing produced water, therefore non-ionic and zwitterionic agents should be alternatives. Experiments show that the zwitterionic bactericide not only effectively controls SRB, but also hardly effect on the ssc of produced water.

1. Introduction

It has been known that developed oilfields in central and eastern China generally have come to a production mode of water injection cycle. Produced water is not only the product of oil recovery, but also the driving source of oil production. Therefore, the treatment quality of produced water determines the production effect, and is also an important guarantee for oil production.

After treatment, the produce d water must meet the corresponding injection standard [1-3]. Among them, the average corrosion rate and bacterial content are two very important indicators, which are related to the corrosion and scaling of pipelines and tank walls, as well as the formation blockage. Adding corrosion inhibitor and bactericide is a very important way to control the average corrosion rate and bacterial content in injection water. Therefore, the adaptability of corrosion inhibitors and bactericides has also been reported. In enhanced oil recovery or tertiary oil recovery technologies, polymer solutions are often prepared with treated produced water. It was found that [4] cationic bactericide can precipitate or flocculate polymer solution and reduce the viscosity of polymer solution. Corrosion inhibitors, bactericides and scale inhibitors, flocculants, demulsifiers must be compatible; otherwise it will decrease the performance of various agents [5].

In this paper the compatibility of polymer-bearing produced water and agents, hereinafter referred to corrosion inhibitor, bactericide has been explored from the angle of suspended solids content (ssc) of reinjection water from water treatment process of a station in Shengli Oilfield, and the direction of exploration in the near future on corrosion inhibitor and bactericide has also been mentioned.

2. Process and Effect of Produced Water Treatment

The produced water treatment in this station adopts two-stage sedimentation combined with filtration process, as shown in figure 1. The main equipment consists of 2 primary deoiling tanks of 2000 m^3, 2 secondary coagulation tanks of 2000 m^3, 2 buffer tanks of 1000 m^3 and 8 walnut shell filters of Φ3.4

m. The daily treatment water volume is 2×10^4 m^3. The total salinity of the produced water is 18457.8 mg/L and the water type is calcium chloride. The polymer content is 15 mg/L. The content of SRB bacteria is higher. The composition of produced water is shown in table 1.

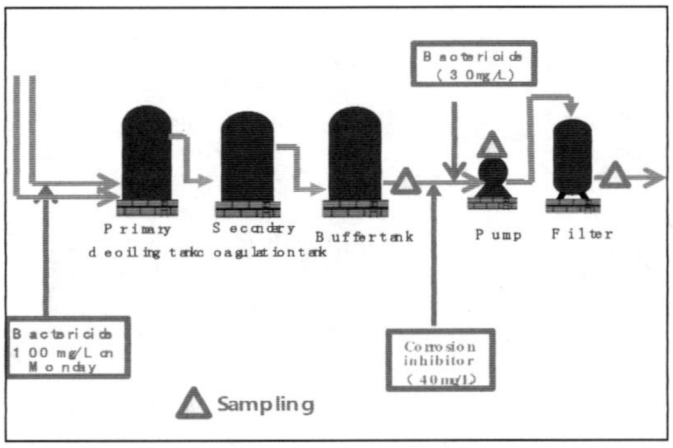

Figure 1. The Produced Water Treatment Process

Table 1. Components and Characterization for Produced Water

Items	Value	Items	Value
Na^+, mg/L	6502.3	Temperature, °C	50
Ca^{2+}, mg/L	615	pH	6.71
Mg^{2+}, mg/L	63.2	Dissolved oxygen, mg/L	0.03
Fe^{2+}, mg/L	4.8	Free carbon dioxide, mg/L	70.4
Cl^-, mg/L	10812.3	Sulfate, mg/L	0.214
HCO_3^-, mg/L	460.2	SRB, /mL	6.0×10^4
Salinity, mg/L	18457.8	FB, /mL	0
Polymer, mg/L	15	TGB, /mL	0.6

According to the screening results in 2014, the imidazoline corrosion inhibitor and the double quaternary amine salt with 40 mg/L dosage respectively had the best performance for this produced water, then corrosion inhibitor and bactericide were designed to continuously pump into the produced water through two dosing points of 10 meter-distance along the pipeline before the buffer tank and the pump. Furthermore, the dosage of 100 mg/L bactericide was being pumped into the produced water for 8 hours on Monday through a dosing point on the pipeline before the primary deoiling tanks.

The ssc was analyzed for the produced water, which was sampled every hour at different nodes of the treatment process. According to standard [1], microporous nitrocellulose membrane (Merck Millpore in Ireland) with a pore size of 0.45 um was applied. The results are shown in figure 2.

There is no treatment between pump and buffer tank except for the dosage of chemicals. Technically, the ssc should keep unchanged. However, on average, the ssc of water sample from pump outlet was averagely 20 mg/L higher than that of the outlet of buffer tank.

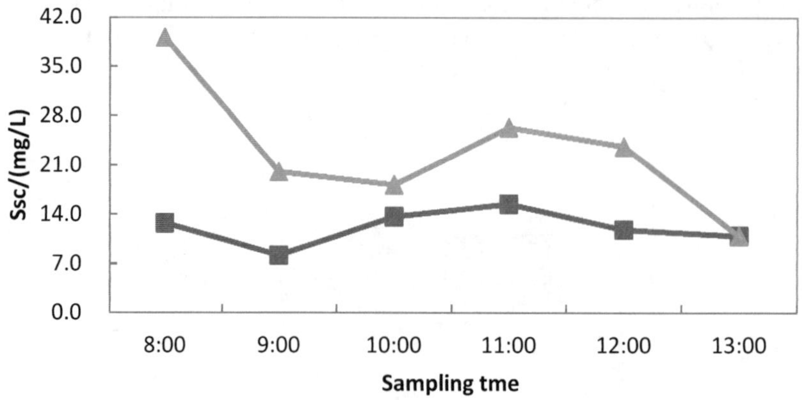

Figure 2. The ssc of Produced Water from Nodes along the Treatment Process
▲ Pump Outlet
■ Buffer Tank Outlet

3. Compatibility of Agents and Polymer-bearing Produced Water

The test above proves that the ssc of produced water along water treatment process is increased after the dosage of the agents. In order to evaluate the compatibility of produced water with agents applied, produced water samples without any agents were taken as blank for laboratory evaluation. The blank samples with addition of corrosion inhibitor and the bactericide respectively were taken as test samples. Afterwards the ssc was analyzed for all those samples, and the results are shown in figure 3. It can be seen that compared with the blank water sample, the ssc of water with the addition of 36.8 mg/L of corrosion inhibitor was increased by 18 mg/L averagely, and the ssc of water with the addition of bactericide 36.8 mg/L was increased by 22 mg /L averagely. Laboratory evaluation is consistent with the field test.

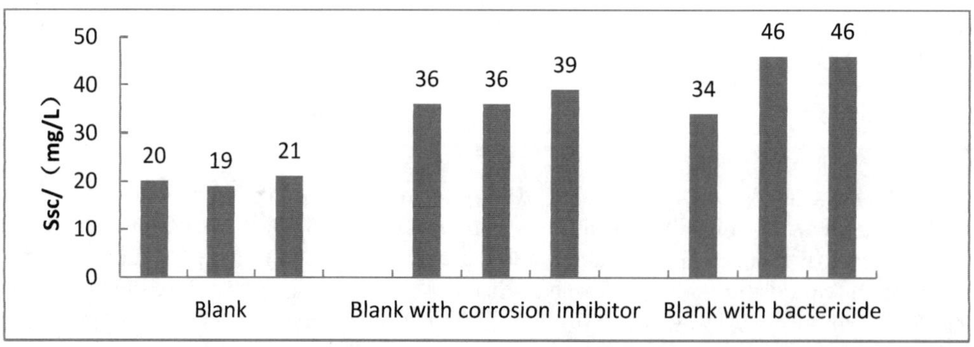

Figure 3. The ssc of Produced Water, of that with the Addition of 36.8 mg/L of Corrosion Inhibitor, and of that with the Addition of Bactericide 36.8 mg/L Respectively

4. Reasons for Incompatibility between agents and Polymer-bearing Produced Water

Corrosion inhibitors are classified into three types: oxidation film, precipitation film and adsorption film. Adsorption film is the most widely used type, and imidazoline compounds are the main ones, see figure 4. Among many bactericides, because bacterial cell walls are usually negatively charged, cationic surfactants are the earliest and most widely used bactericide, such as double quaternary amine salt, as shown in figure 5.

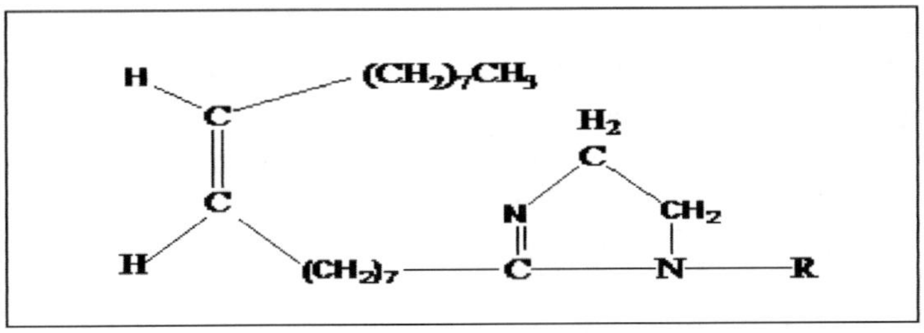

R=-(CH_2CH_2NH)n-$CH_2CH_2NH_2$, n=0, 1, 2......

Figure 4. Imidazoline Corrosion Inhibitor

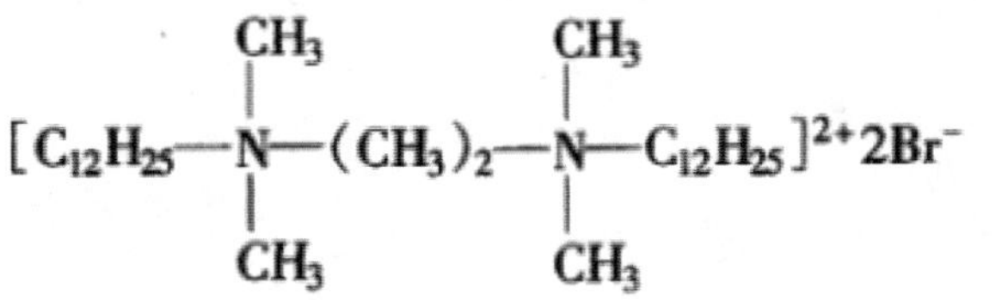

Figure 5. Double Quaternary Amine Salt Bactericide

Double quaternary amine salt is cationic and can ionize into cationic quaternary ammonium group in water. In preparation of imidazoline corrosion inhibitor, tertiary amine cation is formed by adding hydrochloric acid to increase the water solubility of the product. Negatively charged suspended solids and polymer for oil displacement exist in produced water. Polymers for oil displacement are generally anionic partially hydrolyzed polyacrylamide.

Generally, polymer will further degrade and its molecular weight will decrease while migrating and working in the reservoir. According to long-term follow-up analysis, the molecular weight of residual polymer in produced water is generally about 500,000. At the same time, hydrolysis and curl of polymer molecules will occur [6]. Therefore, the polymer in the produced water is negatively charged soft particles.

Polymer soft particles are negative, and the suspended solid particles in produced water are negative, too. They all aggregate with cationic agents because of electrical attraction, that is, cationic agents act as flocculants. There are three reasons for the increase of ssc in the suspended solids test process based on filter membrane filtration. Firstly, cationic agents' aggregate small particles which could pass through 0.45μm filter membrane into larger particles which could not pass through 0.45μm filter membrane and those large particles remain on the filter membrane. Secondly, agents and polymers also remain on the filter membrane. Thirdly, the retention of larger particles, polymers and agents in the filter membrane results in dense plugging of the filter membrane, and pore size of the filter membrane getting smaller and smaller, through which the smaller particles in produced water can't pass. Any one of the above reasons will lead to the increase of ssc.

5. Attempts to Use Zwitterionic Bactericide

Regarding the influence of cationic corrosion inhibitor on suspended solids in produced water, the investigation on preparation, performance and contribution to suspended solids for a non-ionic Imidazoline corrosion inhibitor has been explored in previous work [7]. Since bactericide should contain cationic groups to adsorb to negatively charged bacteria, bactericides won't contribute to suspended solids only when the bactericide molecules are electrically uncharged. Therefore, this study tried to find a type of zwitterionic bactericide, and checked its performance [2] and the effect on the suspended solids in the produced water.

From table 2, it can be seen that the growth index of SRB is 221, the number of sulfate reducing bacteria (SRB) corresponding to the table [1] was 70 /mL, the dilution factor is 10^4, that is, the number of bacteria in the blank sample was 7×10^4 per milliliter, while table 3 and table 4 indicate that the inhibitory concentration for zwitterionic bactericide was 30-50 mg/L, which satisfies the current standard: when SRB numble is 10^4-10^5/mL, the inhibitory concentration is less than 70 mg/L [2].

Table 2. Numbering of SRB in Blank Water Samples

Order	0	1	2	3	4	5	6	7	8	9
Bottle 1	+	+	+	+	+	-	-	-	-	-
Bottle 2	+	+	+	+	+	+	-	-	-	-

Table 3. Numbering of SRB in Blank Water Samples with Zwitterionic Bactericide

Order	Agent concentration mg/L				
	30 mg/L	50 mg/L	70 mg/L	100 mg/L	120 mg/L
0	+ -	- -	- -	- -	- -
1	- -	- -	- -	- -	- -
2	- -	- -	- -	- -	- -
3	- -	- -	- -	- -	- -
4	- -	- -	- -	- -	- -
5	- -	- -	- -	- -	- -

Table 4. Numbering of SRB in Blank Water Samples with Zwitterionic Bactericide

Order	Agent concentration mg/L				
	30 mg/L	50 mg/L	70 mg/L	100 mg/L	120 mg/L
0	- -	- -	- -	- -	- -
1	- -	- -	- -	- -	- -
2	- -	- -	- -	- -	- -
3	- -	- -	- -	- -	- -
4	- -	- -	- -	- -	- -
5	- -	- -	- -	- -	- -

The contribution of the zwitterionic bactericide to ssc of produced water was also tested. The results are shown in the table 5. The ssc of blank water sample was 3.0 mg/L, with the addition of 40 mg/L of double quaternary amine salt bactericide was 19.4 mg/L averagely, and the ssc of water with the addition of zwitterionic bactericide was 4.8 mg/L averagely. Zwitterionic bactericide has little contribution to polymer-bearing produced water.

Table 5. The Contribution of the Zwitterionic Bactericide to ssc in Produced Water

Sample	Agent added	Agent concentration mg/L	Ssc mg/L	Average ssc mg/L
1	None	-	2.5	3.0
2	None	-	3.6	
3	Double quaternary amine salt bactericide	40	17.4	19.4
4	Double quaternary amine salt bactericide	40	21.5	
5	Zwitterionic bactericide	40	2.8	4.8
6	Zwitterionic bactericide	40	6.9	

6. Conclusions and Suggestions

The results show that cationic corrosion inhibitor and bactericide can interact with negatively charged suspended solids and polymers in produced water, resulting in the increase of ssc of produced water. Therefore, the use of cationic agents should be avoided for polymer-bearing produced water, and non-ionic and zwitterionic agents should be used instead. Experiments show that the zwitterionic ion bactericide not only has good bactericidal effect, but also has little effect on the ssc of produced water.

7. Acknowledgments

This research is supported by Shengli Oilfield Branch Company, SINOPEC (Project No. YC1905), and thanks to those who are involved in.

8. References

[1] SY/T5329—2012
[2] Q/SH1020 0688—2016
[3] SY/T5273-2014
[4] Wu S. Study on Efficiency of Bactericide in the Waster Water with Polymer. Thesis of Master Degree, Da Qing Petroleum Institute, 2007.
[5] Shen Zh., Huang ZhY, Li J, He Zh, Wu XY. Compatibility of the Main Agent in the Oilfield Water Treatment. Oilfield Chemistry. 2017(4): 688-693
[6] Paul Davison, and Eric Mentzer. Polymer flooding in North Sea Reserviors. Society of petroleum engineers journal, SPE 9300, SPE 55the Annual Technical Conference and Exhibition,Sept.21-24,1980, Dallas TEXAS, USA
[7] Sun LM, Gao CX, Gao WM, Cui J and Xu J H. Adaptability of Polymer-bearing Sewage Corrosion Inhibitor. Journal of Materials Science and Chemical Engineering. 2018(6): 70-80.

Fabrication of Bi_2MoO_6 Photocatalytic Fibers via Wet Spinning and Enhanced Photocatalytic Activity

Jun Sun[1,3], Yan Zhang[1,*], Xiaobing Tian[1,2,3], Meijun Yang[1], Xiaolong Yang[1] and Jianqiang Yu[1,*]

[1] Institute of Green Chemistry and Industrial Catalysis, School of Chemistry and Chemical Engineering, Qingdao University, Qingdao 266071, China;
[2] Resources and Environmental Engineering, Wuhan University of Technology, Wuhan 430070, Hubei, China;
[3] Qingdao Rongchuang Institute of Novel Materials, Qingdao International Academician Park, Qingdao 266000, China.

Abstract. In this paper, Bi_2MoO_6 photocatalytic fibers were fabricated by wet spinning method and the photocatalytic activity was found to be enhanced significantly. The Bi_2MoO_6 powders were prepared by hydrothermal process at different pH (0.4, 6, 8, 10), which is recorded as BMO-0.4, BMO-6, BMO-8, BMO-10. The powders were then mixed with sodium alginate colloid, and fabricated into photocatalytic fibers by wet spinning. The samples were recorded as FB-0.4, FB-6, FB-8, FB-10, respectively. The crystal phase, morphology and photophysical properties of the fiber samples were characterized by X-ray powder diffraction (XRD), scanning electron microscopy (SEM), and laser particle size analysis (LPSA). In order to investigate the photocatalytic performance of the synthesized photocatalysts, RhB and formaldehyde were used as simulated water pollutant and air pollutant, respectively. It showed that the efficiency of the degradation of RhB over BMO-10 was 75%, while that enhanced to 85% over FB-10. It should be noted that the content of photocatalyst in the fiber FB-10 was only 1%. The degradation efficiency of formaldehyde in FB-10 sample reaches 92%.

1. Introduction

With the development of the economy and the progress of society, environmental pollution and energy shortage are two major problems that human beings urgently need to solve. Since Fujishima and Honda reported the use of TiO_2 as a photoanode for UV photolysis of water [1], Bard reported that UV-irradiation of TiO_2 can transform CN^- to OCN^- [2], and more and more scholars have set their sights on it. Among the semiconductor materials for photocatalysis, visible-light semiconductor materials have become attractive photocatalysts due to their excellent light-absorption properties, adjustable band gap and low cost [3]. Bismuth molybdate is one of the representative materials. A large number of scholars have studied the doping of bismuth molybdate to form heterojunctions [4], but the application research on bismuth molybdate-based materials has not been developed.

Regarding the investigation of photocatalytic fibers, the most widely studied is TiO_2 photocatalytic fiber, and the photocatalytic performance of the fiber is studied by attaching TiO_2 to the fiber. For example, Yuan [5] research group connected the TiO_2 and activated carbon fiber with the calcined epoxy resin residue, and explored the relationship between the pore size of the fiber and its photocatalytic performance. Zhong [6] research group used hollow fiber (HOF) to combine with TiO_2 matrix composites to improve the photocatalytic activity. Cai [7] studied the preparation of TiO_2 fibers with cotton fibers as template, and then synthesized Ag_3PO_4/TiO_2 fibers by in-situ growth method on Ag_3PO_4 particles. However, TiO_2 can only be excited by ultraviolet light to generate electrons and holes, and photogenerated electrons and holes are easily recombined, resulting in low utilization of light energy. These problems limit its wide application.

The currently prepared photocatalytic fiber method mainly combines the photocatalyst with the fiber by a suitable binder, or directly adds the photocatalyst to the spinning solution and directly spun it by electrospinning or wet spinning. The alginic acid used in this paper is a linear polymer polyelectrolyte [8] extracted from the marine organism brown algae, which has a wide range of sources, environmental protection, biocompatibility, high hygroscopicity, flame retardancy, gel blocking, high oxygen permeability, high ion adsorption, antibacterial, and many other excellent properties. It has a wide range of applications in the fields of biology, food, textile, medical and health. The photocatalyst is used to control different pH values to prepare different structures of bismuth molybdate material [9], and the synthetic material is stirred and mixed with sodium alginate to form a colloid, and the photocatalytic fiber is directly spun by wet spinning. The degradation properties of different fibers to RhB and formaldehyde were investigated under visible light.

2. Experimental Method

2.1. Material preparation

$(NH_4)_6Mo_7O_{24}\cdot 4H_2O$ (0.72 mmol) was sonicated in 20 mL of H_2O, and $Bi(NO_3)_3\cdot 5H_2O$ (10 mmol) was completely dissolved in 20 mL of HNO_3 solution (2.0 mol·L^{-1}) and stirred for 30 min. The cerium nitrate solution was added dropwise to the ammonium molybdate solution under magnetic stirring. The pH of the mixture was adjusted to 0.4, 6, 8, and 10 using ammonia water. Stir for 30 minutes. The mixture was transferred to a reaction kettle and heated at 150 ° C for 24 hours. The product was taken out, washed with deionized water to neutrality, and then dried at 60 ° C for 8 hours to obtain Bi_2MoO_6 synthesized under different pH conditions. Recorded as BMO-0.4, BMO-6, BMO-8, and BMO-10.

20g $(C_6H_7O_6Na)_n$ was slowly poured into 450g of H_2O, stirred evenly to form a colloid. 5g of the obtained Bi_2MoO_6 was added to the colloid, stirred for 24 hours, placed in a spinning machine for 2 h, and $CaCl_2$ was used as a coagulating liquid. Compared with 1:1.2, spinning for one hour, photocatalytic fibers were obtained. The photocatalytic fibers prepared from Bi_2MoO_6 prepared at different pH values of 0.4, 6.0, 8.0, and 10.0 were recorded as FB-0.4, FB-6, FB-, 8, FB-10.

2.2. Sample Characterization

X-ray diffraction (XRD) spectra of the samples were measured on a Bruker D8-advance X-ray diffractometer. The experiment used Cu Ka target n = 0.15478 nm), and the scanning range was 5 ° to

80 °. Scanning electron microscopy (SEM) characterization was performed on a JSM-6700F field emission scanning electron microscope. Laser particle size analysis (LPSA) was performed on a Winner 2005 laser particle size analyzer.

The degradation of RhB was carried out by irradiating RhB under a 500 W xenon lamp (cut filter λ ≥ 420 nm), and the photocatalytic suspension was subjected to photocatalytic activity. In the degradation experiment, 40 mg of Bi_2MoO_6 photocatalyst was dispersed in 40 mL of RhB dye solution (4.0 $mg{\cdot}L^{-1}$) and stirred in the dark for 60 minutes to achieve desorption-adsorption equilibrium. 500 mg of the photocatalytic fiber was dispersed in 200 mL of RhB dye solution (1.0 $mg{\cdot}L^{-1}$) and stirred in the dark for 60 minutes to reach a desorption-adsorption equilibrium, after which the above suspension was irradiated with a light source at a distance of 10 cm. During the photocatalytic reaction, 4 mL of the liquid was taken out from the reactor at intervals of 30 minutes, and then the photocatalyst was separated by centrifugation. The ultraviolet absorption peak of the catalyst-free RhB solution was measured at 554 nm using a spectrophotometer.

Formaldehyde degradation was carried out in a self-made $1m^3$ reaction chamber. The sample fibers were arranged on the same length of 50×50cm transparent glass, placed in the center of the tank, and the continuous formaldehyde tester was connected to the tank with a leather tube. The internal formaldehyde concentration, using a specific method, so that the concentration of formaldehyde gas in the cabin is $1mg/m^3$, recorded as c_0, open 30w fluorescent lamp for irradiation, record the data once every 0.5h, record as c, for 8h, then the degradation efficiency is $(1 - c/c_0)$.

3. Results and Discussion

3.1. Phase Structure Analysis of Bi_2MoO_6

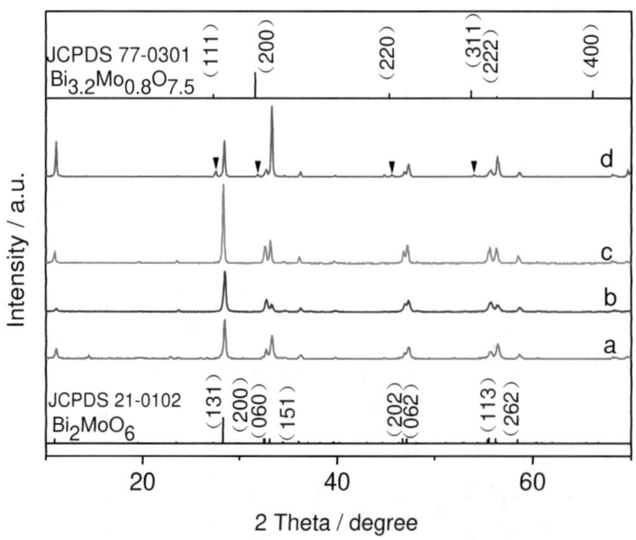

Figure1. XRD Spectrum of Bi_2MoO_6 Prepared under Different pH Conditions (a) pH=0.4; (b) pH=6; (c) ph=8; (d)=10

Figure 1 is a XRD pattern of the product prepared at different pH values. It can be seen from the figure that the diffraction peak intensity of all the samples is large and the peak is sharp, which indicates that the sample prepared by hydrothermal method has good crystallinity. (020), (131), (200), (002), (060), (151), (202), (113), (262) crystal planes appeared in the four groups of samples, indicating that there are γ-Bi2MoO6 in the substance. When the pH value is 0.4, it can be seen from the XRD spectrum that the product is a composite phase of Bi_2MoO_6 and Bi_2O_3. As the pH value increases, the product becomes pure phase Bi_2MoO_6, and the crystallinity of pH 8 is better than the crystallinity of pH 6, and at pH 8, the (111), (200), (220) , (311), (222), (400) crystal plane, indicating that $Bi_{3.2}Mo_{0.8}O_{7.5}$ gradually appeared in the product. When the pH reached 10, it was obvious that orthogonal γ-Bi_2MoO_6 and $Bi_{3.2}Mo_{0.8}O_{7.5}$ existed in the product, it is expected that a $Bi_{3.2}Mo_{0.8}O_{7.5}$ quantum dot modified Bi_2MoO_6 has a heterojunction structure with special properties such as quantum size effect, which is consistent with previous reports.

Figure 2 is a SEM photograph of Bi_2MoO_6 prepared under different pH conditions. It can be seen from the SEM photograph that the four groups of samples are mainly in the form of a sheet, the size of the first three particles are below 1 μm, and the size of the fourth particle is about 3 μm. When the pH reaches 8, spherical particles appear, $Bi_{3.2}Mo_{0.8}O_{7.5}$ appears, which is consistent with the XRD characterization. $Bi_{3.2}Mo_{0.8}O_{7.5}$ forms quantum dots on the surface of γ- Bi_2MoO_6, which is heterogeneous. Junction structure, such a structure helps to improve its photocatalytic efficiency. The synthesized particles are all around 1μm.

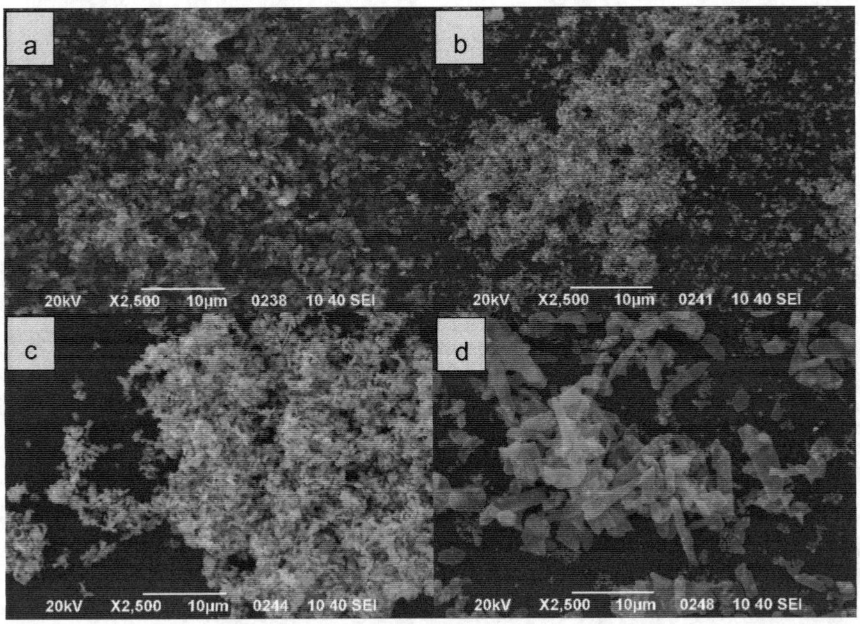

Figure 2. SEM Photograph of Bi_2MoO_6 Prepared under Different pH Conditions (a) BMO-0.4, (b) BMO-6, (c) BMO-8, (d) BMO-10

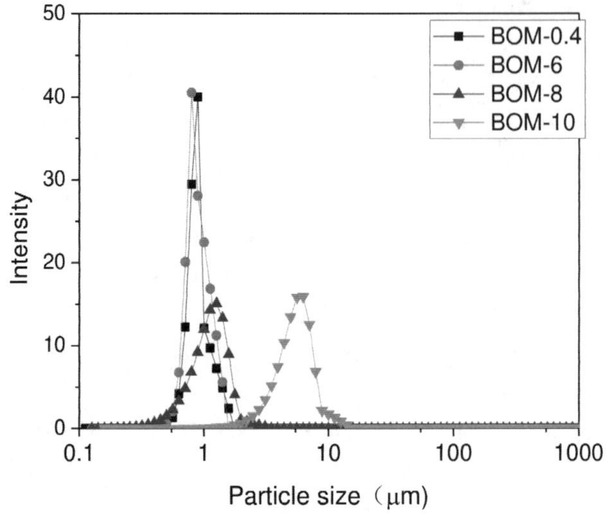

Figure 3. Laser Particle Size Analysis of Samples Prepared at Different pH Values

Figure 3 is a laser particle size spectrum obtained by the test, where in the sample particle size obtained by pH 0.4 is between 0.664-0.907μm, the sample size of pH 6 is between 0.563-0.765μm, and the sample size of pH 8 is The sample size of 0.559-1.320μm and pH 10 was between 2.238 and 4.979μm, which is consistent with the particle size data obtained by SEM. Such particle size ensures the progress of the wet spinning experiment.

3.2. Phase Structure Analysis of Fiber Samples

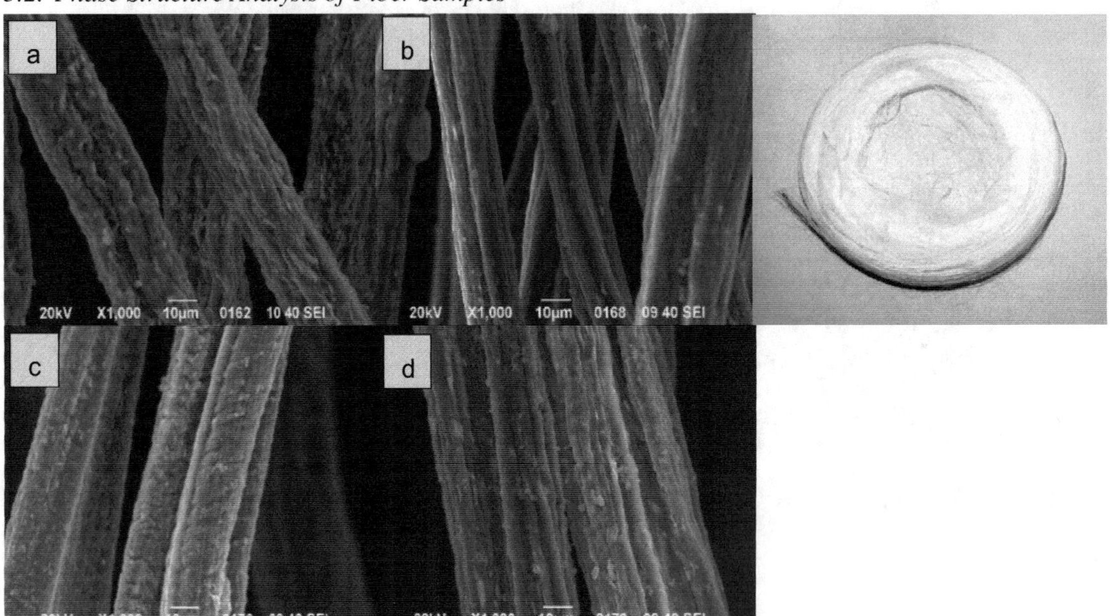

Figure 4. A. SEM Photographs of Fibers Made from Different Samples (a) FB-0.4, (b) FB-6, (c) FB-8, (d) FB-10 B. Apparent Photo of the Fiber Sample

Figure 4 is a SEM photograph of a photocatalytic fiber obtained by a wet spinning method. It can be seen from the picture that the fiber diameter is about 15μm, the aspect ratio is large, the thickness is relatively uniform, the surface is rough, and there are many tiny The particles are distributed on the surface of the fiber, and the distribution is relatively uniform and fine, and no agglomeration occurs. It proves that we successfully woven the photocatalyst into the fiber by the wet spinning method and the morphology is uniform. It is also apparent from the appearance that the fiber sample changes from white to light yellow fiber, and the distribution is fine, the touch is grain-free, and the photocatalyst and the fiber are well connected.

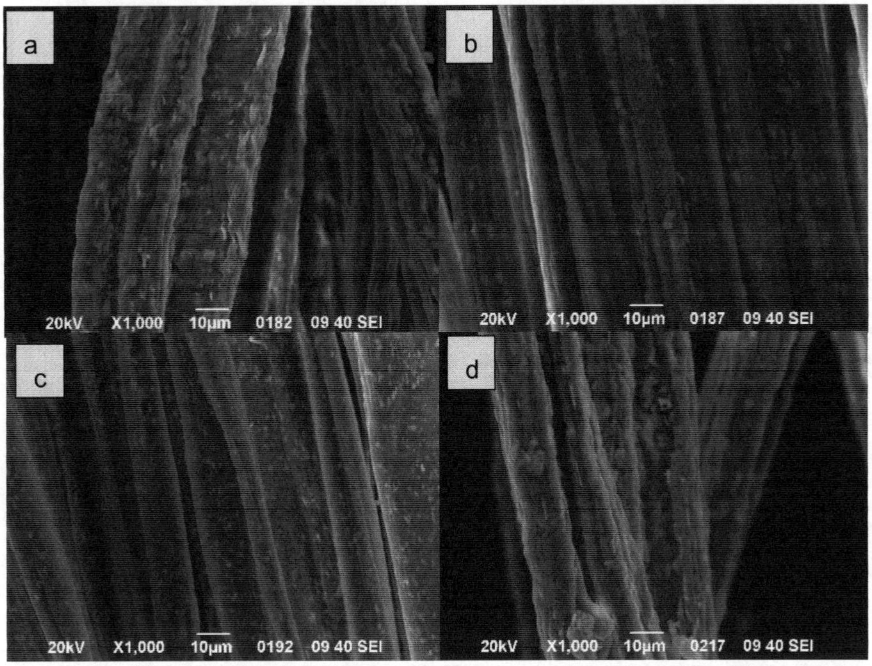

Figure 5. SEM Photograph of Fibers Soaked in Soapy Water for 1 Week

Figure 5 is a SEM picture of the obtained fiber soaked in soapy water for one week. It can be seen from the figure that the surface of the fiber is still attached with many tiny particles, and the distribution is still uniform, which is almost the same as the SEM picture of the fiber before soaking. It can be proved that the photocatalyst in the obtained photocatalytic fiber is relatively strong in connection with the fiber and is not easily broken.

3.3. Photocatalytic Property Investigation

Figure 6 is a graph showing the degradation curves of RhB by powder and photocatalytic fiber under visible light. It can be seen from Figure (a) that the sample prepared by pH 0.4 has a low catalytic performance and can only reach about 10%. The sample prepared at pH 0.6 is pure phase γ- Bi_2MoO_6, and its degradation efficiency can only reach about 25%. With the formation of $Bi_{3.2}Mo_{0.8}O_{7.5}$ and γ-Bi_2MoO_6 heterojunction, the degradation efficiency of the sample is improved. The degradation efficiency of BMO-8 reached 60%, and the degradation efficiency of BMO-10 reached 75%. Figure (b) shows the degradation RhB spectrum of the photocatalytic fiber produced. The degradation efficiency of FB-0.4 and FB-6 samples is very low, only about 10%, and the degradation efficiency of sample FB-8 can reach 60%. The FB-10 sample can reach about 85%. The degradation trend of the powder sample and the fiber sample is the same, in which the powder degrades 4 mg/L of RhB and the photocatalytic fiber degrades 1 mg/L of RhB. It is worth noting that the photocatalyst content in the fiber sample is only 1%. In this way, the photocatalyst adheres to the surface of the fiber, which greatly improves its photocatalytic performance.

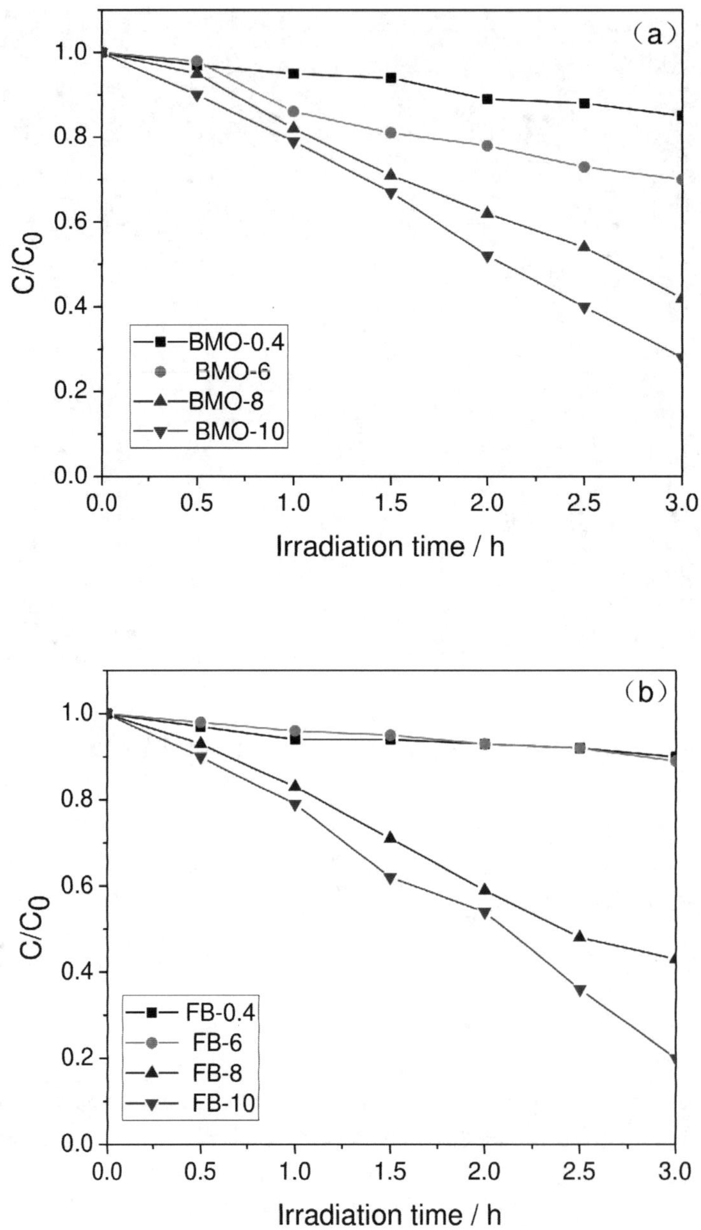

Figure 6. (a) Degradation Curve of RhB in Powder under Visible (b) Degradation of RhB by Photocatalytic Fibers under Visible Light

Figure 7 shows the degradation curve of formaldehyde produced by photocatalytic fiber, which is the same as the degradation of RhB. The degradation efficiency of FB-0.4 and FB-6 to CH2O is only about 15%.with the increase of $Bi_{3.2}Mo_{0.8}O_{7.5}$. The degradation efficiency of the fiber sample prepared by the formation of the heterojunction is greatly improved. The degradation efficiency of the sample prepared at pH 8 is about 70%, and the degradation efficiency of the FB-10 sample is over 92%.

The improvement of photocatalytic performance is due to its uniform distribution on the surface of the fiber, and the surface of the fiber is uneven, which greatly increases the contact area of the material with light and contaminants. The increase in specific surface area makes the degradation efficiency greatly improved.

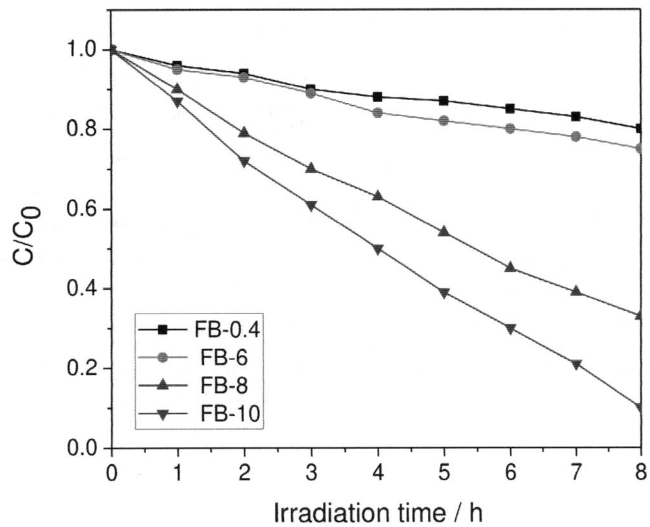

Figure 7. Degradation Curve of Photocatalytic Fiber on CH_2O under Visible Light

4. Conclusion

In this paper, catalyzed thermal and wet spinning methods were used to successfully prepare photocatalytic fibers with high catalytic performance. Among them, γ-Bi_2MoO_6 and $Bi_{3.2}Mo_{0.8}O_{7.5}$ were synthesized under hydrothermal conditions of pH 10,150 °C. The heterojunction structure sample has high catalytic performance. The degradation efficiency of RhB is about 75% at 3h. The photocatalytic fiber produced by this sample has a degradation efficiency of 85% for 3h, and the degradation efficiency is improved. At the same time, only 1% of the light is contained. The catalyst, such that the degradation efficiency of the photocatalyst is greatly improved compared to the powder. Among them, the degradation of formaldehyde by fiber samples reached more than 92%, and it has high degradation efficiency. In the face of the increasingly serious air pollution and water pollution, the fiber produced in this paper can be used for both water pollution treatment and air pollution treatment. It has a wide application prospect. With the deepening of research, it is believed that it is in the environment. Governance can do a lot.

5. References

[1] Fujishima A, Honda K. Electrochemical photolysis of water at a semiconductor electrode [J]. Nature, 1972, 238 (5358): 37-38.

[2] Frank SN, Bard A J. Heterogeneous photocatalytic oxidation of cyanide ion in aqueous solutions at titanium dioxide powder [J]. J Am Chem Soc, 1977, 99(1): 303-304.

[3] Cao S, Zhou P, Yu J. Recent advances in visible light Bi-based photocatalysts [J]. Chinese Journal of Catalysis, 2014, 35(7): 989-1007.

[4] Hao, Z.; Xu, L. Q.; Wei, B.; Fan, L. L.; Liu, Y.; Zhang, M. Q.;Gao, H. Nanosize α-Bi_2O_3Decorated Bi_2MoO_6 via an Alkali Etching Process for Enhanced Photocatalytic Performance. RSC Adv. 2015, 5, 12346-12353.

[5] Rusheng Yuan, Rongbo Guan, Jingtang Zheng. Effect of the pore size of TiO_2-loaded activated carbon fiber on its photocatalytic activity [J]. Scripta Materialia. 2005, 52: 1329-1334.

[6] Nianbing Zhong, Ming Chen, Yihao Luo, Zhengkun Wang, Xin Xin, Bruce E. Rittmann. A novel photocatalytic optical hollow-fiber with high photocatalytic activity for enhancement of 4 chlorophenol degradation [J]. Chemical Engineering Journal.2018, 355: 731-739.

[7] Li Cai, Qiyi Long, Chao Yin. Synthesis and characterization of high photocatalytic activity and Stable Ag_3PO_4/TiO_2 fibers for photocatalytic degradation of black liquor [J]. Applied Surface Science, 2014, 319: 60-67.

[8] LIU L, JIANG L, XU G K, et al. Potential of alginate fibers incorporated with drug-loaded nanocapsules as drug delivery systems [J]. Journal of Materials Chemistry B, 2014, 2(43): 7596-7604.

[9] Yanhua Peng, Qinghua Liu, Jinqiu Zhang, Yan Zhang, Mengjie Geng, and Jianqiang Yu, Enhanced Visible-Light-Driven Photocatalytic Activity by 0D/2D Phase Heterojunction of Quantum Dots/Nanosheets on Bismuth Molybdates [J]. Phys. Chem. C, 2018, 122: 3738-3747.

Preparation and Characterization of Neopentyl Glycol/Epoxy Resin Shape-Stabilized Phase Change Material for Thermal Energy Storage

Duo Meng, Kang Zhao, Anqi Wang and Wei Zhao
School of Civil and Architectural Engineering, Liaoning University of Technology, No.169 Shiying Street, Guta District, Jinzhou, China
Email: mengduo39@163.com

Abstract. Neopentyl glycol is a kind of solid-solid phase change material for thermal energy storage. In order to overcome volatilization weight loss caused by the high saturated vapor pressure of neopentyl glycol, epoxy resin was chosen as the encapsulation matrix to inhibit the volatilization of neopentyl glycol when it happens phase change or works in a relatively high temperature environment. The micro morphology analysis indicated that the neopentyl glycol was encapsulated evenly in the epoxy resin matrix. These two components had no chemical reactions and combined with each other by the physical encapsulation. Phase change temperature and latent heat of the composite was measured as 41.26°C and 57.04 J/g respectively, which proves the thermal heat storage performance of the neopentyl glycol was not influenced by the adding of epoxy resin. Meanwhile, attributed to the effective encapsulation and protection by epoxy resin, volatilization weight loss of the neopentyl glycol in the composite decreased significantly and behaved good thermal stability and reliability. After 100 times thermal cycling test, the composite hardly changed in the thermo-physical properties. Thus, the prepared neopentyl glycol/epoxy resin composite phase change material is supposed to be a potential and promising functional material for the applications in low temperature building energy conservation and solar energy storage.

1. Introduction

Phase change materials (PCMs) are a kind of functional materials which have the ability to absorb and release thermal energy at a certain temperature range near their phase change temperature. PCMs working as energy storage carrier can improve the energy utilization efficiency and alleviate the energy crisis. Attributed to the advantages such as high efficiency, saving energy and environmental protection etc., PCMs have been applied widely in the fields of industrial waste heat utilization, solar energy utilization, building thermal energy storage and so on [1]. The commonly used PCMs include solid-liquid and solid-solid phase transition forms, and there have been many researches and applications about these two kinds of PCMs [2]. However, liquid leakage is always a fatal problem for the liquid-solid PCMs in the process of heat absorption. Therefore, a packaging container [3] or shape-stabilized matrix [4-6] would be needed to overcome this disadvantage. Polyol is kind of solid-solid PCMs which do not generate liquid while absorbing the latent heat. They have the advantages of wider phase transition temperature range, small volume change, non-toxic, non-corrosion and loose requirements for PCM container and operation conditions. The latent heat of polyol is comparative with that of the solid-liquid PCMs and the super-cooling degree is small [7]. So the polyol is suggested to be a kind of potential and promising PCMs in latent thermal energy storage systems [8-10].

The biggest problem of polyol is the higher saturated vapor pressure which can generate the volatilization of polyol in the relatively low working temperature range. The volatilization will cause the weight loss of polyol and its thermal storage capacity will be influenced resultingly. This is profitless to the long-term thermal stability and reliability in engineering application for the polyol PCMs. Taking neopentyl glycol (NPG) as an example, its saturated vapor pressure is about 3 Pascal. In the repeated heating/cooling cycling process or high temperature working environment, the NPG may volatize and leads to the weight loss and finally influence the latent heat storage and service life of NPG servicing as the thermal energy medium. One of the feasible methods is to choose an encapsulation matrix to inhibit the volatilization of NPG. So the objective of this paper was the preparation of NPG/epoxy resin (EP) composite as a thermal-stable composite PCM by incorporation of NPG within the EP to resist the volatilization weight loss of NPG. EP has tight structure and is able to keep solid and rigidity near the solid-solid phase change temperature of NPG. The characterizations of chemical structure, micro morphology, thermo-physical properties and thermal reliability of the NPG/EP composite PCM were examined by FT-IR, SEM and DSC techniques, Service performance. In addition, the influence of the weight loss improvement of the NPG/EP composite PCM was also evaluated experimentally.

2. Experimental

2.1. Materials

Neopentyl glycol (NPG) working as the phase change material was supplied by Tianjin Kemiou Chemical Reagent Co., Ltd. Epoxy resin (EP) of E-44 type which was supposed to be encapsulation matrix to NPG and modified amine curing agent of T-31 type were provided by Wuxi Bluestars Petrochemical Co., Ltd. Anhydrous ethanol and acetone acting as solvent in the preparation of NPG/EP composite PCM was purchased from Tianjin Kemiou Chemical Reagent Co., Ltd.. The main physical parameters of E-44 EP are listed in Table 1.

Table 1. The Main Physical Parameters of EP

Property	Quantity
Appearance	No visible impurities
Epoxy equivalent/ g·eq-1	210~244
Viscosity /Pa·S	6~10
Softening point /°C	12~20

2.2. Preparation of the NPG/EP Composite PCM

The melt blending method was conducted on the preparation of the NPG/EP composite PCM.

1) The NPG was dissolved in anhydrous ethanol firstly and irradiated by the infrared lamp until the anhydrous ethanol volatilized completely. After cooled to the room temperature, the NPG was poured into a mortar and grinded into powders. The dissolved raw material of EP was put into a drying oven for about 2 h at the temperature of 50°C and then cooled to the room temperature for use.

2) A certain amount of NPG was put in the acetone solution and dissolved completely at 50°C. The equal weight of NPG and EP was mixed together uniformly by stirring to be a mixture solution. Then the T-31 modified amine curing agent was dripped in the mixture at the ratio of 20wt% by EP. After agitated evenly, the mixture solution was poured into the stainless steel mold coated with release agent. The mold was vacuumized for 2 h and then cured at the room temperature for 96 h. In order to make the EP cure completely and remove the acetone solvent, the mold filled with NPG and EP was dried in the drying oven at 65°C for 1 h. When cured entirely, the sample was naturally cooled to the room temperature and then taken out of the mold.

3) The anhydrous ethanol was used to flush the composite PCM block for several times until its surface was clear, and the NPG/EP composite PCM was finally obtained after dried in the drying oven to make the anhydrous ethanol volatilize out.

2.3. Instruments and Measurements

Fourier-transform infrared (FT-IR) was carried out on a EQUINX55 spectrophotometer to investigate the compatibility and interaction between NPG and EP matrix. Scanning electronic microscope (SEM) observations were performed for the PCM samples on a JSM-5600LV microscope. Thermal properties such as latent heat and phase change temperature of pure NPG and NPG/EP shape-stabilized PCMs were obtained by using differential scanning calorimetry (DSC) instrument (DSC822^e, Mettler). The amples were heated from 10°C to 70°C at a heating rate of 5°C min^{-1} in a static nitrogen atmosphere. Thermogravimetric (TG) analysis was subjected on a thermo-analyzer instrument (TGA/SDTA851^e, Mettler). The samples were treated in the instrument from atmosphere temperature to 700°C at a heating rate of 10°C min^{-1} in a static nitrogen atmosphere.

2.4. Thermal Cycling Test

For investigating the variation of thermo-physical properties of the NPG/EP composite PCM after enduring many times phase transition, accelerated heating-cooling thermal cycling test was conducted on the composite PCM. In the thermal cycling test, the heating temperature was 60°C, while the cooling temperature was determined to be 0°C implemented in a refrigerator. The heating-cooling process was repeated for 100 times and DSC analysis was used for measuring the phase change temperature and latent heat of the NPG/EP composite PCM after the thermal cycling.

3. Results and Discussion

3.1. Chemical Structure of the Composite PCM

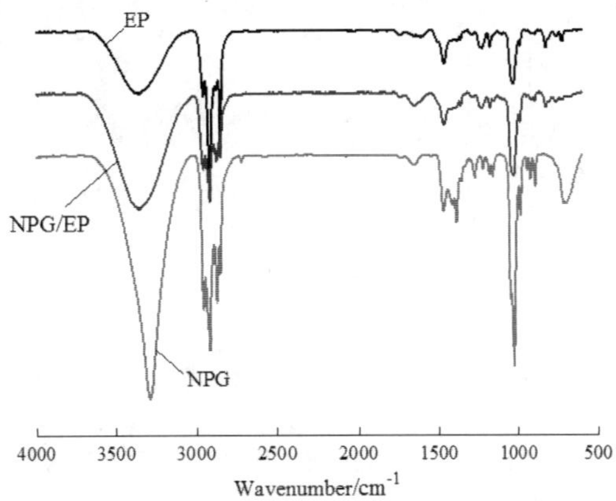

Figure 1. FT-IR Spectra of the Samples

FT-IR spectra of EP and NPG/EP composite PCM are shown in Figure 1. From the infrared absorption spectrum of NPG, it can be seen that there are two obvious absorption peaks in the wavenumber of 2915.84 cm^{-1} and 2848.34 cm^{-1}, which caused by the stretching vibration of C-H group. The overlapping peaks in 1465.64 cm^{-1} and 1390.42 cm^{-1} correspond to the bending vibration of C-H group. The peaks in 1274.72 cm^{-1} and 1224.58 cm^{-1} are the characteristic absorption peaks of -OH groups. As seen in the spectrum of EP in Figure 2, there is an absorption peak at 3355.53 cm^{-1} which is caused by the stretching vibration of N-H. The characteristic peaks of C=C aromatic nucleus appear at the wavenumber of 1461.78 cm^{-1} and 1589.1 cm^{-1}. From the FT-IR spectrum of NPG/EP composite in Figure 2, it is can be found that the typical peak of -OH in NPG has shifted to 1276.6 cm^{-1} and 1226.71 cm^{-1} respectively. The characteristic absorption peaks of C=C and C-H group both have appear, and the rest of the main characteristic peaks just have a little change in location and strength. Moreover, there are no new absorption peaks appear in the composite PCM. It is can be concluded that there are no chemical reactions between NPG and EP in the composite PCM. EP just plays a role in encapsulating NPG to confine its volatilization.

3.2. Morphology Analysis

For investigation the encapsulation effect of EP to NPG, the sectional microstructure of EP and NPG/EP composite PCM were observed by SEM respectively, shown in Figure 1. From the EP morphology of Figure 1(a), it can be seen that the surface of EP is smooth without any pores or particles and the structure is compact. Figure 2(b) shows the section microstructure of NPG/EG composite PCM. Compared with EP, NPG/EP composite is less compact than EP and has a lot of pores ranging from 0.5 μm to 12.5 μm. A certain amount of NPG crystal particles are found in the relatively big pores, which illustrates that most of NPG are encapsulated in the supporting material of EP in the small molecules form. When the environmental temperature reaches to the phase change temperature of NPG, EP can encapsulate the NPG so well that the volatilization of NPG would be restrained effectively. Because the softening point of EP is much lower than the solid-solid phase change temperature of NPG, EP can prevent the weight loss of NPG caused by volatilization in the composite PCM. Thus, the composite PCM can maintain the good thermal storage performance for a long time in service.

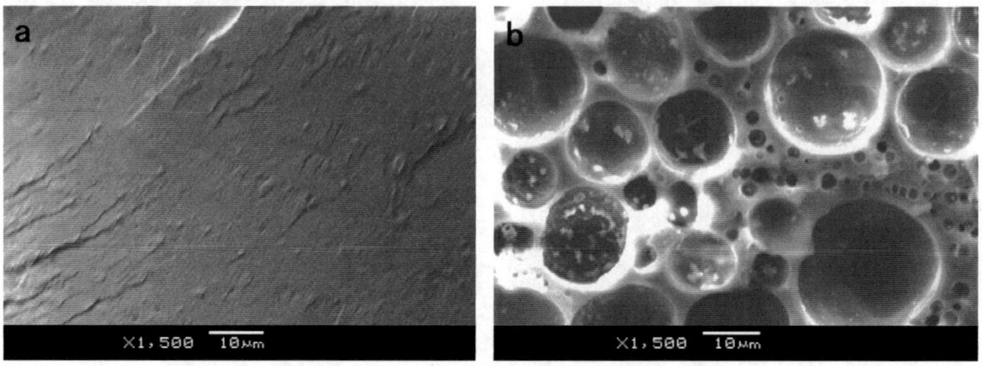

Figure 2. SEM Images: (a) EP; (2) NPG/EP Composite PCM

3.3. Thermo-Physical Properties of the Composite PCM

The important thermo-physical properties of PCMs including phase change temperature and latent heat were measured by DSC technique. Figure 3 illustrates the DSC curves of the NPG and NPG/EP composite PCM. In Figure 3, the DSC curve of NPG shows that the phase change temperature of NPG is 42.88°C and the latent heat is 115.1 J/g. As seen in the DSC curve of NPG/EP composite PCM, the phase change temperature of the composite is 41.26°C which is slightly lower than that of pure NPG. Due to no heat absorption behavior below 70°C of EP, the similar phase change temperature between NPG and NPG/EP composite indicates that the endothermic peak in the composite is only attributed to the phase change heat absorption of NPG. As a result, the latent heat of NPG/EP composite PCM is 57.04 J/g, which means that the mass fraction of NPG in the composite PCM is 49.6% based on Formula (1). The mass fraction of NPG in theory is 50% according to the preparation method, so it can be deduced that NPG has no loss during the reaction process with EP and maintains the intrinsic phase change thermal storage capacity. The FT-IR test results also verify that EP is only a supporting material and have little impact on the latent heat storage properties of NPG.

$$W_{NPG} = \frac{\Delta H_{NPG/EP}}{\Delta H_{NPG}} \tag{1}$$

Where W_{NPG} is the mass fraction of NPG in NPG/EP composite PCM, %; $\Delta H_{NPG/EP}$ is the latent heat of NPG/EP composite PCM, J/g; ΔH_{NPG} is the latent heat of NPG, J/g.

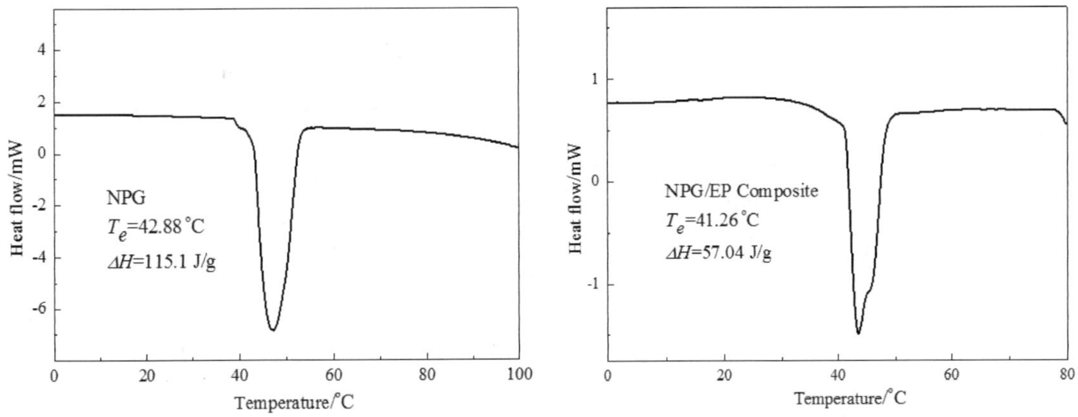

Figure 3. DSC Curves of NPG and NPG/EP Composite PCM

3.4. Volatilization Weight Loss of the Composite PCM

After NPG undergoing the repeated heating-cooling thermal cycling or working under a relatively high temperature environment, NPG may volatilize because of its higher vapor pressure. This will lead to the mass loss of NPG or NPG based composite PCMs, which affects the service performance in latent heat storage application of the PCMs. To examine the volatilization weight loss of the prepared NPG/EP composite PCM and the protection effect of EP to NPG, the weight loss variations were recorded by weighing the tested samples after heated to the temperature of 40°C, 60°C and 100°C, respectively. The weight loss rates of NPG, EP and NPG/EP composite are shown in Table Ⅱ.

Table 2. The Weight Loss of PCMs at Different Temperature

Material	Initial weight /g	Weight loss rate/%		
		40°C	*60°C*	*100°C*
NPG	10	0	0.8	22.1
EP	10	0	0	0
NPG/EP	10	0	0.05	3.25

From the results in Table Ⅱ, it can be known that EP has no weight loss in the three heating processes. Pure NPG has no weight loss either after it was heated to 45°C and happened solid-solid phase transformation. However, when heated to 60°C, the weight of NPG has decreased by 0.8% because of the volatilization. While heated to 100°C, the weight of NPG decreased significantly, by 22.1%. Due to EP having no weight loss below 100 °C, the weight loss of the NPG/EP composite PCM is caused only by NPG. As can be seen, the weight loss rate is only 0.10% at the temperature of 60 °C and 6.5% at 100 °C. By the comparison, it is obvious that the EP has remarkably suppressed the volatilization of NPG and improves the thermal stability of the NPG/EP composite PCM. The prepared NPG/EP composite PCM in our research is supposed to be used in the field of low temperature building energy conservation applications and the designed work temperature is less than 60 °C. While the composite PCM hardly has weight loss below 60°C proved by the test, as a result, the engineering application of this kind of NPG/EP composite PCMs is feasible.

3.5. Thermal Reliability of the Composite PCM

DSC curve of the NPG/EP composite PCM subjected 100 times accelerated heating-cooling thermal cycling test is given in Figure 4. As shown in Figure 4, the phase change temperature of the NPG/EP composite PCM is 41.13°C, being similar to the initial phase change temperature of NPG and the NPG/EP composite. Its latent heat is 55.97 J/g, which is reduced by1.88% than that of NPG/EP composite PCM before thermal cycling. According to the results in 3.4, the NPG/EP composite PCM happened mass loss at the heating temperature of 60°C ascribed to the volatilization of NPG completely. So after the repeated thermal cycling treatment, the heat storage capacity of the composite PCM would decline. On the basis of DSC result in Figure 4, however, the mass loss rate of the composite PCM has no directly proportional relationship to the thermal cycling times on the composite PCM. The latent heat of NPG/EP composite changes small and this change can be neglected in the engineering application. Moreover, the slight mass loss is able to be alleviated even avoided by using a reasonable application means or advanced construction method.

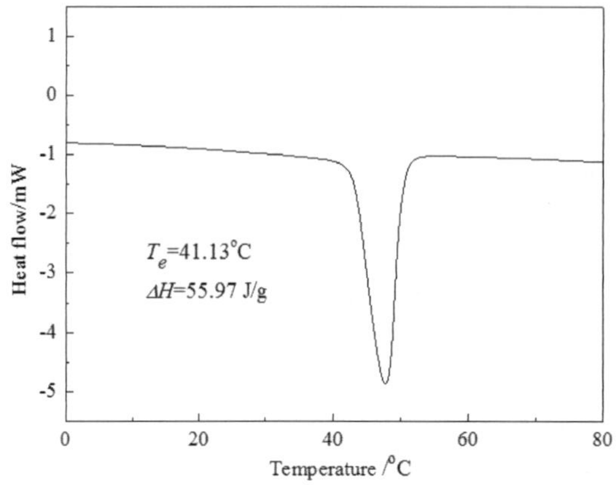

Figure 4. DSC Curve of NPG/EP Composite PCM after 100 Times Thermal Cycling Test

4. Conclusions

The NPG/EP composite PCM in which NPG working as phase change core and E type epoxy resin as encapsulated matrix was successfully prepared through the melt blending method. The NPG/EP composite PCM was characterized by SEM and FT-IR spectroscopy techniques. It is indicated that by curing and molding with EP simultaneously, NPG has been encapsulated into EP well. There were no chemical reactions between NPG and EP. According to the DSC test results, phase change temperature of the NPG/EP composite PCM was 41.26°C which was similar to that of pure NPG. The latent heat was 57.04 J/g, about 49.6% of the pure NPG. Heating experiment on the composite PCM indicated that EP played an important role in inhibiting the volatilization of NPG in the composite. The volatilization weight loss of the NPG/EP composite PCM rate decreased significantly than that of pure NPG. Furthermore, thermal cycling test confirmed the thermal stability and reliability of the NPG/EP composite PCM. The change on phase change temperature and latent heat of the composite PCM after 100 times heating -cooling thermal cycling was so small that can be neglected in the thermal storage applications.

Based on all results, it can be concluded that the NPG/EP composite PCM can be used in the building energy conservation and solar energy utilization etc. as thermal energy material due to absorption of heat in conjunction of solid-solid phase change of PCM.

5. Acknowledgement

The authors thank the testing works provided by the School of Materials Science and Engineering in Dalian University of Technology. This work was supported by the National Natural Science Foundation of China (No. 51308275) and Natural Science Foundation of Liaoning Province (No. SY2016004) is also gratefully acknowledged.

6. References

[1] Tyagi Vineet Veer and Buddhi D, " A review on phase change energy storage: materials and applications," Energy Conversiton and Management, vol. 45, Jun. 2004, pp. 1597-1615, doi:10.1016/j.enconman.2003. 09.015.

[2] N.Soares, J.J. Costa, A.R. Gaspar and P. Santos, "Review of passive PCM latent heat thermal energy storage systems towards buildings' energy efficiency," Energy and Buildings, vol. 59, Apr. 2013, pp. 82-103, doi:10.1016/j.enbuild.2012.12.042.

[3] I. S. Jacobs and C. P. Bean, “Thermal and flow behaviors in heat transportation container using phase change material,” Energy Conversiton and Management, vol. 49, Apr. 2003, pp. 698-706, doi:10.1016/j.enconman.2007.07.022.

[4] M. Li and Z. S. Wu, "A review of intercalation composite phase change material: Preparation, structure and properties," Renewable and Sustainable Energy Reviews, vol. 12, May 2012, pp. 2094-2101, doi.org/10.1016/j.rser.2012.01.016.

[5] D. Meng and L.J. Wang. "Characterization and thermal conductivity of modified graphite/fatty acid eutectic/PMMA shape-stabilized phase change material," Journal of Wuhan University of Technology, vol. 28, Jun. 2013, pp. 586-591, doi:10.1007/s11595-013-0735-0.

[6] T. Khadiran, M. Z. Hussein, Z. Zaina and R. Rusli, "Encapsulation techniques for organic phase change materials as thermal energy storage medium: A review," Solar Energy Materials & Solar Cells, vol. 143, Dec. 2015, pp. 78-98, doi:10.1016/j.solmat. 2015.06.039.

[7] X.W. Wang, E.R. Lu, W.X Lin and C.Z. Wang, "Micromechanism of heat storage in a binary system of two kinds of polyalcohols as a solid-solid phase change material," Energy Conversion and Management.vol. 41, Jan. 2000, pp. 135-144, doi:10.1016/S0196-8904(99)00096-5.

[8] X. W. Wang, E. R. Lu, W. X. Lin, T. Liu, Z. S. Shi, R. S. Tang and C. Z. Wang, "Heat storage performance of the binary systems neopentyl glycol/pentaerythritol and neopentyl glycol/trihydroxy methyl-aminomethane as solid–solid phase change materials," Energy Conversiton and Management, vol. 41, Jan. 2000, pp. 129-134, doi:10.1016/S0196-8904(99)00097-7.

[9] S. Divi, R. Chellappa and D. Chandra, "Heat capacity measurement of organic thermal energy storage materials," J. Chem. Thermodynamics, vol. 38, Mar. 2006, pp. 1312-1326, doi:10.1016 /j.jct.2006.02.005.

[10] Q. Cao and P. S. Liu, “Hyperbranched polyurethane as novel solid–solid phase change material for thermal energy storage,” European Polymer Journal, vol. 42, Jul. 2006, pp. 2931-2939, doi:10.1016/j. eurpolymj.2006.07.020

IOP Conf. Series: Materials Science and Engineering **735** (2020) 012015 doi:10.1088/1757-899X/735/1/012015

Preparation and Electrochemical Performance of Porous Si/SiOx/G Composite Anode for Lithium Ion Batteries

Jing Wang[1,2,3*], Xiaoyan Zhang[1], Ran Wang[1], Guoqiang Tan[1,2], Yuefeng Su[1,2,3] and Feng Wu[1,2,3]

1. School of Materials Science & Engineering, Beijing Institute of Technology, Beijing Key Laboratory of Environmental Science and Engineering, Beijing, 100081, China.
2. National Development Center of High Technology Green Materials, Beijing, 100081, China.
3. Collaborative Innovation Center of Electric Vehicles, Beijing, 100081, China.

[*]e-mail: wangjingbit98@bit.edu.cn

Abstract. Silicon/silicon oxides/graphite (Si/SiOx/G) composite with high capacity and excellent cycling performance has been synthesized via high temperature treatment, HF etching and high-energy ball milling for mixture of silicon oxide and graphite. The chemical composition and morphology of the sample was characterized by XRD, SEM and TEM. Large numbers of nanopores were formed in the as-prepared composite, providing three-dimensional transmission pathways for lithium ions and electrons. After 100 cycles, the electrode kept an excellent reversible capacity of 804.2mAh·g^{-1}. The excellent electrochemical properties are mainly attributed to the uniform distribution of silicon particles in amorphous silicon oxide, the buffering effect of porous structure and the improvement of conductivity of the composite by graphite.

1. Introduction

With the development of a series of new technologies such as electric vehicles and portable electronic devices, the lithium-ion batteries (LIBs) with higher energy density have been in widely demand. However, the theoretical capacity of commercial graphite anodes is only 372 mAh·g^{-1}, which hardly continues to be used as the anode material for the next-generation LIBs [1-3]. Therefore, the silicon-based anode with ultra-high theoretical capacity (3580 mAh·g^{-1}) has caused wide attention in recent years and is considered as the most promising anode material for next-generation LIBs. However, the serious volume change (>300%), which will cause destruction and mechanical pulverization of the material structure, severely limits its commercial application. In addition, the electrical conductivity of crystalline silicon is very low and limits its rate performance [4-7].

Compared with crystalline silicon, silicon monoxide (SiO) anode materials have lower volume expansion and better cycle performance, which is favoured by many researchers [8-10]. However, SiO is unstable at high temperature and is prone to dismutase to nano-silicon and silicon oxide (SiOx) or amorphous silicon dioxide, in which the nano-silicon is surrounded by silicon oxide (SiOx). During the first discharge, SiOx can react with lithium to form lithium oxide (Li_2O) and a series of lithium silicate phases, of which lithium silicate phases have a good buffering effect on the volume expansion of silicon. But, the disproportionation of SiO usually happens at high temperature (>900℃) and this process will take a long annealing time. In addition, the prepared nano-silicon is easy to agglomerate to form large particles [11-15]. In this study, NaOH was used to reduce the temperature of

IOP Conf. Series: Materials Science and Engineering **735** (2020) 012015 doi:10.1088/1757-899X/735/1/012015

disproportionation of SiO, and a porous silicon-based composite material was prepared by etching with HF, followed by high-energy ball milling with graphite to further reduce the size of particles and improve the electrical conductivity of the composite.

2. Experimental Section

2.1. Material Synthesis

The preparation process of the material is shown in Figure 1. First, 2 g of silicon monoxide powder (SiO, 200 meshes) is mixed with sodium hydroxide solution, and stirred at room temperature. When the solution is evaporated to dryness, it is placed in a tube furnace, heat treated at 700℃ for 30 min in an argon atmosphere. When cooling to room temperature, the prepared material was washed with acetic acid solution to remove sodium silicate formed during the reaction, and the product A was obtained. The product A was added to a 10% HF solution, stirred at room temperature for 10 min, then washed with deionised water until neutral, and dried in an oven at 80℃ to obtain a porous material B (labeled as PSS). The product B and the commercial graphite were put into a planetary ball mill (Fritsch Puluerisette 7, Germany) and the final porous Si/SiOx/Graphite composite (labeled as PSS-G) was obtained.

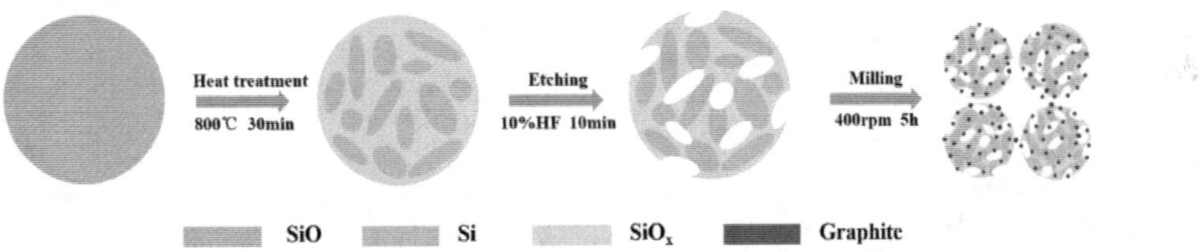

Figure 1. Schematic Illustration of the Synthesis Route of the Si/SiOx/Graphite Composite

2.2. Instruments and Methods

The phase composition of the material was analyzed by X-ray diffractometry (XRD, Ultima IV-185, Japan) with a scanning range of 10° to 80°; the microtopography of the composite was analyzed by scanning electron microscopy (SEM, Hitachi, S-4800) and transmission electron microscopy (TEM); the microstructure of the material was characterized by Raman test (RM, JY Labram HR 800); the surface chemistry of the material was studiedd by X-ray photoelectron spectroscopy (XPS, PHI Quantera).

2.3. Electrochemical Performance

The active material (SiO, PSS, PSS-G), binder (sodium alginate) and acetylene black were mixed at a mass ratio of 70:15:15 to prepare slurry. After casting the slurry on the copper foil, the electrode was dried at 80℃ under vacuum to remove the solvent. Then the film was punched and cut into a circular pole piece with a diameter of 11mm as a working electrode. The coin-type cells were assembled with Li metal as a counter electrode in an Ar-filled glove box, the Celgard 2400 is used as the diaphragm, and the 1 mol/L $LiPF_6$/ (EC+DMC) (volume ratio is 1:1) is used as the electrolyte. Galvanostatic charge/discharge were carried out on a battery testing system(Land CT2100A) within a voltage window of 0.01-2.0 V with current density of $100mA{\cdot}g^{-1}$. Cyclic voltammetry (CV) measurement and impendance testing were performed on a electrochemical workstation (CHI 660E).

3. Results and Discussion

3.1. Structure and Morphology

Figure 2a shows XRD patterns of original SiO, porous material PSS (product B) and PSS-G composite. In the XRD pattern of the original SiO, a diffuse peak appeared between 20° and 30°, indicating that SiO is amorphous. Two sharp peaks appeared at 26.6° and 28.4° respectively, indicating that crystalline SiO_2 and Si phases are present in the original SiO. The XRD pattern of PSS sample shows obvious diffraction peaks of crystalline Si (PDF#27-1402) and cristobalite (PDF#39-1425), which indicates that the disproportionation reaction of SiO occurs after heat treatment at 700 °C. The reaction process is $3SiO + 2NaOH = Na_2SiO_3 + H_2\uparrow + Si + SiO_2$, and the amorphous SiO_2 is converted into a crystalline state of cristobalite in the presence of NaOH [16]. According to the *Scherrer* formula, the average particle sizes of the cristobalite and crystalline Si are 20 nm and 14 nm, respectively [17]. In the XRD pattern of PSS-G, the peaks of graphite appeared at 26.5°, 44.5° and 54.7°, and the diffraction peaks of crystalline Si and cristobalite are still existed, indicating that the ball milling did not destroy the crystal structure of the material. Figure 2b is the Raman spectrum of PSS-G composites, the peaks of D-band and G-band appeared around 1360 cm^{-1} and 1580 cm^{-1} due to the addition of graphite. The weak peak appearing around 520 cm^{-1} proves the presence of crystalline silicon in the PSS-G, which consistents with the XRD results. Figure 2c shows the XPS (Si 2p) spectrum of the PSS-G composite, which contains silicon species in the valence state of Si^{4+}, Si^{3+}, Si^{2+}, Si^{1+} and Si^0, and with the highest content of Si^{3+}, indicating that the silicon in the composite is mainly in the form of silicon oxide, and the appearance of Si^0 also indicates that the crystalline silicon is generated during the disproportionation.

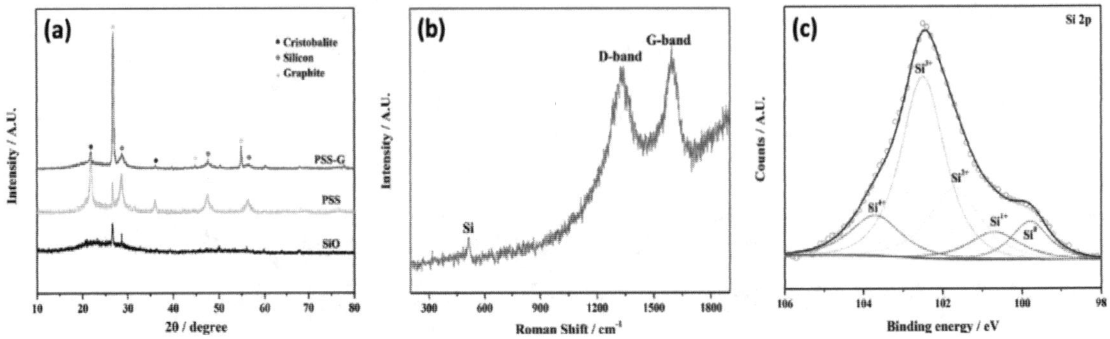

Figure 2. (a) XRD Patterns of Sample SiO, PSS and PSS-G, (b) Roman and (c) XPS Si 2p Spectra of PSS-G Composite

Figure 3 is the morphology images of graphite, SiO and PSS (product B). Figure 3a shows that graphite has a sheet-like structure with a particle size on the level of micronanometers. In Figure 3b, the morphology of SiO is irregular, the surface of the particles is smooth and has sharp edges and corners, and the particle size is about 20-40μm. Figure 3d shows that after high temperature heat treatment and HF etching, the surface of the material particles becomes rough, and many pores appear on the surface to form a porous structure, which can provide a three-dimensional channel for the transport of lithium ions. The material in Figure 3e is flocculated, indicating that a porous structure appears in the material after HF etching. A large number of nanocrystalline particles appear in Figure 3f and are surrounded by amorphous silicon oxide, which illustrating the formation of a crystalline Si and cristobalite phase in the disproportionation reaction, this corresponds to the XRD pattern.

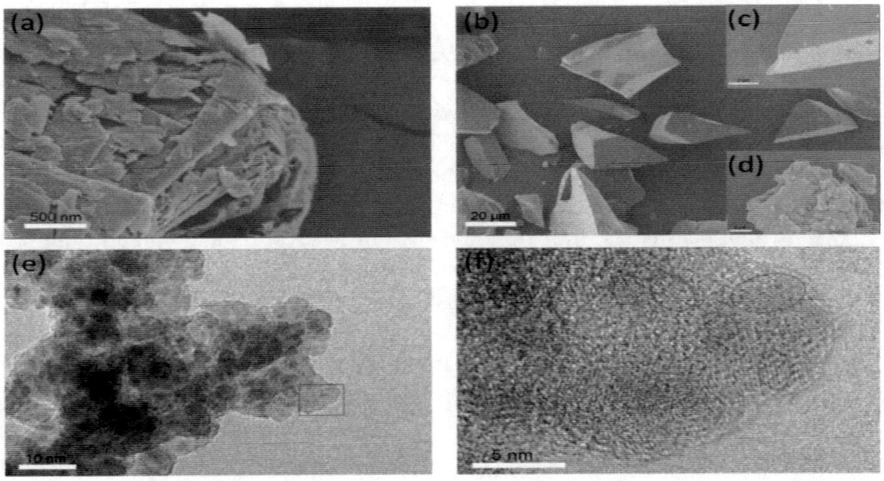

Figure 3. SEM Images of Sample (a) Graphite, (b and c) SiO, (d) PSS and TEM Images of (e and f) PSS Composite

3.2. Electrochemical Properties

Figure 4a is the CV curve of PSS-G composite electrode. In the first lithium intercalation process, there is a broad reduction peak between 0.6-1.0V, corresponding to the decomposition of electrolyte to form the SEI film [18], but the reduction peak disappeared from the second cycle and the curve becomes smoother. There is a sharp reduction peak near 0 V, which corresponds to the intercalation of lithium into bulk silicon. The anode peaks at 0.33V and 0.52V correspond to the delithiation process of SiO [19], and its intensity is continuously enhanced during the subsequent cycles, but the peak position does not move significantly, which indicates that the kinetic enhancement process occurs on the electrode.

Figure 4b shows the charge-discharge curve for the first three cycles of PSS-G electrode. The ICE of the PSS-G electrode is 62.2%, which has a large increase compared to pure SiO (49.8%). But the PSS-G electrode still has a large capacity loss, mainly because the irreversible reaction of SiO during the first cycle, which consumes a large amount of Li^+ to form electrochemically inert Li_2O, Li_3SiO_3, Li_4SiO_4 and so on. The Coulombic efficiency increased rapidly during the next series cycles, which indicating that the SEI film tends to be stable.

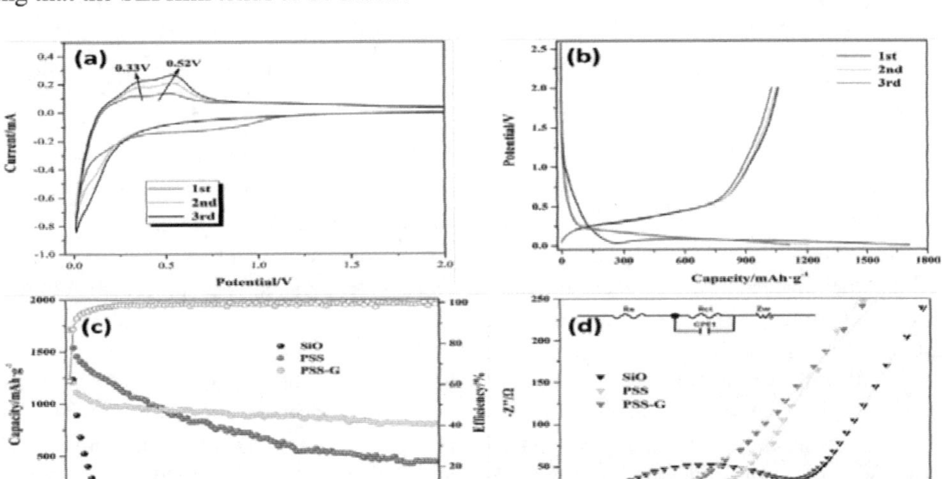

Figure 4. (a) CV and (b) the Charge-discharge Curves of PSS-G Composite Electrode, (c) Cycle Performance and (d) Nyquist Plots of Sample SiO, PSS and PSS-G

Figure 4c shows the cycle performance of sample SiO, PSS and PSS-G. Although the capacity of pure SiO can reach to 2305.5 mAh·g^{-1} in the first cycle, the reversible capacity decreased rapidly as the cycle proceeded. After 17 cycles, the capacity was lower than 100 mAh·g^{-1}. After high temperature treatment and HF etching, the first cycle capacity of the PSS electrode increased to 2535.5 mAh·g^{-1}, and the cycle stability was also improved. This is mainly because the nano-Si particles generated during the disproportionation process are surrounded by the amorphous silicon oxide, which can alleviate the volume effect of nano-Si. Also, partial silicon oxide was etched by HF to form large number of pores, which can reduce the stress change during cycling and can also buffer the volume expansion. However, the capacity of PSS electrode decreased in the subsequent cycles, this is mainly because of the low electrical conductivity of Si and destruction of material structure in cycling. When the PSS and graphite are mixed by ball milling, the cycle stability of the material is further improved. The reason for this is that, on the one hand, the particle size decreases after ball milling, and the stress variation due to volume change during cycling is reduced; On the other hand, graphite has a good conductivity, it can ensure good electrical contact between particles, reducing the possibility of peeling off between active silicon particles for particles. Therefore, the PSS-G composite exhibits a good electrochemical cycle stability, and its reversible capacity can reach 804.2 mAh·g^{-1} after 100 cycles.

Figure 4d shows the Nyquist plots of sample SiO, PSS and PSS-G. Each of the curves has a similar shape which consists of a semicircle and a straight sloping line. The semicirclele indicates the charge-transfer resistance (Rct) and the sloping line indicates the diffusion impedance (Zw) of Li^+ in materials. Obviously, the three curves have significant differences in the electrochemical impedance parameters [20]. Table1 shows the electrochemical impedance parameters of the three electrodes. It can be seen from Table 1 that the PSS-G electrode has the lowest charge transfer impedance, indicating that ball milling and the addition of graphite improves the rate of charge transfer between the interfaces. In addition, there also has a large difference between the ohmic impedances. The data in Table 1 shows that the PSS-G electrode has the lowest ohmic resistance value of 4.0Ω, which indicates that the addition of graphite also improves the electrical contact between the electrode material and the current collector, the structural pulverization of the material and the isolation of the active particles caused by the volume expansion of the material during cycling are avoided, so the cycle stability is improved.

Table 1. Electrochemical Impedance Parameters of the Three Electrodes

Sample	Rs(Ω)	Rct(Ω)
SiO	27.2	102.3
PSS	17.8	78.6
PSS-G	4.0	63.2

4. Conclusion

Porous Si/SiOx/Graphite composite anode materials were prepared by commercial SiO and graphite by high temperature heat treatment, HF etching and high energy ball milling. The nano-Si particles generated by the disproportionate reaction in the composite material are uniformly coated by the silicon oxide substrate. In addition, the material is etched by HF to form a porous structure, which can buffer the volume expansion of silicon, so the cycle performance of SiO was obviously improved. The high-energy ball milling further reduces the size of the particles, the electrical conductivity and cycle performance of the material are further improved with the addition of graphite which acts as a "connection medium" between the active material particles and the current collectors. At a current density of 100mA·g^{-1}, the reversible capacity was 804.2 mAh·g^{-1} after 100 cycles, and no obvious capacity decline was found.

5. Acknowledgments
This work was financially supported by the National Key R&D Program of China(2018YFB0104401).

6. References
[1] J.M. Tarascon, M. Armand, Issues and challenges facing rechargeable lithium batteries, *NATURE*, 414(6861), 359(2001)
[2] M. Armand, J.M. Tarascon, Building better batteries, *NATURE*, 451(7179), 652(2008)
[3] H. Li, Z.X. Wang, L.Q. Chen, X.J. Huang, Research on Advanced Materials for Li-ion Batteries, *Advanced materials*, 21(45), 4593(2009)
[4] C.K. Chan, H.L. Peng, G. Liu, K. Mcilwrath, X.F. Zhang, R.A. Huggins, Y. Cui, High-performance lithium battery anodes using silicon nanowires, *Nature nanotechnology*, 3(1), 31(2008)
[5] Y.S. Hu, R. Demir-Cakan, M.M. Titirici, J.O. Muller, R. Schlogl, M. Antonietti, J. Maier, Superior storage performance of a Si@SiOx/C nanocomposite as anode material for lithium-ion batteries, *Angew. Chem. Int. Ed.*, 47(9), 1645(2008)
[6] H. Gao, L.S. Xiao, I. Plumel, G.L. Xu, Y. Ren, X.B. Zuo, Y.Z. Liu, C. Schulz, H. Wiggers, K. Amine, Z.H. Chen, Parasitic Reactions in Nanosized Silicon Anodes for Lithium-Ion Batteries, *Nano Lett.*, 17(3), 1512(2017)
[7] J. Ryu, D. Hong, M. Shin, S. Park, Multiscale Hyperporous Silicon Flake Anodes for High Initial Coulombic Efficiency and Cycle Stability, *ACS nano*, 10(11), 10589(2016)
[8] C.H. Doh, C.W. Park, H.M. Shin, D.H. Kim, Y.D. Chung, S.I. Moon, B.S. Jin, H.S. Kim, A. Veluchamy, A new SiO/C anode composition for lithium-ion battery, *Journal of Power Sources*, 179(1), 367(2008)
[9] J.H. Yom, S.W. Hwang, S.M. Cho, W.Y. Yoon, Improvement of irreversible behavior of SiO anodes for lithium ion batteries by a solid state reaction at high temperature, *Journal of Power Sources*, 311, 159(2016)
[10] J. Woo, S.H. Baek, A comparative investigation of different chemical treatments on SiO anode materials for lithium-ion batteries: towards long-term stability, *RSC Adv.*, 7(8), 4501(2017)
[11] C.M. Park, W. Choi, Y. Hwa, J.H. Kim, G. Jeong, H.J. Sohn, Characterizations and electrochemical behaviors of disproportionated SiO and its composite for rechargeable Li-ion batteries, *Journal of Materials Chemistry*, 20(23), 4854(2010)
[12] Y.R. Ren, M.Q. Li, Si-SiOx-Cristobalite/Graphite Composite as Anode for Li-ion Batteries, *Electrochimica Acta*, 142, 11(2014)
[13] S.Z. Zeng, X.R. Zeng, L. Huang, H.L. Wu, Y.C. Yao, X.F. Zheng, J.Z. Zou, The formation mechanisms of porous silicon prepared from dense silicon monoxide, *RSC Adv.*, 7(13), 7990(2017)
[14] B.C. Yu, Y. Hwa, C.M. Park, H.J. Sohn, Reaction mechanism and enhancement of cyclability of SiO anodes by surface etching with NaOH for Li-ion batteries, *Journal of Materials Chemistry A*, 1(15), 4820(2013)
[15] J.I. Lee, K.T. Lee, J. Cho, J. Kim, N.S. Choi, S. Park, Chemical-assisted thermal disproportionation of porous silicon monoxide into silicon-based multicomponent systems, *Angewandte Chemie*, 51(11), 2767(2012)
[16] J.I. Lee, N.S. Choi, S. Park, Highly stable Si-based multicomponent anodes for practical use in lithium-ion batteries, *Energy & Environmental Science*, 5(7), 7878(2012)
[17] B.D. Cullity, Elements of X-Ray Diffraction, *Addison-Wesley*, (1978)
[18] H. Tang, J.P. Tu, X.Y. Liu, Y.J. Zhang, S. Huang, W.Z. Li, X.L. Wang, C.D. Gu, Self-assembly of Si/honeycomb reduced graphene oxide composite film as a binder-free and flexible anode for Li-ion batteries, *Journal of Materials Chemistry A*, 2(16), 5834(2014)
[19] J. Wang, M. Zhou, G. Tan, S. Chen, F. Wu, J. Lu, K. Amine, Encapsulating micro-nano Si/SiOx into conjugated nitrogen-doped carbon as binder-free monolithic anodes for advanced lithium ion batteries, *Nanoscale*, 7(17), 8023(2015)

IOP Conf. Series: Materials Science and Engineering **735** (2020) 012015 doi:10.1088/1757-899X/735/1/012015

[20] X. Yu, H. Yang, H. Meng, Y. Sun, J. Zheng, D. Ma, X. Xu, Three-Dimensional Conductive Gel Network as an Effective Binder for High-Performance Si Electrodes in Lithium-Ion Batteries, *ACS applied materials & interfaces*, 7(29), 15961(2015)

IOP Conf. Series: Materials Science and Engineering **735** (2020) 012016 doi:10.1088/1757-899X/735/1/012016

Application of Trace Property in Fire Cause Identification

Liantie Wang and Qingshan Meng

Shenyang Fire Research Institute, 110034, Huanggu District, Wenda road 218-20 first, Shenyang, China

Email: wangliantie@syfri.cn

Abstract. In the investigation of fire causes, there are many kinds of weld marks extracted from the fire site, such as the first and second short circuit weld marks. The relationship between the melting property and the cause of fire can be identified according to the different microstructure characteristics of the weld marks formed on the copper and aluminium wires. Therefore, in the determination of the cause of fire, the nature of the weld mark will directly affect the result of the determination, and sometimes a completely different conclusion will be drawn. However, sometimes in the same fire scene can be extracted from more than one weld mark, and the nature of the weld mark is not the same, which will bring certain difficulties to the fire cause identification work, so the organic combination of all kinds of weld mark, reasonable use, in order to better complete the fire scene cause identification work.

1. Introduction

Society is progressing and science and technology are developing. With the continuous expansion of electrical applications, electrical fires have an increasing impact on people. The number of electrical fires has ranked first among all kinds of fires in China. According to the fire department of the ministry of public security "China fire statistics yearbook" data show [1]: in 2008, a total of 13.30,000 fires (refer to statistics month, excluding forest, grassland, army and underground part of mine, the same below), and 40,000 electrical fires, accounting for 30 percent of the total number of fires.1%.In the 15 years from 1993 to 2007, 152 fires broke out across the country.760,000, including 37 electrical fires.370,000 fires, or 24 percent of the total.5%, which has caused huge loss of life and immeasurable loss of property to the society. Fire scene, however, differ in thousands ways, could even say that no two the same fire scene, this gives the fire cause investigation work bring greater difficulties, which gather relevant evidence, the scene of the fire and scientific analysis, using these material evidence, to provide fire cause investigation is the most direct evidence that is particularly important.

According to the fire rescue bureau of the emergency management department [2], in 2018, a total of 237,000 fires were reported nationwide, resulting in 1,407 deaths and 798 injuries. Direct property losses of 3.675 billion yuan have been counted. Among them, there was one particularly serious fire, one more than the previous year; there were four major fires, one less than the previous year; 67 major fires occurred, two more than the previous year. The fire situation is still not optimistic. Among all the fires, electrical fires still account for a large proportion and the total number of fires caused by the violation of the regulations on electrical installation and use accounts for 34.6%.Major fires have a greater impact, and the number of major fires increased year on year.

2. Extraction of Physical Evidence

Fire cause investigation is a multidisciplinary comprehensive technology. Physical evidence to be extracted on site can be divided into physical evidence, chemical evidence and electronic evidence, etc. Only by mastering multidisciplinary knowledge can the most valuable fire material evidence be extracted to provide strong evidence for the investigation of fire cause. The identification of fire material evidence plays a very important role in fire investigation. The accuracy of material evidence extraction can ensure the accuracy of cause investigation and greatly improve the efficiency of fire investigation.

Copper, aluminum conductor short-circuit arc formation of weld mark are sometimes caused by high temperature melting, sometimes is formed by the fire, but no matter how, the nature of weld mark are can represent the characteristics of the atmosphere at that time, due to the different environment atmosphere in the whole process of the welding scar formation and microstructure of the present also has the characteristics of each are not identical, so keep the weld mark formed their respective characteristics. Therefore, from the extraction of physical evidence at the scene of fire, to the analysis of the property of the material evidence, and then to the application of the property of the weld mark, each step has an internal relationship.

3. Trace Metallographic Analysis

In 2016, when a fire broke out in a comprehensive shopping mall, fire investigators examined the site and extracted the remnants of the fire site, including several molten beads, cable tray and residual wires. Professional and technical personnel carried out technical appraisal on the samples submitted for inspection. The equipment used includes digital camera, grinding machine, metallographic microscope, etc.

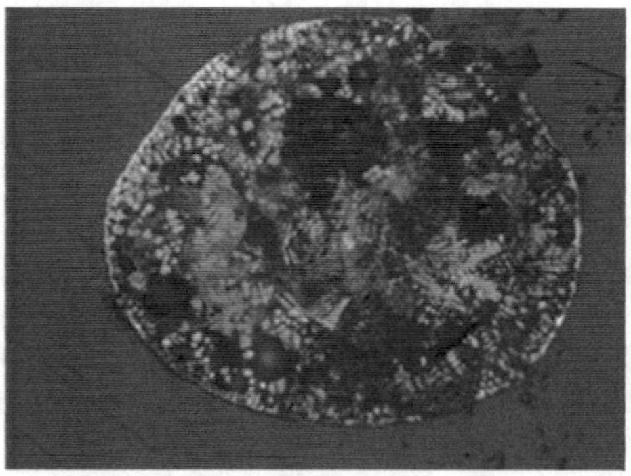

Figure 1. Metallurgic structure of short circuit splashbead

Figure 2. Metallurgic structure of fire bead

3.1. Sample Metallographic Analysis

After metallographic analysis, the sample of # 1 molten bead sent for inspection contains short-circuit splashing molten bead and fire bead (the metallographic photos of one of them are attached respectively, see Fig.1and 2;2# molten beads are all fired beads , see Fig.1;3# melted mark is secondary short circuit melted mark ,see Fig. 4.

First of all, let me show you the fire scars. As the name implies, it is the mark formed by melting after receiving the high temperature of the flame, which is one of the consequences of the fire rather than the cause of the fire.2# weld mark,3# melted mark and4# melted mark do not support that the cause of fire is related to electricity.

3.2. The Relationship between Microstructure Characteristics of Fusion Marks

Secondary short circuit melted mark with a short circuit melted mark belongs to the instantaneous electric arc of high temperature melting, cooling speed, the characteristics of the melting range is small, but they have differences, secondary short circuit melted mark [3], secondary short circuited melted mark refers to the copper and aluminum wire charged, under the effect of external fire or high temperature, the insulation failure leads to short circuit after traces of residue. Secondary short circuit occurred in the fireworks and the temperature of the atmosphere [4-5], and is often a good fire hot melting trace, their time, temperature and different from a short circuit, it has long time temperature, large range of fire, melting temperature is lower than the short-circuit arc temperature, so the organization grain is relatively bulky, according to this characteristic, which can determine the secondary short circuit mark of molten is formed after the fire. And a short circuit occurs in the normal atmosphere, is the short circuit before the fire, but in this case there is no short circuit weld mark, so do not make too much explanation. The short circuit splashes and thermal-electric crevices have similar properties, both of which may be formed before and after fire. The fusion mark formed by electrothermal action is the melting mark formed under the action of electric current or arc, including: leakage fusion mark, local overheat fusion mark formed by electric action, short-circuit fusion mark between charged body and other metal with different potential, etc. Short circuit splash bead is a kind of bead melting trace which is adhered to other carriers when copper and aluminum wires have short circuit fault. In this case, the short circuit scattering bead is separated from the body of the generated bead and exists independently. The short circuit fusion mark formed by the contact between the electrically charged body and other metal cable tray with different potential produces arc.

Therefore, the short circuit splash bead in sample 1# and the secondary short circuit fusion mark in sample 6# can indicate that the trace is formed under the charged condition, but it cannot directly prove that the cause of fire is related to the electrical.

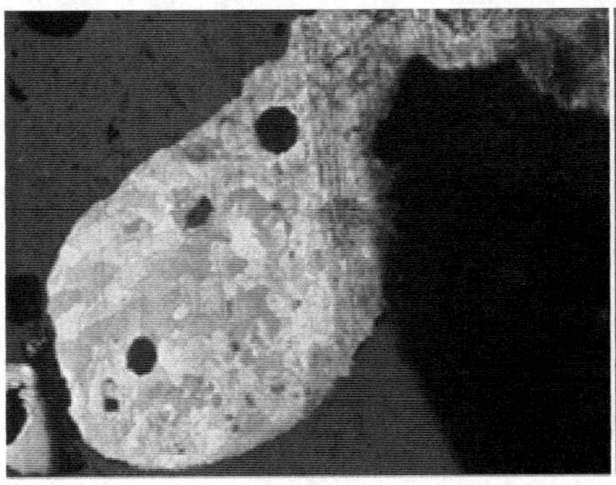

Figure 3. Metallurgic structure of fire melted mark

Figure 4.Metallurgic structure ofsecondary short circuited melted mark

4. Analysis of Causes

Based on the above analysis, none of the weld marks obtained in this case can directly prove that the fire was electrical related. However, if field inspection can prove that the material evidence is within the range of the fire site, and can exclude the cause of fire other than electrical, then the above mentioned short circuit splashes and electric hot melt are the most direct and powerful evidence, which can prove that the fire is caused by electrical fault.

5. Conclusion

The fire scene is complex and changeable, and the application of physical evidence should also be combined with the actual situation, which can be demonstrated from both positive and negative aspects rather than simple comparison. Only in this way can we ensure that everything is foolproof.

6. Reference

[1] Fire department of the ministry of public security. China fire statistics yearbook [M].Beijing: China personnel publishing house, 2016
[2] https://www.jianshu.com/.
[3] Wang xiqing, Han baoyu, Di man, etc. National standards for the People's Republic of China. GB 16840.4 1997
[4] Information on http://www.esafety.com.cn/case/21580.html (In Chinese)
[5] Fire Science and Technology, Vol. 24 (2005) No.4, p.495 (In Chinese)

$CuInS_2$ Quantum Dots-based Off/On Probe for Detection of Cetyltrimethylammonium Bromide and L-Cysteine

Yitong Liu, Zhongkai Guo, Qunxi Tai, Zaiqian Tu, Hanyu Bai and Siyu Liu*
College of Life and Health Sciences, Northeastern University, Shenyang, 110000, China.
Email: Liusiyu@mail.neu.edu.cn

Abstract. We developed a convenient method for cetyltrimethylammonium bromide (CTABr) detection based on the quenching phenomenon of $CuInS_2$ quantum dots. There existed a good linear relationship between the fluorescence intensity of $CuInS_2$ quantum dots and CTABr concentration from 0.08μmol/L to 6.0μmol/L with a detection limit of 25nmol/L. The quenched fluorescence intensity of $CuInS_2$ quantum dots by CTABr could be recovered by L-cysteine, which could be also used to detect L-cysteine from 0.05 to 100μmol/L with a detection limit of 20nmol/L.

1. Introduction

L-Cysteine (Cys) plays a vital role in a series of physiological processes [1]. Multifarious techniques have been utilized for the detection of Cys [2]. In recent two decades, fluorescence assays for Cys based nanosensors have attracted great attentions due to the practicality and convenience for operation [3].

Semiconductor quantum dots (QDs) have attracted more attentions in the past two decades due to their unique physicochemical properties. In recent years, QDs such as CdTe, CdSe, etc, capped by stabilizing agents such as organophosphate or thiols, were generally reported to be utilized in bimolecular recognition, ion detection, DNA and protein analysis [4-6].$CuInS_2$ QDs was a new I-III-VI nanomaterial with low toxicity to organism [7]. According to previous report, mercaptopropionic acid (MPA)-capped $CuInS_2$ QDs were prepared [8].

Cetyltrimethylammonium bromide (CTABr), as cationic surfactants, can promote nucleic acid and acid polysaccharide precipitation from the low ionic strength of solution, so it could be used in the separation and purification of DNA and polysaccharide. During the past two decades, CTABr was also used in the synthesis of mesoporous materials. Although CTABr has been applied in many fields, it was also reported as a pollutant in water due to its high toxicity to human body and animals.

Herein, we used $CuInS_2$ QDs for the fluorescent detection of CTABr and Cys. The fluorescence of $CuInS_2$ QDs would be obviously quenched by CTABr, and then recovered in the presence of Cys. Based on the fluorescent turn off/on process, we proposed a convenient method for determination of CTABr and Cys with high selectivity and sensitivity.

2. Experiment Section

2.1. Synthesis of $CuInS_2$ QDs

MPA-capped $CuInS_2$ QDs was prepared on the basis of previous report [8].

2.2. Fluorescent Detection of CTABr

$CuInS_2$ QDs solution (0.3mL, 1.5mmol/L), PBS (0.2mL, 0.1 mol/L pH 9.0) and different amount of CTABr solution were successively added into a 2 ml test tube, diluted to the mark with deionized water, and then incubated for 5 minutes.

2.3. Fluorescent Detection of Cys

$CuInS_2$ QDs solution (0.3mL, 1.5mmol/L), PBS (0.2mL, 0.1mol/L pH 9.0) and a certain amount of CTABr solution were added into a 2 ml test tube, and then equilibrated for 5 minutes. Different amount of Cys were added into the test tube, diluted to the mark with deionized water. The fluorescence spectra were recorded after 15 minutes incubations.

3. Results and Discussions

3.1. Fluorescence Quenching by CTABr

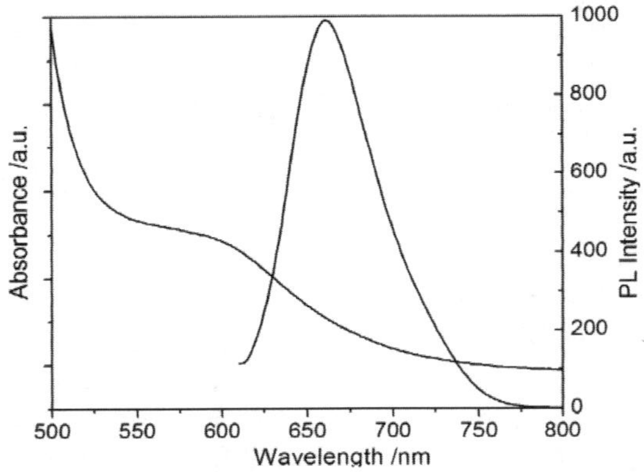

Figure 1. The Fluorescence Emission and UV-Vis Absorption Spectra of the $CuInS_2$ QDs Solution

As shown in Figure 1, the prepared $CuInS_2$ QDs displayed a symmetrical fluorescence emission peak at 660 nm and absorption peak around 570nm, which were consistent with previous report [8].

In recent years, a lot of works have reported the effect of cationic surfactant on the PL of QDs. Herein, we investigated systematically the effects of CTABr on $CuInS_2$ QDs. Figure 2ACurve a showed the temporal evolution of the fluorescence intensity of $CuInS_2$ QDs with 6μmol/L CTABr. The fluorescence intensity sharply reduced after the addition of CTABr in 5 minutes, and reached equilibrium.

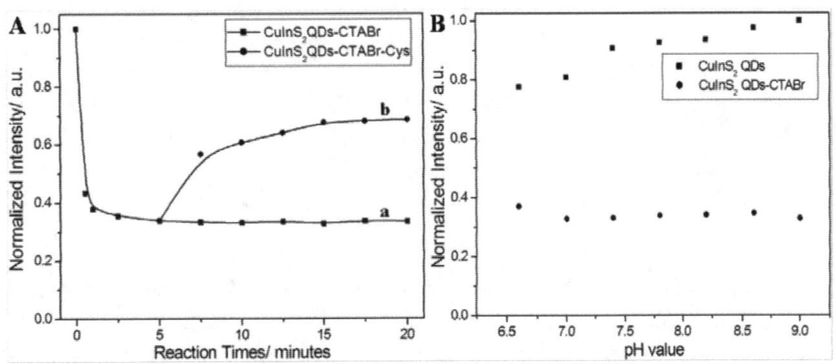

Figure 2. (A) (a) Fluorescence quenching of $CuInS_2$ QDs by 6μmol/L CTABr at different incubation time. (b) Fluorescence recovery of $CuInS_2$ QDs-CTABr system (incubating 5 minutes) by addition of 100μmol/L Cys with increasing incubation times. (B) Fluorescence intensity of $CuInS_2$ QDs without or with 6μmol/L CTABr in different pH environments

We studied the effects of pH on the fluorescence intensity of $CuInS_2$ QDs in the range of 6.6-9.0. Figure 2B showed the fluorescence intensity of $CuInS_2$ QDs solution increased along with the increase of pH from 6.6 to 9.0, which was because that the thiol (-SH) of the stabilizer MPA was easy to be proton in acid medium. The $CuInS_2$ QDs can remain stable in alkali environments, and the pH has a little influence on the fluorescence of $CuInS_2$ QDs-CTABr in the range of 6.6-9.0. Considering the stronger electrostatic interactions between the surfaces of $CuInS_2$ QDs and CTABr in alkali environment, pH 9.0 PBS was used in the next experiments.

Figure 3A showed the fluorescence emission spectra of $CuInS_2$ QDs with various CTABr concentrations. The fluorescence intensity of $CuInS_2$ QDs gradually reduced with CTABr concentration enrichment. Figure 3AInset showed that there was a good linear relationship between $(F_0\text{-}F)/F_0$ (F_0 or F was the fluorescence intensity of the $CuInS_2$ QDs without or with CTABr, respectively) and CTABr concentration in the range of 0.08-6.0μmol/L, which can be described as follow:

$$(F_0\text{-}F)/F_0=0.088+0.099[\text{CTABr}],\ \mu\text{mol/L} \quad (1)$$

The regression coefficient is 0.998 with 0.02μmol/L mol/L detection limit for CTABr. The cationic surfactant CTABr can bind to the surface of $CuInS_2$QDs via electrostatic interactions, which would induce the aggregation and the fluorescence quenching of $CuInS_2$QDs.

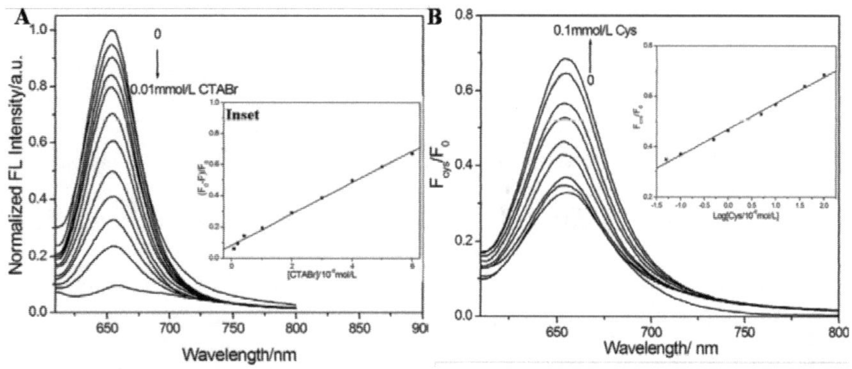

Figure 3. (A) Fluorescence spectra of $CuInS_2$ QDs upon the addition of different concentration of CTABr from 0 to 10μmol/L. Inset: The plot of FL intensity versus the concentration of CTABr. (B)Fluorescence spectra of $CuInS_2$ QDs-6μmol/L CTABr system upon the addition of different concentration of Cys from 0 to 0.1mmol/L. Inset: The plot of F_{cys}/F_0 versus the logarithm of Cys concentration. PBS: 10mmol/ L phosphate buffer solution (pH 9.0)

3.2. Effect of Foreign Substance on the Fluorescence Quenching By CTABr

We investigated the influence of various metal ions and foreign molecule on the determination of CTABr. As shown in Table 1, high concentration of metal cation Na^+, K^+, Ca^{2+}, Mg^{2+}, Ba^{2+} have no remarkable effects. The short chain tetraalkylammonium such as L-carnitine and acetylcholine and the common chelating agent tartaric acid and sodium citrate also nearly have no interference on the determination process.

3.3. Fluorescence Recovery by Cys

The quenched fluorescence of $CuInS_2$ QDs by CTABr could be recovered by Cys. Figure 2ACurve b offered the temporal evolution of the fluorescence of $CuInS_2$ QDs quenched by CTABr after the addition of Cys. The FL of $CuInS_2$ QDs was quenched by CTABr for 5minutes, and then the quenched fluorescence recovered after the addition of Cys, and reached equilibrium after 15 minutes. Figure3B showed that the quenched fluorescence by CTABr was sensitive to Cys. The fluorescence intensity was recovered as the Cys concentration increased from 0 to 100μmol/L. The relationship between F_{cys}/F_0 (F_{cys} is the FL of the $CuInS_2$ QDs-6μmol/L CTABr system with different concentration of Cys, and F_0 is the original FL of $CuInS_2$ QDs) and the logarithm of Cys concentration in the range of 0.05-100μmol/L can be described as the following equation:

$$F_{cys}/F_0=0.470+0.103Log[Cys], \mu mol/L \quad (2)$$

The regression coefficient is 0.997 with0.025μmol/L detection limit for Cys.

Table 1. Tolerance of foreign substance (The concentration of CTABr is 6μmol/L)

Coexisting substance	Tolerance ratios	Change of fluorescence(%)
Na^+	1000	-2.1
K^+	1000	+1.9
Mg^{2+}	100	-4.7
Ca^{2+}	100	+2.6
Ba^{2+}	100	+4.3
L-carnitine	500	+4.9
Acetylcholine	100	-1.5
Tartaric acid	1000	-1.7
Sodium citrate	1000	-2.0

Compared with the previous works about fluorescent detection for Cys, our method has the lower detection limit and similar dynamic range. As previous reports, Cys has strong binding ability to QDs. The CTABr adsorbed on MPA-capped $CuInS_2$ QDs would be replaced by Cys, which changed the surface charges of $CuInS_2$ QDs. The repulsion of the same charges and the steric hindrance effect would prevent the CTABr from fluorescence quenching of QDs, which induced the fluorescent recovery of $CuInS_2$ QDs.

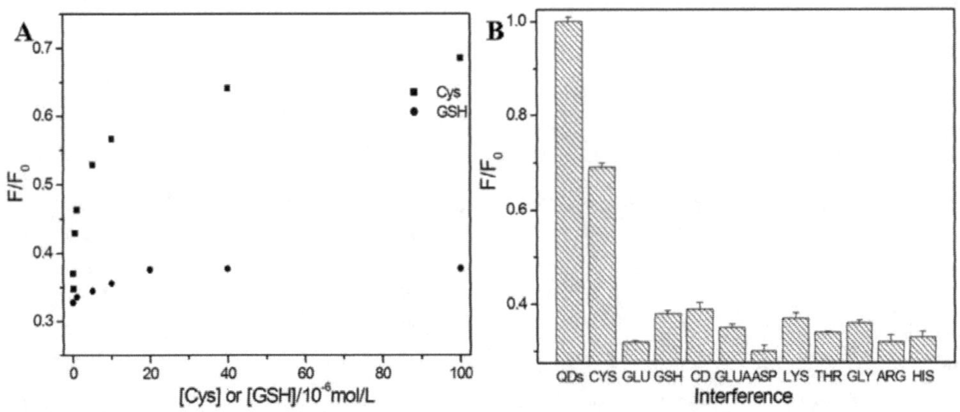

Figure 4. (A) The fluorescence intensity of $CuInS_2$QDs quenched by CTABr with different concentration of L-cysteine and glutathione in the range from 0-0.1mmol/L. (B) Effect of a series of 0.1mmol/L biomolecule such as L-cysteine (Cys), glucose (GLU), glutathione (GSH), cyclodextrin (CD), glutamic acid (GLUA), aspartate (ASP), lysine(LYS), threonine (THR), glycine (GLY), arginine (ARG), histidine (His) on the fluorescence intensity of $CuInS_2$ QDs quenched by CTABr solution

The study on the competition between Cys and glutathione (GSH) was illustrated in Figure4A. The quenched fluorescence by CTABr could restore to 69% of the original fluorescence with 0.1mmol/L Cys. The fluorescence recovery with 0.1mmol/L GSH is not very ideal, which is because that the

CTABr molecules prevent the higher molecule weight GSH molecule from closing to the surface of $CuInS_2$QDs due to the steric hindrance effect. As shown in Figure 4B, the presence of other amino acid only produced small effects on the fluorescence intensity.

4. Conclusions
In summary, the prepared $CuInS_2$ QDs was successfully used to detect CTABr and Cys. This method avoids complicated surface modification and operations, and has simple operation and excellent selectivity.

5. Acknowledgements
This work was supported by the National Training Program of Innovation and Entrepreneurship for Undergraduates (201910145108) and Fundamental Research Funds for Central Universities (N182410001).

6. References
[1] Cebi N, Dogan CE, Develioglu A, Yayla ME and Sagdic O 2017 *Food Chem.* **228**116.
[2] Lee JS, Ulmann PA, Han MS and Mirkin CA 2008 *Nano Lett.***8**529.
[3] Sudeep PK, Shibu Joseph ST and George Thomas K 2005 *J.Am.Chem. Soc.***127**6516.
[4] Wang HF, Li Y, Wu YY, He Y, Yan XP 2010 *Chem. Eur. J.***16**12988.
[5] Mulrooney RC, Singh N, Kaur N and Callan JF 2009 *Chem. Commun.***6** 686.
[6] Gill R, Willner I, Shweky I and Banin U 2005*J. Phys. Chem. B.***109** 23715.
[7] Nakamura H, Kato W, Uehara M, Nose K, Mata T, Otsuka-Yao-Matsuo S, Miyazaki M and Maeda H 2006 *Chem. Mater.***18**3330.
[8] Liu SY, Zhang H, Qiao Y and Su XG 2012 *RSC. Adv.* **2**819.

IOP Conf. Series: Materials Science and Engineering **735** (2020) 012018 doi:10.1088/1757-899X/735/1/012018

Study on Preparation and Properties of Modified Functional Mortar Mixed with Reduced Iron Powder

Yingjian Lu[1,2], Jing Huang[2] and Yi Gao[2*]

1 Naval University of Engineering, Wuhan 430000, China
2 Naval Logistics Academy, Tianjin 300450, China
Corresponding author e-mail: gaoyi5777@163.com
Corresponding author: Yi Gao

Abstract. The effect of reduced iron powder on the performance of functional mortar was studied. Aimed at the construction performance, the experiments of consistency and water retention were carried out. Aimed at the mechanical properties, tensile bond strength, compressive strength and flexural strength experiment were carried out. The results show that the content of powder has little effect on mortar consistency. It can improve water retention. The 14d tensile bond strength is 36.2% higher than the reference mortar, which can meet the requirements of plaster-mortar. The compressive strength of 7d mortar was 24.2% higher than that of reference mortar. Compared with the reference mortar, the 28d compressive strength increased 13.6%, the 7d flexural strength increased 36.6%, and the 28d increased 25.8%.

1. Introduction

Cement-based materials have the advantages of low price and strong applicability. In the face of increasingly complex space requirements, materials prepared by adding special components have been studied a lot. Some scholars have prepared cement-based materials by adding metal powder, and concluded that the materials have good electrical conductivity [1]. Taking graphite powder and carbonyl iron powder as admixtures, the absorption effect of materials in high frequency band was studied [2]. A cement-based material was prepared with carbon nanotubes, and its astuteness of the structure was tested [3]. Aluminate cement matrix composite was prepared with iron powder as admixture, and its thermal properties at high temperature were investigated [4]. Some studies have prepared concrete slab with reduced iron powder and conducted relevant studies on its mechanical properties [5]. As a kind of ultrafine metal powder, reduced iron powder has good electrical conductivity [6]. The plastering mortar prepared by adding reducing iron powder in cement mortar can provide a choice for reconstruction of old engineering with special needs. Mechanical properties, as the most obvious performance of cement-based materials in engineering applications. The thesis studies the effect of reduced iron powder on the properties of mortar, which lays a foundation for the study of its construction and application.

2. Experiment

2.1. Raw Materials

Reduced iron powder: 200 mesh secondary reduced iron powder; Cement: grade 42.5 ordinary Portland cement; Sand: medium river sand; 1250 mesh fly ash: first class fly ash.

2.2. Mix Proportion
According to the proper standard, the proportion of plastered mortar is selected and the strength grade of test match is adjusted appropriately. The content of reduced iron powder is respectively 0kg/m^3, 100kg/m^3, 200kg/m^3, 300kg/m^3, 400kg/m^3 and 500kg/m^3. The specific proportion is shown in table 1.

Table 1. Mix proportion

Specimen number	Content (kg/m^3)				
	Reduced iron powder	Cement	Sand	Fly ash	Water
1	0	450	1498	50	300
2	100	450	1473	50	300
3	200	450	1448	50	300
4	300	450	1423	50	300
5	400	450	1398	50	300
6	500	450	1373	50	300

2.3. Preparation Methods
According to the selected mix proportion, weigh the quality of each group. The cement and reduced iron powder were put into the mixer for 60s dry mixing at a slow speed, followed by 60s dry mixing at a fast speed, so that they were fully mixed. Sand and fly ash were added at a uniform velocity, and water was added at the same time. The time was no less than 180s.

2.4. Method of Experiment
The consistency and water retention of functional mortar were tested. The specimens were cured to 28d under the specified environment, and the cube compressive strength was tested. The specimens were cured to 14d under the specified environment, and the tensile bond strength was tested. The specimens were cured to 7d and 28d respectively under the specified environment, and the compressive and flexural strength were tested.

3. Experimental Results

3.1. Effect of Reduced Iron Powder on Consistency of Mortar
With the continuous increase of reduced iron powder, the consistency shows the law of decreasing, and the fluidity becomes worse. At this time, the consistency of mortar can meet the basic requirements of plastered mortar. When the content is 400kg/m^3 and 500kg/m^3, the consistency decreases obviously, which cannot meet the requirements of plastering mortar. The effect is shown in figure 1.

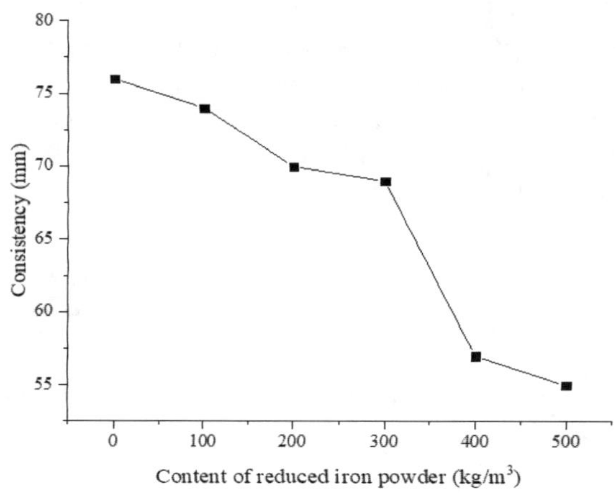

Figure 1. Effect of Reduced Iron Powder on the Consistency of Functional Mortar

Through analysis, the main reasons for the above phenomenon has the following two aspects: on the one hand, the reduced iron powder has a larger specific surface area, also, there are bulk powder particles adhered to the surface, which can make it combined with water molecules easily, the powder can reduce the water which is to participate in the hydration reaction, which cause mortar to thicken. On the other hand, the powder belongs to super fine powder, there are more contact surfaces between particles, and the relative friction force increases, which reduces the fluidity of mortar.

3.2. Effect of Reducing Iron Powder on Water Retention of Mortar

When the content is less than 400kg/m^3, with the increase of the content, the water retention rate shows a linear trend of increase. When the content is 400kg/m^3, the rate reaches a peak of 95.8%. Compared with the reference mortar, the rate increases by 1.7%. When the content was increased to 500kg/m^3, the rate decreased significantly. The effect is shown in figure 2.

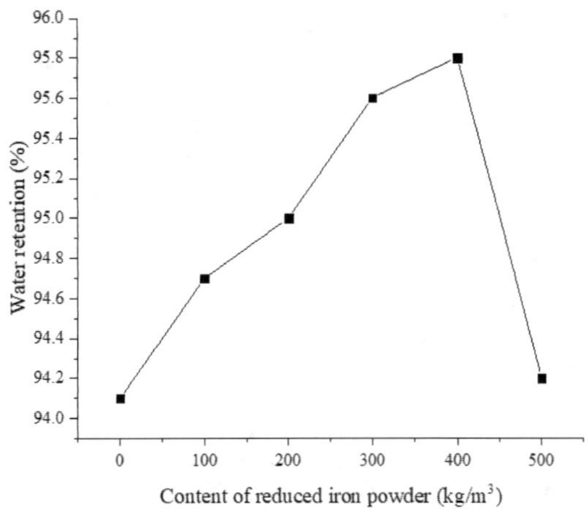

Figure 2. Effect of Reduced Iron Powder on the Water Retention of Functional Mortar

Through the analysis, the reduced iron powder has a porous sponge structure, which is easy to combine with the free water in the mortar, making it difficult to separate out. When the right amount of reducing iron powder is added, it can improve the water-retention performance of the mortar .When the content of reduced iron powder increases to 500kg/m^3, because the density of reduced iron powder is larger, when the content is too much, settlement will occur in the mortar, the rate decreases obviously.

3.3. Effect of Reduced Iron Powder on Tensile Bond Strength

According to the experiment, when the content of reduced iron powder is less than 300kg/m^3, the tensile bond strength shows an increasing trend. When the content is 300kg/m^3, the strength reaches the peak value of 0.79MPa, which is 36.2% higher than the reference mortar. When added to 500kg/m^3, the tensile bond strength of mortar specimens decreases gradually. The effect is shown in figure 3.

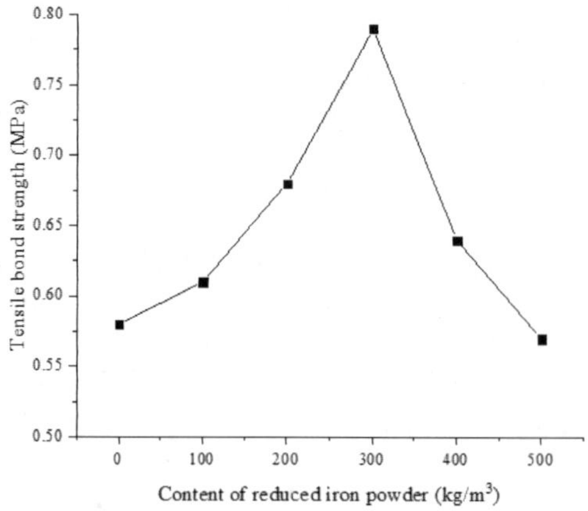

Figure 3. Effect of Reduced Iron Powder on the Tensile Bond Strength of Functional Mortar

Through the analysis, reduced iron powder belongs to super fine powder, when mixed with mortar, it has a physical filling effect [7], the interface between the cement particles is more compact, the workability has improved, which can improve the bonding strength. When the content added to 500 kg/m^3, the strength decreases, on the one hand, the powder has a large density, when the content is too much, it will subside, the physical filling effect decrease, causing strength reduced; On the other hand, the powder has a large specific surface area, when the content is too much, it will lead to the increase of the porosity, and the insufficient cement slurry used to bond decrease, thus reducing the strength.

3.4. Effect of Reduced Iron Powder on Compressive Strength

According to the experiment, when the content of reduced iron powder is less than 300kg/m^3, the compressive strength of the specimens increases gradually with the increase of the content. The 7d strength reaches the peak strength when the content is 300kg/m^3. When the content of powder increases to 400kg/m^3, the 28d strength reaches its peak. The strength of the cube reached a peak of 34.2MPa, 13.6% higher than that of the reference, and the strength of the prismatic specimen reached a peak of 31.3MPa, 7.2% higher than that of the reference. When the content of powder increased to 500kg/m^3, the strength decreased significantly. The effect is shown in figure 4.

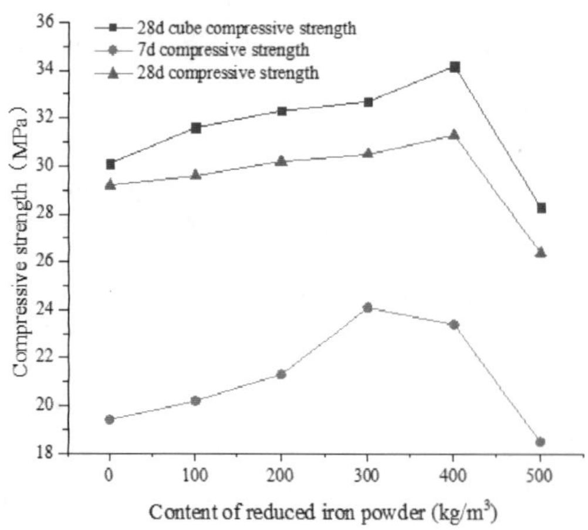

Figure 4. Effect of Reduced Iron Powder on the Compressive Strength of Functional Mortar

Through the analysis, reduced iron powder has a great effect on the early strength of specimen. On the one hand, the powder has a degree of water absorption, which can reduce the free water in the mortar, and the internal foam is more likely to burst, so that the pores are reduced, the structure is more compact, and the strength is increased. On the other hand, the powder belongs to super fine powder. When it is mixed with mortar, it will play a physical filling role between cement and aggregate, improve the pore structure and particle gradation inside mortar specimen, making the interior structure more compact.

The main reasons that the strength reduced are as follows: on the one hand, the hydration reaction is an important factor of strength increasing, the powder has a certain water absorption [8], when the content is too much, it will make a water shortage, leading to the decrease of strength of cement hydration reaction, the filling effect of powder have been unable to offset the lack of hydration reaction. On the other hand, the density of powder is large, and it will subside when the content is too much, and it cannot be evenly dispersed in mortar, leading to the increase of the weak points of specimens. By comparing the two groups of 28d specimens, it can be seen that the strength of the cube is relatively high. When the powder is not added, the compressive strength has a small difference. When the content of powder is 400kg/m^3, the difference is 2.9MPa. The cube compressive strength specimen is prepared by manual tamping, while the prism one is formed by mechanical vibration. Due to the large density of reduced iron powder, mechanical vibration is more likely to cause the settlement than manual vibration, which leads to the reduction of its filling effect. This indicates that choosing proper mixing and vibrating methods is beneficial to reduce the settlement of powder in mortar and make the best use of it.

3.5. Effect of Reduced Iron Powder on Flexural Strength

It was founded that when the content of reduced iron powder was less than 400kg/m^3, the flexural strength of the two groups gradually increased with the increase of the content. When the content was 400kg/m^3, the specimen reached the peak strength, the 7d specimen reaches the peak strength of 5.6MPa, which was 36.6% higher than that of the reference mortar. The 28d specimen reached the peak strength of 7.8MPa, which was 25.8% higher than the reference mortar. When the content increased to 500kg/m^3, the flexural strength decreased obviously. The effect is shown in figure 5.

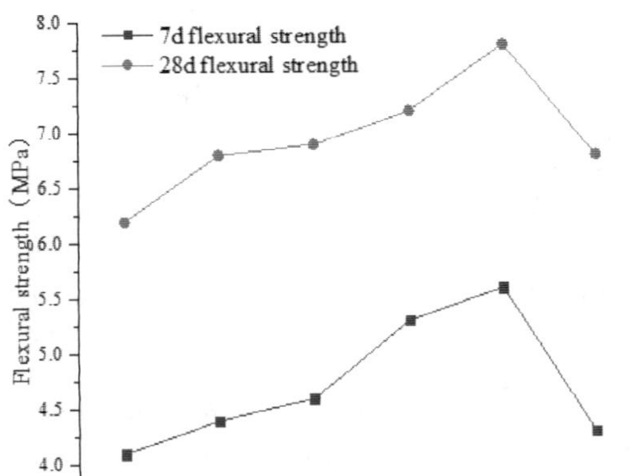

Figure 5. Effect of Reduced Iron Powder on the Flexural Strength of Functional Mortar

Through the analysis, the main reasons for the above phenomenon are as follows: on the one hand, the interface area between cement slurry and aggregate is the weak point, and a good interface area structure can effectively improve the flexural strength. When proper amount of reduced iron powder is added, its good physical filling effect can significantly improve the interface structure and improve the flexural strength. On the other hand, as a super fine powder, the reduced iron powder has a high strength and elastic modulus. When the specimen is subjected to flexural failure, the powder can buffer the stress with its bending deformation [9], delaying the cracking, so as to improve the strength. When the content powder is increased to 500kg/m^3, the strength decreases, when adding reduced iron powder. When the content is too large, on the one hand, it is easy to cause the settlement phenomenon in mortar, which leads to the increase of the weakness of specimen and the decrease the bending strength. On the other hand, it will also lead to a decrease of sand that plays the role of "skeleton", which will increase the brittleness of mortar specimens and reduce the flexural strength.

4. Conclusion

(1) Reduced iron powder will reduce the consistency of mortar. When the content is less than 300kg/m^3, the mortar consistency can meet the basic requirements of plastered mortar construction.

(2) An appropriate content of reduced iron powder can improve the water retention of mortar and meet the construction requirements of plastered mortar. When the content is less than 400kg/m^3, with the increase of content, the rate gradually increases, and reaches the peak when the content is 400kg/m^3.

(3) An appropriate content of reduced iron powder can effectively improve the tensile bond strength and meet the construction requirements of plastered mortar. When the content is less than 300kg/m^3, the strength gradually increases with the increase of the content. When the content is 300kg/m^3, the strength reaches the peak.

(4) An appropriate content of reduced iron powder can improve the compressive strength of mortar specimens. The powder has a great effect on the early strength of the specimen. When the content was 300kg/m^3, the strength of the specimen on 7d reached its peak. When the content increased to 400kg/m^3, the compressive strength of 28d specimens reached the peak. By comparing the compressive strength of the two groups of 28d specimens, it can be seen that the selection of appropriate mixing and vibrating methods is conducive to preventing excessive settlement.

(5) An appropriate content of reduced iron powder can improve the flexural strength of functional mortar specimens. When the content is 400kg/m^3, it reaches the peak strength.

5. References

[1] Chen Yangru, Xiong Guoxuan, Zhang Zhibin. The application study of shielding media in cement-based materials, J. New Building Materials, 2010, 37 (10) 80-82.

[2] Han bin, Ji Zhi jiang, Zhang Zhong lun, et al. Research on Absorbing Properties of Cement-based Coating Material J. Materials Review, 2009, 23 (z1) 370-373.

[3] Li Yunfeng, Sun Jing, Wang Quanxiang. Monitoring of carbon nanotubes based on structural cementitious composites sensor J. New Building Materials, 2015, 42 (10) 34-37, 53.

[4] Fu Quanhai, Liao Wei, Ni Yaru, et al. Properties of Iron Powders Incorporate with Aluminate Cementitious Thermal Storage Materials J. Bulletin of the Chinese Ceramic Society, 2015,34 (6) 1559-1564.

[5] Zheng Guozhi, Chen Bin, Zhang Zehai, et al. Reflection and transmission performances of concrete slabs mixed with reduced iron powder J. High Power Laser and Particle Beams, 2015,27 (4) 151-156.

[6] Xu Fangxing, Zeng Guoxun, Zhang Haiyan, et al. Preparation and electromagnetic properties of flaky reduced iron powders J. Electronic components and materials, 2014, (12) 41-48.

[7] Wan Wei, Yue Zhihua, Chen Fu, et al. Effect of Grain Size of Hematite on High-Density Cement Slurry Performance J. Natural Gas Exploration and Development, 2008, (1) 51-59.

[8] Liu Donghua, Qian Xiaotai, Liang Yi, et al. Microstructure and Performance of Iron Powders Prepared by Reduction-grinding and Seperation J. Metal Materials and Metallurgy Engineering, 2013,41 (1) 8-11.

[9] Shi Mingxia, Long Guangcheng, Xie Youjun. Effects of Various Particles on Mechanical Properties of Very High Strength Cement-based Composites J. Materials Review, 2009, 23 408-411.

Development Trend of Low Thermal Conductivity Ceramic Materials

Xiqian Luo, Bohan Guo, Xiaoqiang Wang and Zeyu Zhang

School of Materials and Metallurgy, Wuhan University of Science and Technology, Wuhan, Hubei 43000, China
Email:1076678956@qq.com

Abstract. In recent years, high entropy ceramics have become one of the research hot spots in the ceramic field due to its superior high entropy characteristics, and the use of blast furnace slag as high entropy material is also in line with the general trend of energy conservation and emission reduction in today's society. This paper will briefly introduce the preparation and properties of several common porous ceramics by studying the low thermal conductivity mechanism of traditional ceramic materials. The high entropy ceramic materials are further introduced. By introducing its low thermal conductivity principle, preparation process and its low thermal conductivity related experiments, the superior performance of low thermal conductivity high entropy ceramic materials compared with traditional porous materials is demonstrated, the development trend of low thermal conductivity ceramic materials is analyzed, and finally the application prospect of blast furnace slag in high entropy materials is prospected.

1. Introduction

In recent years, materials with low thermal conductivity, especially oxide ceramics, have attracted much attention due to their potential as thermal barrier coatings for high-temperature engines. Traditional ceramic materials use porous structures to reduce thermal conductivity, while new high entropy ceramics use their high entropy characteristics to achieve low conductivity. This article will analyze the development trend of low thermal conductivity ceramic materials from the related introduction of traditional porous ceramics to low thermal conductivity and high entropy ceramics.

2. Development of Traditional Porous Ceramics with Low Thermal Conductivity

When it comes to low thermal conductivity ceramic materials, we have to talk about its low thermal conductivity. Reducing the thermal conductivity of ceramic materials is to make the materials have better thermal insulation performance and can be applied to the required fields. This article will explore the low thermal conductivity mechanism of traditional porous ceramic materials and briefly introduce several porous ceramic materials.

2.1. Low Thermal Conductivity Mechanism of Porous Ceramic Materials

In this paper, the low thermal conductivity principle of traditional ceramic materials will be explained by the microstructure and room temperature thermal conductivity of porous yttria stabilized zirconia ceramic (YSZ) with thermal conductivity close to the lower thermal conductivity limit (0.06 W/m·K) [1].

Firstly, we have to measure the pore diameter distribution and pore diameter ratio of the sample to determine its density, and then compare it with the density of the fully dense ceramic of the material, wherein the pore diameter distribution can be obtained by mercury ion implantation method, and the porosity can be calculated by measuring the volume and mass, and the density of the fully dense ceramic of the material is known as 6.0g/cm^3.

From the details of pore morphology and intergranular bonding in the sintered ceramic of Figure 1, it can be seen that as the sintering temperature increases from 1450°C to 1550°C, the crystal grains gradually grow and the porosity decreases.

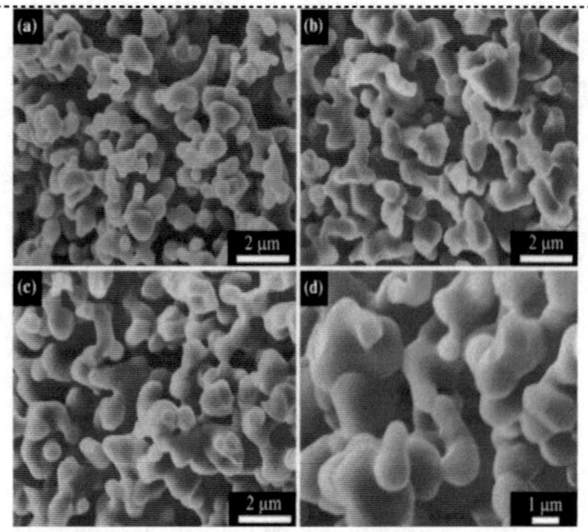

Figure 1. Porous Image of YSZ Sample at Different Temperatures: Interconnection of a 1450 °C, B 1500 °C, C 1550 °C and D typical grains.

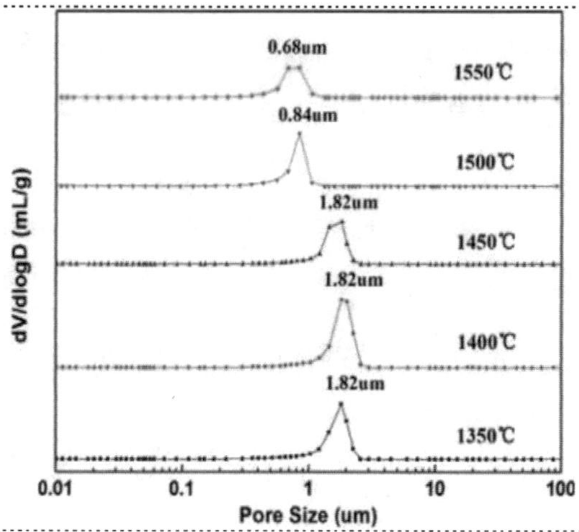

Figure 2. Pore Size Distribution of Porous YSZ Samples with Solid Content of 10% by Volume and Different Sintering Temperature.

Figure 2 shows the pore size distribution of the sample at different sintering temperatures. From the figure one narrow half-peak width single peak can be seen, indicating that the pore size distribution is uniform. Among them, the total volume fraction of the sample pores is reflected by the area range under the curve, and due to the increase of temperature, the crystal grains grow rapidly, resulting in the total volume fraction and average pore diameter of the sample pores decrease with the increase of sintering temperature.

Figure 3 and Figure 4 show the change trend of porosity and room temperature thermal conductivity of the material with solid loading of 10vol% and 15vol% respectively as the temperature increases from 1450°C to 1550°C.

When studying the microstructure of porous yttria stabilized zirconia ceramic, we found that the pores trapped a large amount of air, and the thermal conductivity of air was lower than that of the material, which better blocked the heat transfer. In addition, a large number of voids are connected with each other, which makes the ceramic particles no longer continuous, thus leading to the reduction of heat conduction level. Meanwhile, a large number of pores and micro-sized interfaces provide significant phonon and photon scattering. Therefore, the lower thermal conductivity of the material is caused by the porous structure.

IOP Conf. Series: Materials Science and Engineering **735** (2020) 012019 doi:10.1088/1757-899X/735/1/012019

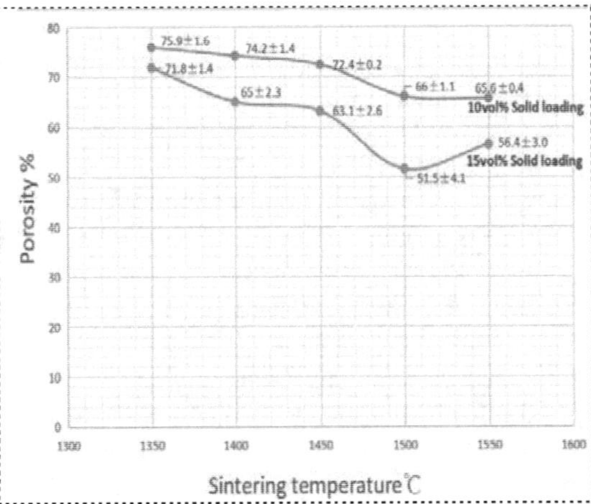

Figure 3

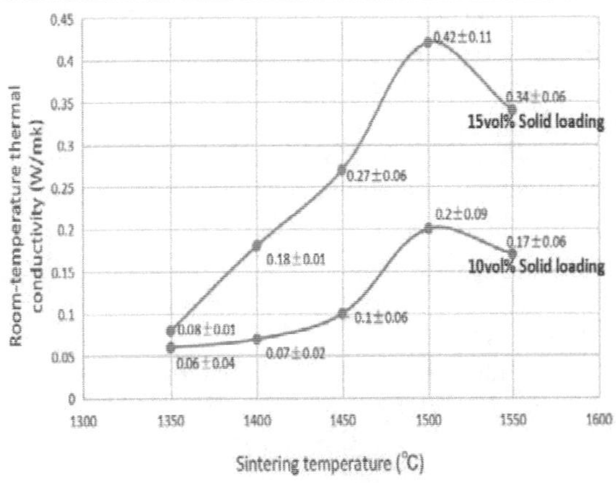

Figure 4

2.2. *Porous Ceramic Materials*

2.2.1. *Porous Anorthite Ceramics with Low Thermal Conductivity*

Ye Li [2] et al. have successfully prepared porous anorthite ceramics with low thermal conductivity by direct foaming and grouting with fly ash and gypsum as raw materials, and found that after adding 10% gypsum and 0.8% sodium hexametaphosphate, the porosity of the prepared anorthite ceramics is as high as nearly 95% and the thermal conductivity is as low as 0.042W/m·K.

2.2.2. *Porous Anorthite Ceramics with Low Thermal Conductivity*

Yao Han [3] et al. used alumina, calcium carbonate and silicon powder as raw materials to prepare porous anorthite ceramics with low thermal conductivity, as low as 0.018 W/m·k. Then the microstructure of the ceramic samples was analyzed by scanning electron microscope and XRD. The results show that porosity, pore size, pore structure and particle size are all the reasons that lead to the decrease of thermal conductivity.

2.2.3. *Porous Mullite Ceramic*

Lunun Gong [4] et al. prepared porous mullite ceramic from industrial mullite powder by foaming and starch curing, and characterized its thermal conductivity and studied the effects of suspension solid content and sintering temperature on porosity and compressive strength. The results show that its thermal conductivity is as low as 0.09 W/m·K, the porosity of sintered mullite ceramic is 86%-73% of volume, and the compressive strength is 1-22MPa. The porous has a multi-modal microstructure and the inner wall has large spherical pores and small pores.

2.2.4. *Porous Y_2SiO_5 Ceramics*

Duanyang Li [5] et al. have successfully prepared porous silica ceramics with neat straight pore structure and low thermal conductivity by TBA-based solidification casting using tert-butyl alcohol as solvent. It is found that the porosity of sintered silica ceramics can be adjusted by adjusting the initial solid load, and when the initial silica content decreases from 20% to 10% by volume, the porosity increases linearly from 45% to 71%.

The above is the principle of traditional low thermal conductivity ceramic materials and introduces some common porous ceramic materials. The following will introduce a new way to reduce the thermal conductivity of ceramic materials, namely synthesizing high entropy ceramic materials.

3. Low Thermal Conductivity and High Entropy Ceramic Materials

Correspondingly to low entropy, high entropy refers to the state where the system is highly chaotic or disordered, compared with traditional materials, high entropy materials are superior, so they have received considerable attention in materials science and engineering. In recent years, high entropy ceramics have been widely studied because of their interesting properties, i.e. low thermal conductivity, good high temperature stability, slow diffusion rate and serious lattice distortion.

3.1. Principle of Low Heat Conduction of High Entropy Ceramics

Previously, the research focus of low thermal conductivity materials focused on pore regulation, while high entropy ceramics (HEC) are characterized by the inherent atomic disorder of metal elements at cationic positions, resulting in complex compositions and significant lattice distortion. The following is an example of high entropy carbide ceramics to explore the low heat conduction principle of high entropy ceramics.

Xue-liang Yan et al. synthesized high entropy carbonized ceramic $(Hf_{0.2}Zr_{0.2}Ta_{0.2}Nb_{0.2}Ti_{0.2})C$ by spark plasma sintering, in which HfC, ZrC, TaC, NbC and TiC powder are equivalent and mixed. The results show that $(Hf_{0.2}Zr_{0.2}Ta_{0.2}Nb_{0.2}Ti_{0.2})$ C has a very low thermal diffusivity (3.6mm^2/s) at 29.5°C, which is 57% of the lowest value (6.29mm^2/s) of the five binary carbides at 25°C. With the increase of temperature, the thermal conductivity of this material decreases to 5.42W/m·K at 70.8°C.

The low thermal conductivity of $(Hf_{0.2}Zr_{0.2}Ta_{0.2}Nb_{0.2}Ti_{0.2})C$ may be caused by the high entropy effect. The main carrier of solid heat transfer is phonon and electron. For transition metal carbide, the contribution of electron at room temperature is about 25% of the total heat transfer. According to WiedemannFranz's law, it is estimated that about 75% of the phonon contribution is left [6]. When HfC, ZrC, TaC, NbC and TiC form a high entropy phase $(Hf_{0.2}Zr_{0.2}Ta_{0.2}Nb_{0.2}Ti_{0.2})$ C, the contribution of phonon changes greatly when the electronic contribution is not changed. It is found that the five independent metal atoms in the cation position have different bonding strength and size, which leads to the deformation of the atomic scale lattice of the sample, resulting in low thermal conductivity.

3.2. High Entropy Ceramic Materials

3.2.1. Ultra-high Hardness Dense High Entropy Boride Ceramics

Yan Zhang et al [7] prepared ultra-high hardness high density high entropy boride ceramics by spark plasma sintering self-synthesis method. Ultrafine high entropy boride powders $(Hf_{0.2}Zr_{0.2}Ta_{0.2}Cr_{0.2}Ti_{0.2})$ B2, $(Hf_{0.2}Mo_{0.2}Zr_{0.2}Nb_{0.2}Ti_{0.2})$ B_2 and $(Hf_{0.2}Mo_{0.2}Ta_{0.2}Nb_{0.2}Ti_{0.2})$ B2 were obtained by boron thermal reduction method. The measured yields were 89.9%, 89.6% and 97.3%, respectively. After solid phase extraction at 2000°C, it is found that high entropy boride ceramics have higher target phase yield.

3.2.2. Reactive Flash Discharge Plasma Sintering of High Entropy Ultra High Temperature Ceramics

Joshua Gild et al [8] synthesized high entropy metal diboride $(Hf_{0.2}Zr_{0.2}Ta_{0.2}Nb_{0.2}Ti_{0.2})B$ and high entropy metal carbide $(Hf_{0.2}Zr_{0.2}Ta_{0.2}Nb_{0.2}Ti_{0.2})C$ by reactive flash discharge plasma sintering . It is found that the addition of a small amount of carbon can synthesize and homogenize the high entropy metal borides without severe mechanical alloying, so as to avoid serious pollution and oxidation problems.

3.2.3. $(Hf_{0.2}Zr_{0.2}Ta_{0.2}Nb_{0.2}Ti_{0.2})C$ low Thermal Conductivity High Entropy Ceramics

Xue-liang Yan et al [9] characterized the phase composition and crystal structure of a high entropy carbide ceramic $(Hf_{0.2}Zr_{0.2}Ta_{0.2}Nb_{0.2}Ti_{0.2})$ C synthesized by spark plasma sintering with X-ray diffraction. By comparing its XRD spectrum, lattice parameters and thermal properties, it was confirmed that the single-phase halite structure of five metal elements (Hf, Zr, Ta, Nb, Ti) with cation positions, and its low thermal conductivity coefficient may be due to phonon scattering on its twisted anion sublayer.

3.2.4. High Entropy Ceramic $He(La_{0.2}\,Ce_{0.2}\,Nd_{0.2}\,Sm_{0.2}\,Eu_{0.2})_2Zr_2O_7$

Zi-fan Zhao et al. [10] designed and successfully synthesized a new high entropy rare earth zirconate $He(La_{0.2}Ce_{0.2}Nd_{0.2}Sm_{0.2}Eu_{0.2})_2Zr_2O_7$ for the first time.XRD and SEM analysis showed that the thermal conductivity of high entropy solid solution was as low as 0.76Wm^{-1}K^{-1} at room temperature due to its slow diffusion effect. The grain growth rate of the material is slower than that of $La_2Zr_2O_7$, making it hopeful to be used as thermal energy barrier coating material.

With the rapid development of high entropy ceramic materials, they have virtual superior performance compared with traditional low heat conduction porous ceramic materials, which makes them more widely used, and makes them hopeful to replace traditional porous ceramics in more fields and become better low heat conduction materials.

4. Application Prospect of Blast Furnace Slag

As a by-product of steel production, the world's annual blast furnace slag production is a huge number. Different from other European and American countries, China's slag utilization rate has been very low. For example, in 2016, its utilization rate will be less than 30%, and more than 3 million tons of slag will not be utilized. This not only causes environmental pollution, but also is a great waste of resources. The author believes that if blast furnace slag is used as raw material to prepare low thermal conductivity high entropy ceramics, it not only realizes the reuse of waste slag, but also has great potential for carbon dioxide and energy conservation, and is a promising application of blast furnace slag.

5. Summary

This paper mainly introduces the traditional porous ceramics and new high entropy ceramics, reflects the development trend of low thermal conductivity ceramic materials, and forecasts the application prospect of blast furnace slag in high entropy materials.

6. References

[1] Liangfa Hu, Chang-An Wang et al. Porous yttria-stabilized zirconia ceramics with ultra-low thermal conductivity. The Journal of Materials Science [J/OL] 2011, 46: 623–628.

[2] Ye Li, Xudong Cheng et al. Fabrication and characterization of anorthite foam ceramics having low thermal conductivity. Journal of the European Ceramic Society. [J/OL] 2015, 35: 267–275.

[3] Yao Han, Cuiwei Li et al. Porous anorthite ceramics with ultra-low thermal conductivity. Journal of the European Ceramic Society [J/OL], 2013, 33: 2573–2578.

[4] Lunlun Gong et al. Porous mullite ceramics with low thermal conductivity prepared by foaming and starch consolidation. Journal of Porous Materials. [J/OL] 2014, 21: 15–21.

[5] Duanyang Li1, Meishuan Li. Porous Y2SiO5 Ceramic with Low Thermal Conductivity. Journal of materials science & technology. [J/OL] 2012, 28(9): 799–802.

[6] Morelli DT. Thermal conductivity and thermoelectric power of titanium carbide single crystals. Phys Rev B. 1991; 44:5453.

[7] Yan Zhang, Wei-Ming Guo et al. *Dense high-entropy boride ceramics with ultra-high hardness, Scripta Materialia.* [J/OL] 164 (2019) 135–139.

[8] Joshua Gild, Kevin Kaufmann, Kenneth Vecchio et al. *Reactive flash spark plasma sintering of high-entropy ultrahigh temperature ceramics. Scripta Materia* 170(2019)106110.

[9] Xueliang Yan, et al. $(Hf_{0.2}Zr_{0.2}Ta_{0.2}Nb_{0.2}Ti_{0.2})C$ high-entropy ceramics with low thermal conductivity. Journal of the America ceramic society.[J/OL], 2018, 101:4486–4491.[2019.8.1]

[10] Zi-fan Zhao, Hui-min Xiang et al. *A novel high-entropy ceramic with low thermal conductivity and sluggish grain growth rate. Journal of Materials Science and amp* [J/OL] S1005-0302(19)30193-8.

IOP Conf. Series: Materials Science and Engineering **735** (2020) 012020 doi:10.1088/1757-899X/735/1/012020

Microstructures, Optical Properties and Degradation Efficiency of Nitrogen-doped Titanium Dioxide to Automobile Exhaust

Zhihui Hu[1, 2], Wei Liu[1, 3], Tao Xu[1, 2*] and Xiaoyan Li[1, 3]

1 National Engineering Laboratory for Advanced Road Materials, 2200 Chengxin Avenue, Nanjing 211112, Jiangsu, China
2 College of Civil Engineering, Nanjing Forestry University, 159 Longpan Road, Nanjing 210037, Jiangsu, China
3 JSTI Group, 2200 Chengxin Avenue, Nanjing 211112, Jiangsu, China
*Corresponding author E-mail: seuxt@163.com

Abstract. To understand the influence of nitrogen (N) dopant on phase structure, morphology, optical property, photocatalytic degradation of TiO_2 to automobile exhaust (AE) under visible light irradiation, the N-doped TiO_2 was first prepared, and then X-ray diffraction (XRD), transmission electron microscopy (TEM), ultraviolet-visible light diffuse reflectance spectra (UV-vis DRS) and photocatalytic degradation tests were conducted. Results indicated that only anatase phase was found in both pure TiO_2 and N-doped TiO_2, and the N doping improved the dispersity of TiO_2. Also, the N doping increased the absorbance of TiO_2 under UV and visible light, respectively. An obvious red shift was observed on UV-Vis DRS of pure and N-doped TiO_2 samples, indicating that the spectral response range of TiO_2 was expanded to visible light region, and the utilization efficiency of solar irradiance was increased. The prepared N-doped TiO_2 nanoparticles with an N dopant content of 3.0% showed better optical properties. Finally, the N doping improved the photocatalytic activity of TiO_2 to AE in the visible light region. Furthermore, the N-doped TiO_2 presented lower degradation efficiency to CO than that to CO_2 under visible light radiation. Thus, it is found that the degradation efficiency of TiO_2 to AE can be increased by the N doping under visible light radiation.

1. Introduction

Automobile exhaust (AE) including NO_x, HC, CO_2, CO, etc. causes serious environmental pollution [1]. AE brings great harms to human health and ecological environments so that some photocatalytic technologies have been developed to purify AE. Among them, TiO_2, as one of important photocatalysts, is utilized to chemically degrade AE and other air pollutants due to its advantages of high photoactivity, good stability, non-toxicity, low cost, etc [2].

However, owing to the high band gap energy of TiO_2, it can only be excited by ultraviolet (UV) light in sunlight to exert the property of photocatalytic degradation. However, the UV light onlu occupies about 4% in the total solar radiation [3]. This not only causes the insufficient use of solar energy but also lowers the photocatalytic efficiency of TiO_2. Another disadvantage of pure TiO_2 is that the recombination of photo-generated carriers decreases the photocatalytic activity of TiO_2 [4]. Therefore, it is meaningful to increase the photocatalytic efficiency of TiO_2 under visible light which occupies about 45% of total solar radiation.

Currently, the transition metal or nonmetal doping can not only broaden the photoresponse range of

TiO_2, but also promote the generation of electron holes. It has become one of the important methods to improve the photocatalytic efficiency of TiO_2 [5]. Among them, the N-doped TiO_2 has been paid more efforts due to its lower preparation cost and high photocatalytic activity under visible light, and controllable synthesis [6]. The photocatalytic performances of N-doped TiO_2 to methylene blue under the irradiation of visible light were discussed by Ramchiary et al. [7]. Powell [8] used the N doping method of TiO_2 to increase the absorbance of visible light, improving the photocatalytic activity of TiO_2. Todorova [9] reported the N-doped TiO_2 showed higher degradation efficiency to NO_x than pure TiO_2. Triantis et al. [10] claimed that energy states in N doping configuration were formed in the TiO_2 band gap.

Additionally, it is noted that the photocatalytic degradation mechanism of N-doped TiO_2 under the radiation of visible light has not been well understood. Also, few researches have been focused on photocatalytic degradation efficiency of N-doped TiO_2 nanoparticles to AE. The purpose of this study is to use the N element as a dopant for doping TiO_2, improving the degradation efficiency of TiO_2 to AE under visible light. For this, in this study, the N-doped TiO_2 nanoparticles were synthesized using the sol-gel method. Then the phase microstructure of prepared TiO_2 nanoparticles were tested using X-ray diffraction (XRD), and the micromorphology and dispersity of N-doped TiO_2 nanoparticles were characterized by the transmission electron microscopy (TEM). In addition, the optical performance of TiO_2 nanoparticles were analyzed by ultraviolet-visible light diffuse reflectance spectra (UV-vis DRS). Finally, the degradation tests were conducted to validate the photocatalytic activity of N-doped TiO_2 to AE under visible light.

2. Materials and Method

2.1. Synthesis of TiO_2 Samples

N-doped TiO_2 nanoparticles were prepared using the sol-gel method. The tetra-n-butyl titanate and carbamide were utilized as titanium and N precursor, respectively. Firstly, 17 mL tetra-n-butyl titanate and 30 mL absolute ethanol were blended for 30 min at room temperature, preparing the solution A. Secondly, 8 mL deionized water, 20 mL acetic acid and 28 mL anhydrous alcohol were stirred, and then a calculated amount of carbamide (the mole ratio of N/Ti was 1.0%, 2.0%, 3.0% and 4.0%, respectively.) were dropped and fully mixed, preparing the solution B.

Then, the solution B was slowly dropped into solution A, and was stirred using a magnetic stirrer at 30 °C, thus the transparent and uniform sol was obtained, and the sol was matured at room temperature for 7 days to prepare the gel. After that, the prepared gel was dried at 80 °C for 1 day, and was milled into powders and calcined at 400 °C for 2.5 h. The N-doped TiO_2 nanoparticles were obtained. The prepared samples were marked as TiO_2+NX (the X was the N dopant content). Based on the same approach, N-doped TiO_2 samples containing various N dopant contents were synthesized. However, the pure TiO_2 nanoparticles were synthesized without the carbamide in the mixed liquor in the second step.

2.2. Characterization Methods

The phase structures of prepared TiO_2 samples were tested by XRD (Ultima IV type, Rigaku, Japan) with a Cu-$K\alpha$ radiation (λ=0.15418 nm). XRD patterns were recorded in the 2θ range of 10° - 80°at a rate of 2°/min. The applied current was 30 mA, and the accelerating voltage was 40 kV.

A field emission TEM (JEM-2100 (HR) type, JEOL, Japan) was utilized to characterized microstructures and morphology of TiO_2. Its lattice resolution was 0.14 nm, and the amplification times were 50-1500000, and the point resolution was 0.23 nm.

The DRS of TiO_2 samples were recorded by a UV-vis spectrophotometer (Lambda 950 type, PE, USA) coupled with an integrating sphere. The scanning range was 200–800 nm at a data interval of 1nm. The reference sample for baseline correction was $BaSO_4$ standard.

2.3. Photocatalytic Activity Tests

As shown in Fig.1, an experimental setup was used to validate the photocatalytic degradation efficiency of prepared TiO_2 to AE under visible light. The prepared TiO_2 samples were first evenly scattered on a plate with 300 mm in both length and width, and then the plate was placed in the closed non-transparent reactor. The height of incandescent lamp was about 20 cm from the plate. AE was produced by a gasoline engine generator. Its exhaust pipe was connected with the closed reaction reactor. The degradation tests of TiO_2 to four constituents of AE were carried out under visible light. A vacuum pump connected to the reaction reactor was utilized to control the initial pressure of the closed reaction reactor.

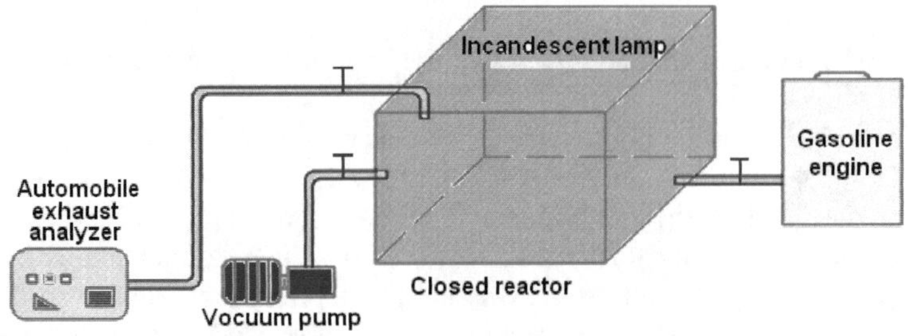

Figure 1. Schematic Diagram of Test Setup of Photocatalytic Degradation

Firstly, the vacuum pump started to run for 60 s. The gasoline engine was then started to produce AE until it run steadily. The intake valve was opened, and AE was introduced into the closed reaction reactor. The intake valve was closed when the initial content of each constituent of AE approached a stable value.

After that, the incandescent lamp was turned on to validate the photocatalytic activity of TiO_2 to each constituent of AE. The content change of each constituent of AE vs. the radiation time was recorded for 60 min by an AE analyzer (HPC506 type, Huapeng, China). In the radiation process of incandescent lamp, the contents of NO_x, CO, HC and CO_2 were tested at the interval of 10 min.

Finally, the degradation efficiency of TiO_2 to each constituent of AE was taken as the evaluating index calculated by Eq. (1).

$$\eta = (C_0\text{-}C_{end})/C_0 \tag{1}$$

Where, C_0 was the initial content of each constituent, and η was the degradation efficiency, and C_{end} was the residual content of each constituent at the end of degradation reaction.

3. Results and Discussion

3.1. Phase Structure

XRD patterns of pure and N-doped TiO_2 are presented in Figure 2. As illustrated in Figure 2, the N-doped and pure TiO_2 samples show similar phase structures. The representative diffraction peaks are observed at 62.7°, 55.0°, 53.9°, 48.0°, 37.8° and 25.3°, which correspond to six crystal planes of (204), (211), (105), (200), (004) and (101), respectively. It is noticed that the typical diffraction peaks of the rutile phase of TiO_2 are not observed, and all N-doped TiO_2 samples show anatase phase based on the comparison with the JCPDS NO: 21-1272. This suggests that the synthesized TiO_2 samples have high purity and crystallinity.

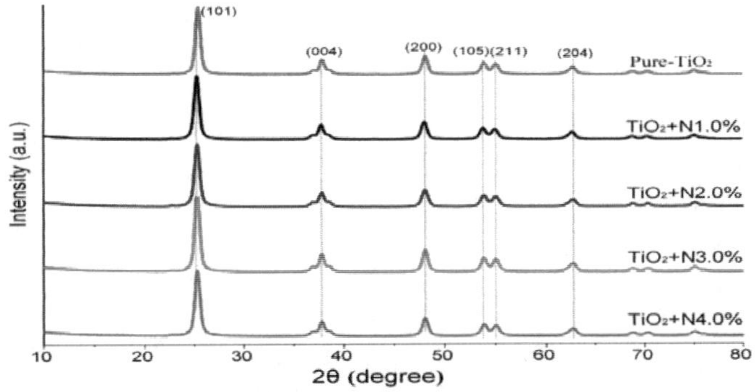

Figure 2. Typical XRD Patterns of N-doped and Pure TiO_2 Samples

By comparison, the phase compositions of N-doped TiO_2 samples are almost unchanged when compared with that of pure TiO_2 sample. Also, it is observed that the peak positions of major diffraction peaks of N-doped TiO_2 samples are similar with those of pure TiO_2 nanoparticles, but the above diffraction intensities are slightly changed. Moreover, the diffraction peaks of N impurities in the XRD diffraction patterns are not found. This suggests that the N dopants are evenly dispersed in the crystal lattices of TiO_2 samples, and its effects on the microstructures of TiO_2 are ignorable. Therefore, the crystal phase microstructures of TiO_2 nanoparticles are not changed by the N dopant.

3.2. Morphology Changes

The influences of the N dopant on the morphology and dispersity of TiO_2 are analyzed by TEM images of pure and N-doped TiO_2 as illustrated in Figure 3. From Figure 3, the particle morphologies of pure and N-doped TiO_2 present no obvious changes, and still show the spherical nanoparticles. The crystallite sizes of pure and N-doped TiO_2 are in the range between 13 nm to 16 nm. This is consistent with the XRD test results. Additionally, the dispersity of TiO_2 nanoparticles is increased when the N dopant content is raised from 0 % to 4.0 %. This is because the N-doping lowers the TiO_2 agglomeration in the heat treatment process to increase the dispersibility of TiO_2 nanoparticles. Also, this improves the specific surface area of TiO_2 nanoparticles, which is beneficial to raise the photocatalytic degradation efficiency of N-doped TiO_2.

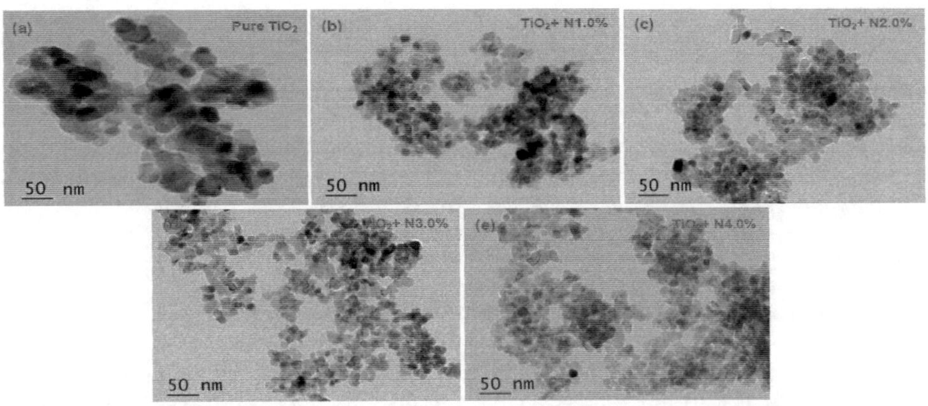

Figure 3. TEM Images of Pure and N-doped TiO_2 Nanoparticles

3.3. Optical Property

In order to discuss effects of different N dopant contents on optical performances of TiO_2 nanoparticles, the test results of UV-vis DRS of N-doped and pure TiO_2 samples are provided in Figure 4.

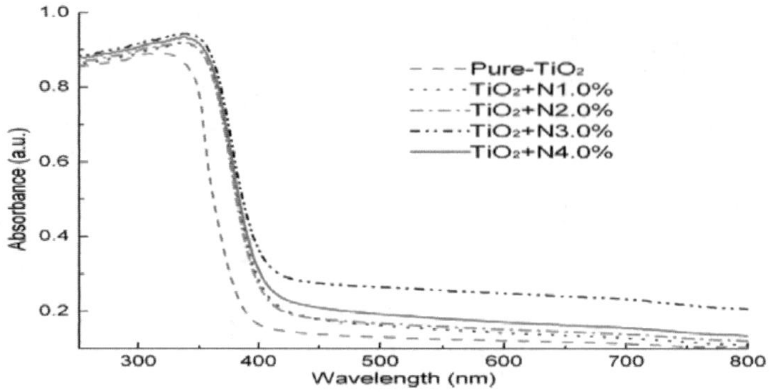

Figure 4. Test Results of UV-Vis DRS of N-doped and Pure TiO_2 Nanoparticles

It is seen from Figure 4 that the absorbance of N-doped TiO_2 is obviously improved under visible light radiation as compared with that of pure TiO_2 samples. Moreover, the optical absorption edges of N-doped TiO_2 present a red shift toward high wavelength, and the TiO_2+N3.0% shows the most obvious red shift toward high wavelength. This implies that the N-doping expands the optical absorption edge of TiO_2 from UV light to visible light region. This is conductive to increase the photocatalytic degradation efficiency and extend the utilization range of TiO_2 nanoparticles. The reason for this is that the local intermediate band gap (N2p) energy level is formed by the introduced N dopant on the top of O2p valence band in the TiO_2 nanoparticles [11]. Electrons of the doping energy level can be activated to the conduction band of TiO_2, resulting in the long wave photons to be absorbed [12]. Therefore, the absorption spectrum range of TiO_2 is expanded to the visible light region and the absorption performance in the visible light region is increased.

3.4. Photocatalytic Degradation Efficiency to AE under Visible Light

Based on the above analysis results, it is found that TiO_2+N3.0% sample has better optical properties. Therefore, we only discussed the photocatalytic activity of TiO_2+N3.0% and pure TiO_2 to each constituent of AE under visible light. The changes in NO_x and HC concentration are shown in Figure 5 during the photocatalytic degradation of TiO_2+N3.0% and pure TiO_2.

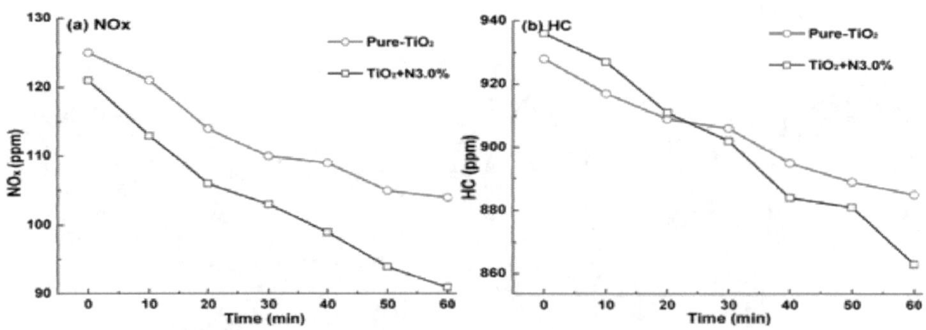

Figure 5. Concentration Changes of NO_x and HC under Visible Light

It is found from Figure 5 (a) that the photocatalytic degradation of TiO_2+N3.0% and pure TiO_2 to NO_x are similar in the first 30 min, and then the degradation effects of TiO_2+N3.0% is much higher than that of pure TiO_2. As presented in Fig.5 (b), TiO_2+N3.0% exhibits a better photocatalytic degradation performance to HC in the first 10 min, and then degradation effects of the two TiO_2 samples tend to be equal in the next 20 min, but TiO_2+N3.0% sample shows a better photocatalytic degradation performance in the last 30 min. This indicates that the N doping actually improves the photocatalytic degradation efficiency of TiO_2 to NO_x and HC under visible light.

This is because the impurity levels in TiO_2 are produced due to the substitution of doped N for partial O element. The electronegativity differences between N and O element form oxygen vacancies around the substituted position, leading to the crystal lattice distortion in TiO_2 [13]. This causes some defects to generate on the surface of crystal, thereby the recombination of photo-generated electrons and holes is inhibited and the photocatalytic active centers are increased simultaneously [14]. All these are beneficial to increase the photocatalytic degradation efficiency of TiO_2 to AE under visible light.

The changes in CO and CO_2 concentration are shown in Figure 6 during the photocatalytic degradation of TiO_2+N3.0% and pure TiO_2.

It is noted from Figure 6 (a) that the photocatalytic degradation efficiencies of TiO_2+N3.0% and pure TiO_2 to CO are not obviously improved on the whole, but the N-doped TiO_2 sample still presents a slightly higher photocatalytic degradation property to CO than that of pure TiO_2 sample. As illustrated in Figure 6 (b), the photocatalytic degradation efficiencies of TiO_2+N3.0% and pure TiO_2 to CO_2 are almost equal each other in the first 30 min, but the pure TiO_2 shows a lower degradation efficiency than that of TiO_2+N3.0% sample in the next 30 min.

This suggests that the photocatalytic activity of TiO_2+N3.0% to CO and CO_2 is higher than that of pure TiO_2 although the photocatalytic degradation efficiencies of TiO_2+N3.0% and pure TiO_2 samples to CO_2 and CO are not obviously increased. All these indicate that the N doping can increase the photocatalytic degradation efficiencies of TiO_2 to CO_2 and CO under visible light. The reasons for this are the band gap of TiO_2 is narrowed owing to the N doping, and the absorption range of TiO_2 to visible light is broadened to absorb more visible light energy and generate more electron holes [15]. Also, the electrons in activated state are more likely to jump and accelerate the transfer rate of electron interface, improving the photocatalytic degradation efficiency under visible light.

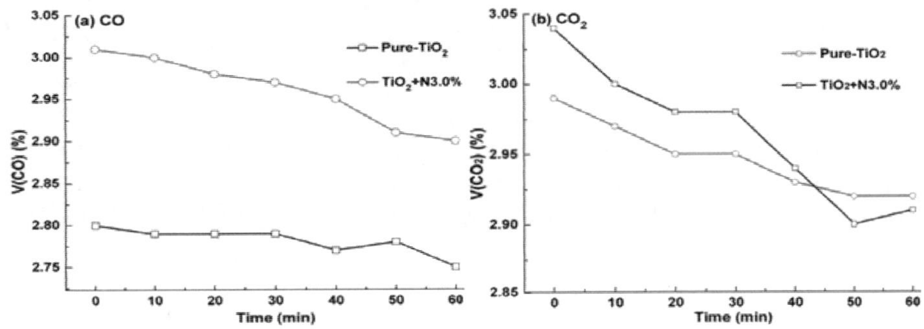

Figure 6. Concentration Changes of CO and CO_2 under Visible Light Irradiation

Table 1 provides the calculated results of photocatalytic degradation efficiencies of TiO_2+N3.0% and pure TiO_2 to each constituent of AE under visible light.

Table 1. Calculated Results of Photocatalytic Degradation Efficiencies of TiO_2+N3.0% and Pure TiO_2 to Each Constituent of AE

TiO_2 samples	Photocatalytic degradation efficiency (%)			
	CO	CO_2	HC	NO_x
Pure-TiO_2	1.5	2.1	4.6	16.7
TiO_2+N3.0%	3.7	4.3	7.8	25.2

From Table 1, it is found that the pure TiO_2 has limited photocatalytic degradation efficiencies to different constituents of AE under visible light. However, the photocatalytic degradation efficiencies of N-doped TiO_2 to different constituents of AE under visible light are distinctly increased. It is concluded that the N doping is a feasible approach to increase the photocatalytic degradation efficiency of TiO_2 to AE under visible light.

4. Conclusions

(1) Only contain anatase phase is found in the prepared N-doped nanoparticles with a slight

distortion. As the N dopant content is raised from 0% to 4%, the dispersity of TiO_2 nanoparticles is increased, improving the specific surface area and photocatalytic degradation efficiency of TiO_2. The grain size range of TiO_2 sample is 13-16 nm with a spherical shape.

(2) N-doped TiO_2 show higher optical absorptions than pure TiO_2 in UV light and visible light regions. A red shift to higher wavelength is seen on UV-Vis DRS due to the N doping of TiO_2, leading to the decrease in band gap of TiO_2. This causes TiO_2 to absorb more visible light and increase the utilization efficiency of solar irradiance, increasing the degradation efficiency of TiO_2.

(3) The N doping improves the degradation efficiency and photocatalytic activity of TiO_2 to AE under visible light. The N-doped TiO_2 shows lower degradation efficiencies to CO_2 and CO than those to NO_x and HC of AE. Also, the N-doped TiO_2 shows lower photocatalytic activity to CO than to CO_2.

(4) It is concluded that the N dopant improves the photocatalytic degradation efficiency of TiO_2 to each constituent of AE under visible light for air purification, and expands the application range of TiO2. These are beneficial to further develop the pollutant purification technology using N-doped TiO_2 nanoparticles as raw materials of building coating, road pavement, etc. along the street.

5. Acknowledgements

Authors would like to thank the financial support from Open Research Fund of National Engineering Laboratory for Advanced Road Materials, China (No. NLARMORF-2018-02), and China International Research Cooperation Projects (No. 2016YFE0108200), Research Project of Department of Transportation of Jiangsu, China (No. 2016 Y01, 2016Y03 and 2018Y25). Also, we would like to thank Advanced Analysis &Testing Center of Nanjing Forestry University for the assistance in experiments.

6. References

[1] Jin, J., Xiao, T., Tan, Y.Q. Effects of TiO_2 pillared montmorillonite nanocomposites on the properties of asphalt with exhaust catalytic capacity. J. Clean. Produt. 2018, 205: 339-349.

[2] Zhang, W.G., Zhang, Y.X., Jia, Z.R. Test method and material design of asphalt mixture with the function of photocatalytic decomposition of automobile exhaust. Constr. Build. Mater. 2019, 215: 298-309.

[3] Veluru, J.B., Appukuttan, S.N., Zhu, P.N. Synthesis and characterization of rice grains like Nitrogen-doped TiO_2 nanostructures by electrospinning–photocatalysis. Mater. Lett. 2011, 65 (19- 20): 3064.

[4] Daghrir, R., Drogui, P., Robert, D. Modified TiO_2 for environmental photocatalytic applications: a review. Ind. Eng. Chem. Res. 2013, 52 (10): 3581.

[5] Hoang, S., Guo, S., Hahn, N.T. Visible light driven photoelectrochemical water oxidation on nitrogen–modified TiO_2 nanowires. Nano Lett. 2012, 12(1): 26–32.

[6] Mamane, H., Horovitz, I., Lozzi, L., Camillo. The role of physical and operational parameters in photocatalysis by N–doped TiO_2 sol–gel thin films. Chem. Eng. J. 2014, 257(6): 159–169.

[7] Ramchiary, A. and Samdarshi, S.K.. Hydrogenation based disorder–engineered visible active N–doped mixed phase titania. Sol. Energy Mater. Sol. Cells 2015, 134: 381 –388.

[8] Powell, M.J., Dunnill, C.W., Parkin, I. P. N–doped TiO_2 visible light photocatalyst films via a sol–gel route using TMEDA as the nitrogen source. J. Photoch. Photobio. A 2014, 281(5): 27–34.

[9] Todorova, N., Hishita, S., and Boukos, N. N and N, S–doped TiO_2 photocatalysts and their activity in NOx oxidation. Catal. Today 2013, 209 (12): 41–46.

[10] Triantis, T.M. and Fotiou, T. Photocatalytic degradation and mineralization of microcystin–LR under UV–A, solar and visible light using nanostructured nitrogen doped TiO_2. J. Hazard. Mater. 2012, S (211–212)(2): 196–202.

[11] Wang, W.N., Ramalingam, B., and Mukherjee, S. Size and structure matter: enhanced CO_2 photoreduction efficiency by size–resolved ultrafine Pt nanoparticles on TiO_2 single crystals. J. Am. Chem. Soc. 2012, 134(27): 11276–11281.

[12] Burda, C., Lou, Y., Chen, X. Enhanced nitrogen doping in TiO_2 nanoparticles. Nano Lett. 2003, 3, (8): 1049–1051.

[13] Niu, P.J., Yan, J.L., Xu, C.Y. First-principles study of nitrogen doping and oxygen vacancy in cubic $PbTiO_3$. Comput. Mater. Sci. 2015, 98: 10–14.
[14] Tang, B., Liu, X., Cao, X. Experimental and theoretical studies on photocatalytic activity of N–doped TiO_2 with different concentration. J. Synth. Cryst. 2015, 44(7): 1885–1889.
[15] Pan, H., Zhang, Y.W., Shenoy, V.B., Gao, H. Effects of H–, N–, and (H, N) –doping on the photocatalytic activity of TiO_2. J. Phys. Chem. C 2011, 115: 12224–12231.

IOP Conf. Series: Materials Science and Engineering **735** (2020) 012022 doi:10.1088/1757-899X/735/1/012022

Magnetism of Undoped ZrO_2 Nanoparticles Deposited by Plasma-Gas-Condensation Technology

Wenxin Zhou, Kangqi Lin, Laisen Wang*, Qing Luo and Dongliang Peng

College of Materials, Xiamen University, Xiamen 361005, P. R. China
Email: wangls@xmu.edu.cn

Abstract. The magnetic properties of undoped ZrO_2 nanoparticles (NPs) deposited by plasma-gas-condensation technology were studied. The as-prepared ZrO_2 NPs with controllable particle sizes (6-13 nm) presents tetragonal phase which was hardly acquired by traditional methods at room temperature. The as-deposited ZrO_2 NPs were annealed at 300 to 1000 °C in air or mixed gas of Ar (95%) + H_2 (5%) for 3 h, respectively. XRD results show that the phase of ZrO_2 NPs gradually transformed from tetragonal to monoclinic with increasing the annealing temperature. The magnetic results of ZrO_2 NPs under all annealing treatments were paramagnetic, which are different with the diamagnetic or ferromagnetic ones, also different from magnetic property changed after annealing in some other works.

1. Introduction

Although unanticipated room-temperature ferromagnetism was found in many nonmagnetic metal oxide nanostructural materials without any doping [1-10], the origin of ferromagnetism still maintained debatable. Thanks to the wide band gap, thermally stability and high dielectric constant [11], ZrO_2 has a promising future in spintronics material and was being studied as the gate-dielectric in field effect transistors [12, 13]. Therefore, studying on the magnetic properties of undoped ZrO_2 is of great academic and practical significance.

Due to the possible undetected ferromagnetic impurities, the controversy in the ferromagnetism of nonmagnetic metal oxide nanostructural materials mainly embodied on whether the ferromagnetic was intrinsic [14-16]. For example, Hadacek *et al.* [14] and Abraham *et al.* [15] consider that ferromagnetism in such materials arise from the ferromagnetism impurity contamination. However, the coercivity detected in such materials is too small to distinguish the ferromagnetic is intrinsic or extrinsic. Meanwhile, creation mechanism of this ferromagnetism in such undoped materials was diversiform without a consensus. For example, Rao *et al.* [3, 6] consider that the exchange interactions of unpaired electron spins resulting from superficial oxygen vacancies generate the ferromagnetism. Coey *et al.* [1, 2] consider that the necessary ferromagnetic coupling is obtained from the polarized impurity band formed by defect-related electrons. Sanvito *et al.* [5] and Gallego *et al.* [17] consider that cation vacancy is the origin of this ferromagnetism by means of *ab initio* calculation. Therefore, for clarifying the origin of ferromagnetism in undoped nonmagnetic metal oxide nanostructural materials, a preparation method which can synthesize high purity nanosacle samples and systematically study on effects of defect on magnetic properties of such materials are necessary.

In this work, the plasma-gas-condensation technology [18] was used to synthesize ZrO_2 nanoparticles with uniform morphology and controllable particle diameters. The as-prepared undoped ZrO_2 NPs are surface clean and contamination free thanks to the vacuum condition and organic-free in the gas-phase synthesis process. Therefore, the preparation method in our work is very suitable for studying the magnetic properties of such undoped nonmagnetic metal oxide nanostructural materials because the impurity interference can be ignored. In this work, we not only ensure the purity of samples in preparation but also focus on all possible pollution in handling or measurement procedures, especially ensure that there is no contamination on account of using iron based tools in any step of sample processing. As done by many studies, systematic and various annealing treatment were applied to our ZrO_2 NPs for observing changes of magnetic properties.

2. Experimental

2.1. Deposition of ZrO_2 Nanoparticles

During the process of gas-phase synthesis, high density Zr vapor was produced by two Zr targets (99.99%) fixed oppositely with a dc sputtering source. For obtaining nanoparticles with various morphology, the different Ar gas flow rate were imported into sputtering chamber with the fixed sputtering current 0.8 A. The sputtered Zr atoms nucleated and grew by colliding each other and then flew into deposition chamber, finally landed onto the substrates and grew into nanoparticles. The deposition rate of ZrO_2 nanoparticles was measured by the quartz oscillator thickness monitor. The equivalent thickness of ZrO_2 particles assembling films can be estimated by the deposition time and deposition rate. The Zr nanoparticles with controllable sizes were obtained by changing the flow rate of Ar gas. Due to high surface activity, the as-prepared Zr NPs at initial deposition stage completely oxidized to tetragonal ZrO_2 when removed from the deposition device. For studying the magnetic performance of ZrO_2 NPs with different crystal structure and defect concentration, the as-deposited samples were annealed at 300 to 1000 ^oC in air and mixed gas of Ar(95%)+H_2(5%) for 3 h, respectively.

2.2. Measurement of ZrO_2 NPs Properties

Transmission electron microscopy (TEM) of JEOL-2100 was employed to studied ZrO_2 NPs with thickness of about 2 nm deposited on TEM copper grids. Surface morphology of NPs with thickness of 400 nm was tested by scanning electron microscope (SEM) LEO-1530FE. The crystalline structures of ZrO_2 NPs on quartz glasses before and after annealing were investigated by a Panalytical X'pert PRO X-ray diffractometer. The MPMS was used to investigate the magnetic hysteresis loops of ZrO_2 NPs with effective thickness of about 400 nm at 5K.

3. Results and Discussion

The TEM images of ZrO_2 NPs prepared with Ar gas flow rate of 200 to 500 sccm deposited on TEM copper grids with effective thickness about 2 nm are shown in **Figure 1(a)**. The corresponding statistical results of size distribution and average particle diameter (*d*) estimated by image-analysis software are shown in **Figure 1(b)**. It can be seen that ZrO_2 NPs prepared at different R_{Ar} had a narrow size distribution and the average diameters were controlled between 6.3 to 12.7 nm. The average particle diameter *d* increases with increasing the flow rate of R_{Ar}. In the synthesis process, Ar gas is used as sputtering and condensation gas at the same time. As the increasing of R_{Ar}, quantity of Zr vapor and Zr-Ar collision probability increases and then the growth of ZrO_2 NPs are promoted. If the ZrO_2 NPs continuously deposited onto the silicon, the nanoparticles assembling films were obtained. It can be seen from the SEM image (Figure. 2) that ZrO_2 nanoparticles-assembled thin films deposited on Si wafer with effective of about 400 nm are accumulated continually by lots of tiny ZrO_2 particles.

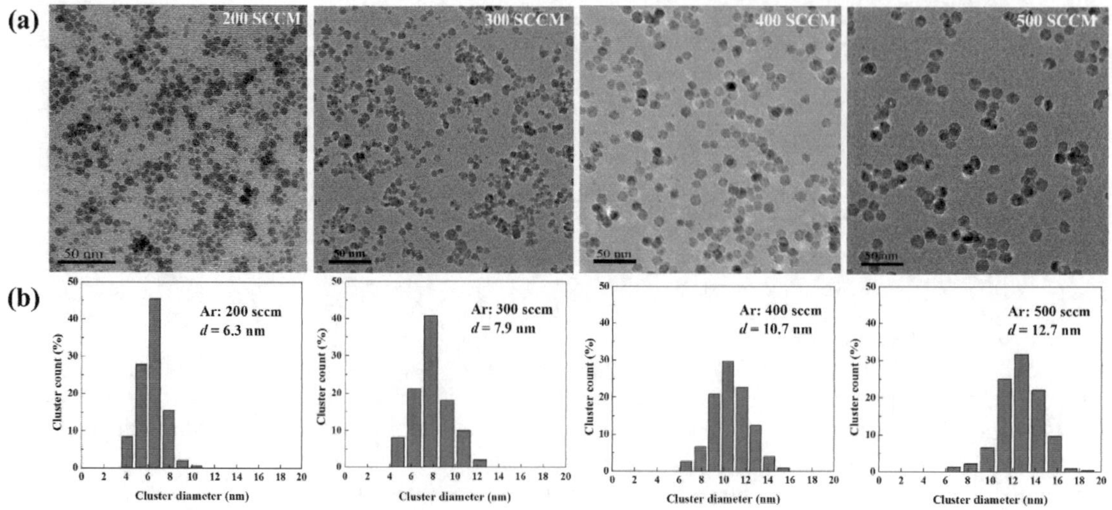

Figure 1. The TEM Images (a) and Corresponding Size Distribution (b) of ZrO_2 Nanoparticles Obtained with Different R_{Ar}

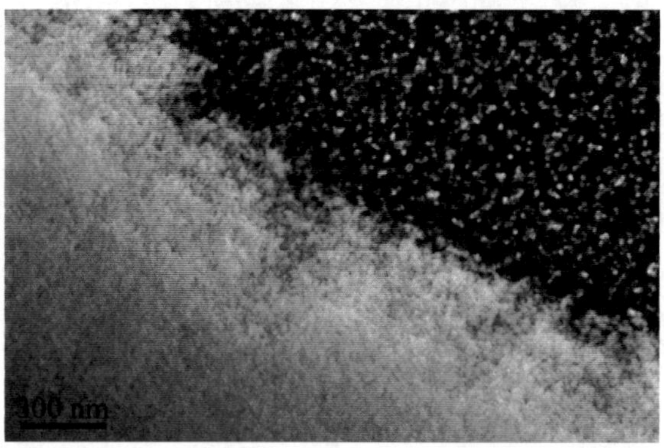

Figure 2. SEM Image of ZrO_2 NPs Assembling Films with Effective Thickness of 400 nm Prepared at R_{Ar} = 200 sccm and I = 0.8 A.

Figure 3(a) presents the XRD pattern of as-prepared ZrO_2 NPs (d = 6.3 nm) deposited on glass with thickness of 400 nm. Compared with the pattern of pure quartz glass (blue line in Figure3 (a)), the ZrO_2 emerges the peak of tetragonal ZrO_2 (101) crystal plane. Because of small size, high defect concentration and poor crystallinity the diffraction peak broadens seriously, but this condition can be improved well by annealing. In normal case, bulk ZrO_2 is stable with monoclinic phase instead of tetragonal phase below 1170 ^{o}C under normal atmospheric conditions [19]. Owing to dimension effect, the as-prepared ZrO_2 NPs possess high defect concentration and surface energy. For reducing the system energy of NPs, ZrO_2 revealed the high temperature phase (tetragonal) at ambient pressures and temperatures [20-23]. Thus, the undoped tetragonal ZrO_2 which was difficult to obtain at room temperature via traditional synthetic methods can be synthesized by plasma-gas-condensation. **Figure 3(b)** shows the magnetization curve of ZrO_2 NPs on Si substrate. It can be seen that pure Si substrate reveals obvious diamagnetic (red line in Figure 3(b)). After eliminating the diamagnetic signal of Si substrate, the magnetization curve of as-prepared ZrO_2 NPs is almost linear. This indicates that the tetragonal ZrO_2 NPs with average diameter of 6.3 nm is paramagnetic. This result is not in agreement with ferromagnetism [1-10] or diamagnetism [15] found by many similar studies.

In the study of diluted magnetic oxides, the magnetic properties are closely related to the crystal structure and defects [5, 24]. According some reports, the ferromagnetism nanoscale HfO_2 or ZrO_2 usually present monoclinic phase [2] rather than tetragonal phase in our ZrO_2 NPs. Therefore, for studying the effects of crystal structure, crystallinity and oxygen vacancy on magnetic properties, as-prepared ZrO_2 NPs were annealed in air from 300 to 1000 °C for 3 h, respectively. The XRD patterns of ZrO_2 NPs annealed with various temperatures are shown in **Figure 3(c)**. This indicated that the crystallinity of tetragonal ZrO_2 NPs was greatly improved as annealing temperature increased. At the annealing temperature of 800 °C, ZrO_2 NPs still maintain tetragonal phase so that the pure tetragonal ZrO_2 at low temperature can be study by plasma-gas-condensation technology. After annealing at 1000 °C in air for 3 h, part of ZrO_2 transformed into monoclinic phase from tetragonal phase. The corresponding magnetic properties of ZrO_2 NPs annealed at various temperatures in air were measured by SQUID at 5 K (**Figure 3(d)**). All the magnetic results indicate that ZrO_2 NPs are paramagnetic after all annealing treatments despite it is tetragonal or monoclinic phase, poor or good crystallinity. Furthermore, defect concentration especially oxygen vacancy decreased with the increase of annealing temperature in air (oxidizing atmosphere). Therefore, air-annealing couldn't change the magnetic property of ZrO_2 NPs which is not in agreement with some similar studies [1, 4, 7, 8, 25].

Some researchers also consider that for achieving ferromagnetism in nanoscale oxide, the vacancy concentration of the system should be high [26-28]. For further study on effects of high oxygen vacancy on magnetic properties, ZrO_2 NPs were annealed in reducing atmosphere (mixed gas of Ar(95%) and H_2(5%)) from 300 to 1000 °C for 3 h, respectively. A similar XRD results are shown in **Figure 3(e)**. As annealing temperature increased, crystallinity of tetragonal ZrO_2 NPs was greatly improved and transformed into monoclinic phase at the annealing temperature of 1000 °C. **Figure 3(f)** shows the corresponding magnetization curves. It can be seen that the magnetic properties still maintain paramagnetic although the paramagnetic moment decreased slightly. Therefore, the increasing oxygen vacancy obtained by annealing in reducing atmosphere couldn't change paramagnetic result to ferromagnetism.

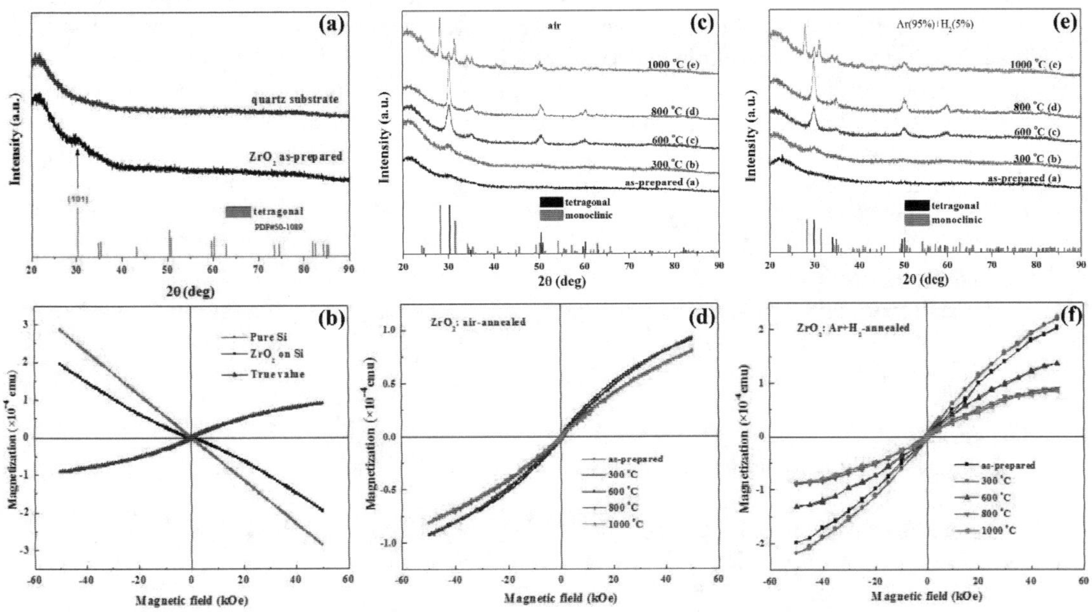

Figure 3. (a) XRD patterns and (b) magnetization curves of as-prepared ZrO_2 NPs deposited on quartz glass and pure Si substrates, respectively. (c) XRD patterns and (d) magnetization curves of ZrO_2 NPs before and after annealing in air atmosphere for 3 h at various temperatures, respectively. (e) XRD patterns and (f) magnetization curves of ZrO_2 NPs before and after annealing in mixed gas of Ar (95%) and H_2 (5%) for 3 h at various temperatures, respectively.

As described above, the ZrO_2 NPs with effective thickness of 400 nm and average diameter of 6.3 nm deposited on Si substrate presents paramagnetic instead of diamagnetism or ferromagnetism reported in recent research. Similar paramagnetic behavior has also been found in HfO_2 by Tirosh *et al.* [25] and copper-doped ZrO_2 by Dutta *et al.* [29], in which this result was regarded as defect or vacancy related. The extended defects might create delocalized and spin-polarized states and then generate magnetic moment [30]. As a result of size-effect, there is a large aggregate superficial area and defect concentration in ZrO_2 nanoparticles. Therefore, we think that the observed paramagnetic behavior in ZrO_2 NPs should be also defect related. However, the paramagnetic result rather than ferromagnetism might be because magnetic properties of undoped nonmagnetic metal oxide nanostructural materials are related with their preparation method, and ferromagnetic results might have stringent demands on morphology, defect types, defect concentration and distribution in such nanostructural materials.

The paramagnetic was maintained even when crystal structure, crystallinity and oxygen vacancy concentration changed after annealing at different temperatures and atmosphere (see Figure 3). This indicates that the paramagnetic of undoped ZrO_2 NPs prepared by plasma-gas-condensation method is stable and rarely affected by the change of crystal structure, crystallinity and oxygen vacancy concentration. This is not the same with magnetism changed from ferromagnetic or paramagnetic to antimagnetic after annealing treatments detected by Coey and Tirosh [1, 25], respectively. Therefore, further work on the clarification of the diamagnetic behavior in such nanoscale material synthesized by gas-phase method is underway.

4. Conclusions

In our work, tetragonal ZrO_2 nanoparticles with uniform morphology and controllable particle diameters (6-13 nm) through modifying the flow rate of Ar gas were prepared by plasma-gas-condensation technology. After annealing at different temperatures in oxidizing or reducing atmosphere for 3h, respectively, tetragonal ZrO_2 partly converted into monoclinic phase at the annealing temperature of 1000 ^{o}C and their magnetic properties remained paramagnetic despite the change of crystal structure, crystallinity and oxygen vacancy concentration. The paramagnetic result in our work is different with the ferromagnetic or diamagnetic one, also different from magnetic property changed after annealing treatments in some reports. However, whether this experimental phenomenon also applies to every such nanoscaled materials with different morphologies and dimensions, this interesting result provides a new experimental phenomenon and basis for the study on magnetic properties of undoped nonmagnetic metal oxide materials.

5. Acknowledgements

This work was supported by the National Natural Science Foundation of China (Grant No. 51771157, 51571167 and 51901197), the Fundamental Research Funds for the Central Universities of China (Xiamen University: Nos. 20720190007).

6. References

[1] Coey J.M.D., Venkatesan M., Stamenov P., Fitzgerald C.B. and Dorneles L.S. 2005 *Phys. Rev. B* **72** 024450.
[2] Venkatesan M., Fitzgerald C.B. and Coey J.M.D. 2004 *Nature* **430** 630.
[3] Sundaresan A., Bhargavi R., Rangarajan N., Siddesh U. and Rao C.N.R. 2006 *Phys. Rev. B* **74** 161306.
[4] Ma N.G., Zhang Y.J., Cao E.S., Sun L., Hao W.T. and Yang Z. *Chem. Phys. Lett.* 2019 **727** 6-104.
[5] Pemmaraju C.D. and Sanvito S. 2005 *Phys. Rev. Lett.* **94** 217205.
[6] Sundaresan A. and Rao C.N.R. 2009 *Nano Today* **4** 96-106.
[7] Yi J.B., Pan H., Lin J.Y., Ding J., Feng Y.P., Thongmee S., Liu T., Gong H. and Wang L. 2008 *Adv. Mater.* **20** 1170-1174.
[8] Nandy A., Pal U. and Pradhan S.K. 2019 *J. Alloy. Compd.* **793** 220-231.
[9] Esmaeily A.S., Venkatesan M., Sen S. and Coey J.M.D. 2018 *Phys, Rev. Mater.* **2** 054405.

[10] Gao Q., Chen H.L., Li K.F. and Liu Q.Z. 2018 *ACS Appl. Mater. Inter.* **10** 27503-27509.
[11] Seema K. and Kumar R. 2015 *J. Supercond. Nov. Magn.* **28** 2735-2742.
[12] Ostanin, S. *et al.*, 2007 *Phys. Rev. Lett.* **98** 016101.
[13] Konda R.B., White C., Smak J., Mundle R., Bahoura M. and Pradhan A.K. 2013 *Chem. Phys. Lett.* **583** 74.
[14] Hadacek N., Nosov A., Ranno L., Strobel P and Gal´era R-M 2007 *J. Phys. Condens. Matter* **19** 486206 14.
[15] Abraham D.W., Frank M.M. and Guha S. 2005 *Appl. Phys. Lett.* **87** 252502.
[16] Jo Y., Hwang I.R., Park B.H., Lee K.J., Lee S.I. and Jung M.H. 2009 *Appl. Phys. Lett.* **95** 263504
[17] Gallego S., Beltr´an J.I., Cerd´a J and Mu˜noz M.C. 2005 *J. Phys. Condens. Matter* **17** 451–457.
[18] Wang L.S., Wen R.T., Chen Y., Yue G.H. and Peng D.L. 2011 *Appl. Phys. A* **103** 1015.
[19] Yoshimura M. 1988 Am. Ceram. Soc. Bull. **67** 1950.
[20] Garvie R.C. and Goss M.F. 1986 J. Mater. Sci. **21** 1253.
[21] Bokhimi X., Morales A., Novaro O., Portilla M., López T., Tzompantzi F. and Gómez R. 1998 *J. Solid. State. Chem.* **135** 28-35.
[22] Duwez P., Odell F. and Brown F.H. 1952 *J. Am. Ceram. Soc.* **35** 107.
[23] Duwez P., Brown F.H. and Odell F. 1951 *J. Electrochem. Soc.* **98** 356.
[24] Ye L.H. and Freeman A. J. 2006 *Phys. Rev. B* **73** 081304.
[25] Tirosh E. and Markovich G. 2007 *Adv. Mater.* **19** 2608-2612.
[26] Maoz B.M., Tirosh E., Sadan M.B. and Markovich G. 2011 *Phys. Rev. B* **83** 161201.
[27] Zunger A., Lany S. and Raebiger H. 2010 *Physics* **3** 53.
[28] De Souza A.O., Ivashita Flávio F., Biondo V., Paesano A. and Mosca D.H. 2016 *J. Alloy. Compd.* **680** 701-710.
[29] Dutta P., Seehra M. S., Zhang Y. and Wender I. 2008 *J. Appl. Phys.* **103** 07D10428.
[30] Schoenhalz A.L., Arantes J.T., Fazzio A. and Dalpian G.M. 2009 *Appl. Phys. Lett.* **94** 162503 30.

IOP Conf. Series: Materials Science and Engineering **735** (2020) 012023 doi:10.1088/1757-899X/735/1/012023

A Refined Model for Predicting the 31-Coupling Effect of Piezoelectric Structural Fiber Composites

Yaoling Xu and Lishuang Zheng

Key Laboratory of Mechanical Reliability for Heavy Equipments and Large Structures of Hebei Province, Yanshan University, Qinhuangdao, China.
Email: xylysu@163.com

Abstract. A refined model is proposed to predict the 31-coupling effect of a piezoelectric structural fiber (PSF) and PSF composites. The PSF model consists of a piezoelectric shell, inner and outer electrode layers, and a piezoelectric inactive core fiber. The inner and outer electrodes provide a radial electric field passing through the piezoelectric shell, which leads to the axial deformation of the piezoelectric shell, the piezoelectric inactive core fiber provides the mechanical support to the brittle piezoelectric shell. Analytical expressions of the effective piezoelectric coefficient d_{31} of the single PSF and the PSF composites are derived, the influences of aspect ratio of the piezoelectric shell, thickness of the electrode layer and PSF volume fraction, on the effective piezoelectric coefficient d_{31} are demonstrated. The numerical results show that the model without electrode layers may lead to a significant error.

1. Introduction

The piezoelectric materials have been widely used in sensing and actuation [1-4], fragile nature of the monolithic piezoelectric materials limits their applications in many situations, so the piezoelectric fiber composites have been developed, in which the polymer matrix can provide improved robustness by protecting the fragile fibers and allow the composites to more easily conform to the curved surfaces found in more realistic industrial applications. A common form of piezoelectric fiber composites is based on embedded solid piezoelectric fibers activated by an external interdigitated electrode pattern. Although this type of composite is very useful for anisotropic activation of composites, it requires the electric field to pass through the composite matrix (the electrodes are placed on the matrix surface), which leads to the requirement of high voltages for actuation due to significant electric field losses. Additionally, this type of composite is constrained to non-conductive matrix materials and constructed usually in the form of a patch of material which can be bonded to the surface of a structure or laid-up as active layers along with conventional fiber-reinforced composites, and it is still generally separate from the structural components and not intended to provide any load bearing functionality [5].

In order to overcome the drawbacks of the solid piezoelectric fiber composites, the hollow piezoelectric fiber composite was proposed [6-8], and the feasibility of producing hollow piezoelectric fiber has been investigated [9]. In the hollow piezoelectric fiber composite, the hollow piezoelectric fibers are individually electroded on both the inside and outside surfaces of the hollow fibers, and activated by an electric field applied directly across the wall of the hollow fiber, which generate longitudinal strain due to the piezoelectric coefficient d_{31}. Even though the longitudinal strain of the hollow piezoelectric fiber composite is decreased by approximately half by using d_{31} versus the d_{33} mode used in solid piezoelectric fiber composites, the required voltage can be decreased by a factor of ten or more since the electric field is applied only across the wall of the hollow piezoelectric fiber instead of through the matrix, thereby electric field losses are eliminated. Moreover, the hollow piezoelectric fibers isolate the inner electrode from the matrix, which enable the fiber to be embedded in electrically conductive matrices.

However, hollow fibers are prone to cracking and failure under mechanical loading due to the fragile nature of the fiber walls. Recently, the piezoelectric structural fiber (PSF), which consists of a piezoelectric inactive core fiber coated with a piezoelectric material, has been developed [10], the piezoelectric inactive core fiber provides the mechanical support to the brittle piezoelectric shell and carries part of the mechanical loading, an electrical potential difference between the inner and outer electrodes gives rise to an electrical polarization and an actuating electric field in the radial direction of the piezoelectric shell, corresponding axial deformation of the fiber is induced by the 31-coupling effect of the piezoelectric materials. The piezoelectric inactive core can be nonconductive materials such as glass [11] or conductive materials such as metals [12, 13], Carbon and Silicon Carbonate [14]. The conductive core can serve as an electrode, while for nonconductive cores, an inner electrode must be placed between the core fiber and the piezoelectric shell.

The single PSF and PSF composites in which PSFs are deployed unidirectionally into the polymer matrix have many practical applications in sensing and actuating, the effective piezoelectric coefficients are key parameters of the PSF and the PSF composites. Brei and Cannon [8] evaluated the effective d_{31} of the hollow piezoelectric fiber and hollow piezoelectric fiber composites, and investigated the effects of the key design parameters (matrix/fiber Young's moduli, aspect ratio of the individual fibers, and the overall active composite volume fraction) on the performance, manufacturing and reliability of the hollow piezoelectric fiber and hollow piezoelectric fiber composites. Beckert et al. [11] predicted effective properties of PSF composites with the glass core by three different methods (finite element method, effective field method, and homogeneous filed approach). Lin and Sodano [5, 14] developed a one-dimensional micromechanics model to obtain the effective d_{31} of the PSF and the PSF composites.

In the works mentioned above, the electrode layers are neglected in the models due to its small thickness. In some cases, the electrode layer should be considered in the models based on the following reasons: (1) very thin electrode layers are prone to damage, which will leads to the failures of the PSF fiber and the PSF composites due to losing electric connectivity, so a thick electrode layer is beneficial to maintain the integrity of the electrode; (2) hollow piezoelectric fibers can be fabricated in very small sizes, even the wall thickness of the hollow piezoelectric fiber maybe on the order of a grain size. Compared with the wall thickness, the electrode layer cannot be neglected in size; (3) the stiffness of the metal (typical platinum and wolfram) is usually several times larger than that of the piezoelectric ceramics, so significant errors may occur due to neglecting the electrode layer in the model. This paper proposes a refined PSF model, in which inner and outer electrode layers are considered. Analytical expressions of the effective d_{31} of the single PSF and the PSF composites are derived, the influences of aspect ratio of the piezoelectric shell, thickness of the electrode layers, and PSF volume fraction, on the effective piezoelectric coefficient d_{31} are investigated.

2. Model and Effective Piezoelectric Coefficients of the Single PSF and the PSF Composites

A PSF and its cross section are shown in figure 1. The proposed model consists of a piezoelectric shell which is polarized along the radial direction (r or x_3 direction), a piezoelectric inactive core fiber which provides the mechanical support to the piezoelectric shell, and inner and outer electrode layers. An electrical potential difference between the inner and outer electrodes gives rise to a radial electrical field which is used to drive the piezoelectric shell, and a corresponding deformation in axial direction (z or x_1 direction) of the PSF is induced by the piezoelectric 31-coupling effect.

The behavior of piezoelectric materials is described by the following constitutive equation:

$$\begin{aligned} \varepsilon_{ij} &= s^E_{ijmn}\sigma_{mn} + d_{nij}E_n \\ D_i &= d_{imn}\sigma_{mn} + \kappa^{\sigma}_{in}E_n \end{aligned} \tag{1}$$

where ε_{ij} , σ_{mn} , D_i and E_n are the strain, stress, electric displacement, and electric field, respectively; s^E_{ijmn} , d_{nij} and κ^{σ}_{in} are elastic compliance tensor (measured in a constant electric field), piezoelectric field-strain tensor (measured in a constant electric field or at a constant stress), and dielectric tensor (measured at a constant stress), respectively.

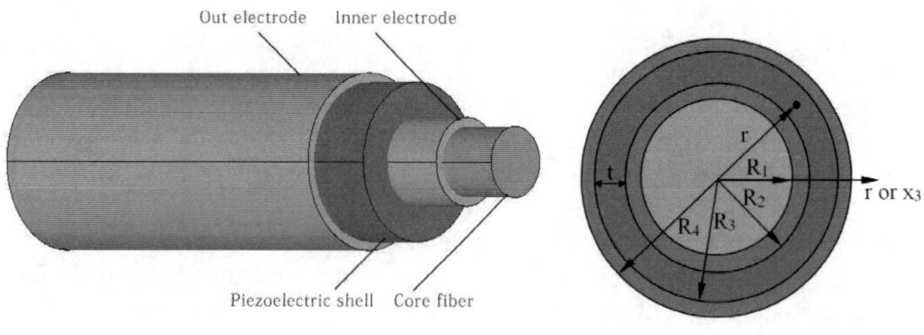

Figure1. Piezoelectric Structural Fiber and Its Cross Section

For the piezoelectric shell, if the material is polarized along the radial direction, it exhibits transversely isotropic behavior in the x_1-x_2 (z - θ) plane, and the constitutive equation (1) can be rewritten as the following matrix form:

$$\begin{pmatrix} \varepsilon_{11} \\ \varepsilon_{22} \\ \varepsilon_{33} \\ \varepsilon_{23} \\ \varepsilon_{31} \\ \varepsilon_{12} \\ D_1 \\ D_2 \\ D_3 \end{pmatrix} = \begin{bmatrix} s^E_{11} & s^E_{12} & s^E_{13} & 0 & 0 & 0 & 0 & 0 & d_{31} \\ s^E_{12} & s^E_{11} & s^E_{13} & 0 & 0 & 0 & 0 & 0 & d_{31} \\ s^E_{13} & s^E_{13} & s^E_{33} & 0 & 0 & 0 & 0 & 0 & d_{33} \\ 0 & 0 & 0 & s^E_{44} & 0 & 0 & 0 & d_{15} & 0 \\ 0 & 0 & 0 & 0 & s^E_{44} & 0 & d_{15} & 0 & 0 \\ 0 & 0 & 0 & 0 & & s^E_{66} & 0 & 0 & 0 \\ 0 & 0 & 0 & 0 & d_{15} & 0 & \kappa^{\sigma}_{11} & 0 & 0 \\ 0 & 0 & 0 & d_{15} & 0 & 0 & 0 & \kappa^{\sigma}_{11} & 0 \\ d_{31} & d_{31} & d_{33} & 0 & 0 & 0 & 0 & 0 & \kappa^{\sigma}_{33} \end{bmatrix} \begin{pmatrix} \sigma_{11} \\ \sigma_{22} \\ \sigma_{33} \\ \sigma_{23} \\ \sigma_{31} \\ \sigma_{12} \\ E_1 \\ E_2 \\ E_3 \end{pmatrix} \tag{2}$$

An electrical potential difference between the inner and outer surfaces of the piezoelectric shell gives rise to an actuating electric field, the expression for the electric field E at a distance r from the geometric center of the piezoelectric shell is given as follows [7]:

$$E(r) = -\frac{V}{r\ln(1-\alpha)} \tag{3}$$

Where t is the wall thickness of the piezoelectric shell, $\alpha = t/R_3$ is the aspect ratio, V is the voltage applied across the piezoelectric shell.

The average electric field in the piezoelectric shell is

$$E_{ave} = \frac{\int_{R_2}^{R_3}\int_0^{2\pi} E(r)rdrd\theta}{\pi(R_3^2 - R_2^2)} = -\frac{E_{tw}}{(1/\alpha - 0.5)\ln(1-\alpha)} \tag{4}$$

Where $E_{tw} = V/t$.

IOP Conf. Series: Materials Science and Engineering **735** (2020) 012023 doi:10.1088/1757-899X/735/1/012023

Compared with the length of the PSF, the diameter of the PSF is very small, so the one-dimensional model is usually used to study the 31-coupling effect of the PSF [2], the axial strain in the piezoelectric shell is derived from equation (2) as follows:

$$\varepsilon_{11}(r) = d_{31}E_3(r) \tag{5}$$

The averaged axial strain of the piezoelectric shell is

$$\varepsilon_{11,ave}^{pie} = \frac{\int_{R_2}^{R_3}\int_0^{2\pi}\varepsilon_{11}(r)rdrd\theta}{\pi(R_3^2 - R_2^2)} = \left(-\frac{d_{31}}{(1/\alpha - 0.5)\ln(1-\alpha)}\right)E_{tw} = d_{31,eff}^{pie}E_{tw} \tag{6}$$

Where $d_{31,eff}^{pie}$ is defined as the effective d_{31} of the piezoelectric shell.

Due to the restriction of the electrode layers and the core fiber of the PSF, the piezoelectric shell cannot deform to $\varepsilon_{11,ave}^{pie}$ freely, its practical averaged axial strain is denoted by $\varepsilon_{11,fact}^{pie}$, the averaged axial strains of the electrode layers and the core fiber can also be approximated as $\varepsilon_{11,fact}^{pie}$. Because the resultant axial force on the cross section of the PSF is zero, we have

$$\frac{(\varepsilon_{11,ave}^{pie} - \varepsilon_{11,fact}^{pie})}{s_{11}^E}\pi(R_3^2 - R_2^2) = \frac{\varepsilon_{11,fact}^{pie}}{s_{11}^{out}}\pi(R_4^2 - R_3^2) + \frac{\varepsilon_{11,fact}^{pie}}{s_{11}^{in}}\pi(R_2^2 - R_1^2) + \frac{\varepsilon_{11,fact}^{pie}}{s_{11}^{core}}\pi R_1^2 \tag{7}$$

Where s_{11}^E, s_{11}^{out}, s_{11}^{in} and s_{11}^{core} are the elastic compliance coefficients of the piezoelectric shell, the outer electrode, inner electrode and the core fiber, respectively.

From equations (6) and (7), $\varepsilon_{11,fact}^{pie}$ is derived as follows:

$$\varepsilon_{11,fact}^{pie} = \left(\frac{\frac{(R_3^2 - R_2^2)}{s_{11}^E}d_{31,eff}^{pie}}{\frac{(R_4^2 - R_3^2)}{s_{11}^{out}} + \frac{(R_3^2 - R_2^2)}{s_{11}^E} + \frac{(R_2^2 - R_1^2)}{s_{11}^{in}} + \frac{R_1^2}{s_{11}^{core}}}\right)E_{tw} = d_{31,eff}^{PSF}E_{tw} \tag{8}$$

Where $d_{31,eff}^{PSF}$ is defined as the effective d_{31} of the PSF.

To predict the effective d_{31} of the composites embedded with the PSFs, an annular which represents the matrix is attached to the outside of the outer electrode layer of the PSF model, and the area of the annular is determined by the volume fraction of the matrix. Similarly, the effective d_{31} of the PSF composites can then be derived as:

$$d_{11,eff}^{com} = \left(\frac{\frac{(R_3^2 - R_2^2)}{s_{11}^E}d_{31,eff}^{pie}}{\frac{(R_5^2 - R_4^2)}{s_{11}^{matrix}} + \frac{(R_4^2 - R_3^2)}{s_{11}^{out}} + \frac{(R_3^2 - R_2^2)}{s_{11}^E} + \frac{(R_2^2 - R_1^2)}{s_{11}^{in}} + \frac{R_1^2}{s_{11}^{core}}}\right)E_{tw} = d_{31,eff}^{com}E_{tw} \tag{9}$$

IOP Conf. Series: Materials Science and Engineering **735** (2020) 012023 doi:10.1088/1757-899X/735/1/012023

Where $d_{31,eff}^{\text{com}}$ is defined as the effective d_{31} of the PSF composites, R_5 is the outer radius of the matrix annular, s_{11}^{matrix} is the compliance coefficient of the matrix material.

3. Examples

Piezoelectric shells with different aspect ratios have been manufactured through microfabrication by coextrusion technique [9], their diameters are about several hundred micrometers. A PSF consists of a PZT-5A shell, inner and outer wolfram electrode layers, and a glass core fiber is considered. The outer radius of the PZT-5A shell is equal to $300\mu m$, the inner and outer electrode layers have the same thickness. The material properties are listed in Table 1.

Table 1. Material Properties

	PZT-5A	Wolfram	Glass	Epoxy
Elastic compliance coefficient, $\times 10^{-12} m^2 N^{-1}$	$s_{11}^E = 16.4$	2.90	14.3	292.8
Piezoelectric field-strain coefficient, $\times 10^{-12} CN^{-1}$	$d_{31} = -171$	0	0	0

The normalized effective d_{31} of the PSF ($d_{31,eff}^{PSF} / d_{31}$) with respect to the aspect ratio and the normalized effective d_{31} of the PSF composites ($d_{31,eff}^{com} / d_{31}$) with respect to the PSF volume fraction λ for $\alpha = 0.5$ are plotted in figure 2 for several electrode thicknesses (t=0, 1, 3, and $5\mu m$). It is seen from figure 2 that the effective d_{31} of the PSF shows a non-monotonous relationship with the increasing aspect ratio, the maximum value is obtained at approximately $\alpha = 0.74$. The effective d_{31} of the PSF composites shows a sharp increase when λ is small, and then a slow increase is observed with the increasing PSF volume fraction. The electrode layers have significant effects on the effective d_{31} of both the PSF and the PSF composites, the effective d_{31} decreases obviously with the increasing electrode thickness, which indicate that the model without electrode may cause significant errors. It is worth pointing out that the electrode layer leads to the decrease of the effective d_{31}, but a thick electrode is beneficial to prevent the electrode from damage, and improve the mechanical stability of the piezoelectric shell.

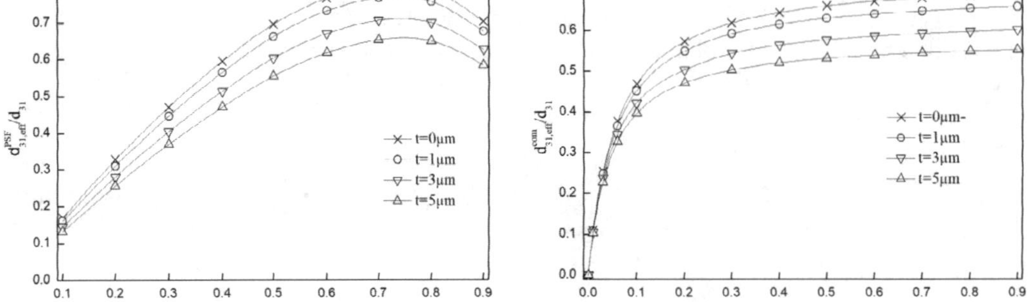

Figure 2. Normalized Effective d_{31} (a) PSF (b) PSF Composites ($\alpha = 0.5$)

4. Conclusions

A refined model is proposed to predict the 31-coupling effect of the PSF and the PSF composites, analytical expressions of the effective piezoelectric coefficient d_{31} of the single PSF and the PSF composites are derived, and the influences of aspect ratio of the piezoelectric shell, thickness of the electrode layer, and PSF volume fraction on the effective d_{31} are demonstrated. The results show that the electrode layers have significant effect on the effective d_{31} of both the PSF and the PSF composites, the model without electrode layers may produce a significant error.

5. References

[1] Bent A A, Hagood N W and Rodgers J P 1995 *J. Intel. Mat. Syst. Str.* **6** 338-49
[2] Niezrecki C, Brei D, Balakrishnan S and Moskalik A 2001 *Shock and Vibration Digest* **33** 269-80
[3] Sodano H A, Park G and Inman D J 2005 *J. Spacecraft Rockets* **42** 370-73
[4] Liu J, Qiu J H, Chang W J, Ji H L and Zhu K J 2010 *Proc. Conf. on Smart Materials, Adaptive Structures and Intelligent Systems* (Philadelphia)
[5] Lin Y and Sodano H A 2008 *Compos. Sci. Technol.* **68** 1911-18
[6] Zhang Q, Wang H and Cross L E 1993 *J. Mater. Sci.* **28** 3962-68
[7] Fernandez J F, Dogan A, Zhang Q M, Tressler J F and Newnham R E 1995 *Sens. Actuator A-Phys.* **51** 183-92
[8] Brei D and Cannon B J 2004 *Compos. Sci. Technol.* **64** 245-61
[9] Cannon B J and Brei D 2000 *Proc. Int. Symposium on Smart Structures and Materials* (San Diego)
[10] Qiu J, Tani J, Yamada N and Takahashi H 2003 *Proc. Int. Symposium on Smart Structure and Materials* (San Diego)
[11] Beckert W, Kreher W, Braue W and Ante M 2001 *J. Eur. Ceram. Soc.* **21** 1455-58
[12] Sebald G, Qiu J, Guyomar Dand Hoshi D 2005 *Japan-Soc. Appl. Phys.* **44** 6136-56
[13] Takagi K, Sato H and Saigo M 2006 *Adv. Compos. Mater.* **15** 403-17
[14] Lin Y and Sodano H A 2009 *Compos. Sci. Technol.* **69** 1825-30.
[15] Dai Q L and Ng K 2012 *Mech. Mater.* **53** 29-46.
[16] Halliday D and Resnick R 1988 *Fundamentals of Physics* (New York: John Wiley & Sons).
[17] Gael S, Abdelmjid B, Jinhao Q, Guiffard B and Guyomar D 2006 *J. Appl. Phys.* **100** 054106-1-6-6

IOP Conf. Series: Materials Science and Engineering **735** (2020) 012024 doi:10.1088/1757-899X/735/1/012024

Study on Hot Deformation Behavior and Constitutive Model of TNTZ Alloy

Jing Cheng, Kelu Wang, Xin Li and Jianjun Liu

School of Aerospace Manufacturing Engineering, Nanchang Hangkong University, Nanchang 330063,China

E-mail:ChagallCheng@163.com; Tel: 157-9775-7281

Abstract. The isothermal constant strain rate compression experiment of Ti-29Nb-13Ta-4.6Zr (TNTZ) alloy with deformation temperature of 700~900°C and strain rate of 0.001~1s^{-1} was carried out by Gleeble-3500 thermal simulator. The high temperature deformation behavior of the alloy was discussed and the thermal deformation activation energy was calculated. The hyperbolic sinusoidal constitutive model of TNTZ alloy was constructed. The results show that the flow stress of the alloy tends to be steady at high temperature and low strain rate. The constitutive model established by the Arrhenius model and modified by Zener-Hollomon parameter, which its correlation coefficient R and the average relative error E is 0.971 and 5.92%. The results predicted value for high temperature deformation behavior and agrees well with the experimental value of the alloy.

1. Introduction

TNTZ alloy is one of the new medical β-type titanium alloy Ti-Nb-Zr-Ta series and its wear resistance, biocompatibility and its elastic mode is significantly lower than that of α+β type titanium alloy, which has significantly effect in biomedical applications, such as bone repair and artificial joints [1]. In recent years, many scholars have carried out research on the properties of the alloy. Huang et al [2] prepared a TNTZ alloy with good comprehensive performance by spark plasma sintering, and studied the sintering temperature of the alloy. The effects of structure and mechanical properties. Ye [3] studied the effects of rolling annealing process on the high temperature deformation behavior and hot working process of TNTZ and determined that the rolling treatment can effectively reduce the elastic modulus and improve the strength. In addition, the annealing time, rolling ratio and the influence of $(Ti,Zr)_2Si$ make the elastic modulus and plasticity of the alloy significantly improved and the elastic modulus (24.4-28.9GPa) matching the human skeleton can be obtained. Therefore, TNTZ is widely used in human prostheses and joints.

In this paper, the true stress-true strain curve characteristics of the TNTZ alloy at a temperature of 700°C to 900°C and a strain rate of 0.001s^{-1} to 1s^{-1} is analyzed and the traditional Arrhenius model was established to study the constitutive relation. By comparing the predicted stress value and the experimental stress value, the applicability of the model was corrected, which could provide a theoretical basis for the subsequent thermal processing.

2. Materials and Methods

The experimental material is TNTZ alloy and its nominal composition is Ti-29Nb-13Ta-4.6Zr (at %), the α+β→β phase transformation temperature of the alloy is 850°C. The compressed sample was a cylinder of Φ8 mm×12 mm and an isothermal constant strain rate compression experiment was performed using a Gleeble-3500 thermal simulator. The experimental schemes are: deformation temperatures of 700, 750, 800, 850, 900°C and strain rates of 0.001, 0.01, 0.1, $1.0s^{-1}$. The height reduction ratio is 70% and the corresponding true strain is about 1.2. The sample was heated to a deformation temperature at a rate of 10°C /s and then held for 300s to make the temperature uniform. After the end of compression, spray water to cool to room temperature immediately. During compression, the device automatically records true stress-true strain data.

3. Results

3.1. Flow Stress Analysis

Figure 1 shows the true stress of TNTZ alloy about a deformation temperature of 700°C at a strain rate of $0.001s^{-1}$ to $1.0s^{-1}$ and a strain rate of $1.0s^{-1}$ at a deformation temperature of 700~900°C true strain curve. It can be seen from Figure 1(a) that when the deformation temperature is constant, the increase of the strain rate causes an increase in the flow stress. From Figure 1(b), when the strain rate is constant, the increase of the deformation temperature causes the decline of the flow stress. The main reason is when the deformation temperature increases, the atomic diffusion rate of the metal increases and the deformation energy storage effect of the metal causes dynamic softening, thereby reducing the flow stress [4].

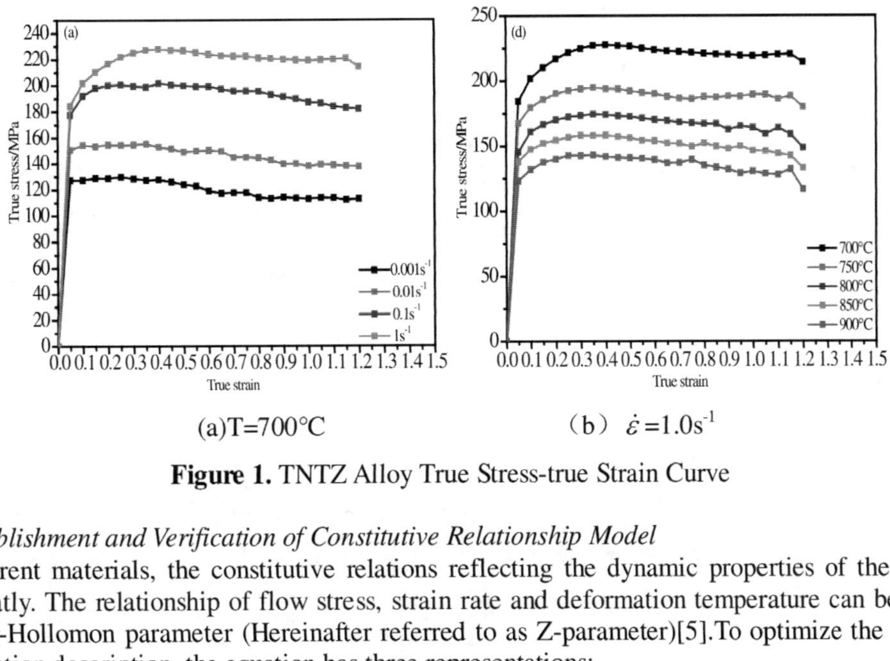

(a)T=700°C (b) $\dot{\varepsilon}$=$1.0s^{-1}$

Figure 1. TNTZ Alloy True Stress-true Strain Curve

3.2. Establishment and Verification of Constitutive Relationship Model

For different materials, the constitutive relations reflecting the dynamic properties of the materials vary greatly. The relationship of flow stress, strain rate and deformation temperature can be obtained by Zener-Hollomon parameter (Hereinafter referred to as Z-parameter)[5].To optimize the Arrhenius type equation description, the equation has three representations:

$$\dot{\varepsilon}\exp(\frac{Q}{RT}) = A_1\exp(\beta\sigma) \quad (1)$$

$$\dot{\varepsilon}\exp(\frac{Q}{RT}) = A_2\sigma^{n_1} \quad (2)$$

$$\dot{\varepsilon}\exp(\frac{Q}{RT}) = A_3[\sinh(\alpha\sigma^{n})] \quad (3)$$

Where σ is the flow stress(MPa); $\dot{\varepsilon}$ is the strain rate(s^{-1}); Q is the deformation activation energy(kJ/mol); T is the absolute deformation temperature(K); A_1, A_2, A_3, α, β are material constants; n, n_1 is a work hardening constant, where $\alpha = \beta/n_1$. The above equations (1) to (3) are exponential function equations, power function equations and hyperbolic sine function equations. The deformation process of the material can be well described [6].

Logarithmically on both sides of equations (1), (2) and (3) can be derived, as shown in Figure 2 the fitting slope $1/n_1$, $1/\beta$ and get α=0.008841448242728. Equation (3) can be transformed into the following:

$$n_2 = \frac{\partial \ln \dot{\varepsilon}}{\partial \ln[\sinh(\alpha\sigma)]} \tag{4}$$

$$k = \frac{\partial[\sinh(\alpha\sigma)]}{\partial(1/T)} = \frac{Q}{nR} \tag{5}$$

$$Q = n \cdot k \cdot R \tag{6}$$

When the deformation temperature T is constant, it can be obtained by combining the equation (6) and Figure 2(c), n=5.8. When the strain rate is constant, the linear relationship is obtained by the equation (5) in combination with figure 4(d), k=5938.699.When the strain reaches the peak value, the deformation activation energy Q=286.075kJ/mol, after optimization by Z-parameter, n_2=5.794 and the intercept $\ln A_3$=27.260 and the above constant is obtained. The Arrhenius constitutive equation of the TNTZ alloy at a strain of 0.3 is obtained by calculation:

$$\dot{\varepsilon} = \exp(27.26)[\sinh(8.841\times10^{-3}\sigma)]^n \exp(\frac{-286.075}{8.314T}) \tag{7}$$

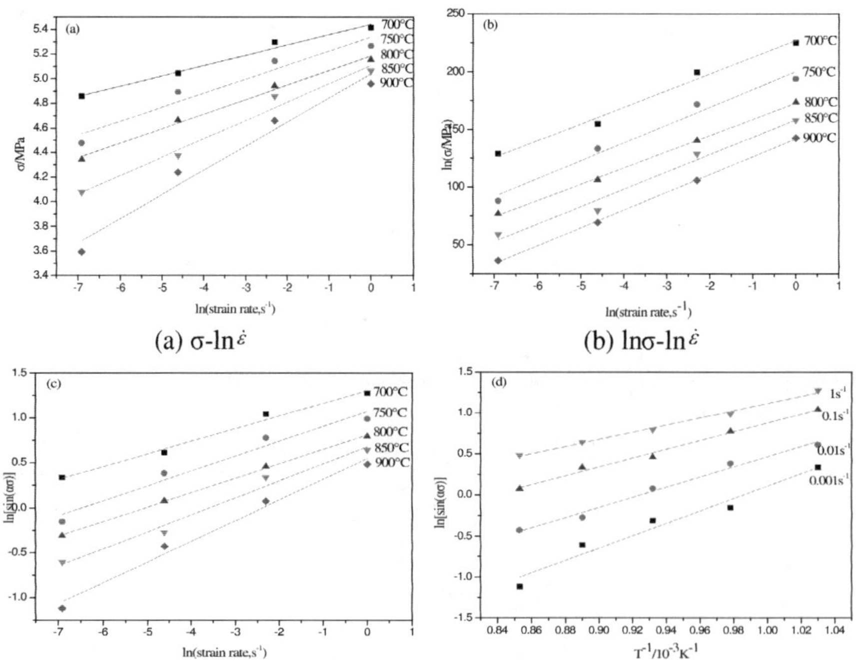

(a) σ-ln$\dot{\varepsilon}$ (b) lnσ-ln$\dot{\varepsilon}$

(c) ln[sinh(ασ)]-ln$\dot{\varepsilon}$ (d) ln[sinh(ασ)]-1/T

Figure 2. Relationship between Flow Stress, Strain Rate and Deformation Temperature of TNTZ Alloy

The Z-parameter can be used to express the relationship between deformation temperature and strain rate of plastic deformation [7, 8] and the activation energy Q value is substituted, as shown in equation (8):

$$Z = \dot{\varepsilon}\exp(\frac{Q}{RT}) = \dot{\varepsilon}\exp(\frac{286.075}{RT}) \tag{8}$$

According to the equation of the hyperbolic sine function, the flow stress can be expressed as a function of the Z-parameter [9]:

$$\sigma = \frac{1}{\alpha}\ln\{(\frac{Z}{A})^{\frac{1}{n}} + [(\frac{Z}{A})^{\frac{2}{n}} + 1]^{\frac{1}{2}}\} \tag{9}$$

Substituting the experimental data into equation (9), the constitutive model of the peak stress, strain rate and temperature T represented by the Z-parameter is obtained as follows:

$$\sigma = \frac{1}{0.008841}\ln\{(\frac{Z}{6.9\times10^{11}})^{\frac{1}{5.794}} + [(\frac{Z}{6.9\times10^{11}})^{\frac{2}{5.794}} + 1]^{\frac{1}{2}}\} \tag{10}$$

3.3. Verifying the Constitutive Model

The experimental deformation temperature and strain rate were substituted into the above constitutive model to obtain a predicted value and compared with the experimental value. According to the prediction results, the mapping is performed for accuracy analysis and the model correlation coefficient R and the average relative error E are used to analyze the accuracy of the constitutive model. The expression is as follows:

$$R = \frac{\sum_{i}^{N}(C_i - \bar{C})(T_i - \bar{T})}{\sqrt{\sum_{i=1}^{N}(C_i - \bar{C})^2}\sqrt{\sum_{i=1}^{N}(T_i - \bar{T})^2}} \tag{11}$$

$$E = \frac{\sum_{i=1}^{N}\left|\frac{C_i - T_i}{C_i}\right| \times 100\%}{N} \tag{12}$$

Where C is the experimental value; T is the predicted value; N is the number of data points. according to equation (11) with (12), comparing the experimental values and experimental values, the results are shown in Figure 3. The correlation coefficient R of the values is 0.9714. The experimental value is basically consistent with the predicted value, that is, the constitutive model has higher precision.

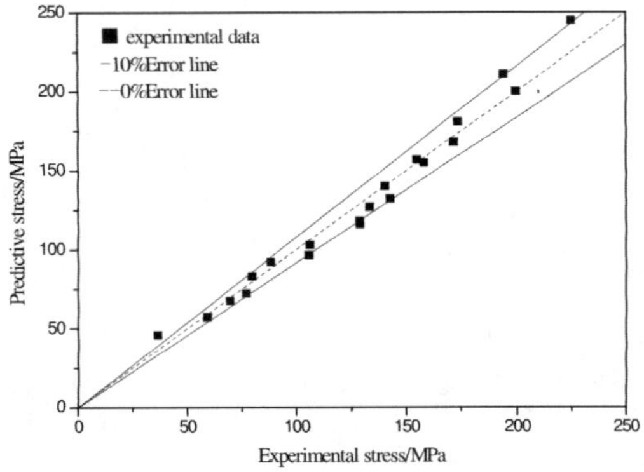

Figure 3. Flow Stress Results and Predicted Values of TNTZ Alloy

4. Discussion

The true stress-true strain of the TNTZ alloy decreases as the deformation temperature increases and the strain rate decreases. When the temperature is lower and the strain rate is higher, the curve shows a flow softening phenomenon; when the temperature is higher and the strain rate is lower, the curve shows a steady flow phenomenon.

Using the hyperbolic sinusoidal function to establish the stress-strain constitutive equation for TNTZ alloy, the activation energy is calculated to be 286.075kJ/mol and the constitutive equation as follow:

$$\sigma = \frac{1}{0.008841}\ln\{(\frac{Z}{6.9\times10^{11}})^{\frac{1}{5.794}} + [(\frac{Z}{6.9\times10^{11}})^{\frac{2}{5.794}} + 1]^{\frac{1}{2}}\}$$

Error analysis shows that the constitutive model has higher precision.

5. Acknowledgments

This study was financially supported by the National Natural Science Foundation of China (No. 51464035).

6. References

[1] Y.Z Wu and A.H Guo 2010 *DEV APPL MATER* vol 2 pp 81-85.
[2] H.G Huang, Y.Q Zhang, Y.H Jiang, et al 2013 *RARE METAL MAT ENG* vol 6 pp 1173-1177.
[3] W Ye 2017 Effect of rolling annealing process on texture evolution and mechanical properties of medical β-type titanium alloy *Diss.*
[4] Takematsu E , Cho K , Hieda J , et al 2016 *J MECH BEHAV BIOMED* vol 61 pp 174 -181.
[5] ZENER C and Hollomon J H 1944 *J APPL PHYS* vol 15chapter1 pp 22-32.
[6] MCQUEEN H J 1993 *J MATER PROCESS TECH* vol 37 chapter 1-4 pp 3-36.
[7] SELLARS C M Mctegart W J 1966 *Acta Metall* vol 14 chapter 9 pp 1136-1138.
[8] J.X Li, S.L Guo, S.Y Liu, et al 2018 *Journal of Netshape Forming Engineering* vol 10 chapter pp 132-138.
[9] LIU Q and Nash P 2011 *INTERMETALLICS* vol 19 chapter 9 pp 1282-1290.

IOP Conf. Series: Materials Science and Engineering **735** (2020) 012025 doi:10.1088/1757-899X/735/1/012025

Effect of Rolling Processing on Mechanical Properties of Material Cr12

Yichen Liu[1], Taoyuan Li[2], Haiquan Zhang[1] and Peiquan Guo[1]*

1. School of Mechanical Engineering, University of Jinan, Jinan 250022, China.
2. Department of Mechanical Engineering, Zaozhuang Vocational College of Science & Technology, Tengzhou 277599, China.

*Corresponding author. Email: oss_guopq@ujn.edu.cn

Abstract. The rolling process changes the mechanical properties of the specimen's surface material, and it also directly affect the fatigue strength and service life of the workpiece. The effects of rolling parameters on surface residual stress, microhardness, surface roughness and surface microstructure are studied by rolling test of Cr12 specimens. After rolling, the residual stress on the surface of Cr12 specimens changes from tensile stress to compressive stress, and the maximum compressive stress is -216MPa. With the increase of contact stress and rolling passes, the residual compressive stress increases at first and then remains stable. Rolling processing can effectively improve the surface hardness of specimens up to 30%, and the surface roughness can be reduced by 25%-67%. The fine grains aligned with the rolling direction, and the 50-60 μm grain refinement layers were observed, which resulted in the changes in the surface mechanical properties of the specimens.

1. Introduction

The generation and growth of fatigue cracks are mainly concentrated on the materials' surface. By applying strengthening treatment, increasing the strength or changing the prestressing state of surface materials can effectively restrain the fatigue cracks, thereby improving the working performance and service life of parts [1,2]. Zheng et al.[3] demonstrated the influence of optimal processing parameters on surface roughness, microhardness and residual stress. Jiang et al.[4] reported the effects of different rolling times on surface morphology, roughness, residual stress and microhardness. When the load reaches a certain value, elastic and plastic deformations occur on the surface of the workpiece, which results in dislocation configuration and phase changes. At the same time, subgrains of the rolling layer are refined to form gradient nanocrystalline layer [5,6]. After rolling, the roughness of the workpiece surface is reduced[7,8], while the work hardening phenomenon[9] and the residual compressive stress are introduced on the workpiece surface[10,11], which can prevent the occurrence of fatigue failure and improve the fatigue strength and the fatigue life of the workpiece [12-14].

2. Experimental Procedure

2.1. Experimental Material

The testing material is Cr12 (the chemical composition is shown in Table 1), and it is reshaped into specimens with a diameter of 40 mm and a length of 300 mm by turning after being quenched and tempered.

Table 1. The chemical constituents of Cr12

Element	C	Cr	Mn	Si	P	S	Others
Content (wt.%)	2.00-2.30	11.50-13.00	≤0.40	≤0.40	≤0.030	≤0.030	≤0.025

2.2. Experimental Method

In the experiment, the iXRD stress analyzer produced by Proto Company of Canada was used to measure the residual stress of the specimen. A digital microhardness tester measured the surface microhardness of the rolled specimens. The surface roughness Ra of the specimen is measured by Mitutoyo SJ-410 surface roughness tester. Microstructure along the depth direction was observed using a scanning electron microscope (SEM, EVO/LS15).

In this work, the two-factor test was designed, where contact stress and rolling passes were selected as test variables. The set forces were 300N, 400N, 500N, 700N and 1000N, and the maximum contact stresses were 555 MPa, 641 MPa, 717 MPa, 848 MPa and 1013 MPa, respectively. In addition, rolling passes were selected as one, two, three, five and eight times. Meanwhile, the rolling test was carried out at a fixed speed of 44 r/min.

3. Results and Discussion

3.1. Surface Residual Stress

The surface residual stress of the machined specimens is 108MPa.After rolling, the surface residual stress of Cr12 specimens has changed significantly from tensile stress to compressive stress. As shown in Figure 1. During the change of rolling passes and contact stress, a certain change rule can be observed. After rolling, the residual stress on the surface of Cr12 specimens changes from tensile stress to compressive stress, and the maximum can reach -216 MPa. With the increase of contact stress and rolling passes, the residual compressive stress increases first and then keeps stable.

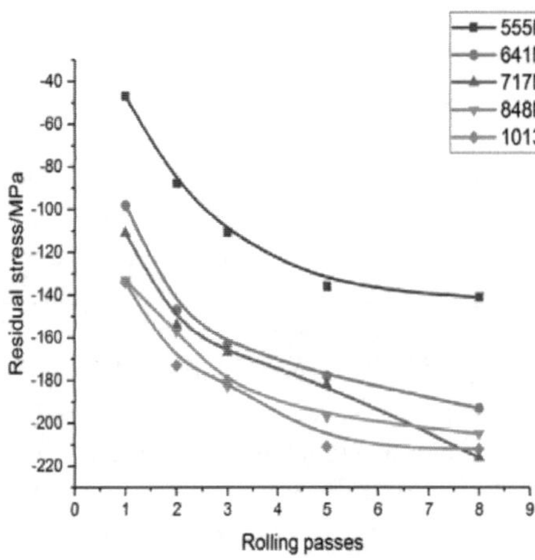

(a)

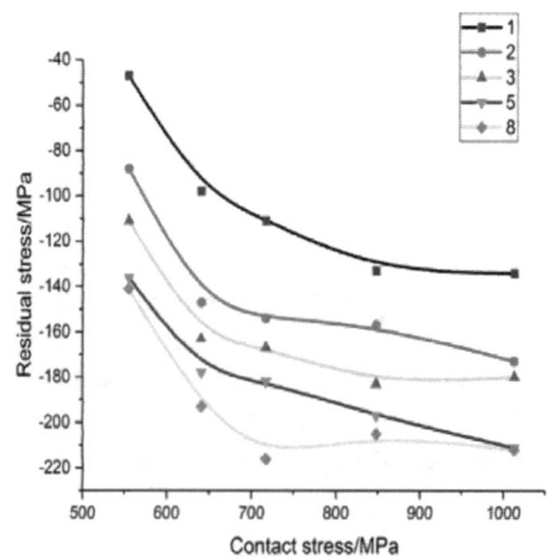

(b)

Figure 1. Surface residual stress of the specimens at different contact stress and rolling passes. (a)influenced by contact stress. (b)influenced by rolling passes.

3.2. Microhardness

The surface microhardness of the non-rolling specimens is $270HV_{0.2}$. After rolling, the surface microhardness of the specimens increases in varying degrees. As shown in Figure 2.With the increase of contact stress and rolling passes, the surface microhardness of the specimens reveals improving tendencies. By comparison, the parameters of rolling passes have a stronger impact on the microhardness of the material. In particular, the microhardness increases significantly before the third rolling pass, and then gradually reaches out to stabilization.

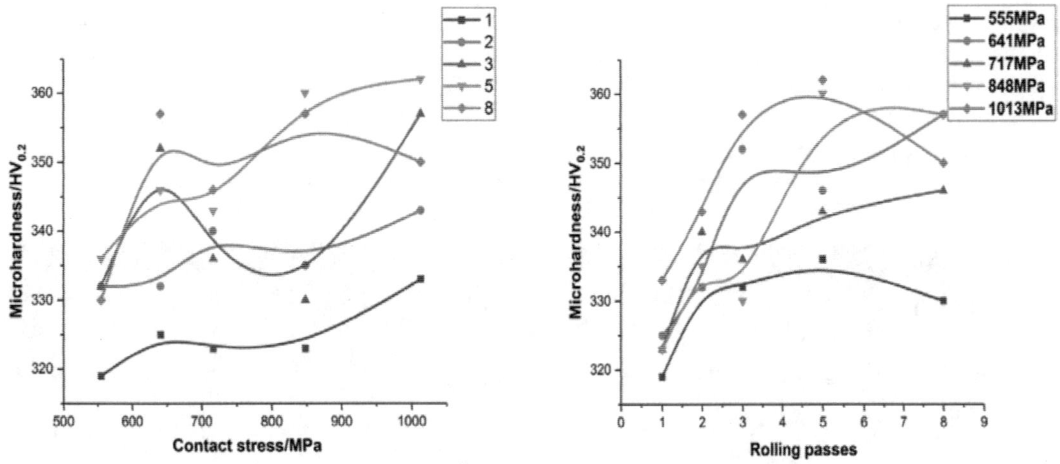

(a) (b)

Figure 2. Microhardness of the specimens at different contact stress and rolling passes. (a) Influenced by contact stress. (b) Influenced by rolling passes

3.3. Surface Roughness

After rolling, the specimen surface produces elastic-plastic deformation, which will cause the change of surface roughness. The initial value of surface roughness of the specimen is 3.079 μm and that of the roller is 0.413 μm. As shown in Figure 3. It can be seen that the surface roughness indicates a decreasing tendency with the increasing rolling passes. Especially, when the contact stress is 1013 MPa, it is clear see that the minimum surface roughness is 1.019μm, and the roughness decreases by 67%.

During the rolling process, the main reason for the reduction of surface roughness is that the peak of surface roughness is filled into the trough by plastic flow. After rolling, the surface morphology of the roller will be reproduced on the surface of the specimen to a certain extent, that is, the surface roughness of the specimen will gradually reach to the surface roughness of the roller.

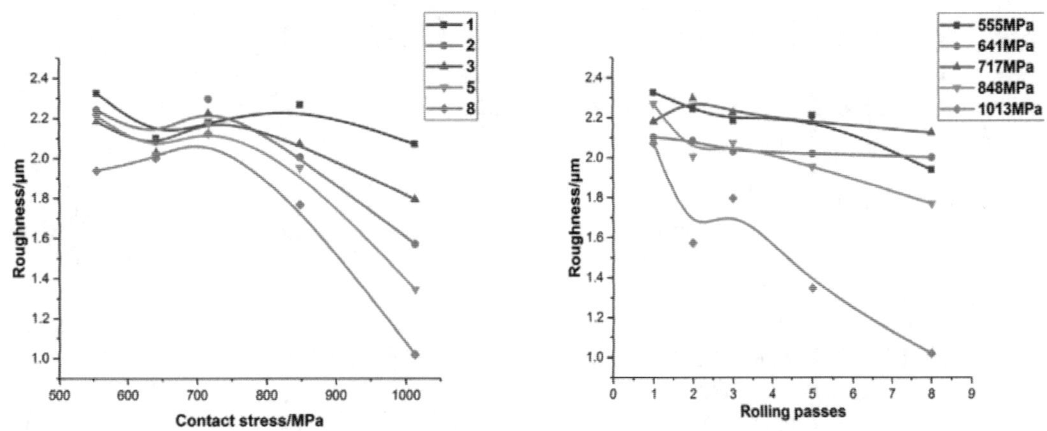

(a) (b)

Figure 3. Surface roughness of the specimens at different contact stress and rolling passes. (a)influenced by contact stress. (b)influenced by rolling passes.

3.4. Microstructure

In order to explore the intrinsic mechanism of rolling technology, the surface microstructures of the rolled surface of specimens were observed and studied. Samples with contact stress of 1013MPa and rolling passes of 8 times were selected to observe SEM images along the depth direction. Figure 4 shows that the surface morphology of sample surface becomes smooth and flat, and the grains are crushed into fine grains and ultra-fine grains, which flow along the direction of rolling. With the increase of depth along the surface of the specimen, grain refinement layers of about 50-60 μm are observed, and the grain size gradually increased.This is the main reason for improving the surface performance.It is clear see that the degree of grain refinement is the most serious when the grain refinement layer is 20 microns to 30 microns away from the surface layer.

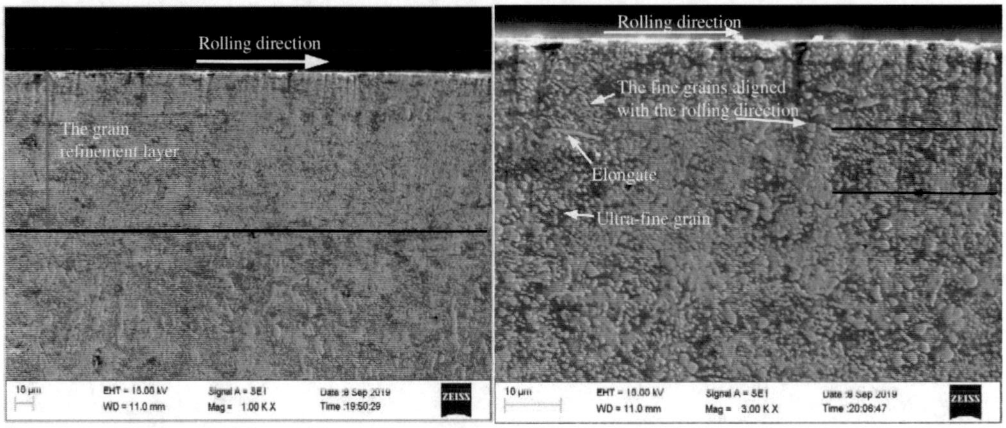

(a) (b)

Figure 4. (a) Microstructure of specimen along the depth direction (b) Partial enlarged view of the surface layer

4. Conclusions

In this work, the main conclusions can be drawn as follows:

1) After rolling, the residual stress on the surface of Cr12 specimens changes from tensile stress to compressive stress, and the maximum compressive stress is -216MPa.It is found that the influence depth of residual compressive stress is about 0.2 mm.
2) Rolling processing can effectively improve the surface microhardness of the specimens and reduce the surface roughness of the parts. The experimental results show that the maximum surface microhardness increased by 33% and the maximum reduction of surface roughness is 67%.
3) After rolling processing, the surface of the specimens undergoes plastic deformation, and the grains are elongated and refined. It can be clear found that the degree of grain refinement is the most serious when it is 20 microns to 30 microns away from the surface layer.

5. Acknowledgments

This work was supported by Key R&D Project of Shandong Province (Grant No.2019GGX104096).

6. References

[1] Revankar G D, Shetty R, Rao S S, et al. Analysis of surface roughness and hardness in ball burnishing of titanium alloy[J]. *Measurement*, 2014, 58: 256-268.

[2] Tsuji N, Tanaka S, Takasugi T. Effect of combined plasma-carburizing and deep-rolling on notch fatigue property of Ti-6Al-4V alloy[J]. *Materials Science and Engineering: A*, 2009, 499(1-2):482-488.

[3] Zheng J X, Ren Y C. Comprehensive assessment of surface integrity in two dimensional ultrasonic rolling 7050 aluminum alloys[J]. *China mechanical engineering*, 2018, 29(13):1622-1626.

[4] Jiang S X, Zheng J X. Microstructure and mechanical properties of 7050 aluminum alloy treated by two dimensional ultrasonic rolling[J]. *Heat Treatment of Metals*, 2018, 43(5):116-119.

[5] Zhao J, Xia W, Li N, et al. A gradient nano/micro-structured surface layer on copper induced by severe plasticity roller burnishing[J]. *Transactions of Nonferrous Metals Society of China*, 2014, 24(2):441-448.

[6] Pu Z, Yang S, Song G L, et al. Ultrafine-grained surface layer on Mg–Al–Zn alloy produced by cryogenic burnishing for enhanced corrosion resistance[J]. *Scripta Materialia*, 2011, 65(6):520-523.

[7] Korzynski M. Modeling and experimental validation of the force–surface roughness relation for smoothing burnishing with a spherical tool[J]. *International Journal of Machine Tools and Manufacture*, 2007, 47(12-13):1956-1964.

[8] Hiegemann L, Weddeling C, Khalifa N B, et al. Prediction of roughness after ball burnishing of thermally coated surfaces [J]. *Journal of Materials Processing Technology*, 2015, 217:193-201.

[9] Rodriguez A, Lopez de Lacalle L.N, Celaya A,et al. Surface improvement of shafts by the deep ball-burnishing technique[J]. *Surface & Coatings Technology*, 2012, 206(11-12):2817-2824.

[10] Meyer D. Cryogenic deep rolling-An energy based approach for enhanced cold surface hardening[J]. *CIRP Annals - Manufacturing Technology*, 2012, 61(1):543-546.

[11] Juijerm P and Altenberger I. Effect of temperature on cyclic deformation behavior and residual stress relaxation of deep rolled under-aged aluminium alloy AA6110[J]. *Materials Science and Engineering: A*, 2007, 452-453:475-482.

[12] Avilés R, Albizuri J and Rodríguez A. Influence of low-plasticity ball burnishing on the high-cycle fatigue strength of medium carbon AISI 1045 steel[J]. *International Journal of Fatigue*, 2013, 55:230-244.

[13] Yekta R T, Ghahremani K and Walbridge S. Effect of quality control parameter variations on the fatigue performance of ultrasonic impact treated welds[J]. *International Journal of Fatigue*, 2013, 55(Complete):245-256.

[14] Zhang P, Lindemann J, Ding W J, et al. Effect of roller burnishing on fatigue properties of the hot-rolled Mg-12Gd-3Y magnesium alloy[J]. *Materials Chemistry and Physics*, 2010, 124(1):835-840.

IOP Conf. Series: Materials Science and Engineering **735** (2020) 012026 doi:10.1088/1757-899X/735/1/012026

Synthesis and Adsorption Properties of Insoluble Starch Xanthate (ISX)

Haihua Li[1,2], Lihui Zhang[1], Yuhua Gao[2] and Zhenfa Liu[1]

[1] Institute of Energy Sources, Hebei Academy of Sciences, Shijiazhuang Hebei 050081, China;
[2] Hebei Engineering Research Center Water Saving in Industry, Shijiazhuang Hebei 050081, China
Email: 873756647@qq.com

Abstract. Starch xanthate was synthesized from corn starch and CS_2 under alkaline condition. The effects of catalyst dosage, CS_2 dosage and reaction temperature on sulfur content and copper ion removal rate were studied by single factor experiments, and the structure of the product was characterized by FTIR, SEM and XRD. The results showed that the optimum synthetic condition of ISX was NaOH dosage 16 mL, CS_2 dosage 2.5 mL, reaction temperature 35 °C and reaction time 2 h. The sulfur content of ISX synthesized under this condition was 7.54%, the removal rate of copper ion was 90.5%, and the adsorption capacity of ISX was 90.0 mg/g.

1. Introduction

Nowadays, all kinds of industrial enterprises in the world still emit a large number of heavy metal ions to the environment, such as cadmium, mercury, lead, arsenic, copper, zinc, nickel and so on. Because these heavy metals can not be degraded by microorganisms, they become persistent pollutants, which greatly endanger the ecological environment and people's health [1-3]. For the treatment of wastewater containing heavy metal ions, chemical precipitation, solvent extraction, membrane separation, electrolysis and ion exchange methods are commonly used [4]. However, these methods generally have the problems of high cost, complex operation and secondary pollution. Polymer heavy metal traps can combine heavy metal ions in a wide range of pH and ambient temperature to produce stable and insoluble mental chelates with high reaction efficiency and good selectivity, have obvious advantages in the treatment of heavy metal wastewater with low concentration (100 mg/L). They have become the key method for the prevention and control of heavy metals in water [5-7]. Synthetic organic macromolecule heavy metal collectors are mainly from petroleum products, and their prices are increasing year by year. Moreover, the toxicity of residual monomers limits their application in water treatment. Compared with the synthetic organic polymer, the modified natural macromolecule heavy metal collector has the advantages of high selectivity, non-toxicity and low price [8, 9].

Among many natural macromolecule heavy metal traps, the research and development of starch modification is particularly noticeable. Starch derivative flocculants have been used in wastewater treatment in the United States, Japan, Britain and other countries. Starch xanthate (ISX) is a starch derivative obtained by xanthation of starch with carbon disulfide after cross-linking under alkaline conditions [10]. Compared with other heavy metal traps, ISX is an economic and environmentally friendly adsorbent with simple synthetic process, good removal effect of heavy metal ions and low production cost.

In this paper, the synthesis conditions of ISX were systematically studied, and the structure of the product was characterized. Copper sulfate solution was used to simulate heavy metal wastewater to study the adsorption performance of ISX for copper ions.

2. Materials and Methods

2.1. Reagents
Corn Starch, Reagent Grade, was purchased from Shanghai Aladdin Chemical Co., Ltd., China. CS_2, was purchased from Shanghai Aladdin Chemical Co., Ltd., China. NaOH, Acetone and $CuSO_4 \cdot 5H_2O$ were purchased from Tianjin Fuchen Chemical Reagent Factory, China. All the above ingredients except corn starch were of analytical grades.

2.2. Synthesis of Cross-linked Starch [11, 12]
15 g corn starch was dispersed in 50 mL of the concentration of 1% NaCl solution and the resulting slurry was placed in a water bath preset at 30°C. Then, 15% NaOH solution 4mL and epoxy chloropropane (EPI) 1mL were added dropwise into the reaction system under continuous stirrig, and reacted at constant temperature for 5 h.

The synthesized product was separated by filtration. The precipitate was dispersed in distilled water and adjusted pH by HCl to 7-8. Then, the product was washed by a lot of water and ethanol, finally, dried to constant weight in a vacuum oven at a 60 °C.

2.3. Synthesis of Insoluble Xanthate Starch (ISX)
5 g cross-linked starch was dispersed in a flask with three necks with 40mL deionized water placing in a 30°C water bath. The system was alkalized for 0.5 h under stirring by adding 16mL of the concentration of 15% NaOH solution. After alkalized, 2.5mL CS_2 was added in the flask under stirring and reacted for 2 h. Then, 25mL $MgSO_4$ solution (60 g/L) was added to the system and continued to react for 15 minutes. After the reaction, the cake obtained by filtering was washed with deionized water until its pH=7-8, and then washed with acetone three times to remove water from the product. The filter cake was vacuum dried for 2 h at 40°C, and the light yellow powder was ISX. The reaction processes were as follows:

$$S{=}C{=}S + NaOH \longrightarrow HO{-}\overset{\overset{\displaystyle S}{\|}}{C}{-}SNa \quad (1)$$

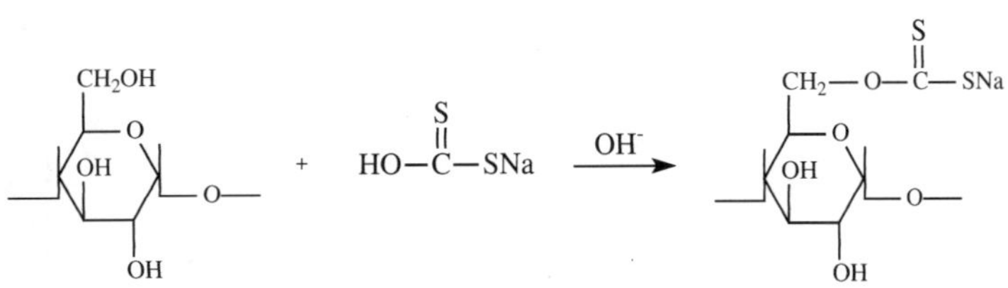

(2)

2.4. Characterization of Corn and ISX
Infrared spectra of corn starch and ISX were measured via a SP100 Fourier-transform infrared (FTIR) spectrometer (PerkinElmer, USA.) using the potassium bromide pellet method. The wavenumber range was 4000-400 cm^{-1}. XRD patterns of the samples were obtained using a Ultima Ⅳ X-ray Diffractometer (Rigaku, Japan) equipped with a Cu-ka radiation source operated at 40 kV and 40 mA, with the diffraction angle scanned at a speed of 4°/min over a 2θ range of 10-60°. SEM images were observed on a Inspect S50 scanning microscope (FEI, USA). The acceleration voltage was 5 kV and the magnification was 2000 times. The content of sulfur of ISX was determined by a vario mICRO cube element analyzer (Elementar, Germany).

2.5. Adsorption Experiments
The reserve solution of copper ions (1 g/L) was prepared by dissolving 1 g $CuSO_4 \cdot 5H_2O$ in 1000mL deionized water. The water sample to be tested of copper ions (25mg/L) was prepared by removing

5mL reserve solution into 200mL deionized water. A certain amount of ISX was accurately weighed and put into the sample. The sample was stirred by magnetic force for 40 min at 500r/min, and then precipitated for 4 h. The concentration of Cu^{2+} in water samples was determined by sodium diethylaminodithiocarbonate spectrophotometry via a TU-1900 Dual-beam Ultraviolet-Visible Spectrophotometer (Beijing General Analytical Instrument Co., Ltd., China). According to the following formula, the removal rate of Cu^{2+} ($P\%$) was calculated.

$$P\% = \frac{c_0 - c_1}{c_0} \times 100\% \tag{3}$$

In the formula, c_0 and c_1 were the initial and final concentrations of Cu^{2+} in water sample, respectively, mg/L.

3. Results and Discussion

3.1. Synthetic Conditions of ISX

The sedimentation volume of the cross-linked starch was 19 mL. The synthetic conditions of ISX were screened based on the content of sulfur (S%) and the removal rate of Cu^{2+} (P%). The dosage of ISX was 250 mg/L.

3.1.1. Effect of NaOH Dosage. In xanthogen acidification of cross-linked starch, alkaline conditions are necessary, and the amount of alkali has a great influence on the degree of xanthogen acidification. The effect of NaOH dosage on the content of sulfur and the removal rate of Cu^{2+} were investigated under the conditions that CS_2 dosage was 2.5 mL, reaction temperature was 35 °C and reaction time was 2 h. As shown in figure 1, the sulfur content of ISX increased with the increase of NaOH dosage. When NaOH dosage was less than 16mL, the sulfur content of the product increased significantly. While continuing to increase the dosage of NaOH, the increase of S% slowed down. The tendency of the removal rate of Cu^{2+} by ISX was similar to that of S%. This was mainly due to the sulfhydryl group as the functional group in ISX. With the increase of S%, the content of sulfhydryl group in the product increased, and the removal rate of Cu^{2+} also increased. But the more alkali was used, the more deionized water was needed to wash and remove excess alkali, which results in waste of water. Therefore, the alkali dosage we choosing was 16mL.

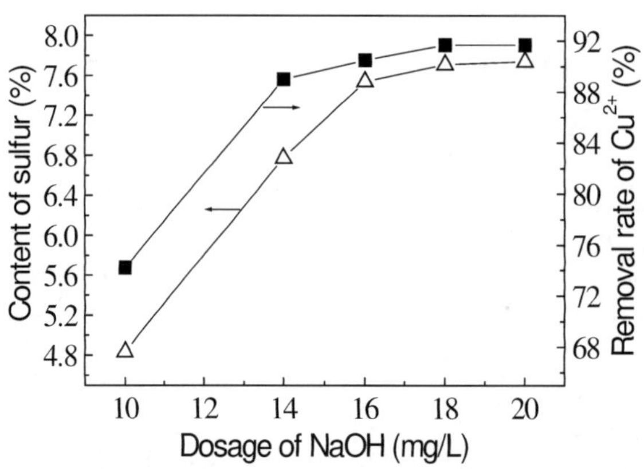

Figure 1. Effect of NaOH Dosage on Sulfur Content and Cu^{2+} Removal Rate

3.1.2. Effect of CS_2 dosage. As shown in figure 2, the S% increased with the increase of CS_2 dosage. When the dosage of CS_2 exceeded 2.5 mL, the sulfur content of the product decreased. This was mainly due to the fact that excessive CS_2 inhibited the opening of starch ring and reduced the reaction rate.

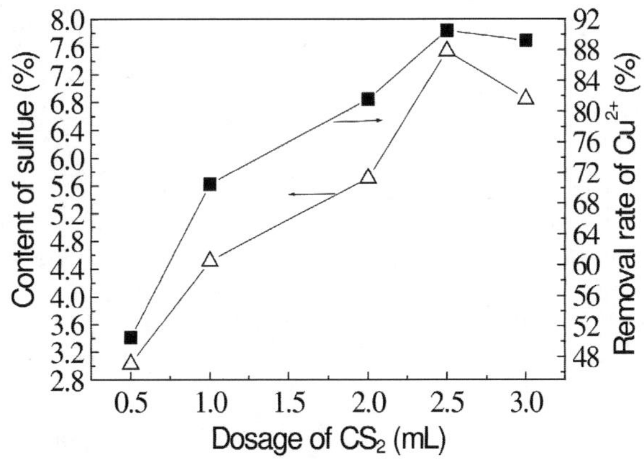

Figure 2. Effect of CS_2 Dosage on Sulfur Content and Cu^{2+} Ion Removal Rate

3.1.3. Effect of Temperature. It was reported that the temperature of xanthate acidification is generally 20-40°C, which mainly because the boiling point of CS_2 is very low (46-47 °C), and the water solubility of CS_2 is not good, and the reaction rate of xanthate with alkali is slow. Xanthan acidification was carried out at different temperatures to investigate the effect of temperature on the product. As shown in figure 3, when temperature was below to 35°C , S% gradually increased with the temperature increasing. While when the temperature was above 35°C, S% decreased rapidly. This was mainly because the reaction system of CS_2 and NaOH was heterogeneous, including solid-phase cross-linked starch, liquid-phase CS_2 and water, and gas-phase CS_2 and steam. high reaction temperature would cause the volatilization of CS_2, resulting in inadequate xanthate acidification reaction and the adverse side reaction. And the reaction was exothermic, and low temperature was conducive to the positive direction of the reaction. But the lower the reaction temperature, the longer the reaction time, so the temperature we choosing was 35°C.

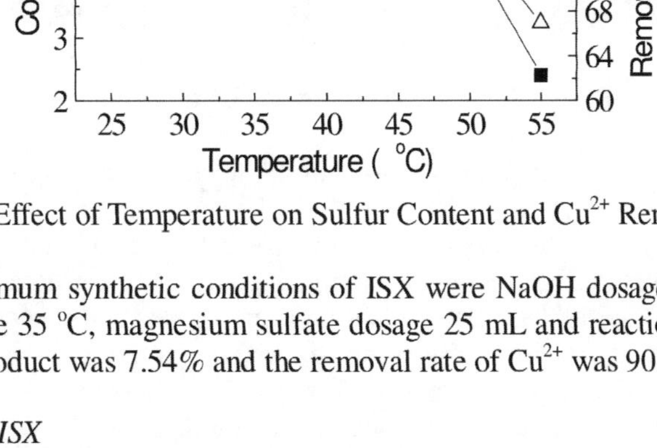

Figure 3. Effect of Temperature on Sulfur Content and Cu^{2+} Removal Rate

In summary, the optimum synthetic conditions of ISX were NaOH dosage 16mL, CS_2 dosage 2.5 mL, reaction temperature 35 °C, magnesium sulfate dosage 25 mL and reaction time 2 h. Under these conditions, S% of the product was 7.54% and the removal rate of Cu^{2+} was 90.5%.

3.2. Characterization of ISX

3.2.1. FTIR analysis. The FTIR spectra of corn starch and ISX were shown in figure 4. Compared with the FTIR spectrum of corn starch, the peak of ISX at 3700 cm^{-1} was the absorption peak of $Mg(OH)_2$, the absorption peak at 1702 cm^{-1} was the stretching vibration absorption peak of C=S, the characteristic absorption peak at 1517 cm^{-1} was the thiol group, and the deformation vibration peak at 853 cm^{-1} was the deformation vibration peak of -CSS- functional group. The emergence of a series of new peaks illustrated that xanthate group was introduced into starch molecular chain.

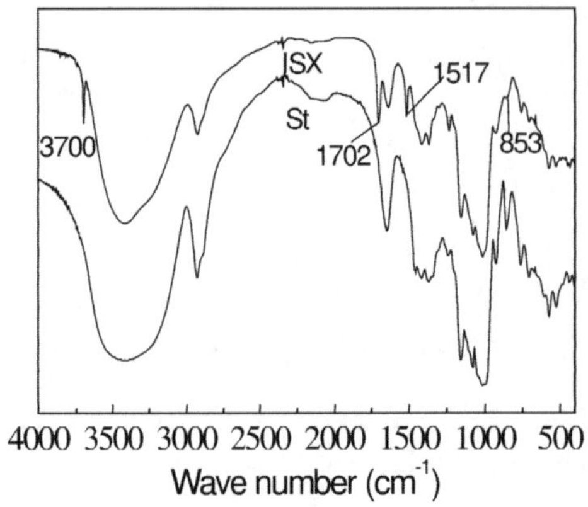

Figure 4. Infrared Spectra of Corn Starch and ISX

3.2.2. SEM Analysis. The surface morphology of corn starch and ISX particles was observed by SEM. As shown in figure 5, the granule surface of corn starch was smooth and oval. This was due to the interweaving of amylose and branched chains in starch molecules, which keeps starch granules intact (a). After xanthate acidification, the granule of ISX exhibited a rough surface with pits or holes (b), which was conducive to the adsorption of heavy metal ions.

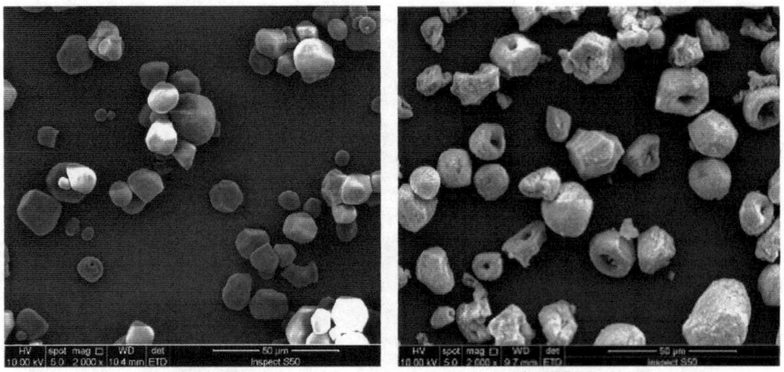

(a) St (b) ISX

Figure 5. SEM Images of Corn Starch and ISX

3.2.3 XRD Analysis. The XRD patterns of corn starch and ISX were shown in figure 6. There were obvious diffraction peaks near 15°, 17°, 18° and 23° in the spectrogram of corn starch, which indicated that corn starch had complete microcrystalline region and belonged to typical A-type crystal. While the peaks disappeared completely at the same diffraction angle in the spectrogram of ISX, which might be due to the weakening of intermolecular forces of the starch molecular chains and the hydrogen bonds of hydroxyl groups. In the xanthate acidification, the crystal structure of starch was destroyed, while the activity and the porosity of starch increased which conductive to absorb heavy metal ions. In addition, the pattern of ISX showed two obvious diffraction peaks at 37.88° and 58.86°, the two peaks attributed to Mg(OH).

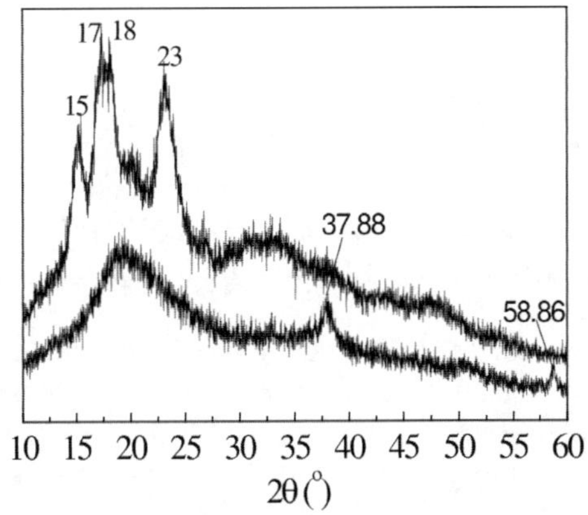

Figure 6. XRD Patterns of Corn Starch and ISX

3.3. ISX Adsorption of Copper Ions

The effect of dosage on the removal rate of Cu^{2+} was tested at room temperature using the simulated wastewater with Cu^{2+} concentration of 25 mg/L. As shown in figure 7, when the dosage of ISX was less than 250 mg/L, the removal rate of Cu^{2+} increased rapidly with the dosage increasing. While when the dosage was more than 250 mg/L, the increase of Cu^{2+} removal rate slowed down. When the dosage was 250 mg/L, the removal rate of Cu^{2+} was 90.5%, and the adsorption capacity of ISX was 90.0 mg/g.

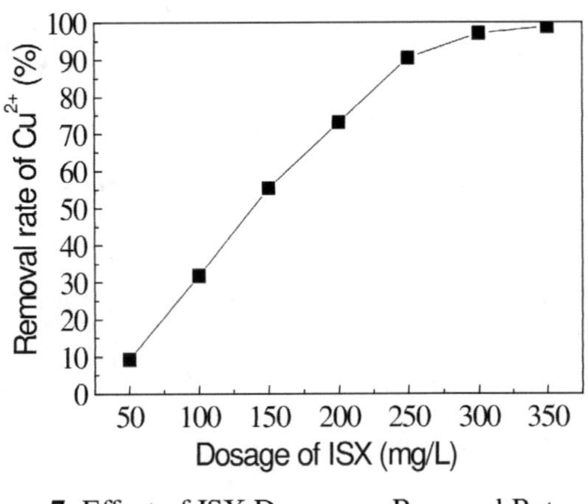

Figure 7. Effect of ISX Dosage on Removal Rate of Cu^{2+}

4. Conclusions

4.1. The synthetic conditions of ISX were screened by single factor experiments with the sulfur content of the product and the removal rate of copper ions as the evaluation criteria. The results showed that the optimum synthetic conditions were NaOH 16 mL, CS_2 2.5 mL, reaction temperature 35 °C and reaction time 2 h. Under these conditions, S% of the product was 7.54%, and the removal rate of copper ion was 90.5%.

4.2. The effect of ISX dosage on the removal rate of copper ion was investigated. The results showed that when the dosage of ISX was 250 mg/L, the removal rate of copper ion was 90.5%, and the adsorption capacity of ISX was 90.0 mg/g.

5. Acknowledgements

The research was sponsored by the National Natural Science Foundation of China (21376062), the Natural Science Foundation of Hebei Province (18964005D) and the Science and Technology Project of Hebei Academy of Sciences (19701).

6. References

[1] Kim Y J, Choi J D, and Kim E S 2019 Removal of Heavy Metal Ion in Wastewater by Dendritic Nano-Structured Complex Material. *Journal of Nanoscience and Nanotechnology*, **19**(2):.1010-1013.

[2] Wang J, Liu M, Duan, Sun J P, and Xu Y W 2019 Preparation and characterization of cellulose-based adsorbent and its application in heavy metal ions removal. *Carbohydrate Polymers*, **206**: 837-843.

[3] Chang Q 2015 New research area of flocculation in water treatment-macromolecule flocculant with the function of trapping heavy metal. *Acta Scientiae Circumstantiae*, **35**(1): 1-11.

[4] Zhang Y M, Yunus R F, Nanayakkara K N, and Chen J P 2012 Electrochemical decoloration of synthetic wastewater containing rhodamine 6G: behaviors and mechanism. *Industrial & Engineering Chemistry Research*, **51**(17): 5953-5960.

[5] Gao L L, Li Q R, Han L Hu X H, and Wang X P 2016 Preparation and Pb^{2+} adsorption property of Mg Al Layered double hydroxides (LDHs) based onbiomorphic manganese oxide. *Environmental chemistry*, **35**(9):1961-1967.

[6] Sun Y K, Guo X Z, Li D Y, and Yang H 2019 Selective adsorption of heavy metal ions by $MgFe_2O_4$@PDA@PAMAM nanocomposites. *Journal of the Chinese Ceramic Society*, **47**(5): 664-669.

[7] Zhao D M, Bai L, Liu Y, Zang L L, and Li J 2018 Preparation and research of core shell fiber adsorbing heavy metal ions. *Journal of Functional Materials*, **49**(12): 12107-12112.

[8] Li T Q, and Hu F 2018 Characterization and trapping efficiency of starch based heavy metal trapping materials. *Fine Chemicals*, **35**(3): 463-468, 524.

[9] Cheng A H, Qian D P and Yao M Y 2018 Preparation of Ca^{2+} imprinted cross-linked modified chitosan and its adsorption propertied for Cd^{2+}. *Fine Chemicals*, **35**(3): 482-488.

[10] Wei X Y, Liu B S, Ourang Y Y, Sun P, and Tan J 2015 The study on the adsorption mechanism of insoluble xanthates for the removal of copper ions from wastewater. *Journal of Jiaxing University*, **27**(6): 76-82.

[11] Joana B, Edita L K, and Ramune R 2018 Biodegradation of cross-linked and cationic starches [J]. *International Journal of Biological Macromolecules*, **119**.

[12] Zhang H, Fan X Y, Wang J K, Guo J, and Liang K 2017 Synthesis of cross-linked carboxymethyl starch and its adsorption properties on heavy mental ions. *Chemical Industry and Engineering Progress*, **36**(7): 2554-2561.

Regenerated Bamboo-Derived Cellulose Fibers/RGO-Based Composite for High-Performance Supercapacitor Electrodes

Tianhao Wang,[1] Wentao Zhang,[1] Shujuan Yang,[1] Weiqian Tian[2] and Liping Zhang[*1]

1 Department of Material Science and Technology, MOE Key Laboratory of Wooden Material Science and Application, Beijing Forestry University. No.35, Tsinghua East Road, Haidian District, Beijing, China.
2 Department of Fibre and Polymer Technology, KTH Royal Institute of Technology, Teknikringen 56, 10044 Stockholm, Sweden
E-mail address: Zhanglp418@163.com (L.P. Zhang)

Abstract. Bamboo-derived cellulose fibers/RGO carbon aerogel composite was prepared by using a facile aerogel-based method, in which bamboo pulp fibers were dissolved and incorporated in an ionic liquid system, and RGO was introduced by thermal reduction approach. The obtained bamboo-derived cellulose fibers/RGO carbon aerogel composite shows a large specific surface area, and excellent electrochemical performance. When the GO content was 2.5 wt%, the obtained composite showed a high specific surface area of 1957 m^2/g, and high specific capacitance of 351 F/g in 6 M KOH electrolyte solution even with a more than 90% capacitance retention at a high scan rate of 200 mV/s. The bamboo-derived cellulose fibers/RGO composite electrodes show the low equivalent series resistance of 5.0 Ω and small charge transfer resistance of 0.30 Ω which further demonstrate the excellent electrochemical behaviors.

1. Introduction

Shortages of fossil fuels and climate change are serious problems associated with environmental pollution; hence, there is an urgent need for highly efficient and low-cost sustainable energy storage devices. Among energy storage systems, electrochemical capacitors (supercapacitors) have good cycle stability, high power densities, and environment-friendly [1]. In recent years, supercapacitors have drawn interest in the applications for hybrid vehicles and portable electronic devices [2]. While increasing the specific surface area of the electrode material, increasing the conductivity of the material can effectively improve the performance of the supercapacitor [3]. Carbon nanomaterials can well meet these requirements. Among carbon nanomaterials, graphene has the characteristics of good electrical conductivity, high specific surface area, high heat resistance and good chemical stability. It is the most commonly used electrode material of supercapacitor at present [4].

However, due to the high preparation cost, easy agglomeration and low reaction capacity of graphene. Therefore, reduced graphene oxide (RGO) is often used as an additive to the electrode material of supercapacitors, and graphene alone is rarely used as the electrode material [5]. After a simple carbonization process, the biomass porous carbon material can have a high specific surface area, which is very helpful to improve the specific surface area of the supercapacitor electrode. Meanwhile, in this process, the pore diameter and pore distribution of biomass porous carbon materials can be adjusted by the conditions of carbonization process. In this way, the optimal pore structure and specific surface area can be obtained, which is conducive to improving the capacity of supercapacitors to store electrical energy [6]. Among the numerous biomass porous carbon materials, bamboo has the advantages of wide distribution, fast growth cycle and high economic effect. Therefore, we prepared the supercapacitor electrode with bamboo cellulose fibers as the main material.

Then, in order to improve the specific surface area of the composites, we prepared porous carbon composites by using the strategy based on carbon aerogel and high temperature activation method. Carbon aerogel has large specific surface area, good chemical stability and high porosity. Low density, three-dimensional network structure. These characteristics enable electrolytes to enter mpterial interior for energy storage [7].

To sum up, the experiment we here dissolved bamboo cellulose in a quaternary ammonium salt ionic liquid, which disrupted the structure of the cellulose to release the internal hydrogen-bonded molecular chains. We used this solution to produce carbon aerogels with a porous structure to realize the high specific surface area of composite materials, and added oxidized graphene to improve the conductivity of the composite materials. We examined the effects of different mass loading of RGO onto the performance as a supercapacitor electrode material in 6 M KOH aqueous electrolyte.

2. Experimental

2.1. Materials

Bamboo pulp (constituents listed in Table 1), was provided by Jilin Chemical Fibers Co., Ltd., Jilin, China. Tetrabutylammonium acetate (IL) was purchased from Tokyo Chemical Industry Co., Ltd., Shanghai, China. Graphene oxide was provided by Jining Lite Nanotechnology Co., Ltd., Shandong, China. Dimethyl sulfoxide (DMSO), polyvinylidene fluoride, N-methyl pyrrolidone, and hydrochloric acid were purchased from Beijing Chemical Works, Beijing, China. Potassium hydroxide was purchased from Xilong Chemical Co., Ltd., Hubei, China. Deionized water was used throughout the experiments.

Table 1. Compositions of Bamboo Pulp Fibers

Sample	Degree of polymerization	α-Cellulose (wt%)	Moisture (wt%)	Cellulose (wt%)	Hemicellulose (wt%)	Lignin (wt%)
Pulp Fibers	728	82.6	5.6	92.8	2.3	4.5

2.2. Preparation of Cllulose Fbers/GO Arogel

A cellulose solvent was prepared by mixing the tetrabutylammonium acetate with the cosolvent of DMSO, and then different proportions of GO (i.e., 1.5 wt%, 2.5 wt%, and 3.5 wt%) were added into it. 5 wt% bamboo pulp fibers were dissolved in the cellulose solvent at room temperature over 48 h. After that, A regenerated cellulose alcohol gel was obtained by solvent exchange with deionized water and tert-butanol. The gel was frozen in liquid nitrogen for 5 min before freeze-drying (-60°C, 40 Pa) for 48 h. Finally regenerated cellulose/GO aerogel obtained, was denoted as RA_x, where x refers to the content of GO.

2.3. Carbonization and Ativation

The regenerated cellulose/GO aerogels were carbonized in a tuber furnace under N_2 atmosphere from room temperature at a rate of 5 °C/min up to 900 °C for 1 h. The samples were then cooled to room temperature, and cellulose/GO-derived carbon aerogels were obtained as CRA_x, where x refers to the different proportion of GO. In the subsequent activation process, we used KOH as an activator with a 1:2 ratio of carbon aerogel and KOH.

In brief, the carbon aerogels were immersed in KOH solution and then were heat at 100 °C in an oven to evaporate the excess water. The resulting mixture was transferred to a tuber furnace and heated in N_2 atmosphere from room temperature to 900 °C at a rate of 5 °C/min and maintained at that temperature for 1 h before cooling to room temperature. We used 1 M HCl and deionized water to wash the sample until the pH up to neutral. We used the bamboo-derived cellulose fibers/RGO carbon aerogel composite, denoted as A-CRA_x, where x represents the different proportions of GO added.

2.4. Characterization

The morphology of the prepared sample was investigated with a scanning electron microscope (SEM, S-3000n, Hitachi, Japan). X-ray diffractometer (XRD) patterns were obtained from a powder X-ray diffractometer (Bruker D8 Advance) with Cu-Ka radiation at 40 kV and 40 mA. XRD data were collected in continuous scan mode from 10° to 50°, with a step size of 0.01°. FT-IR spectra were acquired on a Bruker Vector 33 spectrometer with KBr pellets. The reflectance spectra were measured over the energy range of 400 to 4000 cm^{-1} at a spectral resolution of 4 cm^{-1} with 32 scans. Thermogravimetric analysis (TGA) was studied in a TGA-600 analyzer (Shimadzu, Japan). Samples (5-10 mg) were heated up to 600 °C at a rate of 20 °C/min, and the flow rate of nitrogen was 20 mL/min. The porous features were determined by using a Micromeritics analyzer (ASAP 2460) at 77 K, with the degassing time of 10 h and the degassing temperature at 200 °C. The specific surface area was calculated by the Brunauer-Emmett-Teller method (BET) method based on adsorption data in the relative pressure (P/P0) range from 0.06 to 0.20 and total pore volume was obtained at a relative pressure (P/P0) of 0.99 from the amount of adsorbed nitrogen. The pore size distributions (PSDs) were calculated by using non-local density functional theory (NLDFT) method.

2.5. Electrochemical Aalysis

We used a three-electrode system to characterize the electrochemical properties of the sample. In three-electrode system, Pt metal and saturated Ag/AgCl calomel electrodes were used as the counter and reference electrodes, respectively. The electrolyte used was 6 M KOH solution. The work electrode was prepared as follows: the sample was ground in an agate mortar, and was then mixed with PVDF, acetylene black in NMP to obtain slurry which was pressed onto the nickel foam as thin slices and dried in an oven.

The electrochemical performances of the materials were evaluated with a CHI 760D electrochemical station (CH Instruments, USA) by Cyclic voltammetry (CV), galvanostatic charge/discharge (GCD) measurements, and electrochemical impedance spectrogram (EIS) techniques. CV curves were measured at different potential sweep rates of 10-200 mV/s within the potential window of 0-0.9 V. The gravimetric specific capacitances were calculated from the CV curves according to the equation $C = \int IdV/(\upsilon m\Delta V)$, where I is the current (A), ΔV is the working voltage window (V), m is the mass of active material (g), and v is the scan rate (mV/s). GCD w performed at a different current densities in the range of 0.5-10 A/g (with a potential window of 0-0.9 V). The specific capacitances were calculated from the discharge curves by the equation of $C = I\Delta t/(m\Delta V)$, where I is the discharge current (A), Δt is the discharge time(s), ΔV is the working voltage window (V). Finally, EIS was conducted in the frequency range from 0.01 Hz to 100 kHz.

3. Results and Discussion

3.1. Synthesis of Prous A-CFA

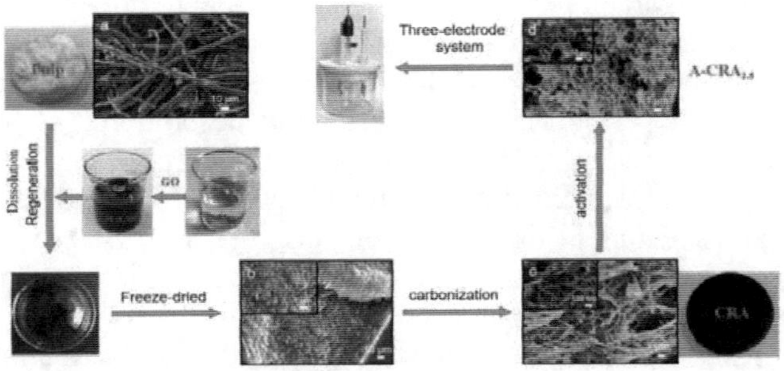

Figure 1. Schematic Ilustration of the Fbrication Pocess of A-CFA: FE-SEM of Pristine Bamboo Pulp Fibers (a), RA (b), CRA (c) and A-$CRA_{2.5}$ (d)

Figure 1 shows the schematic of the fabrication of A-CFA and the SEM images of A-CRA and A-CRA$_{2.5}$. The bamboo pulp fibers showed a rod-like structure with a diameter of approximately 6-8 μm (Figure 1a). After the bamboo pulp fibers were added to the tetrabutylammonium acetate/DMSO ionic liquid system, they swelled and intra- and inter-molecular hydrogen bonding was partially disrupted. In the process of regeneration, the broken cellulose molecular chains became unstable and rearranged to a stable crystalline structure.

SEM images show that many micropores were generated on the surface of the bamboo fibers surface after the activation treatment. Similar to other biomass-derived carbon products obtained by KOH activation [8], A-CRA$_{2.5}$ had disordered and wormlike micropores, other than that we can see RGO was implanted in the structure of A-CRA. Thus, the chemical activation effectively increased the porosity of the walls of the fiber. The surface of the bamboo pulp fibers was uneven, and microfibers aggregates were visible. In the regenerated aerogel (Figure 1b), a large number of pore structures appeared on the surface of the aerogel, and large voids appeared in the structure.

These features provided a large specific surface area, which improved the power storage capacity. After carbonization, carbon aerogels were obtained (Figure 1c). The surface of the carbon aerogel developed a network pore structure, and a large number of three-dimensional nanopores formed, which greatly improved the specific surface area. Figure 1d shows the carbon aerogel materials with 2.5 wt% GO. The surface of the material had a loose sponge-like structure, it means RGO was exactly implanted in the bamboo-derived cellulose fibers/RGO carbon aerogel composite as we expected and the graphene nanosheet layer was dispersed in the three-dimensional network pores of the carbon aerogel, which contributed to the conductivity of the composite material.

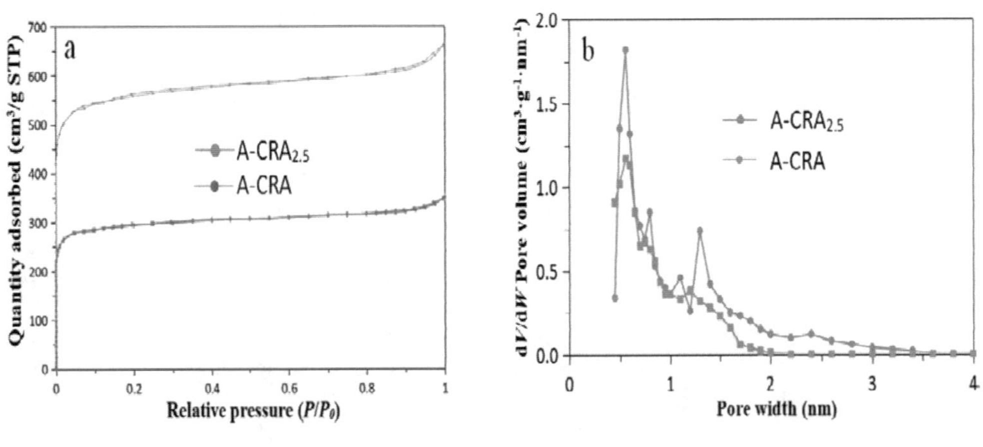

Figure 2. N_2 Adsorption Isotherms of A-CRA and A-CRA$_{2.5}$ (a) and Pore Size Distributions (b) of A-CRA and A-CRA$_{2.5}$

The SEM images of the carbon aerogel showed a large number of pore structures ranging from micro- to nanometer size. These macropores were mainly produced in the gel and regeneration process, which was controlled by the lyophilization and carbonization process. To characterize the micro- and mesopores in the material, we performed nitrogen adsorption/desorption tests on the sample A-CRA$_{2.5}$. Figure 2 shows the nitrogen adsorption/desorption isotherm and pore size distribution of A-CRA$_{2.5}$. The nitrogen adsorption of A-CRA increased sharply at relative pressures (P/P_0) less than 0.1, indicating the richer pore structure of the carbon aerogels after the KOH activation. Furthermore, the increase of the pressure curve saturation indicates that the material contained a large number of microporous structures, a hysteresis appeared in the range of $0.5 < P/P_0 < 1.0$, indicating that mesopores and macropores existed in A-CRA$_{2.5}$ [9]. The pore size distributions for the A-CRA and A-CRA$_{2.5}$ (Figure 2b) show that the A-CRA and A-CRA$_{2.5}$ have pores in the range of 0.5-2.0 nm and peaks appeared at 0.56, 0.8, and 1.3 nm, indicating A-CRA$_{2.5}$ with hierarchical pore structures. During the KOH etching process, elemental potassium can embed into the carbon skeleton and creat voids with two times as large as the radius of potassium atoms (2.78 nm) is retained after elemental potassium is removed by HCl [10]. This result is consistent with the nitrogen adsorption/desorption isotherms. Moreover, the mesoporous and micropores increase the specific surface area of the composites from 1563 m^2/g for A-CRA to 1957 m^2/g for A-$CRA_{2.5}$. The loading of GO markedly increased the specific surface area and changed the pore distribution of carbon aerogels.

3.2. Chemical Characteristics of the Material

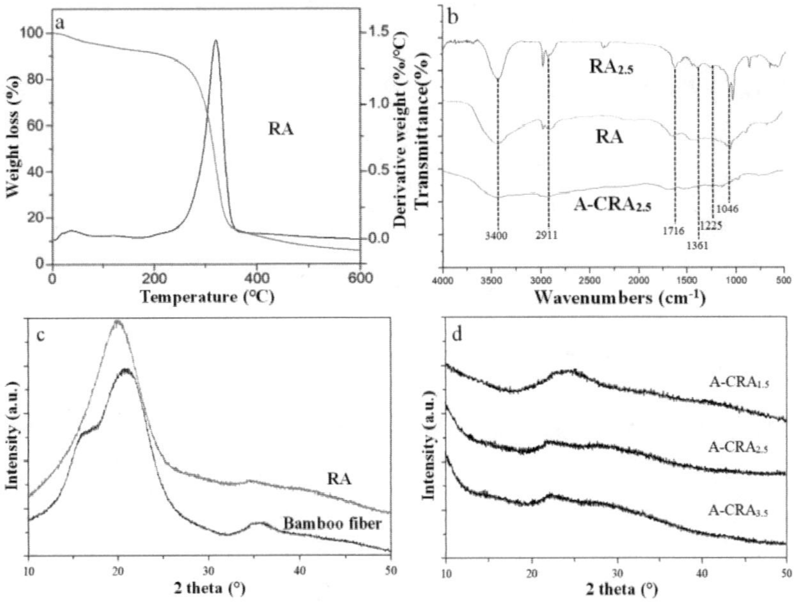

Figure 3. TG, DTG Curves for RA (a) and FTIR Spectra of RA, $RA_{2.5}$, A-$CRA_{2.5}$ (b) and X-ray Diffraction Patterns of Bamboo Fibers and RA (c), A-CRA_x (d)

We used thermogravimetric analysis (Figure 3a) to discover the carbonization temperature of bamboo cellulose fibers/GO aerogels. The mass loss of the bamboo cellulose aerogels over the range of 30-250 °C was only 5%, derived from moisture and carbon dioxide [11]. The aerogel pyrolysis temperature range was 250-350 °C and the mass loss rate was as high as 85%. The thermal decomposition rate of the largest peak temperature was 320 °C, and the corresponding mass loss rate was 60%. This stage is mainly the result of cellulose glycosidic molecular chain decomposition, and a small amount of hemicellulose and lignin degradation/polycondensation reactions leading to pyrolysis and aromatic ring formation. Between 400–600 °C the mass loss is approximately 4%, which is attributable to a small amount of lignin degradation. These results further illustrate that the main constituent of the bamboo pulp fibers is cellulose. This is also consistent with the results in Table 1.

Figure 3b shows the FTIR spectra of RA, $RA_{2.5}$, and A-$CRA_{2.5}$. Characteristic peaks of cellulose appeared in the RA at 3400, 2911, 1466, 1361, and 1046 cm^{-1}, which are attributed to stretching vibration of O-H, CH_2, H-C-H and H-O-C bending vibration, and H-C-C, H-C-O, H-O-C bending vibration. In the spectrum of $RA_{2.5}$, characteristic peaks of GO appeared at 1716, 1361, and 1225 cm^{-1} respectively [12], corresponding to C=O stretching vibration of carboxyl group, C-H bending, and C-O stretching vibrations. By comparing the infrared spectra of RA and $RA_{2.5}$, we can see that the carbonyl absorption peak of the composite was increased significantly after adding GO. It can also further validate the XRD results that GO were implanted in the composite. After carbonization, the carbon aerogels showed no characteristic peaks of cellulose and only a small amount of inorganic H_2O and CO_2 absorption peaks. It shows that the basic structure of cellulose was destroyed and the oxygen-containing functional groups on GO were reduced.

IOP Conf. Series: Materials Science and Engineering **735** (2020) 012027 doi:10.1088/1757-899X/735/1/012027

Figure 3c shows XRD spectra of the bamboo pulp fibers and bamboo regenerated cellulose aerogel. Two characteristic peaks for cellulose appeared at 2θ = 16.3° and 22.0° in the bamboo pulp fibers, assignable to the (101) and (200) planes of type I cellulose respectively [13]. After regeneration, the bamboo regenerated cellulose aerogel was obtained. The peak at 16.3° disappeared and the peak width at 22.0° increased, indicating that the cellulose type I structure changed and converted to a cellulose type II structure. Figure 3d shows XRD results for the bamboo-derived cellulose fibers/RGO carbon aerogel composite with different contents of GO. All the patterns show a characteristic peak of graphene at approximately 11°, indicating that the GO was reduced during carbonization. Broad peak appeared at 22° shifted to higher angles as the increase of graphene content, indicating a higher degree of graphitization in the materials [14].

3.3. Electrochemical Performance

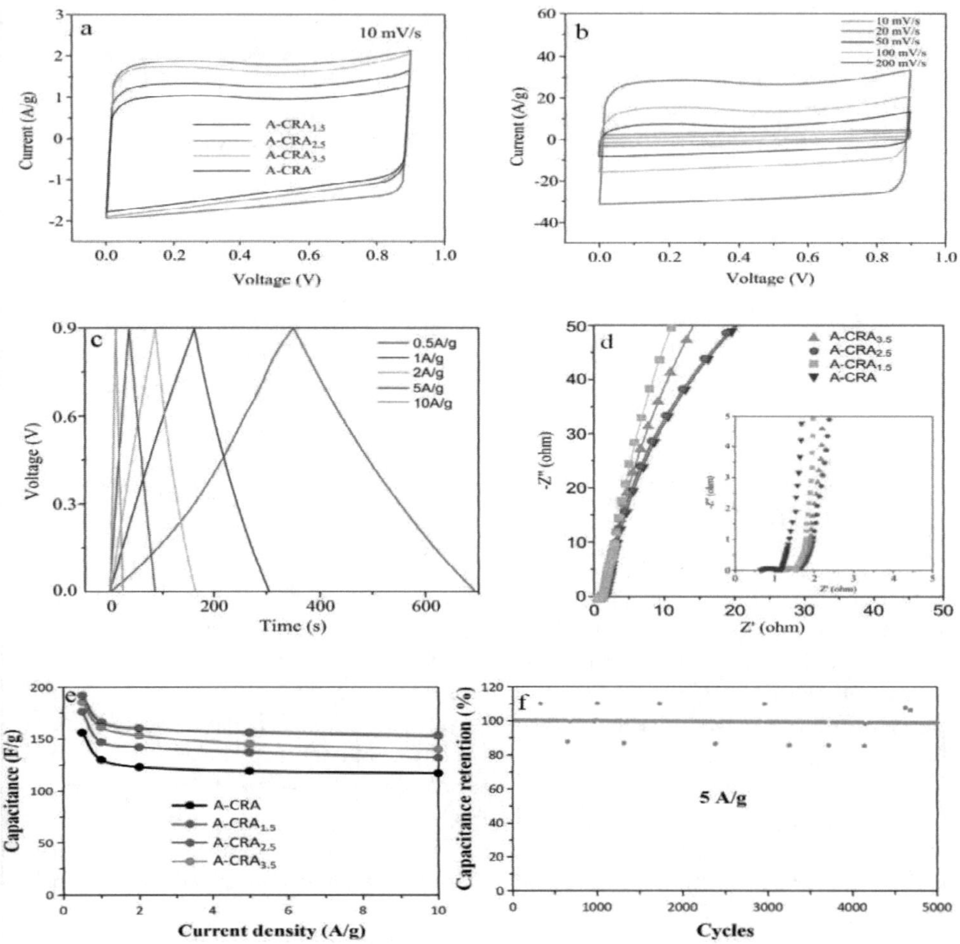

Figure 4. Electrochemical Properties of A-CRA$_x$ Samples in Three-Electrode Cell: CV Curves at Various Scan Rate of 10 mV/s (a); CV Curves of A-CRA$_{2.5}$ at Different Scan Rates (b); GCD Curves of A-CRA$_{2.5}$ Samples at Current Densities of 0.5-10 A/g (c); Nyquist Plots of A-CRA$_{2.5}$ (d); Capacitances at Different Current Densities (e); Cyclability for A-CRA$_{2.5}$ (f)

Figure 4a shows the CV curves of CRA and A-CRAx electrodes at a scanning rate of 10mV/s and a potential range of 0-0.9 V. The CV curve of CRA shows a deformed rectangle with a lower capacitance of 156 F/g. But the CV curve of A-CRAx showed a rectangular shape, indicating that a perfect electrical double layer (reversible adsorption and desorption of ions) formed throughout the carbon network. The specific capacities of the A-CRA$_{1.5}$, A-CRA$_{2.5}$, and A-CRA$_{3.5}$ electrodes were 325, 351, and 334 F/g, respectively, at a scan rate of 10 mV/s and these values were 14%-24% higher than that of the A-CRA. The A-CRAx electrode maintained a good rectangular CV profile (Figure 4b) with the increase of scan rate. The A-CRA$_{2.5}$ sample had the maximum corresponding current and area under the CV curve (Figure 4a), indicating that it has the maximum specific capacitance, followed by that of the A-CRA$_{3.5}$ sample, and A-CRA$_{1.5}$ sample. Thus, the specific capacitance of carbon aerogels can be improved by the addition of GO. As the amount of GO was increased, the specific capacitance tended to increase. Figure 5b shows CV curves of A-CRA$_{2.5}$ at different scan rates which showed good performance of an electric double-layer capacitor. Even at a high scan rate of 200mV/s, the material maintained a symmetric rectangular shape suggesting that the material has good ratio performance [15].

As observed from the above results, the increase of graphene content in the composite resulted in an increase of the specific capacitance. However, as the GO content is increased up to a certain value, the specific capacitance of the composite material will decrease again. It probably because the graphene sheets would re-agglomerate, which reduces the overall specific surface area of the composite.

Figure 4c shows the GCD curves of A-CRA$_{2.5}$ under different current densities with all symmetrical triangles, indicating that the A-CRA$_{2.5}$ has high charge-discharge efficiency good electrolyte diffusion in the pore structures [16]. When the current density increased from 0.5 to 10 A/g, the specific capacitance of all the samples decreased, which was consistent with the CV data. When the current density was 0.5 A/g, the A-CRA$_{2.5}$ shows a capacitance of 192 F/g. When the current density increased to 10 A/g, the A-CRA$_{2.5}$ capacitance was maintained at 153 F/g. These results indicate that A-CRA$_{2.5}$ owns a good rate performance [17].

The resistance inside the electrode (Rs) reflects the power loss caused by itself. The smaller the internal resistance, the smaller the voltage drop caused by the internal resistance, indicating the small power loss, which is very important for the electrode material. In addition, supercapacitors also have a charge transfer resistance (Rct), which is a reflection of the resistance to charge movement. The smaller the charge transfer resistance, the better the electrical conductivity of the material. Figure 4d shows the EIS of A-CRAx. All samples show, almost perpendicular to the coordinate axis in the low-frequency region. The A-CRA$_{1.5}$ has the steepest curve, indicating a low diffusion resistance of electrolyte ions in its porous structure. In the high frequency region, the intercept between the curve and the coordinate axis is the resistance inside the electrode (Rs), which reflects a small charge transfer resistance (Rct) and the high current charge-discharge performance of the sample [18]. The Rct results for A-CRA$_{1.5}$, A-CRA$_{2.5}$, A-CRA$_{3.5}$, and A-CRA were 0.46, 0.30, 0.37, and 0.34 Ω. The test results show that the bamboo-derived cellulose fibers/RGO carbon aerogel composite has a small internal resistance and charge transfer resistance, indicating that the bamboo-derived cellulose fibers/RGO carbon aerogel composite has a small energy consumption and well electrical conductivity; it is completely suitable for supercapacitor electrode materials.

Figure 4eshows the capacitances of different samples at different current densities. It can be seen that when the current density increases from 0.5 A/g to 10 A/g, the specific capacitance of all samples presents a downward trend, and it declines rapidly in the range of low current density (0.5 A/g~1 A/g) and slowly in the range of high current density (2 A/g~10 A/g). At the same current density, the specific capacitance of A-CRA$_{2.5}$ is the largest, which is consistent with the detection result of CV. It is related to its large pore volume, wide pore distribution, smaller average pore and smaller internal resistance.

In order to characterize the cyclic stability of A-CRA$_{2.5}$, 5000 charge and discharge cycles were performed on the samples at the current of 5 A/g (Figure 4f). The specific capacitance retention rate is basically maintained above 99%, the results show that the material has good cyclic stability.

As we expected, we can see from the characterization of electrochemical properties that bamboo-derived cellulose fibers/RGO carbon aerogel composite have larger specific capacitance, well multiplier performance and rate performance, as well as excellent cyclic stability, these performances allow it to be used in the field of supercapacitor electrode materials.

4. Conclusions

In this experiment, we used tetrabutylammonium acetate/DMSO system to dissolve bamboo pulp fibers, the electrochemical properties of the materials were improved by doping RGO, and the porous structure of the materials were achieved by preparing the carbon aerogels. Finally, a porous bamboo-derived cellulose fibers/RGO carbon aerogel composite was obtained by freeze-drying and carbonization, which was then used as the electrode material for supercapacitors. The microstructures show a cross-linked fibers skeleton doped with porous nano-graphene sheets, with a specific surface area of 1957 m^2/g. Owing to these structural advantages, the material has a large specific capacitance (351 F/g). Lightweight, porous, and stable carbon aerogels show great promise as electrode materials for flexible ultracapacitors in the future. In addition, the preparation of green and environment-friendly regenerated cellulose carbon aerogel is a good way of applying biomass resources. Thus, these findings may also help to find new value-added applications for biomass materials.

5. Acknowledgments

This research was supported by the China Ministry of Science and Technology (2017YFD0600804).

6. References

[1] Dobashi A, Maruyama J, Shen Y, Nandi M, Uyama and Hiroshi 2018 Activated carbon monoliths derived from bacterial cellulose/polyacrylonitrile composite as new generation electrode materials in EDLC. *Carbohyd. Polym.* **200** 381-90

[2] Pérez-Madrigal M, Edo M, Saborío M, Estrany F and Alemán C 2018 Pastes and hydrogels from carboxymethyl cellulose sodium salt as supporting electrolyte of solid electrochemical supercapacitors. *Carbohyd. Polym.* **200** 456-67.

[3] Liu N, Shen J and Liu D 2013 Activated high specific surface area carbon aerogels for EDLCs. *Micropor. Mesopor. Mat.* **167** 176–81.

[4] Liang H, Guan Q, Chen L, Zhu Z, Zhang W and Yu S 2012 Macroscopic-Scale Template Synthesis of Robust Carbonaceous Nanofiber Hydrogels and Aerogels and Their Applications. *Angew. Chem. Int. Edit.* **51(21)** 5101–05.

[5] Ren F, Li Z, Tan W, Liu X, Sun Z, Ren P and Yan D 2018 Facile preparation of 3D regenerated cellulose/graphene oxide composite aerogel with high-efficiency adsorption towards Methylene blue. *J. Colloid. Interf. Sci.* **532** 58-67.

[6] Zu G, Shen J, Zou L, Wang F, Wang X, Zhang Y and Yao X 2016 Nanocellulose-derived highly porous carbon aerogels for supercapacitors. *Carbon.* **99** 203–11.

[7] Liu H, Wang A, Xu X, Wang M, Shang S, Liu S and Song J 2016 Porous aerogels prepared by crosslinking of cellulose with 1, 4-butanediol diglycidyl ether in NaOH/urea solution. *Rsc. Adv.* **6(49)** 42854-62.

[8] Sevilla M, Ferrero G and Fuertes A 2017 Beyond KOH activation for the synthesis of superactivated carbons from hydrochar. *Carbon.* **114** 50-8.

[9] Mi H, Jing X, Politowicz A, Chen E, Huang H and Turng L 2018 Highly compressible ultra-light anisotropic cellulose/graphene aerogel fabricated by bidirectional freeze drying for selective oil absorption. *Carbon.* **132** 199–209.

[10] Yang X, Fei B, Ma J, Liu X, Yang S, Tian G and Jiang Z 2018 Porous nanoplatelets wrapped carbon aerogels by pyrolysis of regenerated bamboo cellulose aerogels as supercapacitor electrodes. *Carbohyd. Polym.* **180** 385–92.

[11] Javadi A, Zheng Q, Payen F, Javadi A, Altin Y, Cai Z and Gong S 2013 Polyvinyl alcohol-cellulose nanofibrils-graphene oxide hybrid organic aerogels. *Acs. Appl. Mater. Inter.* **5(13)** 5969-75.

[12] Li C and Shi G 2014 Functional Gels Based on Chemically Modified Graphenes. *Adv. Mater.* **26(24)** 3992–4012.

[13] Hong K, Qie L, Zeng R, Yi Z, Zhang W, Wang D and Huang Y 2014 Biomass derived hard carbon used as a high performance anode material for sodium ion batteries. *J. Mater.Chem. A.* **2(32)** 12733.

[14] Hao P, Zhao Z, Tian J, Li H, Sang Y, Yu G, Cai H, Liu H, Wong C and Umar A 2014 Hierarchical porous carbon aerogel derived from bagasse for high performance supercapacitor electrode. *Nanoscale.* **6** 12120-9.

[15] Xu L, Guo M, Liu S and Bian S 2015 Graphene/cotton composite fabrics as flexible electrode materials for electrochemical capacitors. *Rsc. Adv.* **5(32)** 25244–9.

[16] Oschatz M, Boukhalfa S, Nickel W, Hofmann J, Fischer C, Yushin G and Kaskel S 2017 Carbide-derived carbon aerogels with tunable pore structure as versatile electrode material in high power supercapacitors. *Carbon.* **113** 283–91.

[17] Gao S, Li X, Li L and Wei X 2017 A versatile biomass derived carbon material for oxygen reduction reaction, supercapacitors and oil/water separation. *Nano. Energy.* **33** 334–42.

[18] Zhang Q, Wang Y, Zhang B, Zhao K, He P and Huang B 2018 3D superelastic graphene aerogel-nanosheet hybrid hierarchical nanostructures as high-performance supercapacitor electrodes. *Carbon.* **127** 449–58.

IOP Conf. Series: Materials Science and Engineering **735** (2020) 012028 doi:10.1088/1757-899X/735/1/012028

Comparison in Chemical Composition of Lime Binders between Several Typical Ancient Buildings in China

Peng Zhao[a], Xiuzhi Zhang[b], Lei Qin[a], Yunsheng Zhang[c], Hongwei Ren[d], Yinghui Lv[a] and Gengle Liu[a,*]

[a] School of Civil Engineering and Architecture, University of Jinan, Jinan 250022, People's Republic of China;
[b] School of Materials Science And Engineering, University of Jinan, Jinan 250022, People's Republic of China;
[c] Jiangsu Key Laboratory for Construction Materials, Southeast University, Nanjing 211189, People's Republic of China;
[d] School of Electrical Engineering, University of Jinan, Jinan 250022, People's Republic of China;
* Corresponding author: Gengle Liu, 155900251@163.com

Abstract. This paper aims to compare the chemical and phase compositions of lime binders in ancient buildings. To this end, X-ray diffractometer (XRD) and FT-IR testing were used to analyse the representative samples by sampling the lime-based binding materials of several typical ancient buildings in China. The test results showed that the lime binder contained residual glutinous rice amylopectin; just because of this inorganic and organic synergistic effect of glutinous rice and lime can the lime binders have excellent properties. This provides a theoretical basis for the subsequent study.

1. Introduction

In China's ancient buildings, the lime-based binders were used for a long period. With the progress of history, such binding material gradually evolves from the original single-component mortar into a complex system containing multiple components [1, 2]. It has been found that China started to use an organic-inorganic composite material long before, that is, a traditional Chinese lime mortar combining the inorganic materials such as lime and the organic materials such as glutinous rice etc [3, 4]. Using modern technology to explore the physical properties, chemical and phase composition, and ingredient proportion of this organic-inorganic composite lime binder, as well as the interaction mechanism between key components can not only promote Chinese culture, but also absorb and utilize the advantages of traditional technology for current use. In particular, it is of great practical significance for guiding and promoting the protection of ancient buildings in the future [5, 6]. Therefore, taking the lime binder materials of several ancient architectural sites as samples, this paper analyses the chemical composition of the representative samples by X-ray diffractometry (XRD) and FT-IR test, which provides a theoretical basis for the subsequent study.

2. Test of Ancient Lime Samples

2.1. *Sampling Location*

According to archaeological records, when the Great Wall was built in Qin Dynasty, people learned to use a composite binding material composed of glutinous rice and lime.; in the northern and southern dynasties, a composite material containing starch was added to the lime; in the Ming Dynasty, the glutinous rice-lime composite binder was the most frequently used. In this study, nine ancient architectural buildings were sampled, from the Ming Dynasty to the Qing Dynasty. Table 1.1 lists the sample information.

Table 1. The Sample Information

Sample	Sample source	Date
#1	The Gate of China, Nanjing City	1369-1375 CE (Ming dynasty)
#2	Ming Tomb, Nanjing City	1381-1405 CE (Ming dynasty)
#3	Scholar Zhu lane, Nanjing City	1573-1620 CE (Ming dynasty)
#4	Tunxi Old Street, Huangshan City	Before 1548 CE (Ming dynasty)
#5	YangliuVillage, Nanjing City	1579 CE (Qing dynasty)
#6	Zhangtan, Huangshan City	1832 CE (Qing dynasty)
#7	Zhouzhuang Town, Suzhou City	1720 CE (Qing dynasty)
#8	Hongcun Town, Huangshan City	Before 1620 CE (Qing dynasty)
#9	Yu Garden, Nanjing City JinXi	1876 CE (Qing dynasty)

2.2. *Test Methods*

After crushing each sample, some small pieces in size 2.5mm-5.0mm were taken. Then, removing the sandstone contained, they were ground to less than 0.08mm in an agate mortar, and dried at 40°C for 24 hours. Next, the dried samples were kept in the dryer for XRD and FT-IR testing.

3. Results and Discussion

3.1. *XRD Analysis*

Figure 1 shows the XRD pattern of the lime binder samples. It can be seen from the XRD line that a main peak appeared at a 2θ angle close to 30°, which corresponded to a characteristic peak of calcium carbonate in the calcite crystal form. In addition, in the 3#, 4#, 8#, and 9# samples, another significant main peak appeared at a 2θ angle of approximately 27°, which corresponded to a characteristic peak of the quartz crystal form. Also, the XRD pattern analysis found that the main components of the 1#, 2#, 5#, 6#, and 7# samples were calcite crystal calcium carbonate in which a small amount of clay mineral was doped, while the 3#, 4#, 8#, and 9# samples contained not only calcite crystal calcium carbonate, but also the significant characteristic peaks of quartz crystal form, which indicates that these four samples are mainly composed of lime and clay, similar to the described trinity mixture fill in the relevant literature.

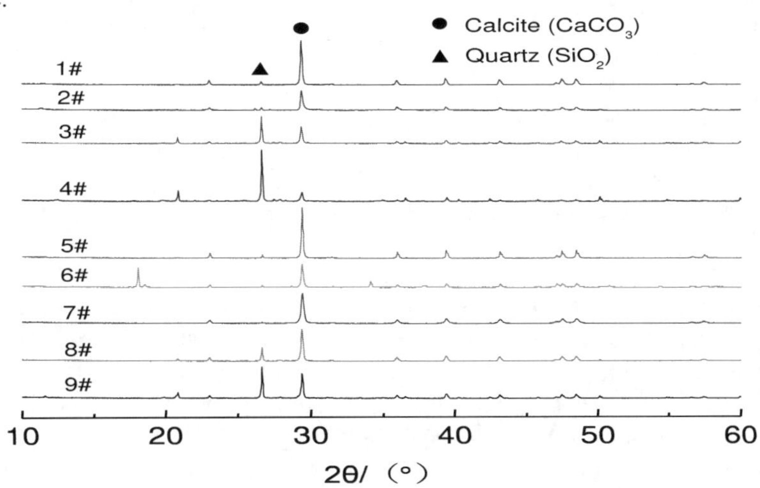

Figure 1. XRD Patterns of Lime Binders

3.2. FT-IR Analysis

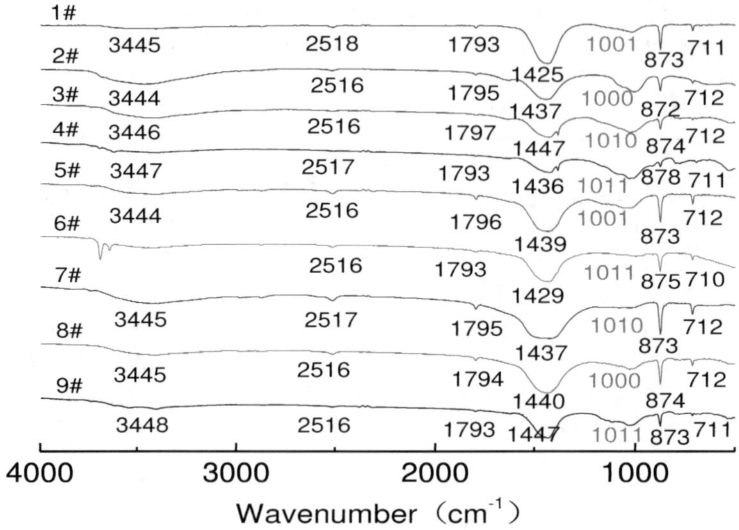

Figure 2. FT-IR Patterns of Lime Binders

Figure 2 shows the results of the FT-IR test. It can be seen from the figure that the absorption peaks near the wavenumbers of 712, 876, 1429, 1794, and 2513 cm-1 were attributed to the characteristic peak of calcite, so it can be inferred that the calcium carbonate contained in samples mainly used a calcite crystal form; also, the absorption peak (a.1000cm-1, c.1010cm-1, d.1011cm-1) near wavenumber 1018cm-1 corresponded to the characteristic peak of a constituent polysaccharide compound in the amylopectin of glutinous rice, which is caused by stretching vibration of a large number of C-O(H) groups [130]. Through the relative intensity of the main strong peak of amylopectin, rapid mass identification can be carried out to determine the amount of amylopectin added in different representative samples. Among them, the amount of amylopectin added in samples 2#, 3# and 4# was the most, followed by the samples 8#, 9#, and samples 6#, 7# had the least proportion of amylopectin.

3.3. Mechanism Explanation

Studies have shown that as a key component in traditional mortars, glutinous rice and lime play a major role in improving the bond strength of mortar, while soil and sand are inert components, and they basically function only as fillers. Besides, glutinous rice has a regulating effect on the carbonation of lime. Glutinous rice is a kind of polysaccharide matter. Modern scientific research shows that bio-polysaccharides can often be used as a template agent for biomineralization process to constrain and regulate the size, morphology and structure of crystalline particles formed by inorganic ions during crystallization. Therefore, added with this agent-glutinous rice, the calcium carbonate formed by the mineralization reaction of Ca(OH)2 solution is a nanometre-scale fine particle of calcite crystal form, which is much finer and denser than that without glutinous rice. This fine structure can greatly improve the compressive strength and surface hardness of the mortar.

4. Conclusions

The hardening process of the quicklime mortar involving the glutinous rice is actually the biomineralization process of calcium carbonate added with the natural bio-polysaccharide (glutinous rice). In this process, glutinous rice not only plays a role in regulating the crystallization process and microstructure of calcium carbonate, but also closely combines with the generated calcium carbonate to form a composite structure in which organic matter and inorganic matter are combined and densely filled. This should be the microscopic basis for excellent mechanical properties of glutinous rice such as high mortar strength and toughness.

5. Acknowledgements

This work was supported by the Key R & D project of Shandong Province (2019GSF109119), the Doctoral Foundation of the Shandong Natural Science Foundation (ZR2018BEE038), the National Key Research and Development Program of China (2017YFC0703100), and the National Natural Science Foundation of China (51778269, 51678277).

6. References

[1] Wang X, Lam C C, Iu V P, Experimental investigation of in-plane shear behaviour of grey clay brick masonry panels strengthened with SRG, Eng. Struct. 162 (2018) 84-96.
[2] Pan G G, A History of Chinese Architecture (in chinese), China Architecture & Building Press, Beijing, 2004.
[3] Liu Y, Brief Excavation Report of the Eastern Zhou Dynasty Rouyou site in Shanxi province (in chinese), Archaeol. Rel. 31(1991) 9-10.
[4] Shan C R, The report on archaeological excavations of the Chenhan tomb site in Shanxi province (in chinese), Shanxi people's publishing house, Shanxi, 2006.
[5] Shen Y, The study of ancient Chinese eaves tiles (in chinese), Cultural relics publishing house, Shanxi, 2006.
[6] Wu J, Wu J H, Wang H S, Lu X K, Wu J M, The technical research on the huge glazed brick from the relic of Nanyue Kingdom Palace, Sci. China. Ser. E. 51(1) (2008) 16-24.

IOP Conf. Series: Materials Science and Engineering **735** (2020) 012029 doi:10.1088/1757-899X/735/1/012029

Design and Experiment of Dynamic Measurement Method for Bulk Material of Large Volume Belt Conveyor Based on Laser Triangulation Method

Fusong Min[1], Andong Lou[2,3] and Qun Wei[1]

1Nanjing Coal Science &Technology Research Co., Ltd., Nanjing, Jiangsu 210018, C hina

2 Luoyang Mining Machinery Engineering Design and Research Institute Co., Ltd., Luoyang, Henan 471039, China

3 Corresponding author and e-mail: Andong Lou, citichma_louad@163.com

Abstract. In order to improve the measurement accuracy and efficiency of bulk material transported by large-capacity belt conveyor, an online dynamic non-contact metering system based on laser triangulation is designed on the belt conveyor platform. The system includes: laser Scanners, industrial camera, signal processing software, mechanical assemblies. High-precision basic measurement data is an important guarantee for the accuracy of bulk flow measurement. The laser scanner is used as a line source, and the scanning frequency is set to 300HZ, which is installed on the belt conveyor. Industrial camera accepts images produced by a line source projected onto the bulk surface. The signal processing software system analyzes the images, and uses the triangulation method directly calculate the three-dimensional contour information and volume flow of the bulk material. The flow filtering algorithm filters the instantaneous flow, smoothing the fluctuation of the flow data to reflect the real changes of the dry and wet degree of the bulk material on the belt, the degrees of surface cracking and shapes. Establish a mathematical model based on neural network and apply the belt shaking elimination algorithm to correct the previously calculated flow data to reduce the measurement error. In order to restore the actual belt conveyor bulk transportation conditions, a test belt with length 20m and speed 2m/s was produced according to the actual size. The physical simulation test of bulk material for potassium salt was carried out, the results show that the on-line measurement method based on laser triangular method has high efficiency and stable operation, and eliminates the influence of bulk density. The measurement error is ≤3.

1. Introduction

The three-dimensional contour measurement method was contact measurement in the early stage. In recent years, this method has been developed unprecedentedly [1]. Currently, there are many types of non-contact contour measurement methods, including: time-of-flight method [2], holographic interferometry. [3], speckle interferometry [4], confocal microscopy [5], phase method [6], binocular stereo method [7], and laser triangulation method [8] and so on.

At present, in the production process of factories and mines, the electronic belt scale and the nuclear belt scale are the two mainstream metering devices for dynamic weighing. The electronic belt scale is a contact type measurement. It uses the pressure sensor installed under the conveyor belt to sense the mass of the bulk material on the belt. However, for the belt under motion, the measurement result is inevitably subject to the tension, rigidity and weight of the belt and so on. The nuclear scale is non-contact measurement and is made using the principle that substances absorb γ-ray. Due to the ever-changing shape of the bulk materials on the conveyor belt, the ionization chamber is often covered by the material falling off the belt, which causes the accuracy of the nuclear scale to decrease, and the application of the radioactive source is cumbersome, and there are certain safety hazards in the use and storage of the radioactive source.

2. Measuring Principle and Process of Metering System

In this paper, a new concept based on laser triangulation is designed to measure the material flow of conveyor belt is designed. The online measurement method of bulk materials is attributed to the online measurement of bulk materials as two problems: online measurement of bulk volume and measurement of bulk density. The laser scanner emits a fan-shaped laser beam as a line source and shoots it vertically downwards to the belt. Using laser image analysis technology, a series of material surface contour coordinates can be obtained, and then triangulation can be used to calculate the flow over the belt per unit time. The bulk volume of the material is measured online, and the bulk density of the material can be determined by physical calibration. The volume and bulk density of the material on the conveyor belt are obtained, and the material flow rate is determined. This paper introduces the following three processes of the measurement.

2.1. Find the Surface Height of the Bulk Material

First, the line source is vertically irradiated on the surface of the calibration object. By adjusting the height of the calibration object, the industrial camera acquires the image information, and then the software performs image processing to complete the calibration of the relationship between the actual height h of the object and the distance d between the source coordinates.

The line source is illuminated on the surface of the mineral material, the red line (Figure. 1) in the schematic diagram of the laser stripe on the surface of the mineral material. Take the photo, and obtain the coordinate value of the line source by image processing. Using the relationship between h and d calibrated in the previous step, the actual height of the line source, that is, the height of the surface of the mineral material. The actual height of the entire line source is determined by n height values. The yellow dot in the figure below is a schematic diagram of the n feature points on the selected line source. The larger the value of n, the more accurately the real height of the mineral material required to be taken Real height.

Figure 1. Laser Triangulation Method for Measuring Surface Height

2.2. Find the Cross-sectional Area of the Bulk Material

Using the values between the point i and the point i+1 (i=1, 2,..., n-1) height points, the cross-sectional area is obtained by linear interpolation, The cross-sectional area of the irregular object, the sum of all the cross-sectional areas is the cross-sectional area of the mineral material.

2.3. Find the Flow of Bulk Material

Take a image information (picture) every **Δt** time. If the distance traveled by the material is **k** in **Δt** time, multiply the cross-sectional area obtained in the previous step by **k**, which is the volume of the material passing through **Δt** time. The volume of materials obtained during the time is all added up. If **t** time contains **m** times **Δt** times, the flow rate of the requested material is Q:

IOP Conf. Series: Materials Science and Engineering **735** (2020) 012029 doi:10.1088/1757-899X/735/1/012029

$$Q = \frac{\sum_{j=1}^{m} V_j}{m\Delta t}$$

3. Key Technologies for the Metering Systems

3.1. High-frequency High-precision Outdoor Laser Scanning Instrument

High-precision basic measurement data is an important guarantee for material flow measurement accuracy. In order to improve the adaptability and stability of the metering system and improve its measurement accuracy, the main technical parameters of the selected high frequency and high precision laser scanner are shown in the following table1.

Table 1. Main Technical Parameters of Laser Scanner

Laser scanner technical parameters		
Number	Name	Parameter
1	Data points	1280
2	Z-axis linearity	0.04%
3	Z-axis resolution	0.092-0.448 mm
4	Z-axis repeatability	12μm
5	X-axis resolution	0.375-1.1mm
6	Measuring distance	350mm
7	Measuring range	800mm
8	Scanning range	390-1260mm
9	Input voltage	24-48V
10	Scanning frequency	170-5000HZ
11	Protection level	IP67
12	Temperature	0-50°C

According to the actual working conditions, the scanning frequency of the scanner is set to 300HZ. According to the average belt speed of 1.65m/s, the surface contour line spacing of the scanned materials is 8mm, and the X-direction (vertical belt moving direction) scanning point spacing is 0.375mm. The accuracy is less than 0.5mm. The scanner itself has IP67 protection and can be used in outdoor working environment. The operating temperature range is 0-50 °C, which can meet the normal working conditions of the instrument in relatively low temperature and high temperature environment, so the instrument has the ability to work around the clock.

3.2. Triangulation Method to Calculate Volume Flow

There are two ways to perform triangulation of spatial points: one is to directly triangulate points in space, and the other is to map points in space to plane and triangulate points in two-dimension. And then back map into 3D space. The latter method is relatively simple. The latter method is used to triangulate feature points in a two-dimensional image by triangulation algorithm (Shewehuk, 1995), as shown in Figure3.and then due to three-dimensionality in space. The points are calculated by the feature points, so according to this mapping relationship, the two-dimensional points of the split are mapped into the three-dimensional space, and the triangulation of the spatial points are realized as shown in Figure 4.

The laser is projected onto the surface of the material on the belt, which results in a series of scan point coordinates that form a material surface profileon the XZ coordinate plane. As shown in Figure 2.

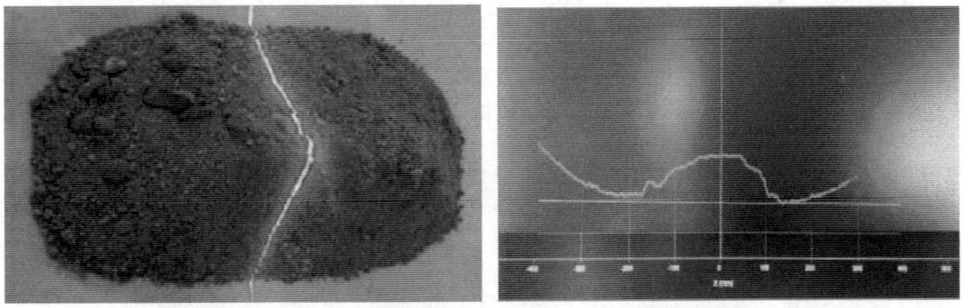

Figure 2. Material Profile Contour Laser Contour

Assuming the number of scanning points **N** on the contour line, two adjacent contour lines triangulate from the minimum to the maximum along the X coordinate, and **2 (N-1)** bevel triangular prisms can be obtained. The distance of the contour line in the belt movement direction (Y direction) is the product of the belt speed and the time interval between the contour lines. By accumulating the volumes of these **2 (N-1)** bevel triangular prisms, the volume of materials between the two contours can be obtained. The triangulation method is shown in Figure 2.

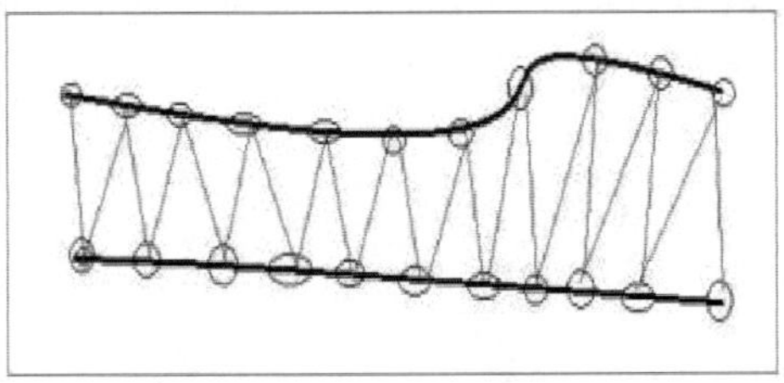

Figure 3. Interface Triangulation Diagra

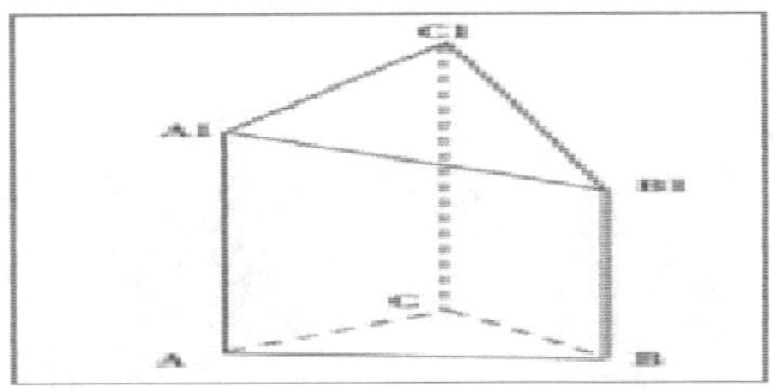

Figure 4. Beveled Triangular Prism

Assume that in Figure 3, the i-th beveled triangular prism has a volume of:

$$dV_i = \frac{1}{3} S_{\Delta ABC} * (Z_{A1} + Z_{B1} + Z_{C1})$$

The volume between the two contour lines is:

$$V = \sum_{i=1}^{2(N-1)} dVi$$

Assuming that the time interval between the two contour lines is **Δt,** the instantaneous volumetric flow rate of the material on the belt is:

$$F = \frac{V}{\Delta t}$$

As shown in Figure 5, the feature points acquired by the high-frequency scanning picture set are combined, and the information of all the feature points obtained by the picture set is integrated to obtain the reconstructed three-dimensional contour figure by triangulation, thereby calculating the volume flow rate.

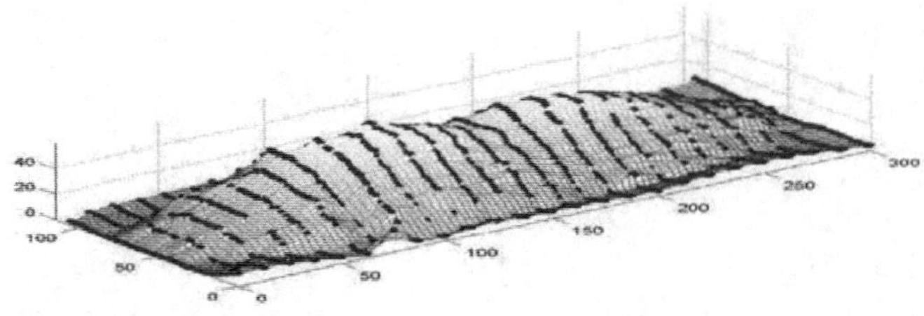

Figure 5. Reconstructed 3D Contour Graphics

The instantaneous weight flow of the material is obtained by multiplying the instantaneous volume flow by the bulk density of the material.

3.3. Traffic Filtering Algorithm

Due to the change of the dryness and wetness of the material during the conveying process of the belt conveyor, when the material falls onto the belt, the surface cracking degree is different and the shape is also ever-changing, resulting in a large fluctuation of the measured instantaneous flow. As shown in Figure6. The purpose of filtering the instantaneous flow is to smooth the fluctuation of the flow data and reflect the real change of the material on the belt in real time.

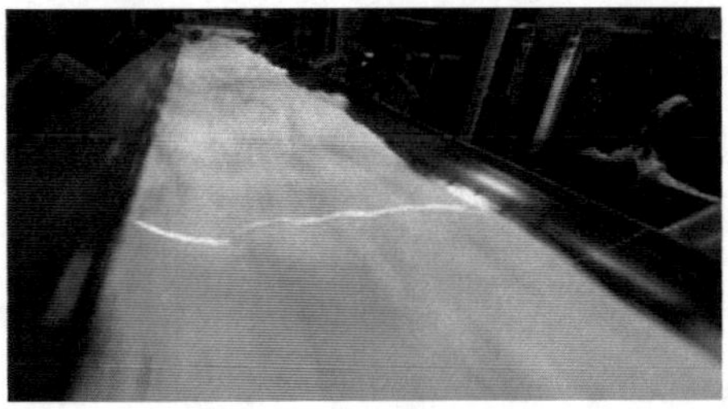

Figure 6. Surface Profile (potassium salt) on the Belt

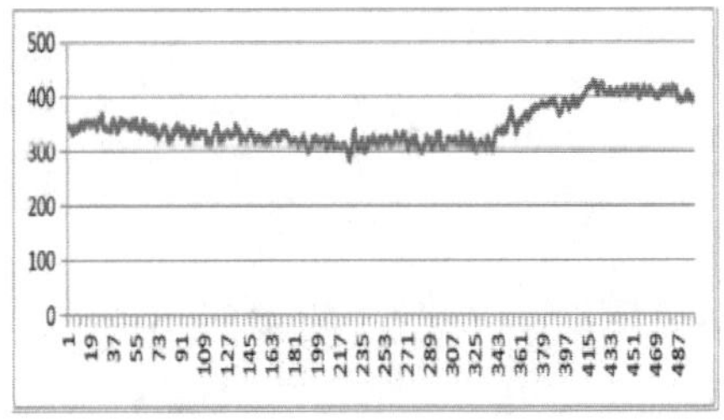

Figure 7. Filtered Flow Curve

First of all, the instantaneous flow data is accumulated in unit time Δ **t**, and one actual flow data is calculated per unit time Δ **t**, which can reduce the output frequency of the actual flow data and reduce the fluctuation. Assuming that the number of scanners in per unit time is **M** scanning contour lines, two adjacent contour lines calculate a instantaneous flow data, and the total number is **M-1**, then the volume flow in Δ **t** is:

$$F = \frac{\sum_{i=1}^{M-1} Vi}{\Delta t}$$

Secondly, the flow data calculated in unit time is subjected to five-point method smoothing filtering. The filtering formula is:

$$FR_k = \alpha_{k-2}F_{k-2} + \alpha_{k-1}F_{k-1} + \alpha_k F_k + \alpha_{k+1}F_{k+1} + \alpha_{k+2}F_{k+2}$$

The real flow values F_{k-2}, F_{k-1}, F_k, F_{k+1}, and F_{k+2} of the **k**-th cycle of FR_k are the volume flow values calculated in units of 5 consecutive cycles. α_{k-2}, α_{k-1}, α_k, α_{k+1}, α_{k+2} are smoothing coefficients, which are calculated according to the Hamming function, and are taken as 0.04, 0.24, 0.44, 0.24, and 0.04, respectively.

Figure 7 shows the filtered flow data curve for a certain period of time. The burr-like value jump on the curve is reduced, which can also reflect the material change on the belt.

3.4. Belt Jitter Effect Elimination Algorithm Based on Neural Network

When there is no material transport on the belt, the laser source projects a bright band formed on the belt, which is called "background bright band". Since the bright band is composed of a series of spots, the geometric coordinates of each spot can be obtained through a series of transformations. In the case of materials on the belt, the scanner continuously ingests the contour lines of the material surface at a certain time interval, that is, the bright lines, and calculates their geometric coordinates. By comparing with the background bright bands, a series of material cross sections are obtained. Shape and calculate its area. As shown in Figure 8. The background bright band coordinates data of the laser scanning, so the contour data of the empty belt is an important basic factor affecting the measurement accuracy.

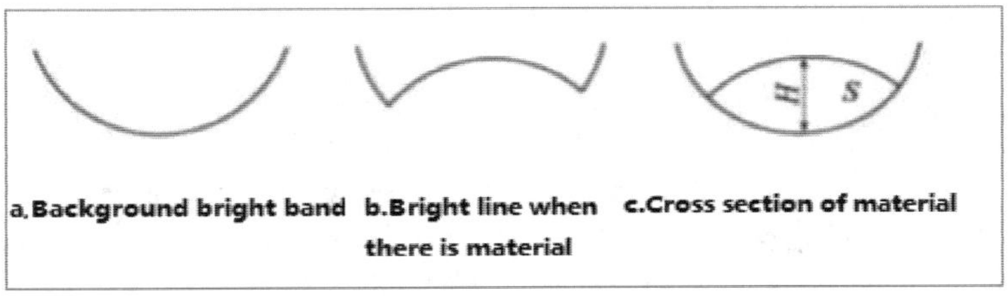

Figure 8. Background Bright Band

When the belt moving at high speed passes by the roller, there will be obvious up and down jitter, which will cause the Zcoordinate of the contour of the background light to move up and down. When calculating the cross section of the material, deviation will occur, which will cause measurement error. When there are materials on the belt, the materials cover the bottom of the belt, so the coordinates of the background light strip cannot be accurately obtained. By adjusting the field of view of the scanner, the scanner can always scan the marked area where the edge of the belt is never covered by the materials. At the time the belt is empty, a neural network model between the marker region coordinates and the center point Zcoordinate on the background light band is established. When the belt is covered by the materials, the coordinate change of the marked area is obtained, and the neural network model can be used to predict the change of the Zcoordinate of the center point of the background light band, thereby eliminating the influence of the belt jitter effect.

The neural network is recognized as an effective model building method that can be used to establish a mathematical model between the coordinates of the marker area and the Zcoordinate of the center point on the background light strip. In this way, during the measurement process, the change of the coordinate of the background bright band contour can be calculated according to the coordinate change of the mark area, which is used to correct the data of the flow calculated in the foregoing, and reduce the measurement error.

RBF network (radial basis function neural network) is a very popular modeling method. Consider an RBF network: it has n inputs, M hidden nodes and an output, and uses Gaussian radial basis functions. Before using the RBF network, you must first determine the hidden node data center (including: number, value, expansion constant) and a corresponding set of weights.

According to the on-site calibration result, the Z coordinate of 10 consecutive scanning points in the marking area is used to form the input vector. When the belt is empty, the Z coordinate of the center point is the target value, the belt is idling for 5 minutes (the belt runs about two times), and 15000 sets of data are collected for Neural network training. The training process is as follows:

1) First, 50 sets of data are randomly selected as the data center vector, and the data center distance threshold δ is set.

2) Cross-calculate the Euclidean distance between 50 sets of data center vectors, and eliminate vectors with high similarity.

3) Calculate the Euclidean distance of the remaining 14950 data sets, comparing the remaining data center vectors in 2), and add the vector with the distance greater than δ into the data center vector until the number of vectors reaches 50, and the number of hidden nodes in the neural network is 50.

4) Training a neural network uses a two-step approach. Idea: The first step is to determine the data center of h hidden nodes in the RBF network by unsupervised learning method, and determine the

expansion constant of the hidden node according to the distance between each data center. The second step is to use supervised learning (gradient method) training. The output weight of each hidden node.

After determining the extension constant of the data center and the output weight of the hidden node, the neural network model f(X) is determined. In the actual measurement process, the input vector **X** is the Z coordinate of 10 consecutive scanning points in the marking area, and the input neural network f(X) to the network output value Zx, Zx is the center point Z coordinates of the neural network prediction. Zx is subtracted from The Z coordinate of all the scanning points on the contour line, and then the triangulation, volume calculation and filtering budget are performed to obtain real flow data.

4. Overall Design of the Metering System

4.1. Structure of the Metering System

The metering system is designed not only to measure the flow volume measurement of the fine particles on the transport belt but also to meet the requirements of measuring the volume flow of large materials. When measuring large materials, it can be adjusted by increasing the power of the laser source. The parameters of the motion control module of the system realize the measurement and calculation of the contour and volume of large materials. As shown in Figure 9.

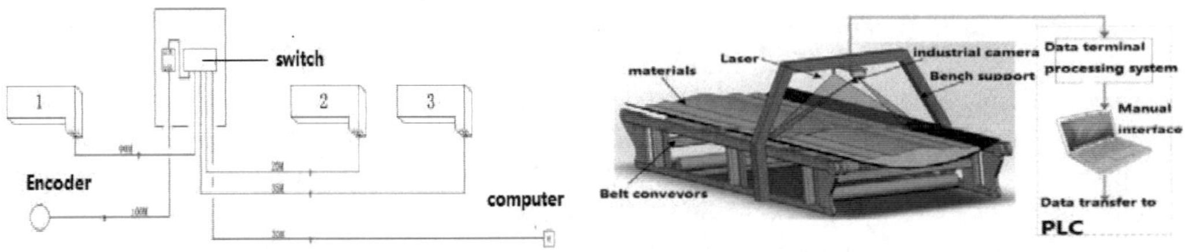

Figure 9. Test System Structure

4.2. Computer Data Information Processing Software

According to the application measurement requirements and marketing of the measurement system for actual working conditions, a dedicated measurement and test system software has been developed , can set the system technical parameters according to the production requirements.

5 .Test Results and Analysis

5.1. Laboratory Physical Simulation Test

The metering system installation and simulation tests were carried out in the laboratory. According to the size specification of the 105 belt, a belt with a length of about 20 meters was produced, and the belt speed was 2m/s. Physical simulation tests and verification were carried out. Install the branch support above the belt conveyor to install the equipment in the normal working form. The installation height of the equipment takes into account the measuring range. And then the calibration and scanning area settings are performed according to the equipment usage specifications. The entire laboratory test environment is shown in Figures 10.

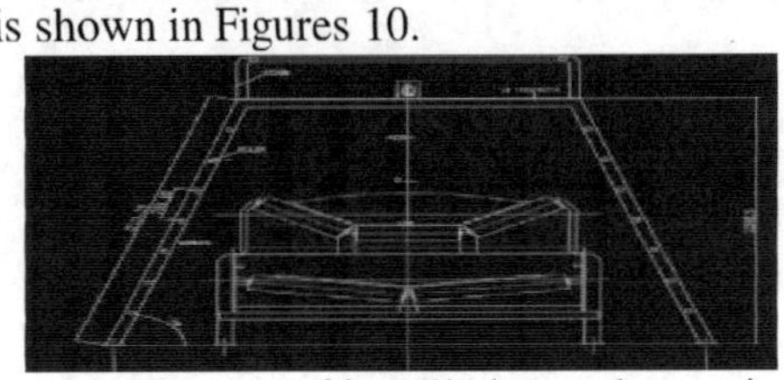

(a) Schematic diagram of installation and scanning range

(b) Device initialization settings

Figure10. Physical Simulation Test Scenario

In order to verify the measurement accuracy of the system, two laboratory tests were carried out: carton volume measurement and potassium salt volume measurement. The fixed volume of the carton and the potassium salt are stacked on the belt, the belt is started, and the volume and actual volume are measured by the device to calculate the measurement error. As shown in Figure 11, the measurement results are shown in Table 2 and Table 3.

Figure11. Laboratory Simulation Test

Table 2. Carton Volume Measurement Results

Carton volume	Number	Measuring volume	Actual volume	error
0.01584m^3	10	0.16074m^3	0.158m^3	1.47%

Table 3. Volumetric Measurements of Potassium Salts on the Belt

Number	Salt volume	Measure volume	Error
1	0.03654m^3	0.03765m^3	0.00111m^3
2	0.03654m^3	0.03564m^3	-0.00090m^3
3	0.03654m^3	0.03544m^3	-0.00110m^3
4	0.03654m^3	0.03550m^3	-0.00104m^3
5	0.03654m^3	0.03735m^3	0.00081m^3
6	0.03654m^3	0.03560m^3	-0.00094m^3
7	0.03654m^3	0.03760m^3	0.00106m^3
8	0.03654m^3	0.03736m^3	0.00082m^3
9	0.03654m^3	0.03591m^3	-0.00063m^3
10	0.03654m^3	0.03770m^3	0.00116m^3

The accuracy of the measuring instrument is expressed by the standard deviation σ:

$$\sigma = \sqrt{\frac{\delta_1^2 + \delta_2^2 + \ldots + \delta_n^2}{n}} = \sqrt{\frac{\sum_{i=1}^{n} \delta_i^2}{n}}$$

σ--standard deviation
Δ--measurement error
n--Number of measurements
Result: σ=0.00097m^3

Measurement error: $f = \frac{0.00097}{0.03654} = 2.65\%$

6. Conclusion

The online dynamic non-contact metering system based on laser triangulation method has some advantages, such as simple structure, high degree of automation and good stability. The system uses a line laser source in the measurement process, which improves the measurement system compared with the traditional point laser source. In the actual working conditions of the factory belt conveyor to transport bulk material, the metering volume flow error 2.65% ≤ 3%, the metering system designed in this paper meets the real-time measurement requirements of the factory belt conveyor, which is a good application prospect in the practice of fine industrial metering.

7. Acknowledgement

Support from the special project of Science and Technology Innovation and Entrepreneurship Fund of China Coal Technology & Engineering Group Corp(Grant No. 2018-2-QN005) is acknowledg ed.

8. References

[1] Javier Garcia, Jose J. Esteve-Taboada, Jose J. Valles. Detection of Three Dimensional Objects Based on Phase Encoded Range Images [J]. Proc SPIE Vol.5477, 2002,158 ~167

[2] Markus-Christian Amann, Thierry Bosch Risto Myllyla, et.a1. Laser ranging: A critical review of usual technlques or distance measurement[J].Opt Eng,2001,40(l):10 ~19

[3] Goodman J W, Lwrence R W. Digital image formulation from electronically detected hologram. Appl phys letter, 1967,11(3):77 ~ 99

[4] Le Kaiduan et al. Large dynamic range deformation measurement [J]. Acta Photonica Sinica, 1998, 27(6): 6Wilson T, C J R Sheppard. Theory and practice of scanning optical microscopy [M].London Academic Press, London. 1984. 1~ 65

[5] Huang J, Zhang X B. 3D measuring techniques based on image processing[J].Chinese Journal of Electron Devices, 2002, 25(4): 364 ~ 368

[6] Coodall A J, Burton D R .The future of 3D range image measurement using binaary-encoded pattern projection [J]. Opt. Loers Eng.,1994, 21(l-2): 99 ~ 110

[7] Huang J, Zhang X B. 3D measuring techniques based on image processing[J].Chinese Journal of Electron Devices， 2002， 25(4):364～368

[8] Coodall A J, Burton D R .The future of 3D range image measurement using binaary-encoded pattern projection[J]. Opt.Loers Eng.,1994,21(l-2):99～110

Volumetric Displacement in Large Deflection of Symmetrically Layered Piezoelectric Plate under Pretension

Chunfu Chen* and Hsiangjuei Yu+

*Professor, +Graduate Student, Department of Mechanical Engineering,
Chung Hua University, Hsin Chu. Taiwan 30067.
cfchen@chu.edu.tw

Abstract. The volumetric displacement of a simply supported and symmetrically layered piezoelectric circular plate under lateral load and pretension in large deflection is studied. The approach extends von Karman's large deflection theory to a layered case and accounts for the associated piezoelectric force terms. The nonlinear governing equations based on the equilibrium were derived, but the arising nonlinear terms in governing equations were dropped, to have an analytical investigation. The solutions were found to be expressible with modified Bessel or standard Bessel functions, depending on whether it is a mechanical-load dominating or piezoelectric-load dominating condition. The previously derived expressions for lateral slopes and deflections are further manipulated to derive the analytical expressions for the volumetric displacements, following strictly the related recurrence relations. The presented approaches are implemented with typical silicon-based layer materials used in a sensing or actuating device. Various dimensions, especially the relative thickness of the piezoelectric layer, as well as different applied voltages and pretensions are considered to have an extensive parametric study. The results indicate that, piezoelectric effect is only apparent in a low pretension condition. Upon reaching a moderate pretension, the pretension effect may always be dominant over the effects of varying the applied voltage and relative thickness.

1. Introduction

Piezoelectric devices have long been utilized as micro transducers including sensors and actuators, especially micro pumps in practical applications. This type of devices has shown many advantages in characteristics such as rapid response, high-energy capacity, and excellent efficiency in power consumption [1]. In the recent years, similar application has been broadened and extended to the field of medical industry. In this area, a comparatively new application is to make use of the property of piezoelectricity of a piezoelectric material in pumping drug fluid as a micro pump for drug delivery [2]. The same pumping principle appears to already have a very industrialized application in automobile field for pumping fuel flow for fuel delivery [3]. Similar piezoelectric actuating mechanisms have also been employed in PZT energy transducers to harvest energy of sound wave [4]. For the consideration of exploiting sustainable energy and resource, energy harvesting floors based on similar PZT materials have been developed and commercialized as well [5]. These kinds of devices are commonly designed and fabricated in a layered configuration, or maybe bossed with a proof mass. Devices with such configuration have demonstrated better performances such as improved sensitivity and linearity over a conventional flat member without a center boss [6]. In practical applications, however, they may often undergo a large deflection condition. Thus, a more advanced approach beyond classical plate theory that bases on Kirchhoff hypothesis is required to evaluate their structural responses. In literature, there have been available quite a few studies dealing with this type of problems [7-8]. In practical fabrication, they are often made by simply depositing a piezoelectric sensing or actuating layer onto a substrate

[9-11]. Thus, a shallow-patched structure is quite common for this type of devices.

Another key issue in designing and fabricating these types of members appears to be the arising of initial tension, on the other hand. A severe pretension may cause a possible warpage in early stage and premature failure. The pressure sensitivity may thus be degraded for the case a pressure sensor [12]. A worth-to-note fact is, however, pretension induced buckling is also the most reliable mechanism in evaluating the residual stresses [13]. Apparently, problems of large deflection of a patched plate under pretension are of practical importance. In literature, however, related available studies often consider the cases of clamped-ended edges and the center boss is usually taken as a rigid part, thus neglecting the possible elastic deformation of the center boss. For a layered piezoelectric transducer such as a micro pump, a center boss may be of a thickness comparable to the annular region. The possible elastic deformation should be included in problem formulation in assessing their structural behaviors. On the other hand, a simply supported problem is deemed to represent the other limited case regarding end support for practical application, though it has been considered in less. Solutions for a simply supported case should be of definite significance as it may serve as an opposite bound in structural characterization. It can likely be of more practical merit in simulating a real application for piezoelectric miniaturized devices.

This study is thus motivated to investigate the volumetric displacement in large deflection of a simply supported and piezoelectric layered plate under pretension. The approach extends von Karman's large deflection theory to consider a symmetrically layered piezoelectric plate. Thus, the associated piezoelectric force term will be included in formulating the problem. For a primary insight, however, the thus derived nonlinear governing equations were reduced to linear ones, by neglecting the arising nonlinear terms. This gives rise to a modified Bessel equation or standard Bessel equation for the lateral slope, depending on the relative magnitude of piezoelectric force term. The corresponding analytical solutions were developed by considering the boundary condition of the problem. The solutions for various geometrical responses, including lateral curvatures and deflections have been illustrated by the present authors in our previous work [14], all expressible in terms of either modified Bessel or standard Bessel functions. Following the solution of lateral deflections, one more tractable and feasible step is conducted in the present study to find the volumetric displacement, by integrating the solution of lateral defection over the entire plate. The significance of extending such analytical solutions is to provide a clue for detecting the quasi-static dosage for a single stroke in drug or fuel delivery applications [3], or maybe piezoelectric actuators operating alike. In real application, the importance and merit of finding the solution of volumetric displacement should worth further note, as it may also represent the key element for evaluating the acoustic compliance for an electro-acoustic device [15]. The developed approach will be implemented with typical silicon-based layer materials. A parametric study will be conducted to evaluate the effects of the ratio in thickness between the center piezoelectric layer and the substrate layer, as well as the magnitude of pretension and electric voltage applied across the piezoelectric layers. A recent work presented by Fox, et. al. [16] had ever evaluated the effect of dimension (radius and thickness) of an annular piezoelectric actuator film upon the behavior of lateral deflection of a circular plate, for different kinds of boundary conditions. However, the influence due to initial tension was not considered.

2. Problem Description and Solution Method

In order to simulate a more practical situation compared to our previous investigation [14], a simply-supported isotropic circular plate symmetrically sandwiched between two piezoelectric outer layers polarized in the thickness direction is considered. It is subjected to an initial in-plane tension, N_0, a uniform lateral load, p_z=p_0, and an applied voltage V, across the piezoelectric layers polarized in the thickness direction, as shown in Figure 1. The governing equations based on force and moment equilibrium can be expressed by lateral slope, $w_{,r}$, in-plane force resultants, N_r and N_θ, and the moment resultants, M_r and M_θ, such that,

$$(rN_r)_{,r} - N_\theta = 0, \quad (rQ_r)_{,r} + (rN_r w_{,r})_{,r} = -p_0 r, \quad (rM_r)_{,r} - M_\theta - rQ_r = 0, \tag{1}$$

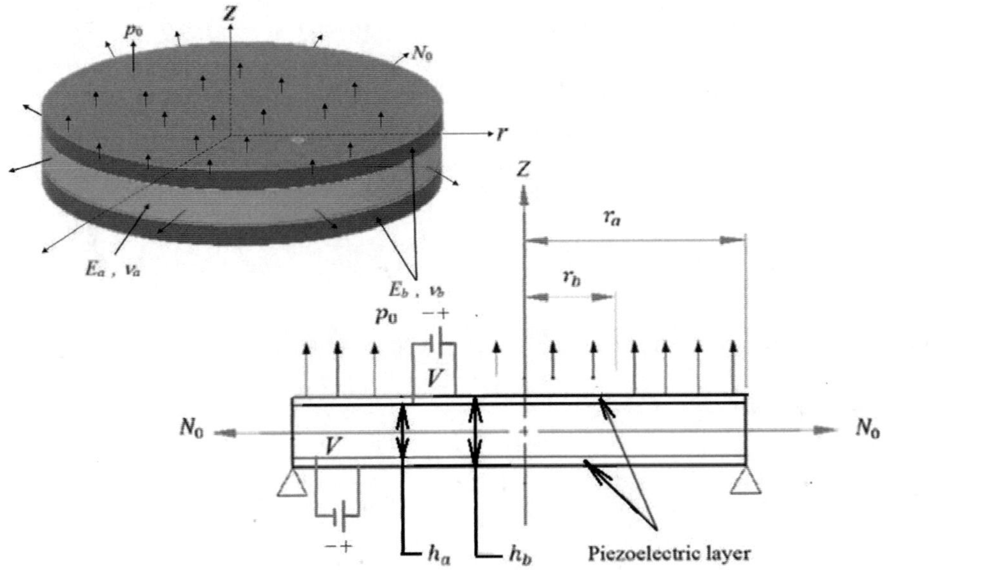

Figure 1. Simply Supported and Sandwiched Piezoelectric Layered Plate under Pretension in Lateral Load

Where Q_r is the transverse shear force resultant, and Ns and Ms are the force resultants and moment resultants, respectively, with the following common definitions based on laminate constitutive laws,

$$\begin{Bmatrix} N_r \\ N_\theta \end{Bmatrix} = \begin{bmatrix} A_l & A_t \\ A_t & A_l \end{bmatrix} \begin{Bmatrix} \varepsilon_r^\circ \\ \varepsilon_\theta^\circ \end{Bmatrix}; \quad \begin{Bmatrix} M_r \\ M_\theta \end{Bmatrix} = \begin{bmatrix} D_l & D_t \\ D_t & D_l \end{bmatrix} \begin{Bmatrix} \kappa_r \\ \kappa_\theta \end{Bmatrix} \tag{2}$$

In the above, As and Ds are the elements of the extensional and bending stiffness matrix, respectively, and both are expressible in terms of moduli and thickness of the composted layers of the plate. The subscript r (θ) represents a radial (circumferential) component, and the subscript l (t) for stiffness components represents a diagonal (off-diagonal) component. In addition, ε_α° and κ_α ($\alpha = r, \theta$) are the mid-plane strain and curvature components defined by the radial displacement, u, and the lateral deflection, w, such that,

$$\varepsilon_r^\circ = u,_r + (w,_r)^2/2, \quad \varepsilon_\theta^\circ = u/r; \quad \kappa_r = -w,_{rr}, \quad \kappa_\theta = -w,_r/r. \tag{3}$$

2.1. Nonlinear Governing Equations

For the problem of lateral loading after in-plane pretension, an incremental form for the equilibrium equations can be derived, following the approach presented previously [14]. For completeness, the previously illustrated equations are cited as follow,

$$\begin{cases} \hat{N}_r,_r + \dfrac{\left(\hat{N}_r - \hat{N}_\theta\right)}{r} = 0; \\ w,_{rrr} + \dfrac{w,_{rr}}{r} - \dfrac{w,_r}{r^2} - \dfrac{1}{D_l}\left(\hat{N}_r - N_r^P + N_0\right)w,_r = \dfrac{p_0 r}{2D_l} \\ r\hat{N}_\theta,_r + \left(\hat{N}_\theta - \hat{N}_r\right) + A_q\left(N_\theta^P - N_r^P\right) + \dfrac{1}{2\bar{A}_l}(w,_r)^2 = 0 \end{cases} . \tag{4}$$

Where a caped superscript and the superscript "P" represent incremental force and piezoelectric force terms, respectively. In additions, $A_q = 1 - \overline{A}_t / \overline{A}_l$ and $\overline{A}_t = -A_t \overline{A}_l / A_l$ are all expressible in terms of extensional stiffness of the layered plate. Employing a non-dimensional scheme similar to that defined by Sheplak and Dugundji [8], these equations can further be simplified and merged to read.

$$\begin{cases} \xi^2\theta'' + \xi\theta' - [\xi^2(k^2 + 12S_r - N_r^P) + 1]\theta = 6P\xi^3 \\ \xi^2 S_r'' + 3\xi S_r' = -\dfrac{a^2 A_q}{D_l}(N_\theta^P - N_r^P) - \dfrac{h^2\theta^2}{2\overline{A}_l D_l' \xi} \end{cases} \tag{5}$$

The related non-dimensional quantities are defined as follow:

$$\xi = \frac{r}{a},\quad W = \frac{w}{h},\quad U = \frac{u}{h},\quad P = \frac{p_0 a^4}{hD_l'},\quad \theta = \frac{dW}{d\xi} = \frac{aw_{,r}}{h},\quad \Psi = \frac{d\theta}{d\xi} = \frac{a^2 w_{,rr}}{h},$$

$$\theta' = \frac{d\theta}{d\xi},\ \ldots \text{etc},\quad k = \sqrt{\frac{N_0 a^2}{D_l}},\quad (S_r, S_\theta) = (\hat{N}_r, \hat{N}_\theta)\frac{a^2}{D_l'},\quad (D_l', D_t') = 12(D_l, D_t).$$

2.2. Linear Problem and Analytical Geometrical Responses

Due to the presence of nonlinear terms including θ^2 and $S_r * \theta$, apparently, trying to solve the above nonlinear equations is formidable. For a preliminary insight, a simplified linear study, i. e., the case of small deflection is considered. The non-dimensional force resultant S_r may be small as well, so that the nonlinear term of product, $S_r * \theta$, can be negligible. In this manner, the first of previous equations for lateral slope can be reduced to the following linear one, i. e.

$$\xi^2\theta'' + \xi\theta' - [\xi^2(k^2 - N_r^P) + 1]\theta = 6P\xi^3, \tag{6}$$

Where ξ is the normalized radial coordinate. Depending upon the sign of the term, $k^2 - N_r^P$, this equation can be reformed as either one of the following two cases :

(i) For $(k^2 - N_r^P = k_m^2 > 0)$, i. e., **mechanical-load dominating** case, the previous equation may clearly be cast in a form as a modified Bessel equation, i. e.,

$$\xi^2\theta'' + \xi\,\theta' - [k_m^2\xi^2 + 1]\,\theta = 6P\xi^3. \tag{7}$$

Boundary conditions of the problem to be satisfied include: (i) Free lateral slope at the center of the plate due to axis-symmetry, i.e, $\xi = 0$: $\theta = 0$; and (ii) Free radial moment resultant at support ends, i. e, $\xi = 1$: $M_r = 0$. Considering the asymptotic behaviors of the modified Bessel functions at infinity, the solution for θ can be derived in terms of modified Bessel functions, i. e.,

$$\theta(\xi) = \frac{6P}{k_m^2}[D_l I_1(k_e\xi) - \xi],\quad D_l = \frac{(D_t + D_l)}{[(D_t - D_l)I_1(k_m) + D_l k_e I_0(k_m)]} \tag{8}$$

Where $I_1(k\xi)$ is the modified Bessel function of the 1st kind. Subsequently, taking another step of integration and differentiation, respectively, the lateral deflection and curvature can be obtained following the recurrence relationships of the modified Bessel functions, i. e.,

$$W(\xi) = \int_1^\xi \theta(\xi)d\xi = \frac{6P}{k_m^2}\left\{\frac{D_l}{k_m}[I_0(k_m\xi) - I_0(k_m)] - \frac{1}{2}(\xi^2 - 1)\right\}, \tag{9}$$

$$\Psi(\xi) = \frac{6P}{k_m^2}\left\{ D_I \left[kI_0(k_m\xi) - \frac{1}{\xi} I_1(k_m\xi) \right] - 1 \right\} \cdot \tag{10}$$

Likewise, the solution of volumetric displacement (V_D) can further be derived, by integrating the lateral deflection over the entire plate based on similar recurrence relations of I_1 and J_1 [18], as well as the definition of (V_D). Thus, for mechanical-load dominating case,

$$V_{D(M)} = \int_0^1 (2\pi\xi) * W(\xi) d\xi = \int_0^1 (2\pi\xi) * \frac{6P}{k_m^2}\left\{ \frac{D_I}{k_m}\left[I_0(k_m\xi) - I_0(k_m) \right] - \frac{1}{2}(\xi^2 - 1) \right\}$$

$$= \frac{12\pi P}{k_m^2}\left\{ \frac{D_I}{k_m}\left[\int_0^1 \xi \bullet I_0(k_m\xi) d\xi - I_0(k_m) \int_0^1 \xi d\xi \right] - \int_0^1 \frac{1}{2}(\xi^3 - \xi) d\xi \right\}$$

Where since $\int_0^1 \xi \bullet I_0(k_m\xi) d\xi = \frac{1}{k_m^2}\int_0^1 (\eta_m) \bullet I_0(\eta_m) d(\eta_m) = \frac{1}{k_m^2}\left[(\eta_m) \bullet I_1(\eta_m) \right]_0^1 = \frac{I_1(k_m)}{k_m}$, we may have

$$V_{D(M)} = \frac{12\pi P}{k_m^2}\left\{ \frac{D_I}{k_m}\left[\frac{I_1(k_m)}{k_m} - \frac{I_0(k_m)}{2} \right] + \frac{1}{8} \right\} \tag{11}$$

(ii) For $N_r^p - k^2 = k_c^2$ **> 0**, or **piezoelectric force dominating** situation, on the other hand, the problem reduces to a standard Bessel equation, i. e.,

$$\eta^2\theta'' + \eta\theta' + [\eta^2 - 1]\theta = 6P\eta^3 / k_c^3 \ , \tag{12}$$

Where $k_c^2 = N_r^p - k^2$; and $\eta = k_c\xi$. The solution for lateral slope can be similarly formulated to read,

$$\theta(\xi) = \frac{6P}{k_c^2}\left[\xi - D_J J_1(k_c\xi) \right], \quad D_J = \frac{(D_t + D_l)}{\left[(D_t - D_l) J_1(k_c) + D_l k_c J_0(k_c) \right]} \tag{13}$$

The corresponding solutions for lateral deflection and curvature can subsequently be derived, i. e.,

$$W(\xi) = \int_1^\xi \theta(\xi) d\xi = \frac{6P}{k_c^2}\left\{ \frac{1}{2}(\xi^2 - 1) + \frac{D_J}{k_c}\left[J_0(k_c\xi) - J_0(k_c) \right] \right\}, \quad \Psi(\xi) = \frac{6P}{k_c^2}\left\{ 1 - D_J \left[k_c J_0(k_c\xi) - \frac{1}{\xi} J_1(k_c\xi) \right] \right\} \tag{14}$$

In the above, $J_0()$ and $J_1()$ are the Bessel functions of the 1st kind with order 0 and 1, respectively. Thus, the volumetric displacement can be similarly derived to take the form as,

$$V_{D(E)} = \int_0^1 (2\pi\xi) * W_{bE}(\xi) d\xi = \int_0^1 (2\pi\xi) * \frac{6P}{k_c^2}\left\{ \frac{1}{2}(\xi^2 - 1) + \frac{D_J}{k_c}\left[J_0(k_c\xi) - J_0(k_c) \right] \right\}$$

$$= \frac{12\pi P}{k_c^2}\left\{ \int_0^1 \frac{1}{2}(\xi^3 - \xi) d\xi + \frac{D_J}{k_c}\left[\int_0^1 \xi \bullet J_0(k_c\xi) d\xi - J_0(k_c) \int_0^1 \xi d\xi \right] \right\}$$

In the last equation, since $\int_0^1 \xi \bullet J_0(k_c\xi) d\xi = \frac{1}{k_c^2}\int_0^1 (\eta_c) \bullet J_0(\eta_c) d(\eta_c) = \frac{1}{k_c^2}\left[(\eta_c) \bullet J_1(\eta_c) \right]_0^1 = \frac{J_1(k_c)}{k_c}$, a few more steps of manipulations can give rise to the expression of volumetric displacement for this case that,

$$V_{D(E)} = \frac{12\pi P}{k_c^2}\left\{\frac{D_J}{k_c}\left[\frac{J_1(k_c)}{k_c} - \frac{J_0(k_c)}{2}\right] - \frac{1}{8}\right\} \quad (15)$$

3. Numerical Remarks

To implement the developed approach, a simply-supported and symmetrically layered plate composed of silicon (Si) substrate layer sandwiched between two piezoelectric outer layers of either poly-silicon (Poly-Si) or silicon-dioxide (SiO_2) were considered during the course of this study. The material properties (Young's moduli and Poisson's ratios) for the layers are listed in Table 1 [18]. The case with poly-silicon (Poly-Si) outer layers simulates a nearly monolithic plate (NMP), while that with the SiO_2 outer layers represents a typically layered plate. An outer radius of 500 (μm) for the plate is considered. The thickness of the substrate layer (h_a) is taken to be 10 (μm) and that of the piezoelectric layer (h_p) is set to be 0.25, 0.5, 1.0 or 2.5 (μm), respectively; to have a total thickness ratio of h_t/h_a = 1.05, 1.1, 1.2, and 1.5 between the entire sandwiched plate and that of the middle substrate layer. In this way, hence, only comparatively thin piezoelectric layers were considered, to avoid the risk that the polarization orientation may be affected as the electric voltage is applied. The pretension parameter with k_a=1, 10, and 50 was implemented, similar to Sheplak and Dugundji [8]. Meanwhile, piezoelectric constant of the symmetric patches is taken to be d_{31}=50 pm/V [16] and the applied voltage is considered to be V=1, 5 and 10 (Volt) [11].

Table 1. Material Properties for the Layers of Piezoelectric Plates

Layer Material	Young's Modulus (Gpa)	Poisson's Ratio
PolySi	170	0.22
Si	165	0.27
SiO_2	75	0.17

For demonstration, however, the afore-mentioned nearly monolithic plate (NMP) was implemented first. In the case of a relatively low applied voltage, V=1 (volt), simulating a nearly pure mechanical loading condition, the solutions of lateral slope and deflection for various initial tensions versus central deflection (W_0) are presented in Figure 2 and 3, respectively. The results following CPT (Classical Plate Theory) obtained by taking the layered plate as a single layer with Young's modulus, E=170 (GPa) and the same Poisson's ratio as the silicon layer without initial tension is included for comparison. Apparently, the present solutions for the case of very low initial tension (k=1) correlate well with the CPT results and thus the present approach is validated.

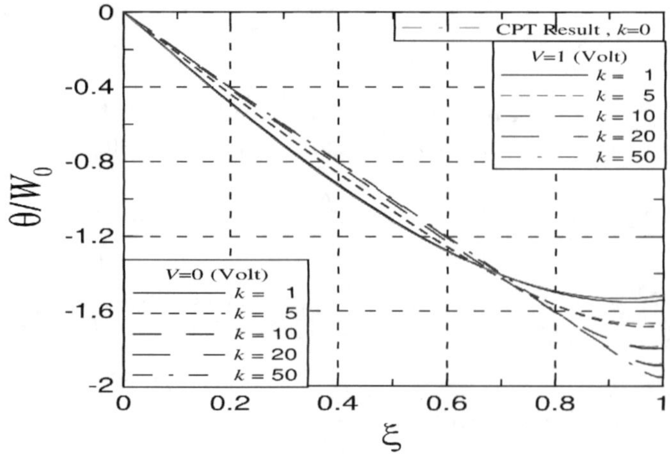

Figure 2. Comparison of Lateral Slope for Nearly Monolithic Plate with CPT Solutions

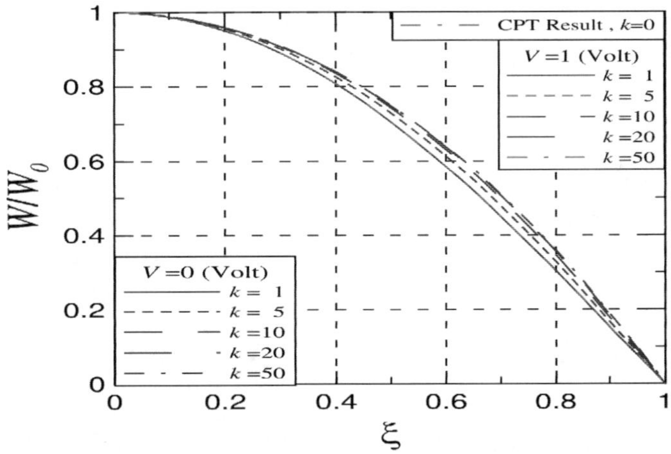

Figure 3. Comparison of Lateral Deflection for Nearly Monolithic Plate with CPT Solutions

Volumetric displacement (V_D) for the same plate under various applied voltages are evaluated for thickness ratios of h_t/h_a = 1.05, 1.1, 1.2 and 1.5 between the total sandwiched plate and the substrate layer. The results are plotted against the magnitude of pretension (k=k_a), as shown in Figure 4 to 7, respectively. Due to the inherent normalization scheme, it should be noted that, the present solutions for (V_D) are also normalized quantities relative to exactly the dimension of structural volume of the considered sandwiched plates. The obtained solutions show that V_D always have the highest magnitude when no pretension (k=k_a=0) is involved. The thinner the piezoelectric layers, the greater the piezoelectric effect is observed. This implies that not only a larger V_D will be obtained at the left end for the shown curves; it also shows a very sensitive change for V_D as the applied voltage (V) is varied. Under a low pretension and applied voltage, on the other hand, varying the thickness of the piezoelectric layers seems to have an apparent influence on the results of (V_D). In a low initial tension, conceivably, raising the applied voltage may rapidly increase the volumetric displacement, especially for a thin piezoelectric layer cases. Yet, as the pretension grows, volumetric displacement (V_D) may decrease apparently. The decreasing of (V_D) relative to pretension further shows that pretension will dominate over both the piezoelectric effect and the geometrical size (layer thickness) effect of the sandwiched plate. The volumetric displacement will eventually vanish, as the pretension keeps increasing to become relatively large.

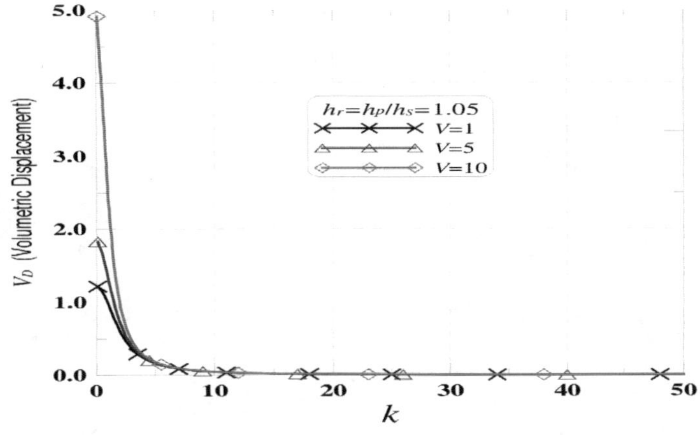

Figure 4. Volumetric Displacement of [Poly-Si/Si/Poly-Si] Plate, h_r=1.05

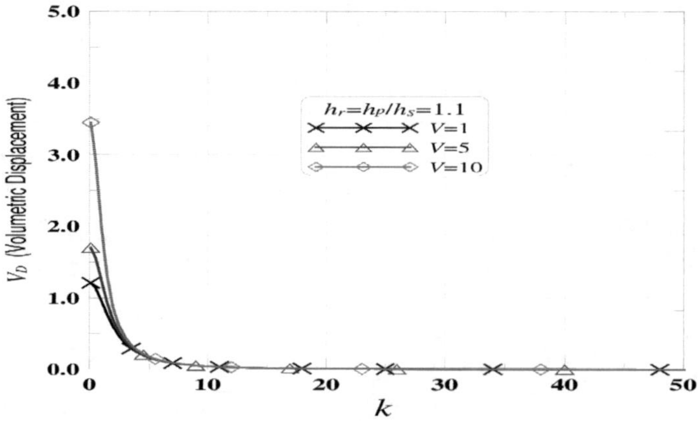

Figure 5. Volumetric Displacement of [Poly-Si/Si/Poly-Si] Plate, h_r=1.1

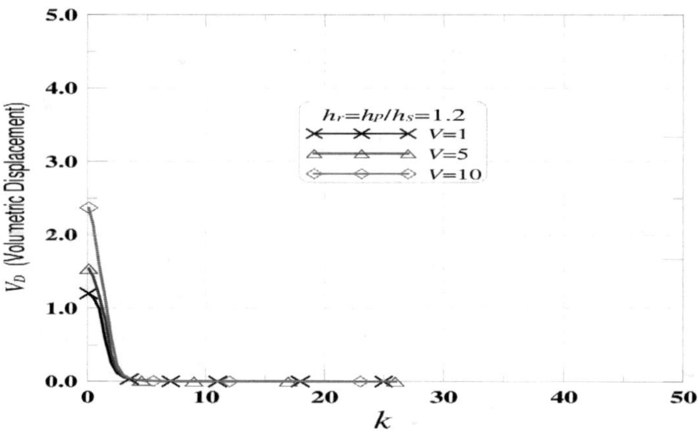

Figure 6. Volumetric Displacement of [Poly-Si/Si/Poly-Si] Plate, h_r=1.2

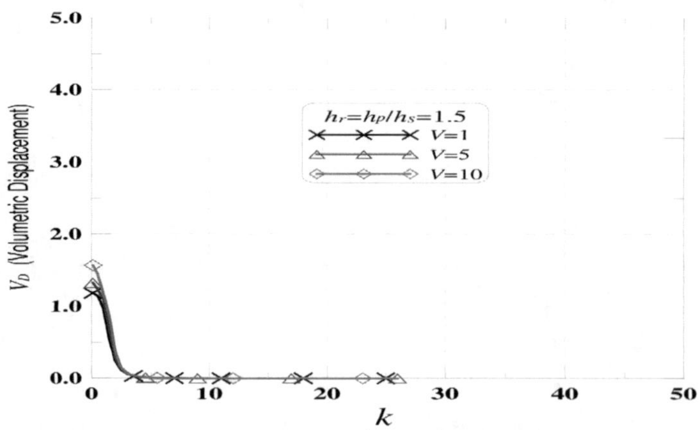

Figure 7. Volumetric Displacement of [Poly-Si/Si/Poly-Si] Plate, h_r=1.5

4. Conclusion

The problem volumetric displacement in large deflection of a pre-stressed piezoelectric and sandwiched plate due to lateral load is studied. For a nearly monolithic plate under a very low applied voltage, the obtained solutions agree with available solutions for the case of pure mechanical loading and thus the presented approach is checked. The solutions for typically layered plate show that, piezoelectric effect and geometric size effect of the layered plate may come into play when it is in a low pretension. Upon reaching a moderate magnitude for the pretension, the pretension effect will dominate over the effects of the applied voltage and relative thickness ratios of the layers.

5. References

[1] S.Trolier-McKinstry and P. Muralt, 2004, "Thin Film Piezo-electrics for MEMS", J. Electro-ceramics, Vol. 12, pp. 7–17.

[2] T.C. Yih, C. Wei, B. Hammad, 2005, "Modeling and Characterization of a Nanoliter Drug-delivery MEMS Micropump with Circular Bossed Membrane", Nanomedicine: Nanotechnology, Biology, and Medicine, Vol. 1, pp. 164–175.

[3] S.Chandika and R. Asokan, 2011, "Design and Analysis of Piezo-actuated Micro-pump for Fuel Delivery in Automobiles", Journal of Scientific & Industrial Research, Vol. 70, pp. 448-454.

[4] L.H. Fang, S. I. S. Hassan, R. A. Rahim, M. Isa, and B. B. Ismail, 2017, "Exploring Piezoelectric for Sound Wave as Energy Harvester", Energy Procedia, Vol. 105, pp. 459–466.

[5] A.M. Elhalwagy, M. Y. M. Ghoneem, and M. Elhadidi, 2017, "Feasibility Study for Using Piezoelectric Energy Harvesting Floor in Buildings' Interior Spaces", Energy Procedia, Vol. 115, pp. 114 – 126.

[6] A.Ettouhami, N. Zahid, and M. Elbelkacemi, 2004, "A Novel Capacitive Pressure Sensor Structure with High Sensitivity and Quasi-linear Response", C. R. Mecanique, Vol. 332, pp. 141-146.

[7] J.A.Voorthuyzen, and P. Bergveld, 1984, "The Influence of Tensile Forces on the Deflection Diaphragms in Pressure Sensors", Sensors Actuators A, Vol. 6, pp. 201-213.

[8] M.Sheplak, and J. Dugundji, 1998, "Large Deflections of Clamped Circular Plates under Initial Tension and Transitions to Membrane Behavior", Trans. ASME J. Appl. Mech., Vol. 65, pp.107-115.

[9] Y.H. Su, K. S. Chen, D. C. Roberts, and S. M. Spearing, 2001, "Large Deflection Analysis of a Pre-stressed Annular Plate with a Rigid Boss under Axisymmetric Loading", J. Micromech. Microeng, Vol. 11, pp. 645-653.

[10] G.-S. Chen, 2002, "Design of Bulk-Micro-machined Mechanical Sensors and Investigation of LIGA- like Technology", Ph. D. Dissertation, Dept. of Mechanical Engineering, NCKU, Tainan, Taiwan, ROC.

[11] G.Wang, B. V. Sankar, L. N. Cattafesta, and M. Sheplak, 2002, "Analysis of a Composite Piezoelectric Circular Plate with Initial Stress for MEMS", Proc. of 2002 ASME Int. Mech. Engng. Congress & Exposition (Paper No. 34337), pp. 1 – 7.

[12] S.T. Cho, K. Najafi, and K. D. Wise, 1992, "Internal Stress Compensation and Scaling in Ultra-sensitive Silicon Pressure Sensors", IEEE Transaction of Electron Devices, Vol. 39, pp. 836-842.

[13] D.-Y. Qiao, W.-Z. Yuan, Y.-T. Yu, Q. Liang, Z.-B. Ma, and X.-Y. Li, 2008, "The Residual Stress-induced Buckling of Annular Thin Plates and Its Application in Residual Stress Measurement of Thin Films", Sensors and Actuators A, Vol. 143, pp. 409–414.

[14] C.-F. Chen and J.-H. Chen, 2010, "Analytical Geometrical Responses in Large Deflection of Simply Supported Piezoelectric Layered Plate under Initial Tension", Journal of the Chinese Institute of Engineers, Vol. 33, No. 7, pp. 1005-1013.

[15] S. A. N. Prasad, B. V. Sankar, L. N. Cattafesta, S. Horowitz, Q. Gallas, and M. Sheplak "Two-Port Electroacoustic Model of an Axisymmetric Piezoelectric Composite Plate", AIAA-2002-1365.

[16] C.H. J. Fox, X. Chen, and S. McWilliam, 2007, "Analysis of the Deflection of a Circular Plate with an Annular Piezoelectric Actuator", Sensors and Actuators A, Vol. 133, pp. 180–194.

[17] M.Abramowitz and I. A. Stegun., Handbook of Mathematical Functions. (http://jonsson.eu/resources /hmf/?page=index)

[18] Zhou, M.-X., Huang, Q.-A., Qin, M., 2005, "Modeling, Design, and Fabrication of a Triple-Layered Capacitive Pressure Sensor", Sensors Actuators A, Vol. 117, No. 1, pp.

IOP Conf. Series: Materials Science and Engineering **735** (2020) 012031 doi:10.1088/1757-899X/735/1/012031

Design and Implementation of Fixed Length Profile Cutting System Based on Mitsubishi Q Series and CC-Link Field Bus Control

Liqing Wang, Yimin Cheng, Xujie Hou and Chang Wang

College of Mechanical and Electrical Engineering of Weifang Vocational College, Shandong Weifang 262737, China

Abstract. In order to achieve the fixed length cutting of profiles, this paper designs a fixed length cutting system using Mitsubishi Q series PLC as the main controller, Fx series PLC as the slave controller and CC-Link as the field bus. Based on the working principle of the equipment, the paper analyzes the operation process of the cutting system. Based on PLC controller, stepping motor and servo motor, the hardware design of the system is carried out. A mathematical model for calculating the initial remaining length was established and the software was designed. The use indicates that the work intensity of the operator is greatly reduced, the yield is improved, the production cost is reduced, and the application prospect is good.

1. Introduction

Profile fixed length cutting system is an automatic control system that can constantly cut the profiles into set lengths per the custom requirements of the enterprises. A few years ago, profiles cutting required workers using sawing machines or using mechanically fixed flying shears to cut the profiles at a fixed state. These processes make the production efficiency low and require the device to be started and stopped frequently, which affect the stability of the device itself and the power supply. This paper designs a fixed length cutting system based on the profile cutting requirements from Shandong Luxing Steel Pipe Ltd, by using Mitsubishi Q series PLC as the main controller, Fx series PLC as the slave controller and CC-Link as the field bus. This cutting system is for the purpose of improving the yield and reducing the production cost.

2. Structure Design of the Fixed Length Cutting System

The structure of the fixed length cutting system designed in this project is shown in Figure 1.

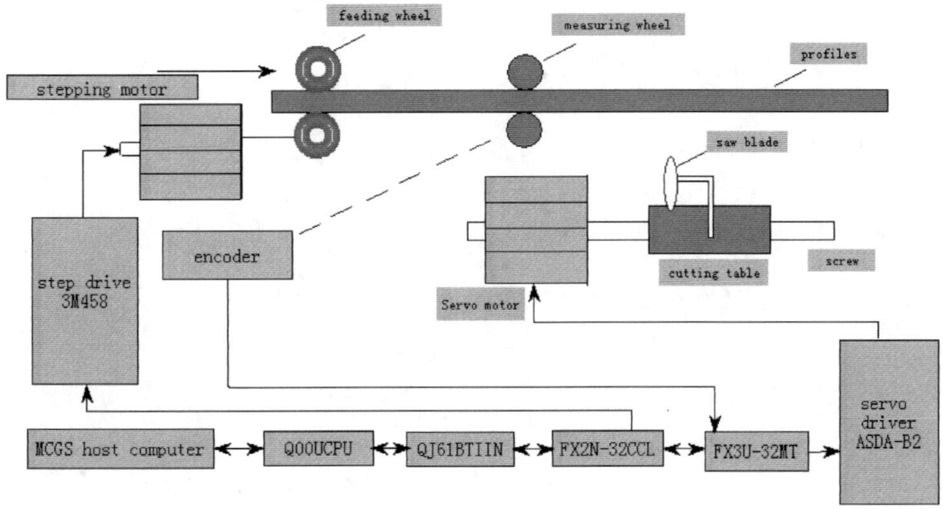

Figure 1. The Structure of the Fixed Length Cutting System

The mechanical structure of the system is mainly composed of servo motor, stepping motor, cutting saw, cutting cylinder and limit switch etc. The electrical control part is mainly composed of Mitsubishi Q series PLC, Mitsubishi Fx series PLC, MCGS touch screen, servo driver, stepping driver and Hengte encoder etc. [1]

According to the working process, the system consists of feeding mechanism, measuring mechanism and cutting mechanism. The feeding mechanism is driven by the stepping motor to push the profile into the cutting mechanism steadily and continuously at a set speed. Them easuring mechanism is powered by the cut profile. When the profile moves forward, it drives the measuring wheel of the encoder to rotate and triggers the encoder to generate pulse, and sends it to the Mitsubishi Fx 3U slave controller in real time. With combining the measurement of the distance traveled by the profile in a circle of rotation, the displacement of the profile is calculated. The cutting mechanism (cutting table) is controlled by the servo motor to make a back-and-forth movement, the hydraulic cylinder coordinates with the rotation and expansion of the saw blade, adjust the speed of the saw blade according to the cutting requirements to achieve the pipe cutting.

3. System Working Process Analysis

3.1. System Working Process

The system is controlled by the MCGS touch screen of the host computer. When the equipment is powered on, parameters such as profile feeding speed, profile cutting length, acceleration of cutting table and profile thickness are set on the start-upinterface. The profile starts moving after the equipment is started up, FX3U PLC receives the pulse transmitted by the encoder, calculates the displacement of the profile, compares it with the initial remaining length, to determines the start time of the cutting table. When all the requirements are met, the cutting table starts accelerating according to the set acceleration. When the speed is the same as the profile's speed, the profile length at the front meets the cutting requirements. At this time, the motor of the saw starts and the saw starts cutting profile. After the cutting is finished, the cutting table returns to the datum point and waits for the next instruction. A completed cutting process is finished. [2]

3.2. Working Process of Cutting Table and Saw

From the working process analysis above, when the profile displacement does not reach the initial remaining length, the cutting table waits at the datum point [3]. When the initial remaining length is

IOP Conf. Series: Materials Science and Engineering **735** (2020) 012031 doi:10.1088/1757-899X/735/1/012031

reached, the cutting table completes a cutting process through the process of "waiting--approaching--synchronizing & executing the cutting--forward deceleration to stop--reverse acceleration--reverse deceleration--return to the datum point to wait for the next cutting" [2]. The change of the motion speed pulse of the cutting table is shown in Figure 2.

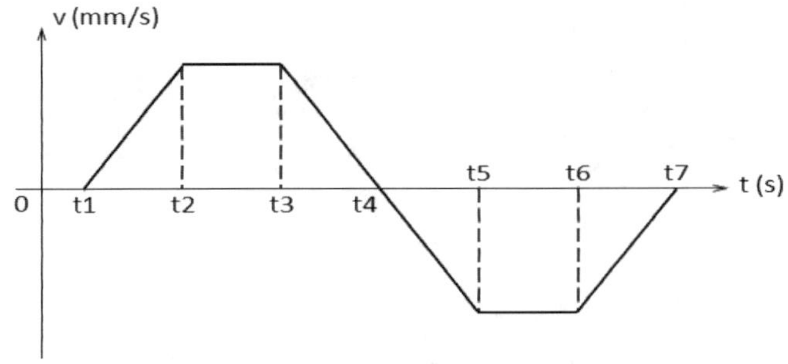

Figure 2. Chart of the Pulse Change

Waiting section 0-t1: start the control system through the touch screen, the profile is pushed forward continuously, and the cutting table is in the waiting section at the starting position.

Forward pursuit section t1-t2: when the profile's displacement is equal to the initial remaining length, Mitsubishi Fx3U slave controller sends tracking signal to the servo driver through the high-speed output outlet. The cutting table accelerates from 0 and reaches the set speed according to the set acceleration time; the set speed is equal to the profile's speed.

Cutting section t2-t3: the Mitsubishi Fx 3U slave controller sends pulses to the servo driver at a fixed frequency, the cutting table moves forward at a constant speed, the hydraulic cylinder of the cutting equipment extends, the sawing blade moves down and starts cutting until completed.

Positive deceleration section t3-t4: after receiving the cutting completed signal, the hydraulic cylinder retracts and the saw blade is reset. Fx series PLC sets the deceleration time to reduce the sending of pulse until the sending pulse is reduced to zero.

Return to the datum point section t4-t5: the Mitsubishi Fx 3U slave controller sends pulses to the servo driver at a fixed frequency to control the cutting table to move toward the datum point. The cutting table accelerates and then moves with a constant speed and then decelerates until the speed reaches zero and the cutting table returns to the datum point at the same time.

4. The design of the System Hardware

4.1. Selection and Design of the Controllers

The system is designed to be able to complete the speed and length measurement of the profile, establish tracking motion curve, precisely control the operation of the servo motor, allow human-machine interaction, and meet the extensibility of the equipment's function at later stages. After comprehensive considerations, Mitsubishi series PLC is selected as the main controller, Mitsubishi Fx 3U series PLC is selected as the controller [4], the connection and the communication between the two controllers is achieved through the CC-Link communication method. Cc-link is an open field bus developed by Mitsubishi Corporation [5], which is based on Mitsubishi PLC and other automatic control systems. It has the advantages of fast communication speed and convenient connection. [6]The input/output terminals of the controller are allocated as shown in Table 1.

Table 1. The Input/Output Terminals of the controller

F_X3U Input		
Terminal	Label	Function
X0	Encoder	Encoder input pulse port
X1	Encoder	Encoder input pulse port
X2	SB1	Start Button
X3	SB2	Stop Button
X4	SQ1	Saw blade limit switch
X5	SQ2	Datum point limit switch
X6	FR1	Saw blade motor thermal protection
F_X3U Output		
Terminal	Label	Function
Y0	CW	Servo pulse input
Y1	CCW	Servo Direction
Y2	PLS-	Step-by-step pulse
Y3	DIR-	Step-by-step Direction
Y4	KM1	Saw motor control coil
Y5	Electromagnetic Valve	Control electromagnetic valve

4.2. Selection and Design of Servo System

The servo system is used to move the cutting table. The servo driver receives the pulse [7] sent by the Mitsubishi Fx3U slave controller's high-speed pulse port to control the operation of the servo motor [8]. This system selects Delta servo driver and Delta ASDA-B2 servo motor that adapt the servo driver [9].

4.3. Selection and Design of Stepping Motor and Encoder

The stepping motor drives the feeding wheel to rotate, and the profile is returned to the cutting mechanism at a constant speed. Meanwhile [10], the friction force generated by the forward moving profile drives the speed measuring wheel to rotate, triggering the encoder to send pulse to Fx3U slave controller to calculate the displacement of the profile [11]. In this system, HETKJ Hengte HTK4808 incremental encoder is selected, the encoder sends out 1000 pulses in one complete speed wheel rotation.

The electrical schematic diagram of the system is shown in Figure 3.

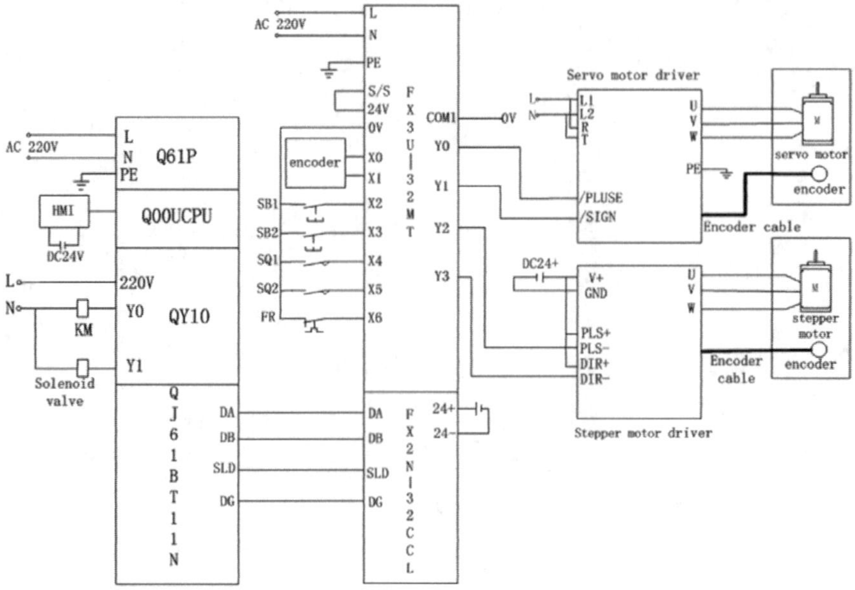

Figure 3. The Electrical Schematic Diagram of the System

4.4. Host Computer Selection and Interface Design

In order to understand the real-time data of the profile cutting system directly and to facilitate the setting of relevant parameters such as profile cutting length and acceleration of the cutting table, MCGS touch screen is selected for the host computer [12]. The touch screen is connected to PLC serial port of Mitsubishi Q series main station through RS232 module to complete data exchange [13]. Host computer interface design shows profile speed, profile cutting length, chasing acceleration, profile thickness and other settings; it also has start and stop control buttons. The interface is shown in Figure 4.

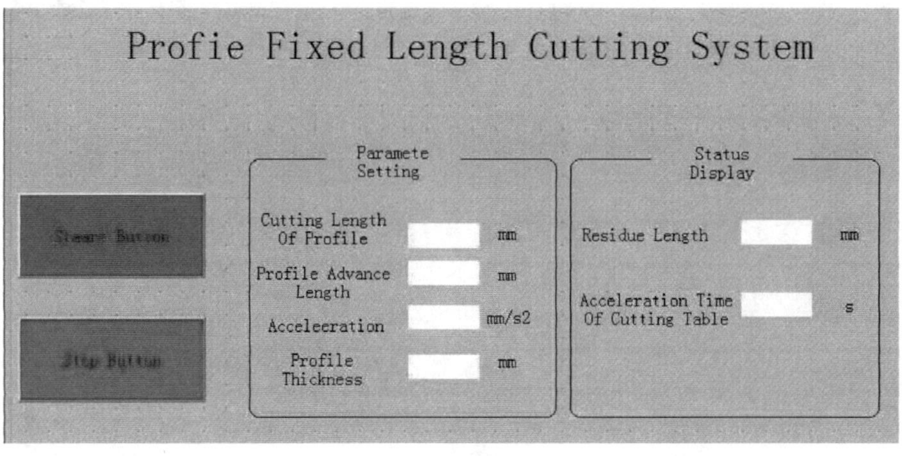

Figure 4. Interaction Simulation Interface

5. Software Design

5.1. Calculation of Initial Remaining Length

To complete the cutting process, the key point is to determine when to start the cutting table [14]. The design takes the forward displacement of the front end of the profile from the Datum point as the reference, the cutting table began to accelerate to chase when the profile displacement reached the initial remaining length. Once the cutting table's speed and the profile's speed are equal, the profile's length at the front of the saw blade is S0, which is the length set on the interface.

Set C as the initial remaining length and V0 as the speed of the profile. When C meets the starting conditions, the cutting table accelerates from zero with acceleration a, and the profile continues to be pushed forward with V0. The two speeds are equal after time t, the mathematical model is shown in Figure 5.

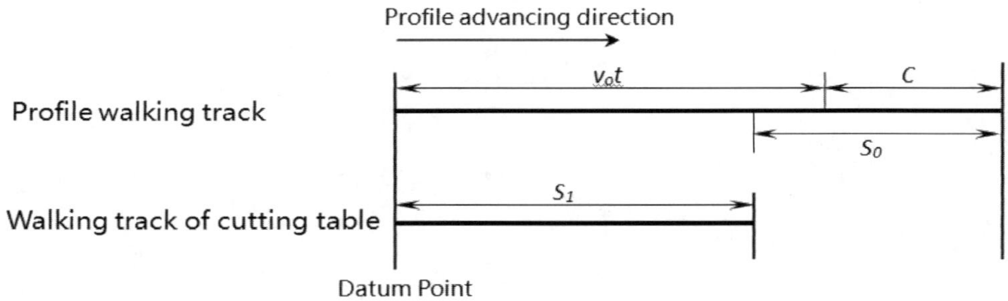

Figure 5. Mathematical Mode for Initial Remaining Length Calculation

At this point

$$t = \frac{V_0}{a}$$

The distance traveled by the cutting table

$$S1 = \frac{V_0^2}{2a}$$

According to the above formula, the Initial remaining length can be obtained.

$$C = S_0 + S_1 - V_0 t = S_0 - \frac{V0^2}{2a} \quad [15]$$

5.2. System Flowchart

According to the control requirements, the work flow of cutting table and saw blade is shown in Figure 6.

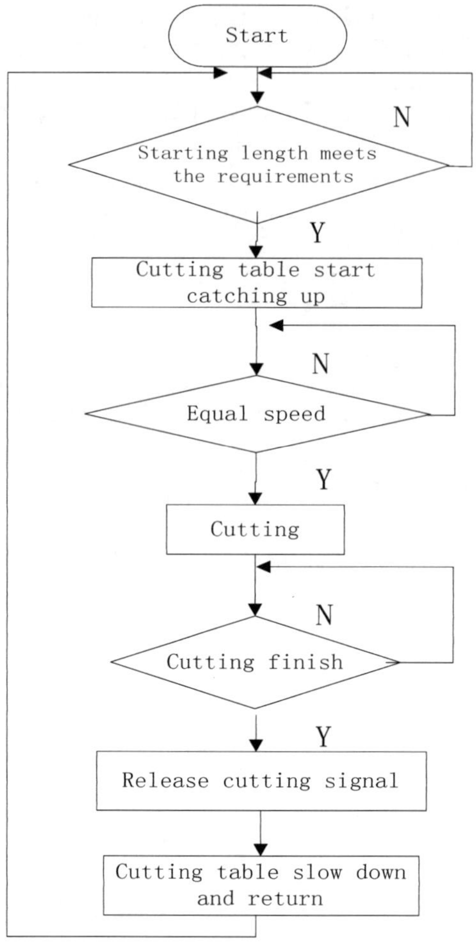

Figure 6. The Work Flow of Cutting Table and Saw Blade

6. Conclusion

The system adopts Mitsubishi Q series PLC as the main station and Mitsubishi Fx3U series PLC as the slave station, the stations are connected by Mitsubishi cc-link field bus. An automatic fixed-length profile cutting production line is established with stepping motor feeding and servo motor driving cutting table. The results show that the system is intelligent and easy to maintain, which greatly reduces the work intensity of the operators, improves the yield and reduces the production cost.

7. References

[1] Linkun Li. Design of press mounting system of automobile coupling based on PLC [D]. Shanghai: Donghua University, 2014.
[2] Huijie Lei, Yantao Chen, Yanwei Zhang. Application of PLC in servo control system of steel pipe production line [J]. Manufacturing automation, 2015, 4:120-133.
[3] Hao Wang, Qin Wang. Realization of servo control of two axis system based on S7-200 smart PLC [J]. Electrical technology, 2016, 17 (5): 85-90.
[4] Haitao Wang. Application of PLC network control system in automobile axle welding production line [D]. Chengdu, Sichuan province: Xihua University, 2011.
[5] Guochen Ma. Application of Mitsubishi PLC technology in modern industrial electrical automatic control system [J]. Internal combustion engine and accessories, 2017, (17): 134-135.
[6] Tao Wang, Ruibin Kuang. Design of automatic filling production line control system based on cc-link network [J]. Electrical automation, 2017, 6: 94-95, 98.

[7] Bin Yao. Prospective control algorithm and device for glass bevel grinding speed [D]. Hangzhou city, Zhejiang, Zhejiang University of Science and Technology, 2017.
[8] Zhichao Chen, Wei Qi, Wenju Ning. Design of motor stator drying control system based on PLC [J]. Automation and instrumentation, 2018, (9).
[9] Haibin Song. Research on human skull capture technology based on monocular machine vision [D]. Luoyang, Henan, Henan University of Science and Technology, 2013.
[10] Hongye Han. Development of synchronous following cutting machine for seam spacing surface material film [D]. Dalian, Liaoning: Dalian University of Technology, 2013.
[11] Liehuai Wang, Qiaoling Xu. Research on the algorithm of automatic production line kit sorting based on PLC control [J]. Journal of Wanxi University, 2015, 31(2).
[12] Yang Xiang, Yang Fu. Control system design of servo motor testing platform based on PLC [J]. Science and information, 2017, (35): 46, 49.
[13] Yu Fu. Design of reconfigurable control system for wrench production line based on cc-link [D]. Zhenjiang, Jiangsu, Jiangsu University of Science and Technology, 2015.
[14] Jiewen Lin, Yifeng Wu. Design of servo motor motion control system based on PLC [J]. Electromechanical Technology, 2015, (5):20-23, 26.
[15] Changqing Liu, Minghai Li, Wei Xi, etc. Stepwise start-stop control algorithm for multiple cooling devices based on PLC [J]. Manufacturing automation, 2015, 4:109-110,122.

Study on the Corrosion Inhibition of Benzotriazole in the Blast Furnace Gas Condensate Water

Yuhua Gao[1,2], Haihua Li[1,2], Lihui Zhang[1,2] and Zhenfa Liu[1,2,*]

[1] Institute of Energy Resources, Hebei Academy of Sciences, Shijiazhuang 050081, China
[2] Hebei Engineering Research Center for Water Saving in Industry, Shijiazhuang 050081, China
Email: lzf63@sohu.com

Abstract. Preparation of the corrosion inhibitor was carried out by adding the benzotriazole in sodium hydroxide solution in view of the corrosion problem of blast furnace gas condensate in iron and steel enterprises. Real-time corrosion rate of blast furnace gas condensate water with and without the corrosion inhibitor was monitored throug corrosion rate tester. The average real time corrosion rates of the blast furnace gas condensed water contacting the various dosage of the corrosion inhibitor were lower than the average real time corrosion rates without adding the corrosion inhibitor. Corrosion inhibition behavior of the corrosion inhibitor on carbon steel in the blast furnace gas condensed water using weight loss measurement method.

1. Introduction

With the continuous development of China's iron and steel enterprises, efforts to reduce the cost of blast furnace smelting and enhance the intensity of blast furnace smelting have become one of the goals pursued by various iron and steel enterprises. As a widely used technology to enhance the blast furnace smelting intensity and reduce the cost of blast furnace smelting, such as oxygen-enriched pulverized coal injection technology and sinter spraying $CaCl_2$ solution technology, which all can led to the content of acid gases (such as SO_2, SO_3, H_2S, HCl, etc.) in blast furnace gas higher[1-3]. Especially after the dry gas dedusting technology is widely used in iron and steel enterprises, almost all of the above acid gases are retained in the gas and cannot be discharged. As a result, the pH value of the condensate water of blast furnace gas decreases obviously, and the acid corrosion of blast furnace gas gradually appears, which has become the main factor affecting the safe operation of blast furnace gas pipelines and their auxiliary equipment [4]. Therefore, how to effectively prevent and control the acid corrosion of blast furnace gas is an important issue facing iron and steel enterprises.

Wang Xueming et al. [5] determined the idea of adjusting pH by spraying pesticide through hanging piece simulation experiment in view of the rapid corrosion of blast furnace gas pipeline network of Shouqin Company in 2011. Ammonia was determined as the preferred chemical through comparative test and analysis of NaOH, Na_2CO_3 and ammonia, which dosage was 10000 mg/L. Ammonia could be used as a reagent to adapt to the great change of condensate pH in the pipeline network. When the reagent is excessive, it will not have a negative impact on the pipeline, and has a protective effect on the pipeline. The ammonia spraying method needs less dosage, and all the condensate water is used for steel-making slagging treatment, which will not cause environmental pollution.

Chen Chao et al. [6] summarized and discussed the preventive measures for corrosion of gas pipelines, such as spraying NaOH solution on the rear pipeline of TRT in Shougang Group Company to control acid corrosion of gas condensed water. Its advantage is that it could better control the acidity of blast furnace gas. But it has many disadvantages, such as taking up much room, severe corrosion when alkali is prepared and transported, the larger maintenance workload, and so on. The addition of $CaCl_2$ reinforcer in sinter was controlled to control the acidity of blast furnace gas in Jigang Company. Its advantage is that there is no investment in equipment and facilities, but its disadvantage is that the fluctuation of $CaCl_2$ additives will affect the strength control of sinter feeding, which may affect the blast furnace operation states and is not conducive to the stable operation of production. The alkali adding device is adopted in the spray equipment of dry gas dedusting blast furnace gas in the steel plant of Shougang and Henggang, which turns the original neutral water mist into the alkaline water mist. The purpose is to neutralize the sour gas in the gas, and reduce the acidity of the condensed water in the pipeline. That is to alleviate the corrosion phenomenon of the gas pipeline.

We preparated the corrosion inhibitor by adding the benzotriazole in sodium hydroxide solution to effectively alleviate the corrosion problem of blast furnace gas in view of the corrosion problem of blast furnace gas condensate in iron and steel enterprises. If the corrosion inhibitor was adopted in the blast furnace gas, which will prolong the service life of gas pipelines and guarantee the safe and stable operation of blast furnace gas system. That will promote the development of iron and steel enterprises in China, and have significant economic and social benefits.

2. Experimental

2.1. Preparation of the Corrosion Inhibitor
Preparation of the corrosion inhibitor was carried out by adding the benzotriazole in sodium hydroxide solution (mass content was 20%). Then stir with a magnetic stirrer until benzotriazole dissolved completely. And then dilute the corrosion inhibitor with deionized water to about 10 mg/mL. Keep the prepared corrosion inhibitor in the refrigerator.

2.2. Real-time Corrosion Rate Monitoring
A QYFS-I type corrosion rate tester equipped with two measuring electrodes and the associated testing software was used to monitor the real-time corrosion of A3 carbon steel in the blast furnace gas condensed water. The following testing steps were adopted for studying the corrosion inhibition performance: the corrosion inhibitor was added to a beaker containing 200mL of the blast furnace gas condensed water. The two electrodes used for corrosion measurements were rinsed with absolute alcohol and dried, before their installations into the test system. After the corrosion detector was placed into the beaker containing the blast furnace gas condensed water with or without corrosion inhibitor, corrosion monitoring commenced. The average corrosion rate was recorded at 11 min intervals for a total testing period of 72 h. Composition and properties of the blast furnace gas condensed water are given in table 1.

Table 1. Composition and Properties of the Blast Furnace Gas Condensed Water

pH	Cl^- (mg/L)	Ca^{2+} (mg/L)	Mg^{2+} (mg/L)	OH^- (mg/L)	Conductivity (μs/cm)
5.68	72.3	5.0	5.0	608.1	1710

2.3. Weight Loss Measurement of the Rotating Hanging Coupons Test
Corrosion inhibition behavior of the corrosion inhibitor on carbon steel in the blast furnace gas condensed water using weight loss measurement were assessed in accordance with Chinese Standard GB/T 18175-2000. A XYZK-A type rotating hanging coupon test device was used for measuring weight loss. The following procedures were adopted: The corrosion inhibitor was added in the 1800 mL of the blast furnace gas condensed water. Then the beaker containing experimental water was

placed into the water bath of the rotating hanging coupon test device and the experimental temperature set to 45±1°C. Carbon steel samples (50mm×25mm×2mm) were then suspended in the blast furnace gas condensed water with and without corrosion inhibitor. The rotating speed of the rotating axis was 75 rpm. The experimental period lasted 72 h. The corrosion rate of the steel sample was calculated with equation (1).

$$X = \frac{8760 \times 10 \times (W_0 - W)}{A \times D \times T} = \frac{87600 \times (W_0 - W)}{28 \times 7.85 \times 72} \quad (1)$$

Where X is the average corrosion rate of carbon steel in blast furnace gas condensed water (mm/y); the 8760 and 10 are conversion factors for number of hours in a year and dimensional (cm to mm) conversion respectively; W_0 and W are the weights (g) of the test samples prior to and after testing; A is the steel sample surface area (28 cm^2); D is the density of the steel samples (7.85 g/cm^3); T is test time (72 h).

Corrosion inhibition efficiencie of the corrosion inhibitor was calculated from equation (2).

$$\eta = \frac{(X_0 - X_1)}{X_0} \times 100\% \quad (2)$$

Where η is the corrosion inhibition efficiency of the corrosion inhibitor (%); X_0 is the annual corrosion rate of carbon steel in the absence of blast furnace gas condensed water (mm/y); X_1 is the iannual corrosion rate of carbon steel in the presence of the blast furnace gas condensed water (mm/y).

3. Results and Discussion

3.1. Real-time Monitoring of Corrosion Rate

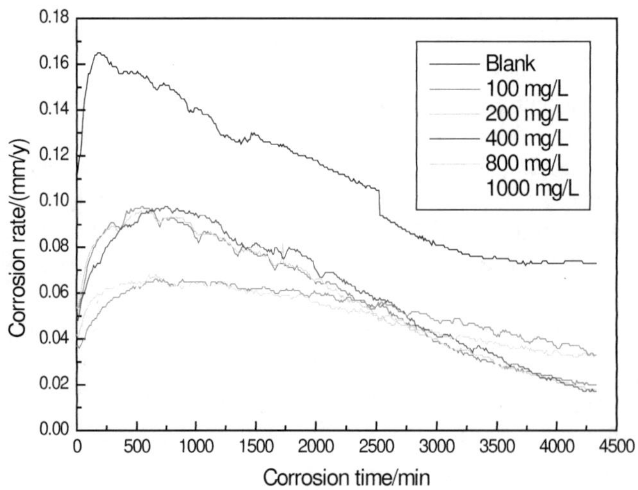

Figure 1. Corrosion Rate of Various Dosage of the Corrosion Inhibitor with Time

The real time corrosion rates of carbon steel in the uninhibited and inhibited blast furnace gas condensed water environments are shown in figure 1. The average real time corrosion rates of the blast furnace gas condensed water contacting the various dosage of the corrosion inhibitor were lower than the average real time corrosion rates without adding the corrosion inhibitor. For the uninhibited system, the corrosion rate increased sharply initially with corrosion time, reaching a maximum of ~0.165

mm/y at about 200 min. Further corrosion time was accompanied by a gradual decrease in corrosion rate and tended to be stable at about 2500 min attaining a final value of ~0.087 mm/y. The average corrosion rate of the blast furnace gas condensed water with 100 mg/L, 200 mg/L or 400 mg/L inhibitor all showed better inhibitory performance, with a maximum of ~0.10 mm/y at about 500 min, and then a gradual decrease in corrosion rate and attaining a value of ~0.05 mm/y at 2500 min. The average corrosion rate of the blast furnace gas condensed water with 800 mg/L or 1000 mg/L inhibitor increased with increasing time during the early stages of the testing, which to achieve constant corrosion rate values with ~0.06 mm/y at about 500 min, ~0.05 mm/y at 2500 min and 0.03 mm/y at the end of the test period. While the average corrosion rate of the blast furnace gas condensed water with 100 mg/L, 200 mg/L or 400 mg/L inhibitor was 0.02 mm/y at the end of the test period. So we can add 800 mg/L corrosion inhibitor in the blast furnace gas condensed water and require supplementary inhibitor treatment when corrosion time to 2500 min to prevent corrosion in practical application.

3.2. Corrosion Inhibition Efficiency Evaluated With Weight Loss Method

The corrosion rate and the corrosion inhibition efficiency of corrosion inhibition test results at the different dosages are presented in table 2.

Table 2. The Corrosion Rates and Corrosion Inhibition Behaviour of the Corrosion Inhibitor

Dosage	Corrosion rate (mm/y)	Corrosion inhibition efficiency (%)
Blank	1.624	/
100 mg/L	1.132	30.3
200 mg/L	1.086	33.1
400 mg/L	0.865	46.7
800 mg/L	0.438	73.0
1000 mg/L	0.227	86.0

The corrosion rate decreased with the dosage of the corrosion inhibitor creased from 100 mg/L to 1000 mg/L. The corrosion rate was 0.227 mm/y (corrosion inhibition efficiency was 86.0%) when 1000 mg/L corrosion inhibitor was added, which significantly less than without corrosion inhibitor in the blast furnace gas condensed water. These results implied that the corrosion inhibitor was affective for the corrosion of the blast furnace gas condensed water. This maybe because benzotriazole contained benzene ring and three nitrogen atoms, bonded and lone pair electrons, which were easily adsorbed on the surfaces of various metals and their alloys. A multilayer protective film was formed on the surface of metals, which was hard to dissolve and separate, thus achieving the corrosion inhibition effect of segregation between metal surface and corrosive medium, making the surface of metal without oxidation-reduction, producing no hydrogen, and playing an anti-corrosion role[7].

4. Conclusions

Preparation of the corrosion inhibitor was carried out by adding the benzotriazole in sodium hydroxide solution. The average real time corrosion rates of the blast furnace gas condensed water contacting the various dosage of the corrosion inhibitor were lower than the average real time corrosion rates without adding the corrosion inhibitor. The average corrosion rate achieved constant corrosion rate values with ~0.06 mm/y at about 500 min, ~0.05 mm/y at 2500 min and 0.03 mm/y at the end of the test period under 800 mg/L or 1000 mg/L inhibitor. The corrosion rate was 0.227 mm/y (corrosion inhibition efficiency was 86.0%) when 1000 mg/L corrosion inhibitor was added through weight loss method, which significantly less than without corrosion inhibitor in the blast furnace gas condensed water.

5. Acknowledgments
The authors would like to acknowledge the Fundamental Research Project of Hebei Province (18964005D), the Science and Technology Plan Project of Hebei Province (17393604D), the Science and Technology Plan Project of Hebei Academy of Sciences (19705) and the National Natural Science Foundation of China (21376062), for providing financial support for this project.

6. References
[1] Xiyan Guan, Zuofang Ma and Shulin Xiong 2011 Causes of corrosion of blast furnace gas pipe and anticorrosive measures *Metallurgical Power* **148** 25-9
[2] Xin Jiang, Ping Wang and Bingyu Zhang 2010 Study and analysis on blockage in gas pipeline of 5# coke oven in Baotou Steel *Science and Technology of Baotou Steel* **36** 7-10
[3] Zuhua Zuo 2014 Causes of corrosion of coking gas pipes and preventive measures *Metallurgical Power* **169** 23-5 mechanism of benzotriazole as corrosion inhibitor of copper *Journal of Materials Protection* **46** 45-8
[4] Zhiyuan Gao, Liping Liang and Zhiqiang Liu 2008 Research on corrosion mechanism of pipe for blast furnace dusting by dry dusting method and protection measures *Metallurgical Power* **128** 28-30
[5] Xueming Wang, Tingjun Chen, Xuming Liu, Junjie Zhao, Ji Zhang and Zhenbo Li 2011 The acid treatment of blast-furnace gas condensed water *8th China Iron and Steel Annual Conf.* (Beijing) pp 2121-7
[6] Chao Chen and Cuijiao Ding 2015 Research on corrosion mechanism and anticorrosive measures of gas piping *Industrial Heating* **44** 41-3
[7] Wenzhou Xie, Hesheng Li, Zhilin Li and Yu Yang 2013 Research progress of inhibition mechanism of benzotriazole as corrosion inhibitor of copper *Journal of Materials Protection* **46** 45-8

Micro-mechanism on the Metal Materials with the Fatigue Propagation

Molan Li

School of Materials Science and Engineering,Tianjin University of Technology, Tianjin 300384, China
Email: litj2016@163.com

Abstract. Low stress failure due to fatigue is one of the most common and common forms of component failure. Especially for aerospace and Aeronautical products, the system structure is complex, the environmental conditions are complex, the application of new technologies and materials is significant, and the investment of a single product is high. Therefore, the abnormal low stress damage of individual structures often causes catastrophic losses. In practice, most of the components with defects are not easy to find because of the stress damage caused by the continuous crack growth under fatigue load. Therefore, the study of fatigue crack and growth rate has great practical significance. In this paper, the general law and microscopic mechanism of crack growth of metal structural materials under normal temperature and transverse fatigue load are studied.

1. Introduction

When metal materials are subjected to fatigue alternating load stress, they will show cracking and other phenomenon, at the same time, metal materials will be deformed from plastic to brittle zone directly. At present, according to, metal materialsfatigue in the engineering industry can be divided into the following categories the external environmental conditions: high-cycle fatigue [1-3], low-cycle fatigue [4], Thermal fatigue [5], corrosion fatigue [6], Contact fatigue [7]. High cycle fatigue denote fatigue with a stress cycle of more than 100,000 weeks under low stress conditions (as we all known the Working stress is usually sated by under the yield limit of the materials, or even below the elastic limit). Low cycle fatigue refers to fatigue with a stress cycle of less than 10,000 to 100,000 weeks under high stress (the working stress is close to the yield limit of the material) or high strain conditions. The main mechanism of fatigue under this cycle load is the plastic strain. Thermal fatigue refers to the repeated effects of thermal stress due to temperature changes, resulting in fatigue damage. Corrosion fatigue refers to the fatigue damage caused by machine parts under the joint action of alternating loads and corrosive media (such as acids, bases, seawater, active gases, etc.) Contact fatigue refers to the contact surface of the machine parts in the repeated action of contact stress, the hemp point peeling or the surface crushing and peeling occurs, resulting in the failure and destruction of the machine parts.

The study on the metal materials' influence mechanism directly converted from plastic zone to brittle material under fatigue loading has also become the focus of attention in recent years. Among them, plasticity refers to the ability of metal materials to produce plastic deformation (permanent deformation) under loading. The plasticity of the polycrystalline material will be improved with the decrease of the grain. In addition, due to the inverse ratio of the grain boundary diffusion creep rate to the cubic size of the grain size, especially during the expansion of the high cyclic fatigue crack, due to the expansion of the grain boundary. The change in creep deformation rate, this leads to an increase in the rate of crack propagation. In addition, the dislocation movement of the grain boundary is also the reason for the change of the crack propagation rate.

In addition, due to the particularity of the interaction between the twin interface and dislocation, the common lattice boundary also has a special contribution to the strength and plasticity of the material compared to the ordinary bighorn crystal boundary. For example, nano-twinning copper containing high-density growth twins exhibits high tensile strength and good plasticity at the same time. This is mainly due to: On the one hand, the twinning interface [8] Can be as effective as the bighorn crystal boundary to block dislocation through, improve the strength; On the other hand, the twinning interface itself, as a sliding surface, also allows dislocation to slip on it, thereby increasing plasticity. Under cyclic loading conditions, the twin interface also shows different characteristics from the ordinary bighorn crystal boundary. The stress concentration of elastic Anisotropy on both sides of the twin interface makes it easy to start fatigue crack under high cycle fatigue load. Under the low cycle fatigue load, fatigue cracking is more difficult to occur in the larger angular crystal boundary of the twin interface. The understanding of resistance to low cycle fatigue cracking at the higher interface of twin crystal is still unclear. Shenyang Materials Science National (Joint) Laboratory[9-11] the low cycle fatigue cracking behavior of twin interface in face-centered cubic metal was systematically studied by Zhefeng Zhang researcher group of material fatigue and fracture research department. The previous study found that with the reduction of the fault energy of the alloy layer, the tendency of fatigue crack initiation at the twin interface increased. Recently, they also studied the effects of crystal orientation, layer error energy, and slip patterns on the fatigue cracking behavior of twin interfaces in Cu and Cu-Zn and Cu-Al alloys. The results show that under the same fatigue loading conditions, fatigue cracks in several alloy materials can be generated either along the slip zone or along the interface of the twin crystal. This shows that whether fatigue cracks are generated at the interface of the twin crystal depends not only on the external loading. And the material itself, and it is closely related to the specific load conditions of each grain.

In addition, the most commonly used method for the simulation analysis of the fatigue crack propagation process of metal materials is the molecular dynamics method, such as:

Based on studying the special crystallographic relationship between the interface of twinning crystal and the starting sliding system, the most common method for the study of the micro-influence mechanism of metal material in the process of fatigue crack propagation is to carry out the method of determining the grain orientation directly through the sliding morphology. To accurately describe the effect of grain orientation on the fatigue cracking behavior of the twinning crystal interface, the parameter Schmid factor difference(ΔΩ) was introduced to describe the relationship between shear stress and dislocation density during plastic deformation. In addition, to accurately describe the number of dislocation of the product at the surface of the twin boundary due to stress incoordination on both sides of the twinning interface, combined with Schmid's law, the number of dislocation is determined to be

$$n_{\Delta\Omega} = \frac{L(\sigma-\sigma_0)}{KGb}\Delta\Omega \quad (1)$$

The critical stress required for dislocation operation is approximately equal to the yield strength. On the other hand, the extension dislocation needs to be clustered to a certain critical width (Rc) when passing through the twin interface, and the extension width of this dislocation is determined by the layer error energy. Therefore, according to the definition of layer fault energy and the property of dislocation plug product, the relationship between the number of dislocation product and layer fault energy at the twin interface can be determined:

$$n_{RSE} = \frac{M/R_C-\gamma}{\tau_0} \quad (2)$$

M is the repulsive force of two extended dislocations, andτ0 is the piston force provided by a fault in a stop-fault group.

In addition, can also according to the above about the crystal orientation and dislocations that can place dislocation of the twin boundary surface accumulation of objective influence, coupled with the dislocation slip way influence on twin boundary cracking, they established differential Schmid factor

(ΔΩ), fault (gamma), sliding way and the twin boundary surface cracking critical criterion semi-quantitative equation:

$$\Delta\Omega = \frac{KGb}{L(\sigma-\sigma_0)}\left(\frac{n_c}{\eta} - \frac{M/R_c-\gamma}{\tau_0}\right) \tag{3}$$

The nc for the twin boundary surface cracking need critical product number of dislocation, the approximate constant can be thought of a material, η to consider the parameters of the dislocation sliding way and introduce, K, G, b, M for material constant, and L, Rc, σ0andτ0 to a certain group of materials also haven't changed much. Boundary cracking zone facing the above analysis shows that the twin boundary is Δ Ω and piecewise linear equations of eta: with the change of the dislocation sliding mode (η), Δ Ω and gamma of the linear relationship between the slope will also change. Based on the above model, they use 6 kinds of alloy with different fault can (γ) material in the statistical analysis of a large number of Schmid factor poor relations with the twin boundary surface fatigue cracking or not experimental evidence, the results show that with the wrong alloy material layer can increase and the decrease of the twin boundary surface Schmid factor difference on both sides, twin boundary surface fatigue cracking is becoming more and more difficult, and with the dislocation slip way gradually by Cu wavy slip into Cu-16% et. Al plane slip, the influence of the two is more and more obvious, as shown in figure 1.

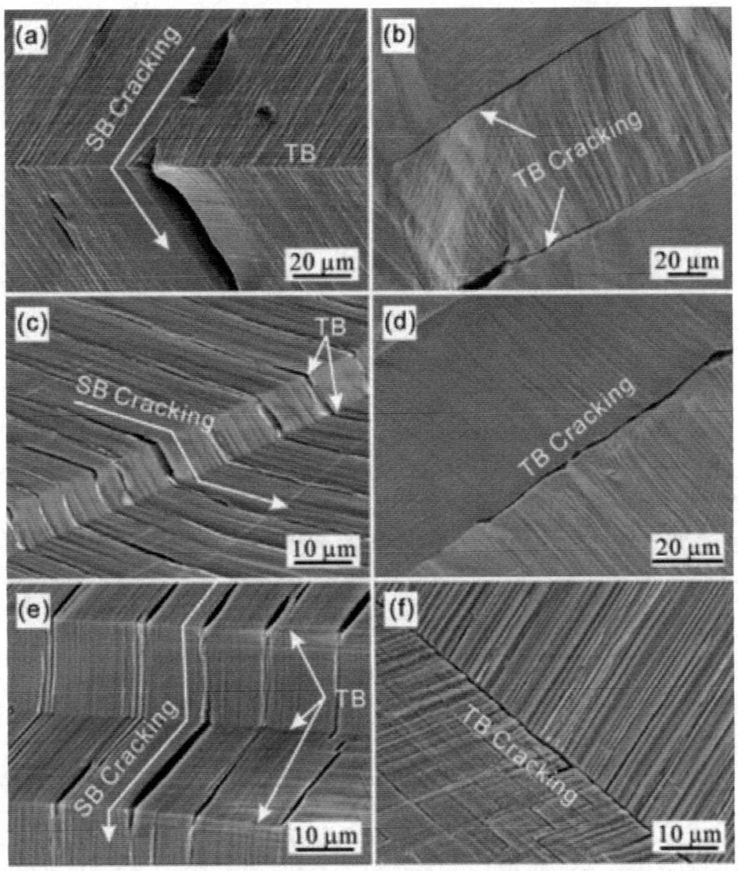

Fig.1 Slip Band Cracking and Twin Boundary Cracking Behavior of different materials Cu (a,b), Cu-10at. %Zn (c,d) &Cu-32at.%Zn (e,f)

2. Molecular Dynamics Simulation Method

Molecular dynamics method is a computer simulation experimental method. This technique can not only obtain the trajectory of atoms, but also observe various microscopic details of the process of atomic motion. It is a powerful supplement to theoretical calculation and experiment. Because molecular dynamics research is not limited by sample preparation and testing techniques, it is helpful to understand the relationship between microstructure and properties of materials. It is widely used in material science, biophysics and drug design.

In 1957, some research firstly used molecular dynamics to study the equation of state of gas and liquid under the hard ball model, which set a precedent for studying the macroscopic properties of matter by molecular dynamics simulation. In 1972, the method and extended the non-equilibrium system with velocity gradient has been developed. Andersen et al. created the method of constant pressure molecular dynamics in 1980. Method to non-equilibrium systems with temperature gradient is developed. Andthus, forming a molecular dynamics method system for non-equilibrium systems. The creation of thermostatic molecular dynamics is developed. In 1985, the first principle molecular dynamics method, which organically combines electron theory with molecular dynamics method, was proposed for semiconductor and metal Car, etc., which are difficult to model potential function. In 1991, a molecular dynamics method of giant regular systems for the treatment of adsorption problems has been further proposed. In the late 1980s, the rapid development of computer technology, coupled with the proposal and development of multi-body potential function, is a further development of molecular dynamics simulation technology.

The particle motion of the treatment system in MD obeys Newton equation is that:

$$\vec{F_i}(t) = m_i \vec{a_i}(t)$$

In the formula, the force acting on particle $\vec{F_i}(t)$ is the mass of particle m_i and the acceleration of particle $\vec{a_i}(t)$. The force acting on particle $\vec{F_i}(t)$ is the first derivative of the potential energy function to the coordinates, that is, the mass of particle m_i and the acceleration of particle $\vec{a_i}(t)$.

$$\vec{F_i}(t) = -\frac{\partial u}{\partial r_i}$$

$$m_i \frac{\partial \overrightarrow{v_i}}{\partial t} = \vec{F} = -\frac{\partial U}{\partial \overrightarrow{r_i}} + \cdots \frac{\partial U}{\partial \overrightarrow{r_n}}$$

$$\vec{r}(t) = \vec{v}(t)$$

Generally, numerical solutions of these equations need to be carried out by numerical methods. These numerical solutions generate a series of positions and velocities of$\{x_n, v_n\}$, n represents a series of discrete times, $t = n\,\Delta t$, Δt represents the time step. To solve the equations, the initial coordinates and velocities of each particle must be given.

3. Conclusion

In this paper, the propagation process of preset micro cracks in alloys is simulated by molecular dynamics method, and the fatigue crack growth behavior of different synthetic materials is studied. According to the basic principle of molecular dynamics, the expression of interaction force between atoms is deduced, and the process of tension, shear and bending of single crystal copper is simulated by establishing molecular dynamics model. The experimental results show that the cracks affect the mechanical behavior of the strength and modulus of elasticity of single crystal copper plate.

4. Reference

[1] Petit, J., and C. Sarrazinbaudoux. "An overview on the influence of the atmosphere environment on ultra-high-cycle fatigue and ultra-slow fatigue crack propagation." International Journal of Fatigue 28.11(2006):1471-1478.

[2] Li, Wei, et al. "Surface vs. interior failure behaviors in a structural steel under gigacycle fatigue: Failure analysis and life prediction." International Journal of Fatigue 64.14(2014):42-53.

[3] Kemsley, D. S. . "The effects of water vapour content and temperature of an air environment on the fatigue behaviour of an ultra-high strength steel." ScriptaMetallurgica 10.5(1976):0-424.

[4] Henaff, G, J. Petit, and B. Bouchet. "Environmental influence on the near-threshold fatigue crack propagation behaviour of a high-strength steel." International Journal of Fatigue 14.4(1992):211-218.
[5] Cotterill, P. J . "The Influence of a Coal Gasfier Atmosphere on Fatigue Crack Growth Rates in BS 4360 Steel." International Journal of Fatigue 15.1(1993):27-30.
[6] Zhang, J. M., et al. "Influence of inclusion size on fatigue behavior of high strength steels in the gigacycle fatigue regime." International Journal of Fatigue 29.4(2007):765-771.
[7] Furuya, Yoshiyuki . "Size effects in gigacycle fatigue of high-strength steel under ultrasonic fatigue testing." Procedia Engineering2.1(2010):485-490.
[8] Furuya, Yoshiyuki, H. Hirukawa, and M. Hayakawa. "Gigacycle Fatigue Properties of Hydrogen-Charged JIS-SCM440 Low-Alloy Steel Under Ultrasonic Fatigue Testing." Metallurgical & Materials Transactions A 41.9(2010):2248-2256.
[9] Furuya, Y , et al. "Gigacycle fatigue properties for high-strength low-alloy steel at 100 Hz, 600 Hz, and 20 kHz." Scripta Materialia 46.2(2002):157-162.
[10] Wang, Q. Y., et al. "Effect of inclusion on subsurface crack initiation and gigacycle fatigue strength." International Journal of Fatigue 24.12(2002):1269-1274.
[11] Lambrighs, Kasper , et al. "Influence of non-metallic inclusions on the fatigue properties of heavily cold drawn steel wires." Procedia Engineering 2.1(2010):173-181.

Design of A Drive Scheme for Untethered Magnetic Manipulation Polishing

Xiaoqin Zhou, Haocheng Wang, Baoquan Sun, Yifei Wang and Xiaoyang Wu

School of Mechanical and Aerospace Engineering, Jilin University, Changchun 130022, China
wanghcjlu@163.com

Abstract. Based on the untethered magnetic manipulation polishing device, this paper proposes a reasonable driving scheme, which makes the teleoperation process of the device more stable and easy to implement. The change of the position of the external permanent magnet can drive the internal polishing tool to produce corresponding changes. The comparison is made between the axial and radial driving modes, and finally the axial driving mode is selected for control.

1. Introduction

The existing polishing method is applied to the free-form surface of aluminum alloy. The non-contact driving of magnetic field is applied to the free-surface polishing. Based on the theory of dynamic magnetic field drive, the principle of polishing method for untethered magnetic manipulation polishing device is proposed. As shown in **Figure1**, the external permanent magnet is driven by the mechanical structure to realize the movement of multiple degrees of freedom and the rotation of the shaft to generate a dynamic magnetic field. The polishing tool is rotated by the dynamic magnetic field and rotates in the same direction as the external permanent magnet, and the magnetic pressure is pressed. The magnetic polishing liquid inside the workbench is also subjected to a magnetic field. The magnetic particles in the polishing liquid form a chain structure along the direction of the magnetic field, and the coated abrasive grains move in the direction of the magnetic field gradient, gather around the polishing tool and the polishing area, and generate with the dynamic magnetic field. The vortex flows and flows dynamically around the polishing zone; the polishing abrasive particles in the vortex flow are pressed against the workpiece by the magnetic force F of the polishing tool, and the material of the polishing region is taken away from the surface of the workpiece to realize the removal of the surface material of the workpiece.

Figure 1. Schematic Diagram of Untethered Magnetic Manipulation Polishing

The remote steering drive process of the device should be stable and easy to implement, and the change of the position of the external permanent magnet can drive the internal polishing tool to produce corresponding changes. When the external polishing tool is in the axial or radial position of the external permanent magnet, the rotation axis of the polishing tool is parallel to the external permanent magnet, that is, w=1, and the control of the polishing tool is the simplest, so the polishing tool is used in two The torque and force under driving are analyzed. The relative positions of the external permanent magnets and polishing tools in the two drive modes are shown in **Figure2** and **Figure3**.

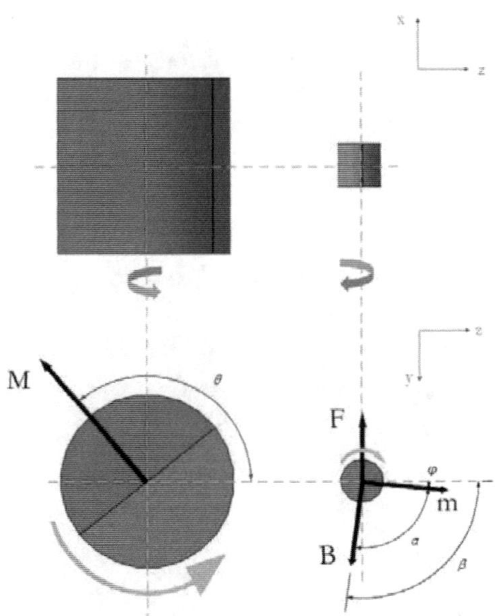

Figure 2. Radial Drive Schematic

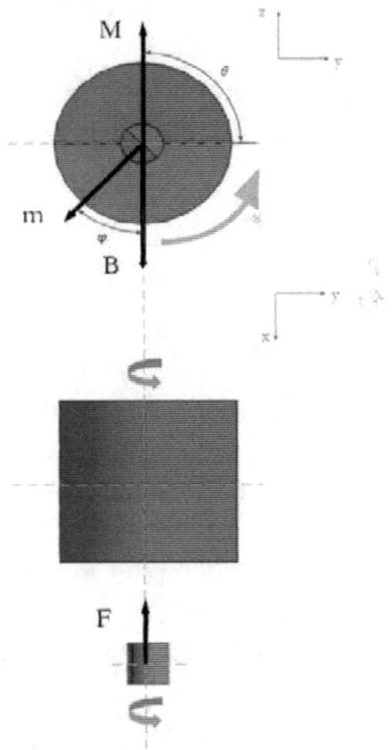

Figure3. Axial Drive Schematic

2. Analysis of Force and Torque of Polishing Tools in Radial Drive

When the polishing tool is placed in the radial direction of the outer permanent magnet, the polishing tool rotates in the opposite direction to the outer permanent magnet, as shown in Figure2. It is assumed that the polishing tool is limited to the positive z-axis direction of the external permanent magnet, that is, the unit position vector $\mathbf{p}^T = (0,0,1)$, and the magnetic formula is simplified and obtained:

$$\mathbf{B} = \frac{\mu_l|\mathbf{M}|}{4\pi|\mathbf{p}|^3}\begin{bmatrix} 0 \\ \sin\theta \\ 2\cos\theta \end{bmatrix} \tag{1}$$

Where θ is the angle between the magnetic moment M of the external permanent magnet and the positive direction of the z-axis.

During the operation of the device, the axial direction of the control external permanent magnet is always parallel to the x-axis direction of the space rectangular coordinate system, and the polishing tool is rotated by the torque applied by the dynamic magnetic field generated by the external permanent magnet. The magnetic field torque τ_M to which the polishing tool is subjected is:

$$\tau_M = |\mathbf{m}||\mathbf{B}_l|\sin\alpha = \frac{\mu_l|\mathbf{m}||\mathbf{M}|\sin\alpha\sqrt{1+3\cos^2\theta}}{4\pi|\mathbf{p}|^3} \tag{2}$$

Where $\mathbf{m}$ is the dipole moment of the permanent magnet inside the polishing tool, α is the angle between the magnetic moment $\mathbf{m}$ of the polishing tool and its magnetic induction $\mathbf{B}$.

The polishing fluid near the polishing tool produces a resistive torque τ_f for the polishing tool, which is linearly related to the rotational speed of the polishing tool:

$$\tau_f = -c\dot{\varphi} \tag{3}$$

Where c is the coefficient of drag of the polishing fluid and $\dot{\varphi}$ is the rotational angular velocity of the polishing tool.

If inertia is neglected, when the system is in steady state, the polishing tool rotates at a constant speed, and the received magnetic field torque τ_M is equal to the resistance torque τ_f:

$$\tau_M + \tau_f = |\mathbf{m}||\mathbf{B}_l|\sin\alpha - c\dot{\varphi} = 0 \tag{4}$$

In order to maximize the angular velocity $\dot{\varphi}$ of the polishing tool in equilibrium, it is necessary to always maintain $\alpha = 90°$, which has the maximum polishing efficiency. From the geometric relationship $\beta = \alpha + \varphi$, we know that $\dot{\beta} = \dot{\varphi}$. And $\tan\beta = \frac{\sin\theta}{2\cos\theta} = \frac{1}{2}\tan\theta$, so:

$$\dot{\beta} = \frac{1}{1+(\frac{1}{2}\tan\theta)^2} \times \frac{1}{2\cos^2\theta} \times \dot{\theta} = \frac{2}{1+3\cos^2\theta} \times \dot{\theta} \tag{5}$$

Substituting $\alpha = 90°$ into equation (11):

$$|\tau| = |\mathbf{m}||\mathbf{B}_l|\sin\alpha = |\mathbf{m}|\frac{\mu_l|\mathbf{M}|}{4\pi|\mathbf{p}|^3}\sqrt{1+3\cos^2\theta} \tag{6}$$

Finishing out the driving equation for the external permanent magnet is:

$$\dot{\theta} = \frac{\mu_l|\mathbf{m}||\mathbf{M}|}{8\pi|\mathbf{p}|^3 c}(1+3\cos^2\theta)^{\frac{3}{2}} \tag{7}$$

The relationship between the rotational speed and the rotation angle is shown in **Figure4**. The abscissa is the rotation angle θ and the ordinate is the rotational speed $\dot{\theta}$. In this drive mode, the polishing tool maintains the fastest and most stable motion. In this state of motion, according to the polishing tool in the magnetic field, the force is $\mathbf{F}=(\mathbf{m}\cdot\nabla)\mathbf{B}$, and the magnetic force of the polishing tool can be obtained by substituting the parameters:

$$\mathbf{F}=\frac{3\mu_l|\mathbf{m}||\mathbf{M}|}{4\pi|\mathbf{p}|^4}\left(\frac{1+\cos^2\theta}{\sqrt{1+3\cos^2\theta}}\right)\begin{bmatrix}0\\-1\\0\end{bmatrix} \tag{8}$$

When $|\mathbf{p}|=5\text{cm}$, the relationship between the magnetic force $\mathbf{F}$ and the rotation angle θ is shown in **Figure5**. In the figure, the abscissa is the force $\mathbf{F}$ of the polishing tool and the ordinate is the outer permanent magnet rotation angle θ. The magnetic force of the polishing tool is always in the negative direction of the y-axis, and its value changes with the change of θ. The variation range of $\mathbf{F}$ fluctuates with the rotation angle θ between 94%-100%. In this driving mode, the instantaneous acceleration of the permanent magnet is large, and the polishing tool is unstable, and it is difficult to realize the driving method.

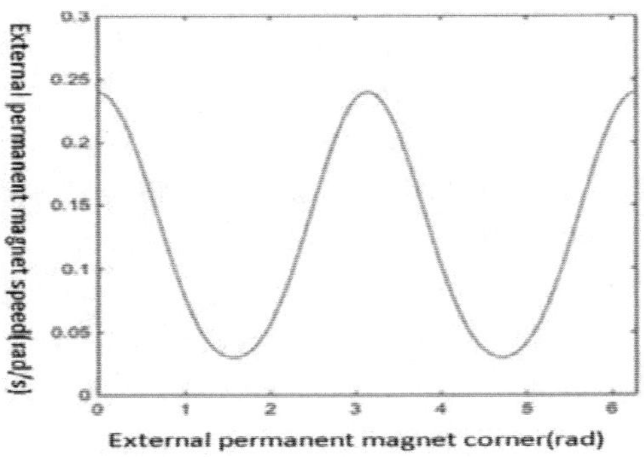

Figure 4. Relationship between Permanent Magnet Rotation Speed and Rotation Angle

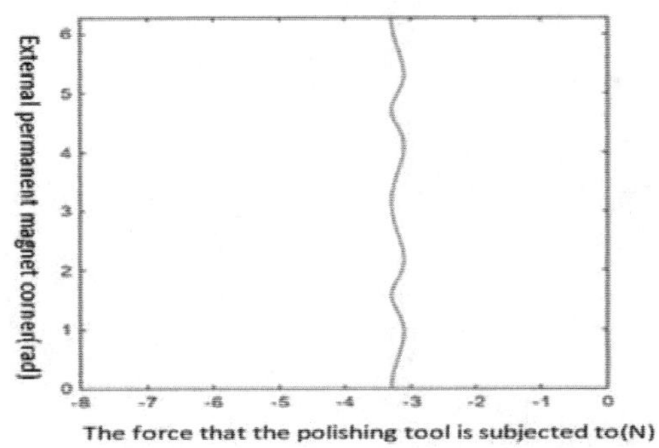

Figure 5. Polishing Tool Force and Angle Relationship

3. Analysis of Force and Torque of Polishing Tools in Axial Drive

When the polishing tool is placed in the radial direction of the outer permanent magnet, the polishing tool is opposite to the direction of rotation of the outer permanent magnet, as shown in **Figure3**. It is assumed that the polishing tool is limited to the positive z-axis direction of the external permanent magnet, that is, the unit position vector $\mathbf{p}^T=(0,0,1)$, which can be obtained:

$$\mathbf{B}=\frac{\mu_l|\mathbf{M}|}{4\pi|\mathbf{p}|^3}\begin{bmatrix}0\\\sin\theta\\2\cos\theta\end{bmatrix} \tag{9}$$

In a certain range of speeds, the polishing tool and the external permanent magnet can be rotated synchronously, and the retardation angle of the polishing tool dipole moment $\mathbf{m}$ and the magnetic induction intensity $\mathbf{B}$ it receives is φ. At this time, the polishing tool dipole moment $\mathbf{m}$ is:

$$\mathbf{m}=|\mathbf{m}|\begin{bmatrix}0\\-\cos(\theta+\varphi)\\-\sin(\theta+\varphi)\end{bmatrix} \tag{10}$$

The axially driven lower polishing tool is also subjected to the magnetic field torque τ_M and the resistive torque τ_f, which are equal when the constant speed is stably rotated, as shown by equation (12). At this time, the magnitude of the torque τ_M to which the polishing tool is subjected is:

$$\tau_M = |\mathbf{m}||\mathbf{B}_l|\sin\varphi = \frac{\mu_l |\mathbf{m}||\mathbf{M}|\sin\varphi}{4\pi|\mathbf{p}|^3} \tag{11}$$

The magnetic force $\mathbf{F}$ obtained by the finishing tool can be:

$$\mathbf{F} = \frac{3\mu_l |\mathbf{m}||\mathbf{M}|\cos\varphi}{4\pi|\mathbf{p}|^4}\begin{bmatrix}-1\\0\\0\end{bmatrix} \tag{12}$$

It can be seen from the formula (12) that the direction of the magnetic force $\mathbf{F}$ received by the polishing tool is always in the negative direction of the x-axis.

The magnitude of the magnetic field torque τ_M and the magnetic force $\mathbf{F}$ of the polishing tool depend on the hysteresis angle φ. When $\varphi = 0$, the magnetic force is the largest of $\varphi = 90°$, and the magnetic field torque is 0; when the magnetic force is zero, the magnetic field torque is the largest. During the actual polishing process, the magnitude of the hysteresis angle φ is related to the external permanent magnet and the rotational speed of the polishing tool. When the rotational speed is kept constant, the size of the polishing tool can be kept constant.

When $\varphi = 45°$, the relationship between the magnetic force $\mathbf{F}$ and the distance $|\mathbf{p}|$ of the polishing tool is as shown in **Figure6**. When the distance $|\mathbf{p}|$ is less than 5 cm, the polishing tool is subjected to a significant magnetic force; when the distance is greater than 10 cm, the polishing tool is subjected to a magnetic force of almost zero. Therefore, in order to subject the polishing tool to a sufficient magnetic force, the distance between the center of the outer permanent magnet and the center of the polishing tool should be kept as close as possible.

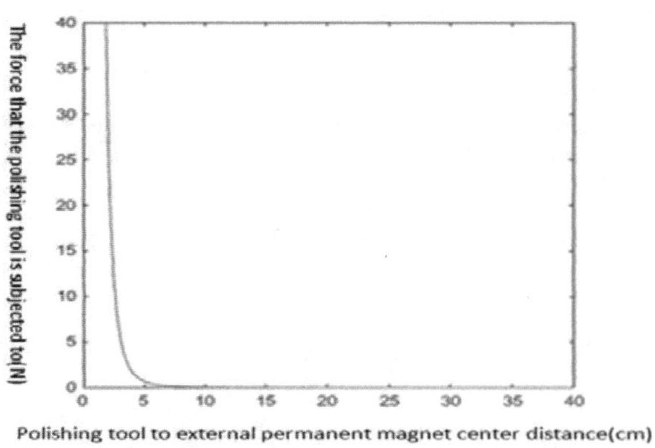

Figure 6. Polishing Tool Magnetic Force and Distance Diagram

4. Conclusion

The two driving schemes of radial driving and axial driving are analyzed. In the radial driving mode, in order to ensure synchronous rotation, the external permanent magnet is controlled to rotate non-uniformly, and the polishing tool is periodically floated; the axial driving mode is polished. The tool is magnetically attractive, more controllable and stable. Therefore, this paper chooses the axial driving method for the polishing tool.

5. Acknowledgments
This research has been supported by the Department of Science and Technology of Jilin Province (grant no. 20170414005GH).

6. References
[1] Choi J, Choi H, Cha K, et al. Two-dimensional locomotive permanent magnet using electromagnetic actuation system with two pairs stationary coils[C]// IEEE International Conference on Robotics and Biomimetics. IEEE, 2009:1166-1171.

[2] Yu C. Novel Electromagnetic Actuation System for Three-Dimensional Locomotion and Drilling of Intravascular Microrobot[J]. Sensors & Actuators A Physical, 2010, 161(1):297-304.

[3] Kummer M P, Abbott J J, Kratochvil B E, et al. OctoMag: An electromagnetic system for 5-DOF wireless micromanipulation[C]// IEEE International Conference on Robotics and Automation. IEEE, 2010:1610-1616.

[4] Kummer M P, Abbott J J, Kratochvil B E, et al. OctoMag: an electromagnetic system for 5-DOF wireless micromanipulation[J]. IEEE Transactions on Robotics, 2010, 26(6):1006-1017.

[5] Kratochvil B E, Kummer M P, Erni S, et al. MiniMag: A Hemispherical Electromagnetic System for 5-DOF Wireless Micromanipulation[M]// Experimental Robotics. Springer Berlin Heidelberg, 2014:317-329.

[6] Schuerle S, Erni S, Flink M, et al. Three-Dimensional Magnetic Manipulation of Micro- and Nanostructures for Applications in Life Sciences[J]. IEEE Transactions on Magnetics, 2012, 49(1):321-330.

[7] Mahoney A W, Abbott J J. Managing magnetic force applied to a magnetic device by a rotating dipole field[J]. Applied Physics Letters, 2011, 99(13):55.

[8] Mahoney A W, Cowan D L, Miller K M, et al. Control of untethered magnetically actuated tools using a rotating permanent magnet in any position[C]// IEEE International Conference on Robotics and Automation. IEEE, 2012:3375-3380.

Effects of Cooling Conditions on Surface Integrity in Turning of TC4 Alloy

Jianming Li[1], Taoyuan Li[2], Xiangyu Wang[1], Yang Qiao[1] and Peiquan Guo[1]*

1. School of Mechanical Engineering, University of Jinan, Jinan 250022, China.
2. Department of Mechanical Engineering, Zaozhuang Vocational College of Science & Technology, Tengzhou 277599, China.
*Corresponding author. Email: oss_guopq@ujn.edu.cn

Abstract. Machined induced surface integrity of titanium alloy is of great importance due to its strong correlation with the functional performance and service life. In this paper, the effects of different cooling conditions on the surface integrity of TC4 alloy were studied experimentally. The results showed that the values of surface roughness and surface residual stress under cryogenic condition were lower than those under dry condition and emulsion cooling condition with the same cutting parameters, while the influence of cooling condition on the microhardness of the machined surface is not obvious. Cryogenic cutting is an effective way to machine titanium alloy.

1. Introduction

TC4 is widely used in aerospace industry, chemical industry and health care industry due to its excellent inoxidability, thermal stability and specific strength[1]. To satisfy the instant requirements of good surface integrity for engineering practice, high efficiency and clean machining of TC4 is an inevitable trend.

However, due to its excellent mechanical and physical properties, TC4 is a typical difficult-to-machine material. During the machining process, TC4 is often accompanied by high cutting temperatures, severe tool wear[2]. These adverse processing characteristics inevitably have certain impacts on the surface integrity of TC4 component, thereby affecting its performance[3].

In recent years, with the development of cryogenic machining technology, it offers a potential processing technology to machine TC4 and other difficult-to-machine materials[4]. Cryogenic cooling can effectively reduce the cutting temperature during the machining process, change the thermal deformation form of the machining area, and effectively improve the machining quality[5]. In addition, cryogenic machining is a clean and sustainable processing technology. This paper carried out the single-factor comparison test of machining TC4 to explore the influence of cooling conditions on surface integrity.

2. Experimental Procedure

2.1. Experimental Material

The material used in the test was TC4 bar material, whose chemical composition is shown in Table 1. The material was machined into a dish workpiece of φ 110mm × 4mm by WEDM. After being fixed with a fixture, the material was subjected to conduct a radial turning test.

Table 1. The chemical composition of TC4 (The mass fraction, %)

Al	V	Fe	Si	C	N	H	O	Ti
5.6	3.86	0.18	0.01	0.02	0.023	0.01	0.17	allowance

2.2. Experimental Method
The turning test was carried out on CAK3665 CNC lathe, uncoated carbide tool (Kennametal grade: K313, inserts TPGN220408 geometry) was selected in this experiment, the rake angle of the tool after installation was 3° and the back angle was 8°. In the turning test, the TC4 was machined using radial turning, and the test parameters were shown in Table 2. Liquid nitrogen has good cooling effect and no pollution, so liquid nitrogen was used as coolant in the process of cryogenic cutting. In the turning test, a circular liquid nitrogen nozzle was adopted, and the injection area was the workpiece in front of the cutting area. During the cutting process, the pressure in the nitrogen canister was always kept at 1.5 MPa. Figure 1 shows the operating conditions of cryogenic cutting.

Table 2. Cutting test parameters

cooling condition	v_c (m/min)	f_z (mm/r)	a_p (mm)
dry cutting emulsion cutting cryogenic cutting	80	0.1	0.8

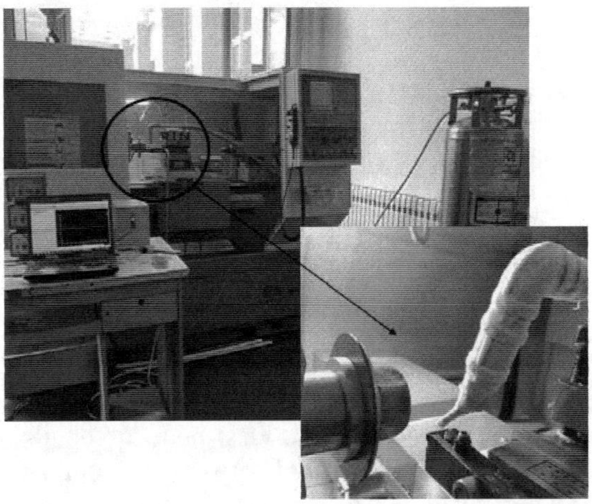

Figure 1. Testing apparatus of cryogenic cutting

After the cutting test, the surface roughness, surface residual stress and microhardness of the machined workpiece were measured. The surface roughness was tested by Mitutoyo Sj-410 profilometer. The surface roughness was measured along the cutting feed rate direction at the middle position of the machined surface. Each parameter was measured for 5 times and the average value was calculated as the final measurement result. In order to further study the variation of machined surface, Usp-sigma white light interferometer was used to take photos of the 3D morphology of the machined surface. The surface residual stress was measured using PROTO iXRD-Combo residual stress instrument, and the residual stress of the workpiece along the cutting feed rate direction was measured. The target material selected for residual stress test was Cu-Kα, the diffraction crystal plane was {213}, and the Bragg Angle was 140°. The microhardness was tested by 402MVD microhardness tester. The loading force was 0.5N and the dwelling time was 10s. Each workpiece was measured 5 times and its average value was taken as the measurement result.

3. Results and Discussion

3.1. Surface Roughness

Surface roughness refers to the roughness of the machined surface with small spacing and small peaks and valleys[6]. And it is an important index to measure the quality of workpiece after cutting process. Figure 2 shows the test results of the effects of cooling conditions on surface roughness.

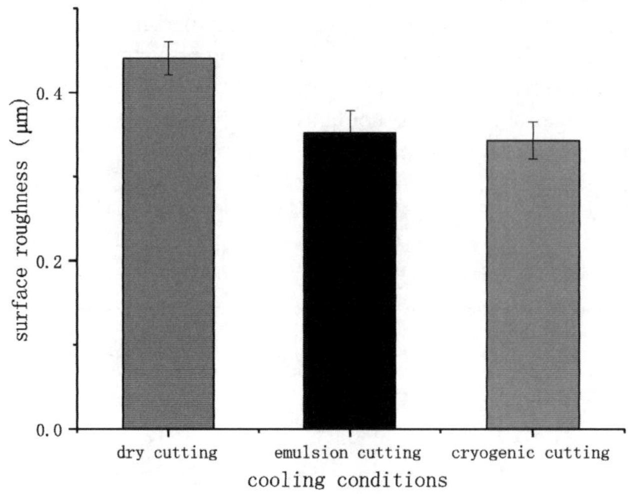

Figure 2. Surface roughness under different cooling conditions

By comparing the influence of different cooling conditions on the surface roughness in Figure 2, it was observed that at same cutting parameters, both the cryogenic cutting and emulsion cutting showed comparable results. Compared with dry cutting, both of them reduce the surface roughness of the machined surface, and the effect of cryogenic cutting is much better.

In dry cutting conditions, there are problems such as high cutting temperature, large cutting force and serious tool wear. In addition, the chip is not easy to discharge, and these will affect the quality of the machined surface, resulting in rise of surface roughness[7]. It is well known that both emulsion and liquid nitrogen have good cooling performance, which can reduce the temperature of the cutting area and toobtain low tool wear. Therefore, both methods are beneficial to improve the surface roughness. But, compared with emulsion, liquid nitrogen has a stronger cooling effect, so cryogenic cutting can obtain better surface roughness[8]. Besides, both the emulsion and liquid nitrogen can lubricate well, thus reducing the friction coefficient between the tool and the workpiece, which will greatly improve the surface quality of the workpiece. In addition, surface plastic side flow is an important factors which influence the machining induced surface roughness. The metal of workpiece in front of tool nose occur plastic deformation because of the tool extrusion and the coupling of cutting heat[9]. With the decrease of cutting temperature, the plasticity of material decreases and the trend of surface plastic side flow decreases, thereby reducing the surface roughness of the workpiece.

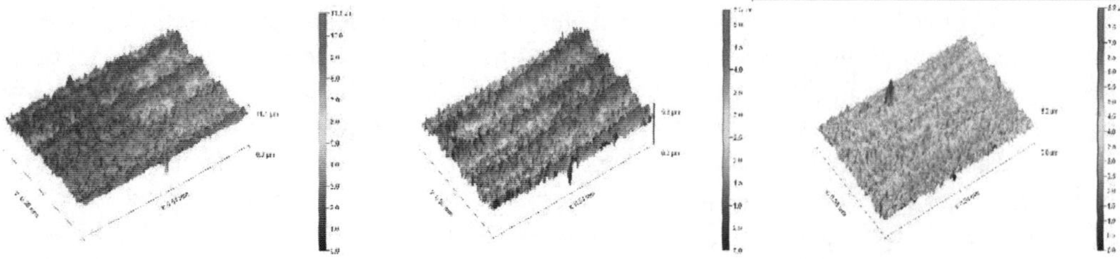

(a)Dry cutting (b) Emulsion cutting (c)Cryogenic cutting

Figure 3. 3D topography of the machined surface under different cooling conditions

Figure 3 shows the 3D topography of the machined surface under different cooling conditions. In the meantime, by observing the three-dimensional topography of the surface under three cooling methods, it can be found that a lot of flash and burrs are generated in the dry cutting conditions, the surface is rough, and the surface quality is poor. In cryogenic cutting and emulsion cutting conditions, the surface quality is very good, and cryogenic cutting produces a highly polished surface.

3.2. Surface Residual Stress

Surface residual stress affects machining accuracy, dimensional stability, static strength, fatigue strength and corrosion cracking of components. During the machining, the combination of mechanical plastic deformation, thermal plastic deformation and phase transition produces the residual stress on the machined surface[10]. Figure 4 shows the influence of different cooling conditions on surface residual stress.

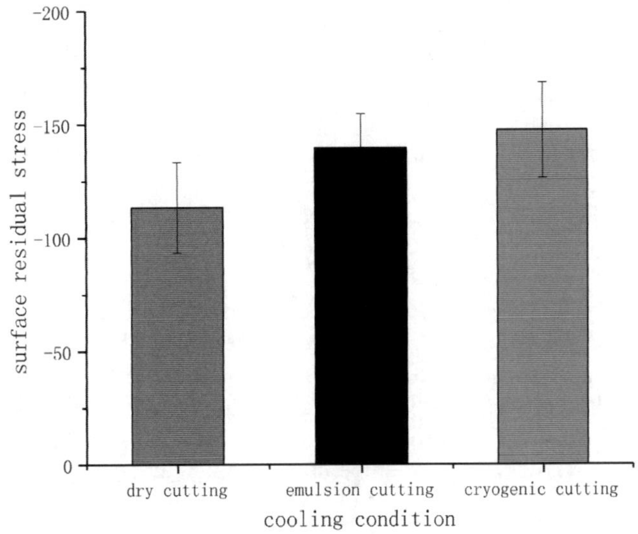

Figure 4. Surface residual stress under different cooling conditions

It can be seen from Figure 4 that all the residual stress generated by three cooling conditions is residual compressive stress, and the absolute value of residual stress generated by emulsion cutting and cryogenic cutting is greater than the absolute value of residual stress generated by dry cutting. The reason is that the cooling effect of emulsion and liquid nitrogen greatly reduces the cutting temperature, that is, reduces the tendency of thermal effect to produce residual tensile stress, so as to obtain better residual compressive stress. The cooling effect of liquid nitrogen is greater than that of emulsion, so in most cases, the inhibition effect of liquid nitrogen on residual tensile stress is greater than that of emulsion.

3.3. Microhardness

In the cutting process, because the cutting edge is not absolutely sharp, the surface material of the workpiece is plastically deformed by the effect of extrusion and friction, which results in a phenomenon that the surface layer hardness of the workpiece is strikingly improved, which is called work hardening[11]. At the same time, a certain extent of thermal softening is caused by the influence of cutting heat. Therefore, the microhardness of the surface is the result of the coupling of mechanical loads and thermal loads. Figure 5 shows the influence of different cooling conditions on microhardness[12].

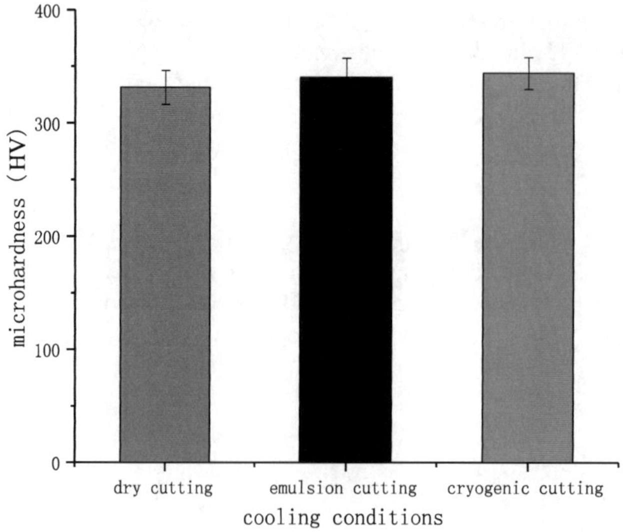

Figure 5. Microhardness under different cooling conditions

It can be found form Figure 5 that: the difference of micro-hardness of machined surface is not so big under the three cooling methods. Micro hardness produced by cryogenic cutting and emulsion cutting is slightly bigger than that produced by dry cutting, with a difference of about 10HV. The main reason is that a large amount of cutting heat is generated during dry cutting, resulting in the high cutting temperature, and the heat generated accumulates near the cutting edge, enhancing the phenomenon of machining heat softening, leading to the decrease of hardness. The cutting temperature can be reduced under the cryogenic cutting and emulsion cutting, which will inhibit the adverse influence of machining heat softening effect, and increase the micro hardness. At the same time, the literature of some scholars shows that low temperature cutting can help to keep the smaller grain size after dynamic recrystallization of materials, refine the grains on the machined surface, and improve the hardness.

4. Conclusions

According to the cutting test of TC4, the effects of cooling conditions on surface integrity under the same cutting parameters were discussed, and the following conclusions can be drawn:

1) Both cryogenic cutting and emulsion cutting reduce surface roughness, but cryogenic cutting has better surface roughness as its better cooling effect.
2) Cryogenic cutting can achieve better residual stress with its excellent cooling effect.
3) Cryogenic cutting and emulsion cutting effectively suppress the thermal softening effect of temperature and obtain higher microhardness.

In summary, cryogenic cutting can achieve a better machining induced surface integrity of TC4, which certainly could enhance its functional performance. Furthermore, cryogenic machining is more cleaner and sustainable, so it offers an excellent alternative processing technology for TC4 alloy.

5. Acknowledgments

This work was supported by Key R&D Program of Shandong Province (Grant No. 2018GGX03010 and Grant No. 2019GGX104096), and Shandong Provincial Natural Science Foundation (Grant No. ZR2019QEE032).

6. References

[1] Li G R, Li Y M, Wang F F, et al. Microstructure and performance of solid TC4 titanium alloy subjected to the high pulsed magnetic field treatment[J]. *Journal of Alloys and Compounds,* 2015, 644:750-756.

[2] Pervaiz S, Rashid A, Deiab I, et al. An experimental investigation on effect of minimum quantity cooling lubrication (MQCL) in machining titanium alloy (Ti6Al4V) [J]. *International Journal of Advanced Manufacturing Technology,* 2016, 87(5-8):1371-1386.

[3] Yang S C, Cui X Y, Zhang Y H, et al. Effect of tool wear on surface qualities in milling of TC4[J]. *Materials Science Forum,* 2016, 836-837:132-138.

[4] Safari H, Sharif S, Izman S, et al. Cutting force and surface roughness characterization in cryogenic high-speed end milling of Ti–6Al-4V ELI[J]. *Materials & Manufacturing Processes,* 2014, 29(3):350-356.

[5] Jawahir I S, Attia H, Biermann D, et al. Cryogenic manufacturing processes[J]. *CIRP Annals - Manufacturing Technology,* 2016, 65(2):713-736.

[6] Mia M, Dhar N R. Optimization of surface roughness and cutting temperature in high-pressure coolant-assisted hard turning using Taguchi method[J]. *International Journal of Advanced Manufacturing Technology,* 2016, 88(1):1-15.

[7] Deng J, Li Y, Song W. Diffusion wear in dry cutting of Ti–6Al–4V with WC/Co carbide tools[J]. *Wear,* 2008, 265(11):1776-1783.

[8] Dinesh S, Senthilkumar V, Asokan P. Experimental studies on cryogenic machining of biodegradable ZK60 Mg alloy using micro-textured tools[J]. *Materials and Manufacturing Processes,* 32(9):979-987.

[9] Xu F, Fang F, Zhang X. Side flow effect on surface generation in nano cutting[J]. *Nanoscale Research Letters,* 2017, 12(1):359-364.

[10] Dhar N R, Kamruzzaman M. Cutting temperature, tool wear, surface roughness and dimensional deviation in turning AISI-4037 steel under cryogenic condition[J]. *International Journal of Machine Tools & Manufacture,* 2007, 47(5):754-759.

[11] Ginting A, Nouari M. Surface integrity of dry machined titanium alloys[J]. *International Journal of Machine Tools and Manufacture,* 2009, 49(3-4):325-332.

[12] Wang F, Zhao J, Li A, et al. Experimental study on cutting forces and surface integrity in high-speed side milling of Ti-6Al-4V titanium alloy[J]. *Machining Science and Technology,* 2014, 18(3):448-463.

Experimental Study on Basic Mechanical Properties of Calcium Sulfate Whisker-Reinforced Cement

Xin Li, Liang Li, Chi Zhang and Yanyan Zheng

School of Resources and Civil Engineering, Northeastern University, NO. 3-11, Wenhua Road, Heping District, Shenyang, P. R. China.
Email: lixinedu@mail.neu.edu.cn

Abstract. In order to study the effect of calcium sulfate whisker on the mechanical characteristics of Portland cement, the fluidity, initial setting time, final setting time, flexural strength and compressive strength of calcium sulfate whisker-reinforced cement with 0~1.0% (mass fraction) were studied using experimental method. The results show that calcium sulfate whisker has little effect on the fluidity, initial setting time and final setting time. For the 28-day flexural strength and compressive strength of calcium sulfate whisker-reinforced cement, the optimum whisker content is 0.4% and 1.0% and the corresponding strength is increased by 19.54% and 35.84%, respectively. Compared with calcium carbonate whisker, the reinforcement effect of calcium sulfate whisker on the strength of Portland cement is better.

1. Introduction

Ordinary Portland cement is widely used in civil engineering. However, it has low tensile strength, small tensile strain, easy cracking, low permeability and other shortcomings. Researchers use some fiber materials to improve the performance of cement-based materials, such as steel fiber, carbon fiber, glass fiber, polypropylene fiber and so on [1, 2]. But these fiber materials are not perfect, such as steel fiber is easy to agglomerate in construction, the price of carbon fiber with high tensile strength and elastic modulus is expensive and it has poor dispersion, glass fiber will be eroded by the high alkaline hydration products of Portland cement [3, 4]. Therefore, it is necessary to look for and study a fiber reinforced material with excellent performance, reasonable price and sustainable development.

Calcium sulfate whisker is also called gypsum whisker, which is fibrous single crystal of calcium sulfate. Compared with traditional fibers, it is an ideal fiber substitute. Calcium sulfate whisker has the particle size of granular fillers and the length-diameter ratio of short fibrous fillers, and has the advantages of stable size, high strength, good toughness, high temperature resistance and chemical corrosion resistance [5]. The price of calcium sulfate whiskers is much lower than that of other high performance fibers. Calcium sulfate whiskers are mainly prepared from gypsum. At present, China's proven gypsum reserves are about 100 billion tons, ranking first in the world. In addition, there are other raw materials, such as ammonia alkali waste $CaCl_2$, phosphate rock, calcium salts and sulfates, converter slag and tin tailings, rare earth tailings and so on, so sources of raw materials are extensive and using them are conducive to environmental protection and sustainable development [6, 7].

The properties of calcium sulfate whiskers-reinforced cement are studied using experimental method in this paper. Compared with pure cement materials, calcium sulfate whisker has little influence on the working performance of cement-based materials and improves obviously the flexural strength and compressive strength.

2. Working Performance Test

2.1. Test Material and Method

The working performance test of calcium sulfate whisker cement-based material includes fluidity, initial and final setting time. Ordinary Portland cement with strength grade of 42.5 is used. Calcium sulfate whisker (SEM photo as shown in figure 1) is produced by Northeastern University, China and the technical index is shown in table 1.

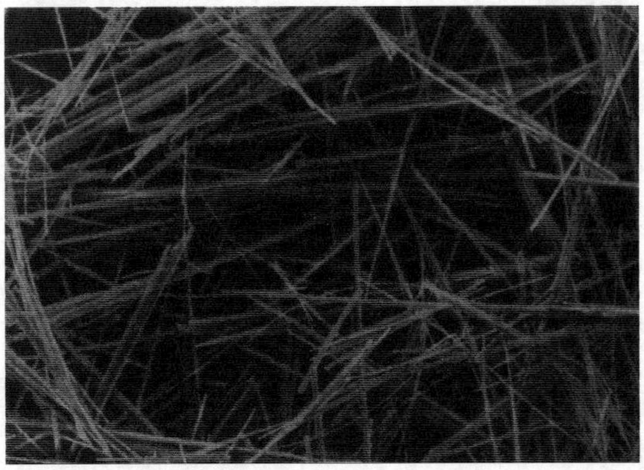

Figure 1. SEM Photograph of CaSO4 Whiskers

Table 1. The Technical Index of Calcium Sulfate Whisker

Index	Value	Index	Value
Density (g/cm^3)	2.96	Melting point (°C)	1450
Bulk density (g/cm^3)	0.1~0.4	Water-soluble (22°C) (ppm)	<1200
Length (μm)	50~200	Moh's hardness	3~4
Diameter (μm)	0.5~4	$CaSO_4$ content (%)	≥96
Tensile strength (GPa)	20.5	Water content (%)	<1.2
Tensile modulus (GPa)	178	pH	7~8

As shown in figure 2, for the measurement of fluidity, the glass plate with the size of 400mm×400mm×5mm is firstly placed on the horizontal position, and the glass plate, agitator, truncated cone die and mixing pot are uniformly rubbed with a wet cloth to make the surface wet without driving water stains. Then put the truncated cone die (upper diameter 36 mm, lower diameter 60 mm, height 60 mm) on the center of the glass plate, and put the mix the cement slurry into the die quickly and flatten it with a scraper. Lift the truncated cone die in the vertical direction and start timing with a stopwatch. When the cement slurry flows for 30 seconds on the glass plate, measure the maximum diameters of the two perpendicular directions of the flowing slurry. Finally, the fluidity of cement slurry is the average value of the two maximum diameters. As shown in figure 3, Vicat apparatus is used to measure the initial setting time (figure 3 (a)) and the final setting time (figure 3 (b)). When the test needle sinks to (4±1) mm from the bottom plate, the cement reaches the initial setting state; when the needle sinks 0.5 mm into the cement body, the cement reaches the final setting state.

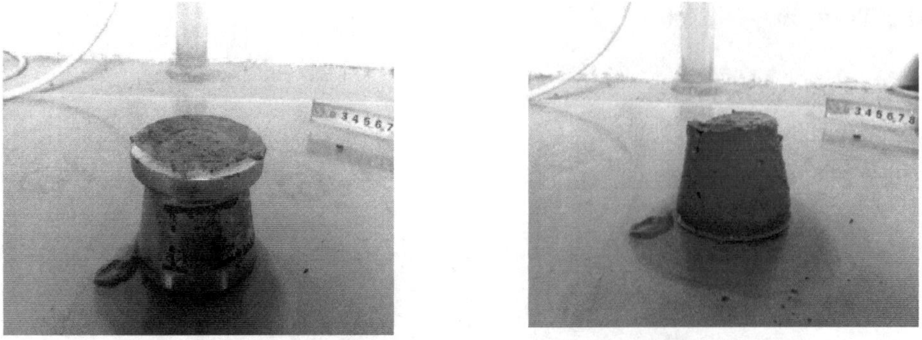

(a) Truncated Cone Die (b) Flowing Cement Slurry

Figure 2. The Test of Fluidity

(a) The Initial Setting Time (b) The Final Setting Time

Figure 3. The Test of the Initial and Final Setting Time

2.2. Test Results and Analysis

The test results of fluidity and initial and final setting time are shown in table 2.

Table 2. The Test Results of Fluidity and Initial and Final Setting Time

Mixed content (%)	Fluidity (mm)	Initial setting time (min)	Final setting time (min)
0	65.17	270	330
0.2	64.83	270	315
0.4	63.50	270	330
0.6	63.50	270	315
0.8	63.50	265	315
1.0	64.00	275	315

According to the data in table 2, when 0.2%~1.0% (mass fraction) calcium sulfate whisker is added, the fluidity of cement slurry decreases, but the extent of decrease is small. Because calcium sulfate whisker is fibrous single crystal that is a microfiber and has no bending deformation, the phenomenon of agglomeration and winding in steel fiber and polypropylene fiber will not occur when a small amount of calcium sulfate whisker is added [8]. In addition, the diameter of the calcium sulfate whisker is relatively small compared to other fibers, so the cement slurry is not affected by fiber blocking effect [9, 10]. At the same time, mixing 0.2%~1.0% calcium sulfate whisker had little effect on the initial and final setting time of cement slurry.

IOP Conf. Series: Materials Science and Engineering **735** (2020) 012036 doi:10.1088/1757-899X/735/1/012036

3. Bending Test

3.1. Test Instrument and Method

YAW-1000 automatic pressure testing machine is used in the test, and the size of test block is 40mm×40mm×160mm, as shown in figure 4. 3-day, 7-day, 14-day and 28-day flexural strength tests are performed on cement samples with calcium sulfate whisker mass fraction of 0, 0.2%, 0.4%, 0.6%, 0.8% and 1.0%, respectively. The loading rate is (50±10) N/s.

The flexural strength R_f is calculated according to the following formula:

$$R_f = 1.5FL / b^3 \tag{1}$$

Where: F is the failure load (N), L is the center distance of the supporting cylinder, L=100mm, and b is the side length of the square section of the test block, b=40mm.

The test results of flexural strength of each group are the arithmetic mean of the measured values of the three test blocks. When any of the three flexural strength values is plus or minus 10% above the mean, this measured data should be eliminated. The final flexural strength test results are averaged with the remaining two flexural strength test values.

(a) Test Instrument

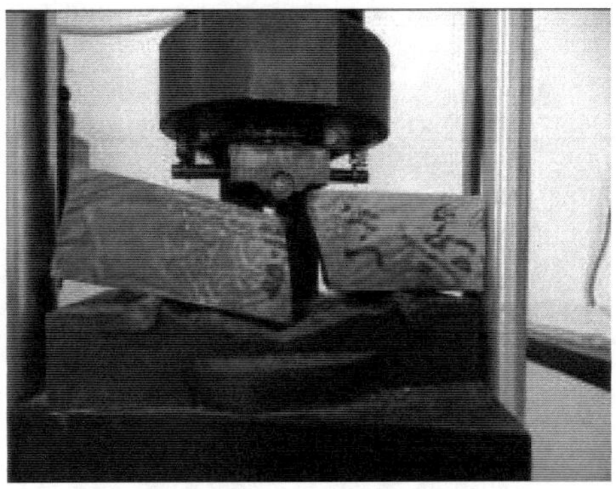

(b) Flexural Test Block

Figure 4. Test Instrument and Flexural Test Block

3.2. Test Results and Analysis

The bending test results are shown in table 3.

Table 3. Flexural Strength of $CaSO_4$ Whisker Cement-based Materials

Mixed content (%)	Flexural strength (MPa)			
	3 days	7 days	14 days	28 days
0	7.55	11.31	14.24	15.10
0.2	8.20	13.59	14.59	15.40
0.4	6.67	13.23	16.05	18.05
0.6	7.46	12.72	16.11	16.26
0.8	6.31	10.50	16.03	16.20
1.0	7.64	11.83	14.58	15.29

It can be seen from table 3 that the mixture of calcium sulfate whiskers improves the flexural strength of cement materials in general. When the curing time is 3 days and 7 days, the flexural strength is improved less because the properties of cement are more unstable. When curing time is 14 days, the flexural strength of cement test block is maximum with the mass fraction of calcium sulfate whisker of 0.6%. When curing time is 28 days, the flexural strength of cement test block is maximum (i.e. 18.05 MPa) with the mass fraction of calcium sulfate whisker of 0.4% and is increased by 19.54% compared with ordinary Portland cement test block without calcium sulfate whisker.

4. Compression Test

4.1. Test Method

The compression test blocks are the fractured prismatic blocks in the bending test. The loading rate is (2400±200) N/s and the loading area of the test block is 40mm×40mm. There are 6 test blocks for one group and the test compressive strength of each group is the arithmetic mean of the compressive strength test values of six test blocks in the same group. When one of the six measured compressive strength values in a group is plus or minus 10% above the mean, this measured data should be eliminated. The final compressive strength test results are averaged with the remaining five compressive strength test values. If any of the remaining five compressive strength values is plus or minus 10% above the new mean value, the test results of this group will be invalidated.

4.2. Test Results and Analysis

The compression test results are shown in table 4.

Table 4. Compressive Strength of $CaSO_4$ Whisker Cement-based Materials

Mixed content (%)	Compressive strength (MPa)			
	3 days	7 days	14 days	28 days
0	49.42	61.16	66.26	72.99
0.2	40.79	64.17	76.02	91.21
0.4	42.61	61.18	73.03	85.87
0.6	42.45	67.77	67.70	78.89
0.8	42.29	66.56	67.34	96.60
1.0	49.47	63.09	68.88	99.15

As can be seen from table 4, the effect of calcium sulfate whisker on early compressive strength of cement-based materials is not obvious. However, with the increase of curing time, the enhancement effect of calcium sulfate whisker increases gradually. When the mass fraction of mixed calcium sulfate whisker is 1.0%, the 28-day compressive strength reaches the maximum value, i.e. 99.15 MPa. Compared with the 28-day compressive strength of pure Portland cement test block, the strength is increased by 35.84%.

5. Comparison of Test Results Between $CaSO_4$ and $CaCO_3$ Whisker

According to reference [3], considering the mass fractions of calcium carbonate whisker with 0%, 2%, 3% and 4%, the bending and compressive tests of calcium carbonate whisker cement-based test blocks are carried out, respectively. The comparison test results of cement-based materials with calcium sulfate whisker and calcium carbonate are shown in table 5.

Table 5. Comparison Test Results between Calcium Sulfate Whisker and Calcium Carbonate

Item	$CaSO_4$ whisker	$CaCO_3$ whisker
Optimum content for flexural strength	0.4%	3.0%
Increased percentage for flexural strength	19.5%	12.1%
Optimum content for compressive strength	1.0%	2.0%
Increased percentage for compressive strength	35.8%	12.5%

Table 5 shows that for the flexural strength and compressive strength of whisker cement-based materials, the optimum content of calcium sulfate whisker is 2/15 and 1/2 of that of calcium carbonate whisker, respectively. However, the increased percentage for flexural strength and compressive strength of calcium sulfate whisker cement-based material is 1.61 times and 2.86 times than that of calcium carbonate whisker cement-based material, respectively. Therefore, compared with calcium carbonate whisker, calcium sulfate whisker has better enhancement effect on the flexural strength and compressive strength of cement-based materials.

6. Conclusion

When the content of calcium sulfate whisker is 0.2%~1.0%, it has little effect on the fluidity, initial and final setting time of cement slurry.

For the 28-day flexural strength, the optimum fiber content of calcium sulfate whisker is 0.4%, and the flexural strength is increased by 19.54% compared with ordinary Portland cement test block without calcium sulfate whisker.

For the 28-day compressive strength, the optimum fiber content of calcium sulfate whisker is 1.0%, and the compressive strength is increased by 35.84% compared with ordinary Portland cement test block without calcium sulfate whisker.

Compared with calcium carbonate whisker, calcium sulfate whisker has better enhancement effect on the flexural strength and compressive strength of cement-based materials.

7. References

[1] Yao W and Zhong W H 2007 *J. of Wuhan University of Technology-Mater. Sci. Ed.* **22** 52–55
[2] Bentur A, Diamond S and Mindess S 1985 *Journal of Materials Science* **20** 3610–20
[3] Cao M L, Wei J Q and Wang L J 2011 *J. of Wuhan University of Technology-Mater. Sci. Ed.* **26** 749–53
[4] Ma Y P, Zhu B R and Tan M H 2005 *Cement and Concrete Research* **35** 296–300
[5] Ma J H 2005 *Preparation of calcium sulphate whisker and its application study in the paving asphalt* (Shandong: Shandong University of Science and Technology)
[6] Zhang L H 2010 *Research progress on preparation of calcium sulfate whiskers and its application* (Shenyang: Northeastern University)
[7] Yuan Z T, Wang X L, Han Y X and Yin W Z 2008 *Journal of Northeastern University (Natural Science)* **29** 573–76
[8] Zhang X H, Xu L, Han W B, Weng L, Han J C and Du S Y 2009 *Solid State Sciences* **11** 156–61
[9] Muller F A, Gbureck U, Kasuga T, Mizutani Y, Barralet J E and Lohbauer U 2007 *Journal of the American Ceramic Society* **90** 3694–97
[10] Faber K T and Evans A G 1983 *Acta Metallurgica* **31** 565–76

IOP Conf. Series: Materials Science and Engineering **735** (2020) 012037 doi:10.1088/1757-899X/735/1/012037

Research on Preparation of Nano-porous Lithium Iron Phosphate for Lithium-ion Battery Electrode Materials

Quanmao Yu[1,*], Tao Ren[1], Xiaoxia Li[1], Zhong Chen[2], Qiurong Liu[1], Zhenyu Chen[1] and Zhengyi Huang[1]

(1. Institute of Functional Materials, School of Software and Internet of Things Engineering, Jiangxi University of Finance and Economics, Nanchang 330013, PR China
2. School of Materials and Mechanical and Electrical Engineering, Jiangxi Science and Technology Normal University, Nanchang 330013, PR China)
* Corresponding author. Email: bsyqm@163.com

Abstract. Citric acid, nitric acid, ferrous phosphate and lithium carbonate as raw materials, the precursor $Fe_3(PO_4)_2$ were synthesized by precipitation method, and nano-porous lithium iron phosphate ($LiFePO_4$) was prepared by modified sol-gel method. The influence of pH to purity and yield of precursor $Fe_3(PO_4)_2$ and sintering temperature to purity of $LiFePO_4$ were studied. X-ray diffraction (XRD), field emission scanning electron microscopy (FESEM) and transmission electron microscopy techniques (TEM) were used to character the phase and morphology of nano-porous lithium iron phosphate ($LiFePO_4$).

1. Introduction

$LiFePO_4$ has been largely used as electrode materials for lithium-ion batteries due to its environmentally friendly, low cost, high safety, excellent charge and discharge, reversible performance [1–3]. However, $LiFePO_4$ has considerable low lithium-ion diffusion coefficient (10^{-14} $cm^2\,s^{-1}$) [4], which make batteries have longer charging time and smaller specific capacity and batter energy density. These short comes limit its practical applications in high energy density batteries using in electric vehicles (EVs) and other high power equipment [5].

Much work has been done to improve lithium-ion diffusion coefficient or 'effective Li^+ density' in $LiFePO_4$ electrode materials lithium-ion batteries to improve the specific capacity and batter energy density [6-10]. Among them, 'decreasing the $LiFePO_4$ particles size to improve effective Li^+ density' is one of theoretical ideal method [8–10]. Because Li^+ can only passed through a few nanometres thickness of the $FePO_4$/$LiFePO_4$ electrode material interface at the charging and discharging process of the batteries, the 'effective Li^+ concentration' of the large crystal grain (such as particle size >1μm) $LiFePO_4$ electrode material is not high. On the contrary, Nano-size $LiFePO_4$ can provide more 'effective Li^+' due to the larger surface portion volume fraction. The large specific surface and porous structure improve the diffusion of Li^+ extraction and embedding during charging and discharging of the battery, thereby increasing the capacity of the battery. Zhang [11] synthesized $LiFePO_4$ quantum dots, the first discharge capacity at 0.1C rate to 197 $mAhg^{-1}$, which is much higher capacity than that of $LiFePO_4$ materials synthesized by traditional methods.

In this paper, we synthesized ferrous phosphate precursor material by precipitation method. Then used low costs, green and low temperature Modified Sol-Gel method synthesize Nano-porous $LiFePO_4$.

IOP Conf. Series: Materials Science and Engineering **735** (2020) 012037 doi:10.1088/1757-899X/735/1/012037

2. Experimental

2.1. Materials

All chemicals and materials used in this research were analytical-grade reagents (A.R.) and purchased from Sinopharm Chemical Reagent Co., Ltd. They were Ammonium Dihydrogen Phosphate ($NH_4H_2PO_4$), Potassium Dihydrogen Phosphate (KH_2PO_4), Citric Acid ($C_6H_8O_7$), Lithium Carbonate ($LiCO_3$), Ferrous Sulfate ($FeSO_4$), Urea (CH_4N_2O), Nitric acid (HNO_3) and Sodium Hydroxide (NaOH). $NH_4H_2PO_4$, KH_2PO_4, $LiCO_3$ and $FeSO_4$ were the source materials for supplying P, Li and Fe, respectively.

2.2. Synthesis

The preparation process has precursor $Fe_3(PO_4)_2$ synthesized by precipitation method and nano-porous lithium iron phosphate ($LiFePO_4$) prepared by modified sol-gel method. KH_2PO_4 is dissolved in deionization water, and then the solution pH is adjusted by NaOH. $FeSO_4$ is dissolved in deionization water. And then fully mix the solutions till the precipitate is repeatedly. Then wash the precipitate with deionization water. Finally, pure $Fe_3(PO4)_2$ is got after the mixture is dried at 80°C in a vacuum.

Figure 1 shows the Nano-porous $LiFePO_4$ synthese process. $C_6H_8O_7$, $NH_4H_2PO_4$ and $LiCO_3$ are mixed and dissolved in deionization water. Then mixes the solution with $Fe_3(PO_4)_2$ solution. Add dilute HNO_3 into the solution till $Fe_3(PO_4)_2$ is completely dissolved. Adjust the pH by Urea. Dry the solution at 80°C till it become colloid. Finally the colloid is sintered in nitrogen gas.

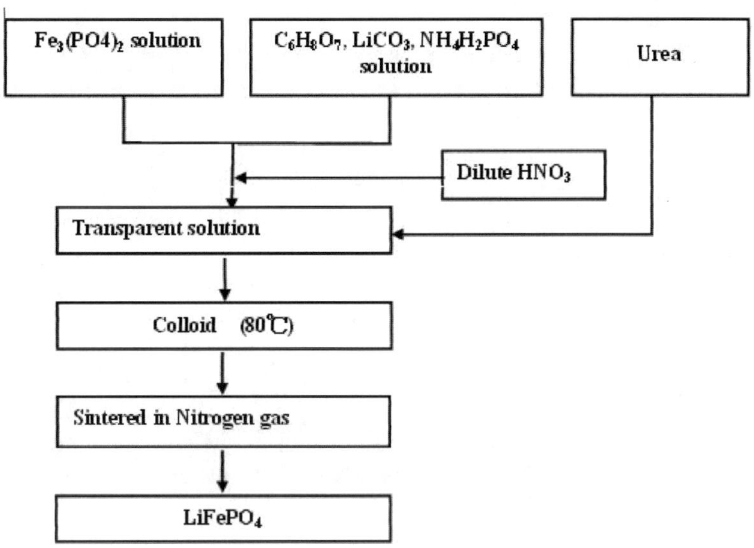

Figure 1. Schematic Diagram of Preparation Process of Nano-porous $LiFePO_4$

2.3. Characterization

The phases were characterized by X-ray diffraction (XRD) Rigaku D/max-rB (Rigaku Co., Japan, Cu-Kα1). The powders size, morphology and structure were characterized by field emission scanning electron microscopy (FESEM, S-3000H, Hitachi Co., Japan) and transmission electron microscopy (TEM, H-7000, Hitachi Co., Japan).

IOP Conf. Series: Materials Science and Engineering **735** (2020) 012037 doi:10.1088/1757-899X/735/1/012037

3. Results and discussion

3.1. Influence of pH to Purity and Yield of $Fe_3(PO_4)_2$

Figure 2 shows XRD patterns of $Fe_3(PO_4)_2{\cdot}8H_2O$ prepared in different pH (from 4.0 to 6.0). It indicates that the $Fe_3(PO_4)_2{\cdot}8H_2O$ phase is very pure by precipitation method in different pH. But our research results indicate that the pH value effects the precursor $Fe_3(PO_4)_2{\cdot}8H_2O$ yield when the molar ratio of KH_2PO_4 to $FeSO_4{\cdot}7H_2O$ is 1:1. The $Fe_3(PO_4)_2{\cdot}8H_2O$ yield is 70.0% (wt.) when the solution pH keeps at around 4.0. The $Fe_3(PO_4)_2{\cdot}8H_2O$ yield is 94.0% (wt.) when the solution pH keeps at around 5.0. The $Fe_3(PO_4)_2{\cdot}8H_2O$ yield is 98.0% (wt.) when the solution pH keeps at around 6.0. However, there will be yellow-white precipitate (Fe^{2+} are oxidized to Fe^{3+}) when the pH above 6.0.

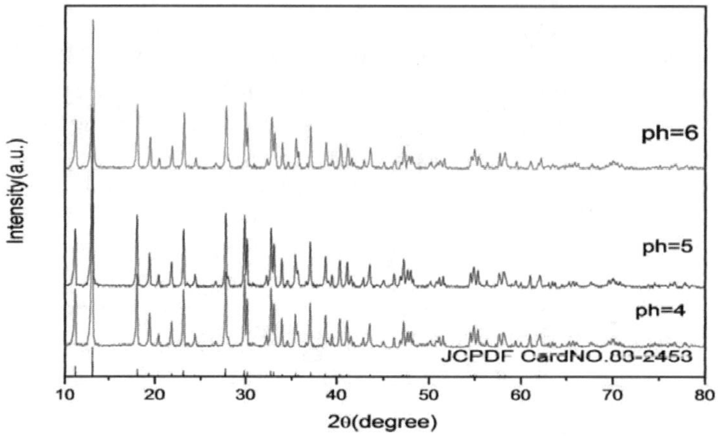

Figure 2. XRD Patterns of $Fe_3(PO4)_2{\cdot}8H_2O$ Precipitate at Different pH

3.2. Influence of Sintering Temperature to Purity of $LiFePO_4$

Figure 3 shows the XRD pattern of $LiFePO_4$ prepared at different sintering temperatures. The diffraction peaks of $LiFePO_4$ by sol-gel method match well with the standard card (40-1499). But the heights of corresponding peaks rise with the sintering temperature arise from 500°C to 600°C. However, the corresponding peaks strength at 450°C are stronger than those of at 500°C, mainly because the amount of $C_6H_8O_7{\cdot}H_2O$ added at 450°C is more than those of at 500°C, which resulting in higher temperature caused by intense instantaneous combustion of citric acid nitrate. (However, if the amount of $C_6H_8O_7{\cdot}H_2O$ added at 450°C was same as 500°C, the red Fe^{3+} compound would appear.)

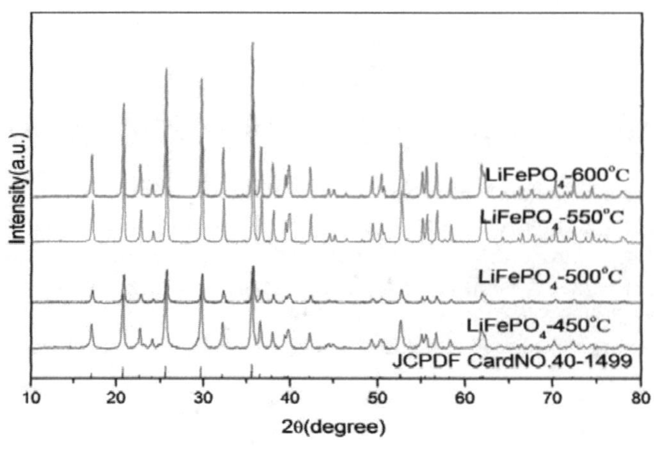

Figure 3. XRD Patterns of $LiFePO_4$ Sintering at Different Temperature

3.3. Morphology Characterization of $LiFePO_4$

Figure 4 shows the FESEM image of $LiFePO_4$ sintered at 500°C for 30minutes. The $LiFePO_4$ particles are homogeneously distributed spherical-like. The $LiFePO_4$ powders loosely agglomerate together forming porous constructions.

Figure 5 shows the TEM image of $LiFePO_4$ sintered at 500°C for 30minutes. The $LiFePO_4$ particles have the sizes of about 100 nm and shape of sphere or ellipsoid.

Figure 4. FESEM Image of Nanoporous $LiFePO_4$

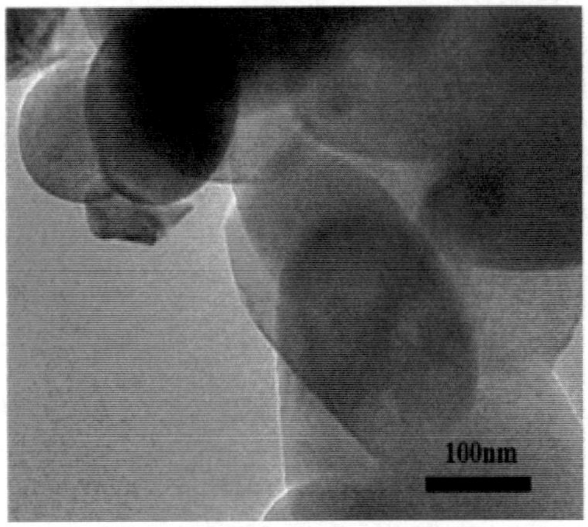

Figure 5. The TEM Image of $LiFePO_4$ Powers

4. Conclusion

Nano-porous $LiFePO_4$ with particle size about 100 nm for Lithium-ion Battery Electrode Materials was synthesized by Citric-Urea-Nitrate Combustion aid modified Sol-Gel method. The pure precursor $Fe_3(PO_4)_2 \cdot 8H_2O$ synthesized by precipitation method had the yield of 98.0% (wt.) when the precipitation solution pH was 6.0 with molar ratio of KH_2PO_4 to $FeSO_4 \cdot 7H_2O$ is 1:1. The Nano-porous pure $LiFePO_4$ with the particle size of about 100 nm is prepared sintering 30 minutes at 500°C in nitrogen gas.

5. Acknowledgements

This work was supported by the Key Science Project of Jiangxi Provincial Department of Education (No. GJJ180246).

6. Reference

[1] Padhi A K, Nanjundaswamy K S and Goodenough J B. Phosphoolivines as positive-electrode materials for rechargeable lithium batteries 1997 *J Electrochem Soc.* **144**: 1188-1194.
[2] Wang J J and Sun X L. Olivine $LiFePO_4$: the remaining challenges for future energy storage 2015 *Energ Environ Sci.* **8**: 1110-1138.
[3] Giorgetti M, Berrettoni M, Scaccia S and Passerini S. Characterization of sol-gel-synthesized $LiFePO_4$ by multiple scattering XAFS 2006 *J. Inorg. Chem.* **45**: 2750-2757.
[4] Prosini P P, Lisi M, Zane D and Mauro P. Determination of the chemical diffusion coefficient of lithium in $LiFePO_4$ 2002 *Solid State Ionics.* **1481**: 45-51.
[5] Lei X L, Zhang H Y, Chen Y M, Wang W G and Ye Y P. A three-dimensional $LiFePO_4$/carbon nanotubes/graphene composite as a cathode material for lithium-ion batteries with superior high-rate performance 2015 *J. Alloy Compd.* **626**: 280-286.
[6] Guan Y B, Shen J R, Wei X F, Qizhen Z, Xiaohui Z and Shuqin X. $LiFePO_4$/activated carbon/graphene composite with capacitive-battery characteristics for superior high-rate lithium-ion storage 2019 *Electrochimical Acta.* **294**: 148-155.
[7] Jing J D, Ying L, Han W Z, Hai B Y, Lei W and Zhou H D. Three-Dimensional Architecture

Reduced Graphene Oxide-$LiFePO_4$ Composite: Preparation and Excellent Microwave Absorption Performance 2019 *Inorg Chem.* **58**(3): 2031-2041.
[8] Kim H J, Bae G H, Lee S M, Ahn J H and Kim J K. Properties of lithium iron phosphate prepared by biomass-derived carbon coating for flexible lithium ion batteries 2019 *Electrochimica Acta.* **300**: 18-25.
[9] Mollazadeh M, Habibi B. $LiFePO_4$/Carbon/Reduced Graphene Oxide Nanostructured Composite as a High Capacity and Fast Rate Cathode Material for Rechargeable Lithium Ion Battery, 2019 *Catalysis Letters.* **148**: 7-18.
[10] Paolella A, Bertoni G, Marras S, Enrico D, Massimo C, Mirko P, Andreas R, Mauro P, Alberto A, Karim Z, Liberato M and Chandramohan G. Etched colloidal $LiFePO_4$ nanoplatelets toward high-rate capable Li-Ion battery electrodes 2014 *Nano Lett.* **14**(12): 6828-35.
[11] Zhang X D, Bi Z Y, He W and Yang G. Fabricating high-energy quantum dots in ultra-thin $LiFePO_4$ nanosheets using a multifunctional high-energy biomolecule-ATP 2014 *Energ Environ Sci.* **7**(7):2285-2294.

Hybrids of Nickel-histidine Functionalized Graphene Quantum Dots: Fabrication and Non-enzymatic Glucose Sensing

Yongqiang Yang, [1*†] Wei Chen, [2†] Qinsheng Wang[1] and Qun Wang[1]

1 National Graphene Products Quality Supervision and Inspection Center (Jiangsu), Jiangsu Province Special Equipment Safety Supervision Inspection Institute·Branch of Wuxi, Yanxin Road 330, Wuxi 214174, China
2 School of Chemical and Material Engineering, Jiangnan University, Wuxi 214122, China
[*] E-mail: yqyang@wxtjy.com
[†] These two authors contributed equally to the work.

Abstract. In this work, a method of in-situ recombination of Ni^{2+} with histidine functionalized graphene quantum dots (His-GQDs) was constructed a three-dimensional network porous structure. The collected precipitate complex precursor was oxidized and the NiO-His-GODs were obtained. The NiO-His-GQDs compound was subjected to high-temperature thermal reduction by inert gas to form Ni-His-GQDs. On the one hand, the intimate contact between Ni nanoparticles and His-GQD greatly shortens the gap between the two complexes, resulting in faster electron migration speed; On the other hand, the conductive (Ni)/semiconductor (His-GQDs) catalytic interface can produce Stokes diode-like structures, accelerate the migration rate of holes carriers in the semiconductor, and exhibit good electrocatalytic activity. A Ni-His-GQDs modified electrode as non-enzymatic glucose electrochemical sensor was shown good electrocatalytic activity and anti-interference property. The concentration of glucose was detected by amperometric analysis, and showed a good linear relation in the concentration range of 5.0×10^{-6}~2.0×10^{-3} M. The detection limit (S/N = 3) is 1.7×10^{-6} M. In addition, interference of sodium chloride (NaCl), uric acid (UA), dopamine (DA) and ascorbic acid (AA) was less than 5.0 % in the same conditions indicating that the hybrids modified electrode sensor could be used for the sensitive and selective detection of glucose.

1. Introduction

Glucose is widely distributed in nature and played an important position as the main functional materials of biology. So the detection of glucose is also important in lots of field, especially in medical treatment field. [1-4] Up to now, the biosensor technology was often used for the detection of glucose, especially obtained the information of glucose by glucose oxidase (GOD). Because GOD can quickly catalyze glucose to gluconic acid and H_2O_2, thereby further identify the information of glucose. [5, 6] However, the GOD activity often is influenced by temperature, humidity, etc. [7, 8] Hence, there are interest to fabricated glucose biosensor which can detect glucose through the direct oxidation of glucose. At present, a part of metallic, metal oxide and alloy nanoparticles have found the ability as electrocatalyst materials, especially the Ni-based materials. [1, 9-17] Because the Ni-based materials could form a high-valent, oxyhydroxide species [NiOOH] in the oxidation processes, and could is obtained in large numbers at low cost. [14, 17]

Graphene, a single atomic layer that consisted of sp2 carbon atoms covalently bonded in a hexagonal manner, has excellent electrical, thermal, mechanical and other physical and chemical properties.[14-18] Moreover, graphene have large surface which afforded the place to disperse metal nanoparticles and obtained nanocomposites. The graphene materials could enhance the electron transport and keep the active of metals nanoparticles in the composites, which is beneficial to the composites as electrochemical sensing materials for the detection of glucose.[13-17] Different from common graphene, graphene quantum dots (GQDs) is less than 100 nm in lateral sizes, have several unique properties besides properties of graphene materials.[19-20] The studies have revealed that GQDs as better electron transporter and acceptor could using an all-right candidate for electrochemical sensing materials.[20, 21]

In this work, the hybrids of Ni-His-GQDs hybrids were obtained and served as electrochemical electrode material for glucose electrochemical detection. As electrochemical electrode, Ni-His-GQD hybrids modified electrodes exhibited high sensitivity, low detection limit, and excellent selectivity in non-enzymatic glucose electrochemical detection.

2. Experimental

2.1. Materials

D-glucose, citric acid, histidine and nickel chloride hexahydrate ($NiCl_2 \cdot 6H_2O$) were bought from Sinopharm Chemical Reagent Co., Ltd (Shanghai, China). His-GQD was generated as described in Ref. [22] Phosphate-buffered saline (PBS, pH 7.0, 0.01 M) was used as the supporting electrolyte. Ultrapure water (18.2 M cm) generated by a Milli-Q purification system was used for all the experiment. All other reagents were analytical purity grade and were bought from Shanghai Chemical Company (Shanghai, China).

2.2. Synthesis of Ni-His-GQDs hybrids

A $NiCl_2 \cdot 6H_2O$ aqueous solution (0.25 M, 40 mL) was added into the His-GQDs dispersion (50 mg mL^{-1}, 20 mL, pH 7) with stirring. The precipitate was collected and then further oxidized at 300 °C for 3 h in air and the NiO-His-GQD was obtained. Finally, Ni-His-GQDs hybrids were obtained after thermal reduction processing at 600 °C for 0.5 h in N_2 atmosphere. Meanwhile, for comparison, NiO-His-GQD and His-GQD were also treated using the same processing.

2.3. Electrochemical Characterization

The electrochemical experiments were performed with a CHI660D electrochemical workstation (Shanghai, China) using a three electrode system. The Ni-His-GQD, NiO-His-GQD and His-GQD modified or bare GCEs (Φ = 1.0 mm) were employed as work electrode, an Ag/AgCl electrode (saturated KCI) as reference electrode, a platinum wire as counter electrode. The GCEs were modified with different materials to obtain different sensors by the same procedure: Ni-His-GQD, NiO-His-GQD and His-GQD dispersing in the chitosan solution (5 mg mL^{-1}) and the suspension (2 mg mL^{-1}) was obtained, respectively. Then these suspensions (10 mL) were dropwise on the GCEs surface, and were dried in air for using, respectively. All measurements experimental solution was degassed by nitrogen for at least 15 min under room temperature.

3. Results and Discussion

3.1. Synthesis and Characterization of Material

Because His-GQDs contains multiple imidazole groups, which can coordinated with Ni^{2+} and rapidly extend outward with Ni-N bonds to form the three-dimensional network precursor. So, first, His-GQDs reacted with Ni^{2+} to form Ni-His-GQD precursor. Second, the precursor was further oxidized at 300 °C in air atmosphere; thereby gain NiO-His-GQDs hybrids. In order to reduce the NiO and

IOP Conf. Series: Materials Science and Engineering **735** (2020) 012038 doi:10.1088/1757-899X/735/1/012038

improve the crystallinity of the His-GODs, the NiO-His-GQDs hybrids were further treated at 600 °C in N_2 atmosphere. Finally, Ni-His-GQDs hybrids were obtained.

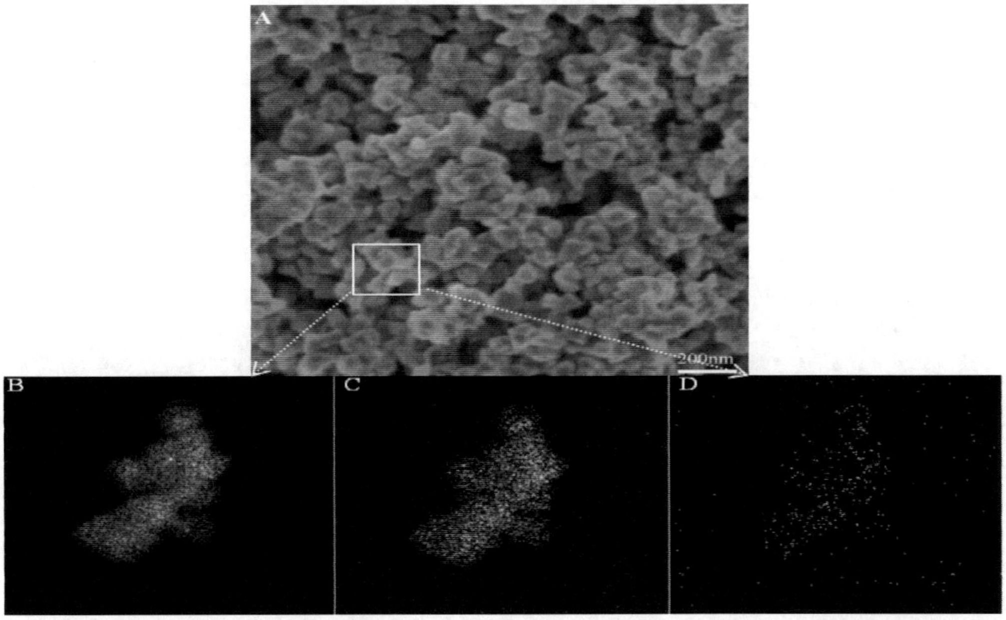

Figure 1. The SEM (A), Nickel Element (B), Carbon Element (C), Nitrogen Element (D) Mapping of the Ni-His-GODs Hybrids

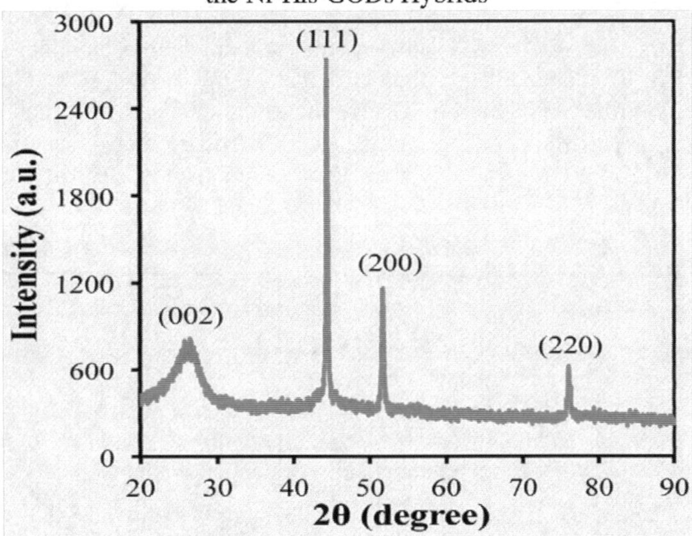

Figure 2. The XRD Spectrum of the Ni-His-GODs Hybrids

The morphology and chemical composition of Ni-His-GQD hybrids were examined by SEM and EDX, and the results were shown in Figure 1. The Ni-His-GQD hybrids show hierarchical architecture (Figure 1A). Nevertheless, the architecture morphologies of the hybrid was preserved properly after the annealing process when the NiO-His-GQD complex conversing to Ni-His-GQD hybrids. Moreover, the size dimension of the Ni-His-GQD hybrids particles were about 50 nm. The results of EDX analysis revealed that the hybrids contained of nickel (Ni), carbon (C) and nitrogen (N) elements (Figure 1C). The nitrogen element could only come from the imidazole groups of His-GODs.

In the Ni-His-GQD hybrids XRD pattern (Figure 2), the diffraction peaks corresponding to the diffractions of (111), (200), (220) of the Ni (fcc), respectively (JCPDS 04-0805). A broad diffraction peaks at ~26.0° is corresponding to the diffraction of (002) of His-GQDs materials, suggesting the His-GQD successful reduction in thermal reduction processing.

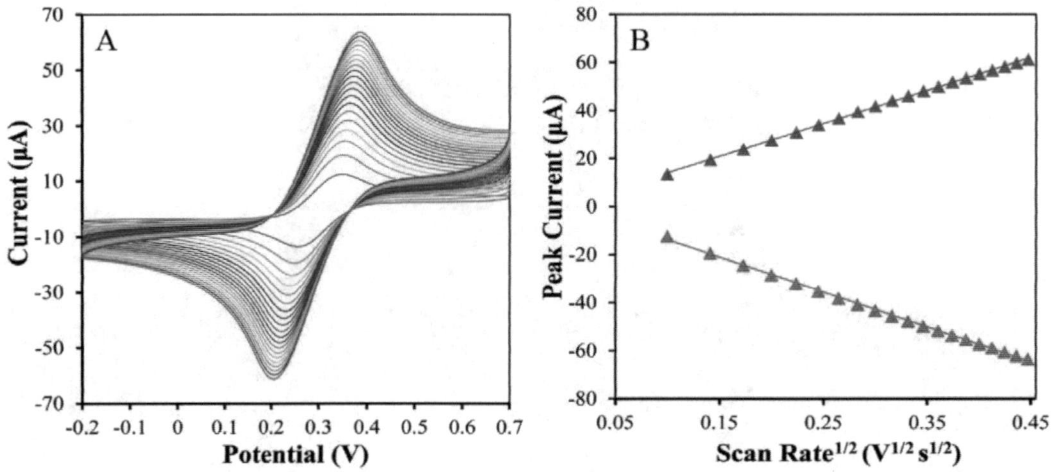

Figure 3. A: CV Curve of Ni -His-GQDs Hybrids Modified Electrode in a PBS (pH 7.0, 1.0 mM $K_3[Fe(CN)_6]$) at Various Scan Rates. B: Ip vs. Square Root of Scan Rate

To investigate the electrocatalytic activity of Ni-His-GQDs hybrids modified electrode, the cyclic voltammograms (CV) were measured at different scan rates (Figure 3). The results revealed that the anodic peak occur a significantly positive shift and the cathodic peak occur a negatively shift with the increase of scan rates. This phenomenon indicated that the Ni-His-GQD hybrids-modified GCE is rapid and reversible. Furthermore, Figure 3B reveals anodic and cathodic peak currents of Ni-His-GQDs hybrids modified electrode is linearly scale with the square root of scan rates, suggesting that the redox reaction of $K_3[Fe(CN)_6]$ is a typical diffusion-controlled electrochemical process at the electrode surface.[13,14]

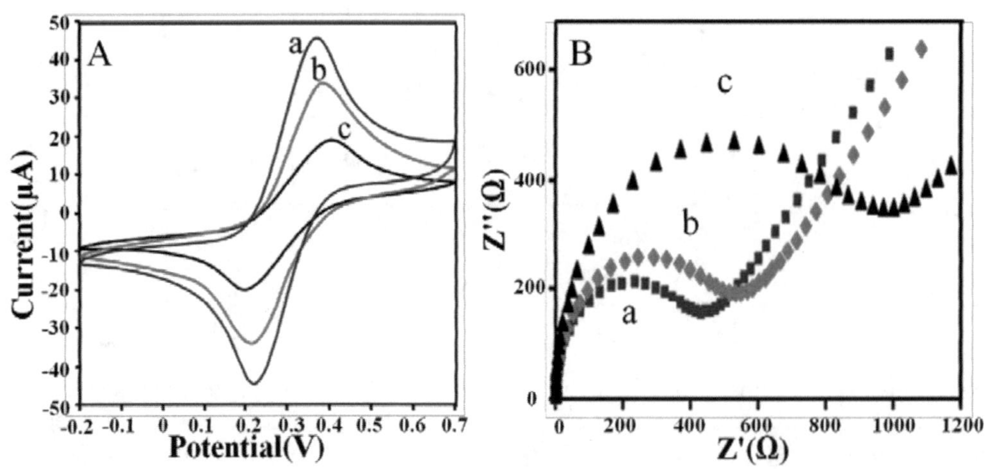

Figure 4. CV (A) and EIS (B) Curve of Ni -His-GQDs (a), NiO-His-GQDs (b) and His-GQDs (c) Hybrids Modified Electrode in a PBS (pH=7, 1.0 mM $K_3[Fe(CN)_6]$) at the Scan Rate of 100 mV s^{-1}

To compare the electrocatalytic activity, their CV behaviours of NiO-His-GQDs, His-GQDs and Ni-His-GQDs hybrids modified electrodes were studied in PBS (pH 7.0, 1.0 mM $K_2Fe(CN)_6$) at the scan rate of 100 mV s^{-1}(Figure 4A). All CV curves of the hybrids electrode shown obvious redox peak, displaying the oxidation and reduction reactions of $K_2Fe(CN)_6$ in electrodes surface. Moreover, the Ni-His-GQDs hybrids modified electrode exhibited the highest peak currents. This is mainly due to the Ni crystals nanoparticles in hybrids have better the electrical conductivity and electrochemical properties compared with NiO crystals nanoparticles, resulting to an evidently improved electrocatalytic activity.

EIS is used to investigate the charge transfer ability of the different electrodes surface (Figure 4B). The Nyquist complex plane impedance plots included a semicircle and a linear portion. The high-frequencies semicircle region corresponds to the electron transfer limited process, and the low-frequencies region linear portion is assigned to the diffusion process. The electron transfer resistance (Rct) of the electrodes surface can be quantified through the diameter of the semicircle diameter. The Rct of Ni-His-GQDs hybrids sensor was much smaller than that of His-GQDs and Ni-His-GQDs hybrids sensors, indicating Ni nanoparticles are beneficial for electron transfers in the hybrids and probably generate synergistic effects. Therefore, Ni-His-GQDs hybrids could dramatically improve the electron transfer efficiency between the electrolyte and electrode interface.

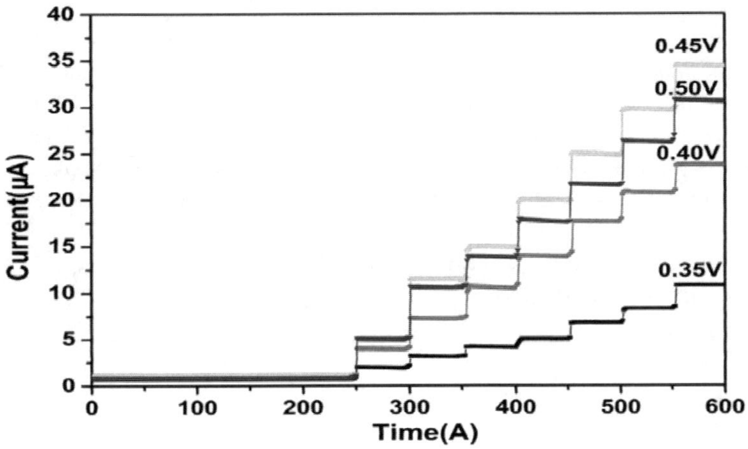

Figure 5. Amperometric Current Response of Ni-His-GQDs Hybirds Sensor at Different Applied Potentials with a Dropwise Addition of 2 mM Glucose into the 0.1 M NaOH

To obtain better detection potential on amperometric current response of sensors, the effect of the applied potential was further investigated from +0.35 V to +0.50 V. The result revealed that the current response is gradually increase from 0.35 V to 0.45 V, and then begin to decrease (Figure 5). It indicated that the applied potential of +0.45 V has the maximum current response and is the optimized suitable working potential in the subsequent studies. It is mainly because of the high potential activated interfering substances and generated many intermediates, which would interact with the electrode, thereby leading to the lowered current. In the meanwhile, the irreversible absorption of intermediates could further harm the electrode and hindered the glucose oxidation. [1]

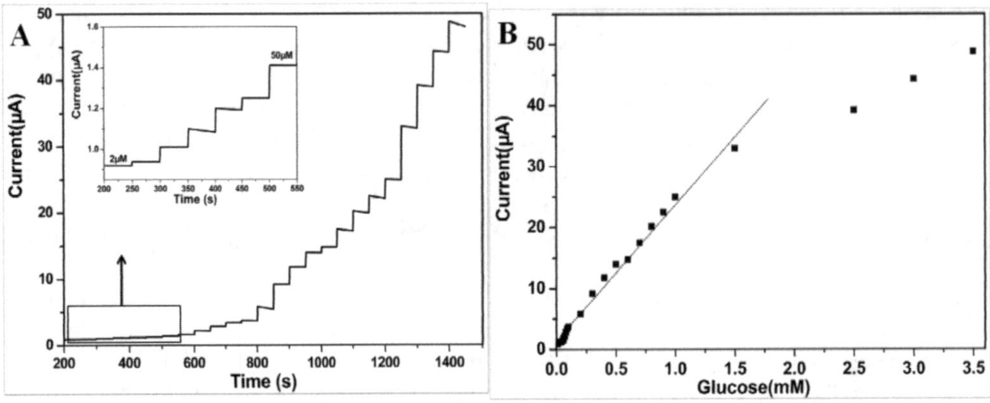

Figure 6. A: Amperometric current response of Ni-His-GQDs hybrids modified electrode on the successive addition of glucose into NaOH (0.1 M) at 0.45 V. The inset shows a magnification of the current response in a low concentration of glucose; B: Current response versus glucose concentration

To confirm the amperometric responses steady state of the Ni-His-GQDs hybrids modified electrode, the dynamic steady state current response of the interface was measured through successive additions of glucose. Figure 6A displayed a current-time plot of the Ni-His-GQDs hybrids modified electrode interface on successive additions of glucose at the +0.45 V potential. The result shows a very quickly response to glucose. The inset results revealed that the current response could begin to the addition of 2 μM glucose (inset in Figure 6A). In addition, the sensor exhibits a broad linear detection range of 5.0×10^{-6}~2.0×10^{-3} M (Figure 6B). The linear regression equation is IP (μA) = 20.89 C (mM) + 1.7713. Based on a signal-to-noise ratio of 3, the detection limit is 1.7×10^{-6} M, suggesting the extremely fast and high sensitivity of the electrochemical electrode.

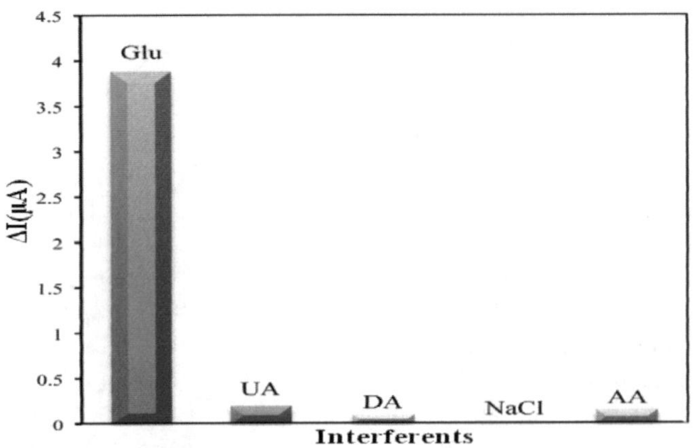

Figure 7. The Bar Diagram of Normalized Amperometric Current against Interferents

Besides the above factors, other substances probably simultaneous oxidize at the same applied potential which is also one challenge in detection field of glucose. In addition, chloride ions could also poison electrodes in some cases. So, the effect of other four kinds of interferential compounds, NaCl, UA, DA and AA were investigated. The result presented in Figure 7, the current response of these four kinds interferential were all less than 5.0%, suggesting that the Ni-His-GQDs hybrids modified electrode can be used in the sensitive and selective detection of glucose, and could neglect the interference of NaCl, UA, DA and AA, etc.

4. Conclusions

In summary, the Ni-His-GQDs hybrids were obtained through coordination between Ni^{2+} and His-GQDs, oxidation and reduction processing. The good electrical contacts between Ni and reduced His-GQD afford synergistically exceptional electrochemical properties in the Ni-His-GQDs hybrids which could obviously enhance catalytic activity. The Ni-His-GQDs hybrids modified electrode exhibited high sensitivity, quickly response and excellent stability for glucose detection. It will be the most promising composites for fabricating Ni-His-GQD hybrids for practical nonenzymatic glucose sensors in biotechnology and clinical diagnosis.

5. Acknowledgment

This work was supported by the Science Foundation of Jiangsu Province Special Equipment Safety Supervision Inspection Institute (KJ(Y) 2017014).

6. References

[1] S. Fu, G. Fan, L. Yang, F. Li, Electrochimi. Acta, 2015,152 146–154
[2] C. Chen, Q. Xie, D. Yang, H. Xiao, Y. Fu, Y. Tan, S. Yao, RSC Adv., 2013, 3(14): 4473-4491
[3] K. J. Chen, C. F. Lee, J. Rick, S. H. Wang, C. C. Liu, B. J. Hwang, Biosens. Bioelectron., 2012, 33 (1): 75-81
[4] A. Heller, B. Feldman, Chem. Rev., 2008, 108, 2482-2505
[5] Y. Liu, M. K. Wang, F. Zhao, Z. Xu, S. J. Dong, Biosens. Bioelectron., 2005, 21, 984-988.
[6] P. Si, S. J. Ding, J. Yuan, X. W. Lou, D. H. Kim. ACS Nano, 2011, 5, 7617-7626.
[7] T. W. Tsai, G. Heckert, L. F. Neves, Y. Tan, D. Y. Kao, R. G. Harrison, D. E. Resasco, D. W. Schmidtke, Anal. Chem., 2009, 81(19): 7917-7925
[8] Z. G. Zhu, L. Garcia-Gancedo, A. J. Flewitt, H. Q. Xie, F. Moussy, W. I. Milne, Sensors, 2012, 12, 5996.
[9] J. Zhang, X. Zhu, H. Dong, X. Zhang, W. Wang, Z. Chen, Electrochim. Acta 105 (2013) 433–438.
[10] K. Li, G. Fan, L. Yang, F. Li, Sensor. Actuat. B-Chem., 199 (2014) 175–182.
[11] S. Liu, B. Yu, T. Zhang, Electrochim. Acta, 102 (2013) 104–107.
[12] A. Ciszewski, I. Stepniak, Electrochimi. Acta, 111 (2013) 185–191.
[13] A. Zhan, C. Liu, H. Chen, H. Shi, L. Wang, P. Chen, W. Huang, X. Dong, Nanoscale, 2014, 6(13): 7424-7429
[14] P. Subramanian, J. Niedziolka-Jonsson, A. Lesniewski, Q. Wang, M. Li, R. Boukherroub, S. Szunerits, J. Mater. Chem. A, 2014, 2, 5525-5533
[15] G. Li, H. Huo, C. Xu, J. Mater. Chem. A, 2015, 3, 4922-4930
[16] Z. Wang, Y. Hu, W. Yang, M. Zhou, X. Hu. Sensors, 2012, 12, 4860-4869
[17] J. Yang, J. H. Yu, J. R. Strickler, W. J. Chang, S. Gunasekaran, Biosens. Bioelectron., 2013, 47, 530–538
[18] Y. Yang, R. Pang, X. Zhou, Y. Zhang, H. Wu, S. Guo, J. Materi. Chem., 2012, 22, 23194-23200
[19] Y. Yang, Q. Wang, Z. Liu, W. Liu, Y. Wen, L. Jin, Y. Liu, B. Ou, Adv. Eng. Res., 2018, 120, 1611-1615
[20] W. Chen, R. Li, Z. Li, Y. Yang, H. Zhu, J. Liu, J. Alloy. Compd., 2019, 777, 1001-1009
[21] H. Chu, J. Hu, Z. Li, R. Li, Y. Yang, X. Sun, Sensor. Actuat. B-Chem., 2019, 298, 126866
[22] X.Y. Zhou, R.Y. Li, Z.J. Li, G.Z. Gu, G.L. Wang, RSC Adv., 2016, 6, 42751-42755
[23] Z. Chen, Y. Zhang, J. Zhang, J. Zhou, J. Electrochem. Soc, 166 (2019) B787-B792.

Simulation Study of the Effect of Projectile Length-Diameter Ratio on Anti-Penetration Performance of Composite Armor

Yuanhao Zhang[1,2], Zhongqing Cheng[2], Yanru Li[2] and Ruijie Han[1,2]

1. College of Surface Ship and Ocean, Naval University of Engineering, 717 Jiefang Road, Wuhan, Hubei, China.
2.Department of Coast Defence Engineering, Naval Logistics College of PLA, 1Hebei Road, Binhai, Tianjin, China.
Email: 158241904@qq.com

Abstract. We studied the failure modes of 55 g cylindrical projectile penetrating ceramic/titanium alloy/gap/polyethylene/gap/921A steel composite armor structure by numerical simulation, obtained the influence of projectile length-diameter ratio on the anti-penetration performance of structure by analysis. The results showed that, ceramic breaks in-plane and forms ceramic cone, titanium alloy target shows shear failure and bending deformation, 921A steel target shows adiabatic shear failure mode when it is penetrated, bulging when it is not penetrated, the front polyethylene target shows compressive shear deformation, the rear polyethylene target shows bending deformation, and fiber tensile failure is the main failure mode, the impact fiber layer has obvious delamination and degumming phenomenon. Under the same mass condition, the larger the length-diameter ratio of the projectile, the smaller the size of the breach and the greater the penetration depth. With the further increase of the length-diameter ratio, the impact of the projectile on the penetration effect is weakened.

1. Introduction

Anti-ship missiles are developing all over the world. Semi-armour-piercing anti-ship missiles invade and explode inside the ship. The high-speed fragments produced will seriously affect the important cabins and personnel of the ship. Therefore, the anti-armour-piercing problem of the ship bulkhead has attracted wide attention. In order to improve the vitality of warships, the armor of warships has gradually evolved from single homogeneous metal armor to composite armor. Ceramic armor has the characteristics of low density, high hardness and high compressive strength, so it is suitable for the bullet-proof panel of composite armor [1], however, because of its poor plasticity and low fracture strength, ceramics cannot be used as protective armor alone, so it needs rigid backplane support [2]. With the development of bullet-proof armor, fiber reinforced composites (FRC) will be widely used in naval armor structures because of their high specific strength, specific stiffness and good impact resistance. Yuxin Xu [3-4] carried out experimental and numerical simulation studies on sandwich panels with different ratios of aramid FRC laminates as sandwich materials under high-speed impact of 10 g FSP. It was concluded that aramid laminates as sandwich materials have better elastic resistance. Masta et al. [5] designed four structural models of polyethylene (PE) with a total thickness of 11.7 mm before and after wrapping 31.6 mm thick aluminium alloy plate. High-speed ballistic experiments were carried out with 12.7 mm spherical projectile. It was concluded that the anti-penetration performance of the rear wrapping structure is the best.

In this paper, finite element analysis software ANSYS/LS-DYNA was used to study the failure mode of 55g cylindrical projectile penetrating ceramic/titanium alloy/gap/polyethylene/gap/921A steel composite armor structure. The influence of the length-diameter ratio of projectile on the elastic performance of the structure was obtained through analysis.

2. Finite Element Simulation

The finite element software ANSYS/LS-DYNA was used to simulate the penetration of high-speed cylindrical projectile into ceramic/titanium alloy/gap/polyethylene/gap/921A steel composite armor structure. The projectile bodies adopted 55g cylindrical projectile with sizes of *d* (diameter) 12.2mm and *l* (length) 60mm, *d*13.4mm and *l*50mm, *d*15mm and *l*40mm, *d*17.4mm and *l*30mm, *d*21.2mm and *l*20mm. In the structure, the size of ceramics is 100 mm x 100 mm x 10 mm, the size of titanium alloy plate is 200 mm x 200 mm x 6 mm, the bullet-proof layer of polyethylene is made of four polyethylene laminates, the size of single polyethylene laminate is 200 mm x 200 mm x 20 mm, the size of 921A steel plate is 200 mm x 200 mm x 8 mm, and the structural gap is 20 mm. Solid 164 three-dimensional solid elements with 8 nodes and Lagrange mesh were used to model the projectile, ceramic, titanium alloy, polyethylene and 921A steel plates. The projectile body was divided into 1 mm 1 mesh along the diameter and length direction, and the ballistic layer was divided into 1/2 mm uniform mesh along the thickness direction. The in-plane mesh was densified in the penetration area and transited to the surrounding sparse. The numerical model was based on cm-g-us unit system. The finite element model (projectile *d*15mm, *l*40mm) is shown in figure 1 and 2.

Figure 1. The Finite Element Integral Model

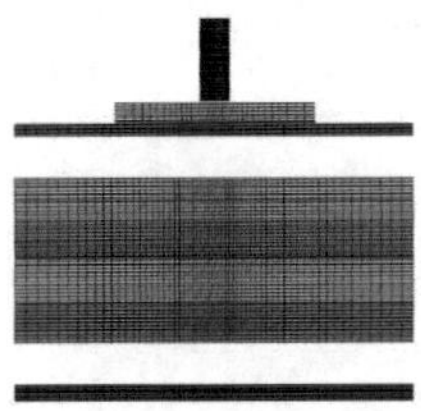

Figure 2. Cutaway of Finite Element Model

The projectile material is 35CrMnSiA low alloy ultra-high strength steel. The projectile and 921A steel adopt MAT_PLASTIC_KINEMATIC bilinear elastic-plastic constitutive model. The strain rate effect of the model is described by Cowper-Symonds model. The material parameters are shown in table 1.

Table 1. Mechanical Properties of Steel

Material	ρ/g·cm^{-3}	*E*/Gpa	μ	σ_y/Mpa	E^T/Mpa	*C*	*P*
35CrMnSiA	7.85	210	0.3	1280	525	40	5
921A	7.80	210	0.3	590	450	40	5

SiC ceramics are described by Johnson-Holmquist_ceramics material model and Gruneisen equation of state. The material parameters are shown in table 2.

Table 2. Material Parameters of SiC Ceramics

Parameter	Figure	Parameter	Figure	Parameter	Figure
ρ/g cm^{-3}	3.10	σ_{HEL}/GPa	6.7	D_2	0.45
G/GPa	192	P_{HEL}/GPa	5.9	K_1	2.05
A/GPa	0.96	*T*/GPa	1.37	K_2	0
B/GPa	0.35	*m*	1	K_3	0
C	0.0	*n*	0.65	*FS*	1.6
EPSI	1×10-6	D_1	0.005		

Titanium alloys are described by Johnson-cook material model and Gruneisen equation of state. The material parameters are shown in table 3.

Table 3. Material Parameters of Titanium Alloys

Parameter	Figure	Parameter	Figure	Parameter	Figure
ρ/g cm^{-3}	4.51	m	1.1	D_3	0
A/GPa	0.922	T_m/K	1941	D_4	0
B/GPa	0.7	T_0/K	300	D_5	0
n	0.93	D_1	0.8		
c	0.014	D_2	0		

Mat_Composite_Damage material model is used to define the brittle failure criterion of polyethylene laminates by Chang-Chang. The material parameters are shown in table 4.

Table 4. Material Parameters of Polyethylene

Parameter	ρ/g cm^{-3}	E_1/GPa	E_2/GPa	E_3/GPa	υ_{12}	υ_{13}	υ_{23}	G_{12}/GPa	G_{13}/GPa	G_{23}/GPa	K_{fail}/GPa
Figure	0.97	30.7	30.7	30.7	0.008	0.044	0.044	0.85	0.67	0.67	2.2
Parameter	AOPT	MACF	S_c/GPa	X_t/GPa	Y_t/GPa	Y_c/GPa	α	S_n/GPa	S_{yz}/GPa	S_{xz}/GPa	
Figure	0	1	0.48	3.5	3.5	2.5	0.5	0.95	0.95	0.95	

3. Numerical Calculation and Analysis

3.1. Effect of Length-Diameter Ratio of Projectile on Structural Failure Modes

In order to study the influence of projectile length-diameter ratio on the failure modes of composite armor structures, five kinds of projectile sizes were adopted: d (diameter) 12.2mm and l (length) 60mm, d13.4mm and l50mm, d15mm and l40mm, d17.4mm and l30mm, d21.2mm and l20mm, the initial velocity of the projectile was defined as 1600 m/s. Through numerical simulation, the deformation of the composite structure under penetration by different length-diameter ratios of projectiles is obtained. The concrete results are shown in figure 3.

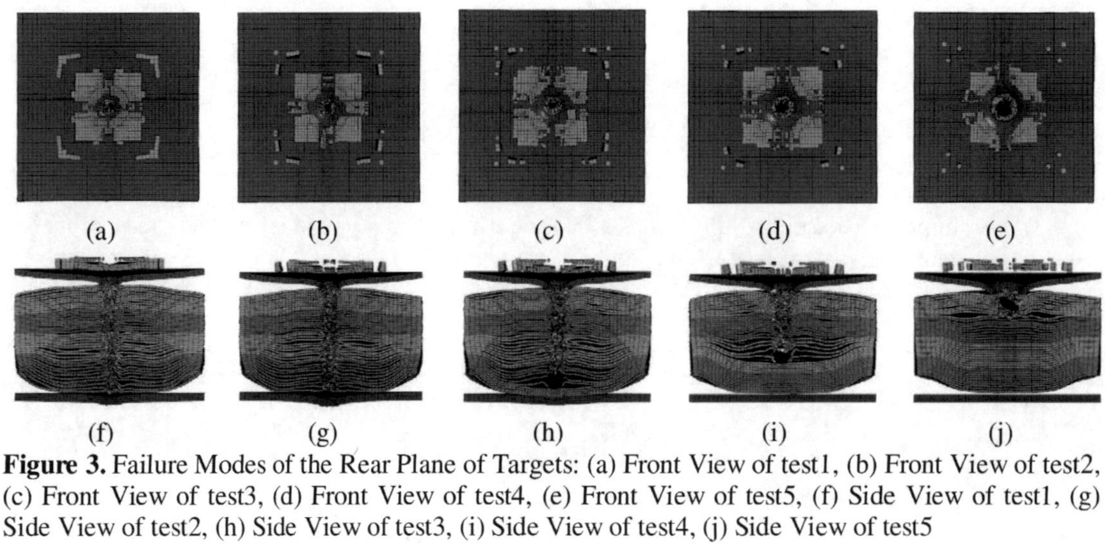

Figure 3. Failure Modes of the Rear Plane of Targets: (a) Front View of test1, (b) Front View of test2, (c) Front View of test3, (d) Front View of test4, (e) Front View of test5, (f) Side View of test1, (g) Side View of test2, (h) Side View of test3, (i) Side View of test4, (j) Side View of test5

Figure 3 shows the failure morphology of the front and side view of the structure in test 1-5. From the observation of figure 3 (a) ~ (e), it can be seen that due to the non-uniformity of ceramics, irregular cracks are formed in the panel after impact by the projectile. The compression wave formed when the projectile contacts the ceramic panel reflects to the tensile wave through the boundary, and the microcracks in the plane propagate under the action of the tensile wave, which ultimately leads to the failure of the ceramic panel. Comparing with figure 3 (a) ~ (e), it is found that the size of ceramic crack is much larger than the diameter of projectile, and the smaller the length-diameter ratio of projectile, the larger the size of crack, the more serious the cracking degree of ceramic panel. As the length-diameter ratio decreases, the diameter of projectile body increases, and the cross-sectional area of projectile cavity increases correspondingly when the projectile penetrates the structure. From the observation of figure 3 (f) ~ (j), it can be seen that the titanium alloy target plate shows shear failure accompanied by bending deformation; the front polyethylene target plate shows compressive shear deformation, the rear polyethylene target plate shows bending deformation, and tensile failure of fibers is the main failure mode, and the impacted fiber layer is accompanied by obvious delamination and degumming phenomenon. The 921A steel target plate shows adiabatic shear failure mode when it is penetrated, it will form bulge when it is extruded by the rear polyethylene plate. Compared with figure 3 (f) ~ (j), it is found that with the increase of the length-diameter ratio of projectile, the contact section between the projectile and the target decreases, the resistance suffered during the penetration process decreases correspondingly, and the penetration depth increases.

3.2. Effect of Length-Diameter Ratio of Projectile on Anti-penetration Performance of Structure

In order to study the impact of the projectile length-diameter ratio on the anti-penetration performance of composite armor structure, the initial velocity of the projectile body was defined as 1000 m/s, increasing 100 m/s one by one until 1300 m/s. The above 20 conditions were calculated by numerical simulation, and the effects of different projectile length-diameter ratios on the penetration depth of polyethylene bullet-proof plate and the break size of titanium alloy target plate in composite structures were obtained. The results are shown in table 5.

Table 5. Numerical Results of Structural Anti-penetrating

Number	Initial velocity (m/s)	Length-diameter ratio	Penetration depth (h_s/mm)	Break size (d_s/mm)	Number	Initial velocity (m/s)	Length-diameter ratio	Penetration depth (h_s/mm)	Break size (d_s/mm)
1		0.94	4	28.5	11		0.94	8	28.7
2		1.72	14	22.6	12		1.72	26	22.9
3	1000	2.67	26	19.2	13	1200	2.67	44	19.5
4		3.73	40	17.7	14		3.73	66	17.3
5		4.92	50	14.7	15		4.92	80	15.3
6		0.94	8	28.5	16		0.94	10	29.3
7		1.72	20	22.9	17		1.72	32	23.2
8	1100	2.67	34	19.2	18	1300	2.67	50	19.7
9		3.73	52	15.9	19		3.73	76	17.9
10		4.92	66	15	20		4.92	80	15.4

According to the data in table 5, the curves of penetration depth h_s of PE bullet-proof plate and break size d_s of titanium alloy target plate with the different projectile length-diameter ratios are drawn, as shown in figure 4 and 5.

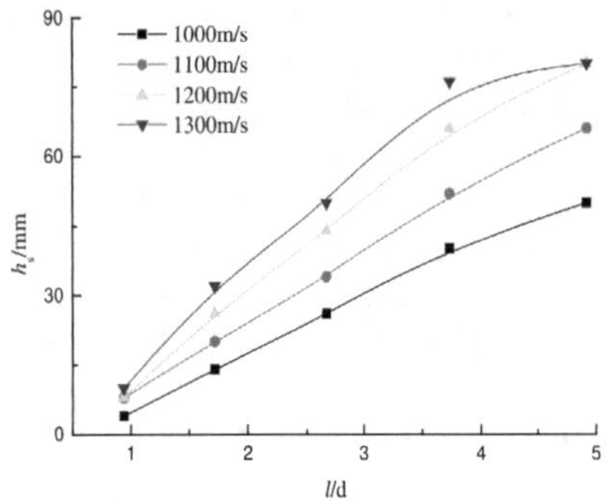

Figure 4. Curve of Penetration Depth with Length-diameter Ratio

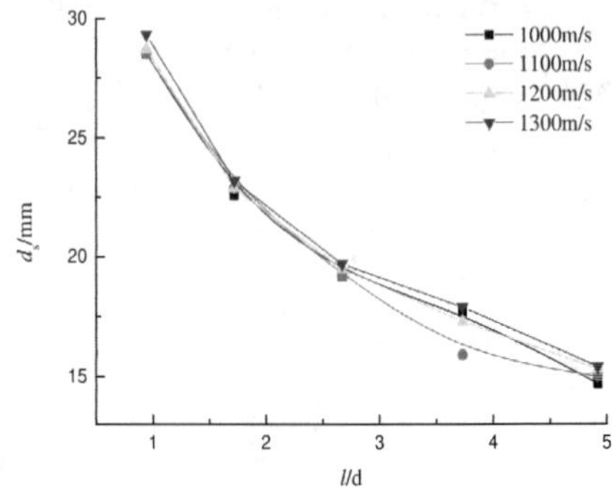

Figure 5. Curve of Break Size with Length-diameter Ratio

Observing figure 4 and 5, it is found that under the same mass condition, the larger the length-diameter ratio of the projectile, the smaller the size of the breach of the titanium alloy target, and the greater the penetration depth of the polyethylene target. This is because the larger the length-diameter ratio of the projectile, the smaller the contact force surface, and the smaller the pit opening range of the structure on the projectile face, which results in the decrease of the resistance of the projectile and the deeper penetration depth per unit weight. With the further increase of length-diameter ratio, the impact of projectile on penetration effect decreases.

4. Conclusion

In this paper, the finite element analysis software ANSYS/LS-DYNA was used for numerical simulation. The failure modes of 55 g cylindrical projectile penetrating ceramic/titanium alloy/gap/polyethylene/gap/921A steel composite armor structures were studied. The influence of the length-diameter ratio of the projectile on the anti-penetration performance of the structure was discussed. The main conclusions are as follows:

(1) Ceramics break in-plane and form ceramic cones. Titanium alloy target shows shear failure and bending deformation. 921A steel target plate shows adiabatic shear failure mode when it is penetrated, it will form bulge when it is extruded by the rear polyethylene plate.

(2) The front polyethylene target shows compression shear deformation, the rear polyethylene target shows bending deformation, and fiber tensile failure is the main failure mode, the impact fiber layer accompanied by obvious delamination degumming phenomenon.

(3) Under the same mass condition, the larger the length-diameter ratio of the projectile, the smaller the size of the breach and the greater the penetration depth. With the further increase of the length-diameter ratio, the impact of the projectile on the penetration effect is weakened.

5. References

[1] Zhijie S, Yan W and Zuoguang Z 2000 *J. Aerospace Materials & Technology* **17** 541
[2] Xiaoqiang W, Xi Z and Zhiyuan M 2011 *J. Journal of Harbin Engineering University* **32** 556
[3] Yuxin X, Shushan W and Wenkang Y 2012 *J. Acta Materiae Compositae Sinica* **29** 72
[4] Yuxin X, Wenxi W and Shushan W 2012 *J. Journal of Vibration and Shock* **33** 134
[5] O'Masta M R, Deshpande V S and Wadley H N G 2014 *J. International Journal of Impact Engineering* **74** 16

Seismic Risk Probability Assessment of Frame Structures Based on Incremental Dynamic Analysis

Beibei Yang, Zhitao Wang and Donghui Ma

College of Architecture and Civil Engineering, Beijing University of Technology, Beijing 100124, China
Email: yangbb@emails.bjut.edu.cn

Abstract. To assess the seismic risk probability of frameworks in different seismic fortification intensity zones in China, three fortification intensity framework models-specifically, 6 (0.05 g), 7 (0.1 g), and 8 (0.2 g)-were established as the research objects. The incremental dynamic analysis method was used to conduct a structure seismic fragility assessment, in which the fragility curves as well as the damage probabilities under certain peak ground accelerations were obtained. The risk probabilities of all the seismic damage levels of the frameworks in the next 50 years were calculated using the Monte Carlo method based on a study of the seismic hazard of different fortification intensity zones. The results indicate that, in the next 50 years, the risk probabilities of fortification intensity framework models 6, 7, and 8 exceeding "serious damage" were 1.92, 4.25, and 8.28%, respectively. Additionally, the risk probabilities of the models having a damage degree of "collapse" were 0.14, 0.43, and 1.07%, respectively. Based on the risk probability of the structure, the risk level is classified. This study forms the basis for the comprehensive evaluation of building seismic damage risk and can be used as a reference for the development of building seismic disaster prevention countermeasures.

1. Introduction

An earthquake is a type of extremely destructive disaster that results in significant economic and social losses in earthquake-prone countries all over the world, threatening the safety of people's lives and property [1]. Adopting seismic fortification for building structures is a fundamental means to reduce the loss caused by building seismic disasters. However, in preparing such fortifications, high-intensity earthquakes that exceed the fortification intensity are often not considered. For example, in many earthquakes, such as the Tangshan (1978) and Wenchuan earthquakes (2008) in China, the fortification intensity was exceeded by the high-intensity earthquake [2], which caused widespread building collapses and displaced millions of people. Therefore, it is crucial to conduct a seismic risk assessment and consider corresponding disaster prevention measures for buildings before the disaster occurs. Conventional seismic fragility analysis is predominantly conducted by analyzing the possibility of different damage grades of structures under specific seismic actions, which ignores the uncertainty of earthquake actions, making it difficult to properly reflect the potential seismic risk. Seismic risk probability assessment considers both the seismic hazard as well as the structural vulnerability [3] and represents the probability of the structure exceeding or reaching a certain earthquake damage state in a certain period in the future. Seismic risk assessment can provide clear measurements of the possible seismic risk level of buildings in the future, scientific basis for decision-making regarding disaster prevention beforehand, earthquake emergency preparedness, and post-disaster rescue work. Further, it can provide an invaluable paradigm shift from the idea of disaster loss reduction to that of disaster risk reduction [4].

In the long-term practical exploration, various scholars worldwide have conducted studies that have provided valuable insights into structural damage seismic risk. For example, Melani et al. [5] conducted incremental dynamic analysis (IDA) on a low-rise reinforced concrete frame structure, plotted its vulnerability curve, and evaluated the possible economic loss from seismic damage. Berto et al. [6] analyzed the time-varying characteristics of the seismic vulnerability of a steel reinforced concrete structure and conducted seismic risk assessment based on the whole life cycle of the structure. Gu et al. [7] developed a 3D finite element model of a typical low-tower cable-stayed bridge with OpenSees software, performed IDA analysis on the transverse and longitudinal directions of the bridge, and calculated the risk probability of each seismic hazard level of a typical low-tower cable-stayed bridge based on the Monte Carlo method. Lü et al. [8] summarized the general methods of seismic risk assessment from the two aspects of site seismic hazard and seismic vulnerability, analyzed the impact of essential uncertainty and knowledge uncertainty on the structural seismic risk assessment results, and derived relevant formulas. However, the majority of these studies focused on structural vulnerability; there are few reports on the seismic risk probability of frame structures in different seismic fortification intensity zones. To this end, this study focuses on establishing a framework structure for different seismic fortification intensity zones.

Earthquakes in China are characterized by wide distribution, high frequency, high intensity, and shallow source. According to the fifth China ground motion parameter zoning map [9], 84.4% of the 2,860 cities and towns at or above the county level are located in regions with fortification intensities 6, 7 (0.1 g), and 8 (0.2 g). Considering this fact, in this study, three structural models in accordance with the 6, 7 (0.1 g), and 8 (0.2 g) fortification intensity requirements were established as the research objects, and seismic fragility and probability analysis of different seismic actions were conducted on the models based on the IDA and Monte Carlo methods, respectively. On this basis, the seismic risk level was classified into three grades, and the risk level of the frame structure evaluated to reflect its earthquake damage risk degree in different fortification intensity zones. Thus, this study can provide a basis for the formulation of earthquake disaster countermeasures.

2. Materials and Methods

2.1. Incremental Dynamic Analysis Method

Based on the differences between the influence areas of seismicity, seismic damage data acquisition and calculation methods and earthquake fragility evaluation methods primarily consist of empirical and theoretical methods [10]. Theoretical methods reflect the probability relationship between the damage states of the structure and the ground motion by numerical calculation. They include damage condition determination, structural model determination, select and input ground motion, IDA analysis, and fragility curve production. Currently, the IDA method is one of the most commonly used theoretical methods to evaluate structural seismic responses and predict structural damage in performance-based earthquake engineering [11-13]. This method can analyze the entire process of the structure from elasticity to elastoplasticity to collapse, and it can better reflect the change rule of the seismic capacity of the structure under different earthquake actions in the future. Therefore, in this study, the IDA method is used to analyze structural vulnerability. Considering that the reinforced concrete frame structure is the most important structural form in China, a multi-story frame structure is considered as an example in this study.

2.1.1. Selection of IDA Indicators In this example we can see that there are footnotes after each author name and only 5 addresses; the 6th footnote might say, for example, 'Author to whom any correspondence should be addressed.' In addition, acknowledgment of grants or funding, temporary addresses etc might also be indicated by footnotes.

2.1.2. Classification of Damage States For seismic damage prediction, HAZUS [15] has classified the damage status of buildings into five levels: nearly intact, minor damage, moderate damage, serious damage, and collapse. In accordance with the code [16], the relationship between θ_{max} and the seismic damage degree is presented in Table 1.

Table 1.

Damage degree	Nearly intact	Minor damage	Moderate damage	Serious damage	Collapse
θ_{max}	θ_{max} <1/400	1/400≤ θ_{max} <1/250	1/250≤ θ_{max} C1/125	1/125≤ θ_{max} <1/50	θ_{max} ≥1/50

2.1.3. Model Information In comparison with the uncertainty of the ground motion, the uncertainty of the structural model has little influence on the analysis results [17]. In accordance with the relevant Chinese codes [18-19], three frame structures with three spans and 10 floors with seismic fortification intensities of 6 (0.05 g), 7 (0.1 g), and 8 (0.2 g) were established. The height of the first layer of the structure was 4.5 m, and the height of each of the other layers was 3.6 m. The standard value of the constant load was 4.6 kN/m2, and the standard value of the live load was 0.5 kN/m2 for an unoccupied roof and 2.5 kN/m2 for the floor. The site classification was class II and the classification of the design earthquake was group II. The characteristic period of the site (T\-g) was 0.4 s. The frame structure plan is shown in Figure 1.

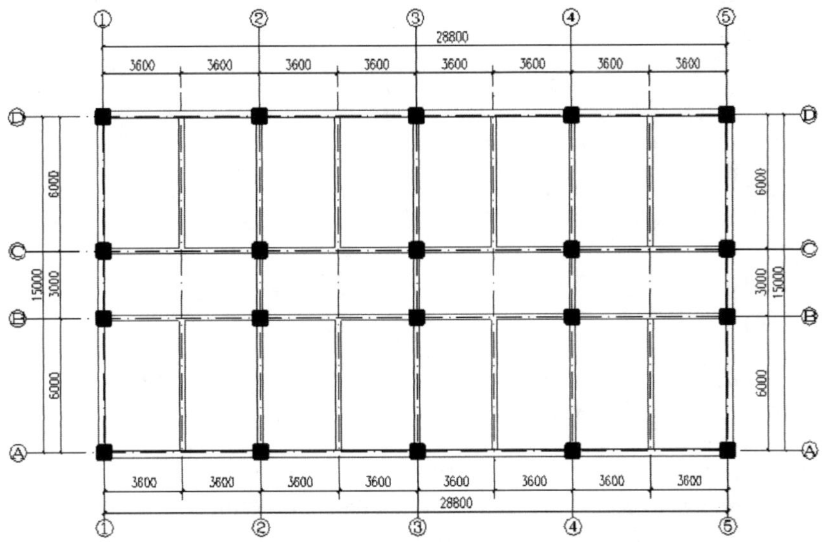

Figure 1. Frame Structure Plan (Unit: mm)

2.1.4. Select and Input Ground Motion Different ground motions have different earthquake effect values, which can differ by up to a factor of ten [20]. To reduce the uncertainty of ground motion on the structure, in this study, 30 seismic waves (more than 20 suggested by FEMA) suitable for the structure were selected from the earthquake ground motion database of the Pacific earthquake center of the United States in accordance with the IDA method requirements. The PGA amplitude modulation was used as the ground motion input for the structural analysis. The selected partial seismic records are listed in Table 2.

IOP Conf. Series: Materials Science and Engineering **735** (2020) 012040 doi:10.1088/1757-899X/735/1/012040

Table 2. Ground Motion Records

Number	Name	Time	Station	PGA (g)
1	Livermore-02	1980	San Ramon - Eastman Kodak	0.191
2	Westmorland	1981	Parachute Test Site	0.219
3	Imperial Valley-02	1940	El Centro Array #9	0.341
4	Imperial Valley-06	1979	Delta	0.284
5	Northridge-06	1994	LA - Century City CC North	0.123
6	Whittier Narrows-01	1987	Downey - Co Maint Bldg	0.177
7	Whittier Narrows-01	1987	Brea Dam (Downstream)	0.231
8	Whittier Narrows-02	1987	LA - 116th St School	0.157

2.1.5. Fragility analysis After the IDA analysis of the structure, the corresponding relationship between θ_{max} and PGA can be obtained. The probability function of the structural reaction can be obtained by considering the logarithms of θ_{max} and PGA [21]:

$$\ln\theta_{max} = a + b\ln PGA, \tag{1}$$

Where a and b are constants, which can be obtained by fitting the results of the structural IDA analysis. It is assumed that the probability function of the structural response D and structural capability parameter C all follow lognormal distribution [22], i.e.,

$$D \sim \ln(\hat{D}, \beta_d), C \sim \ln(\hat{C}, \beta_c), \tag{2}$$

Where

$\hat{D}$ represents the values of the structural responses, $\hat{C}$ indicates the structural capacity parameter, β_d is the logarithmic standard deviation of the structural response, and β_c is the logarithmic standard deviation of the structural capacity.

The fragility curves represent the conditional probability that the response of the structure under different ground motions exceeds the capability parameter defined in a damage level, namely, structural failure probability P_f:

$$P_f = P(C/D \le 1) \tag{3}$$

According to Equation (2), both parameters are known to follow a lognormal distribution. Equation (3) for failure probability can be expressed as follows:

$$P_f = \Phi[\frac{(\ln(D/C))}{\sqrt{\beta_c^2 + \beta_d^2}})], \tag{4}$$

By substituting the fitting result of Equation (1) into Equation (4), the failure probability of the structure can be represented as

$$P_f = \Phi[\frac{(\ln(e^a(PGA)^b/C))}{\sqrt{\beta^2{}_c + \beta^2{}_d}})], \tag{5}$$

Where

e is the base of the natural logarithm, HAZUS99 [23] when the structural fragility curves take PGA as the independent variable, set to 0.5; here $\sqrt{\beta_c^2 + \beta_d^2} = 0.5$. $\Phi(\cdot)$ is the standard normal distribution function.

2.2. Seismic Hazard Analysis Method

The earthquake intensity of a zone in the next 50 years obeys the extremum III type distribution in accordance with Gao's [24] results for his seismic risk analysis of 45 cities in North, Northwest, and Southwest China. The distribution function is as follows:

$$F(x) = \exp[-(\frac{\omega - x}{\omega - \varepsilon})^K] \tag{6}$$

The derivative of the distribution function is considered and the probability density function is obtained, which represents the probability density of the earthquake intensity occurring at a certain site in the next 50 years:

$$F(x) = \frac{K(\omega - x)^{K-1}}{(\omega - \varepsilon)^K} \exp[-(\frac{\omega - x}{\omega - \varepsilon})^K] \tag{7}$$

Where x represents the seismic intensity and is a discrete variable in the range 1–12;ω is the upper limit of intensity, which is 12 in this case, according to the current common seismic intensity division method in China; ε represents a "multi-value strength" earthquake, the seismic intensity with a probability of exceeding 63.2% in the next 50 years; K is the shape parameter, which is related to the basic design intensity.

According to Lv [25], the value of shape parameter K is presented in Table 3. The seismic probability density curve drawn by Equation (7) is depicted in Figure 2.

Table 3. Value of Shape Parameter K

Basic intensity	6	7	8
K	9.7932	8.3339	6.8713

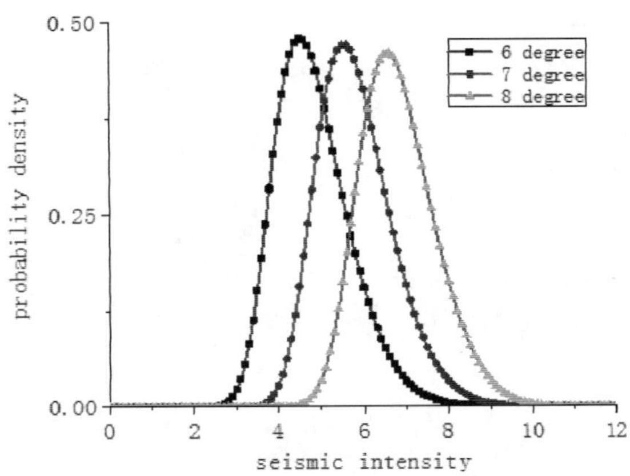

Figure 2. Probabilistic Density of Seismic Intensity in Different Seismic Fortification Zones

2.3. Seismic Risk Analysis Method

The structural seismic risk can be defined as the possibility of various damage degrees of the structure based on the site seismic hazard. The structural seismic risk is numerically equal to the convolution of the site hazard and seismic fragility. The calculation formula is as follows:

$$\mathbf{P} = \sum_{all,m_i} P(C \leq D|PGA = x_i)P(PGA = x_i), \quad (8)$$

In the formula, x_i is the intensity of the earthquake that may occur in the future. As the value of PGA is continuous, Equation (8) can be written as

$$\mathbf{P} = \int P(C \leq D|PGA = x_i)f(x_i)dx, \quad (9)$$

Where f(x_i) is the probability density function of PGA, which is consistent with the probability density function of seismic intensity.

2.4. Monte Carlo Method

The Monte Carlo method can generate a large number of random numbers in line with the probability density of earthquakes to simulate the earthquake intensity that the site may encounter in the future. Considering the accuracy of the integral calculation, the seismic intensity is taken as a continuous variable, and the Monte Carlo method is adopted to sample seismic events in accordance with Equation (7). If the sampling number is n, then the probability of each seismic event is 1/n. The larger the sampling number is, the closer the result is to the real value, and here we consider n=50000.

The seismic intensity obtained after sampling is converted into PGA, according to the formula proposed by Liu [26], as follows:

$$\mathbf{PGA} = 10^{(I \cdot \lg 2 - 0.01)} \quad (10)$$

According to the law of large numbers, Equation (11) can be written as

$$\mathbf{P} = \frac{1}{n}\sum_{i=1}^{50000} P(C \leq D|PGA = x_i) \quad (11)$$

The risk probability of the structure exceeding all degrees of damage is

$$\mathbf{P} = \frac{1}{n}\sum_{i=1}^{50000} P(C \leq D|PGA = x_i) \quad (12)$$

Where j=1, 2, 3, 4, 5 represents the damage degrees, namely nearly intact, minor damage, moderate damage, serious damage, and collapse, respectively.

Similarly, the risk of structural damage at all degrees is as follows:

$$\mathbf{C}_j = \begin{Bmatrix} \boldsymbol{P_j - P_{j+1}} & \boldsymbol{j \leq 4} \\ \boldsymbol{P_j} & \boldsymbol{j \leq 5} \end{Bmatrix} \quad (13)$$

3. Seismic Risk Probability Assessment

3.1. Vulnerability Analysis Results

After the IDA analysis of the structural model, the fragility curves can be plotted according to the above calculation, as shown in Figures 3–5.

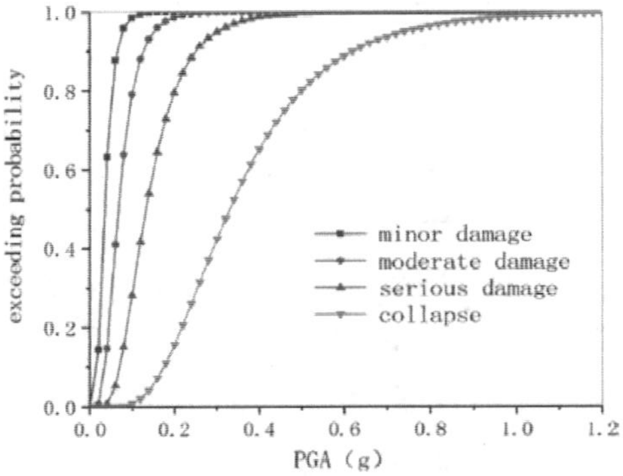

Figure 3. Fragility Curves of a 6-degree Fortification Frame Structure

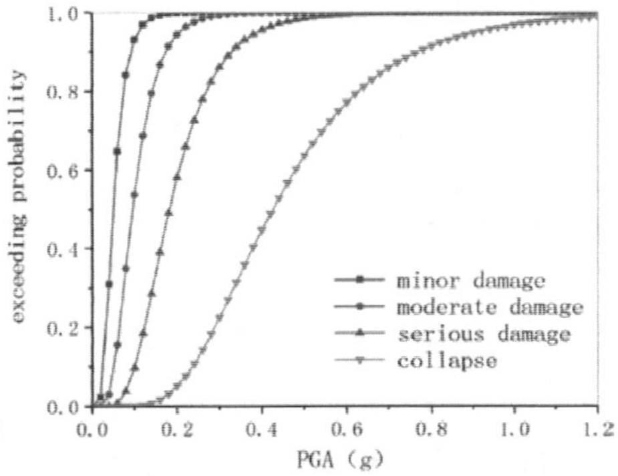

Figure 4. Fragility Curves of a 7-degree Fortification Frame Structure

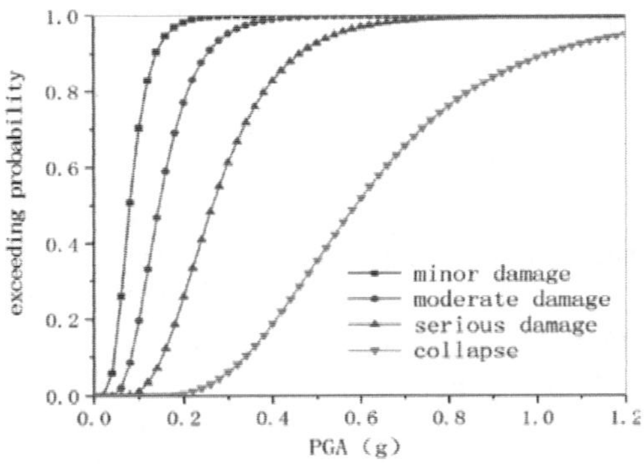

Figure 5. Fragility Curves of an 8-degree Fortification Frame Structure

As can be seen from Figures 2–4, the probability cumulative curves of various damage degrees exhibit an upward trend with the increase of PGA. The larger the PGA is, the more severe the structural damage is. For the probability cumulative curves of the same damage degree, the fragility curve of the 8-degree fortification structure exhibits the slowest rising trend. Under an action of the same seismic intensity, the probability of serious damage and collapse of the 8-degree fortification structure is lower. This indicates that the 8-degree fortification structure has better seismic resistance than the 6- and 7-degree fortification structures.

3.2. Seismic Risk Results

According to the above analysis model, the results of the seismic risk analysis for the next 50 years with different fortification intensity frameworks are calculated as presented in Tables 4–6.

Table 4. Risk Probability Assessment of 6-degree Fortification Frame Structure

Risk probability	Nearly intact	Minor damage	Moderate damage	Serious damage	Collapse
Exceeding probability	100%	29.04%	8.91%	1.92%	0.14%
Occurrence probability	70.96%	20.13%	6.99%	1.78%	0.14%

Table 5. Risk Probability Assessment of 7-degree Fortification Frame Structure

Risk probability	Nearly intact	Minor damage	Moderate damage	Serious damage	Collapse
Exceeding probability	100%	42.34%	16.09%	4.25%	0.43%
Occurrence probability	57.66%	26.25%	11.84%	3.82%	0.43%

Table 6. Risk Probability Assessment of 8-degree Fortification Frame Structure

Risk probability	Nearly intact	Minor damage	Moderate damage	Serious damage	Collapse
Exceeding probability	100%	55.20%	25.64%	8.28%	1.07%
Occurrence probability	44.80%	29.56%	17.36%	7.21%	1.07%

As can be seen from Tables 4–6, the risk occurrence probability of the structure having a damage degree of "nearly intact" and "minor damage" is relatively high at 91.09%, 83.91%, and 74.36%, respectively. The risk occurrence probability of a structure having a damage degree of "serious damage" and "collapse" is relatively low. Through the seismic risk occurrence probability comparison of the 6-, 7-, and 8-degree seismic fortification frameworks, it can be seen that the 8-degree fortification framework's occurrence probabilities of "serious damage" and "collapse" are the largest, with values of 7.21% and 1.07%, respectively. The 7-degree fortification framework's occurrence probabilities of "serious damage" and "collapse" are 3.28% and 0.43%, respectively. The 6-degree fortification framework's occurrence probabilities of "serious damage" and "collapse" are the lowest, with values of 1.78% and 0.14%, respectively. The results of the comprehensive seismic fragility of structures also demonstrate that the seismic risk factors of the seismic damage are related to the seismic fortification area in addition to the seismic capability of the structures.

3.3. Seismic Risk Level Classification

Risk acceptability is a scale that measures whether a certain type of risk is acceptable or not; it is also the basis for formulating risk management countermeasures and schemes. Generally, the risk acceptability is subjective and is based on the public's opinions. If the risk level of a building is "High", it is considered to exceed that of public expectations, and effective measures should be taken to reduce the risk until it is at a level that is accepted by the public.

According to the risk probability of the "collapse" damage in the next 50 years, this study classifies the risk level into three levels, namely high risk, medium risk, and low risk. The risk levels are listed in Table 7.

Table 7. Risk Grade Classification

Risk level	Exceeding collapse damage probability (%)	Risk acceptability
High	[2,100)	Unacceptable risk
Medium	[1,2)	Nearly acceptable risk
Low	[0,1)	Acceptable risk

In urban planning, according to the principle of "priority first, then general, then serious, then light," the countermeasures of seismic strengthening, reconstruction, or demolition of buildings must be formulated based on the principle of "gradual improvement in stages and batches combined with urban reconstruction." For buildings with a low risk level, the construction risks can be ignored. Buildings with a medium risk level must weigh the costs and benefits of risk management measures. If increased investment does not contribute significantly to risk reduction, risks are allowed. Buildings with a high risk level must take certain measures to reduce risks, such as seismic reinforcement, seismic transformation, demolition, and reconstruction, until the risks are acceptable. According to Table 7, combined with the seismic risk assessment results of buildings from Tables 4–6, it can be seen that the risk probabilities of frame structures fortified in degrees 6, 7, and 8 exceeding "collapse damage" are 0.14%, 0.43%, and 1.07%, respectively. The risk level of frame structures fortified in degrees 6 and 7 are both "Low" and within the acceptable risk range. The risk level of the frame structure fortified in degree 8 is "Medium." We recommend that further consideration should be given in the formulation of urban disaster prevention planning.

4. Conclusions

In this study, the IDA and Monte Carlo methods were used to systematically analyze the seismic risk of frame structures with different fortification intensities. The main conclusions of the effort in the realm of seismic risk assessment can be outlined as follows:

·Considering the seismic fragility and seismic risk of buildings, the risk level can be quantified and evaluated to reflect the potential seismic disaster risk of buildings in the future. Thus, this study can be used as a reference for earthquake disaster prevention planning and risk level classification in different cities.

·Through the seismic risk assessment of frame structures with different fortification intensities, the risk levels of structures with fortification intensities 6, 7, and 8 were preliminarily discussed, and it was found that the risk levels of structures with a fortification intensity of 8 were higher than those of structures with fortification intensities 6 and 7. Therefore, in urban disaster prevention planning, high-intensity zones need to pay additional attention to risk prevention.

·This study only considers a typical framework as an example. Owing to the lack of sample adequacy and representativeness and the different construction ages in cities, the evaluation results of this study are still limited. The uncertainties in the building model, seismic event, and cognition, and the influence of other factors on the risk need to be investigated in a future study. The first paragraph after a section or subsection heading should not be indented; subsequent paragraphs should be indented by 5 mm.

5. Acknowledgments

Authors wishing to acknowledge assistance or encouragement from colleagues, special work by technical staff or financial support from organizations should do so in an unnumbered Acknowledgments section immediately following the last numbered section of the paper.

6. Appendices

Technical detail that it is necessary to include, but that interrupts the flow of the article, may be consigned to an appendix. Any appendices should be included at the end of the main text of the paper, after the acknowledgments section (if any) but before the reference list. If there are two or more appendices they should be called appendix A, appendix B, etc. Numbered equations should be in the form (A.1), (A.2), etc, figures should appear as figure A1, figure B1, etc and tables as table A1, table B1, etc.

7. References

[1] Ö. Necmioggĺu. " Design and challenges for a tsunami early warning system in the MarmaraSea." Earth, Planets and Space, vol. 68, no. 1, p. 13, 2016.

[2] Z. -Y. Cheng, " Postscruip: Lessons from Wenchuan earthquake and strategies for post-earthquake reconstruction and retrofit. Wenchuan earthquake: Report of Building Seismic Damage Survey and Post earthquake Reconstruction Analysis, " Beijing: China Architecture and Building Press, 2008. (in Chinese).

[3] Preciado, A. Ramirez-Gaytan, R. A. Salido-Ruiz, et al. " Earthquake risk assessment methods of unreinforced masonry structures: Hazard and vulnerability, " Journal of Earthquakes and Structures, vol. 9, no. 4, pp. 719-733, 2015.

[4] S. B. Manyena, E. Mavhura, C. Muzenda, et al. " Disaster risk reduction legislations: Is there a move from events to processes? " Global Environmental Change, vol. 23, no. 6, pp. 1786-1794, 2013.
A. Melani, R. Khare, R. P. Dhakal, et al., " Seismic risk assessment of low rise RC frame structure, " Structures, vol. 5, pp. 13-22, 2016.
[5] L. Berto, A. Saetta, P. Simioni, " Structural risk assessment of corroding RC structures under seismic excitation, " Construction and Building Materials, vol. 30, no. 5, pp. 803-813, 2012.
[6] Y. Gu, H. Zhong, W.-D. Zhuo, " Lower tower cable stayed bridge seismic vulnerability analysis, " China Civil Engineering Journal, vol. 45 (S1), pp. 218-222, 2012 (in Chinese).
[7] D. -G. Lü, X. -H. Yu, " Theoretical study of probabilistic seismic fragility ", Journal of Building Structures, vol. 10, pp. 41-48, 2013 (in Chinese).
[8] GB 18306—2015, China Ground Motion Parameter Zoning Map.
[9] D. Wu, S. Tesfamariam, S. F. Stiemer, et al., " Seismic fragility assessment of RC frame structure designed according to modern Chinese code for seismic design of buildings, "Earthquake Engineering and Engineering Vibration, vol. 11, no. 3, pp. 331-342, 2012.
[10] E. Brunesi, R. Nascimbene, F. Parisi, et al., " Progressive collapse fragility of reinforced concrete framed structures through incremental dynamic analysis, " Engineering Structures, vol. 104, pp. 65-79, 2015.
[11] M. R. Sheidaii, M. Tahamouliroudsari, M. Gordini. " Investigation of the nonlinear seismic behavior of knee braced frames using the incremental dynamic analysis method. " International Journal of Advanced Structural Engineering, vol. 8, no. 2, pp. 151-159, 2016.
[12] J. Kiani, M. Khanmohammadi. " New Approach for Selection of Real Input Ground Motion Records for Incremental Dynamic Analysis (IDA). " Journal of Earthquake Engineering, vol. 19, no. 4, pp. 592-623, 2015.
[13] P. Zarfam, M. Mofid, " On the modal incremental dynamic analysis of reinforced concrete structures, using a trilinear idealization model, " Engineering Structures, vol. 33, no. 4, pp. 1117-1122, 2011.
[14] " Seismic risk assessment of low rise RC frame structure, " Structures, vol. 5, pp. 13-22, 2016.
[15] D. -H. Ma, X. -D. Guo, Z. -T. Wang, " Implementation guide to standard for urban planning on earthquake resistance and hazardous prevention, " Beijing: China Building Industry Press, vol. 107, 2007 (in Chinese).
[16] M. Avital, R. Kamai, M. Davis, et al. "The effect of alternative seismotectonic models on PSHA results - a sensitivity study for two sites in Israel." Natural Hazards & Earth System Sciences, vol. 18, no. 2, pp. 499-514, 2017.
[17] GB 50011—2010, Code for Seismic Design of Buildings.
[18] GB 50009-2012, Load Code for the Design of Building Structures.
[19] J. Kiani, M. Khanmohammadi. " New Approach for Selection of Real Input Ground Motion Records for Incremental Dynamic Analysis (IDA). " Journal of Earthquake Engineering, vol. 19, no. 4, pp. 592-623, 2015.
[20] F. M. Nazri, N. A. Alexander. " Exploring the relationship between earthquake intensity and building damage using single and multi-degree of freedom models. " Canadian Journal of Civil Engineering, vol. 41, no. 4, pp. 343-356, 2014.
C, Celik, B. R. Ellingwood, " Seismic fragilities for non-ductile reinforced concrete frames – Role of aleatoric and epistemic uncertainties, " Structural Safety, vol. 32, no. 1, pp. 1-12, 2010.
[21] HAZUS99, User's Manual, Washington D C: Federal Emergency Management Agency, 1999.
[22] X. -W. Gao, A. -B. Bao, " Probabilistic model and its statistical parameters for seismic load, " Earthquake Engineering and Engineering Vibration, vol. 5, no. 1, pp. 13-22, 1985 (in Chinese).
[23] D. -G. Lv, " Research on intelligent optimization design of disaster resistant structure based on optimal fortification load and performance, " PhD thesis of Harbin Jianzhu University, Harbin: library of Harbin Jianzhu University, pp. 9-11, 1999 (in Chinese).

[24] H. -X. Liu, R. -J. Lu, D. -S. Chen, et al., "A proposal to revise the old seismic intensity table. Seismological engineering research reports of the institute of engineering mechanics," Chinese academy of sciences, Beijing: science press, pp. 1-13, 1981 (in Chinese).

IOP Conf. Series: Materials Science and Engineering **735** (2020) 012041 doi:10.1088/1757-899X/735/1/012041

Effects of Cu^{2+} Doping on Structure, Morphology and Photocatalytic Hydrogen Production Performance of Porous $CdIn_2S_4$ Microsphere

Xuefeng Bai [1,2], Wei Wu [2] and Hongfei Lv [1]

1 Institute of Petrochemistry, Heilongjiang Academy of Sciences, Harbin 150040, China
2 School of Chemistry and Material Sciences, Heilongjiang University, Harbin 150080, China
Email: tommybai@126.com

Abstract. The Cu^{2+}doped porous $CdIn_2S_4$ microspheres ($Cu@CdIn_2S_4$) with better photocatalytic performance were prepared by sodium dodecyl sulphate (SDS)-assisted hydrothermal technology. The $Cu@CdIn_2S_4$ were characterized by SEM, XPS, XRD and DRS. The effects of Cu^{2+} doping amount on the morphology, structure and photocatalytic hydrogen production performance of $CdIn_2S_4$ were studied. The porous $Cu@CdIn_2S_4$ microspheres were flower shape of a average diameter of 3-5 μm and their maximum absorption edge was up to 700 nm. The Cu^{2+} doping was good for the improvement of photocatalytic performance. $Cu@CdIn_2S_4$ with 0.3wt% Cu^{2+}doping amount possessed the highest hydrogen production rate of 1248.1 μmol/ (h·g), much more than that of pure $CdIn_2S_4$.

1. Introduction

Since the photocatalytic splitting of water on TiO_2 electrodes was found by Fujishima and Honda in 1972 [1], a great many researchers have been carried out in theory and application of the photocatalytic reaction. At present, the environmental pollution has become an obstacle to sustainable development. Hydrogen as a clean fuel is the best way to solve environmental pollution [2].

Photocatalytic decomposition of water and hydrogen sulfide to produce hydrogen has attracted more and more attention from scientists in recent years, which provides the most suitable environmentally friendly process route of utilizing solar energy for valuable hydrogen production [3-4]. The development of efficient and cheap photocatalysts is the key to the research.

$CdIn_2S_4$, as a visible light response semiconductor material, had many unique optoelectronic and excellent photocatalytic properties and had been widely used in the field of solar cells, photocatalysts, and light-emitting diodes. $CdIn_2S_4$ with cubic spinel structure has stable chemical properties and narrow band gap suitable for absorbing less than 600 nm of light [5-6].

Metal ion doping into photoconductors not only acted as the trapping sites for benefiting to limit the recombination of e^--h^+,but also formed impure energy levels in the forbidden band for red shifting the absorption side, which is regarded as the most effective way for improving the performance of photocatalytic hydrogen production [7-9].

In this paper, Cu^{2+} doped porous $CdIn_2S_4$ microspheres ($Cu@CdIn_2S_4$)with the better photocatalytic performance was prepared by sodium dodecyl sulfate (SDS)-assisted hydrothermal technology. The effect of Cu^{2+} doping on the structure, morphology and photocatalytic hydrogen performance of $Cu@CdIn_2S_4$ was discussed in detail.

2. Experimental

2.1. Preparation of Cu@CdIn$_2$S$_4$

All chemicals used are chemically pure. The Cu^{2+} doped porous $CdIn_2S_4$ microspheres were synthesized by sodium dodecyl sulfate (SDS)-assisted hydrothermal technology. 2 mmol Cd $(NO_3)_2 \cdot 6H_2O$,4 mmol $In(NO_3)_3 \cdot 4H_2O$, 8 mmol thioacetamide (TAA) and a certain amounts of SDS were adding and mixing in 75 ml of distilled water. Then $CuCl_2 \cdot 2H_2O$ was added into as-above solution according to Cu^{2+}doping amount. The as-prepared solution was well mixed, and then transferred into Teflon-lined autoclave (100 ml). The hydrothermal synthesis of Cu@$CdIn_2S_4$ was completed at 160℃ for 12 h. The final product was obtained after cooling, filtering, washing and drying at 80℃ for 4 h, named as Cu@$CdIn_2S_4$.

2.2 .Characterization of Cu@CdIn$_2$S$_4$

The X-ray diffraction spectrum (XRD) was obtained on a Bruker D8 Advance X-ray diffractometer using Cu $K\alpha_1$ irradiation. The patterns were recorded from 10° to 70° (2θ). The UV-Vis diffuse reflectance spectra (DRS) were carried out on Shimadzu UV-2450 spectrophotometer in the scanning range from 250 to 700 nm. The experimental data were obtained by transferring from reflection to absorption by Kubelka-Munk method. The scanning electron microscopy (SEM) was characterized on a HITACHI S-4800 instrument operated at 230 kV. The X-ray photoelectron spectroscopy analysis (XPS) was completed on a Thermo Electron instrument at 10^{-7} Pa with Al kα radiation.

2.3. Evaluation of Photocatalytic Hydrogen Production Performance

Photocatalytic experiments were carried out in an inner-irradiation quartz reactor. A 250 W high-pressure lamp was been used as light source. In order to get visible light, 1 M $NaNO_2$ solution was added into an inner circulating system to remove the UV part of the light with a wavelength less than 400 nm. 0.1 g Cu@$CdIn_2S_4$ was added into 300 ml aqueous solution of 0.25 M Na_2SO_3/0.35 M Na_2S. Before opening light source, N_2 was purged through as-above prepared solution to remove oxygen. The photocatalytic reaction temperature was controlled at 30 ± 5℃.Photocatalytic hydrogen production performance was evaluated by measuring the amount of hydrogen evolution by drainage extraction method. Hydrogen purity analysis was conducted on GC with thermal conductivity detector.

3. Results and Discussion

3.1. Structure

Figure 1(A&B) showed the typical XRD patterns of porous Cu@$CdIn_2S_4$ with different Cu^{2+} doping amount. It was shown in Figure 1(A) that all Cu@$CdIn_2S_4$ belongs to pure cubic spinel phase, the same as $CdIn_2S_4$. With the increasing of Cu^{2+} doping amount, the position of peak (511) shown in Figure 1(B) moves gradually towards a large diffraction angle. This is due to Cu^{2+} partly occupying the Cd^{2+} site, incorporating into the lattice or inserting the interspace of $CdIn_2S_4$.The XRD analysis indicated Cu^{2+}doping has an effect on crystal structure of $CdIn_2S_4$.

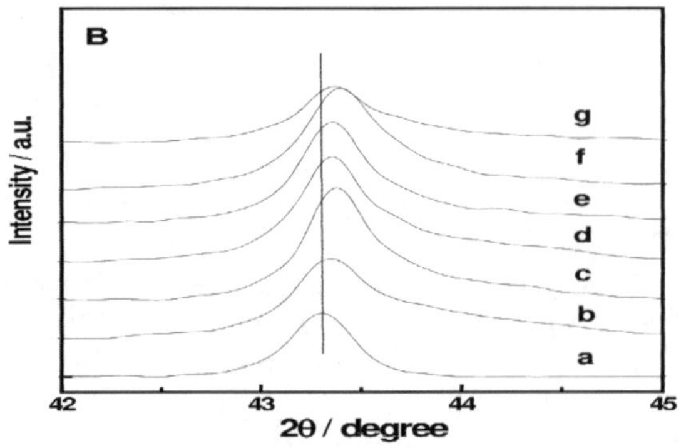

Figure 1. The XRD Patterns of Cu@$CdIn_2S_4$ with Different Doping Amount (A) and the Position of (115) Reflection (B)

3.2. Morphology

Figure 2(a~f) showed that SEM images of Cu@$CdIn_2S_4$ with different Cu^{2+}doping amount. It was seen from Figure2 that the porous Cu@$CdIn_2S_4$ microspheres with unique flower shape of an average diameter of 3-5 μm could be obtained by hydrothermal synthesis technology. The Cu^{2+} doping had great effect on the morphology of Cu@$CdIn_2S_4$ microspheres. It was seen in Figure 2 (a) that the pure $CdIn_2S_4$ without Cu^{2+} doping showed loose flower-like microspheres with numerous sheets. The sheets were interconnected with each other to form many slit-like pores inside of petals and among the curved nanosheets. However, as seen in Figure 2 (b f), with Cu^{2+} as dopant, abundant floss-like spheres with rough surface were mangled partly under this condition. As the concentration of Cu^{2+} dopant increased to 0.7%, mass rule less dollops without pore structure were produced, and obvious aggregation of particles could be seen in Figure 2 (e). When the doping is greater than 1.0 %, the spherical structure is destroyed. This phenomenon is due to that Cu^{2+} doping hindering the self-assembly growth of crystallites into porous microspheres during hydrothermal synthesis process.

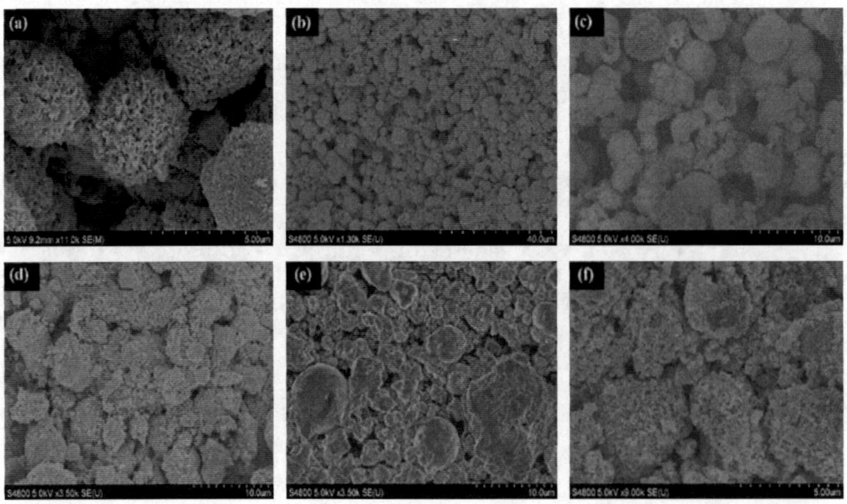

Figure 2. SEM Images of Cu@$CdIn_2S_4$ with Different Doping Amount: (a) 0.0%, (b) 0.1%, (c) 0.3%, (d) 0.5%, (e) 0.7%, (f) 1.0%

3.3. Surface State Analysis

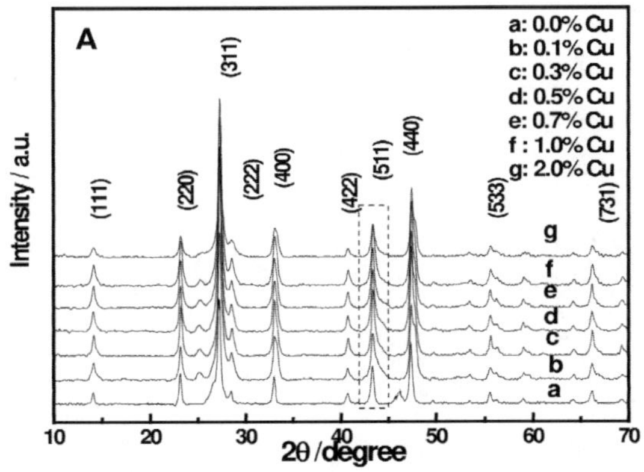

Figure 3. XPS Spectra of $Cu@CdIn_2S_4$ with 0.3wt% Doping Amount: (a) $Cu@CdIn_2S_4$;(b)Cd3d;(c)In3d;(d) S2p;(e)Cu 2p

3.4. Optical Properties

Figure4 showed the DRS spectra of $Cu@CdIn_2S_4$ with different Cu^{2+}doping amount. With the increasing of Cu^{2+} doping amount from 0 to 2 wt%, the maximum absorption edge of $Cu@CdIn_2S_4$ gradually shifted red from 600 nm(band gap of 2.08 eV) without doping to 625 nm~700 nm(band gap of 1.98 eV~1.77 eV) . The reason for this phenomenon is due to the formation of impure energy levels by the Cu^{2+} doping in the forbidden band, reducing band gaps to enhance utilization of visible-light. In addition, the impure energy levels of $Cu@CdIn_2S_4$ could improve the diffusion and transfer of the photo excited charge carriers, and limit the recombination of e^--h^+ .

3.5. Photocatalytic Hydrogen Production Performance

Figure5 showed the photocatalytic hydrogen production rate of $Cu@CdIn_2S_4$ with different Cu^{2+} doping amount. It was seen from Figure5 that $Cu@CdIn_2S_4$ exhibited the higher hydrogen production performance in visible spectra than pure $CdIn_2S_4$. With the increasing of Cu^{2+} doping amount, H_2 production rate showed a trend of increasing first and then decreasing. $Cu@CdIn_2S_4$ with 0.3wt% Cu^{2+}doping amount exhibited the highest rate of 1248.1 μmol/ (h·g), much more than that of pure $CdIn_2S_4$.

Cu^{2+} doping into $CdIn_2S_4$ acted as the trapping sites for benefiting to limit the recombination of e^--h^+. The number of trapping sites was increased with the Cu^{2+} doping amount at low-level. However, when Cu^{2+} doping amount increase at a high addition, a mass of trapping sites acting as recombination centers could make photo excited charge carriers inactive. It has been reported that suitable metal ion doping amount greatly improved the photocatalytic performance [10].

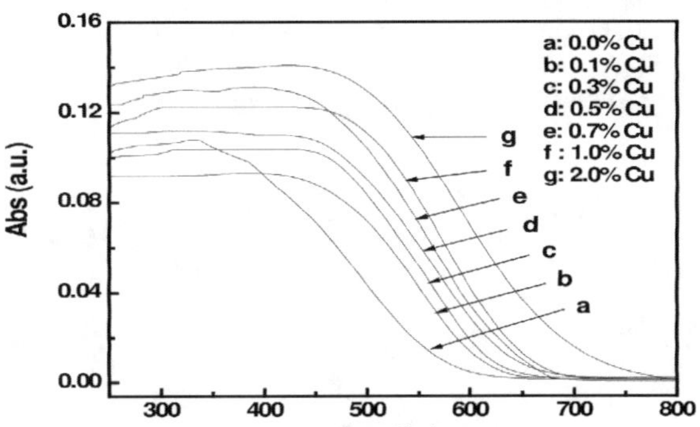

Figure 4. DSR Spectra of $Cu@CdIn_2S_4$ with Different Doping Amount Photocatalysts

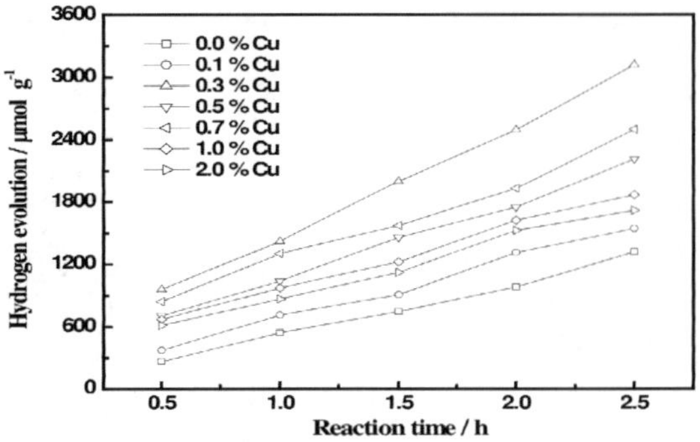

Figure 5. The H_2 Production over Cu@$CdIn_2S_4$ with Different Doping Amount Photocatalysts

4. Conclusions

The Cu^{2+} doped porous $CdIn_2S_4$ microspheres (Cu@$CdIn_2S_4$)with the better photocatalytic performance was prepared by sodium dodecyl sulfate (SDS)-assisted hydrothermal technology. It was indicated from characterization results that the porous Cu@$CdIn_2S_4$ microspheres was unique flower shape of a average diameter of 3-5 μm, and Cu^{2+} partly occupied the Cd^{2+} site and incorporated into the lattice or inserted the interspace of $CdIn_2S_4$.The maximum absorption edge of Cu@$CdIn_2S_4$ gradually shifted red, up to 700 nm.The photocatalytic hydrogen production performance over Cu@ $CdIn_2S_4$ was greatly enhanced by Cu^{2+} doping. With the increasing of Cu^{2+} doping amount, hydrogen production rate showed a trend of increasing first and then decreasing. Cu@$CdIn_2S_4$ with 0.3wt% Cu^{2+}doping amount exhibited the highest rate of 1248.1 μmol/ (h·g), much more than that of pure $CdIn_2S_4$.

5. Acknowledgements

This work was funded by Project of Heilongjiang Science and technology (ZY17A06).

6. References

[1] A Fujishima, K Honda 1972 *Nature*, **238** 37-38
[2] P M Modisha, N M Ouma, R Garidzirai, P Wasserscheid 2019 *Energy & Fuels,doi:10.1021/acs.energyfuels.9b00296*
[3] V Preethi, S Kanmani, 2017 *Int J Hydrogen Energ*, **42** 8997-9002.
[4] S Jeong, K H Chung'H Lee,et al, 2017 *Int J Hydrogen Energ*, **42** 17386-17393
[5] W Jiang, X Yin, F Xin, et al, 2014 *Appl Surf Sci*, **288** 138-142
[6] Q Zhou, Z Zheng,Y Zhang,et al, 2014 *Mater Lett*, **124** 61-63
[7] H R Rajabi, M Farsi2015, *J Mol Cat A:Chem*, **399** 53-61
[8] Z Peng, D Wu, W Wang, et al. 2017 *Powder Technology*, **315** 73-80
[9] K S Siddhapara, D V Shah, 2014 *J Nanosci & Nanotech*, **14** 6337-6341
[10] W Li, J F Li, W Sun, et al. 2008 *J. Chem Phys.* **21** 487-492.

IOP Conf. Series: Materials Science and Engineering **735** (2020) 012042 doi:10.1088/1757-899X/735/1/012042

Research on the Losing Process of the Traditional Gray Bricks and Tiles

Peng Zhao[a], Lei Qin[a], Xiuzhi Zhang[b], Yunsheng Zhang [c], Hongwei Ren[d], Yinghui Lv[a] and Gengle Liu[a,*]

[a] School of Civil Engineering and Architecture, University of Jinan, Jinan 250022, People's Republic of China;
[b] School of Materials Science And Engineering, University of Jinan, Jinan 250022, People's Republic of China;
[c] Jiangsu Key Laboratory for Construction Materials, Southeast University, Nanjing 211189, People's Republic of China;
[d] School of Electrical Engineering, University of Jinan, Jinan 250022, People's Republic of China;
* Corresponding author: Gengle Liu, 455900254@163.com

Abstract. Gray bricks and tiles are important materials of traditional buildings in China. However, the production process of bricks and tiles uses a large amount of clay resources from cultivated land, consumes a large amount of energy such as coal, and at the same time causes a great load to the environment. Therefore, the production process of traditional gray bricks and tiles is facing the risk of loss. This study investigates the traditional brick and tile manufacturers in Jiangsu and Zhejiang provinces and records the manufacturing techniques of traditional brick and tile blanks in detail, providing detailed records for the protection and repair of the architectural heritage.

1. Introduction

Bricks and tiles, as traditional building materials, have been used in architecture for a long time [1, 2]. With the development of industrialization and the shortage of clay resources, the research of traditional brick and tile making and processing technology is gradually reduced. As people raise their understanding of the protection of ancient buildings, more and more ancient buildings get people's attention and protection and repair. However, brick and tile materials are gradually buried in the tide of sustainable development and environmental protection [3, 4]. Adhering to the principle of keeping the original appearance for cultural heritage protection, we should pass down the traditional process of bricks and tiles [5]. Therefore, it is necessary to study the manufacture and application of bricks and tiles in Chinese traditional building construction technology, systematically sort out them with a record, and compare them with modern production technology to form the relevant standard standards for the quality inspection of traditional gray bricks and tiles. At the same time, it is of great significance to rescue traditional processes and protect cultural relics and historic sites.

2. Traditional and Modern Processes of Making Gray Brick and Tile Blanks

The process of making and firing gray bricks and tiles changes due to regional, cultural, folk-custom and other reasons, and it may be elaborated through visiting traditional artists in Suzhou and Jiashan, Zhejiang.

First of all, it's essential to select soil and mud, and good soil quality is the foundation to ensure the solid degree of various brick and tile components. To make bricks and tiles, you must choose loess, which has a greater viscosity and is not easy to break after being fired. There are also several types of loess, such as bean-like soil, clay, and white sand, among which only white sand is suitable for brick and tile production. Generally, the soil layer of 30cm on the surface of the ground shall be removed, because the first layer of the soil on the ground is exposed to the sun for a long period of time and is not sticky enough. The next layer of soil contains more rotten plants or roots and other debris, rich in organic materials and suitable for farming. The third layer contains a lot of minerals, so it's suitable for making bricks and tiles. Due to the quality of the soil, the ground can be excavated to 30 cm to 2 meters from the ground surface. The mud shall then be processed, letting the buffalo tread on the clay mud, which now is gradually changed to use mud mill, and use water pump instead of manual carrying water. The process of processing the mud is shown in Figure 1.

Figure 1. Pug process

The production of traditional brick blank is simple, i.e., filling the mold with mud, smashing, put it on the site, and open the mold, and then the brick blank is finished. Place it for 2 ~ 3 days, after the surface color of the brick blank becomes white, the brick blank can be raised up for drying, and after drying to 70-80%, it can be stored on the rack, to be burned. The manufacturing process of the traditional brick blank is shown in Figure 2.

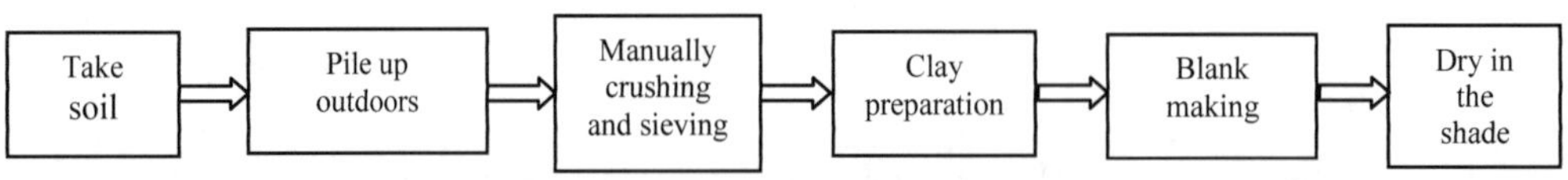

Figure 2. Traditional Production Process of Unburned Brick

The manufacturing process of the tiles is complicated. Firstly, cover the tile on the turntable, and cut the mud evenly with the cutting bow, so that the length of the cut mud blank is the same as the length of the turntable. Place the clay blank on the tile barrel, press and beat with hands, and after the tile barrel and the clay blank are stuck together, the clay barrel and the wood block with the same height and width as the tile will adhere to the clay surface, rotate the turntable, and smooth the surface of the clay blank. Remove the tile from the turntable, place it on the levelled ground, withdraw the tile, and divide the formed clay blank into four pieces with a knife. In the division, only two-thirds of the mud blank needs to be cut, and after the mud blank is 90% dried, the mud blank can be pat open by hands and burned on the rack. The manufacturing process of the tile is shown in Figure 3.

Figure 3. Traditional Production Process of Unburned Tile

Due to the long cycle and low efficiency of the traditional manual process, and the extremely difficult working environment of the workers, the manual process has been gradually replaced by high-efficiency mechanical suppression. In the manufacturing process of modern tile blank, clay is put into a mold and directly pressed into tile blank. In the manufacturing process of modern brick blank, clay is put into the machine for simple mixing then sent by conveyor to the right end and directly cut into brick blank, mainly used for the production of small-volume brick. Or, the prepared brick powder is pressed to the brick blank by 1,000 tons of hydraulic pressure, mainly used for making square brick blank. The manufacturing process of modern gray tile and brick blank is shown in Figure 4.

Figure 4. Modern Production Process of Unburned Tile and Brick

In improving efficiency, the mechanization of manufacturing process of modern gray tile and brick blank has simplified many manufacturing techniques, such as clay preparation, so that the durability, strength and other properties of gray tiles produced now are lower than those of gray tiles produced by traditional process. Compared with the gray tiles made by hands, the surface of the tiles made by machine is more smooth and level. Gray bricks made by machine after firing will often appear the larger thread in the large surface, which is due to the spiral stirring of raw materials. The products produced by the modern manufacturing process are shown in Figure 5.

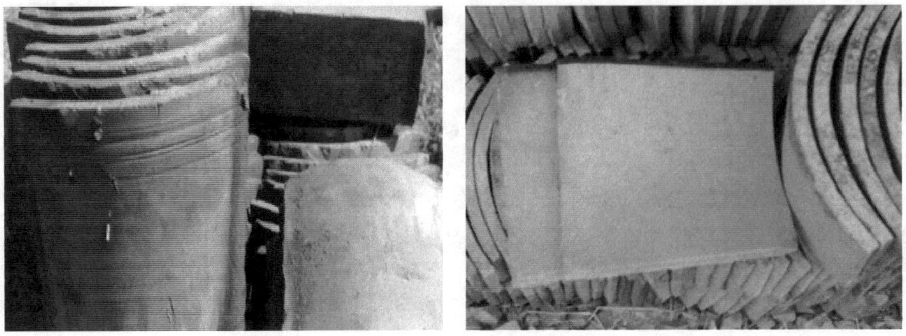

Figure 5. Grey Tile Using Modern Technology

3. Conclusions

This study investigates the brick and tile processing plants in Jiangsu and Zhejiang provinces, and records and collates the manufacturing process of modern antique gray bricks and tiles. The production of brick blank and tile blank goes through such processes as making borrow soil, piling up in the open air, crushing and sifting, preparing clay, making blank and drying in the shade. Modern manufacturing processes have changed with the change of productivity, mainly in the use of machinery to replace the original manual process. Although the mechanical construction has improved the production efficiency, but has changed the traditional manufacturing process, and because of the simplification of the process, the strength and durability of the antique gray bricks and tiles made by modern process have been reduced.

4. Acknowledgements

This work was supported by the Key R & D project of Shandong Province (2019GSF109119), the Doctoral Foundation of the Shandong Natural Science Foundation (ZR2018BEE038), the National Key Research and Development Program of China (2017YFC0703100), and the National Natural Science Foundation of China (51778269, 51678277).

5. References

[1] Lin J S, Lin Z J, Chen J F, The ancient great earthquake and earth-quake-resistance of the ancient buildings (towers, temples and bridges) in Quanzhou city, World. Earthquake. Eng. 21(2) (2005) 159-166.
[2] Ji S J, Some discussions about Nanjing city wall, Palace. Mus. J. 2 (1984) 70-81.
[3] Jiang X F, Linji Zazhi (in chinese), Shanghai Ancient Books Publishing Company, Shanghai, 1988.
[4] Shi X, She W, Zhou H L, Zhang Y S, Shi F, W Chen, Thermal upgrading of Hui-style vernacular dwellings in China using foam concrete, Front. Archit. Res. 1 (2012) 23-33.
[5] Zhao P, Jackson M D, Zhang Y S, Li G Y, Monteiro P J M, Yang L, Material characteristics of ancient Chinese lime binder and experimental reproductions with organic admixtures, Constr. Build. Mater. 84 (2015) 477-488.

Damage Process of High-piled Wharf Subjected to Underwater Explosion

Jinghan Liu[1,2], Ting Tang[2], Zhuobin Wei[2] and Lingfeng Li[1,2]

1. Naval University of Engineering, Wuhan 430033, Hubei, China
2. Naval Logistics Collage of PLA, Tianjin 300450, China
Email: 1226001717@qq.com

Abstract. In order to study the damage process of high-piled wharf subjected to underwater explosion, analysis model of high-piled wharf was established. The load characteristics of piles and panel of high-piled wharf were analysed during shock wave propagation and bubble pulse. The failure mode of wharf is basically completed in the first bubble pulse. The damage effect at the top of the pile is the most serious, the damage at the middle is slight, and the damage at the bottom is the weakest. Concrete damage occurs at the connection between pile and beam, beam and longitudinal beam. As the depth of explosive decreases, the damage effect of piles decreases and the damage effect of wharf superstructure increases. When the explosive explodes near the water surface, the wharf panel is damaged seriously and ultimately unusable.

1. Introduction

The high-piled wharf is one of the most common wharf structures in muddy coasts and estuaries. Underwater explosion is one of the main striking forms to high-piled wharf. It is significant to study the damage process of the high-piled wharf for improving the survivability on the design of high-piled wharf in wartime.

Numerical simulation is more efficient than field test for underwater explosion research. Rajendran R [1] used finite element software LS-DYNA to simulate the dynamic response of circular plate and square plate under underwater explosion. Schiffer [2] considering the cavitation and initial hydrostatic pressure, studied the response of structure subjected to shallow and deep water explosive, and then analysed the cavitation phenomenon and load distribution near composite plate by numerical methods. Based on LS-DYNA simulation, Cardoso [3] analysed and optimized the structural design of plates and shells to improve the protection capability. Flow field boundary has great influences on underwater explosion shock wave and bubble fluctuation.[4] Yan Qiushi [5] used numerical simulation software to analyse the damage effect of reinforced concrete single pile and obtained the anti-explosion area of single pile under specific depth. The study of damage mechanism and failure model of high-piled wharf lays a foundation for further research on anti-explosion performance of high-piled wharf.

2. Numerical Model

Based on the common harbour, the water depth of the harbour basin is 17m and the water bottom is saturated clay. The whole wharf is composed of standard structural sections. The piles, beams and slabs are all reinforced concrete structures. The superstructure is 26.5m wide and contains seven piles, including three vertical piles and two pairs of inclined piles. The distance between vertical pile and inclined pile is 4.5m, and the distance between inclined piles is 1.7m. The cross-sections for both vertical and inclined piles are 600×600 mm. The length of the pile into the soil is 36 m and the length of pile in water is 17m.

The explosive, 117.45 kg TNT, is located between two vertical piles as shown in Figure 1-2. The depth of explosive is 8.5m. Considering the symmetry and calculation efficiency of the finite element model, only the piles near the explosive and the upper structure are established. The 1/2 finite element model is shown in Figure 3. The model includes: concrete, steel, water, soil and explosive. The Lagrange element is used for concrete and steel, and the Euler element is used for air, water, explosive and soil. The coupling between structure and fluid is defined by *Constrained_Lagrange_In_Solid. Concrete model is concrete_Damage_Rel3 model [6], considering the influence of deviator stress invariant on strength failure surface. The concrete strength of pile is 35.0 MPa, and the concrete strength of superstructure is 28.2 MPa, density is 2550 kg/m^3, Poisson's ratio is 0.2. Strain rate effect of concrete is set by the curve of dynamic load increasing coefficient. [7]

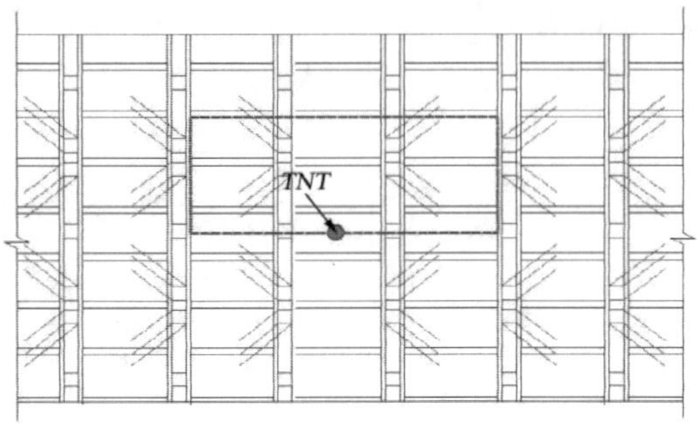

Figure 1. Overhead View

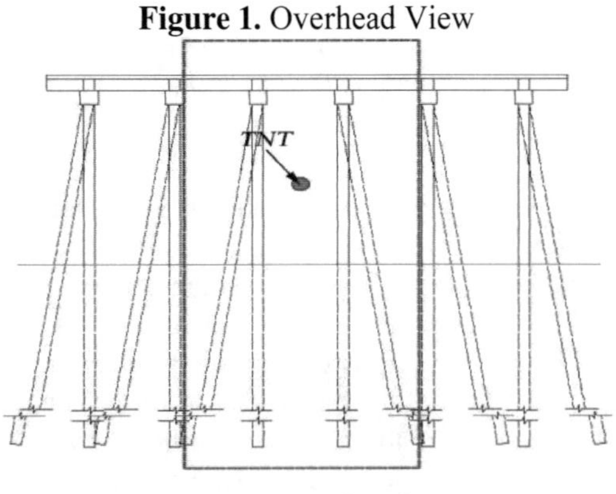

Figure 2. Main View

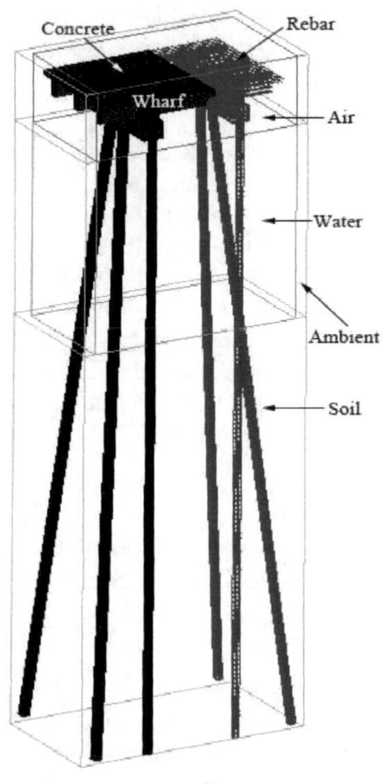

Figure 3. Finite Element Model

The steel model is Mat_Plastic_Kinematic, HRB335. Air is linear polynomial equation of state, C_0~C_6 are parameters of state equation and E is initial internal energy. Water is Gruneisen equation of state, C, S_1~S_3 are parameters of state equation and γ is Gruneisen constant. Explosive is JWL equation, A, B, ω, R_1, R_2 are parameters of state equation. Clay is linear elastic model, E is Elastic modulus and G is shear modulus. Detailed material parameters are shown in Tab 1.

Table.1 Material Parameters of FEM Calculation Model

Material	ρ/(kg·m^{-3})	C_0-C_3	C_4	C_5	C_6	E/(J·kg^{-1})	
Air	1.29	0	0.4	0.4	0	2.5×10^5	
Material	ρ/(kg·m$^{-3)}$	C	S_1	S_2	S_3	γ	
Water	1000	1 480	2.56	-1.986	0.226 8	0.5	
Material	ρ/(kg·m$^{-3)}$	A/GPa	B/GPa		ω	R_1	R_2
TNT	1630	3.74×10^{11}	7.33×10^9		0.3	4.15	0.95
Material	ρ/(kg·m$^{-3)}$		E/MPa			G/MPa	
Soil	1800		22.4			8	

3. Shock Wave Loading

As shown in Figure 4, the initial shock wave propagates spherically and reaches the vertical piles at t=2ms. Because the wave resistance ratio of the concrete is larger than that of the water, the shock wave reflects obviously near piles. During the shock wave passing through the vertical piles (t=2.5ms), the pressure in the front of the pile decreases sharply because the cavitation occurs. The reverse force loads on the piles. When t=3.5ms, the shock wave reaches the inclined piles. Because the inclined pile bears the shock wave load laterally, the reflection of the shock wave is weak and the attenuation of diffracted shock wave is not obvious. Shock wave completely bypasses six piles at t=6ms. It can be found that the reinforcement effect of vertical pile on shock wave is stronger than that of inclined pile and the attenuation of diffracted shock wave near inclined piles is less than that of vertical pile. Therefore, the vertical pile is subjected to stronger shock wave load in the shock wave stage.

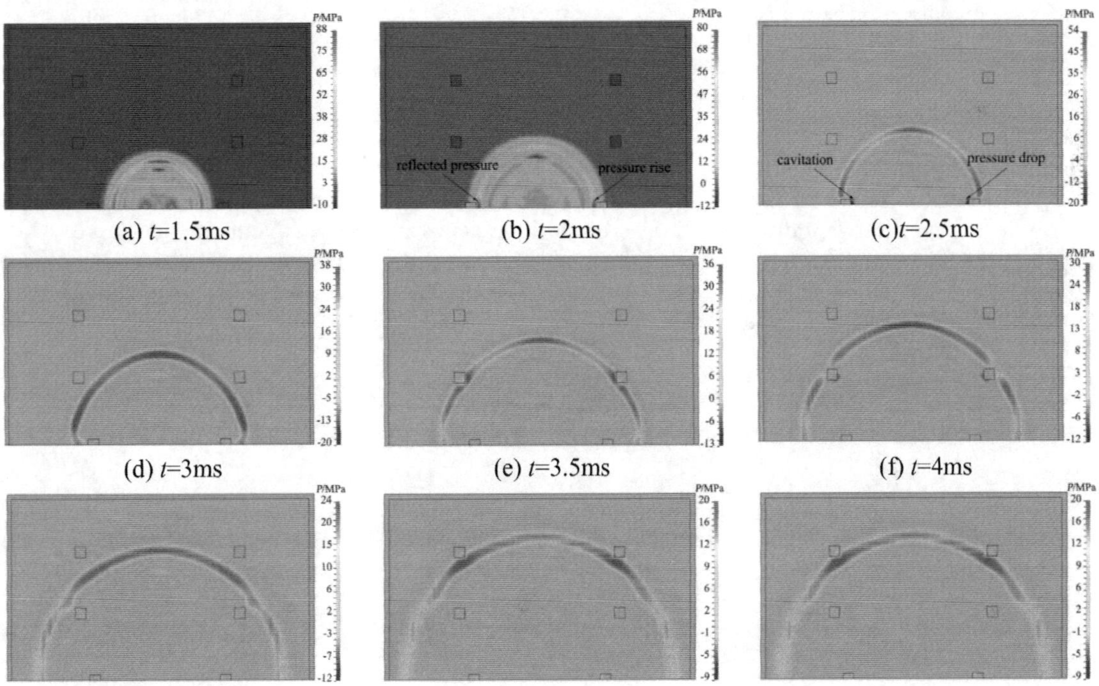

(a) t=1.5ms (b) t=2ms (c) t=2.5ms
(d) t=3ms (e) t=3.5ms (f) t=4ms
(g) t=4.5ms (h) t=5ms (i) t=5.5ms

Figure 4. Propagation of Shock Wave around Piles

4. Bubble pulse

As shown in Figure 5, the bubble expands and extrudes the vertical piles at t=0.07s. The back of the vertical piles are damaged firstly, and then the bottom and top of the piles are damaged. The damage effect at the top of piles is stronger than that at the bottom of the piles because of the buffering effect of clay bottom. Bubble reaches the maximum volume at t=0.27s, then bubble shrinks and the pile is sucked by bubble. At the same time, bubble is affected by Bjerknes force of the piles. The bubble near the pile shrinks slowly and the upper and lower surfaces of bubble shrink fast. It gradually forms bubble jet pointing upward at t=0.62s. The jet breaks through the surface of bubble and forms toroidal bubble. Bubble surface near the piles does not completely shrink to form "tail" near the piles due to Bjerknes force. The toroidal bubble expands in the second bubble pulse process. During the second bubble pulse, the damage phenomenon of high-piled wharf basically remains unchanged.

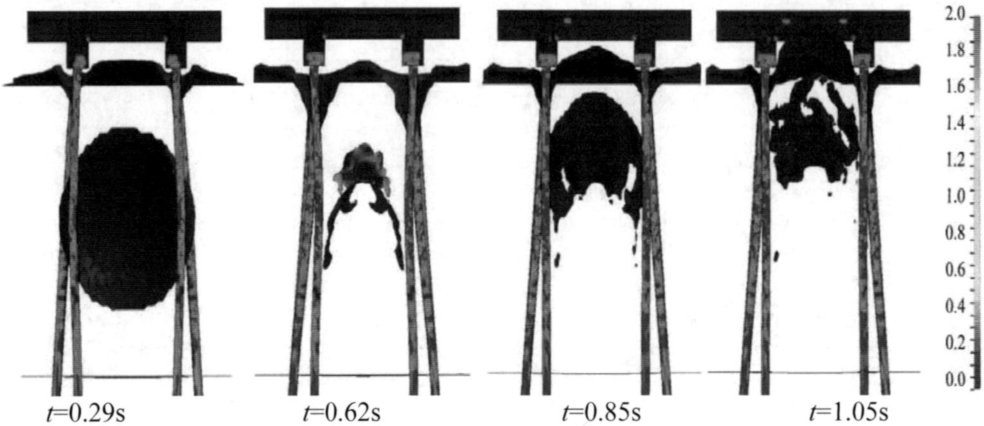

Figure 5. Bubble Pulse Process and the Damage Process of the Wharf

As shown in Figure 6, the first bubble pulse period of underwater explosion is 0.63s, and the maximum radius of bubble is 5.29m at t=0.29s, which is less than the theoretical radius, 6.11m. Piles hinder the expansion of bubble, and bubble collapses earlier in the second bubble pulse period. The maximum bubble radius in the second bubble pulse period reaches only 56.54% of the maximum radius in the first bubble pulse period. There are three peaks in bubble kinetic energy curve, which occur in the initial stage of underwater explosion and at two times of bubble jet. The kinetic energy of two bubble jets accounts for 16.26% and 12.47% of the initial bubble kinetic energy, respectively.

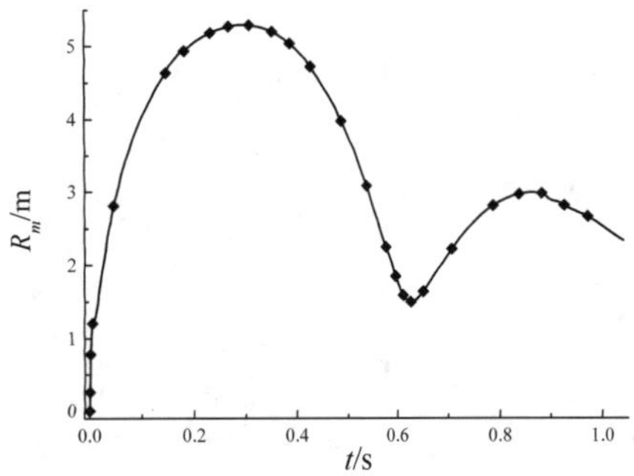

Figure 6. Radius-time of the Bubble

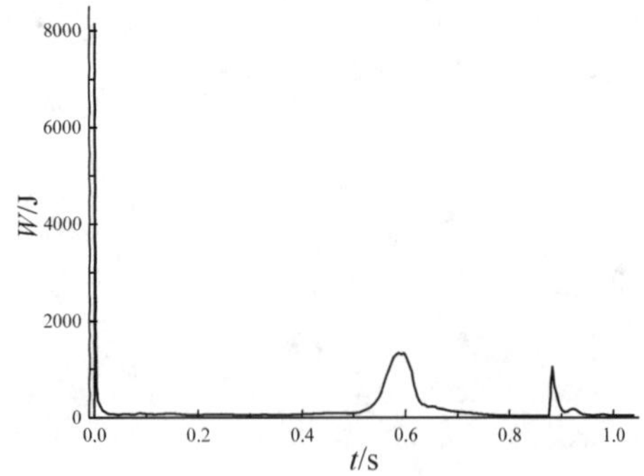

Figure 7. Kinetic Energy-time of the Bubble

(a) t=10ms (b) t=1.05s

Figure 8. The Damage Effect of High-piled Wharf

The Figure 8 shows the damage appearance of high-piled wharf subjected to shock wave loading and bubble pulse. The damage area is mainly at the top, middle and bottom of the piles, the damage of wharf superstructure is slight under shock wave loading. The failure mode of high-piled wharf is mainly formed during bubble pulse. The top and middle of piles are damaged most seriously. Concrete at the connection of piles and beams has been destroyed and failed, and slight damage occurs at the connection of beams and longitudinal beams.

5. Explosive depth

The Figure 9 shows the bubble pulse and the damage process of high-piled wharf when explosive depth is 1m. During the shock wave phase, the shock wave impulse to the pile body is slowed down due to the cut-off effect of water surface. Damage starts to occur in the back of the piles. Subsequently, there is damage effect at the connection of the piles and beams. During the bubble growth, the bubble expands in ellipsoid shape, and the bubble forms a conical water mound on the water surface due to the weak restraint effect of the water surface. As the water mound moves up, bubble directly loads on the wharf panel and longitudinal beam directly. The front face of wharf panel is compressed and the back face is pulled as a result of the bubble load. Because the tensile strength of concrete is far less than the compressive strength, the back of the panel is seriously damaged and cracks are formed in the centre of the panel at t=15ms. Damage effect occurs at the connections of longitudinal and cross beams, longitudinal beams and slabs. The continuous expansion of detonation products loads directly on the wharf panel. The damage area increases obviously in the connections of longitudinal beams, cross beams and wharf panel. After the concrete of the back panel fails, the wharf panel breaks and the steels of the slab are seriously deformed. The bubble collapse because of the wharf superstructure at t=200ms, the explosion product leaks out and turbulence occurs near the water surface. The damage effect of wharf panel is the most serious and the damage of piles decreases gradually as the distance of explosive increases.

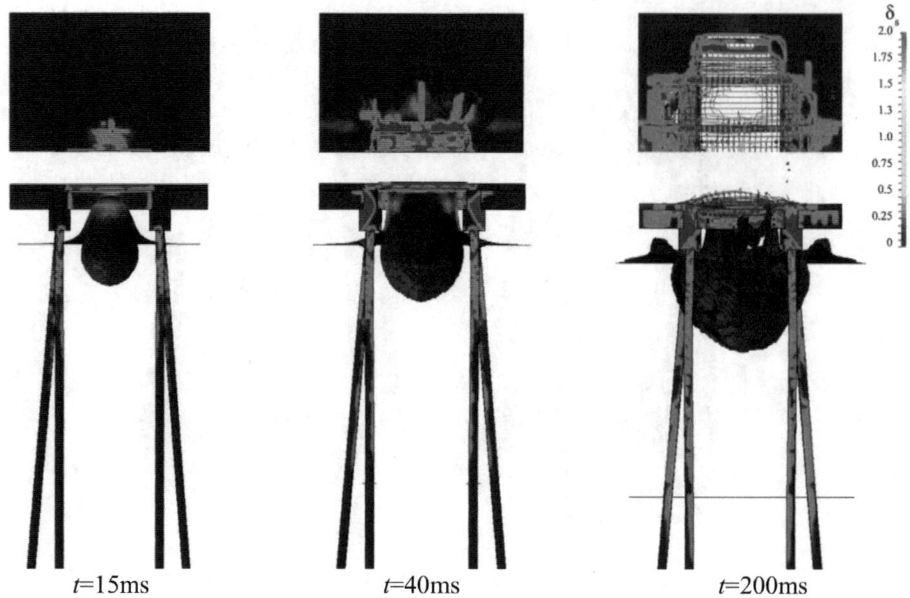

Figure 9. Bubble Pulse Process and the Damage Process of the Wharf (h=1m)

The Figure 10 shows the damage process of high-piled wharf when the depth of explosion is 4.5m. With the increase of explosive depth, the time of bubble forming water mound loading on the wharf panel is delayed. The transverse and oblique cracks are formed on the back of the panel at *t*=0.2s. The bubble breaks up because of the obstruction of the panel at *t*=0.3s. The damage effect of the panel and piles remain basically unchanged after bubble collapse. With the increase of explosive depth, the damage effect of piles increases and the damage effect of wharf panel, crossbeam and longitudinal beam decreases.

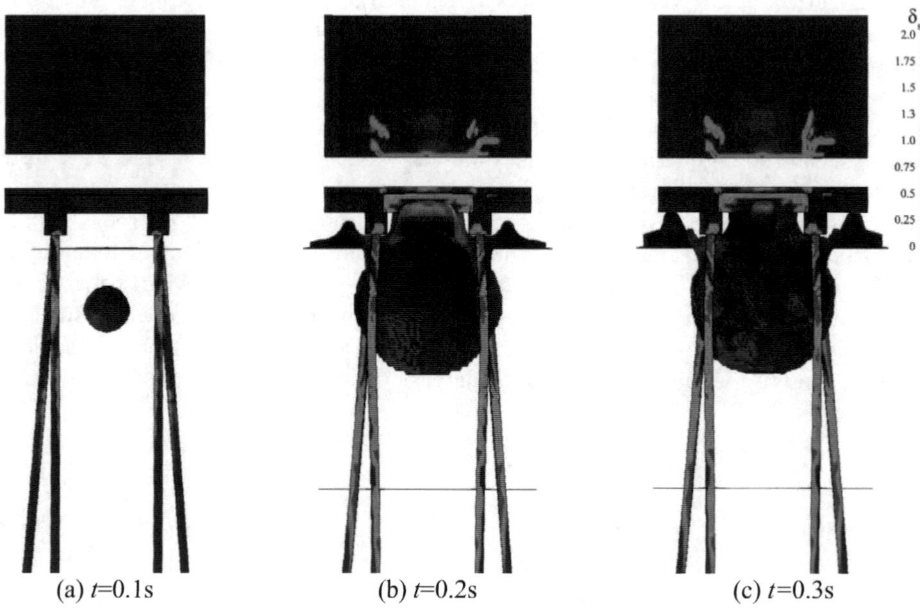

Figure 10. Bubble Pulse Process and the Damage Process of the Wharf (h=4m)

6. Conclusions

In this paper the finite element model of high-piled wharf was established. The damage process of high-piled wharf and bubble pulse were researched. The results presented in this paper lead to the following conclusions.

(1) The damage areas are mainly in the top, middle and bottom of the piles subjected to shock wave loading. The piles are subjected to repeated loading due to the reflection and diffraction of shock wave. The vertical pile is subjected to stronger shock wave load than inclined pile.

(2) The damage appearance of the high-piled wharf basically remains unchanged at the end of the first bubble pulse. The bubble impulsive is the main contributor to the damage of high-piled wharf.

(3) The damage effect of piles decreases and the damage effect of superstructure increases as the depth of explosive decreases. The wharf panel beaks by bubble impulse when the explosive explodes near the water surface.

7. References

[1] Rajendran R. *Numerical simulation of response of plane plates subjected to uniform primary shock loading of non-contact underwater explosion*. 2009. *Materials & Design*. **30**:1000-1007.
[2] Schiffer A, Tagarielli V L. *The response of circular composite plates to underwater blast: Experiments and modelling*. 2015. *Journal of Fluids & Structures*. **52**:130-144.
[3] Cardoso D, Valente R, Paulo R M F. *Numerical simulation and design of extruded integrally stiffened panels (ISP) for aeronautic applications subjected to blast loading*: 2013 *Sensitivity analyses to different stiffener configurations*.
[4] LIU Jinghan, TANG Ting, WEI Zhuobin, etc. *Pressure characteristics of shallow water explosion near the rigid column.2019 Chinese Journal of High Pressure Physics*. **33**(5): 055104.
[5] YAN Qiushi, NING Suyu, DU Xiuli, et al. *Damage effect for a typical reinforced concrete pile under the near field explosion in water*. 2019. *Journal of Beijing University of technology*. **45**(02):55-61.
[6] Tu Z, Lu Y. *Evaluation of typical concrete material models used in hydrocodes for high dynamic response simulations*. 2009. *International Journal of Impact Engineering*. **36**(1):132-146.
[7] Malvar LJ, Ross CA. *Review of strain rate effects for concrete in tension*. 1998. *ACI Materials Journal* **95**(6): 735-739.

Effect of Repeated Cutting on Low-frequency Torsional Vibration Tapping Torque

Wenzhao Li[1], Zhizun Li[1,*] and Fengqi Han[2]

1 Shijiazhuang Campus, Army Engineering University of PLA, Shijiazhuang 050003, China
2 College of Mechanical Engineering, Shijiazhuang Tiedao University, Shijiazhuang 050003, China
E-mail address: jxxylzz@163.com

Abstract. Low-frequency torsional vibration tapping can significantly reduce the tapping torque, which is an effective method to solve the problem of precision machining of small and deep threaded holes on difficult-to-cut materials. In this paper, the mechanism of repeated cutting of low-frequency torsional vibration tapping is studied by comparing continuous tapping with low-frequency torsional vibration tapping on titanium alloy, and the influence law and mechanism of the process parameters such as repeated cutting times and net cutting angle on the tapping torque are studied through the vibration tapping experiments on different materials. The research results show that under the premise of ensuring the processing efficiency, smaller net cutting angle and larger repeated cutting times should be selected to obtain a good process effect.

1. Introduction

Vibration tapping is the application of vibration cutting technology in tapping processing. After years of research and application, it is proved that low-frequency torsional vibration tapping can significantly reduce tapping torque, improve tap life and thread quality, and is especially suitable for the precision machining of small and deep threaded holes in difficult-to-cut materials [1-3].

The difficulty of the application of low-frequency torsional vibration tapping is how to choose the right process parameters for different materials. The main parameters of low-frequency torsional vibration tapping are net cutting angle L_T, backoff angle L_G, and spindle speed V [2, 4]. Net cutting angle and backoff angle determine repeated cutting times and the influence on tapping torque is significant.

In this paper, the effects of repeated cutting on tapping torque and its internal mechanism were studied by vibration tapping experiments of different materials, which provide basis and reference for engineering application of low-frequency torsional vibration tapping.

2. Experiments

2.1. Experiment System

The experiment system is shown in Fig. 1. The torsional vibration of the spindle is realized by the servo motor driven by the control system, and the vibration parameters can be adjusted arbitrarily. The spindle can float freely in the axial direction, and the axial feed is realized by the self-guiding of the tap during the tapping process. Tapping torque is collected by dynamometer.

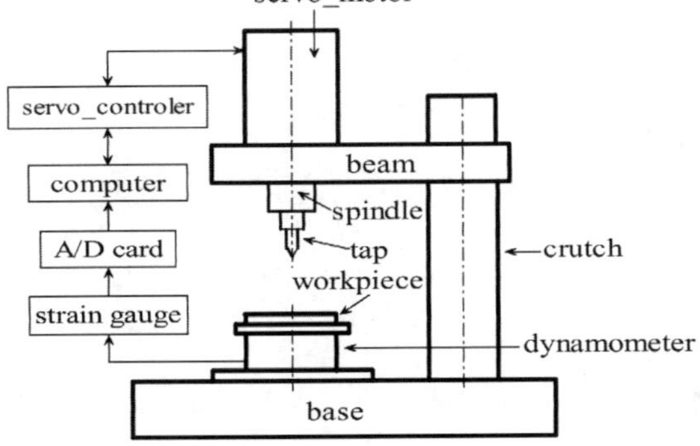

Figure 1. Experiment System

2.2. Experimental Conditions

In order to verify the influence of parameter changes on the tapping torque, difficult-to-cut materials with different processing characteristics were selected as the workpiece. Specific experimental conditions are shown in Table 1.

Table 1. Experimental Conditions

Tap	High speed steel standard machine tap
Workpiece material	High strength steel 30CrMnSi, titanium alloy TC4, superalloy GH4169
Bottom hole	Through hole, depth 15mm
Cutting fluid	Castor oil

2.3. Tapping experiments

In order to study the effect of repeated cutting action of low-frequency torsional vibration tapping, and the influence of parameters on tapping torque, the following experiments were carried out:

(1) Continuous tapping and vibration tapping experiments of M3 threaded holes were performed on titanium alloy TC4 to compare their tapping torque characteristics.

(2) Under the condition that the net cutting angle L_T was constant, backoff angke L_G was changed to perform vibration tapping experiments to study the influence of the change of repeated cutting times on tapping torque.

(3) In order to further study the influence of the contact condition of the rear face on tapping torque, the experiments were carried out by changing the net cutting angle L_T and backoff angle L_G while keeping the repeated cutting times unchanged.

3. Results and Discussion

3.1. Comparison between Low-frequency Torsional Vibration Tapping and Continuous Tapping

When continuous tapping of M3 threaded hole at a speed of 5 rpm were performed on TC4, tapping torque increased with the increase of the tapping depth, and the twist deformation of the tap became more and more obvious. In order to prevent the tap from breaking, the tapping was stopped timely and the tap was retracted, the tapping torque is shown in Fig. 2. It can be seen from the figure that the tapping torque increases rapidly as the tapping depth increases, especially when the tap starts to retreat, the torque has a large negative value, which is the friction torque on the rear face. The torque waveform indicates that the friction torque on the rear face of the tap accounts for a large proportion of the total tapping torque in the continuous tapping process. Therefore, eliminating the friction torque of the rear face is an effective means to reduce the tapping torque.

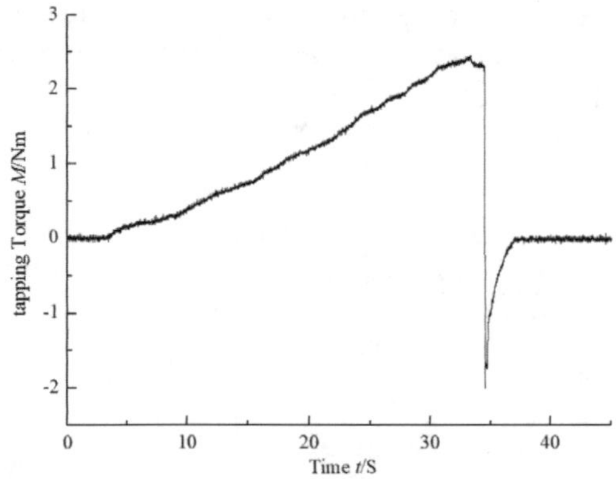

Figure 2. Continuous Tapping Torque Waveform of M3 Threaded Hole of TC4

Low-frequency torsional vibration tapping experiments were carried out on TC4 with different parameters. The experimental results show that during the vibration tapping process, when the cutting part of the tap is completely tapped into the hole, the tapping torque remains basically unchanged, as shown in Fig. 3. The locally amplified vibration tapping torque waveform is shown in Fig. 4. The backoff friction torque directly reflects the friction condition on the rear face of the tap in the tapping process. By comparing Fig. 2 with Fig. 3 and Fig. 4, it can be seen that the peak tapping torque and backoff friction torque of vibration tapping are much smaller than that of continuous tapping.

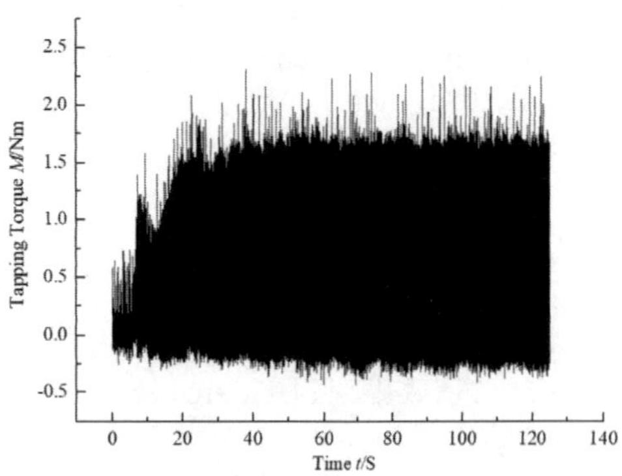

Figure 3. Vibration Tapping Overall Torque Waveform

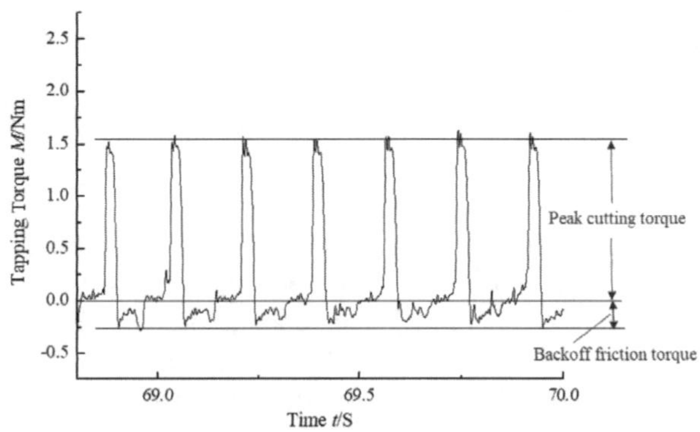

Figure 4. Vibration Tapping Local Torque Waveform

Since the cutting edge is not absolutely sharp, there is always a certain blunt radius, friction is generated in the tool-workpiece contact area, and the springback of the machined surface will also increase the contact area. When the cutting edge and the rear face are worn, the springback amount and contact area will increase, and the rear face may bear the same friction as the rake face [5,6]. Especially when the general tapping method is used to tap on materials with small elastic modulus such as titanium alloy, the amount of springback of the machined surface is large, so the contact area and pressing force between the rear face of the cutter tooth and the workpiece are relatively large, as shown in Fig. 5, the penetration of the cutting fluid into the contact interface between the rear face and the machined surface is very difficult, it is difficult to form an effective boundary lubrication film, so the actual contact area and the cold welding area are large, resulting in a large friction.

Compared with continuous tapping, in the case that the net cutting length is less than the contact length of the rear face and the machined surface, the extrusion of the machined surface to the rear face of the cutter tooth varies in the whole contact range during the cutting stage of the vibration tapping, the amount of springback of the machined surface gradually decreases from the cutting edge to the rear, as shown in Fig. 5, so the extrusion in most areas is relatively light, which is conducive to the penetration of cutting fluid. More importantly, as shown in Fig. 6, after the positive cutting is completed, the machined surface behind the teeth is covered by cutting fluid. In the process of tap retraction, due to the extrusion of the machined surface on the rear face is gradually relieved, the cutting fluid forms a good lubricating film between the rear face and the machined surface under the action of smear and penetration. In addition, the process of tap retraction also provides time for penetration and adsorption of cutting fluid to the newly formed machined surface.

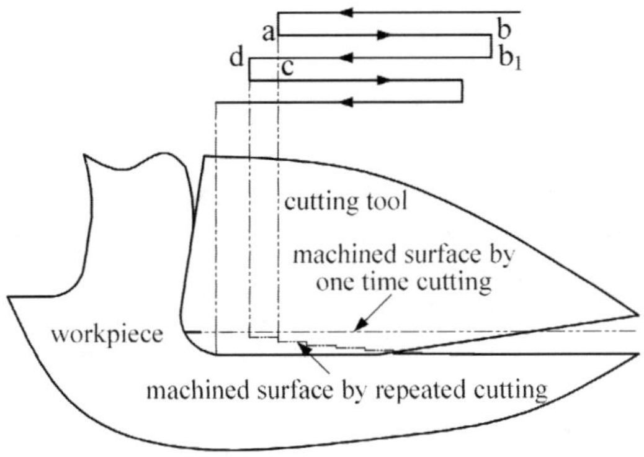

Figure 5. Repeated Cutting Model

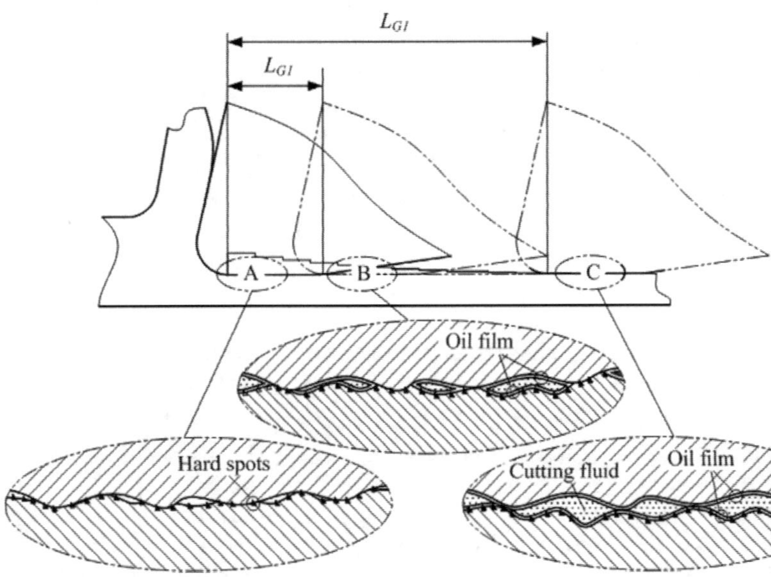

Figure 6. Lubrication of the Rear Face during the Retraction of the Tap

In addition, the formation of thread is a gradual process in the process of tapping, and the shape of any part of thread is formed under the sequential cutting of multiple cutter teeth. If repeated cutting times is enough, the top rear face and side rear face of each cutter tooth can reduce the springback of the machined surface in the current cutting layer to the minimum through sufficient repeated cutting, then the side of the thread is lightly pressed and rubbed against the side rear face of the subsequent teeth, as shown in Fig. 7(a). However, if the repeated cutting times is less, the side rear face of the cuter tooth cannot fully remove the springback of the side of the thread tooth in the current cutting layer, then the newly formed side of the thread will conduct extrusion and friction on the side rear face of the subsequent cutter tooth, and the springback can be gradually removed only after repeated cutting of several subsequent cutter teeth, as shown in Fig. 7(b).

IOP Conf. Series: Materials Science and Engineering **735** (2020) 012044 doi:10.1088/1757-899X/735/1/012044

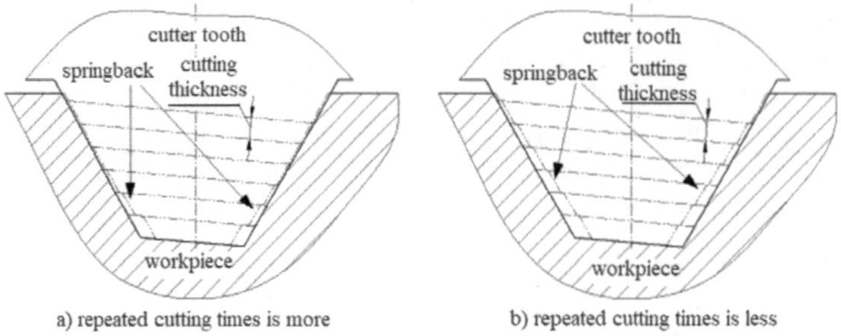

Figure 7. Effect of Repeated Cutting Times on Friction Condition of Side Rear Face of Tap

In a word, under the repeated cutting action of vibration tapping, the contact condition of the rear face has been greatly improved, no matter on the top or both sides of the cutter teeth. On this basis, the periodic tangential separation process between the rear face and the cutting zone greatly improves the boundary lubrication condition of the rear face, which significantly reduces the friction of the rear face during the cutting process, and thus greatly reduces the tapping torque.

3.2. Effect of repeated cutting times on tapping torque

When the spindle speed is 500rpm and the net cutting angle is 1.5°, the law of the backoff friction torque and the peak cutting torque of the M4 threaded holes of the three materials changing with the backoff angle is shown in Fig. 8 and Fig. 9. It can be seen from the figures that the backoff friction torque and the peak cutting torque of the three materials decrease with the increase of the backoff angle. This is because the increase of the bakoff angle increases the repeated cutting times of the machined surface by the cutting edge, thereby reducing the springback of the machined surface and reducing the friction of the rear face. Moreover, since the spindle system, especially the tap, has torsional deformation during the tapping process, the actual reverse angle of the tap cutter teeth is less than the set value. When the backoff angle is small, the separation gap between cutter teeth and workpiece is small, which is not conducive to the smooth entry of castor oil with larger viscosity. The increase of backoff angle can effectively improve this situation. In addition, when the net cutting angle and spindle speed remain unchanged, the increase of backoff angle makes the vibration frequency decrease, and also can fully release the torsional deformation generated by the tap and spindle system during positive and negative rotation, which provides relatively sufficient time and space for castor oil to fully enter the cutting area, which is conducive to reducing tapping torque.

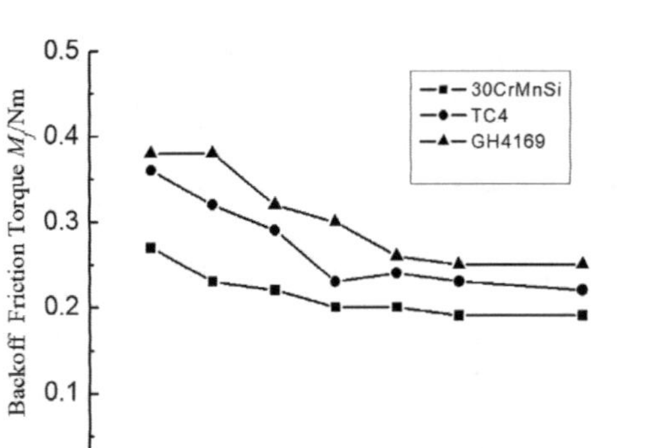

Figure 8. Effect of L_G on Backoff Friction Torque

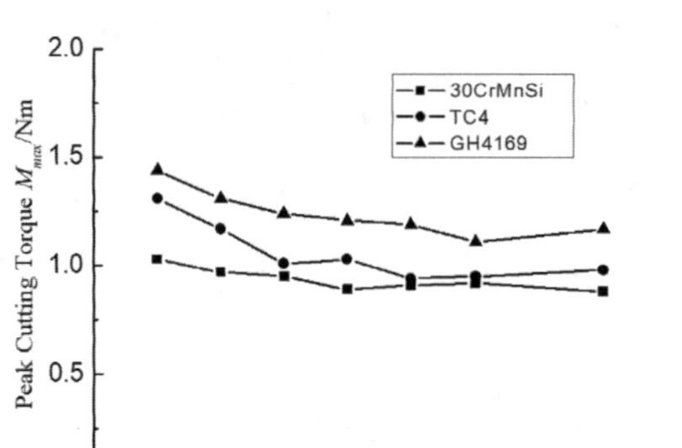

Figure 9. Effect of L_G on Peak Cutting Torque

As can be seen from the above two figures, the vibration tapping torque of the three materials decreases with the increase of backoff angle, and with the increase of backoff angle, its influence on tapping torque gradually weakens. When the backoff angle is greater than 45°, both peak cutting

IOP Conf. Series: Materials Science and Engineering **735** (2020) 012044 doi:10.1088/1757-899X/735/1/012044

torque and backoff friction torque change slightly with the increase of backoff angle. Especially 30CrMnSi, when the backoff angle is greater than 35°, its influence is not obvious.

In addition, it can be seen from the figures that the increase of backoff angle has the greatest impact on the tapping torque of TC4 and the least impact on that of 30CrMnSi. This is because the elastic modulus of TC4 is the smallest, about 1/2 of 30CrMnSi and GH4169, so the effect of repeated cutting is the most obvious. Compared with GH4169, the work hardening of 30CrMnSi is relatively light, and its springback of machined surface can be easily removed during repeated cutting, which can be judged by the fact that castor oil becomes very black during the vibration tapping process, therefore, repeated cutting times has a relatively small influence on the tapping torque.

Therefore, the smaller the elastic modulus of the material, the more repeated cutting times need to be used in the vibration tapping, in order to remove the springback of the machined surface as far as possible, so as to effectively reduce the tapping torque.

3.3. Effect of net cutting angle on tapping torque when repeated cutting times is constant

In the low-frequency torsional vibration tapping, the repeated cutting times is determined by the net cutting angle and the backoff angle, and the net cutting angle determines the cutting stroke of the tap in each torsional vibration period, which has an important influence on the tapping torque. In order to study the influence of net cutting angle, a series of experiments were carried out with the repeated cutting times Q=20 and the spindle speed V=500rpm unchanged. The results are shown in Fig. 10 and Fig. 11. As can be seen from the figures, the backoff friction torque and the peak cutting torque of all three materials increase with the increase of net cutting angle.

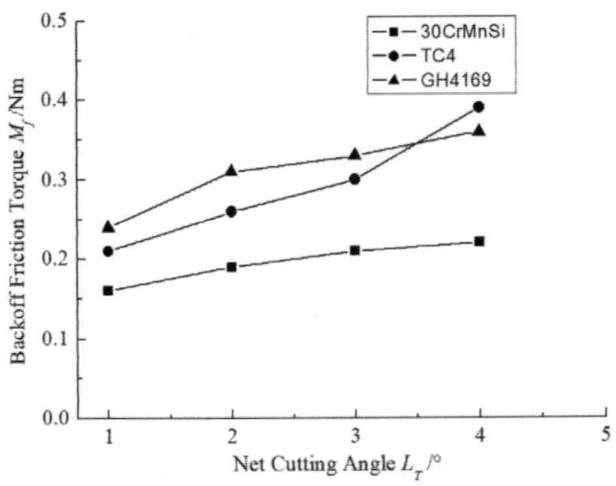

Figure 10. Effect of L_T on Backoff Friction Torque

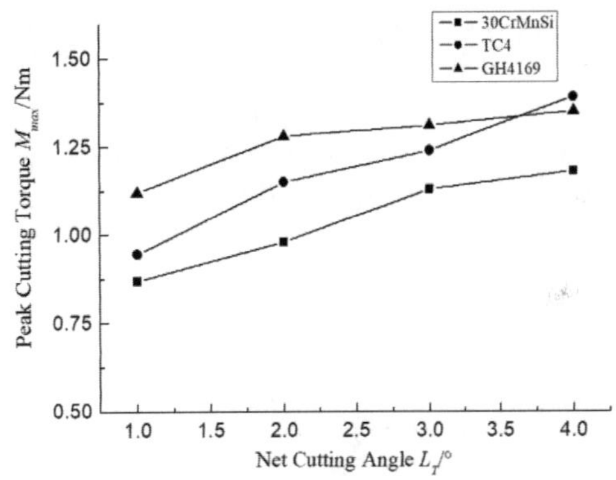

Figure 11. Effect of L_T on Peak Cutting Torque

The extrusion of the machined surface on the rear face of the tap depends on the amount of springback. In the application of vibration tapping, the reverse distance of the tap cutter tooth on the circumference is generally more than 1mm, which is much larger than the width of the wear belt on the rear face of the tap. Therefore, the reverse amount L_G is generally much larger than the contact length between the rear face of the tap cutter tooth and the workpiece in the current cutting layer. In this case, during the cutting stage of each vibration cycle of the tap, the friction condition of the rear face in the current cutting layer of each cutter tooth mainly depends on the net cutting angle L_T, as shown in Fig. 12. The larger the net cutting angle, the greater the length and compressive force of the contact area of the top rear face with the machined surface, and the greater the friction. The friction of the side rear faces in the current cutting layer is equally affected. As mentioned earlier, the magnitude of the effect is related to the properties of the material's elastic modulus and work hardening.

Therefore, in order to reduce tapping torque as far as possible, on the basis of ensuring the necessary repeated cutting times, a smaller net cutting angle should be used.

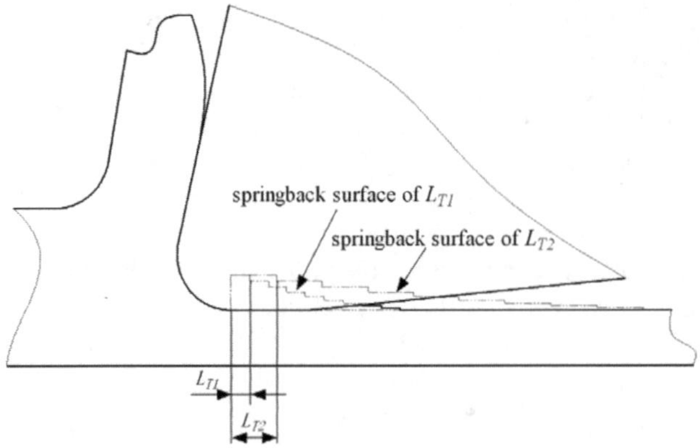

Figure 12. Effect of L_T on the Springback of Machined Surface ($L_{T1}<L_{T2}$)

4. Conclusions
According to the experimental results of low-frequency torsional vibration tapping and the analysis of the friction and lubrication of the rear face, the following conclusions can be drawn:

(1) Low-frequency torsional vibration tapping reduces the extrusion and friction between the machined surface and the rear face of the cutter tooth through repeated cutting, and on this basis, the reciprocating motion of the tap increases the lubrication effect of cutting fluid, effectively reduces the friction torque of the rear face, thus significantly reducing the tapping torque.

(2) Large repeat cutting times should be taken as much as possible to achieve better repeat cutting results.

(3) On the basis of ensuring the machining efficiency, a smaller net cutting angle should be adopted to reduce the extrusion and friction of the machined surface on the rear face of the tap, so as to further improve the technological effect of low-frequency torsional vibration tapping.

5. References
[1] Z.C. Chen: *Studies on the Torsional Vibration Mechanism and Vibration Tapping Technology* (Ph.D. Dissertation, Beijing University of Aeronautics and Astronautics, 2001).
[2] G.J. Li: *Study on the Vibration Tapping and It's Monitoring Technique* (Ph.D. Dissertation, Beijing University of Aeronautics and Astronautics, 2004).
[3] R.D. Han, J. Liu, B. Han 2002 Experiment study of vibration tapping torque of SiCw/Al aluminum composites. *Journal of Harbin Institute of Technology*. **34** 252.
[4] F.Q. Han, Z.Z. Li, D.Y. Zhang 2013 Study on process parameters of low frequency torsional vibration tapping of difficult to cut materials. *Advanced Materials Research*. **712-715** 719.
[5] D.Y. Zhang, D.C. Chen 1998 Relief-face friction in vibration tapping. *Int. J. Mech. Sci.* **40** 1209.
[6] M.S. Shi 1994 *Friction and fluid in cutting*, (China: China railway press) pp 33-97.

IOP Conf. Series: Materials Science and Engineering **735** (2020) 012045 doi:10.1088/1757-899X/735/1/012045

Study on Shearing Characteristics of Circumferential Joint Rebate of Stagger-Jointed Shield Tunnel

Cirong Lu[1,2,*], Huaizhi Zhou[1] and Long Huang[1]

1Shaoxing Rail Transit Group Co., Ltd, China; 2Zhejiang University, China
Email: Lucirong123@outlook.com

Abstract. At present, the design and construction of shield tunnels often adopt stagger-jointed form with rebate, and improving the mechanical properties of rebate is the key to solving the problem of leakage of tunnel structure and other diseases. This paper aimed at circumferential joint rebate of staggered shield tunnel segment, established the local model of discontinuous contact of three-dimensional solid joint to analyse the influence of rebate and the length and inclined angle of it on shear resistance of joint. The result shows five points: (1) When the segment joint is sheared, the greater the axial force of the joint, the stronger the frictional shear resistance. (2) Setting rebate in joints can reduce the number of dislocations in the stage of rapid development of dislocation and increase the shear resistance of joints. (3) The length of rebate has little influence on the shear strength of joints. (4) The angle of inclined plane of rebate has a great influence on the shear resistance of joint. The smaller the angle of inclined plane, the larger the peak shear force and the corresponding dislocation, while the greater the local stress. (5) When the inclined angle increases, the stress of rebate is more uniform, which can prevent the damage of rebate.

1. Introduction

Shield tunnelling is a commonly used tunnel form in urban metro, using assembled lining as a long-term structural carrier. There are two types of lining ring assembling, the through joint and the staggered joint. General segment is usually assembled by staggered joints, which can not only make the stiffness distribution of lining circumferential joints uniform, reduce the deformation of joints and the whole structure, but also be beneficial to waterproof structure and control of shield axis. [1] Therefore, the research on staggered shield tunnel segment is of great significance.

In view of the rebate of circumferential joint assembly shield tunnel segment annular joint, many domestic scholars have carried out relevant research. Wang Tan [2],Zhang Weixi [3], Teng Li [4], Zhu Yaohong et al.[5] studied the staggered shield tunnel segment, which showed that the ability of controlling structural deformation of staggered joint segment was enhanced due to the assembly effect. Zhong Xiaochun [6], Yang Zhao et al. [7] pointed out that the staggered joint assembly method can provide effective constraints on the joint and ensure the long-term waterproof effect of the tunnel. Huang Zhonghui et al. [8] pointed out that the existence of rebate can improve the longitudinal stress state of the tunnel. Zhu Yaohong et al. [10] studied the shear behavior of rebate in Ningbo rail transit shield tunnel. Huang Jun [11] research reduces the breakage of rebate in small radius curves during shield construction in Hangzhou Metro. Jiang Shouchao et al. [12] discussed the performance of rectangular shield joints under the conditions of composite shear resistance, considering installation error and rebate shear resistance alone respectively. In combination with the Shiziyang Tunnel Project of Guangzhou-Shenzhen-Hong Kong High Speed Railway, Feng Kun et al. [13] carried out tests to show the bending stiffness of segment joints under high pressure. In addition, Xiao Mingqing et al. [14] invented a circular rebate shear structure between segments of shield tunnel. Shen Xiufang et al. [15]

designed a concave-convex connection structure between lining segment and inner lining in shield tunnel. Existing studies only focus on staggered shield tunnel segment, but few on the stress characteristics of circumferential joint rebate itself and its size. However, the opening of joints affects the overall force of the segment, so improving the mechanical performance of the circumferential joint rebate is the key to solve the problems of leakage of tunnel structure.

Aiming at the shield tunnel joint with 400 mm segment thickness, a three-dimensional solid discontinuous contact local model of the joint is established, and then the influence of rebate itself and length and inclined angle of it on shear resistance of joint is researched.

2. Finite Element Model of Three-Dimensional Local Joints

The joint structure of shield tunnel is complicated. There are complex structures such as connecting bolts and sealing gaskets, and its mechanical characteristics are obviously non-linear. In order to simulate the real impact of joint components on nearby concrete, a three-dimensional discontinuous contact local model of joint was established by ABAQUS finite element software, which considers the contact between segments and bolts to achieve more accurate simulation. The local model of discontinuous contact of three-dimensional solid joint is shown in Fig 1.

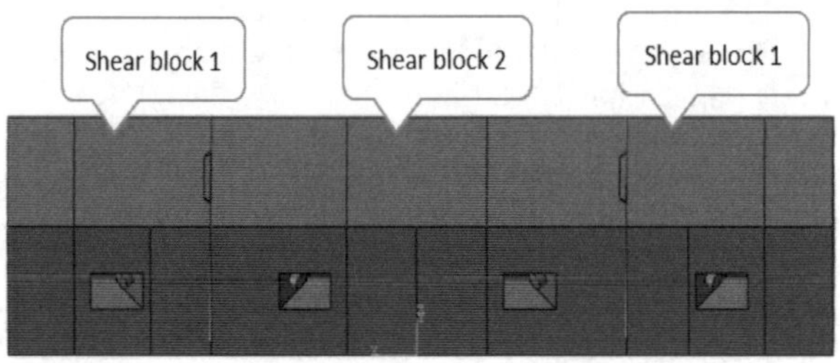

Figure 1. Local Finite Element Model of Joints

3. Geometric Model

Based on the design scheme of shield tunnel segment in a urban rail transit, a geometric model is established, in which the diameter of bolt hole is 36 mm, the diameter of bolt is 30 mm, the thickness of segment is 400 mm, and the length is 500 mm. The geometric structure of rebate is shown in Figure 2, among which the concave tenon length of rebate is 165 mm, and the inclined angle is 55 degrees.

In order to study the change of shear resistance of joints under the change of rebate size, the middle part of rebate was lengthened to 195mm and shortened to 135 mm on the basis of the design scheme. In addition, the inclined plane angles of rebate are transformed to 73 degrees and 33 degrees respectively.

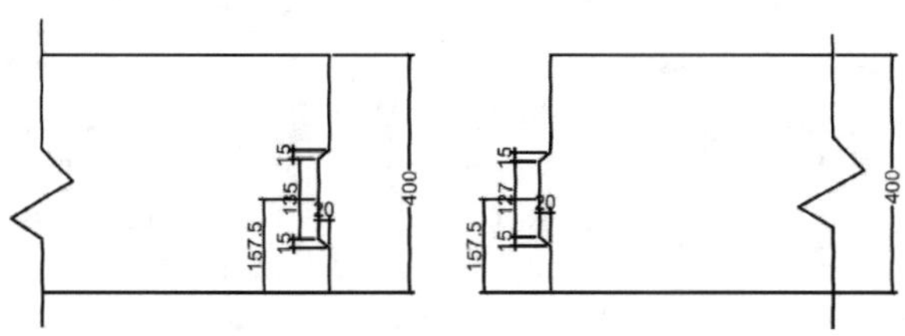

Figure 2. Size Diagram of Rebate

4. Constitutive Properties and Parameters of Materials

The constitutive relationship of concrete is based on the plastic damage model of concrete in ABAQUS, and the elastic damage model of isotropy is used to simulate the inelastic behavior of concrete in combination with isotropic tension or compression plasticity.

In the elastic stage, concrete damaged plasticity model (CDP model for short) uses elastic model to describe the mechanical properties of materials. After entering the damage stage, the relationship between the elastic modulus of the damaged CDP model is as follows:

$$E=(1-d)E_0 \qquad (1)$$

In the formula, E_0 is the initial modulus of elasticity and d is the plastic damage factor d_t or d_c under tension or compression. The range of plastic damage factor d is 0~1. 0 indicates that the material has not been damaged, and 1 indicates that the strength of the material has been completely lost. According to the user-defined nominal stress-strain curves of concrete, they are transformed into the parameters needed in ABAQUS through the principle of energy equivalence. Tensile damage parameters are input into ABAQUS in the form of $\sigma_t - \tilde{\varepsilon}_t^{ck}$ and compressive damage in the form of $\sigma_c - \tilde{\varepsilon}_t^{in}$.

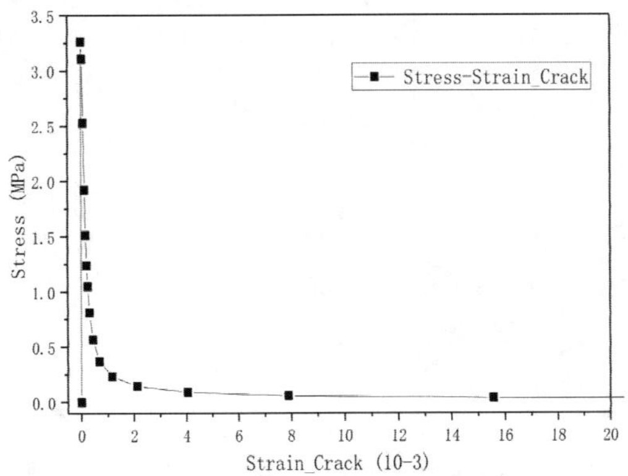

(a) Tensile parametric curve

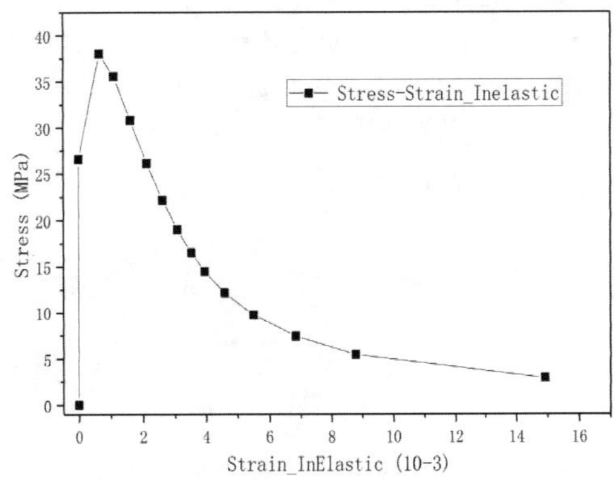

(b) Compression parametric curve

Figure 3. Constitutive Curve of C50 Concrete Material

The values of other constitutive parameters for elastic-plastic damage of concrete are shown in Table 1.

Table 1. Other Constitutive Parameters

dilatancy angle	flow potential skewness	σ_b/σ_c	Kc	viscosity coefficient
38	0.1	1.16	0.6666	0.0005

The bolt constitutive model is simplified to double folded line constitutive model. The elastic modulus in elastic stage is 210 GPa, Poisson's ratio is 0.2, and the elastic modulus after yield is 2.1 GPa. The 5.8 grade bolt material has yield strength of 400 MPa and an ultimate tensile strength of 500 MPa, the constitutive curve is shown in Figure 4.

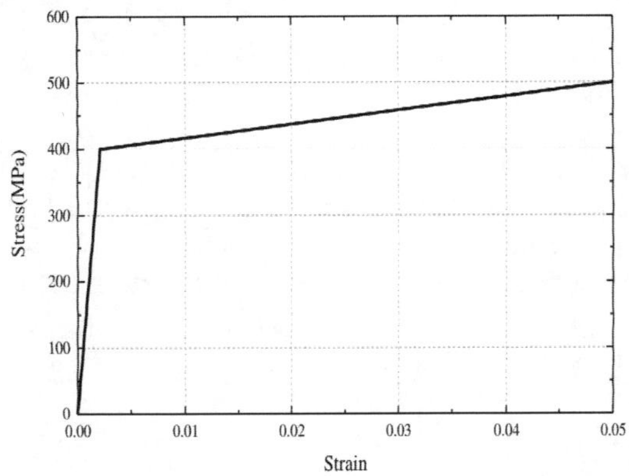

Figure 4. Constitutive Curve of Bolt Material

5. Load and Boundary Conditions

In order to compare and analyze the influence of rebate on the shear properties of joints, the shear properties of two kinds of joints were simulated and calculated: single rebate, rebate and bolt joint. Three working conditions with axial force of 100kN, 300kN and 500kN are selected. Firstly, load-controlled loading mode is used to load horizontal force F on both ends of two shear piece 1 to simulate the axial force at the joint, and then displacement-controlled loading mode is used to load vertical force P on the top of shear piece 2 to simulate the joint being gradually sheared.

When horizontal force F is applied, the displacement of bottom X, Y and Z directions of shear piece 2 and bottom Y and Z directions of shear piece 1 in finite element model are constrained. While the vertical force P is loaded, the displacement in X, Y and Z directions of the bottom of shear piece 1 and the displacement in X and Z directions of the bottom of shear piece 2 are restrained.

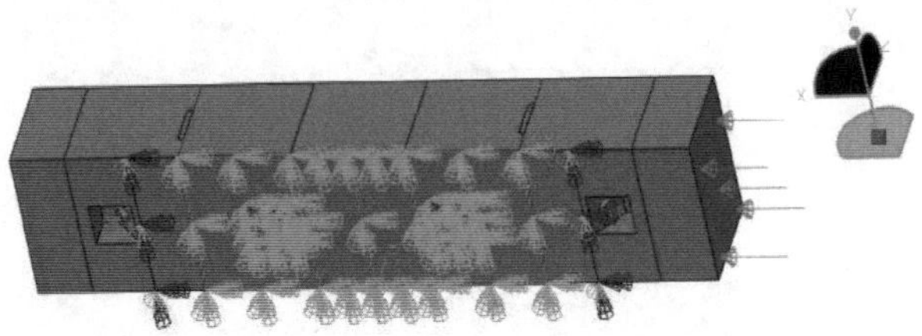

(a) Boundary Conditions for Loading Horizontal Force F

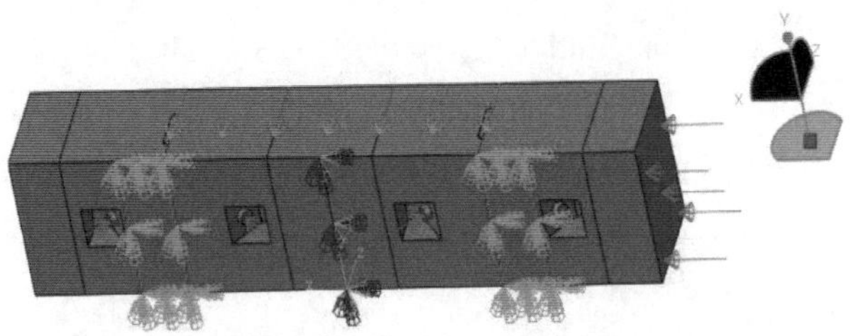

(b) Boundary Conditions for Loading Vertical Force P

Figure 5. Boundary Conditions

Considering the large displacement of joint contact and bolt segment contact in joint test, finite slip is chosen as contact tracking method in this study. In the mechanical behavior of model contact surface, the normal contact behavior is hard touching, and the tangential contact behavior is Coulomb friction contact. Through the compression and shear tests of concrete interface, the static friction coefficient and dynamic friction coefficient of concrete contact surface are 0.5-0.7 and 0.45-0.67 respectively. Therefore, the friction coefficient between the concretes in this model is 0.5, and the friction coefficient between the bolts is 0.3.

6. The Influence of Rebate

The shear-dislocation curve only with bolts is calculated as shown in Fig 6. There are three stages in the curve: the stage of shear resistance of friction, the stage of rapid development of dislocation, and the stage of shear resistance of friction with bolts and rebate. From the graph, it can be seen that the size of the axial force has a significant impact on the shear resistance of the joint. The greater the axial force, the stronger the friction shear resistance.

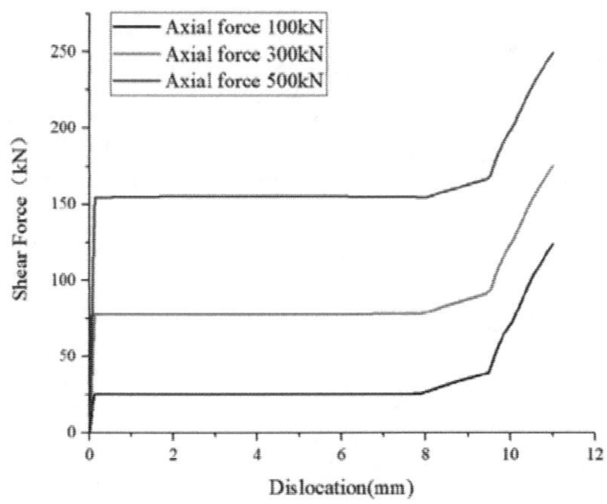

Figure 6. Shear-dislocation Curve of Joint with Bolt

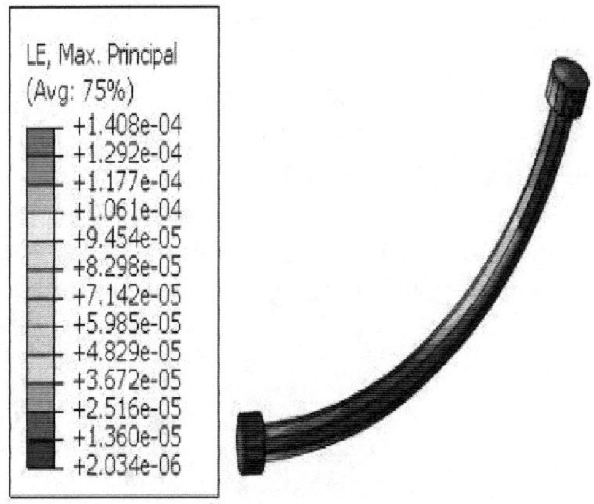

Figure 7. Bolt Strain (300kN)

Under three kinds of axial forces, when the dislocation amounts to 7.74 mm, 7.87 mm and 7.97 mm respectively, the bolt contacts with the bolt hole, and the bolt begins to participate in shear resistance. When the dislocation amounts to 10.4 mm, 10.6 mm and 10.5 mm respectively, the collapsing phenomenon of pore wall concrete has appeared. Taking the case of 300 kN axial force as an example, with the further increase of dislocation, the damage of pore wall concrete continues to develop. The bolts are in the elastic stage during the whole process from dislocation to 11mm. The maximum strain and stress of the bolt are 1.41×10^{-4} and 29.6 MPa when the dislocation is 11 mm, which occur on the arc surface of the bolt, as shown in Figure 7.

The shear-dislocation curves of bending bolt and rebate under different axial load conditions are shown in Fig 7. The curves also have three stages. With the increase of axial force, the maximum shear force at the stage of rapid development of dislocation increases and develops to 4mm. Since then, bolts and rebate have jointly resisted shear, and the shear force increases rapidly with the development of dislocation. In three working conditions, the shear peak values of 1195kN, 1282kN and 1398kN are reached at the dislocation of 4.62mm, 4.75mm and 4.75mm respectively. The increase of axial force can increase the peak shear strength in the third stage, and then enter the softening stage. The smaller the axial force is, the faster the shear strength decreases.

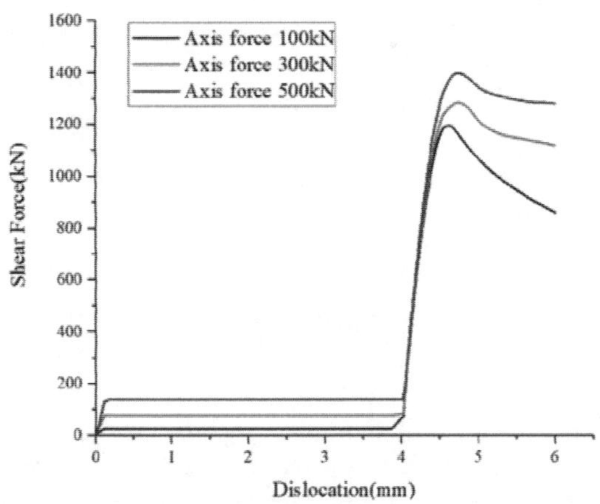

Figure 8. Shear-dislocation Curve of Bending Bolt and Rebate

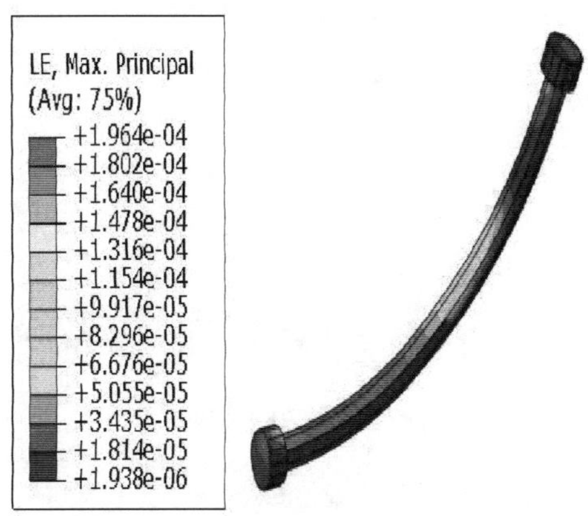

Figure 9. Bolt Strain of Rebate Model (300kN)

When the initial pressure loss of rebate occurs, the dislocation under 100kN, 300kN and 500kN axial force is 4.61mm, 4.57mm and 4.54mm respectively, and the joint shear reaches its peak value.

Taking the 300kN axial force as an example, the bolt is still in the elastic stage when the dislocation develops 6mm. The maximum strain is 1.96×10^{-4} and the maximum stress is 41.2 MPa, which occurs on the inner arc of the bolt shown in Figure 9.

By comparing the shear curves of joints with or without rebate under axial force of 300 kN, it is found that in the second stage, joints without rebate develop rapidly at a lower shear level, and the dislocation can reach 8.1 mm, while in the case of rebate, the staggered stage will enter the third stage when the dislocation reaches 4mm, and the maximum shear force is significantly higher than that of the joint without rebate, as shown in Figure 10.

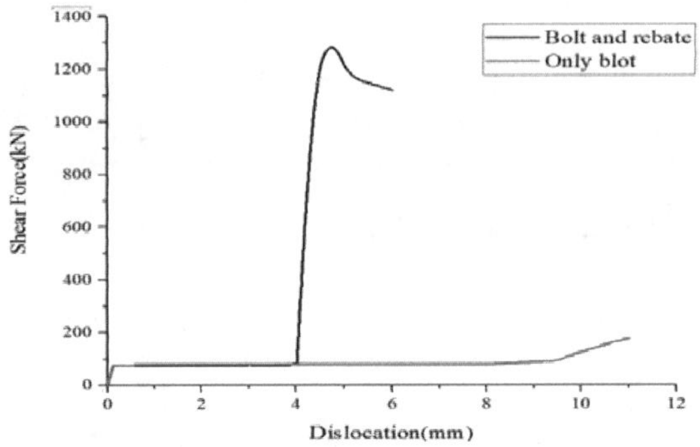

Figure 10. Shear-dislocation Curves of Rebate (300kN)

6.1. The Influence of the Length of Rebate

The shear-dislocation curves calculated by different length of rebate are shown in Figure 11. It can be seen from the graph that changing the length of rebate has little effect on the first and second stages. But the peak value of shear force is slightly different in the third stage that the peak value of rebate model of design length is slightly larger than that of the lengthened and shortened rebate model. In the softening stage, the attenuation of the three models tends to be the same, and the influence of the length of rebate on the curve shape is not significant.

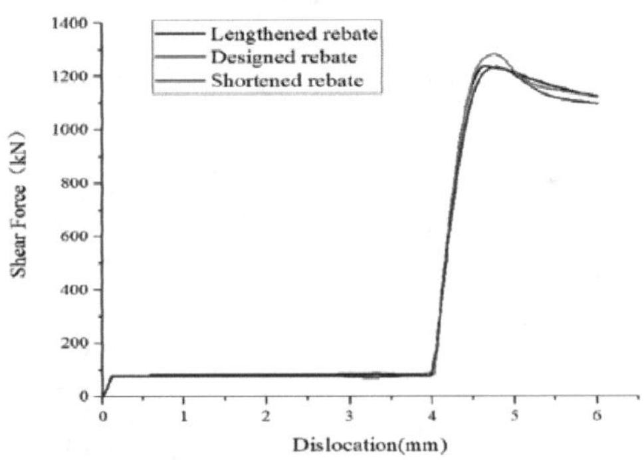

Figure 11. Shear-dislocation Curves of Different Length of Rebate

According to the damage nephogram of rebate, the compressive damage of the lengthened rebate is smaller and distributes along the length direction of rebate, and shortened rebate have large local compression damage and cover the whole tenon, which may damage the whole tenon, while the damage of the designed length model is between the two mentioned above, and its distribution is more uniform but does not penetrate the whole tenon, as shown in Figure12.

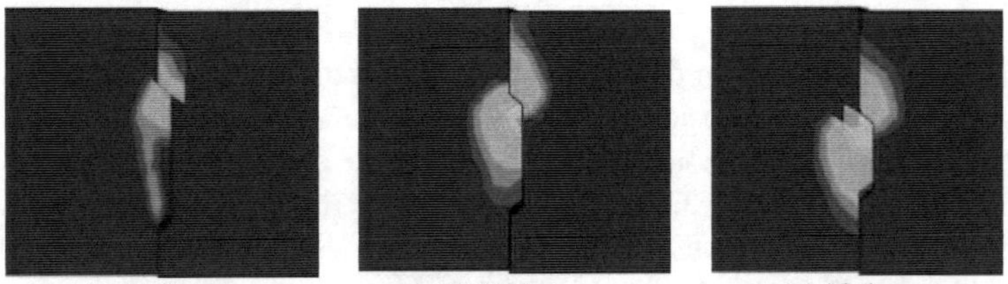

(a)195mm (b)165mm (c)135mm

Figure 12. Initial Compressive Damage Morphology of Rebate with Different Lengths

6.2. The Influence of Inclined Plane Angle of Rebate

The shear-dislocation curves calculated from different inclined angles of rebate are shown in Figure 13. It can be seen from the graph that in the third stage, the peak shear force increases with the increase of inclined angle. The peak shear force at 33, 53 and 73 degrees is 1428kN, 1282kN and 1221kN respectively, and the dislocation peak is 5.01mm, 4.76mm and 4.69mm respectively.

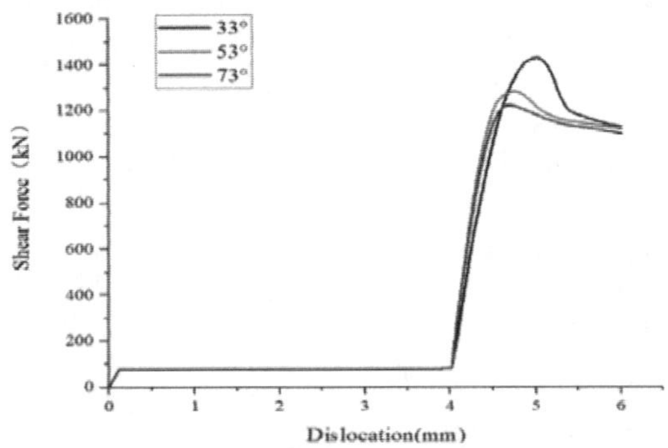

Figure 13. Shear-dislocation Curves of Different Inclined Plane Angles of Rebate

From the damage pattern of rebate, although the peak shear force increases with the decrease of inclined angle, the local damage is relatively increased, which is easy to cause the local damage of rebate. When the inclined angle increases, the range of bearing load concrete at the joint increases and the stress is more uniform, as shown in Figure 14.

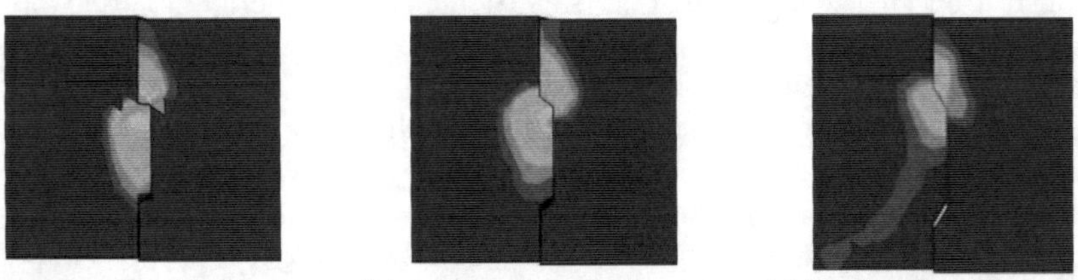

(a) Inclination 33 degrees (b) Inclination 53 degrees (c) Inclination 73 degrees

Figure 14. Initial Pressure Loss Form of Rebate with Different Slope Angles

7. Conclusion

This paper aimed at the circumferential joint rebate of staggered shield tunnel segment, established a three-dimensional solid discontinuous contact local model of joint, studied the effects of rebate and length and inclined angle of it on shear resistance of joint. The following conclusions are drawn:

(1) When segment joints are sheared, whether rebate exist or not, the shear-dislocation curves are divided into three stages: the stage of shear resistance of friction, the stage of rapid development of

dislocation, and the stage of shear resistance of friction with bolts and rebate. The magnitude of the axial force has a significant impact on the shear resistance of the joint. The greater the axial force, the stronger the friction shear resistance.

(2) Compared to joints with only bolts, the setting of rebate can reduce the number of dislocation in the second stage of rapid development of dislocation, and the peak shear force in the third stage increases significantly, which can effectively increase the shear resistance of joint and restrain the development of dislocation.

(3) The length of rebate has little influence on the shear resistance of joint, and the design length is reasonable. The angle of inclined plane of rebate has a great influence on the shear resistance of joint. The smaller the angle of inclined plane, the larger the peak shear force and the corresponding dislocation, but the local stress is greater as well. When the inclined angle increases, the force range of rebate is enlarged and the force is uniform, which can prevent the damage of rebate.

8. References

[1] Yan Jialiang. Research on the behavior of segment joint in shield tunnel[D]. Shanghai: Tongji University, 2006.

[2] Wang Tan, Mo Haihong, et al. The Impact of Staggered Joint and Soil Stiffness on Transverse effective rigidity ratio of shield tunnel [J].Railway Standard Design, 2015, 09: 136-140.

[3] Zhang Weixi, Dong zibo, et al. Experimental study on the force-transfer property of the circumferential joints in quasi-rectangular shield tunnel under the staggered assembling [J]. Journal of Railway Science and Engineering, 2017, 12: 2636-2643.

[4] Teng Li, Lv Jianzhong. Load Test on Typical Segment Joints [J].Journal of Shanghai university (Natural science), 2010, 02: 216-220.

[5] Zhu Yaohong, Xiayang Ganyu, et al. Calculation Parameters of a Shield Tunnel Structure with a Staggered Segmental Tunnel Lining [J]. Modern tunneling technology, 2018, 04: 114-121.

[6] Zhong Xiaochun, Zhu Wei, Qin Jianshe. The comparison of shield tunnel with segment seams in sequence and in stagger [J]. Chinese Journal of Geotechnical Engineering, 2003, 01: 109-112.

[7] Yang Zhao, Pan Xiaoming, Wang Lixin. Research on the Calculation Model of Stagger-jointed Segmental Linings of Shield Tunnel [J]. Journal of Yangtze University (Nat Sci Edit), 2009, 02: 100-103+409.

[8] Huang Zhonghui, Liao Shaoming, Hou Xueyuan. Research on shear model of circumferential joint bolts in stagger-jointed segmental linings [J]. Chinese Journal of Rock Mechanics and Engineering, 2004, 06: 952-958.

[9] Liu Xian, Zhang Weixi, Wang Dongfang. Experimental study on shear performance of longitudinal joint in quasi-rectangular shield tunnel [J]. Journal of Railway Science and Engineering, 2016, 09: 1767-1775.

[10] Zhu Yaohong, Zhang Chen, et al. Experimental study on shear property of circumferential joint in general ring segment under the staggered assembling [J]. Journal of Railway Science and Engineering, 2017, 02: 315-324.

[11] Huang Jun. Fragmentation analysis and Countermeasures of shield segment with concave and convex tenons [J]. Low carbon world, 2017, 04: 226-227.

[12] Jiang Shouchao, Liu Renyong, Liu Xian, et al. Share test study on rectangular shield tunnel segment joint [J]. Building Construction, 2015, 01: 114-117.

[13] Feng Kun, He Chuan, Xiao Mingqing. Bending tests of segment joint with complex interface for shield tunnel under high axial pressure [J]. China civil engineering journal, 2016, 08: 99-110+132.

[14] Xiao Mingqing, Xue Guangqiao, Deng Zhaohui, et al. A Shear Resistance Structure Between Segment Rings of Circular Rebate Shield Tunnel [P]. Chinese patent, CN103924983A, 2014-04-16.

[15] Shen Xiufang, Cao Wenhong, Yang Zhihao, et al. Concave-convex connection method between lining segment and inner lining in Shield Tunnel [P]. Chinese patent, CN101012751A, 2007-02-12.

The Super-hard Cutting Tools' Thermal-cracks Monitoring System Based on Multi-agent Technologies

Caifen Guo and Zhi Dongb
School of mechano-electronic engineering, Suzhou Vocational University, Suzhou, China, 215104
Email: [a]214870053@qq.com, [b]dongzhi@jssvc.edu.cn

Abstract. A thermal crack monitoring system of super-hard cutting tools was set up based on agent technologies. The system structure was designed, together with each agent's workflow, inter-relationship among agents, and their communication mechanisms. If working properly, the system would realize the following functions, data collecting and retrieving, crack conditions monitoring, and resources state inspection and management.

1. Introduction

The working conditions of super-hard cutting tools are very hard, like undergoing loop-impacting loads and elevated temperatures, which would give rise to fluctuating stresses and strains and could eventually result in emergence of tools' thermal fatigue and fractures.

Although researches on thermal fatigue of metal-works have been done thoroughly [1-4], results are all about experiment and qualitive rules, being in short of their mechanism, like the inter-relationship of short cracks, models describing cracks population criterion, especially little about super-hard cutting tools losing effectiveness resulting from thermal fatigue.

Agent is a kind of highly self-government reality, with autonomy, reactivity and initiative. Each agent could perform its work independently, at the same time the whole system objectives would also be carried out through inter-communication and cooperation among agents. Agent can also have perceptual skills to apperceive outside environment changing and infer by the help of knowledge storied in the knowledge system to adjust properties dynamically, so as to make production more practical. Therefore, agent is suitable to be applied to the behavior research of crack individuality and mutual influences among cracks population.

2. System Structure

Agent is used to stand for various physical and logic resources or functional realities in the system. They are organized in mixing hierarchical distribution structures to form a multi-agent system through networks and agent communication protocols. The dispatching agent (Dis-agent for short), administration agent (Adm-agent for short), service agent (Ser-agent for short), cracks agent and resource agent (Res-agent for short) are used to realize monitoring and diagnosing of thermal cracks' conditions in the process of super-hard tools cutting process. The system structure of the cracks monitoring system is shown in Figure 1.

3. System Functions

3.1. Data Acquisition and Retrieving

The data acquisition system is deployed near the cutting tool and acoustic emission signals are collected and kept steady by the BR-ZS1 type of acoustic sensors in the cutting process, which are then transformed into numerical signals by the A/D converter and finally delivered to the FIFO(first in first out, FIFO for short) module. Once accumulated to some kind of amount, data would be carried to the processor and the Res-agent through RS232 bus, like displayed in Figure 2.

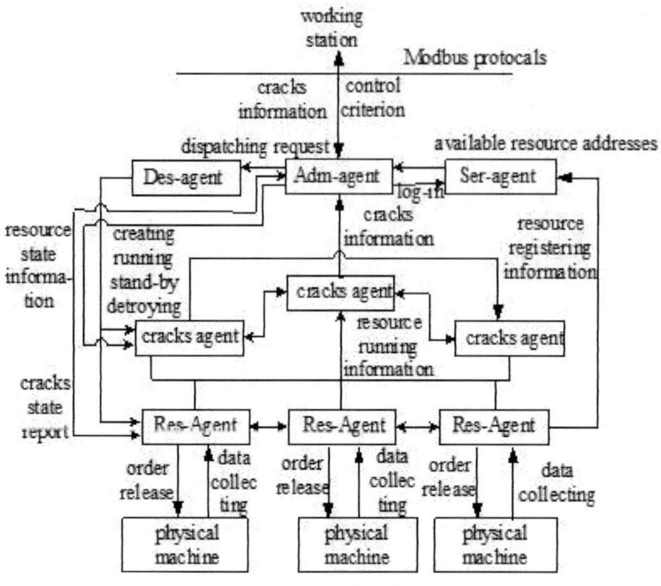

Figure 1. Structure of cracks monitoring system

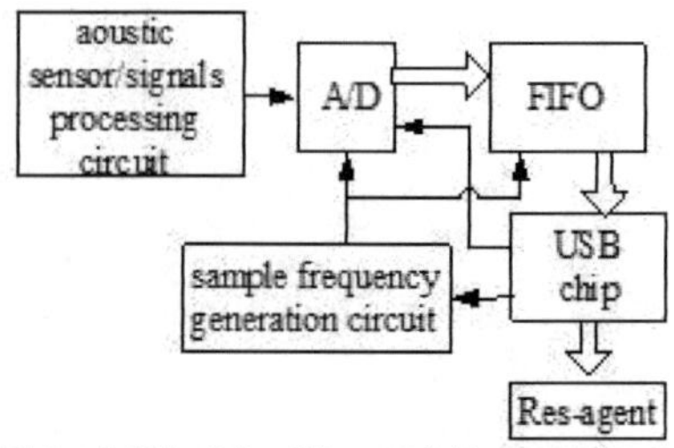

Figure 2. Principle of data aquisition module

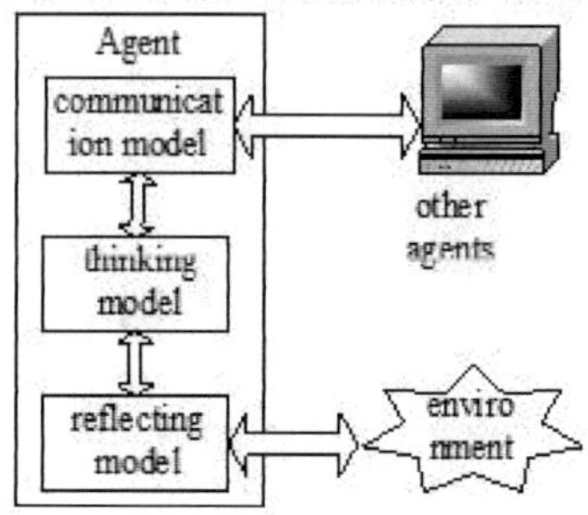

Figure 3. System structure of Agent

3.2. Cracks Monitoring and Tracking

Agent structure is formed by two kinds of modules, one is the thinking module, and the other is the reflecting module, as depicted in Figure 3. Symbol indication and reasoning methods are used to achieve the working goal of the thinking module, while the operation mode is responding directly to the changing of environment in the reflecting module. Once the input is received by agent, the reflecting module starts to work, and then the thinking module begins traditional inferring if no fast reaction or complex inferring is needed.

The Adm-agent receives administration tasks from the cracks monitoring system, then get them analyzed to create the cracks agent. It also interacts with the cracks agent and the Res-agent to grasp the dynamic developing information of cracks. If tools are in good conditions, the Adm-agent would examine whether all production tasks are done, and then finish its working if that is true, otherwise it would check whether tools condition would ensure parts being produced on time, then carry through the production scheme. When tools fail to act actively after taken all the above measures, the cracks monitoring system would dynamically reorganized to respond quickly to the environment and minimize the harmful effects through fast generation and extinction of cracks agent under intervention of Adm-agent. Its workflow is pictured in Figure 4.

Resources allocation optimization is performed by the Dis-agent. After building-up of cracks agent, the Adm-agent requests the Dis-agent to carry out the monitoring process using the enclosed dispatching algorithm, by which the shop production scheme is executed depending upon monitoring task assigned by Adm-agent, some dynamic and static information, such as resource states and abilities. It could interact with other agents to obtain various spots feedback information, through which the monitoring is adjusted. At the same time, the dispatching results are also delivered to the cracks agent and Res-agent. The Dis-agent may operate periodically, or driven by events. Its workflow is shown in Figure 5.

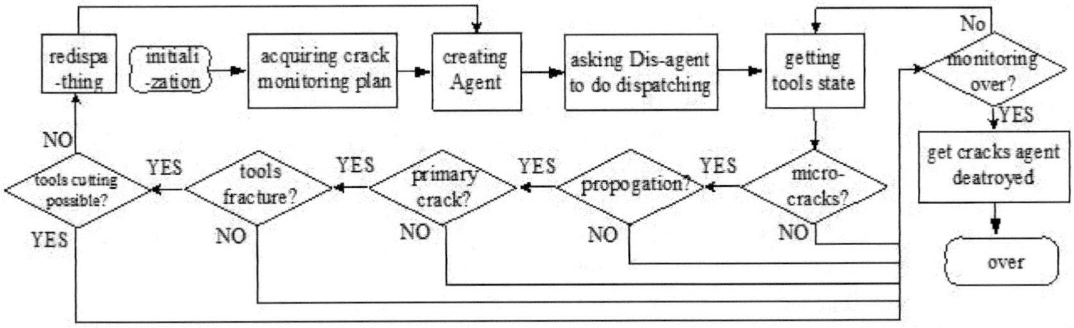

Figure 4. Workflow of Adm-agent

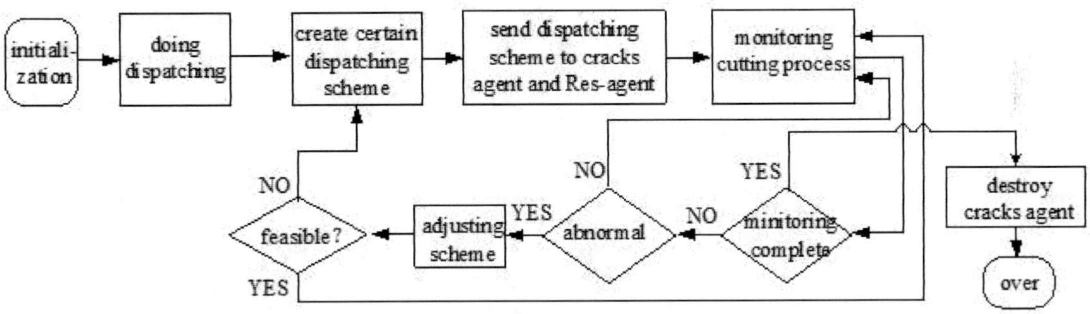

Figure 5. Workflow of Dis-agent

By the task decomposed by Adm-agent the cracks agent is generated. It attaches to the Adm-agent and its lifelong condition is also controlled by it. The cracks agent receives the monitoring sequence generated by the Adm-agent, and by which inquires about the available address information from the Ser-agent, also interacting with it to accomplish tools crack monitoring. The workflow is demonstrated in Figure 6.

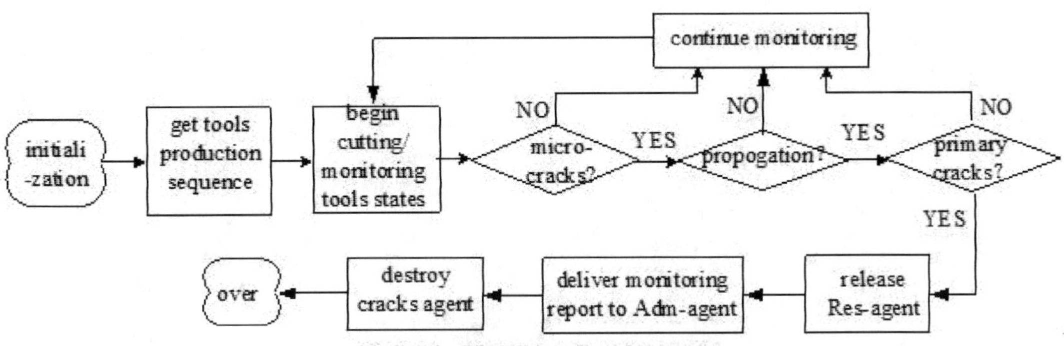

Figure 6. Workflow of cracks agent

3.3. Resources Monitoring and Management

As mapping of physical units in the cracks monitoring system, the Res-agent firstly registers its information to the Ser-agent, like name, address, interest, ability, objective and services provided by its corresponding physical machines. It also can be updated on-line when conditions and functions of physical units changed. The Ser-agent would obtain monitoring tasks by interacting with other agents and drive machines to execute the monitoring tasks. At the same time, the Ser-agent could acquire production data and states of shop-floor machines and deliver them to the cracks agent and the Adm-agent, in order to master monitoring development and resources transformation in time. In case some kind of broken-down happens, the Ser-agent could report it to the Adm-agent and cracks agent without delay and get it maintained on line. Its workflow is shown in Figure 7.

3.4. Communication Mechanism between Multi-agent

The cracks monitoring system is proposed to perform 8 functions, these are, 1) data collecting and acquiring, 2) cracks states monitoring, 3) cracks states tracking, 4) resources condition monitoring, 5) process management, 6) maintenance management, 7) task dispatching, 8) resource management.

From Figure 1 it is known that the cracks monitoring system realize the above functions through mutual coordination, cooperation and negotiation among multi-agent. Essential communication actions and relationships between indicator and receiver of each action are described in Figure 8, in which the figure above each arrow-line is in a consistent one-to-one match with the corresponding function.

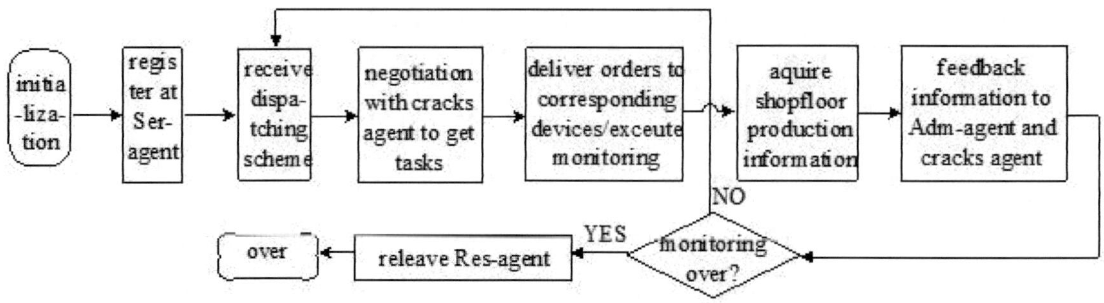

Figure 7. Workflow of Res-agent

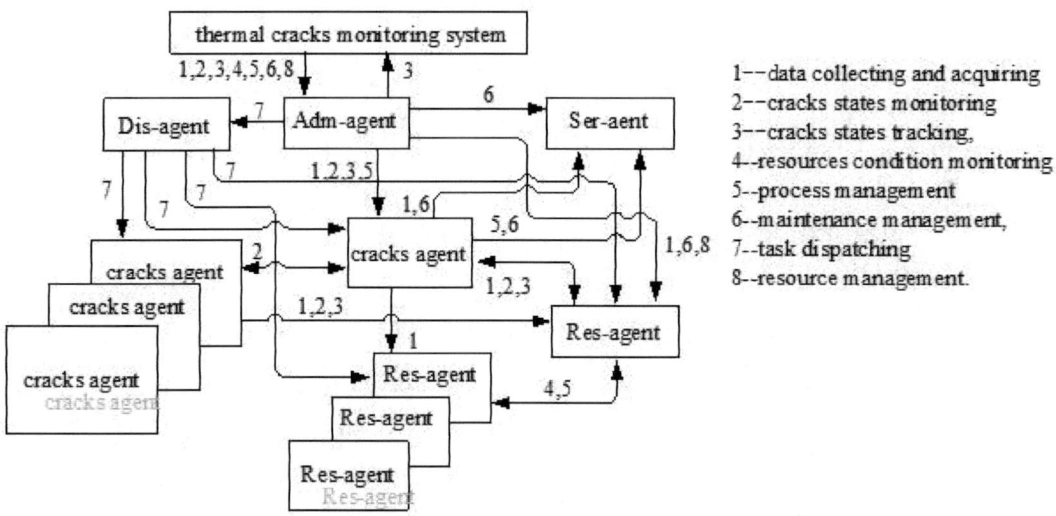

Figure 8. Communication among multi-agent

4. **System Software Realization**

The following crucial problems must be solved during the software actualization process

4.1. Thermal Cracks Typical Behavior Description

Setup cracks condition extension evaluation rules using the extension theory. On the basis of building up cracks evaluation matter-element [5, 6], cracks conditions are identified by calculating matter-element correlation [6, 7].

4.2. Description and Inferring of Thermal Cracks Evolution Rules

Using case-based reasoning methods, combining thermal fatigue tests and experiences [8], cases would be obtained quickly, exactly and practically.

4.3. Designing Agent Thinking-structure and Driven Mechanism

Ensure the agent controlling cracks behavior to be certainly as well as occasionally, the agent thinking structure would adopt the mixing structure combining responding and prudently thinking mode [9].

5. Conclusion

Agent technology is applied to the research of super-hard cutting tools thermal cracks in this paper. The system structure is offered. Such functions like data acquiring and thermal cracks monitoring and tracking, as well as resources real-time monitoring, are realized, by appropriately designing of each agent workflow and their communication relations. The realization of the above research would help to lengthen the tool life, decrease its cost and boost it wider application in various industry fields.

6. Acknowledgement

This research is financially supported by Jiangsu Nature Fund (Project No.BK20171225).

7. References

[1] P. C. Paris, M. P. Gomez, W. P. Anderson. A Rational Analytic Theory of Fatigue [J]. The Trend in Engineering, 1961, 13(1): 9-14.

[2] C. Santus, D.Taylor. Physically Short Crack Propagation in Metals during High Cycle Fatigue [J]. International Journal of Fatigue, 2009, 31(8-9):1356-1365.

[3] J. Polák, T. Kruml, K. Obrtlík, et al. Short Crack Growth in Polycrystalline Materials [J]. Procedia Engineering, 2010, 2(1) :883-892.

[4] Miao.Yu. Research on Short Cracks Initiation and Propagation for Low Cycle Fatigue at High Temperature under Complicated Stress Conditions [D]. Da Lian, China, Dalian University of Technology, 2006, 13-14.

[5] Cai Wen, Wang Guanghua. A New Cross Subject-extenics [J].China Science Foundation, 2004 (5): 268-272.

[6] Dong-min Wu, Yan-nian Rui. Research on the Evaluation of Drill Wear Based on Extension Theory [J]. Modern Manufacturing Engineering, 2012 (4):17-20.

[7] Cai-fen Guo, Zhi Dong, Chang-dong Wan, Yong-kang Zhang. Evaluation of Thermal Fatigue Crack States of Super-hard Cutting Tools Based on the Extension Theory [C]. *MSET2014*, 2014 (936):1894-1899. Jun. 2014, Shanghai, China.

[8] Zhi-wei Ni, Xue-jun Li, Cai-ping Hu, etc. Integration Technique Based on Case-based Reasoning and Rule-based Reasoning [J]. MINI-MICRO SYSTEMS, 2004 (07): 1155-1158.

[9] Si-bo Yang, Min-qiang Li. An Intelligent Control and Decision Making Model Based on Self-organizing Multi-agent System [J].Journal of Tianjin University, 2012 (10) 903-912.

IOP Conf. Series: Materials Science and Engineering **735** (2020) 012047 doi:10.1088/1757-899X/735/1/012047

Dynamic Impact Load Effect on Suspension Crossing Frame at Transmission Tower

Ning Jia, Yongjun Xia, Yong Ma and Rongwang Zhang

China Electric Power Research Institute, Beijing 100055, China
Email: 286488316@qq.com

Abstract. With the increasing number of UHV transmission lines crossing projects, suspension type spans are widely used in cross-construction. Among them, under the condition of construction accidents, the dynamic impact of the bearing cables and towers on the wires has received extensive attention. Based on this problem, this paper studies the impact of dynamic impact loads on transmission towers under construction accident conditions, and proposes reinforcement measures for transmission towers.

1. Introduction

As the company's power grid construction process continues to advance, the situation across construction is increasing. Suspension type spanning frame has the advantages of simple structure, convenient installation, convenient transportation and low construction cost. It is widely used in cross-construction and withstands the dynamic impact of wires under construction accident conditions, thus ensuring the crossing of construction accidents. Under the circumstance of being safe. Under the condition of construction accidents, the load-bearing cable and the tower are subjected to the dynamic impact of the wire. The analysis of the load of the load-bearing cable and the tower under the dynamic impact of the wire is an important basis for the design of the suspension type span and the tower. It is the bearing cable and the tower. The theoretical basis for safety assessment under construction accident conditions.

2. Suspension Spanning Frame and Transmission Tower Introduction

Suspension type spanning frame, also known as no-crossing frame system, means that temporary beams or soft cables are installed as support devices on the towers at both ends of the spanning gear, and the supporting cables are installed between the temporary cross-arms and the insulating nets are installed and the insulating strut is installed. To protect the safety of the crossed objects. It is mainly composed of temporary cross-arm, bearing cable and sealing device. It has the advantages of long distance span, simple structure, convenient installation and small construction difficulty, as shown in

Figure 1.

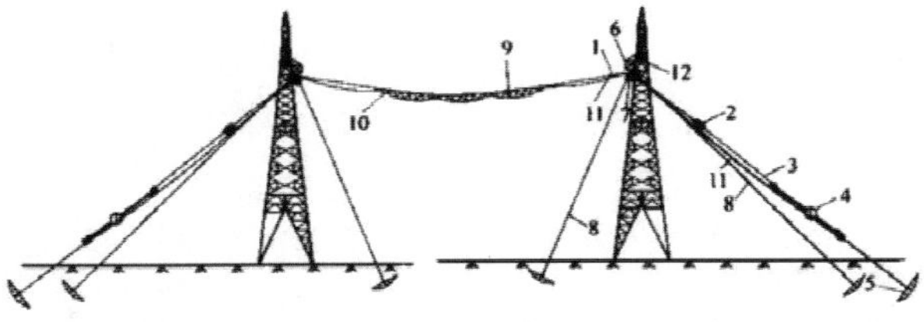

1-Carrier cable (insulation section); 2-Connector; 3-Carrier cable (non-insulated section); 4-Adjustment tool; 5-Ground anchor; 6-Temporary cross arm; 7-Carrier cable pulley; 8-Temporary beam pull; 9-Insulation net; 10-Insulating strut; 11-Anchor rope; 12-Wire take-up pulley

Figure 1. Schematic Crossover System Diagram

A transmission tower is a tower structure used to support and overhead conductors, lightning rods, and other accessories, such that the conductors and conductors, conductors and towers, conductors and lightning conductors, and conductors to the ground or cross-overs maintain a specified safe distance, structure. The tower mainly consists of three parts: the tower head, the tower body and the tower leg, as shown in Figure 2.

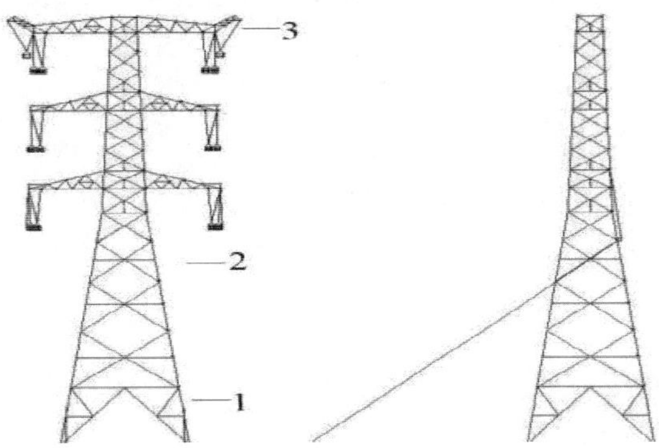

1-Tower leg; 2-Tower body; 3-Tower head

Figure 2. Transmission Tower Schematic

3. Basic Theory of Dynamics

In the course of research, it involves related knowledge of rigid body dynamics, such as the principle of energy conservation and the principle of conservation of momentum.

3.1. Principle of Conservation of Energy

Taking the suspension type spanning frame as the research object, for the running line or disconnection phenomenon in the wire laying line construction, the wire can be calculated in the falling proless, according to the principle of energy conservation, the wire can be calculated before the suspension cable type spanning frame For the speed, take the position of the wire just touching the suspension type span as the zero potential energy reference plane. For the specific relationship, see equation (1):

$$\frac{1}{2}mv_2^2 = \frac{1}{2}mv_1^2 + Q_1 \tag{1}$$

In the middle:

m —Wire quality;

v_1 —Initial speed of the wire;

v_2 —The speed at which the wire just touches the suspension span;

Q_1 —Gravitational potential energy of a wire.

3.2. Principle of Conservation of Momentum

In the process of studying flexible system dynamics modeling method and simulation analysis method, the momentum conservation equation is finite element discrete, and the speed variation is used as the weight function. The solution area is divided into discrete bodies with N1 nodes and N2 units. The derivative of the stress tensor contains the second derivative term of the velocity, and the second derivative term is integrated separately. Combined with the corresponding stress boundary conditions, the momentum conservation discrete form of the flexible system can be derived, as shown in equation (2):

$$\sum_{i=1}^{N_2}\left\{\int_{A_i}\rho\dot{v}_i\delta v_i dA_i + \int_{A_i}\sigma_{ij}\delta v_{i,j}dA_i - \int_{S_i}p_i\delta v_i dS - \int_{A_i}F_i\delta v_i dA_i\right\} = 0 \tag{2}$$

In the middle:

δv_i —Speed variation;

ρ —Flexible body density;

σ_{ij} —Stress tensor in flexible body;

p_i —External force received by the flexible body;

F_i —Physical strength inside the flexible body.

4. Numerical Modeling of Suspension Type Spanning Frame and Transmission Tower

4.1. Creation and Simplification of the Finite Element Model

In this paper, the actual engineering parameters are shown in Table 1, and the finite element model is established. For the dynamic response of the suspension type spanning frame and the transmission tower under the dynamic impact of the conductor, in the unit selection process, the transmission tower adopts the beam element to establish the unit model, and one node has three degrees of freedom, corresponding to the axial direction. The tensile force and bending moment, the force exerted by the node has three components. The suspension type spanning frame and the transmission line are flexible bodies. The finite element model is established by the truss rod unit. One node has two degrees of freedom, and the corresponding one can only bear the axial tensile force, and the tension of the node is decomposed into two components. . The global simulation model established with reference to the engineering site layout is shown in Figure 3.

Table 1. Main Engineering Parameters of Suspension Type Spanning Frame and Transmission Tower

parameter	value
wire	LGJ-630/45
Span	500m
Net length	62m
Net width	12m
Hanging net left and right suspension height	48m
Bearing cable diameter	φ18mm
Wire rope diameter	φ14mm
Number of braided ropes	32
Tray spacing	2m
Tower height	50m
Tower materialQ345	Yield Strength345Mpa

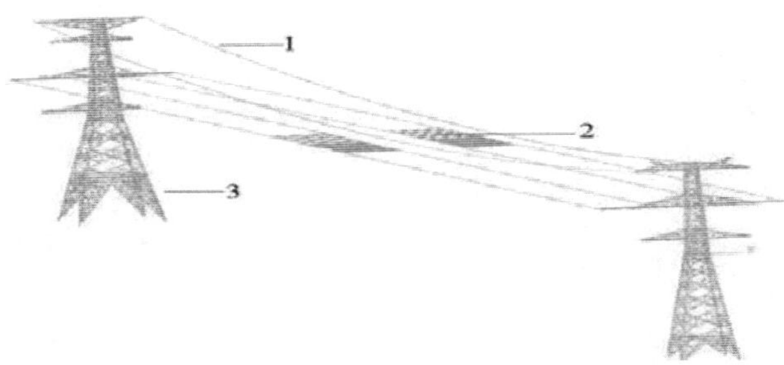

1-wire; 2-Suspension span; 3-Transmission tower

Figure 3. Global Model of Simulation Model for Suspension Type Spanning Frame and Transmission Tower Construction

4.2. Loading Methods and Boundary Conditions

As shown in Figure 4, the gravitational field loading of the wire is achieved by applying a vertical downward gravitational acceleration of 9.8 m/s2 to the wire. A hinge constraint is applied to the left end of the wire and the right end of the wire is released. Set a fixed constraint at the bottom of the tower. A universal contact is provided between the wire and the suspension type span, which has both normal contact force and tangential friction force.

Figure 4. Loading Mode and Boundary Condition Settings

4.3. 630 Wire Impact Spanning Frame Conditions
In order to study the dynamic response of the wire to the suspension type span and the transmission tower, the impact condition of the 630 specification wire at a height of 5 m from the suspension type span is set. To study the dynamic response relationship between the wire and the suspension span and the transmission tower.

5. Numerical Simulation Results
The moment when the wire impact suspension type spanning frame is dropped is shown in Figure 5. The maximum local maximum of the wire impact suspension type spanning frame is 15.7Mpa, which meets the construction requirements. After analysis, it is calculated that the stress on the joint between the temporary cross-arm of the right-hand tower and the tower is 392.5Mpa, which has the risk of failure, as shown in Figure 7.

Figure 5. Wire Impact Suspension Cable Jumper Impact Moment

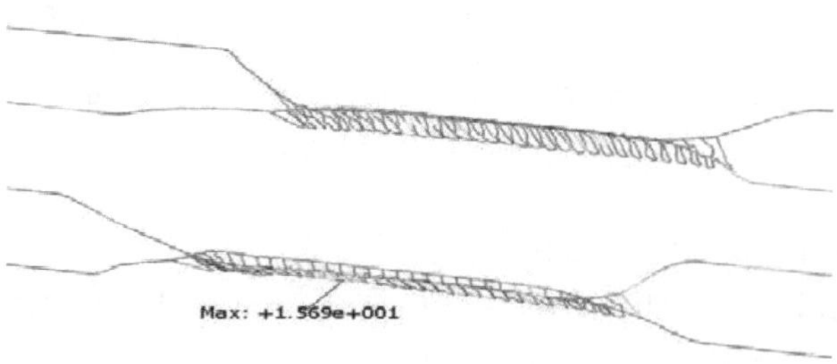

Figure 6. Local Stress Diagram of the Suspension Type Span Frame

(a) Tower maximum stress local map

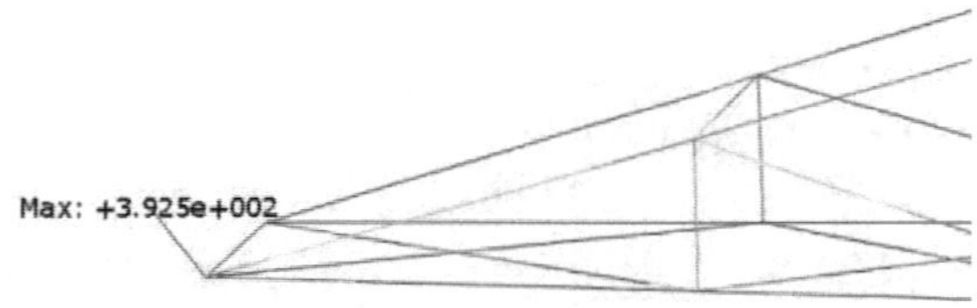

(b) Partial enlargement of the maximum stress of the tower

Figure 7. Tower Stress Map

In view of the failure risk of the tower, a remedy for Φ20 reinforcing cable at the end of the temporary crossarm and the middle portion is taken, as shown in Figure 8. The calculation result is 0.009 MPa as shown in Figure 9. It occurs at the top of the right-hand tower, and it can be seen that the remedial measures are effective.

Figure 8. Strengthening the Cable Installation Diagram

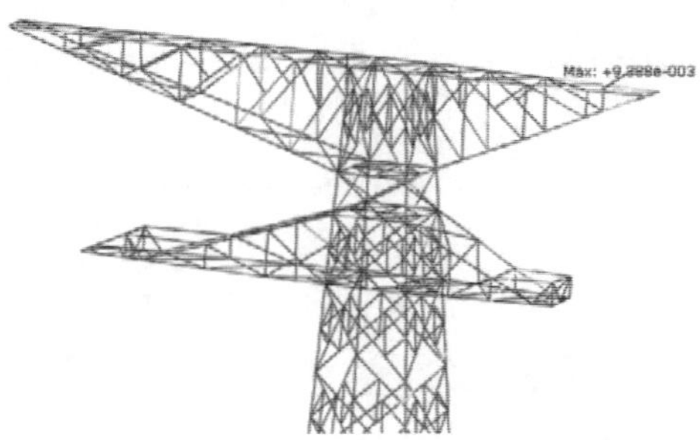

Figure 9. Principal Stress Variation Curve at the Time when the Wire Hits the Right Closed Net Rope at Different Heights from 0 to 0.6 s

6. Conclusion

(1) The 630 wire falls at a height of 5 m, and the impact suspension type span will indirectly cause the risk of failure of the tower.

(2) When the reinforcing tie between the temporary crossarm and the tower is used, the safety of the tower can be significantly guaranteed.

7. Acknowledgement

State Grid Corporation Science and Technology Project Funding (Project Name: Dynamic Impact Load Effect on Suspension Crossing Frame at Transmission Tower, Project No.: GCB11201900108).

8. References

[1] Wei Huang, Hai-xiao Liu, YuShun Lian, Linan Li. Modeling nonlinear time-dependent behaviors of syuthetie fiber ropes under cyclic loading [J]. Ocean Enginearing, **2015.09.**

[2] National Energy Administration overhead transmission line without cross-over frame without power failure spanning line construction process guidelines. DL/T **5301-2013.**

[3] Zhou Lu-yu. Study on UHMWPE weaving process and mechanical properties of fiber rope [D]. Qing Dao University, **2013.10-20.**

[4] Zhang Wei etc. Mechanical Properties and Feasibility Analysis of Ultra High Molecular Weight Polyethylene [J].Engineering plastics application, **2013 (01).**

[5] He Zheng-yang, Pan zhi-juan. Structure and Performance of ultra-high molecular weight polyethylene fiber ropes [J].Modern Science and Technology, **2018, 38 (4).**

IOP Conf. Series: Materials Science and Engineering **735** (2020) 012048 doi:10.1088/1757-899X/735/1/012048

Progress and Challenges Related to Thermal Fatigue Research of Super-hard Cutting Tools

Caifen Guo[a] and Zhi Dong[b]

School of Mechano-electronic Engineering, Suzhou Vocational University, Suzhou, China, 215104.

Email: [a]214870053@qq.com,[b]dongzhi@jssvc.edu.cn

Abstract. A large number of super-hard tools literature has been studied, most of which aimed at tool wear patterns and mechanisms. Researches on thermal lord affecting tools wear are carried out insufficiently and achievements about thermal cracks initiation, propagation are little reported. As existing of elevated temperature and intense heat during cutting processes, super-hard cutting tools thermal fatigue problems are obliged to be faced and some fundamental and unresolved issues need to be answered.

1. Introduction

As one kind of advanced machining technologies, high-speed cutting using super-hard cutting tools (including natural diamond tools, polycrystalline diamond tools (PCD), monolithic polycrystalline cubic boron nitride tools (PCBN), cubic boron nitride (CBN) composite tools) content themselves with high-speed, high-efficiency, low cost and environment friendly machining.

But their working conditions are very hostile, like undergoing loop-impacting loads and elevated temperatures, which would give rise to fluctuating stresses and strains and could eventually result in emergence of tools' thermal fatigue and fracture.

2. Researches Review

2.1. Researches on Tools Wear Patterns and Mechanisms

Researches on super-hard tools' wear patterns and mechanisms have been carried out widely both at home and aboard.

Fatih conducted a research to address wear performance of the CBN and TiN based coated CBN inserts when face milling tool steel. Results showed that macro chipping governing mechanism was responsible for tool wear under all cutting conditions [1].

Zhong investigated micro cutting of pure tungsten with PCD micro slotting tools experimentally. Compared with the flank face, the wear degree of the rake face was more severe. Furthermore, on the rake face micro chipping and flaking emerged, which made up the main wear patterns [2].

With self-developed PCD micro end mills, Bian investigated the tool behaviour in ductile milling of ceramic. During the machining process, diamond particles kept peeling off periodically, which formed the main tool wear pattern [3].

Professor Bai conducted many experiments on 2A12 micro-part with PCD tools. Results showed the predominant features in the damage region of PCD tools were still abrasive, adhesive, and oxidative wear [4].

Priarone compared CBN and PCD tools to carbide tools, indicating that the tool life of the former two kind of tools was clearly longer than that of carbide ones. Because of tool wear, machined surface quality dropped and micro-cracks appeared [5].

Only wear patterns and wear mechanisms, and cutting parameters, mechanical shock and tool material composition effects on tool wear are involved in above researches, little are touched upon influences of cutting temperature and amount of heat on tool wear.

2.2. Studies about Influences of Heat and Thermal Lord

Professor Liew studied the wear features of PCBN tools during machining STAVAX with and without coolant. Results showed that both the tool edge and the rake face were apt to broken. The fracture on the rake face would be resulted from high thermal stresses [6].

Huang investigated lots of CBN tool wear studies and then came to conclusion that abrasion, adhesion, and diffusion would still dominate the CBN tool wear, with each functioning differently, depending on mechanical and thermal load, CBN amount, binder type and quality, chemical stability and parts material, etc. [7].

Professor Ghan did experiments to study CBN tools in hard turning of hardened tool steel. The tool wear, however, under both cutting speeds (υ_c=144.26m/min, and 288.52m/min), was dominated by chipping, which would be alleviated greatly through limiting the heat flow into the tool [8].

With parts possessing both successive and intermittent surfaces, Manu Dogra used CBN and coated carbide tools to finish turning them. Because of high temperature on the tool edge at high cutting speed (190m/min) during turning of continuous surface, the CBN tools tore out quickly, where thermally activated wear was dominant [9].

Doctor Li carried out high speed milling tests on Ti-6Al-4V with PCD tools, finding that the PCD tool possessed kinds of serious chipping and spalling when υ_c was greater than 375m/min. If υ_c was increased still higher, thermal–mechanical impacts would be intensified and the flank wear and breakage became even terrible [10].

With the CBN tool in continuous cutting operations on material Inconel 718, Tatsuya Sugihara did orthogonal machining experiments (υ_c =20~300 m/min) to analyse the tool wear mechanisms. It considered that the trouble lied in the thermal wear originated from the elevated temperature rather than the mechanical wear [11].

Professor Grzesik intended to analyse the changing progress of tool wear using chamfered CBN cutting tools. Results showed that mechanical wear played an important role when υ_c was low, whereas thermal wear when υ_c was high [12].

As far as stresses and temperature were concerned, Li developed and validated evaluation models of PCD tools. What controlled the wear process were still adhesion, abrasion and diffusion, with each wear mechanism functioning at different tribological region [13].

Temperature, Heat quantity, thermal impact and thermal-mechanical coupling are discussed in above documents, but few on initiation, effects and development rules of thermal cracks of super-hard tools.

2.3. Literatures on Thermal Cracks

When milling GG-25 with CBN tools, König discovered continuously enlarged groove wear and thermal cracks at the CBN tool flank face, and there were still chipping and fine grooves distributed on the CBN rake and crater wear faces while cutting Cf53 steel. The latter defect would originated from thermal cracks emerged at the cutting region [14].

Davies discovered that a fine scale attrition governing the wear progress with the low CBN tools. If both CBN and work carbide particle sizes decreased, tool wear rate would be reduced, indicating that the proceeding of micro-fractures was the governing mechanism contained in this kind of tool wear action [15].

Chou investigated microstructure of low CBN content tools in finish hard turning, demonstrating that the key wear reason was fine scale attrition caused by micro-fracture and fatigue [16].

Nabhani employed CBN and PCBN tools to cut titanium alloy and setup a temperature equation, above which the signed adhesion and welding occurred in the contacting surface of tool-part. The failure appeared in the main part of the tool material in all cases, and its pattern was discussed to build the crack propagation route. [17].

In a PCD cutting experiments with laminated flooring with Al_2O_3 overlay, it was found that inter-granular wear and partial cleavage fracture were tool wear mechanisms, among which the micro-cracks played an important role [18].

Professor Cui carried out face milling experiments of AISI H13 steel with CBN tools with high v_c values. It was found that defects resulted from mechanical load changed reversely with v_c , while those of adhesion, oxidation, and thermal cracks boosted by elevated temperature transformed identically [19].

Bushlya presented studies of super-alloy machinability with PCBN tools, on which the crack length was remarkable and stretched beyond the wear region. Experiments told that cracks were not on the surface but extended deep into the tool bulk [20].

From the above papers it is known that the emergence of thermal cracks is one of dominant tool wear mechanisms, but what about cracks initiation characteristics, influence factors and evolution patterns also remain unclear.

3. Challenge

Many researches about super-hard tools mainly concern cutting force, temperature, and tool wear patterns and mechanisms; few are about thermal lord on their wear rules. Literature related to thermal cracks initiation, propagation and evaluation criteria are limited. Many questions remain open

3.1. Cracks Initiation Conditions being not Ascertained

Ding studied the wear behaviour of CBN tools with different grain sizes and various CBN/TiN ratios in ultra-precision cutting of STAVAX. In the machining process, fine-scale cavities appeared on the rake face resulting in cracks formation [21].

Milling of AISI H13 hardened steel with high v_c values, Professor Cui studied the influences of cutting parameters on tool life and wear. Under conditions of elevated temperature and hostile thermal impact, thermal cracks were easily initiated. When v_c=400 m/min, with the following parameter combinations, a_e=15mm, f_z=0.24m/z or a_e=75mm, f_z=0.048m/z, thermal cracks would be found on the rake face with the propagation direction vertical to the cutting edge [22].

Pan studied tool life, tool wear, and causes led to tool failure when end milling of Ti6Al4Vwith PCD tools, indicating that brittle chipping and fatigue were the two controlling failure patterns. One of the possible reasons would be stress concentration [23].

Guangxian Li investigated PCD tools manufactured by electrical discharge grinding (EDG) and common grinding in turning titanium alloy. The initial fragmentation at the cutting stage indicated that interior cracks appeared, which expedited flank wear and tool nose wear and intensified tools fracture [24].

The above researches all attributed tools thermal cracks to stress, temperature, thermal impact and materials hardness, but the individual effect of each factors and their inter-relationships are unknown.

3.2. Thermal Cracks Initiation and Propagation Rules Being Little Stated, Nothing about Thermal Fatigue Evaluation Criteria

Pan established a Finite Element Model to simulate cracks initiation under various loading cycles. Results showed that cracks propagation path was perpendicular to the cutting plane. But whether there existed initiation law, what affected their quantity and growing, how many propagation related factors were there, all related questions waited to be answered [23].

A tool life equation was established by Poulachon when machining hardened steel with CBN tools, which showed that the tool life laid dependence on tool material hardness, cutting parameters (v_c, f, a_p), among which v_c was of the most importance [25].

The above functions are only related to tool life, tool flank and rake wear rates, nothing about thermal fatigue evaluation criteria. If there existed thermal cracks initiation or growth results, they were almost all concerned about metals other than super-hard cutting tool materials.

4. Conclusion

Super-hard cutting tools own a number of benefits over traditional ones. Therefore, they are of great importance to both business and scientific personnel. Towards better understanding and application of super-hard tools, many works need to be done, like thermal cracks initiation, propagation and governing laws, thermal fatigue evaluation criteria, etc.

5. Acknowledgement

This research is financially supported by Jiangsu Nature Fund (Project No.BK20171225).

6. References

[1] Fatih Taylan, Oğuz Colak, Mehmet Cengiz Kayacan. Investigation of TiN Coated CBN and CBN Cutting Tool Performance in Hard Milling Applications [J]. Journal of Mechanical Engineering, 2011, 57(5):417-424.

[2] Li Zhong, Liang Li, Xian Wu, Ning He. Micro Cutting of Pure Tungsten Using Self-developed Polycrystalline Diamond Slotting Tools. Int J Adv Manuf Technol, 2017(89):2435–2445.

[3] Rong Bian, Ning He, Wenzheng Ding, Shuqing Liu. A Study on the Tool Wear of PCD Micro End Mills in Ductile Milling of ZrO 2 Ceramics [J]. Int J Adv Manuf Technol, 2017(92):2197–2206.

[4] Qingshun Bai, Kai Li, Yingchun Liang, Kai Cheng. Wear and Breakage Behaviors of PCD Small-diameter End-mill: A CaseStudy on Machining 2A12 Aluminum Alloy [J]. Int J Adv Manuf Technol, 2015 (77):839–846.

[5] Paolo C. Priarone, Fritz Klocke, Maria Giulia Faga, Dieter Lung, Luca Settineri. Tool life and Surface Integrity when Turning Titanium Aluminides with PCD Tools under Conventional Wet Cutting and Cryogenic Cooling [J]. Int J Adv Manuf Technol, 2016(85):807–816.

[6] W. Y. H. Liew, S. Yuan, B. K. A. Ngoi. Evaluation of Machining Performance of STAVAX with PCBN Tools [J]. Int J Adv Manuf Technol, 2004 (23): 11–19.

[7] Yong Huang, Y. Kevin Chou, Steven Y. Liang. CBN Tool Wear in Hard Turning: A Survey on Research Progresses [J]. Int J Adv Manuf Technol, 2007(35):443–453.

[8] Usman Ghani, Mohammad; Abukhshim, Nuri A.; Sheikh, M. A. An Investigation of Heat Partition and **Tool Wear** in Hard Turning of H13 **Tool** Steel with CBN Cutting **Tools [J]**. International Journal of Advanced Manufacturing Technology, 2009, 39 (9/10):874-888.

[9] Manu Dogra, Vishal S. Sharma, Anish Sachdeva, Narinder Mohan Suri. Finish Hard Turning of Continuous and Interrupted Surfaces with Cubic Boron Nitride (CBN) and Coated Carbide Tools [J]. Materials and Manufacturing Processes, 2012(27): 523–530.

[10] Anhai Li & Jun Zhao & Dong Wang & Jiabang Zhao &Yongwang Dong. Failure Mechanisms of A PCD Tool in High-speed Face Milling of Ti–6Al–4V Alloy [J]. Int J Adv Manuf Technol, 2013 (67):1959–1966.

[11] Tatsuya Sugihara1, Shota Takemura1, Toshiyuki Enomoto1. Study on High-speed Machining of Inconel 718 Focusing on Tool Surface Topography of CBN Cutting Tools [J]. Int J Adv Manuf Technol, 2016 (87):9–17.

[12] Wit Grzesik, Berend Denkena, Krzysztof Żak, Thilo Grove, Benjamin Bergmann. Energy Consumption Characterization in Precision Hard Machining Using CBN Cutting Tools [J]. Int J Adv Manuf Technol, 2016 (85):2839–2845.

[13] Guangxian Li, Shuang Yi, Cuie Wen, Songlin Ding. Wear Mechanism and Modeling of Tribological Behavior of Polycrystalline Diamond Tools When Cutting Ti6Al4V [J]. Journal of Manufacturing Science and Engineering, 2018(140):1-15.

[14] König W, Hochschule T, Komanduri R, Schenectady D, Tönshoff HK. Machining of Hard Materials [J]. Annal CIRP, 1984, 33(2):417–427.

[15] Davies MA, Chou Y, Evans CJ. On Chip Morphology, Tool Wear and Cutting Mechanics in Finish Hard Turning [J]. Annal CIRP, 1996, 45(1):77–82.

[16] Y.Kevin Chou, Chris J.Evans. Tool Wear Mechanism in Continuous Cutting of Hardened Tool Steels [J]. Wear, 1997 (212): 59-65.

[17] F.Nabhani. Wear Mechanisms of Ultra-hard Cutting Tools Materials [J]. Journal of Materials Processing Technology, 2001, 115 (3) :402-412.

[18] Bai QS, Yao YX, Phillip B, Zhang G. Study on Wear Mechanisms and Grain Effects of PCD Tool in Machining Laminated Flooring [J]. Int J Refract Met Hard Mater, 2004(22):111–115.

[19] Xiaobin Cui, Jun Zhao, Xianhua Tian. Cutting Forces, Chip Formation, and Tool Wear in High-speed Face Milling of AISI H13 Steel with CBN Tools [J]. Int J Adv Manuf Technol, 2013(64):1737–1749.

[20] Volodymyr Bushlya, Oleksandr Gutnichenko, Jinming Zhou, Pajazit Avdovic, Jan-Eric Sta˚h. Effects of Cutting Speed when Turning Age Hardened Inconel 718 with PCBN Tools of Binderless and Low-CBN Grades [J]. Machining Science and Tenology, 2013(17):497-523.

[21] X Ding, WYH Liew, BKA Ngoi , JGK Gan, SH Yeo. Wear of CBN Tools in Ultra-precision Machining of STAVAX. Tribology Letters, 2002, 12 (1): 3-12.

[22] Xiaobin Cui, Jun Zhao, Yongwang Dong. The Effects of Cutting Parameters on Tool Life and Wear Mechanisms of CBN Tools in High-speed Face Milling of Hardened Steel [J]. Int J Adv Manuf Technol, 2013 (66):955–964.

[23] Pan WC, Kamaruddin A, Ding SL, Mo J. Experimental Investigation of End Milling of Titanium Alloys with Polycrystalline Diamond Tools[C]. Proc Inst Mech Eng J Eng Manuf, 2014(228):1–13.

[24] Guangxian Li, Mohammad Zulafif Rahim, Songlin Ding, Shoujin Sun. Performance and Wear Analysis of Polycrystalline Diamond (PCD) Tools Manufactured with Different Methods in Turning Titanium Alloy Ti-6Al-4V [J]. Int J Adv Manuf Technol, 2016(85):825–841.

[25] Poulachon G, Moisan A, Jawahir IS. Tool-wear Mechanisms in Hard Turning with Polycrystalline Cubic Boron Nitride Tools [J]. Wear, 2001(250):576–586.

IOP Conf. Series: Materials Science and Engineering **735** (2020) 012049 doi:10.1088/1757-899X/735/1/012049

Effect of the Properties of Basalt Fibers Aging in Salt Solution[1]

Qichang Zhang [1,2], Tan Yan [1,2], Lida Luo [1,2], Qingwei Wang [1,2] and Jin Liu [1,2]
1. State Key Laboratory of Modification of Fiber Materials, Shanghai, 201620, China;
2. Engineering Center of Advanced Glass Manufacturing Technology, Ministry Of Education, Donghua University, Shanghai 201620
Email: 2180271@mail.dhu.edu.cn

Abstract. Basalt fiber is widely used in various fields such as military industry and civil use due to the characteristics of high strength, high temperature resistance and good chemical stability. In this paper, the aging of basalt fiber immersed in different concentrations of NaCl solution at different times was studied. The surface morphology and structure changes of basalt fiber brine after immersion were studied by scanning electron microscopy and Fourier transform infrared spectroscopy. The Fracture strength and weight loss rate of basalt fiber were used to characterize the salt water corrosion resistance of basalt fiber. The results showed that the surface of fiber was slightly corroded and ion exchange occurs in the solution when basalt fiber was soaked in NaCl solutions. The tensile strength of basalt fiber increased first and then decreased with time. It was inferred the ion exchange resulted in surface compressive stress increase the strength of the basalt fiber just like as tempering principle of glass.

1. Introduction

Basalt fiber is a kind of continuous fiber made by basalt ore melting at 1450°C - 1550°C and drawn at high speed through a platinum-iridium alloy drawing and draining plate [1, 2] . It has a series of excellent properties on high temperature resistance, high tensile strength, good chemical stability, as well as the advantage of non-toxic, natural, eco-friendly and inexpensive, At present basalt fiber has been gradually popularized and applied in acoustic thermal insulation materials, chemical filters, electrical insulation materials, composite materials and products with high chemical stability, reinforcing materials, structural materials and other composite products [3-8].

The development of the economic belt in the coastal areas is essential for the development of the whole country, especially the construction of bridges, tunnels and roads, which was called as “The era of marine economy”. Now there are nearly 80% basalt fiber used to strengthen the civil engineering due to the excellent stability, good mechanical properties, eco-friendly and low price [9-12], and it would be vital to explorer the aging properties on the environment of the salt for the application of basalt fiber.

Wei et al. studied the aging properties of basalt and glass fiber reinforced epoxy resin in seawater aqueous solution. The results show that the strength decreases with time [13]. Hu Haitao et al. studied the aging properties of basalt fiber sheets under neutral salt spray conditions, and the aging phenomenon was not obvious [14]. Zhu Ying et al. studied the durability of basalt fiber cloth in alkaline environment, acid environment, salt environment and water environment. The results show that the tensile strength of basalt fiber sheet is obviously reduced after corrosion by alkali solution and acid solution [15]. The salt and water environments have little effect on their tensile strength. Yang Yongxin et al. found that the basalt fiber cloth has a slight decrease in tensile strength and elongation under the environment of 3.5 times concentration and 5 times concentration of artificial seawater. The elastic modulus is improved and the basalt fiber cloth has good resistance. Seawater corrosion performance [16]. Yang Miaomiao et al. found that the tensile strength of basalt fiber and glass fiber monofilament decreased with the increase of soaking temperature and soaking time in seawater immersion conditions, and the tensile strength decreased rapidly in the early stage of immersion, and then the rate slowed down [17]. El Refai et al. studied the aging of basalt fiber reinforced polymer ribs (BFRP) in salt solution. After aging for 70 days in salt solution at room temperature, the fatigue strength of BFRP ribs was reduced by 25 % [18]. Shi et al. studied the aging properties of BFRP in salt solution at different temperatures and times. The results show that the strength of BFRP decreases with increasing temperature and time [19]. Wang et al. studied the aging properties of BFRP in brine. The results show that the static strength of BFRP shows negligible degradation after aging in salt solution, while the aging of salt solution in fatigue load, the adverse effect of brine becomes obvious [20]. At present, most of the researches focus on the durability study of basalt fiber composites, and lack of research on the properties of basalt fiber monofilaments. The aging performance studies on the basalt fiber salinity environment are much less than the acid and alkali resistance studies.

These studies have focused on the materials that the fibers are covered by the matrix. The resistance of the filaments to corrosion is primarily determined by the corrosion resistance of the resin or cement, which is also related to the toughness of the resin or cement. The lack of a complete understanding of the basic mechanisms of fiber damage and degradation is evident.

The purpose of this work was to study the salt solution soak a direct impact on the performance of basalt fiber filaments, excluding the effect of the matrix, and reveal a clear and accurate response to salt corrosion fiber, provide a richer theory for the development of the use of basalt fibers late reference.

2. Experiment

2.1. Preparation of Basalt Fiber Used in the Experiment

The basalt fiber used in this experiment is prepared by laboratory basalt ore. The raw material is selected from the basalt ore in a certain area of Shandong. First, the basalt ore is broken into small pieces and heated to 1280~1350°C by using platinum-plated monofilament furnace. Drawing is started by a fiber monofilament drawing machine. The fiber monofilament is quickly wound onto a lower roller to collect the desired fibers. The diameter of the fiber is controlled by the rotational speed of the wire drawing drum. The composition is shown in the table below. The prepared basalt fibers have an average diameter of 18 μm and an average density of 2.5167g/cm^3.

After the fiber is finished, the sample number is set. About 200 monofilaments are selected for each group of samples, the length is 5 cm, and 20 monofilaments are selected from each group of samples to prepare a fiber tensile strength test sample, which is to be dried after the glue is completely dried. Silk tensile strength test.

Five samples were taken in 3.5 wt% and 15 wt% NaCl solutions, placed in 5 beakers, sealed with plastic wrap, placed in a constant temperature and humidity chamber, and tested at a temperature of 90°C. The reaction time was taken as 1d, 3d, 5d, 7d, and 14d, and a total of 10 samples were collected. After taking out the drying, carry out the diameter test, the strength test, and send the sample to the testing center for further testing by infrared spectroscopy and scanning electron microscopy.

The NaCl is analytically pure, and the experimental water is deionized water. First, a certain amount of basalt fiber is weighed by an analytical balance, placed in a 3.5 wt% and 15 wt% NaCl solutions, and placed in a constant temperature and humidity chamber to set the temperature. It is 90°C. Take 1 day, 3 days, 5 days, 7 days, and 14 days as reaction time, and take them out in order. It is washed several times with deionized water and then dried to obtain basalt fiber after salt treatment.

2.2. Sample Performance and Characterization

The breaking strength of basalt fiber was tested by an electronic single fiber dynamometer. The test standard was in accordance with GB/T7690.3-2001 (method of breaking strength and elongation at break of glass fiber), and the test result was the average value of 20 experiments. The weight loss rate of basalt fiber was tested by testing the weight of the fiber before and after aging of the salt solution.

The apparent morphology changes of basalt fiber before and after treatment were observed by scanning electron microscopy. Characterization of ionic precipitation after immersion of basalt fiber in NaCl solution by ICP. The structure change of basalt fiber surface was characterized by Fourier transform infrared spectroscopy.

3. Results and Discussion

3.1. Effect of Salt Solution on Surface Morphology of Basalt

The SEM image of basalt fibers treated with sodium chloride was shown as Figure 1. It can be seen from Figure 1 that the untreated basalt fiber has a cylindrical shape with a smooth surface and few defects. After the basalt fiber is immersed in a 3.5 wt% NaCl solution at 90°C for 3 days, the surface of the basalt fiber has obvious aging phenomenon. Dense and uniform fiber surface defects appeared. Compared with the NaCl solution with a concentration of 15wt%, the surface of the fiber immersed in a NaCl solution with a concentration of 3.5wt% was more severely corroded. Compared with the fiber immersed for 3 days, the basalt fiber immersed for 14 days has more surface defects and more serious corrosion.

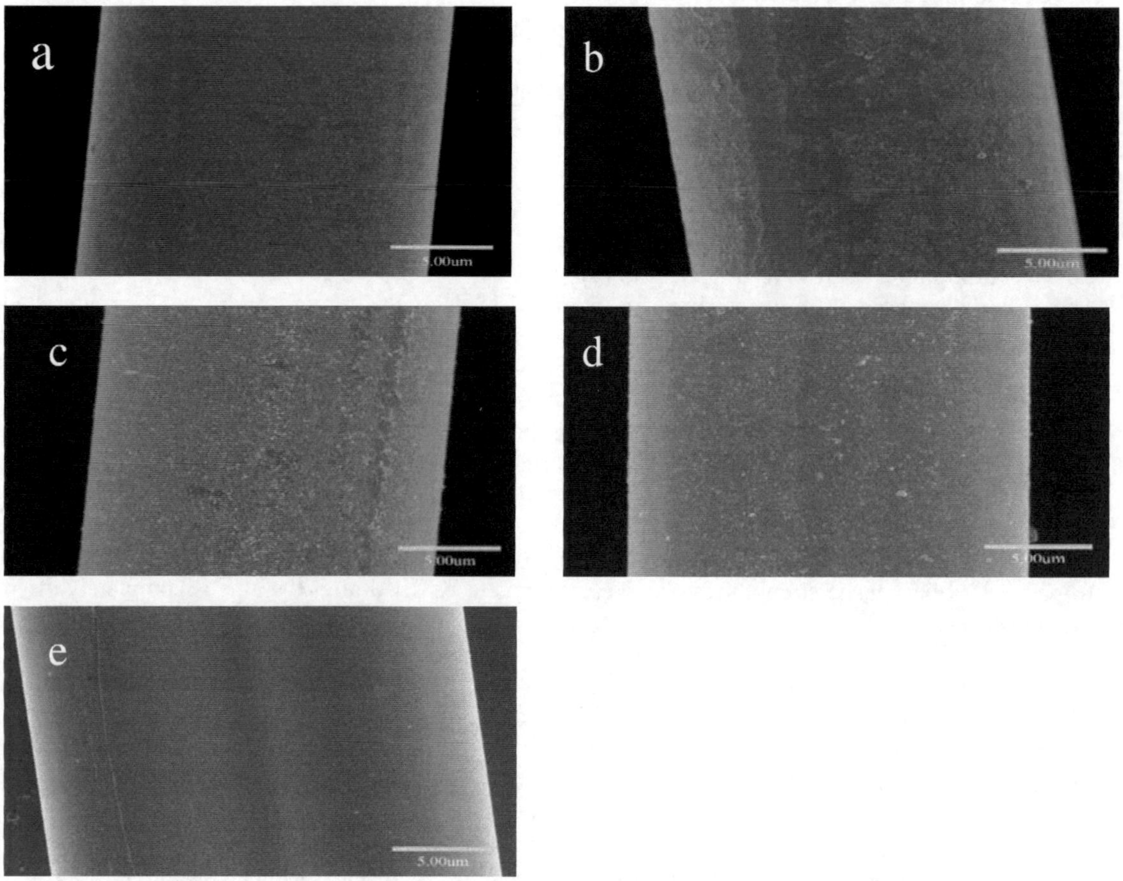

Figure 1. SEM Image of Basalt Fibers Treated with Sodium Chloride (a-3.5wt% 3d; b-15wt% 3d; c-3.5wt% 14d; d-15wt% 14d; e-untreated)

3.2. Analysis of Ion Concentration Precipitation

Figure 2 shows that the basalt fiber has a relatively obvious ion precipitation after immersion aging in 3.5 wt% NaCl solution. In particular, Si^{4+}, Ca^{2+} and K^{+} precipitated 3.87%, 4.04% and 16.94%

respectively after two weeks, and the precipitation content gradually increased with time. It can be seen from Figure 3 that only K^+ is evident after immersion aging of basalt fiber in 15wt% NaCl solution. The precipitation was gradually increased with time, and 39.12% was precipitated after two weeks. It can be seen that during the immersion aging experiment of NaCl solution, the basalt fiber produced a corrosion reaction of ion displacement.

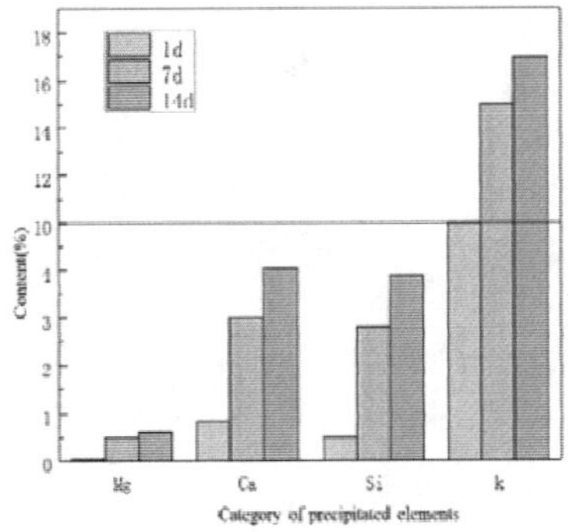

Figure 2. Changes of Elements Precipitated from Basalt Fibers in 3.5wt% NaCl Solution at 90°C

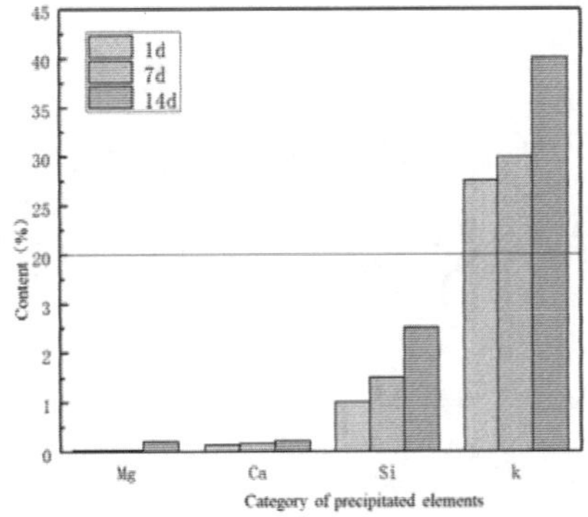

Figure 3. Changes of Elements Precipitated from Basalt Fibers in 15wt% NaCl Solution at 90°C

3.3. Weight Loss Test Results

The weight loss rates were shown in Figure 4 and Figure 5. It can be seen from the figures that after two weeks of aging experiments, the weight loss of basalt fiber clearly increased, and had a gradually increasing trend with time. However, compared with the 3.5wt% NaCl solution, the basalt fiber has less weight loss in the 15wt% solution, and it was inferred the salt concentration is too large to inhibit ion exchange, which makes the basalt fiber weight loss smaller, which is also consistent with the results of the ICP in the previous section.

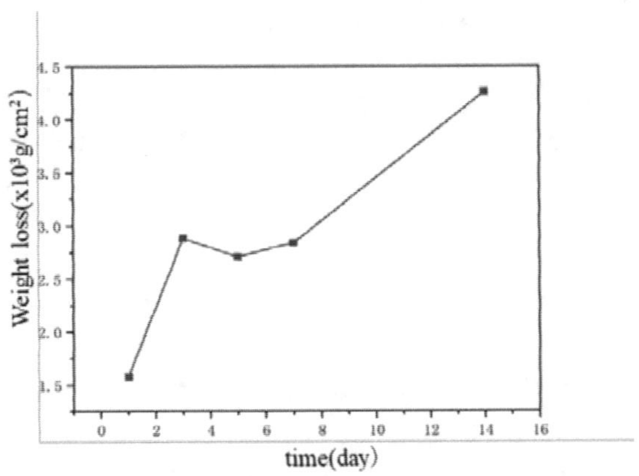

Figure 4. Weight Loss Curve of Basalt Fiber in 3.5wt%NaCl Solution at 90°C

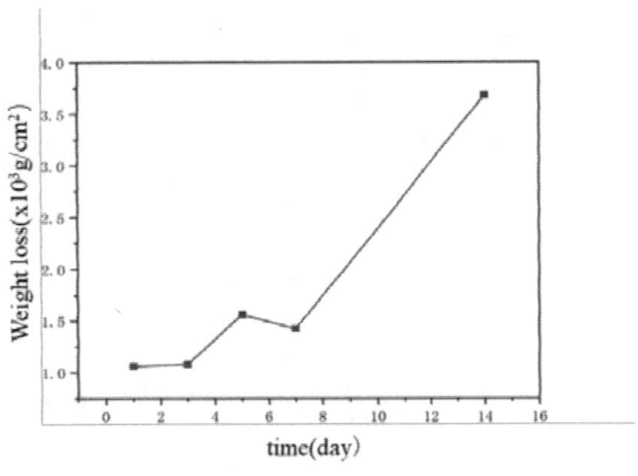

Figure 5. Weight Loss Curve of Basalt Fiber in 15wt% NaCl Solution at 90°C

3.4. Infrared Spectroscopy Test Results

Infrared Spectra of basalt fibers soaked in different concentrations of NaCl solution at 90°C were shown in Figure 6 and Figure 7. Results showed that a broad medium-strong absorption peak appeared at 3290 cm^{-1}. It shows that under the soaking of 3.5wt% NaCl solution, -OH is produced on the surface of basalt fiber. As the reaction progresses, the -OH formed by the reaction further causes a breakdown with the Si-O bond in the fiber skeleton structure, which is also the cause of Si^{4+} precipitation. It can be seen from Figure 8 that the basalt fiber has little change in composition after immersion in a 15 wt% NaCl solution, or the composition change has little effect on the fiber. The absorption peak at 3290 cm^{-1} is small. It is speculated that it may be that a high concentration of NaCl solution inhibits the production of -OH, and the Si-O bond is less destroyed.

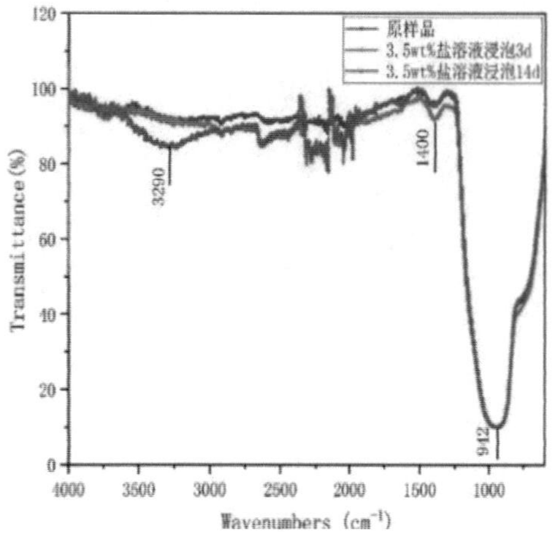

Figure 6. Infrared Spectra of Basalt Fibers in 3.5wt% NaCl Solution at 90°C

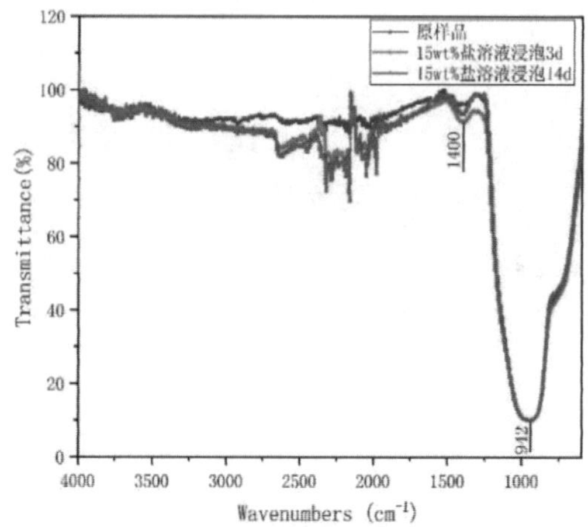

Figure 7. Infrared Spectra of Basalt Fibers in 15wt% NaCl Solution at 90°C

3.5. Tensile Strength Test Results

The Weibull profile of the tensile strength for the basalt fiber in different concentration of NaCl solutions is shown as Figure 8 and Figure 9. The chart evaluates the strength of basalt fiber monofilaments using the Weibull distribution law. There are 20 filaments of samples for each. Results showed that the fracture strength of the basalt fiber in the 3.5wt%, 15wt% NaCl solution was increased first and then decreased. In the first 7 days, the basalt fiber immersed in the 3.5wt% NaCl solution gradually increases the fracture strength, and then the fracture strength decreases. The basalt fiber immersed in the 15 wt% NaCl solution increases in strength on the first day, and then the strength gradually decreases. The strength of the basalt fiber soaked in the 15 wt% NaCl solution is higher than that of the basalt fiber soaked in the 3.5 wt% solution.

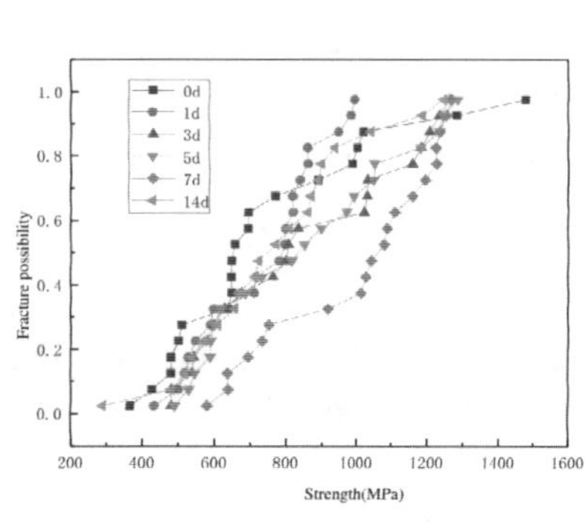

Figure 8. Fracture Strength Change of Basalt Fiber in 3.5wt%NaCl Solution at 90°C

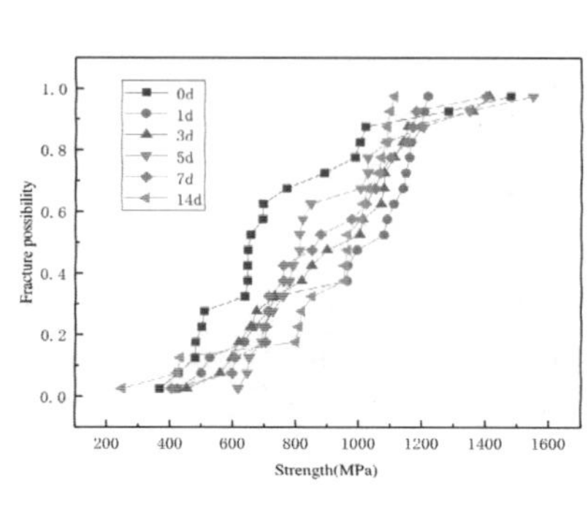

Figure 9. Fracture Strength Change of Basalt Fiber in 15wt%NaCl Solution at 90°C

IOP Conf. Series: Materials Science and Engineering **735** (2020) 012049 doi:10.1088/1757-899X/735/1/012049

3.6. Discussion

3.6.1. Relation of Weight Loss Rate and Ion Precipitation Mechanism. It can be seen from the above experiments that the greater the concentration of sodium chloride solution, the smaller the weight loss of basalt fiber in the salt solution, and the higher the concentration, the inhibition of the production of -OH on the surface of basalt fiber , the decrease of silicon ion precipitation, Si-O bond destruction is reduced. As the Si-O bond is destroyed, the Ca^{2+} and Mg^{2+} precipitates decrease, and only K^+ precipitates most, but at the same time, more Na^+ enters the gap, and in the low concentration sodium chloride solution, not only K^+ Precipitation, silicon ions, calcium ions are also precipitated, especially the precipitation of silicon as a network forming body, the structure will be destroyed, so the smaller the concentration of sodium chloride solution, the greater the weight loss rate.

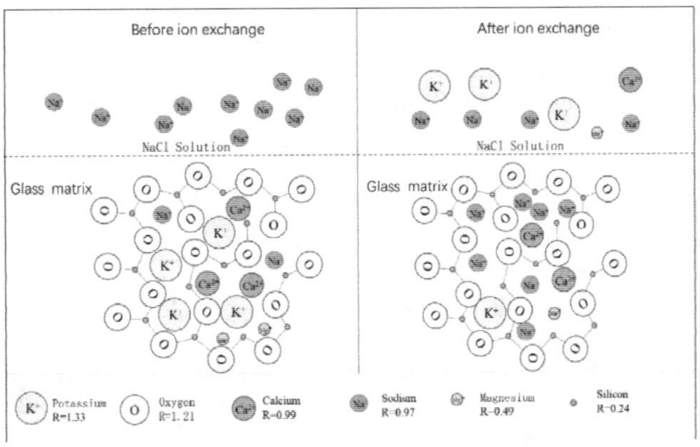

Figure 10. Sketch Map of Ion Exchange of Basalt Fiber in NaCl Solution

3.6.2. Effect of Ion Precipitation on the Strength of Basalt Fiber. As shown in the figure 11, the strength of the basalt fiber after immersion in NaCl solution will increase first and then decrease, because the smaller radius Na^+ enters the vitreous body instead of the larger radius K^+, and the vitreous void is filled with Na^+, which will generate a compressive stress. The strength of basalt fiber is increased, and the mechanism is similar to the principle of glass tempering [21]. However, with the increase of time, the destruction of silicon tetrahedron in the glass body is gradually increased, and the K^+ migration in the glass body is also saturated, and its strength is no longer increased. On the contrary, the strength of the fiber is lowered due to the precipitation of other ions. The higher the concentration of NaCl solution, the faster Na^+ will reach saturation in the vitreous. Therefore, the strength begins to decrease after soaking for 1 day in 15wt% NaCl solution, and the strength begins to decrease after soaking for 3.5 days in 3.5wt% NaCl solution. When immersed in a higher concentration of NaCl solution, more Na^+ will enter the vitreous body, resulting in a larger pressing force, and the strength of the basalt fiber will be higher.

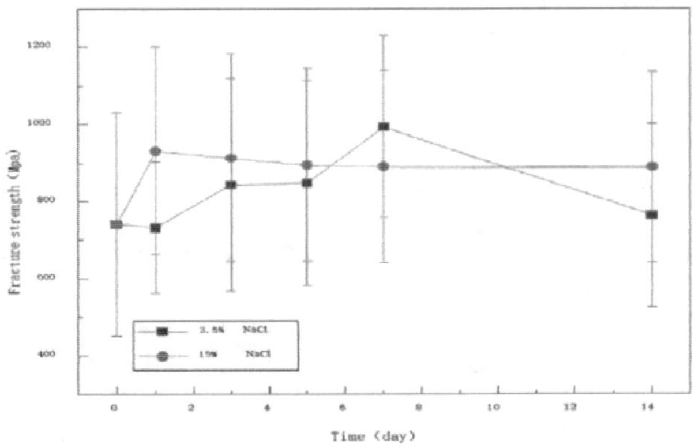

Figure 11. Tensile Strength c Hange of Basalt Fiber Immersed in Different Concentrations of NaCl Solution at 90°C

4. Conclusion

At 90°C, different concentrations of NaCl solution have different effects on the aging properties of basalt fiber. After immersed in 3.5wt% NaCl solution, the surface morphology of basalt fiber is slightly corroded, and a white film is attached to the surface. The strength increases first and then decreases with time. On the 7th day, the strength reaches the maximum value, and then the strength gradually decreases. In the 15wt% NaCl solution, the surface morphology and weight loss are less than 3.5wt% NaCl solution. The intensity of the first day increases gradually with time, then gradually decreases, because the Na^+ in the solution exchanges with K^+ in the glass. The composition of the surface layer of the glass fiber is changed, and a surface compressive stress layer is generated to increase the strength of the basalt fiber [22].

5. Acknowledgment

This work was supported by the National Science Foundation of China (NSFC, No. 51873032), and the Fundamental Research Funds for the Central Universities (223201900003).

6. Reference

[1] Jamshaid H, Mishra R. A green material from rock: basalt fiber–a review [J]. The Journal of the Textile Institute, 2016, 107(7):923-937.

[2] Fiore V, Scalici T, Di Bella G, et al. A review on basalt fibre and its composites [J]. Composites Part B: Engineering, 2015, 74:74-94.

[3] Dong L Q, Chen J F, Guo C M, et al. Application Status and Prospect of Basalt Fiber in Environmental Protection, Contemporary Chemical Industry,2018, 47(2):387-391

[4] Lopresto V, Leone C, De Iorio I. Mechanical characterization of basalt fibre reinforced plastic [J]. Composites Part B: Engineering, 2011, 42(4):717-723.

[5] Sun G Y, Tong S W, Chen D D, et al. Mechanical properties of hybrid composites reinforced by carbon and basalt fibers [J]. International Journal of Mechanical Sciences, 2018, 148:636-651.

[6] Wei B, Cao H L, Song S H. Tensile behavior contrast of basalt and glass fibers after chemical treatment [J]. Materials & Design, 2010, 31(9):4244-4250.

[7] Deák T, Czigány T. Chemical Composition and Mechanical Properties of Basalt and Glass Fibers: A Comparison [J]. Textile Research Journal, 2009, 79(7):645-651

[8] Kizilkanat A B, Kabay N, Akyüncü V, et al. Mechanical properties and fracture behavior of basalt and glass fiber reinforced concrete: An experimental study [J]. Construction and Building Materials, 2015, 100:218-224

[9] Ying X N, Zhou X D. Chemical and thermal resistance of basalt fiber in inclement environments [J]. Journal of Wuhan University of Technology, 2013, 28(3):560-565

[10] Hu X Q, Shen T N. The Applications of the CBF in War Industry & Civil Fields [J]. Hi-Tech Fiber & Application, 2005, 30(6):7-13.
[11] Dhand V, Mittal G, Rhee K Y, et al. A short review on basalt fiber reinforced polymer composites [J]. Composites Part B: Engineering, 2015, 73:166-180.
[12] Wang Y, Yu X, Zhang J J, et al. Development and Application of Basalt Fibers [J]. Fiber Glass, 2017(4):28-31.
[13] Deák T, Czigány T. Chemical Composition and Mechanical Properties of Basalt and Glass Fibers: A Comparison [J]. Textile Research Journal, 2009, 79(7):645-651
[14] Wei B, Cao H L, Song S H. Degradation of basalt fibre and glass fibre/epoxy resin composites in seawater [J]. Corrosion Science, 2011, 53(1):426-431.
[15] Hu H T, Wang X B, Yang Y X. Experimental study on durability of BFRP sheets under salt fog corrosion environment [J]. Industrial Construction, 2010, 40(04):9-12.
[16] Zhu Y, Zhang G C, Wu G. Experimental study on the corrosion resistant performance of
[17] BFRP laminates in corrosive solutions, 2013, 38(1):43-47
[18] Yang Y X, Chen W, Ma M S.Durability of basalt fiber sheets under wet-dry cycling in sea water [J]. Industrial Construction, 2010, 40 (4)
[19] Yang M M. Effects of artificial seawater immersion on the long term properties of basalt and glass fibers reinforced bars [D]. Harbin Institute of Technology, 2014.
[20] El Refai A. Durability and fatigue of basalt fiber-reinforced polymer bars gripped with steel wedge anchors [J]. Journal of Composite for Construction, 2013, 17(6): 04013006
[21] Wang X, Zhao X, Wu Z. Fatigue degradation and life prediction of basalt fiber-reinforced polymer composites after saltwater corrosion [J]. Materials & Design, 2018, 163:107529.
[22] Cheng J S, Zhu L R, Lou X C, et al. Method discussion of glass tempering [J]. Materials Review, 2012, 26(s1):135-137.

High-temperature Creep Behavior of 12Cr1MoVG Steel

Jinping Pan[1*], Chaoming Zhu[1], Xinwei Zhu[1] and Lianjiang Tan[2]

1 Jiaxing Special Equipment Inspection Institute, Jiaxing 314051, China.
2 School of Materials Science and Engineering, Shanghai Institute of Technology, Shanghai 201418, China.
[*] E-mail: jxlpjp@foxmail.com.

Abstract. Heat-resistant steels are key materials for the construction of boilers and high-temperature pipes. Herein, high-temperature creep behavior of 12Cr1MoVG steel has been studied. Creep tests over a long period of time were carried out for 12Cr1MoVG steel samples at varied temperatures. The activation energy for the creep of 12Cr1MoVG steel was evaluated based on the changes of creep rate with temperature when the steel was subject to various stresses. The creep mechanism of the steel has been elucidated by analyzing the relations between the minimum creep rate and the stress. In addition, continuum damage mechanics (CDM) model-based calculation agree well with the strain versus creep time curves obtained from the creep tests. The creep behavior of the 12Cr1MoVG steel at high temperatures has been perfectly described by means of long-time creep tests and mathematical modelling.

1. Introduction

Ultra-super critical (USC) power plants have been developing rapidly due to urgent demands for clean energy and global environment protection [1]. Heat-resistant steels are frequently used for fabricating high-temperature boilers in USC power plants [2, 3]. 12Cr1MoVG steel is a class of high-performance steels frequently used for structural components in boilers. The environment in which the steel works is always harsh. Therefore, a sound understanding of creep behavior of the 12Cr1MoVG steel at high temperatures is fairly important for safe and economical operations of power plants [4, 5]. In this study, the high-temperature creep behavior of the 12Cr1MoVG steel was investigated by creep experiments at selected temperatures and stresses in combination with continuum damage mechanics (CDM) modeling [6]. The creep activation energy of the steel was estimated, and the relevant mechanism of creep was also elucidated.

2. Experimental

2.1. Materials

The 12Cr1MoVG steel samples studied were provided by a power station. The elementary composition was measured using an Optical Emission Spectrometer (ARL 4460, Thermo Scientific). The results are listed in Table 1.

Table 1. Chemical composition of 12Cr1MoVG steel (at. %)

C	Si	Mn	Cr	Mo	V	P	S
0.106	0.255	0.451	1.107	0.286	0.187	0.021	0.012

2.2. Creep Tests
Tensile creep tests at various temperatures were carried out on an electronic high temperature creep & rupture testing machine (GWT2105, MTS) with predetermined loads to examine the creep behavior, using rectangular specimen with the size of 50*15*3 mm.

3. Results and Discussion

3.1. Creep Activation Energy
Creep activation energy plays pivotal roles in the study of creep behavior of materials at high temperatures [7]. Arrhenius equation is often used to calculate the creep activation energy. For most steel materials, Arrhenius equation can be expressed as:

$$\dot{\varepsilon}_m = A\sigma^n \exp\left[-Q_c/(RT)\right] \tag{1}$$

Where $\dot{\varepsilon}_m$ denotes the minimum creep rate; A is a material-dependent constant; σ represents the applied stress (MPa); n is the stress exponent; Q_c is the creep activation energy (kJ/mol); R is the gas constant; T is the absolute temperature (K).

If a constant stress is applied, the creep activation energy can be calculated using the equation as follows:

$$Q_c = -R[\partial ln\dot{\varepsilon}_m/\partial(1/T)]_\sigma \tag{2}$$

The $\dot{\varepsilon}_m$ of 12Cr1MoVG steel can be determined from creep tests under different conditions (changing temperature and stress), where by the creep activation energy was calculated. The results under appropriate conditions are shown in Table 2.

Table 2. Values of Creep Activation Energy for 12Cr1MoVG Steel at Different Temperatures and Stresses

T (℃)	σ (MPa)					
	200	180	160	130	100	80
500-550	815.4	745.6	666.0	580.2	--	--
550-600	--	749.2	685.3	599.7	534.8	--
600-650	--	--	--	616.4	557.5	465.9

In many cases, creep activation energy equals self-diffusion activation energy, as the creep rate is dominated by climbing of dislocations. For most of the heat-resistant alloys, however, such relation does not apply. From Table 2, it can be seen that the creep activation energy increased with reduced enhanced temperature at the same stress level. When the temperature kept constant, the creep activation energy decreased with reduced stress. These results demonstrate the complex creep behavior characteristic of heat-resistant steels. Besides, the following equation is obtained by deriving the equation (1):

$$n = (\partial\ln\dot{\varepsilon}_m/\ \partial\ln\sigma)_T \tag{3}$$

The n value indicates the sensitivity of the material to the applied stress. In the next section, the creep mechanism related to the stress exponent n will be discussed.

3.2. Creep Mechanism
For steel materials strengthened by second-phase particles, the double logarithmic plots of minimum creep rate versus stress can be divided into three regions on the basis of the stress exponent n value [8], which has been defined by the above-mentioned equation (3). It was reported that the n value is near 1 at low stress levels and relatively high temperatures [9]. As the stress grows, the n exceeds 1, and the glide of dislocations dominates the creep [10]. This region is named "power-law-breakdown" (PLB) region where the second-phase particles are bypassed by the Orowan mechanism with n value larger than 7.

The changes of the minimum creep rate with the stress at different temperatures are shown in Figure 1. The experimental data were linearly fit well in the whole stress range. The stress exponent *n* increased with decreasing temperature. At each temperature, the *n* value kept constant despite the changes of stress, the decrease in the *n* value was not attributed to the breakdown of creep strength. Furthermore, the *n* values were larger than 7 in all the cases, implying that the Orowan mechanism accounts for the by-pass of second-phase particles.

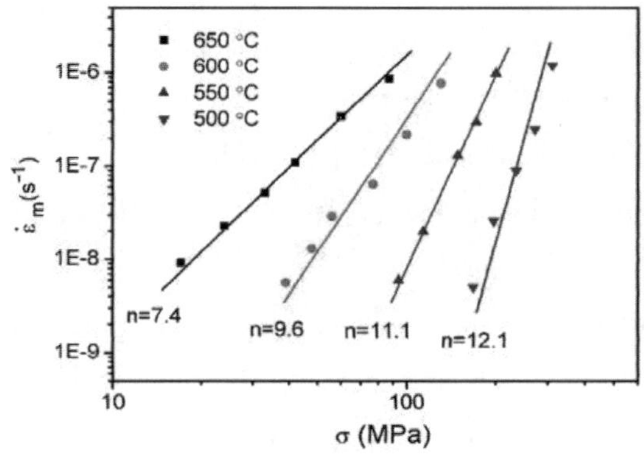

Figure 1. Minimum Creep Rate Versus Applied Stress for 12Cr1MoVG Steel at Different Temperatures. The data is linearly fitted and the *n* values represent the slopes of the fitting lines.

3.3. Strain-creep Time Analysis

The creep curves from the experiments as well as the predicted creep curves by CDM modeling for the 12Cr1MoVG steel under selected testing conditions are demonstrated in Figure 2. A primary creep region can be clearly observed for both experimental curves. Beyond 15000 h, the strain increased more rapidly for both curves. In the data range, the steel sample measured at 200 MPa and 550 °C exhibited larger strains. The creep curves obtained by CDM modeling were in good accordance with the experimental creep data in both cases, showing the high applicability of the CDM model for describing the creep behavior and predicting the creep life of the 12Cr1MoVG steel.

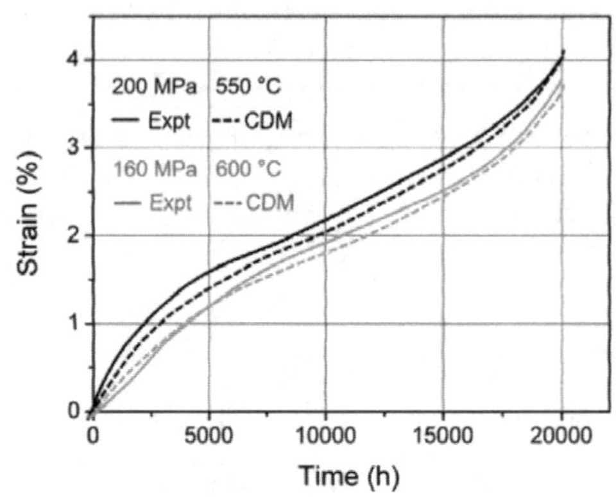

Figure 2. Experimental and CDM Prediction Creep Curves for 12Cr1MoVG Steel under two Testing Conditions

4. Conclusions

To sum up, the high-temperature creep behavior of 12Cr1MoVG steel has been studied. The creep activation energy has been estimated by analyzing the data of creep tests, the results of which show typical creep behavior of a heat-resistant steel. The creep mechanism of the steel proves to be the Orowan mechanism by which dislocations pass by the second-phase particles. Finally, the CDM model has been found to agree well with the experimental creep curves of the 12Cr1MoVG steel over a period of 20000 h, indicating that the CDM model is suitable for describing the creep behavior and predicting the creep life of the 12Cr1MoVG steel.

5. Acknowledgment

This work is financially supported by the Jiaxing Science and Technology Project (2019AD32016) and the AQSIQ Science and Technology Project (Grant No. 2017QK037).

6. References

[1] Aghajani A, Somsen Ch and Eggeler G 2009 *Acta Mater.* **57** 5093.
[2] Rojasa D, Garciab J, Prata O, Carrascoc C, Sauthoffa G and Kaysser-Pyzalla AR 2010 *Mater. Sci. Eng. A* **527** 3864.
[3] Fedoseeva AE, Dudova NR and Kaibyshev RO 2017 *Phys. Metal Metallography* **118** 591.
[4] Vallant JC, Vandenberghe B, Hahn B, Heuser H and Jochum C 2008 *Int. J. Pres. Ves. Pip.* **85** 38.
[5] Bendick W, Cipolla L, Gabrel J and Hald J 2010 *Int. J. Press. Vessel. Pip.* **87** 304.
[6] Kumar M, Deep U and Dixt PM 2017 *Procedia Eng.* **173** 190.
[7] Takeuchi SS and Argon A 1976 *J. Mater. Sci.* **11** 1542.
[8] Dudko VA, Belyakov AN and Kaibyshev RO 2015 *Dokl. Phys. Chem.* **454** 191.
[9] Gh Armaki H, Chen RP, Maruyama K and Igarashi M 2013 *J. Nucl. Mater.* **433** 23.
[10] Pan J, Tu S, Zhu X and Tan L 2019 *Adv. Mech. Eng.* **11**, 1.

IOP Conf. Series: Materials Science and Engineering **735** (2020) 012051 doi:10.1088/1757-899X/735/1/012051

Design of Ionic Fog Filter for AC- 1 Machining Center

Quan Zhou [1, a], Yonghui Xie [2, b] and Juan Chen[3, c]

1 Department of electrical and mechanical engineering, Weifang vocational college, No. 06588, Haian road, science and education innovation park, binhai economic and technological development zone, weifang city, shandong province.
2 Department of electrical and mechanical engineering, Weifang vocational college, No. 06588, Haian road, science and education innovation park, binhai economic and technological development zone, weifang city, shandong province.
3 Department of electrical and mechanical engineering, Weifang vocational college, No. 06588, Haian road, science and education innovation park, binhai economic and technological development zone, weifang city, shandong province.
Email: [a] kyd1981cn@163.com, [b] 16764597@qq.com , [c] kyd1981cn@163.com

Absrtact. In the process of modern NC machine tool processing, a large number of oil fog, smoke, containing polychlorinated biphenyls (PCBs) are produced, which is easy to cause toxic fog and seriously harm the health of staff. This paper introduces the design of an oil mist suction device to avoid the diffusion of toxic fog for a long time, to create a comfortable working environment, to prolong the life of machine tool equipment and to improve the production efficiency. It has been proved by application that it meets the performance requirements of modern NC machining center and obtains good social and economic benefits.

1. Introduction

In air purification, how to collect oil fog gas efficiently and how to purify gas efficiently are the key problems, which directly affect the cleanliness of the purified gas.

2. Design Background and Significance

2.1. Design Background

Because of the late start of China's automation industry, there is little knowledge of the aerosol pollution produced in the process of mechanical processing, and the treatment of the aerosol is not valued, so the research on the aerosol treatment is only in the primary stage, and the theoretical study is lacking. The simple collection and simple treatment of the processed fog can only be carried out by virtue of the accumulated experience, and the blindness and the uncertainty exist to a certain extent. In that domestic market oil mist filter, the purpose of the invention is to realize only the oil mist gas generated by the collection device and to condense and discharge the oil mist gas, and the fine particle solid and the harmful gas cannot be truly recovered and processed.

2.2. Design Significance

In order to meet the national development concept, adhere to green development, promote low-carbon cycle development, and build a clean and low-carbon, safe and efficient modern energy system. The project team independently developed and designed a fully closed ionization aerosol filter, which collected the fog and oil fog produced in the machining process, and then collected it through high voltage electrostatic ionization. The filter layer can be used for a long time, and the filter layer can be replaced regularly to achieve 99.8% purification effect.

3. Design Scheme

The fog filter adopts sealed cabin mechanism, which is mainly composed of oil fog collection tank, front filtration system, electrostatic high voltage ionization system, electrostatic collection system, rear filtration system and electrical control system. According to Coulomb's law, the interaction force between two charged bodies is proportional to the product of the electric quantity and inversely proportional to the square ratio of the distance between them. The electric field intensity of any point on the electric field is equal to the force acting on the unit positive charge at that point in value and direction. The picture shows the effect of electrostatic high voltage aerosol filter.

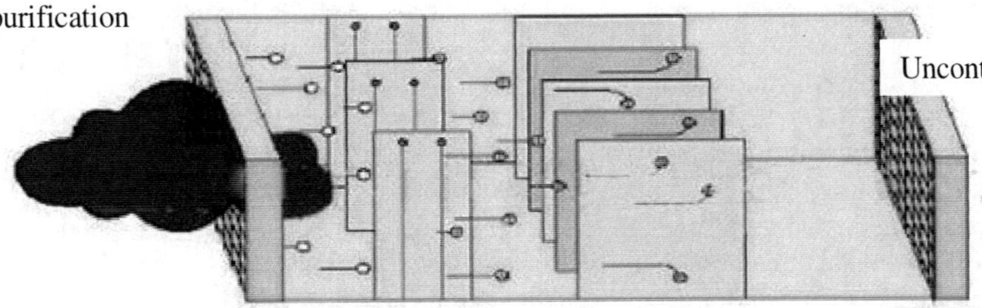

Figure 1. Effect Diagram of Electrostatic High Voltage Aerosol Filter

3.1. Front, Rear Filtering System

In view of the reliable operation and low maintenance cost of the whole filtration system, the collection and treatment bureau of the system adopts separate type, which avoids the erosion of the system by oil fog. In the front filtration system, the centrifugal fan is used to drive the high polishing whirlpool collector to collect the range of oil fog efficiently.

The pre-filter uses stainless steel wire filter to collect most of the large particle oil fog, and different configurations can be selected according to the type and concentration of oil mist on the spot. The rear filter adopts ultra-dense filter and charcoal to form a sandwich filter, and the ionized gas is filtered and purified again to meet the standard discharge requirements. All stainless steel filters and ultra-dense interlayer filters are used for cleaning, maintenance and maintenance in the later stage.

3.2. Electrostatic High Voltage Ionization System

A stable high voltage power supply module is selected in the electrostatic high voltage ionization system. When the polluted oil mist passes through the ionization section, the electric field is formed between the high voltage electrode wire and the electrode plate, and the small particle oil mist is charged at high voltage to carry on the positive charge. When the electric fog passes through the collection section, a multi-layer cathode plate is set up. Because of the interaction between the charged positive charge small particle oil mist and the cathode electrode, all the small particles in the oil fog are adsorbed in the cathode plate. The particles with a certain thickness fall into the ash bucket under the action of self-weight, so as to achieve the purpose of removing the small particles.

3.3. Electric Control System

The ionization fog filter can be connected with the equipment requiring fog treatment as a single device. The electrical control adopts multiple protection, including power supply voltage stabilization monitoring, motor short circuit, overload, overcurrent protection and protective door switch protection, etc., and reserves the signal interaction interface to interact with the equipment that needs fog processing to achieve the automatic start and stop of the filter, alarm output and diagnosis. Realize real-time synchronization with the use of equipment.

4. Working Principle and Performance Analysis

4.1. Operational Principle

The working principle of ionized aerosol filter is to ionize the oil mist by high voltage electric field, and the small particle oil mist charge in the air flow is separated from the air flow under the action of electric field. The positive electrode is made of metal conductors with different section shapes and is called discharge electrode. The negative electrode is made of metal plates of different geometric shapes and is called dust collecting electrode.

The performance of the ionization aerosol filter is affected by the properties of oil mist, the structure of the equipment and the gas flow rate. The specific resistance of oil mist is an index to evaluate the conductivity, which has a direct effect on the efficiency of dust removal. When the specific resistance is too low, it is difficult for the dust particles to remain on the dust collecting electrode, causing it to return to the gas flow. When the specific resistance is too high, the charge of dust particles reaching the dust collecting electrode is not easy to discharge, and a voltage gradient is formed between the dust layers to generate a local breakdown and a discharge phenomenon. These conditions will result in a reduction in the efficiency of the dust removal.

The power supply of ionization fog filter consists of control box, boost transformer and rectifier. The voltage output of the power supply also has a great influence on the ionization efficiency. Therefore, the operating voltage of ionized aerosol filter should be kept above $40 \leq 75$ kV or even above 100 kV.

4.2. Performance Analysis

The ionized aerosol filter adopts advanced design concept, fully considers ergonomics, and adopts some modular designs, including: soft connection of fog collecting port, drawer design of each filter device, overload of ionization system, short circuit protection and electrical control system, which can realize real-time monitoring of the working state of the equipment and fault output and diagnosis of the filter. All the modules of the filter adopt double protection of mechanical structure and electrical structure to realize the protection function of the equipment to the greatest extent.

The experimental results show that the performance of the ionized aerosol filter meets the theoretical design scheme, and good economic benefits are obtained. The electrostatic air purification device has the advantages of high purification efficiency, treatment of flue gas, large fog volume, long service life and low operation cost. It is not only used in grinder, gear machining, large machining center, but also suitable for other equipment that produces a lot of oil fog, dust, oil fume and emulsified fog.

5. Epilogue

With the increasing development of science, technology and industry, people's awareness of the environment, the control of air pollution in operation environment, the control of air pollution in workshop and production site is no longer a luxury. After a lot of experimental tests, the impeller is used to enlarge the air flow as an oil mist suction device in the ionization type air mist filter, which fully meets the demand of oil mist suction in terms of performance. And the protection seal between the impeller and the transmission motor is good, and the oil mist will not be introduced into the transmission device, which reduces the maintenance problem in the later stage.

The ionization method is used as the separation method of oil and gas, which has the advantages of simple structure, convenient installation, convenient maintenance, continuous and reliable work, low cost, easy cleaning and easy automatic control. 1um particles can be collected, energy consumption is low, adsorbed oil fog particles can be accumulated and left behind by oil droplets, so that they can be used better.

At present, there are few corresponding theoretical research reports on the treatment of oil fog in China. The fully closed ionization aerosol filter designed by it has made a better breakthrough in the design of structure and performance. Through the addition of air flow science, it has a better effect on the pre-separation of oil mist cooling. The shape of its fully closed ionization aerosol filter can be well integrated into the airport design style, and the mass production can be realized. In a word, the fully closed ionization aerosol filter has great development opportunities, and its market demand and industrialization prospect are good.

6. References

[1] Sun Zhili, Leng Xingjie, Wei Yangang and other editor-in-chief. Mechanical design Shenyang: northeast University Press 2000.
[2] Liu Tianqi, Qiu Xiaoyan. Power system Analysis Theory Beijing; Science Press 2005.
[3] Study on the Motion Mode of charged particles in Electric Field of Wang Fengming Electrostatic Precipitator Hebei: Hebei University 2006.
[4] GB13223-2001 Air Pollutant Emission Standard of the General Administration of International quality Supervision, Inspection and Quarantine, Ministry of Environmental Protection Beijing: China Environmental Science Press 2011.

Fatigue Performance of T-joint with Complete Weld Details

Xi Liu, Qingyu Meng, Can Wang and Song Huang

Jiangsu XCMG Construction Machinery Research Institute Ltd. Xuzhou, J S 221004, China

Email: 435398417@qq.com

Abstract. The fatigue properties of T-joint of Q345 steel commonly used in construction machinery are studied. The fatigue life of T-joint with complete weld details is analyzed by nominal stress method and the equivalent structural stress method respectively, and compared with the test results. The results show that under the condition of stress ratio R=0.5, the fatigue crack of T-joint with complete weld details often occurs at the weld root of the arc extinguishing and propagate along the weld throat. In production, arc extinguishing should be avoided in key parts. The nominal stress method used to predict the fatigue life of T-joints with complete weld details is dangerous, while the fatigue life of T-joint analyzed by the equivalent structural stress method is basically consistent with the test results.

1. Introduction

T-joint is a typical joint type in welded structures. It is widely used in ships, vehicles, bridge tunnels and other industrial fields. Due to the change of the section form, the stress concentration factor is high. Under cyclic loading, the fatigue fracture often occurs [1].

At present, there are many researches on the fatigue performance of T-joint, but the main object is the fatigue performance of the T-joint eliminated the arc striking and arc extinguishing [2-3]. However, the arc striking and arc extinguishingexist objectively in the actual welding engineering structure, and fatigue cracking often occurs in these parts. At present, the S-N curves given in the fatigue standards of welded structures are mainly based on the fatigue test of joints eliminated the arc-striking and arc extinguishing, and the results may be inaccurate.

Therefore, it is of great significance to study the fatigue performance of welded joint with complete weld details. In this paper, the fatigue test of T-joint with complete weld details is carried out.The nominal stress method and the equivalent structural stress method are used to predict the fatigue life of the T-joint, and compared with the test results to provide a theoretical reference for the selection of fatigue analysis methods for T-joint with complete weld details.

2. Test Method

2.1. Test Materials and Joint Forms

The test material is Q345 steel, the mechanical properties are shown in Table 1. The T-joint form is adopted, and the geometry and dimensions are shown in Figure 1. The welding material is ER50-6 with diameter 1.2mm, and the welding method is gas metal arc welding. The welding process parameters are shown in Table 2. The samples retain the complete weld details, including arc striking and arc extinguishing.

Table 1. Mechanical Properties of Q345Table 2 Welding Process Parameters

Material	Tensile/MPa	Yield/MPa	Elongation/%
Q345	495	396	28

Table 2. Welding Process Parameters

Voltage /V	Current /A	Velocity /mm.min^{-1}	Gas-flow rate/L.min^{-1}
24	225-235	320	15-25

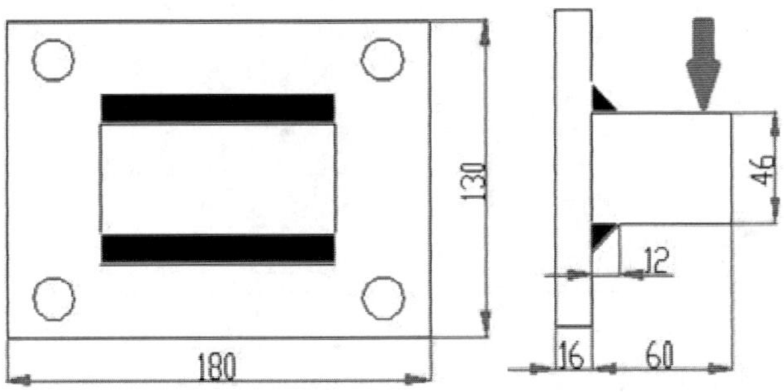

Figure 1. Geometrical Characteristic of Specimen

2.2. Test Method

The test equipment is GPS-200 electromagnetic resonance high frequency fatigue testing machine.The maximum average load is ±200kN, and the maximum alternating load is 100kN. The frequency range is 80~250Hzand the loading error is within ±1%. The fatigue test is carried out under constant amplitude load. The applied load type is pressure-pressure load, and the stress ratio is R=0.5. For the loading form of the T-joint, a special test tool is designed and manufactured.The tooling is fixed on the three-point bending fixture, and the fatigue specimen is fixed on the tooling by bolts,as shown in picture 2.

Figure 2. Fatigue Testing Machine

3. Test Results

The fatigue test results of T-joint specimens with complete weld details are shown in Table 3. The test data are processed according to the statistical method specified by the International Welding Society, and the S-N curves are fitted to 50% survival rate and 95% survival rate respectively, as shown in Figure3. The conditional fatigue strength of the T-joint with complete weld details at 2×106 cycle is shown in Table 4.

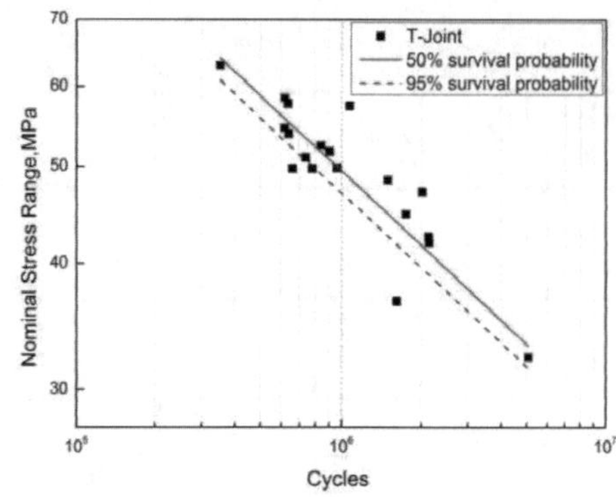

Figure 3. S-N Curve of T-joints

Table 3. Fatigue Testing Results

Number	Stress amplitude /MPa	Fatigue life /10^6	Fracture position
1	32.33	5.07	Weld root
2	36.72	1.61	Weld root
3	42.53	2.13	Weld root
4	49.76	0.78	Weld root
5	54.58	0.62	Weld root
6	58.48	0.62	Weld root
7	62.95	0.35	Weld root
8	41.97	2.14	Weld root
9	47.21	2.02	Weld root
10	49.83	0.96	Weld root
11	52.46	0.84	Weld root
12	51.75	0.90	Weld root
13	57.7	0.64	Weld root
14	57.42	1.07	Weld root
15	49.76	0.66	Weld root
16	51.04	0.74	Weld root
17	53.88	0.64	Weld root
18	48.49	1.49	Weld root
19	44.8	1.75	Weld root

Table 4. Fatigue Strength of T-joint

Specimen type	a	b	$\Delta\sigma 2\times10^6$/MPa
T	3.07494	-0.2297	42.42

Figure 4. A Typical Crack Location of T-joint

The test results show that the fatigue specimens of T-joint with complete weld details cracking at the arc extinguishing position. To observe the inner-crack propagation of the joint, the wire-cutting is used to sample at the crack initiation location of the joint as shown in Figure 4. The fatigue crack growth from the weld root through the weld throat to the surface of the weld until it breaks.

4. Analysis and Discussion

Due to the fatigue failure position of the T-joint with complete weld details is at the weld root, the hot spot stress method cannot be used for fatigue evaluation.In order to investigate the accuracy and applicability of different fatigue analysis methods for the fatigue evaluation of T-joints with complete weld details, the nominal stress method and the equivalent structural stress method are respectively used to evaluate the fatigue of the T-joints with complete weld details, and the results are compared with the fatigue test results.

4.1. The Nominal Stress Method

According to the BS7608 standard [4], the fatigue analysis of the T-joint with complete weld details is carried out. According to the joint type and bearing type, the corresponding joint type number is 7.9 in the BS7608 standard, and the corresponding nominal stress level is G, and the corresponding parameters of the G-class S-N curve as shown in Table 5.The median andthe normal fatigue S-N curves of the G-classare designed and compared with the fatigue test results of the T-joints, as shown in Figure 6.

Table 5. The S-N Parameters of Class G

Class	C0	LgC0	m	STD
G	5.656E11	11.7526	3.0	0.1793

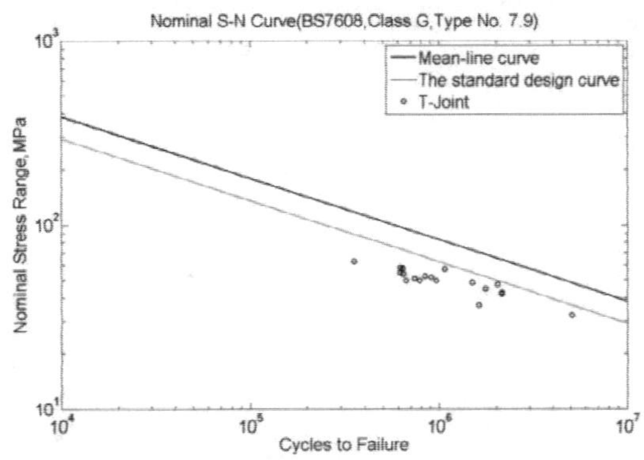

Figure 5. Comparison between S-N Curve (BS7608, Class G) and Test Results of T-joint

As shown in Figure 6, the fatigue test results of the T-joints are lower than the S-N curve of the corresponding joint in the BS7608 standard. So, if the T-joints with complete weld details are predicted using the standard S-N curve, the predicted life is higher than the actual life, and the prediction is dangerous. This shows that the S-N curve in the BS7608 standard is not applicable to T-joints with complete weld details in actual production.

4.2. The Equivalent Structural Stress Method

The fatigue analysis of the T-joint with complete weld details is carried out by the equivalent structural stress method [5-6]. First, the finite element model is established using solid185 unit and the grid size is 4mm.Second, the vertical plate and the bottom plate are only coupled at the root of the weldand the rest of the parts are not coupled in order to establish boundary conditions that match the test conditions and simulate the connection between the actual vertical plate and the bottom plate at the same time.

Figure 6. The Finite Model of T-joint

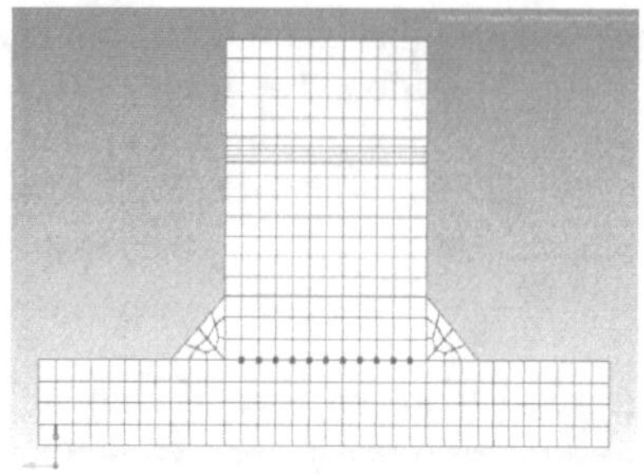

Figure 7. Details of the Nodes Coupling of T-joint

As shown in Figure 9, the equivalent structural stress in the three failure modes is calculated. Figure 10 shows that the distribution of the equivalent structural stress along the weld direction in three failure modes of the T-joint. It can be seen that the maximum equivalent structural stress appears in the root failure mode and is located at the end of the weld, which is consistent with the actual fatigue test results.

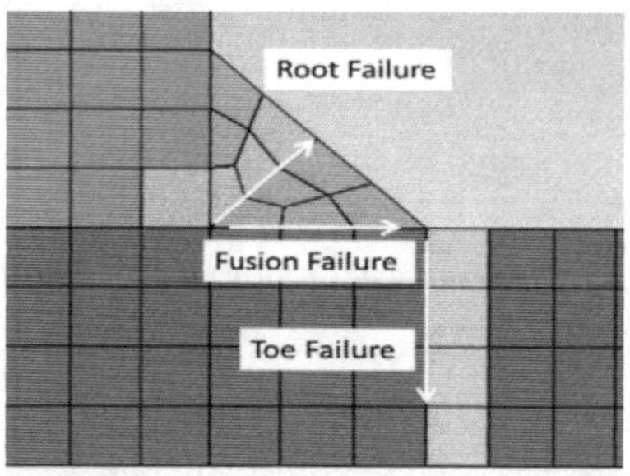

Figure 8. Three Failure Modes of T-joint

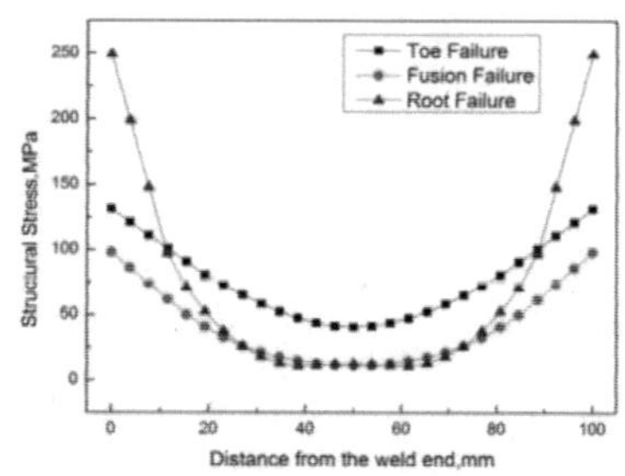

Figure 9. The Comparison of Structure Stress Based on Different Failure Modes

The equivalent structural stress of all T-joints with complete weld details is calculated and compared with the main S-N curve [7] of the root failure, as shown in Figure 11. It can be seen that the test results are basically locate near the median of the main S-N curve. Therefore, the equivalent structural stress method used for fatigue design or prediction of the T-joints with complete weld details that are common in mechanical structures is relatively safe and reliable.

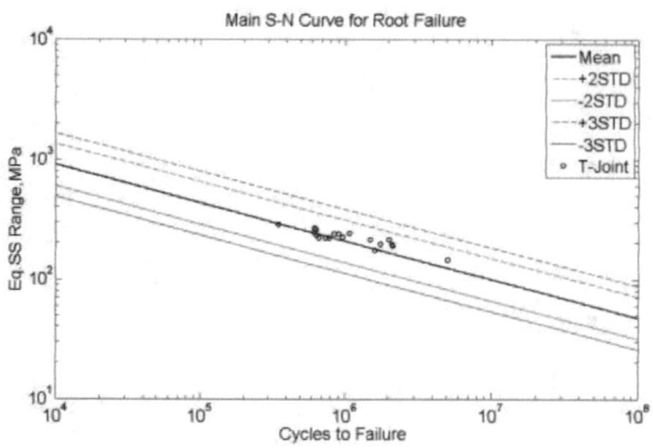

Figure 10. Comparison between Master S-N Curve (for root failure) and Test Results of T-joint

5. Conclusions

(1) Fatigue test is carried out for T-joints with complete weld details. The failure mode is the root cracking at the arc extinguishing position. Therefore, it is recommended to use arc-striking plates or avoid arc extinguishing appearing at key positions for important bearing joints;

(2) The nominal stress method and the equivalent structural stress method are respectively used to evaluate the fatigue life of T-joint with complete weld details. The results show thatthe failure mode and predicted life of the T-joint obtained by the equivalent structural stress method are consistent with the test results. However, the predicted life obtained by the nominal stress method is quite higher and different from the test results.

(3) In practical applications, it is dangerous to use the nominal stress method for fatigue analysis of T-joint in the case of fatigue cracking at the arc-striking or arc-extinguishing position. It is recommended to use the equivalent structural stress method for life prediction and fatigue design of T-joint with complete weld details.

6. References

[1] HuoL X.2000China machine press (Fracture and Assessment of WeldedStructure) p 15
[2] AndreaC, Roberto B and Hans J H 2005International Journal of Fatigue vol 27(Fatigue growth of a surface crack in a welded T-joint)p 59-69
[3] Darko F, Heikki R and Jani R 2011International Journal of Fatigue vol 33(Fatigue assessment of laserstake-welded T-joints)p 102-114
[4] British Standards Institution2014Guide to fatigue design and assessment of steel productsp 1-139
[5] Dong P.2001International Journal of Fatiguevol26 (A Structural Stress Definition and Numerical Implementation for Fatigue Analysis of Welded Joints) p 865–876
[6] Dong P, Hong J Kand Osage D2002 Master S-N curve method for fatigue evaluation of welded Components
[7] Jeong K H 2013Journal of Offshore Mechanics and Arctic Engineeringvol135 (Evaluation of Weld Root Failure Using Battelle Structural Stress Method) 021404-1-7

The Effect of Aluminum Ions on the Acidification and Aging of Paper

Hongwei Zhang, Pengchen Zhai, Qinwen Wang*, Yanqin Li and Aimin Tang

State Key Lab of Pulp and Paper Engineering, South China University of Technology, Guangzhou, China, 510640

Email: qwwang@scut.edu.cn

Abstract. Paper documents provide important records of human civilization and so understanding the causes of paper document aging is very important to delay the process and prolong the lifespan of paper. This study used an anionic rosin as a sizing agent and aluminum sulphate as a fixing agent, to produce paper with different amounts of aluminum sulphate. The effect of different contents of aluminum ions on the acidification and aging of the paper was studied by accelerated aging experiments. The results showed that the presence of aluminum ions in paper could accelerate the acidification and aging process, and the higher the content of aluminum ions, the worse the durability of paper. In the process of accelerated aging, with increasing aluminum ion content, the rate of decline in the strength and whiteness of paper increased whilst the pH value decreased. Accelerated aging under the same conditions for 15 days caused the pH value, whiteness, tensile index and tear index of paper without aluminum ions to decrease by 14.77%, 3.15%, 7.99% and 11.70% respectively. In comparison, paper which had aluminum sulphate added to 30 kg per ton of paper, showed the decreases of pH, whiteness, tensile index and tear index by 22.62%, 4.35%, 24.66% and 31.07% respectively.

1. Introduction

Paper documents provide important records of human civilization, which are affected by time and environmental conditions. Paper aging leads to yellowing, brittleness and even crushing of paper, which seriously affects the lifespan of paper documents[1-2]. Paper aging is significantly affected by air properties[3-5] such as moisture[6-8], temperature[9-10], sunlight[11] and mold[12]. Other factors such as ink[13] and acid substances[14] present in paper can cause acidification and aging. Studies have shown that transition metal ions from inks and pigments, such as Fe^{3+}/Fe^{2+}, Cu^{+}/Cu^{2+}, and Mn^{2+}, catalyze the production of carbonyl or aldehyde groups from cellulose, leading to oxidation reactions and cellulose degradation[15-18]. Ute Henniges et al. [19-20], proved the aging degradation effect of ink on cellulose using carbazole-9-carbonyl oxygen amine (CCOA) notation, and demonstrated the oxidation process using a carbonyl fluorescence labeling method. The study also showed that new carbonyl groups and reductive ends were produced after accelerated aging of the paper samples that contained copper ion pigments. However, since the invention of gum rosin in 1807 by Illing[21], the aluminum-alum sizing system has allowed for internal sizing which has been widely used in large-scale acid paper manufacturing. Few reports have focused on the aging characteristics of this kind of paper. The current study used anionic rosin as sizing agent and aluminum sulfate as fixing agent to produce paper with different amounts of aluminum sulfate. Through an accelerated aging experiment, the influence of aluminum ion content on acidification and aging of paper was studied.

2. Experimental Methods

2.1. Materials

Bleached eucalyptus kraft pulp with a beating degree of 40.5°SR was obtained from Zhuhai Hongta Renheng Packing Co., Ltd.

Anionic rosin with a solid content of 40% was obtained from Guangxi Wuzhou Arakawa Chemical Industries, Ltd. Aluminum sulphate was obtained from, AR, CALMAX Chemical Inc. USA.

2.2. Paper Making

According to ISO 5269/2, paper samples were made in the laboratory with a basis weight of 70 g/m^2, a diameter of 20 cm and an area of 0.0314 m^2. Paper with different amounts of $Al_2(SO_4)_3$ (based on oven dry stock) was made by varying the amount of $Al_2(SO_4)_3$ from 0 kg/ ton to and 30 kg/ ton, whilst keeping the amount of anionic rosin constant at 18 kg/ ton paper. Al2(SO4)3 solution and anionic rosin emulsion were added successively after the pulp (concentration was 0.5%) was digested. After each reagent was added, the reaction was fully stirred for 2 min. The paper was then made using an automatic paper making system (RK3A-KWT, PTI, Austria). After vacuum drying, the paper was placed in a constant temperature and humidity room (relative humidity was 50 ± 5%, temperature was 23 ± 1 °C) to balance moisture for more than 24 hours. The paper was then used for subsequent treatments and performance measurements.

2.3. Accelerated Aging of Paper

According to the standard GB/T 464-2008, paper samples were subjected to a dry-heat accelerated aging test. A certain amount of paper from the different experimental groups was extracted and hung inside the oven. At 105 °C, dry heat aging was carried out. Paper samples were taken every 5 days for performance testing.

2.4. Performance Test and Characterization of Paper

The pH value of the paper samples was determined according to GB/T 450 and GB/T 1545.2-2003. Tensile index was determined according to GB/T 453-2002. The whiteness of the paper was determined according to GB/T 7974-2002.

The degree of polymerization (DP) was determined by weighing a paper sample (with an accuracy of 0.005 g) and then adding the paper sample into 50 ml of polyethylene solution. 2-3 pieces of red copper were then added and a pipette used to absorb 25 ml of deionized water into the solution before the bottle was capped and vigorously shaken. 25 ml of ethylene diketamine hydroxide solution was injected into the bottle and glass beads were added until the solution filled the bottle to eliminate the excess air. The bottle was then placed at a 25 °C constant-temperature shaking table and shaken violently until the sample had completely dissolved. The solution bottle was removed and placed in a thermostat of automatic circulation heat preservation and kept at 25 °C for 10 min. Finally, the outflow time of the solution was determined at 25 ± 1 °C in triplicate and the average value recorded. The relative viscosity was calculated according to the formula ηr=hn·tn. The characteristic viscosity η, and the pulp degree of polymerization were calculated according to the formula DP0.905 = 0.75[η].

The thermostability of the paper samples was determined by weighing 5 mg of the paper samples and placed them onto a platinum crucible. Nitrogen was used as carrier gas with a gas flow rate of 40 mL/min. The temperature rise rate was 20°C /min, and the temperature increases from room temperature to 600 °C. The thermogravimetric (TG) curves of the paper samples were obtained by processing the data collected automatically by the system.

3. Results and Discussion

3.1. Effect of Aluminum Sulphate Content on pH Value of Aged Paper

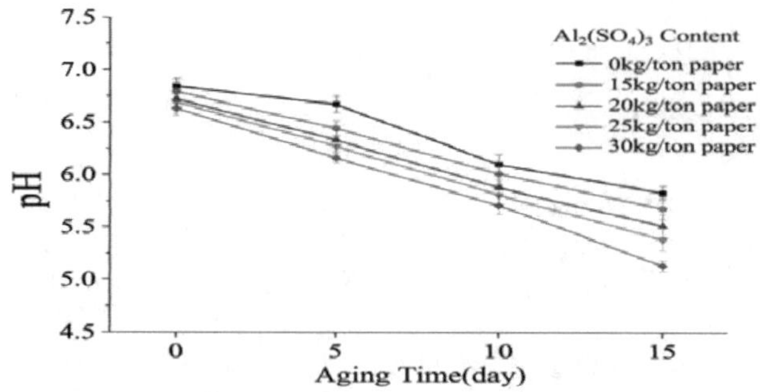

Figure 1. The Effect of $Al_2(SO_4)_3$ Content on the pH Value of Aged Paper

Figure 1 shows the effect of $Al_2(SO_4)_3$ content on the paper pH values after aging for 5, 10 and 15 days. It can be seen that the pH value of all papers manufactured under acidic conditions was greater than 6.5. With increasing aging time, paper acidification and the degree of aging increased, whilst pH value was gradually reduced. The amount of $Al_2(SO_4)_3$ had a major influence on the pH value of the aged paper. At the same aging times, the higher the amount of $Al_2(SO_4)_3$, the higher the content of aluminum ions in the paper, and the more obvious the reduction in pH value. After aging for 15 days, the pH values of the paper manufactured with different contents of $Al_2(SO_4)_3$ (0 kg/ ton paper, 15 kg/ ton paper and 30 kg/ ton paper) decreased by 14.77 %, 16.35% and 22.62 %, respectively. The results show that under the same aging times, the presence of aluminum ions accelerate paper acidification.

3.2. Effect of Aluminum Sulphate Content on the Physical Properties of Aged Paper

The main components of paper are cellulose, hemicellulose and lignin. The aging process is accompanied by the hydrolysis or oxidation of cellulose, hemicellulose and lignin, which consequently alters the mechanical properties of paper. Figure 2 show the effect of $Al_2(SO_4)_3$ content on the tensile index of aged paper and the decline rate of the tensile index of aged paper under different $Al_2(SO_4)_3$ contents. According to these data, as aging time increases, the tensile index of paper decreases to different degrees. The higher the amount of $Al_2(SO_4)_3$, that is, the higher content of aluminum ion in paper, the faster the decline in the tensile index of the aged paper. After 15 days aging, the tensile index of paper without aluminum ions decreased by 7.99%, whilst the paper tensile index when $Al_2(SO_4)_3$ contents are 15 kg/ton paper and 30 kg/ton paper decreased by 11.07% and 24.66%, respectively. These results show the presence of aluminum ions accelerates the aging of paper: the higher the content of aluminum ions, the faster the aging of paper.

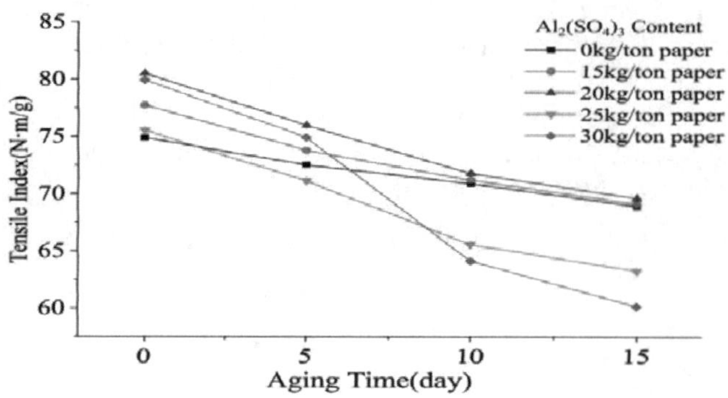

Figure 2. Effect of $Al_2(SO_4)_3$ Content on the Tensile Strength of Aged Paper

3.3. Effect of Aluminum Sulphate Content on the Whiteness of Aged Paper

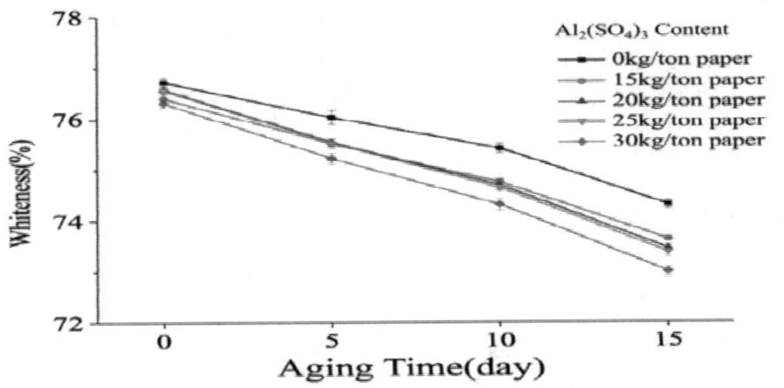

Figure 3. Effect of $Al_2(SO_4)_3$ Content on the Whiteness of Aged Paper

Paper aging is displayed as paper yellowing macroscopically, and the higher degree aging, the more serious the paper yellowing. Changes in paper whiteness are used to measure the yellow degree of paper, which reflect aging and durability. From figure 3, it can be seen that with increasing aging time, the whiteness of the paper gradually decreased. After 15 days of aging, the whiteness of the paper at $Al_2(SO_4)_3$ contents of 0, 15 and 30 kg/ ton paper decreased by 3.15%, 3.64% and 4.35% respectively. The results show the presence of aluminum ions in paper can accelerate the aging and yellowing of paper which worsens the durability of paper.

3.4. Effect of Aluminum Sulphate Content on Polymerization Degree of Aged Paper

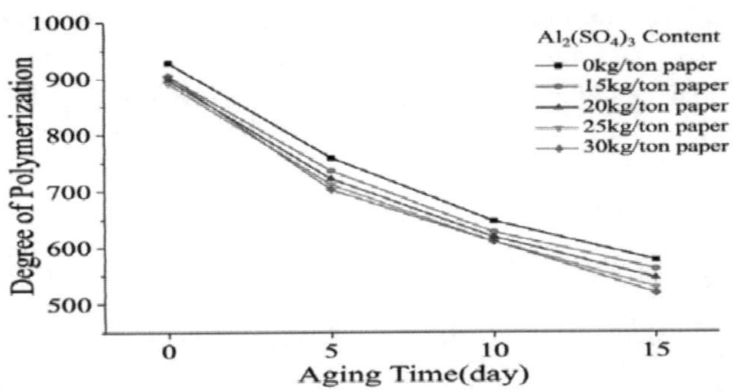

Figure 4. Effect of $Al_2(SO_4)_3$ Content on the Degree of Polymerization of Aged Paper

The average length of cellulose chains could be characterized by the DP in paper. With aging, the cellulose chain broke and the DP decreased. Therefore, the change in the DP of cellulose reflects the degradation of plant fibres during the aging process and can better explain the change in the paper strength properties. Figure 4 shows that the DP of cellulose decreased after aging. Under the same aging times, the paper with higher $Al_2(SO_4)_3$ contents, showed a greater DP decline in cellulose after aging. For example, after 15 days of aging, the DP of cellulose when the content of $Al_2(SO_4)_3$ was 15 kg/ ton paper, reduced by 37.90 % from 903.2 to 560.8, whilst the DP decreased by 42.77% from 905.1 to 518.0 in paper with $Al_2(SO_4)_3$ at 30 kg/ ton. The results showed that the higher the content of aluminum ions in paper, the more seriously the cellulose was degraded during the aging process. Aluminum ions in paper may accelerate the degradation of paper cellulose and so the higher the aluminum content, the more obvious the degradation.

4. Conclusions

The presence of aluminum ions in paper accelerates the acidification and aging of paper: the higher the content of aluminum ions, the worse the durability of paper. During the aging process, with increase content of aluminum ions, the DP of cellulose gradually decreased, and the decline rate in paper strength and whiteness increased, and the pH value is lower. After 15 days of accelerated aging under the same conditions, the pH value, whiteness, tensile index and tear index in the paper without aluminum ions were reduced by 14.77 %, 3.15 %, 7.99 % and 11.70 %, whilst for the paper with an aluminum sulphate content of 30 kg/ ton, the corresponding reductions were 22.62 %, 4.35 %, 24.66 % and 31.07 %.

5. Acknowledgement
Thanks to the support of Science and Technology Plan Special Project of Guangzhou, China (No. GZDD201808).

6. References
[1] Wei L, Sustainable Development and Archives Protection[J]. *Beijing Archives*. **2014,** *5*, 21.
[2] Xia Y Z, Xia S.R, Zhang J P, Zheng D Q, Jiang B. Research onAutomated Deacidification Technology used for Acidified Whole Books[J]. *Science of Conservation and Archaeology*. **2018**, *1*(30), 100-105.
[3] Wang Y P, Han H M. Talking about Paper Chemical Damage andPreservation [J]. *Journal of Puyang Vocational and Technical College*. **2005**, *18*(1), 32-34.
[4] Carter H A. The chemistry of paper preservation. Part 5. Permanent paper [J]. *Journal of Chemical Education*. **2007**, *84*(12), 1937.
[5] Zhao Y.H, Xie M, Analysis on the Preventive Protection of the Rare Ancient Books -Taking the Hubei Provincial Museum for Example[J]. *Paper and Paper Making*. **2019**, *38*(1), 50-54.
[6] Zhang X.M, Su J. Research on consolidation materials for fragile paper[J]. *Science of Conservation and Archaeology*. **2011**, *23*(2), 70-74.
[7] Mussatto S I, Fernandes M, Milagres A M F, Roberto I C. Effect of hemicelluloseand lignin on enzymatic hydrolysis of cellulose from brewer's spent grain[J]. *Enzyme and Microbial Technology*. **2008**, *43*(2), 124-129.
[8] Wang J, Yiu B, Obermeyer J, Filipe C D M, Brennan J, Pelton R. Effects of temperature and relative humidity on the stability of paper-immobilized antibodies[J]. *Biomacromolecules*. **2012**, *13*(2), 559-564.
[9] Emsley A M, Heywood R J, Ali M, Xiao X. Degradation of cellulosic insulationin power transformers. Part 4: Effects of aging on the tensile strength of paper[J]. *IEE Proceedings-Science, Measurement and Technology*. **2000**, *147*(6), 285-290.
[10] Michalski S. Double the life for each five-degree drop, more than double the life for each halving of relative humidity: *Preprints of the ICOM-CC 13th Triennial Meeting*[Z]. *Rio de Janeiro*. **2002**, 66-72.
[11] Weyermann C, Spengler B. The potential of artificial aging for modelling of natural aging processes of ballpoint ink[J]. *Forensic Science International*. **2008**, *180*(1), 24-27.
[12] Sun C, Mei L, Ben S B. Discussion on the Control Measures of Mould Pollution of Ancient Books[J]. *LANTAI WORLD*.**2018**(09), 63-65.
[13] He B, Lin Q, Chang M, Liu C, Fan H M, Ren J L. A new and highly efficient conservation treatment for deacidification and strengthening of aging paper by in-situ quaternization[J]. *Carbohydrate Polymers*. **2019**, *209*, 250-257.
[14] Zhao Y B. A research about influencing factors of ancient books library preservation and measures[J]. *Science & Technology Vision*. **2017**, *27*, 169-170.
[15] Potthast A, Henniges U, Banic G. Iron gall ink-induced corrosion of cellulose: aging, degradation and stabilization. Part 1: model paper studies[J]. *Cellulose*. **2008**, *15*(6), 849-859.
[16] Ahn K, Hofmann C, Horsky M, Pottahast A. How copper corrosion can be retarded–new ways investigating a chronic problem for cellulose in paper[J]. *Carbohydrate Polymers*. **2015**, *134*, 136-143.
[17] Henniges U, Prohaska T, Banik G, Pottahast A. A fluorescence labeling approach to assess the deterioration state of aged papers[J]. *Cellulose*. **2006**, *13*(4), 421-428.

[18] Selih V S, Strlič M, Kolar J, Pihla B. The role of transition metals in oxidative degradation of cellulose[J]. *Polymer Degradation and Stability*. **2007**, *92*(8), 1476-1481.

[19] Henniges U, Bürger U, Banik G, Rosenau T, Potthast A. Copper corrosion: comparison between naturally aged papers and artificially aged model papers[C]. Macromolecular Symposia. Weinheim: WILEY-VCH Verlag. **2006**, *1*(244), 194-203.

[20] Henniges U, Reibke R, Banik G, Huhsmmann E, Hähner U, Prohaska T, Potthast A. Iron gall ink-induced corrosion of cellulose: aging, degradation and stabilization. Part 2: application on historic sample material[J]. *Cellulose*. **2008**, *15*(6), 861-870.

[21] Cheng Z G, Wen Y Q, Jin S H. Analysis of sizing mechanism and development trend of rosin series sizing agent [J]. *Paper chemicals*. **2005**, *1*, 25-28.

IOP Conf. Series: Materials Science and Engineering **735** (2020) 012054 doi:10.1088/1757-899X/735/1/012054

Design and Implementation of a New Circuit Maintenance Training Platform

Zelong Zhou, Xichun Shao, Zhuo Yang and Xiaochun Zhu

Wuhan Mechanical College, Army Engineering University, Wuhan, 430075, China.
Email: zhouzelong2005@163.com

Abstract. This paper designs a circuit maintenance platform for the artillery system. It can read or simulate the important node data of the artillery system in real time online or offline, and provide the basis for the artillery maintenance inspection. It can also be used as a daily teaching operation demonstration platform; analog data detection read and fault demonstration process.

1. Introduction

The artillery can achieve fast, accurate and stable tracking and aiming of moving targets, and can not be separated from the follow-up system. The follow-up system can accept the target elements from the fire-controlled radar in the form of electrical signals, compare the target elements with the muzzle elements, and then perform a logical operation to generate a driving signal, which is driven to amplify the gun position or The high and low execution motors drive the artillery to rotate quickly to continuously track the target. It is not only a signal amplifying circuit for the artillery to automatically track the aiming target, but also an important part of the signal logic processing. The real-time performance state of the follower system plays a decisive role in the normal tracking, aiming and shooting of the artillery.

At present, the system of signal detection and status monitoring of the follow-up system, which is used for equipment inspection and maintenance and teaching training, needs to be designed. In view of this, combined with the actual needs, based on a full understanding of the working principle of the equipment and the logical control relationship, a solution for the maintenance and training platform of the artillery system of the artillery system is proposed. By developing a training system for the artillery follow-up system, it is able to meet the teaching requirements of equipment testing and maintenance, it can also avoid human faults caused by improper operation and save equipment maintenance costs.

2. Circuit Maintenance Training Platform Design Requirements

A new type of artillery tracking system circuit maintenance training platform is a comprehensive system suitable for signal detection and condition monitoring of various types of artillery follower systems. It must meet the following basic conditions: First, the maintenance training platform must have a certain structure. The strength meets the requirements for carrying in outdoor harsh environments under field conditions, and the structural size should be as small as possible to meet the requirements of the electrical interface parameters of the artillery, which is convenient for teaching and training. Second, because the electrical devices of the artillery have different requirements for electrical parameters such as power supply current, voltage, power, etc., and have a certain range, in order to meet the power requirements of different devices, the maintenance training platform must have certain flexibility, that is, according to Different device inputs of different types of guns require adjustment, detection, output and display of electrical parameters within a certain range. Third, the maintenance training platform is required to be modularized, combined installation, easy to carry and store, and suitable for various tasks in different environments.

In order to improve the maintenance efficiency, this paper designs a circuit maintenance platform for the artillery system. It can read or simulate the important node data of the artillery system in real time online or offline, and provide the basis for the artillery maintenance inspection. Daily teaching operation demonstration platform, simulation data detection reading and fault demonstration process.

With the follow-up control box, DC bus measurement and control box, azimuth/high and low receiver, azimuth/high and low drive, bleed box, joystick, angle limiter, azimuth/high and low execution motor, follow-up DC power box, etc. Simulated inspection and maintenance functions for 9 components;

It has the function of setting simulation fault, which is used for maintenance training demonstration and evaluation;

It has the working principle and control simulation function of the follow-up system, and the joint control system has the function of joint control;

It can detect analog signals such as power supply and signal with a multimeter, and can display digital signals such as the position of the gun and the target position;

It has control functions such as starting, stopping, automatic/semi-automatic switching of the follow-up system;

It has the function of detecting the detection hole of each monomer and component and the principle display function of the road; having a protection function for each monomer detection panel;

It has good operability and maintainability;

It has a grounding protection function, which is safe and reliable.

3. Design of the Console Training Platform

The circuit maintenance of a new type of artillery follower system is mainly composed of a system bench, a detection control unit, a detection indicating unit and a connecting cable.

3.1. System Gantry

The system gantry serves as the carrier of the entire maintenance training platform, including the gantry, analog follow-up control box, analog azimuth/high and low receiver, analog azimuth/high and low execution motor, analog azimuth/high and low drive, analog joystick, and analog angle limiter. 9 analog components such as analog DC bus measurement and control box, analog follow-up DC power supply box and analog bleed box. The test pedestal body is constructed of solid wood panels, and the analog component box is processed by aluminum alloy material, and the body is mounted with casters for convenient movement.

The gantry is constructed of solid wood panels, semi-enclosed open design, consisting of countertops, fences, cabinets, etc. The countertops are mainly used to install simulated follow-up control boxes, analog joysticks and other analog components, two bins. Used to install the inspection display computer and control box.

The analog follow-up control box is used to complete the input/output signal test, receiver indication and control functions of the follow-up control box. The analog follow-up control box can receive the position feedback signal of the analog receiver and display it; receive the combined speed signal of the azimuth/high and low analog tachometer and measure in the detection hole; receive the analog power driver signal and perform measurement; can pass the control switch, Control system working methods, etc. The box structure is adopted, and the internal circuit is redesigned and developed according to the working principle and function of the mounting; the panel part leads the detection signal for testing.

Simulated azimuth/high-low receiver device, using the same two-group revolving function as the installation principle, adding a coaxial motor, under the control of the joystick or automatic control signal, driving the rotation through the motor, and electricity The signal is connected to the test panel and the signal is sent via the internal cable to the follow-up control box for display. The azimuth/high and low analog receiver detection panel is shown in Figure 1. The voltage signal of each phase

IOP Conf. Series: Materials Science and Engineering **735** (2020) 012054 doi:10.1088/1757-899X/735/1/012054

winding of the receiver can be tested on the azimuth/high and low analog receiver detection panel, and the static resistance can be detected by switching the switch.

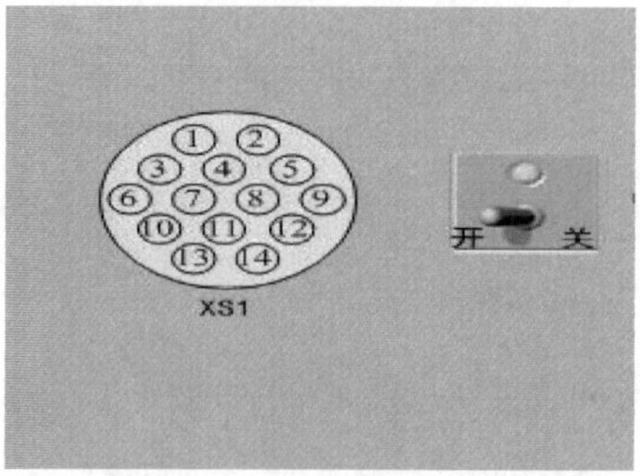

Figure 1. Schematic Diagram of the Receiver Test Panel

The analog azimuth/high-low execution motor uses a stepper motor that rotates under the control of the joystick control signal. Three-phase power signals and control signals can be tested on the test panel. Azimuth/high and low performs the motor detection panel as shown in Figure 2.

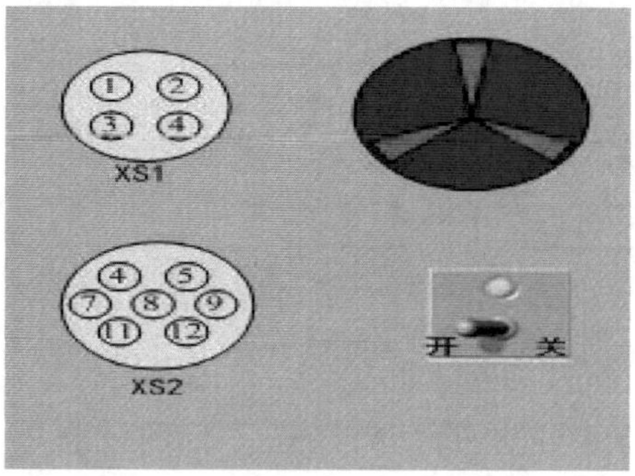

Figure 2. Schematic Diagram of the Executed Motor Detection Panel

The voltage signal of each phase winding of the receiver can be tested on the detection panel, and the static resistance can be detected by switching the switch, which is basically the same from the motor and the high and low motor detection panels.

Analog azimuth/high and low power driver for testing and controlling the input and output signals of the power driver. The box structure is adopted, and the internal circuit is redesigned and developed according to the working principle and function of the mounting; the panel part takes out the detection signal for testing, and the panel design is shown in Fig. 3 and Fig. 4.

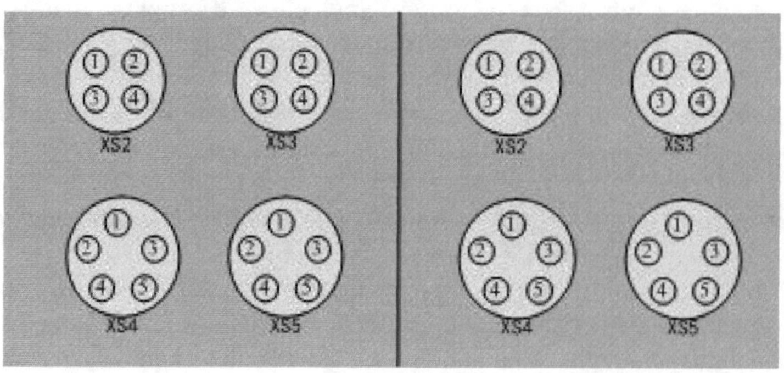

Figure 3. Schematic Diagram of the Analog Azimuth Power Driver

The shape of the analog joystick is designed according to the actual shape and is processed and welded with aluminum alloy material. The joystick body is customized according to the function and operation mode of the original joystick, and its functional shape is consistent with the actual installation. The joystick is developed with imitation components. The control and control functions are similar to the actual installation. By pulling the joystick, the output signal of the joystick can be detected in the detection hole. The analog joystick test panel is shown in Figure 5.

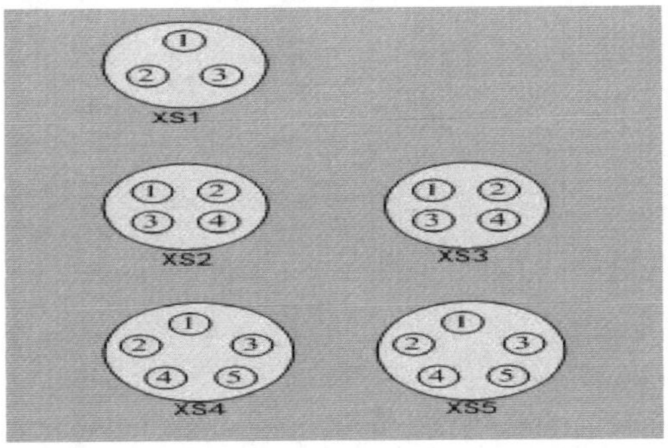

Figure 4. Schematic Diagram of the Analog High and Low Power Driver

Figure 5. Schematic Diagram of the Analog Joystick Test Panel

The analog angle limiter is functionally imitated in the internal structure. The control function is divided into high angle/low angle limit function. The control function is completed by the combination of the mechanical structure part and the control switch. The structure part is coaxially connected to the motor, the motor is controlled by the joystick or automatic signal, and the position feedback is positioned by the receiver to realize the control. Function, while detecting holes can measure input/output signals. The simulated angle limiter is shown in Figure 6.

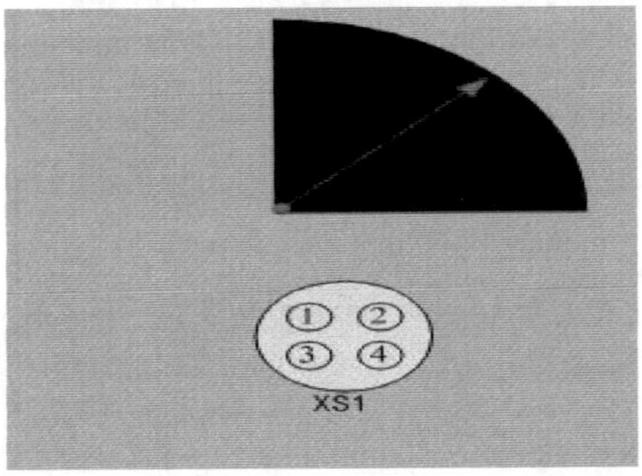

Figure 6. Schematic Diagram of the Analog Angle Limiter

The analog DC bus measurement and control box is used to generate the 325V DC bus voltage and 27V working voltage required for system operation, and adopts switching power supply technology for functional design.

The analog follow-up DC power box is used to generate the DC working voltage required for the operation of the servo system, and the analog power conversion technology is used for functional design.

Analog bleeder resistor box, with built-in resistance, power equivalent resistance, analog impedance test, bleeder current test, etc.

3.2. Detection Control Unit

The detection control unit is used for managing and controlling the correlation between the various analog components, giving the power, control signals and display signals of each unit during operation, simulating the fault setting and the like, so as to realize the simulation between the individual components. Mutual control function. It mainly consists of control box, power supply and signal board, control and display board, management logic board and connecting cable.

3.3. Detection and Control System

The detection control system is the core of the entire maintenance training platform, and completes the power management, control and display, logic management and other functions of the entire detection system. It consists of a cabinet, a power supply and signal board, a control and display panel, a management logic board, and a fault setting unit.

The box body is designed with aluminum alloy structure, and the sub-mother board plug-in structure is adopted inside, and each extension part is installed. The outer shape of the box body is shown in Fig. 7.

Figure 7. Schematic Diagram of the Box Shape

The power supply and signal board are used to convert the mains into 26V/400Hz AC signal power supply, 24VDC, ±15VDC, +5VDC signal power supply for the test platform.

Control and display panels perform various types of signals required to generate a system for impedance matching, amplification, and logic control processing of various power sources and signals. In addition, the circuit board also generates a digital signal that can communicate with a single unit such as a follow-up control box through a single-chip microcomputer system to realize functions such as control and data interaction.

The management logic board is composed of an ARM microcontroller and a logic matrix unit for completing the logic control function of the control system. Fault setting unit As a manual setting function, each fault function can be controlled by the fault selection control. The system sets 1 to 2 fault modes for each analog component.

3.4. Detection Instruction Unit

The detection indication unit is composed of a computer, instruction software, etc., and is mainly used to indicate the circuit structure principle of each single machine component, each input/output channel/detection hole position, signal type, amplitude, etc., to facilitate detection and maintenance training. The detection indication system is mainly used for testing, which is convenient for querying the connection direction of each connector part, signal definition, and the operation principle of the unit and system. It mainly consists of detection indication platform and detection indication software. The detection indicator platform uses a professional touch computer to facilitate query operations. The instruction software is developed by LABVIEW software, which is convenient for manipulation and query. The detection indication system is shown in Figure 8.

The circuit maintenance training platform adds independent motion functions to the azimuth/high-low receiver, azimuth/high-low execution motor, and the angle limiter based on the components of the simulated mounting system, and moves it through the control system. The action is associated with the control signal; the control signal correlation simulation is performed on the follow-up control box, the joystick, the DC bus measurement and control box, and the DC power box of the follower system, thereby realizing the system working state of the simulation installation. In order to facilitate the inspection and maintenance, the detection holes and input/output plug-in signals of each single unit and component are led to the panel for display and detection. At the same time, reference can be made to the detection indication system for the circuit structure principle, pin definition, signal type/amplitude and so on.

The appearance of the circuit maintenance training platform is shown in Figure 9. The specification size is 2000mm × 2000mm × 1500mm, and the total weight is about 200kg. The input voltage is single phase 220V/50Hz or three phase 380V/50Hz. The measurement voltage range is 0 to 325V for DC DC and 0 to 220V for AC. The detection signal coverage is not less than 80%. The number of simulated fault set points is greater than eight. Operating temperature -10 ° C ~ 45 ° C. Continuous working time is greater than 8h.

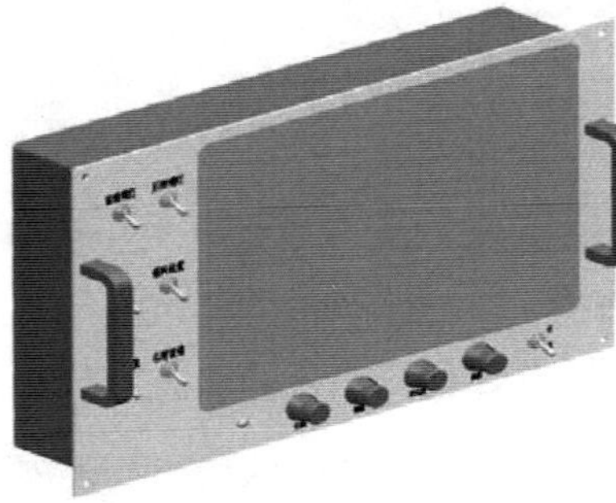

Figure 8. Schematic Diagram of the Detection Indication System

4. Acknowledgments

This research was financially supported by the Equipment Construction Project Foundation of Army Engineering University.

5. References

[1] Zhu Wenbo. Research on Military Training Virtual Driving System [D]. Southwest Jiaotong University, 2017. pp 29–34

[2] Luo Shiliang. Research on PLC-based elevator teaching experiment system [D]. Xiamen University, 2012. pp 56–57

[3] Xu Qinchao, Ouyang Zhonghui, Xiao Fan. Design of Simulation Training Platform for Uninterrupted Operation of Distribution Network Based on Cloud Computing [J]. Journal of Electric Power Science and Technology, 2019, 34(1).

[4] Deng Chenghui, Zhang Zhe, Zhang Qiushi et al. Research on Internet-based simulation training platform for distribution network without power failure [J]. Hydropower Science, 2018, 36(4): 183-186, 174.

[5] Shi Yifeng, Cai Jianwei. Optimized Design and Application of Insulation Cover for Cable Clamping in Distribution Network without Power Failure [J]. Electric Engineering, 2018, (5): 122-123, 126.

[6] Zhang Zhongyuan,Yue Haotian, Wang Bowing, Liu Yunpeng, Luo Shihao. Partial Discharge Ultrasonic Signal Depth Learning Pattern Recognition Method Based on Similar Matrix Blind Source Separation and Convolution Neural Network [J]. Power System Technology, 2019, 43(6).
[7] Li Tianyu, Zhang Yang. Design and Development of Power Line Maintenance Simulation Training System [J]. Information Systems Engineering, 2013, (11).
[8] Zhou Yi, Jiang Jizhen, Chen Xiangjun. Application Research of Substation Simulation Training System [J]. China Electric Power Education, 2012, (15): 25-26.
[9] Zhang Chen Qingyu, etc. Research and Implementation of 3D Demonstration Training System for Traction Substation [J]. Electrical Automation, 2018, (1).

Preparation and Properties of Poly (Silicon-Containing Arylacetylene)/Attapulgite Foams

Jiadong Wu, Shilong Dong and Qiaolong Yuan

Key Laboratory of Specially Functional Polymeric Materials and Related Technology of the Ministry of Education, School of Materials Science and Engineering, East China University of Science and technology, Shanghai, China

Email: qlyuan@ecust.edu.cn

Abstract. Poly(silicon-containing arylacetylene) (PSA) syntactic foams were prepared by thermally expandable microspheres (EM) in the silicon-containing arylacetylene (SA) resin during curing. The morphology, porosity, thermal and mechanical properties of the foams were studied. Furthermore, the PSA foam was reinforced by attapulgite (ATT) nanoparticles to fabricate the PSA/ATT composite foam. The results display that the density, compressive strength and thermal stability of the PSA foam decrease with increasing EM content in the matrix, but the temperature at 5% weight loss in N_2 is still higher than 560°C, and the residual yield at 800°C is larger than 89%. With addition of 10 g ATT in the 100 g matrix, the average specific compressive and flexural strength of PSA/ATT foam are 31.4 MPa $g^{-1}cm^3$ and 33.6 MPa $g^{-1}cm^3$, respectively. The thermal conductivity of the composite foam can decrease to 0.064 W m^{-1} K^{-1} and has a density of 319 kg m^{-3}.

1. Introduction

Lightweight, thermal insulating, high strength and heat-resistant polymeric materials have attracted many attentions among researchers and industries. Polymeric foams are widely used in many applications due to their great features, such as thermal and sound insulation, lightening, cushioning as well as low dielectric constant. Polyarylacetylene (PAA) resin possesses high char yield (over 80%). The PAA foam was prepared by blowing PAA prepolymer with pentane and then carbonized to synthesize carbon foams with high mechanical strength and high strength/density ratio [1].

Poly (silicon-containing arylacetylene) is highly heat- and burn-resistant as ceramic, easy-to-process, and light as thermosetting resin with -C≡C-Si-C≡C-Ar- units. It can be utilized as ablative and heat-protection materials. Itoh and co-workers [2-5] used magnesia to catalyze dehydrogenative coupling polymerization of diethynyl benzenes and hydrosilanes to prepare the poly [(phenylsilylene) ethynylene-1,3-phenyleneethynylene] (called MSP), which has no volatility with cure, high decomposition temperature and high char yield after pyrolysis. Buvat et al [6] prepared a easily processable phenylacetylene terminated poly (silylene ethynylene phenylene ethynylene) (termed BLJ resin) by condensation of dichlorosilane (dichlorophenylsilane) and a mixture of Grignard agents of 1, 3-diethynyl benzene and phenyl acetylene. Huang et al [7] also prepared a silicon-containing arylacetylene resin (SA) by condensation of dichlorodimetylsilane and a mixture of Grignard agents of diethynyl benzene and phenyl acetylene. The resin abbreviated PSA is processable, highly heat resistant and has good mechanical properties as well as stable dielectric property in the range of wide operating temperature and broadband.

The thermally expandable microsphere consists of inert low-boiling liquid hydrocarbon core wrapped by a spherical thermoplastic acrylonitrile copolymer shell to be used as a physical aerating powder [8, 9]. The liquid core begins to evaporate, and meanwhile the thermoplastic shell becomes soft and starts to swell due to the high internal pressure of the gasified blowing agent in the microsphere when the microsphere is heated to above the glass transition temperature (T_g) of the polymer shell. The thermally expandable microspheres (TEMs) are lightweight fillers that can act as physical blowing agent and adjust the size of closed cells when mixed in the resin and subsequently heated to cause expansion within the matrix to fabricate the composite foam. Wang et al [10] prepared the expandable microspheres/epoxy foams with different densities and microstructures, and found that the precuring process and foaming temperatures affected the cell size, the cell size distribution and the density of the foam.

Poly (silicon-containing arylacetylene) foam will be a promising lightweight material which could be applied as adsorbent materials after pyrolysis. Attapulgite (ATT), a kind of natural fibrillar hydrated magnesium-aluminum silicate clay has been used to prepare polymer/ATT nanocomposites, due to its intriguing properties, low cost, and low density. In this paper, the PSA syntactic foams were prepared by mixing with the expandable microspheres (EMs) and then elevating temperature to cure and foam the SA/EMs resins. Furthermore, a nanorod-like ATT nanoparticles were used to reinforce the PSA syntactic foam. The fabrication of PSA foams, the correlation of structure, thermal and mechanical properties of PSA, and PSA/ATT composite foams were attempted to be investigated in detail.

2. Experimental

2.1. Materials

Silicon-containing arylacetylene (SA) resins with ethynyl terminal groups were prepared in our lab according to the procedure shown in reference [7]. The number-average molecular weight ($\overline{M_n}$) and polydispersity index (PDI) of the silicon-containing arylacetylene resins are 2230 and 1.81, respectively. Ethyl acetate, 99+% was purchased from Shanghai Titan Scientific Co., Ltd., China. The expandable microsphere, DU170 (number-average particle size 25.6 μm, density 1.028 g/cm^3) was received from Crerax Science and Technology Co., Ltd., China. Nano attapulgite (ATT) powder was supplied by Jiangsu Jiuchuan Nano Material Technology Co., Ltd., China. All chemicals were used as received.

2.2. Preparation of PSA Syntactic Foam

50 grams of SA and 220 mL of ethyl acetate were added into the 500 mL three-necked flask with mechanical stirrer, condenser and thermometer. The mixtures were heated to 75°C by oil bath, and were agitated until the SA resin was dissolved in ethyl acetate. The certain amount of DU170 powders were poured into the SA melt after the solvent was evaporated under vacuum. The mixture of SA melt and DU170 was stirred until the powders were evenly dispersed in the melt. The melting SA mixtures were poured into the aluminum boxes with sizes of 25 mm×15 mm ×15 mm and 42 mm×42 mm×5 mm. The mixtures of the SA resin and DU170 were foamed and cured in the hot oven in term of the process, 170°C /2 h, 210°C /2 h and 250°C /4 h. SA resin was crosslinked to form poly(silicon-containing arylacetylene) (PSA) during foaming and curing. The added amount of DU170 and the named PSA foams were listed in Table 1.

Table 1. The Named PSA Foams with Different DU170 Content

Coded foam	DU170 content (phr)	Coded foam	DU170 content (phr)
PSA-1.0	1.0	PSA-3.2	3.2
PSA-2.0	2.0	PSA-3.4	3.4
PSA-3.0	3.0	PSA-3.6	3.6
PSA-4.0	4.0	PSA-3.8	3.8

2.3. Preparation of PSA/ATT Composite Foam

In 500 mL round bottom flask, 220 mL ethyl acetate was poured, and then the ATT powders were

IOP Conf. Series: Materials Science and Engineering **735** (2020) 012055 doi:10.1088/1757-899X/735/1/012055

weighed and added with agitation and sonication. After 2 h, the 50 g of SA resin was added into the flask. The dispersion was being agitated and sonicated at room temperature for 1.5 h. And then the dispersion was only heated to 80°C with agitation. The 3.6 parts of DU170 per hundred resin were added into the melted mixture of SA and ATT after the ethyl acetate was partially removed at 80°C for 20 min. The melted mixture was poured into the aluminum box with size of 40 mm × 40 mm × 40 mm until the solvent was evaporated. The mixtures were foamed and cured to achieve the PSA/ATT syntactic foams according to the procedure, 170°C /2 h, 210°C /2 h and 250°C /4 h. With addition of 3.0, 5.0, 7.5, 10.0, 12.5 and 15.0 g of ATT in 100 g of the SA resin, the mixtures were coded as SA/ATT-1, SA/ATT-2, SA/ATT-3, SA/ATT-4, SA/ATT-5 and SA/ATT-6, respectively.

2.4. Characterization
Scanning electron microscopy (SEM, S-4800, Hitachi Company, Japan) was utilized to observe the morphology of the foam samples. Based on Archimedes' principle, the volume density of the PSA foam was tested by drainage method according to the Chinese Standard GB/T 4472-2011. It used the density determination kit for the analytical balance (Mettler Toledo ML204, Switzerland).

The compressive test was done according to Chinese Standard GB/T 8813-2008. The compressive strength of the foams was tested by universal material mechanical testing machine (CMT4204, Shenzhen SANS, China) at room temperature. The testing speed used was 1 mm/min.

Thermal conductivity of the PSA and PSA foams with dimensions of 40 mm ×40 mm ×5 mm was measured by a heat conduction analyzer (TC3000, Xiatech, China).

Thermogravimetric analysis (TGA) was conducted using thermogravimetric analyzer (Mettle-Toledo TGA-DSC1, Switzerland). Nitrogen was used as the purge gas with a flow rate of 50 mL min^{-1}, and a heating rate of 10°C min^{-1} was used.

The flexural properties of the foams were performed on an electronic universal testing machine (CMT4204, Shenzhen SANS, China) at room temperature. The flexural properties of the samples were tested according to Chinese Standard, GB/T 1499-2005. The cross-head speed was 2 mm min^{-1} and the resulting fracture surfaces were used for the SEM observations.

3. Results and Discussions

3.1. PSA Foaming
The morphology of the PSA syntactic foams with different DU170 content displays that the rigid PSA syntactic foam can be obtained during crosslinking and expanding. The most of foams have closed cell structure, and the amount of spherical cells increases with addition of DU170 content in the PSA. The cell wall of the PSA syntactic foam becomes thinner and the cells squeeze and collapse with each other when the 4 parts DU170 per hundred PSA matrix are added. The increase of the cell density can decrease the density of the PSA syntactic foam. Figure 1 shows the fractured surfaces of the PSA syntactic foams with the content of DU170 from 3.2 to 3.8 phr. The spherical cells are evenly dispersed in the PSA matrix. The PSA syntactic foam with high content of closed cells presents the low density and the proper mechanical property.

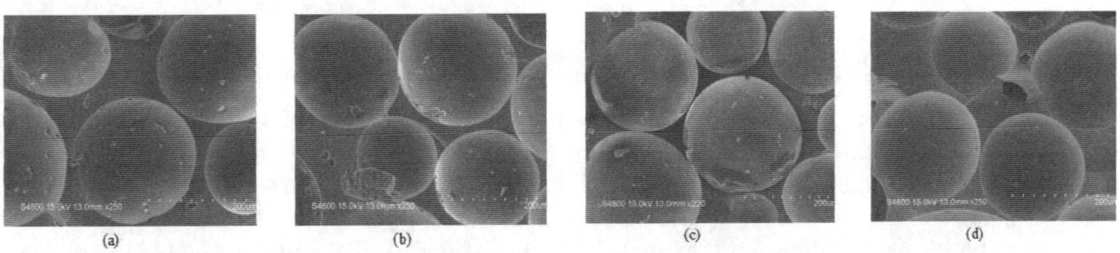

Figure 1. SEM Images of PSA-3.2 (a), 3.4 (b). 3.6 (c) and 3.8 (d) Foams

3.2. Structure and Compressive Strength
The density and compressive strength of the PSA syntactic foams with different DU170 content were

listed in Table 2. The results show that the density of PSA syntactic foam decreases with increase of DU170 content. Meanwhile, the compressive strength of PSA syntactic foam decreases from 10.3 MPa to 4.0 MPa. The decrease of the density of the foam results in the increase of porosity of the foam. The compressive strength of the foam reduces because the wall of the closed cells becomes thinner. However, there is an optimal value, 25.6 MPa g^{-1} cm^3 for the specific compassion strength of the foam when the content of DU170 is 3.6 g in 100 g of PSA matrix. The density of the PSA foam with 4.0 g DU170 in 100 g PSA can reach 296 kg m^{-3}, but the compressive strength of the foam declines to 4.0 MPa since the collapse of the cells occurs in the foam (see Fig. 1(d)).

Table 2. Density and Compressive Strength of PSA Foams with Different DU170 Content

Foam	ρ (g cm^{-3})	σ_c (MPa)	σ/ρ (MPa g^{-1} cm^3)
PSA-3.0	0.511±0.021	10.32±0.22	20.20
PSA-3.2	0.433±0.011	9.92±0.11	22.90
PSA-3.4	0.361±0.006	9.11±0.06	25.23
PSA-3.6	0.341±0.008	8.74±0.09	25.63
PSA-3.8	0.323±0.013	6.07±0.14	18.80
PSA-4.0	0.296±0.004	4.03±0.05	13.62

3.3. Thermal Stability and Conductivity

The effect of DU170 microspheres on the thermal stability of the PSA syntactic foams are shown in Table 3. The results show that the temperature of 5% weight loss (T_{d5}) of the cured SA, namely PSA decreases with addition of DU170 microspheres. However, the T_{d5} of PSA syntactic foams is still higher than 560°C in N_2, and the residual yield at 800°C (Y_{r800}) of the foams and the cured SA are greater than 89%. The PSA syntactic foam still sustains high heat-resistance. The weight loss of the foams is higher than that of the cured SA, and increases with rising temperature since the gases in the cells escape from the foam. There is a first decomposition temperature for the PSA syntactic foams around 330°C. It results from the decomposition of the polymeric shell of DU170 microspheres. The second weight losses are attributed to the decomposition of the PSA. The thermal conductivity of the PSA syntactic foam is in the range of 0.08~0.09 W m^{-1} K^{-1}, which is lower than 0.2 W m^{-1} K^{-1} of the cured PSA.

Table 3. TGA Analysis of the Cured PSA and PSA Foams (in N_2)

DU170 content in 100 g of PSA (g)	T_{d5} (°C)	Y_{r800} (%)
0.0	622.0	89.1
3.0	573.4	90.9
3.2	583.0	91.0
3.4	584.1	90.9
3.6	560.8	89.3
3.8	569.2	89.9
4.0	564.3	89.5

3.4. PSA/ATT Composite Foams

The effect of ATT content on the mechanical properties of the PSA/ATT composite foams are presented in Table 4. The density of the composite foam decreases gently with increasing ATT loading. It probably belongs to the hydrated pristine ATT nanoparticles, which can participate in the foaming with DU170 to reduce the density of the composite foams compared with the neat PSA foam. The nearly free zeolitic water filling up the empty channels which parallel to the ribbons in two silicon-oxygen tetrahedron layers and one octahedral layer of metal cations, escapes firstly during heating dehydration [11, 12]. The escape of water can be gasified and held by the PSA to form the bubble cells in the matrix. The remaining solvents and water resulting from polycondensation reaction were used as blowing agent to prepare polyimide foams [13]. The average specific compressive and flexural strength of PSA/ATT composite foam are higher than that of the neat PSA foam, and increase with increase of ATT loading, and can reach 31.4 MPa g^{-1} cm^3 and 33.6 MPa g^{-1} cm^3, respectively, when the 10 phr of ATT loading are added in the matrix. The epoxy foams expanded by microspheres at 175°C foaming temperature exhibited specific compressive strength of 32.5 MPa g^{-1} cm^3 [10]. The maximum values obtained are attributed to the formation of physical network of ATT nanoparticles in the PSA matrix. The unmodified attapulgite nanosized particles could be well-dispersed in the phenolic (PF) matrix and had the supporting effect in PF foam cell wall to improve the tensile strength and the compressive strength of PF foam nanocomposites [14]. The thermal stability of the PSA/ATT-4 composite foam is also higher than that of the PSA foam (T_{d5} 589°C, Y_{r800} 92.8%). Moreover, the specific strength of the composite foam decreases evidently with addition of higher content of ATT nanoparticles. The fractured surfaces of the PAS/ATT-4 and PSA/ATT-5 composite foams and wall of cells in the foams are shown in Figure 2. The ATT nanoparticles can be evenly dispersed in the cell wall of PSA/ATT-4 composite foam (see white prominent dots in Figure 2(b)). The strength of the PSA/ATT composite foam with higher than 10 phr ATT loading in PSA matrix drops down because of existence of the agglomeration of the filler (as indicated in the black cycle in Figure 2(d)). The flexural modulus of the PSA/ATT composite foam decreases slowly with increasing ATT loading. The decrease of the density of the PSA/ATT composite foam causes the reduction of the thickness of wall of cells which will lead to the decrease of the modulus of the foam. Similarly, the flexural strength of the cured SA was augmented with addition of montmorillonite (MMT) or organically modified montmorillonite (OMMT), but the flexural modulus of the cured SA decreased slightly [15].

Table 4. Mechanical Properties of PSA/ATT Composite Foams with Different ATT Loadings

ATT (phr)	ρ (g cm^{-3})	σ_c (MPa)	E_c (MPa)	σ_f (MPa)	E_f (GPa)
0.0	0.341±0.003	8.74±0.03	284.53±26.31	8.12±0.44	2.45±0.14
3.0	0.337±0.007	9.15±0.18	311.67±19.64	8.27±0.03	2.38±0.11
5.0	0.333±0.010	9.41±0.33	373.37±33.88	8.71±0.12	2.35±0.17
7.5	0.326±0.001	9.87±0.28	447.72±14.67	9.41±0.24	2.25±0.08
10.0	0.319±0.001	10.02±0.78	599.66±97.44	10.72±0.55	2.22±0.10
12.5	0.311± 0.004	8.44±0.34	212.80±26.57	5.88±0.68	1.40±0.22

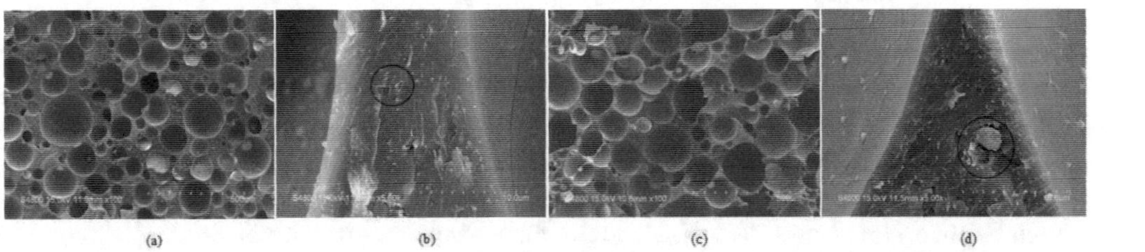

Figure 2. SEM Images of Fractured Surface of PSA/ATT-4 and PSA/ATT-5 Composite Foams

The attapulgite powders could be hot-pressed to form a dehydrated attapulgite with porosity of 45.7%, and exhibited that the thermal conductivity of the attapulgite is as low as 0.34 W m^{-1} K^{-1} at 50°C [16]. This suggests the potential of hot-pressed attapulgite as a candidate for thermal barrier materials. The effect of ATT nanoparticles on the thermal conductivity of the PSA composite foam is presented in Table 5. The results show that the thermal conductivity of the PSA/ATT composite foams are lower than that of the PSA foam, and decreases with increase of ATT loading. It can descend to 0.0642 W m^{-1} K^{-1} when 10 g ATT nanoparticles without premodification are added into 100 g PSA matrix. The thermal conductivity of the microsphere-expanded epoxy foam was about 0.05~0.07 W m^{-1} K^{-1} when the expandable microspheres were added in the range of 3wt%~4wt% [17]. The thermal conductivity of the PSA foam can further be improved by adding ATT nanoparticles in the matrix.

Table 5. Thermal Conductivity of the PSA/ATT Composite Foams

ATT content (phr)	ρ (g cm^{-3})	K (W m^{-1} K^{-1})
0.0	0.341±0.008	0.0821
3.0	0.337±0.007	0.0753
5.0	0.333±0.010	0.0724
7.5	0.326±0.001	0.0688
10.0	0.319±0.001	0.0642

4. Conclusions

Thermally expandable microspheres (EM) were added into the silicon-containing arylacetylene (SA) resin to prepare the SA/EM mixtures The melted mixtures were swollen with expanding of microspheres during curing to form the poly(silicon-containing arylacetylene) (PSA) syntactic foams. Furthermore, the PSA foam was reinforced by attapulgite (ATT) nanoparticles to fabricate the PSA/ATT composite foam. The results show that the foams can form at initial curing temperature of SA resin, and harden to achieve the rigid foams with spherical closed cells after the foams are post-cured. With increasing EM content in the PSA matrix, the density, compressive strength, and thermal stability of the PSA foam decrease, but the T_{d5} of PSA syntactic foam in N_2 is still higher than 560°C, and the Y_{r800} is larger than 89%. The specific compressive strength of the PSA syntactic foam can reach 25.6 MPa g^{-1} cm^3. The thermal conductivity of the foam changes from 0.08 to 0.09 W m^{-1} K^{-1}. With addition of 10 g ATT in the 100 g matrix, the average specific compressive and flexural strength of PSA/ATT foam are 31.4 MPa g^{-1}cm^3 and 33.6 MPa g^{-1}cm^3, respectively. In addition, the thermal conductivity of the PSA syntactic foam can reduce to 0.064 W m^{-1} K^{-1} when the composite foam has a density of 319 kg m^{-3}. A highly heat-resistant poly (silicon-containing arylacetylene) foam is lightweight and heat-insulating syntactic foam for aerospace application, and a promising precursor for preparing carbon foam to be used as an absorbent material.

5. References

[1] Liu M, Gan L, Zhao F, et al 2007 *Carbon* **45** 3055
[2] Itoh M, Mitsuzuka M, Iwata K, Inoue K 1994 *Macromolecules* **27** 7917
[3] Itoh M, Inoue K, Iwata K, Mitsuzuka M, and Kakigano T 1997 *Macromolecules* **30** 694
[4] Itoh M, Iwata K, Ishikawa JI, Sukawa H, Kimura H, Okita K 2001 *J. Polym. Sci.: Part A: Polym. Chem.* **39** 2658
[5] Ogasawara T, Ishikawa T, Yamada T, Yokota R, Itoh M, Nogi S 2002 *J. Compos. Mater.* **36** 143
[6] Buvat P, Jousse F, Delnaud L, Levassort C 2001 *Proc. 46th Int. SAMPE Symp. Exhibition* (California) p134
[7] Wan F, Zhang J, Huang J, Yan H, Huang F, Du L 2006 *Polym. Bull.* **56** 19
[8] Jonsson M, Nordin O, Malmström E, Hammer C 2006 *Polymer* **47** 3315
[9] Kawaguchi Y, Oishi T 2004 *J. Appl. Polym. Sci.* **93** 505
[10] Wang L, Yang X, Zhang J, Zhang C, He L 2014 *Compos. Part B* **56** 724
[11] Haden W L Jr, Schwint I A 1967 *Ind. Eng. Chem.* **59** 58
[12] Post J E, Heaney P J. 2008 *Am. Mineral.* **93** 667
[13] Zhang Y Z, Zhang G C, Sun X F, He Z, Rajaratnam M 2010 *Cell Polym.* **29** 211
[14] Wang H, Wang F, Zheng K, Guo L, Chen L, Zhang X, Tian X 2015 *J. Macromol. Sci. Part A Pure Appl. Chem.* **52** 210
[15] Tong Y, Yuan Q, Huang F 2019 *J. Macromol. Sci. Part B Phys.* **58** 469
[16] Liu Y, Wang X, Wang Y, Tang Z, Makarenko T, Guloy A, Zhang Q, Ren Z 2016 *Appl. Phys. Lett.* **108** 101906
[17] Rheem M J, Jung H, Ha J, Baeck S H, Shim S E 2017 *Colloid Polym. Sci.* **295** 171

IOP Conf. Series: Materials Science and Engineering **735** (2020) 012056 doi:10.1088/1757-899X/735/1/012056

Assembling Construction Technology of Prefabricated Box Culvert Components

Yumei Kang[1], Naiyuan Zhang[1], Tao Wang[2] and Yuhang Song[3]

1 School of Resources and Civil Engineering, Northeastern University, No.3-11, Wenhua Road, Heping District, Shenyang, 110819, China

2 China Railway No.7 Bureau Group No.3 Engineering Co, Ltd., Shanxi, 710043, China

3 School of Architecture & Civil Engineering, Shenyang University of Technology, No.111, Shenliao West Road, Economic & Technological Development Zone Shenyang, 110870, China

Corresponding author's e-mail: 451512696@qq.com

Abstract. Box culverts are common small structures in highway engineering. According to domestic and foreign practical experience in recent years, component factory prefabrication combined with on-site assembly can effectively improve production efficiency and construction quality. It will become the main trend in the box culvert structure and gradually replace the traditional cast-in-place process. Based on the project of 03A design section of Daqing to Guangzhou Expressway, this paper expounds in detail the assembly process and key operation points of prefabricated components. It also compared with the traditional culvert construction technique, which demonstrates the superiority of prefabricated box culvert assembly technology.

1. Introduction

Box culverts are widely used on highways. Its main function is to set up a small drainage structure across the roadbed to vent the surface water flow. The flow of water passes under the road, thus avoiding the impact of accumulated water on the road and detracting from its life [1]. At present, prefabricated concrete box culverts have been widely used in highway engineering in many developed countries. However, prefabricated concrete box culverts are less used, so China's fabricated concrete structure system is still not sound enough [2]. Drawing on the advanced experience of foreign countries, we will gradually form a building industrial chain with Chinese characteristics.

Cast-in-place construction technology is often used in the traditional expressway box culvert. Although the construction is simple and the cost is low [3], with the influence of the age of the concrete, the material operation cycle and the external environment, the construction quality is difficult to guarantee, which brings many difficulties to later management [4]. In order to solve the shortcomings of the traditional construction process, the culvert construction will be developed towards standardization, factorization, and industrialization [5]. Nowadays, the culvert components are prefabricated in the factory and transported to the site for assembly. This not only improves the working efficiency but also effectively reduces the construction cost, which embodies the basic requirements of industrialization construction to the greatest extent [6].

This paper based on the project example of 03A design section of Daqing to Guangzhou Expressway, which researched the assembly and construction technology of prefabricated box culvert components, analyzed the main operational points of the assembly process and compared with the traditional culvert construction. The results show that the centralized operation of the factory has reached the goal of speeding up the project progress and improving labor efficiency.

2. Engineering Situation

The engineering project of the 03A design section of Shuangliao to Taonan section mainly passes through Shuangliao City, Changling County, Tongyu County and Taonan City. The starting point is located at the south side of Hara Road Fortress in Tongyu County, and the stake number is K125+000. The endpoint is located on the southeast side of Changling Village, Tongyu County, the stake number is K150+702.148, the total length is 25.702km, and there are 22 culverts. The project area is located in the Songliao alluvial plain. According to the regional survey geological data, there is no fault structure in the 03A design section, with the thick cover layer, the flat topography, buried depth of water level 1~20m, the better regional stability, and the distribution of saline soil in some paragraphs. The width of the subgrade is 27m, the bearing capacity is 100kPa~150kPa, and the main line of the expressway adopts the standard construction of a four-lane expressway with a design speed of 120km/h.

The project undertakes a total of 60 prefabricated construction tasks for prefabricated reinforced concrete box culverts, including 994 pieces of three structural forms, all of which are made of C40 concrete. Among them, XTI-2×1.5 is prefabricated for the entire section of the whole section, XTI-3×2.2 and XTI-4×3 are a four-component composite structure, which consists of four self-stabilizing members of the roof, floor and two side walls (figure 1). Sizes of prefabricated pipe joints are available in 3m and 1m. The roof of the box culvert is produced by the vertical type, the sidewalls are prefabricated by the inverted type and need to be turned over, the bottom of the floor needs to be pre-embedded, and the hoisting point is cut before assembling. When splicing, the upper roof and sidewall joint are elbow-shaped freely hinged, and then the sling reinforcement and concrete pouring are integrated into one. Taking the four-component composite structure as an example, this paper discusses the construction technology of precast production and assembly of box culvert components.

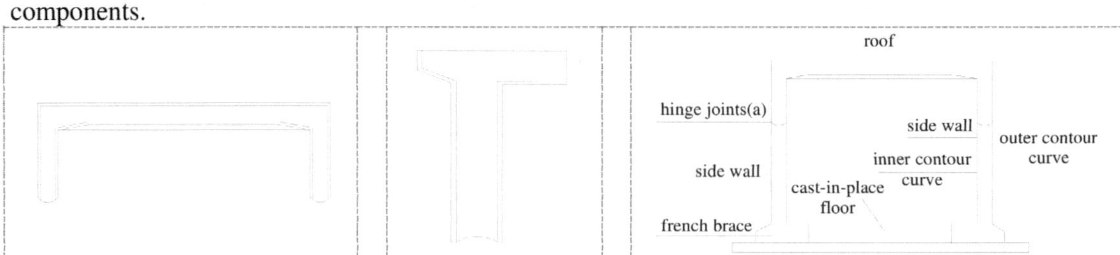

(a) (b) (c)

Figure 1. Four-component Composite Structure of Box Culvert

The design principle of the box culvert is based on the comprehensive consideration of the existing culvert structure problem, which changes traditional concrete on-site pouring to factory centralized prefabrication method to realize the specification factory construction and on-site rapid assembly construction [7]. The structure of the box culvert of this engineering is as follows: (1) The space utilization rate of the box-type culvert is large. In order to exert the force characteristics of the box culvert frame, the cross-center bending moment of the side wall and the top is utilized, which has high rigidity and good integrity. (2) Streamlined prefabricated models, with standard sections of 3m and 1m, further reduce the requirements for on-site lifting capacity and the difficulty of factory prefabrication while ensuring structural stress. (3) The requirements for foundation bearing capacity are low, and it can adapt to various geological conditions. The rectangular section type reduces the space requirements for surrounding embankments. (4) The joint between roof and side wall adopts the free articulated type, and the inner side of the joint head is filled with high-viscosity cement mortar, and the structure is as shown in figure 2(a). At the same time, in order to facilitate the integration of the longitudinal segments, chamfers are provided on the inner and outer sides of the longitudinal joint, and the structure is as shown in figure 2(b).

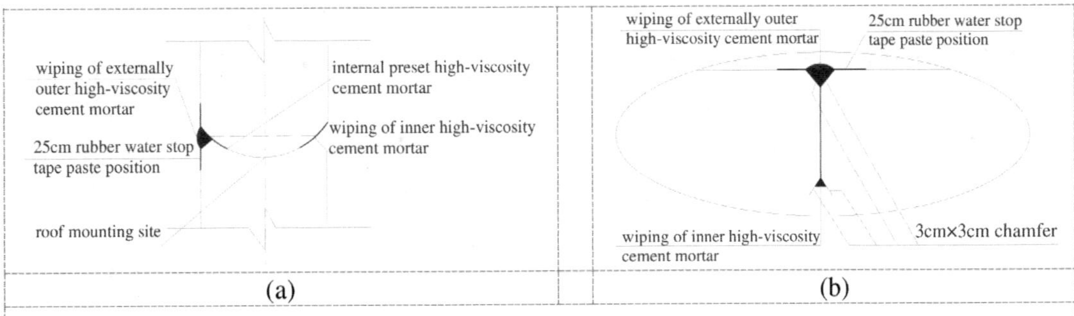

Figure 2. Schematic Diagram of the Construction of the Hinge Joint and the Loop Seam

3. Assembly Technology of Prefabricated Box Culvert

The assembling of the assembled box culvert components shall be carried out through construction preparation, excavation and cushion construction, component assembly, floor and hole cast-in-place, joint and waterproof treatment, backfilling behind abutment and other processes.

3.1. Construction Preparation

The construction preparations are as follows:

(1) Before the construction, comprehensive investigation and verification of culvert location, elevation and geology should be carried out, and geological exploration should be supplemented if necessary. If it is found to be inconsistent with the design, it should be reported to the supervision and design unit immediately and properly resolved as soon as possible.

(2) According to the design requirements, the construction access road is reasonably planned in combination with the terrain characteristics. The prefabricated site selection of the fabricated components is located in the same place as the mixing station and prefabricated beam yard in the S301 provincial highway. The distance between the temporary construction and the station facilities is less than 1km, and the transportation is convenient.

(3) Reasonable arrangement of component prefabrication area, maintenance area, and finished product stacking area is convenient for construction intensive management.

(4) Waterproof and drainage measures shall be taken before excavation of culverts, and construction personnel shall be reasonably arranged and equipped with relevant construction machinery to ensure the continuity of construction.

3.2. Foundation Pit Excavation and Cushion Construction

The foundation pit excavation is the key link of culvert construction. Before excavation, the geological data should be viewed, and the survey personnel should release the four corner piles of the open line of the foundation pit. After the sample is released, it will be reported to the measurement supervision engineer for review. When passing the test, the operator will be arranged to use the white ash to sprinkle the open line in time, and the excavator will enter the site to excavate the foundation pit. The excavation of the foundation pit is shown in figure 3. Excavation is carried out by using 1:1 slope rate for grading and excavation. The foundation pit is excavated according to the working width of the foundation pit width of each structure plus 50cm. Considering the construction of the rainy season and the existence of groundwater, a 50cm×50cm rectangular water collecting well is set at the corner of the foundation pit, the well depth is 60cm, and the water in the collecting well is 50~100m, the head is 2~42m, and the power is 0.75~7.5kW. The submersible sewage pump with a pumping capacity of 7~300m^3/h is collected into the drain and discharged outside the pit.

In order to prevent over-excavation and ensure that the slope is within the reasonable range, when mechanical excavation processed to the design base elevation of 30cm or slope boundary, manual excavation and slope repair should take place. When the foundation pit is excavated to the design elevation, manual leveling is carried out to verify whether the bearing capacity of the foundation meets the design requirements, and the gravel cushion is backfilled after passing the test. Layered and filled according to design and specification requirements, using crushed stone or sand gravel. The cushion layer needs to meet the design plane size and elevation, and the leveling and tamping treatment.

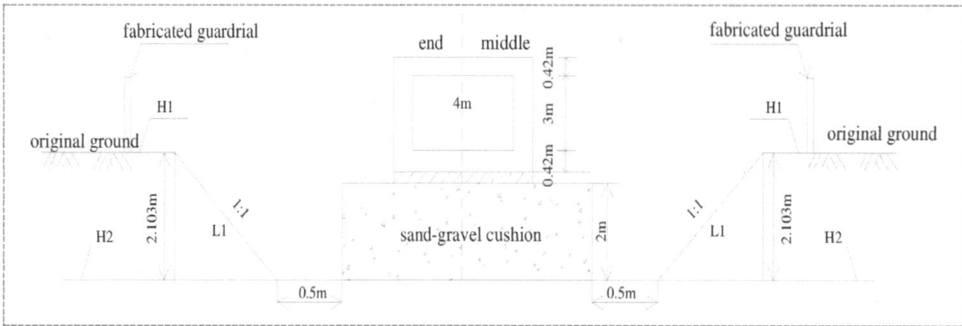

Figure 3. Schematic Diagram of Culvert Foundation Pit Excavation

3.3. Component Assembly

When the prefabricated components are processed in the factory, the concrete strength reaches 90% of the design strength before lifting and transporting. According to the on-site construction access road, two 20t rail flat transport trolleys were used to transport the box and culvert components to the site for assembly. The assembly site is shown in figure 4**.** At the scene, the 50t crane was used to hoist the prefabricated components in place. At the same time, 4 workers, 1 safety technician and 1 technician are arranged to meet the construction needs of the site. Pay attention to the following points when assembling components:

(1) Before assembling the components, the test should be carried out first, and the structural stability and positioning adjustment scheme of the system should be formulated. If necessary, support rods and wire ropes can be used to assist the stability.

(2) When assembling the main body, it should be carried out from the height to the ground, and the same component should be assembled from one end to the other to maintain the stability of the main body.

(3) When assembling the sidewall, ensure the tightness of the base and the cushion, and adjust the lateral position and relative distance at all times.

(4) Before the roof is hoisted, the workers check the position and relative distance of the two walls. Apply viscous M15 cement mortar to sidewall hinge groove to fill it by itself. Use a 3mm chamfer to smooth the cement mortar with an iron trowel.

(5) After sidewall and roof are assembled, the M25 bending bolts are used for connection and fixing, and the assembly should be carried out in an orderly manner according to the design requirements. Finally, check the overall state of the structure to ensure the accurate positioning of the components.

3.4. Floor and Hole Cast-In-Place

When all the components of the box culvert are assembled and accepted, the cast-in-place of floor and hole are made. The transverse steel bars of floor are connected by welding with HRB400 model. The welding quality is strictly implemented in the relevant technical provisions of Technical Specifications for Construction of Highway Bridges and Culverts (JTG/T F50-2011) [8]. Before the concrete is poured, the sidewall and floor joint surface shall be subjected to the brushing treatment, and the height shall be poured into the ground in order. Effective coverage during pouring prevents contamination of side walls. The orthogonal structure models XTI-2×1.5, XTI-3×2.2 and XTI-4×3 hole end segments shall be provided with a splayed wall, and the front view of the entrance hole is shown in figure 5. The eight-character wall and the foundation are cast-in-place, using M15 cement mortar, C30 concrete, 25cm rubber water stop. The four-component structure wing wall needs to be provided with anchor plates.

Figure 4. Box Culvert Roof and Side Wall Assembled Complete

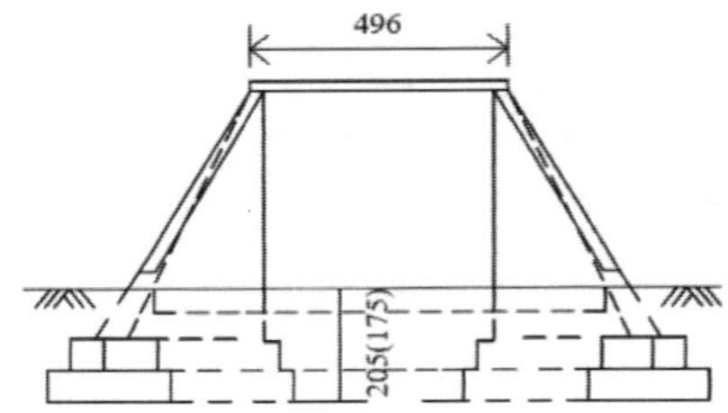

Figure 5. Front View of the Entrance Hole (unit: cm)

3.5. Joint and Waterproof Treatment

The 4cm×6cm groove of the culvert deformation joint is located above roof and floor, and the structure is shown in figure 6 the deformation joint is made of 2cm thick board coated with asphalt paste, and the side is filled with asphalt oakum, which is required to be filled compactness and meet the waterproof requirements [9]. Three layers of asphalt with a width of 50cm and three layers of asphalt and two layers linoleum are set on the outside. The contact part with backfill is coated with three layers of SBS modified emulsified asphalt. The waterproof asphalt is made of No.30 petroleum asphalt. Before the asphalt enters the field, the factory certificate should be inspected and sampled by the tester and sent to the central laboratory for retesting.

When the prefabricated section is waterproofed, the settlement joint and the connection joint shall be set in turn. The settlement joint range shall be set every 4m to 6m, and the inner side shall be attached with 1cm thick asphalt hemp as shown in figure 7. After the seam is fixed, the wire brush is used. The rag will be cleaned, and then coated with a 25cm wide 911 polyurethane waterproof coating. Finally, the special glue will be used to paste the 25cm wide back-mounted rubber water stop. The concrete base surface to which the water stop belt is attached shall be flat and free of cracks, and the allowable deviation of the flatness shall be 3mm. In order to ensure the tightness of the joint section of the joint, the high-strength steel nail may be fixed every 1m.

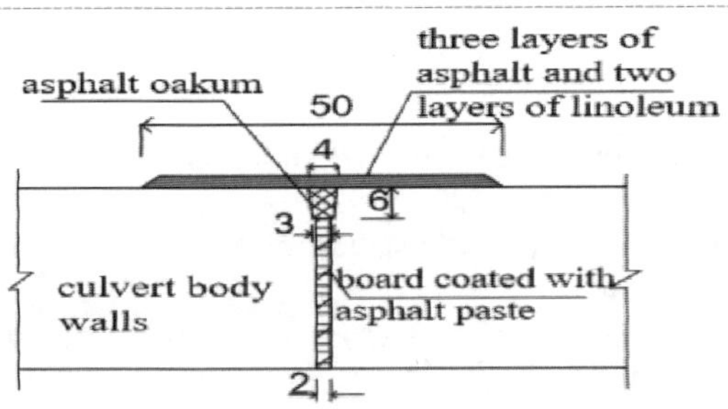

Figure 6. The Structure Diagram of the Deformation Joint of Culvert Body (unit: cm)

Figure 7. Construction of Asphalt Oakum

3.6. Back Filling

The backfilling of box culvert structure shall be carried out after the strength of cast-in-place concrete and mortar reaches 90% of the design strength. The filling shall be compacted in symmetrical layers. The quality of the filling on both sides and top of the box culvert shall be strictly guaranteed to ensure the restraint effect of the soil layer on the box culvert structure [10].

(1) When the culvert is backfilled, the improvement material of cement-soil with good performance should be selected for filling. The compaction thickness of each layer should be less than 20cm and the compaction degree should be less than 96%. When the structure support area is backfilled, the high-density sand can be used for water injection and compaction.

(2) When two side walls are backfilled, it shall be carried out by symmetrical and balanced horizontal layered rolling construction on both sides of the culvert. Manual and small-sized

mechanical compaction shall be used on the side within 1m, so that the backfilling can meet the design requirements of compactness.

(3) When backfilling on the structure roof, the thickness of less than 0.5m should be controlled to use compaction equipment. It is strictly forbidden to use heavy equipment and pile heavy objects, and the operation direction of compaction equipment should always be perpendicular to the structure axis.

3.7. Inspection and Acceptance

In the process of component assembly, the whole assembly process should be inspected and the following basic requirements should be met:

(1) Prefabricated box culvert components must pass the quality inspection of the system and enter the assembly site after passing the quality inspection. The site shall confirm and accept the dimensions, models and quantities of prefabricated components.

(2) During the whole process of component assembly, backfilling construction until completion, the section shape and structural position of the box culvert shall be inspected in three or more different positions.

(3) The construction shall be stopped if the inspection result exceeds the allowable deviation of component installation. The construction shall be continued after the reasons are identified and corresponding measures are taken to confirm the safety and stability of the structure.

4. Conclusion

(1) Relying on the engineering project of 03A design section of Shuangliao to Taonan highway, this paper systematically analyzes the process flow and operational points of assembling prefabricated box culverts components. In the whole assembling process of prefabricated components, the construction requirements of each procedure should be clarified, and accurate connection and coordination should be carried out to ensure assembly efficiency and accuracy.

(2) Compared with the traditional cast-in-place technology, the box culvert of this project adopts the factory centralized operation, which has higher construction efficiency, effectively guaranteed construction quality and enhanced structural durability. At the same time, it can reduce the maintenance cost of the later maintenance, and maximize the superiority of the prefabricated component assembly technology.

5. References

[1] Shi L G, Zhang Y and Li H L. 2017 Construction of precast utility tunnel with upper and lower assembly (Construction Technology) vol 46 18-21
[2] Yang X G. 2013 Design and construction key technology research of the precast reinforced concrete culverts (Central South University)
[3] Wang X X, Liang X L. 2016 Construction technology of factory centralized prefabricated culvert (Journal of Highway and Transportation Research and Development) vol 12 23-5
[4] Zhou W M. 2015 Construction technology of two-component fabricated reinforced concrete culvert (Railway Engineering) vol 02 36-8
[5] Lin Z H, Ma S. 2016 Prefabricated box culvert construction and principle research (Construction Technology) vol 45 329-31
[6] Guo R. 2014 Research on shallow buried soil super-long box culvert jacking key technology of highway (Chang'an University)
[7] Wu X F. 2012 Application of construction technology of fabricated large span arched concrete culvert (Bridge Engineering) vol 30 43-6
[8] CCCC First Highway Engineering Group Co., Ltd, Sichuan Road & Bridge Group Co., Ltd, Hunan Road & Bridge Construction Group Co., Ltd. 2011 (Technical Specification for Construction of Highway Bridge and Culvert)
[9] Xia K, Hu G and Hou Z G. 2012 Prefabricated integrated pipe gallery joint waterproof construction technology (China Concrete and Cement Products) 58-70
[10] Tong X K. 2012 Feasibility analysis of precast box culvert replacing cast-in-place box culvert (China Concrete) vol 38 58-70

Comparisons of Models for Predicting Permeability from Nuclear Magnetic Resonance (NMR) Logging Data in Heavy Oil-bearing Reservoirs

Wei Zhang[1], **Qi Li**[2]

[1] Shenzhen Operating Company of Well-Tech Department, China Oilfield Services Ltd. (COSL), room 4005, block A, CNOOC building, Chuangye Road, Nanshan District, Shenzhen, Guangdong, China
[2] The first oil production plant of petrochina changqing oilfield company, Baota Road, Yan'an, Shaanxi, China

Email: zhangwei53@cosl.com.cn, lqi1_cq@petrochina.com.cn

Abstract. Permeability prediction always faces great challenge, because it cannot be directly measured from any well logging data. The statistical methods had been widely used. The accuracy was too low to be used. The nuclear magnetic resonance (NMR) logging data was considered to be effective in permeability prediction, and many models had been proposed. However, the NMR logging data was heavily affected by the viscosity in heavy oil-bearing formations. The reliability of the permeability estimation model in heavy oil-bearing reservoirs needed to be checked to make them much valuable in field application. In this study, 24 core samples, which were drilled from the heavy oil-bearing reservoirs of the Zhujiangkou Basin in South China Sea, were applied for NMR and mercury injection capillary pressure (MICP) experimental measurements. The classical permeability prediction models of Timur-Coates model and Schlumberger Doll Research (SDR) model were calibrated. Meanwhile, the permeability prediction model based on the Swanson parameter was also established. These three models were applied in field applications to verify the reliability. Comparisons of the predicted permeabilities by using these three models with the core derived results illustrated that the Swanson based permeability model was widely used. Whereas the permeabilities estimated by using the Timur-Coates and SDR models were underestimated. The Swanson based permeability model was given priority to be recommended.

1. Introduction

Permeability is of great importance in formation evaluation, deliverability prediction and validity analysis in any types of reservoirs [1]. Good quality reservoirs always contain high permeability and deliverability. Petrophysicists and geologists tried to predict precise permeability to improve formation evaluation. However, different from the porosity, the permeability cannot be directly estimated from any well logging data [2-3]. Some statistical models were established based on core experimental data to predict permeability from other parameters, such as the porosity, the shaly content and the conventional well logging data [3-4]. These models were empirically established, and were available in specific regions. They cannot be widely used. This made the permeability prediction faced great challenge.

The NMR logging data was considered to be valuable in permeability prediction [5-7]. From the NMR logging data, many available parameters can be acquired, such as the total porosity, the effective porosity, the pore size distribution and the permeability [8-9]. To acquire the permeability from NMR logging data, two classical models were established. These two models were named as the Timur-Coates model and SDR model, and were expressed as follows [10-11]:

$$K = \left(\frac{\varphi}{C}\right)^m \left(\frac{1 - S_{wi}}{S_{wi}}\right)^n \tag{1}$$

$$K = C_1 T_{2lm}^{m_1} \varphi^{n_1} \tag{2}$$

Where, K is the permeability in mD; φ is the porosity in %; T_{2lm} is the NMR T_2 geometric mean in ms; S_{wi} is the irreducible water saturation in fraction. C, m, n, C_1, m_1 and n_1 are the involved parameters in the models. Their values need to be first calibrated by using the NMR experimental data. Once no NMR experimental data was available, their values were defined as 10, 4, 2,10, 4 and 2, separately.

The parameters of φ and T_{2lm} can be directly acquired from the NMR data, the S_{wi} can also be accurately predicted from NMR data once the reasonable $T_{2cutoff}$ is determined [12]. Hence, the permeability can be predicted from NMR data once the values of the involved input parameters were firstly calibrated.

2. Establishment of the Classical Models to Predict Permeability from NMR Data

2.1. Laboratory NMR Experiments

Although the default parameters in equations 1 and 2 can be fixed determined, to make the predicted permeability was much closed to the true value in our target heavy oil-bearing formation in the Zhujiangkou Basin in South China Sea, 24 typical core samples were drilled from the target formations to apply laboratory NMR and MICP experimental measurements. The physical properties of the core samples and the corresponding experimental results were listed in table 1.

IOP Conf. Series: Materials Science and Engineering **735** (2020) 012057 doi:10.1088/1757-899X/735/1/012057

Table 1. The Physical Properties of the Core Samples and the Experimental Results

Core No.	Depth	Porosity	Permeability	Swi	T_{2lm}
	m	v/v	mD	%	ms
1	1333.35	0.23	19.15	66.23	12.19
2	1336.80	0.25	65.49	66.98	12.99
3	1339.35	0.20	17.06	85.95	9.90
4	1354.60	0.25	344.52	97.41	14.87
5	1357.60	0.36	932.71	22.57	24.00
6	1359.10	0.27	68.97	52.57	17.63
7	1363.10	0.27	86.76	52.36	20.79
8	1374.10	0.23	37.44	88.39	13.13
9	1378.35	0.33	457.13	15.90	26.51
10	1385.60	0.30	298.08	29.84	24.57
11	1387.10	0.28	100.99	29.84	27.71
12	1388.85	0.25	87.09	30.93	32.05
13	1393.60	0.21	15.73	61.61	17.18
14	1396.60	0.23	19.63	59.97	22.82
15	1429.90	0.28	2588.71	13.86	21.88
16	1434.35	0.27	154.12	55.13	15.09
17	1436.35	0.33	581.27	35.85	32.89
18	1439.25	0.25	90.88	99.89	11.20
19	1462.10	0.23	24.20	63.93	12.57
20	1467.10	0.22	48.63	67.78	15.89
21	1474.66	0.24	76.77	72.07	44.12
22	1476.60	0.31	573.63	53.99	29.76
23	1479.15	0.26	262.57	95.32	30.95
24	1484.35	0.27	291.69	39.33	42.32

2.2. Calibration of the Involved Model Parameters

By using the NMR experimental results, equations 1 and 2 were calibrated, the involved parameters of C, m, n, C_1, m_1 and n_1 were determined, and these two models were expressed as follows:

$$K=\left(\frac{\phi}{14.823}\right)^{8.395}\times\left(\frac{1-S_{wi}}{S_{wi}}\right)^{-0.064} \tag{3}$$

$$K=1389952\times\left(\frac{\phi}{100}\right)^{7.577}\times T_{2lm}{}^{0.271} \tag{4}$$

Equations 3 and 4 illustrated that the involved input parameters in the classical permeability prediction models in our target formations deviated the empirical values, especially the value of C_1, it was heavily different from the classical values. In addition, the NMR experimental results illustrated that the permeability was negative associated to the ratio of the free fluid volume and bound water content (the value of n lower than 0.0). This might cause by the experimental errors. In the NMR experiment, high centrifugal rotational speed of 9000 round/min was used. The pore structure might be broken, and some free water was through away, this made the determined irreducible water lower than the true value.

2.3. Determination of the Involved Input Parameters

To use equations 3 and 4 to predict permeability from NMR logging data in field applications. The values of φ, T_{2lm} and S_{wi} should be first determined. Although the NMR data can directly offer porosity and T_{2lm}, the results were heavily affected by the viscosity of occupied heavy oil in the pore space, this made the directly extracted porosity lower than true values (figure 1). In this figure, it can be identified that the porosities directly estimated from the NMR logging (TCMR) were lower than the core derived results (PORC) and predicted porosities from conventional well logging (PHIT) in formations with resistivity higher than 1.2 Ω.m. In the other water saturated intervals, these three types of porosities were coincided well with each other. Meanwhile, the S_{wi} prediction is also a difficulty at present. The fixed $T_{2cutoff}$ value was commonly used, and the inaccurate results were acquired [12].

To remove the effect of saturated heavy oil to the parameters acquired from NMR logging data, a proposed technique was used [13]. Based on the proposed technique, the NMR T_2 spectra with oil saturated were corrected as fully water saturated condition. From the corrected NMR T_2 distributions, NMR porosity and T_{2lm} were acquired. Meanwhile, technique of predicting various $T_{2cutoff}$ from NMR T_2 distribution based on the morphological difference was also used in this research to calculate reasonable S_{wi} for permeability prediction [14].

3. Permeability Prediction Based on Formation Pore Structure Characterization

3.1. The Mercury Injection Capillary Pressure (MICP) Curve

Generally, the MICP curve is always displayed in semi-log coordinates, mercury injection saturation (S_{Hg}) is displayed in linear coordinate in X-axis, and mercury injection pressure (P_c) is displayed in logarithmic coordinate in Y-axis [11]. If the MICP curve is displayed in log-log coordinates, it displays as a hyperbolic curve [14]. This hyperbolic curve can be expressed by equation 5:

$$\log_{10}\left(\frac{P_c}{P_d}\right)\times\log_{10}\left(\frac{S_{Hg}}{S_{Hg\infty}}\right)=C \tag{5}$$

Where, P_d is the threshold pressure in MPa; $S_{Hg\infty}$ is the non-wetting phase saturation under infinite mercury injection pressure in % and C is the geometric factor.

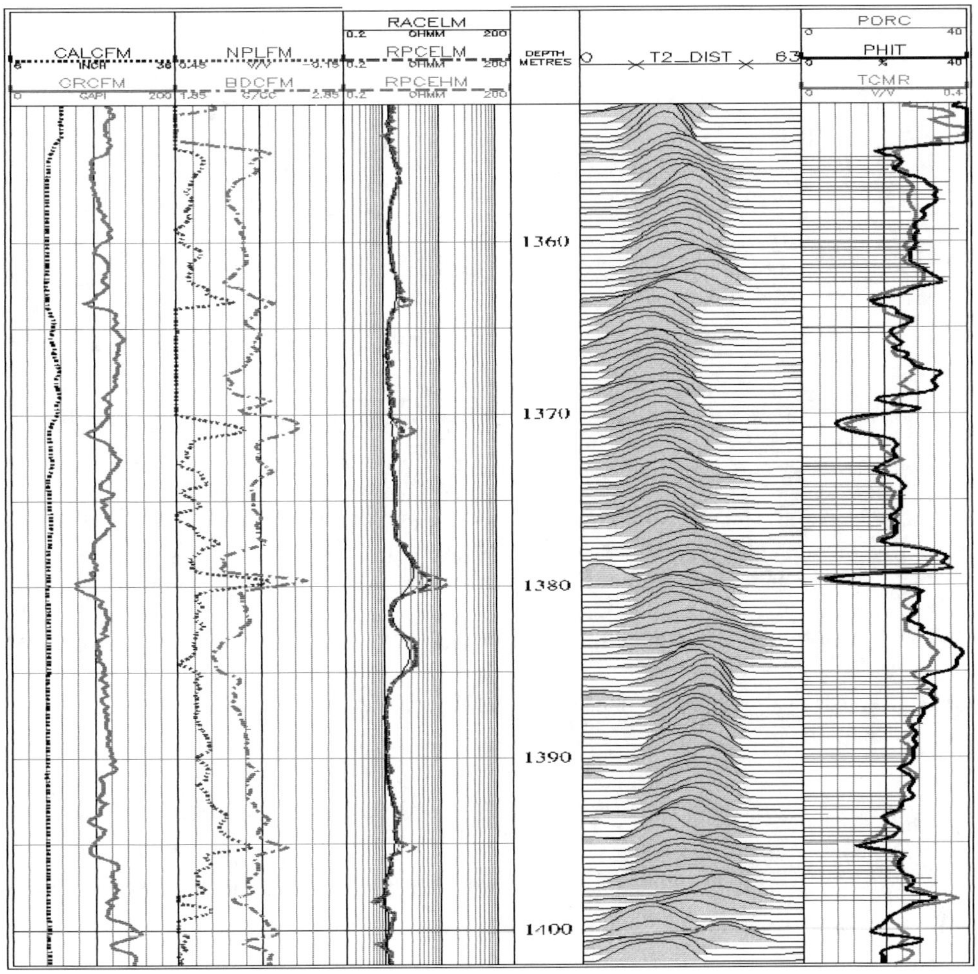

Figure 1. Effect of Saturated Heavy Oil to the NMR T_2 Spectra and NMR Porosity

Based on the analysis of the MICP curves acquired from different types of core samples drilled from common formations, low permeability and tight sandstones, separately, many researchers found that the inflection points of capillary pressure curves were well associated with air permeabilities [14-15]. The physical significance of the inflection point is the mercury injection saturation threshold in the main pore system which primarily controls the fluid flow. If S_{Hg} is displayed in X-axis and the ratio of S_{Hg} and P_c is in Y-axis, the inflection point is located at the apex (figure 2), it was called as the Swanson parameter, and expressed as $(\frac{S_{Hg}}{P_c})_{max}$. the relationship between the Swanson parameter and permeability can be expressed as follows:

$$K = A \times (\frac{S_{Hg}}{P_c})^{B}_{max} \quad (6)$$

Where, $(\frac{S_{Hg}}{P_c})_{max}$ is the Swanson parameter; A and B are the statistical model parameters. For different types of reservoirs, the values of A and B are various, and their values can be calibrated by using the data sets of mercury injection experiment of core samples.

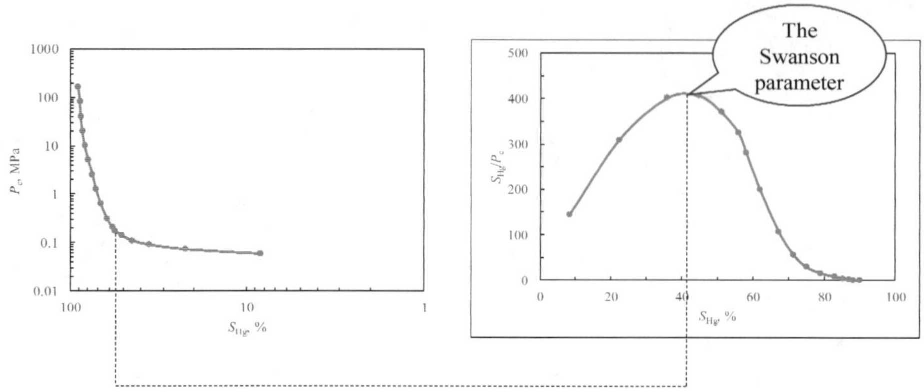

Figure 2. Diagram of Obtaining the Swanson Parameter from MICP Data

3.2. Establishment of Permeability Prediction Based on the Swanson Parameter

Based on above analysis, the MICP curves of 24 core samples in our target formations were reused, the Swanson parameter of every core sample was extracted. We tried to establish the relationship between the Swanson parameter and rock physical properties. Finally, we found that the Swanson parameter was heavily associated with the rock physical parameter of $\sqrt{\frac{K}{\varphi}}$. Hence, we changed the expression of equation 6, and established a permeability prediction model displayed in figure 3 and expressed as equation 7.

$$\sqrt{\frac{K}{\varphi}} = 0.0089\times(\frac{S_{\text{Hg}}}{P_{\text{c}}})_{\max} + 0.1029, \qquad R^2=0.928 \tag{7}$$

Figure 3 illustrated that the Swanson parameter was heavily associated with the rock physical parameter. Based on this equation, once the Swanson parameter and the porosity were first predicted, the permeability can be accurately determined.

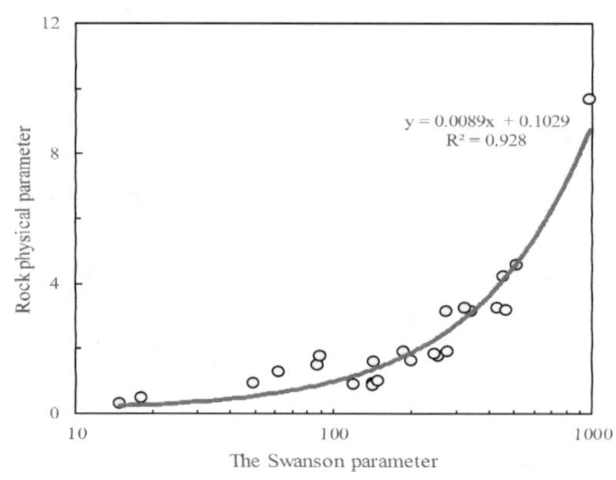

Figure 3. The Relationship between the Swanson Parameter Versus Rock Physical Parameter

4. Comparison of Permeability Prediction Models in Heavy Oil-bearing Reservoirs

To verify the reliability of these three permeability prediction models, we applied them in field applications to process the acquired NMR logging data. The proposed techniques were first used to correct the effect of heavy oil to NMR T_2 spectra [13-14]. Meanwhile, to use equation 7 to consecutively predict permeability, the method of constructing pseudo capillary pressure curves from

NMR logging data was also used [17]. The consecutive pseudo capillary pressure curves and the Swanson parameters were acquired in the intervals with which the NMR logging data was acquired. The permeability estimation models expressed as equations 3, 4 and 7 were used to predict permeabilities. We compared these predicted permeabilities with those of the core derived results, and displayed in figure 4. In the fourth track of figure 4, we compared the corrected porosity (PHIT) from NMR logging data with the core derived porosities (CPOR), good consistency of these two porosities meant well correction of the NMR data. T2_DIST displayed in the fifth track was the corrected NMR T_2 distribution. In the sixth track, the extracted irreducible water saturations (SWICAL) were compared with the core derived results (CSWI). Good consistency of calculated porosity and Swi with the core derived results meant that not any errors would be introduced in permeability prediction. In the last three tracks, we compared the predicted permeabilities from the established models with the core derived results. The results illustrated that the predicted permeability by using the Swanson parameter based model (PERMSWAN) were coincided with the core derived results (CPERM) very well. The estimated permeabilities by using the SDR model (KSDR) and Timur-Coates model (KTIM) were also coincided with the core derived results very well in the majority of intervals. However, in heavy oil-bearing formations (resistivity higher than 1.2 Ω.m), the estimated permeabilities were underestimated; whereas the Swanson parameter based on model was well used.

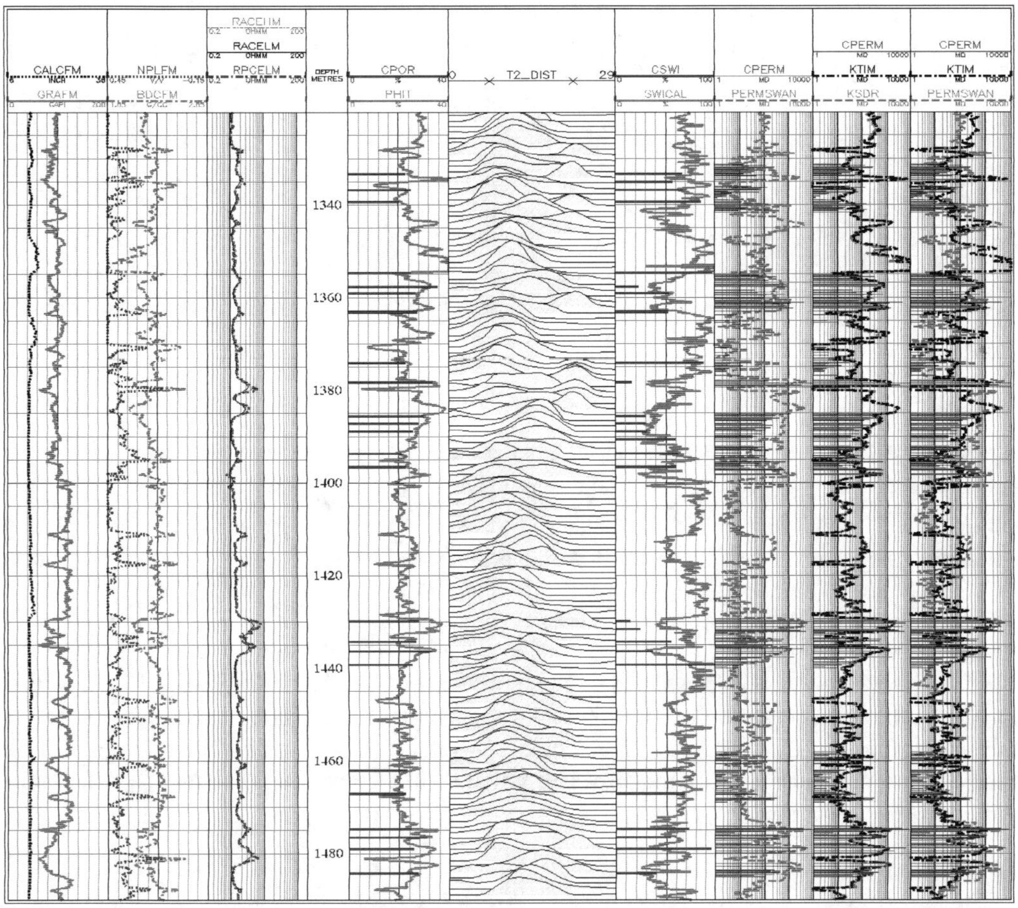

Figure 4. Comparisons of Predicted Permeabilities from Three Models by Using the NMR Logging Data with the Core Derived Results

To verify the wide application of the permeability prediction models, we extended them into the adjacent regions to predict permeability. Figure 5 displayed the comparisons of the predicted permeabilities with the core derived results. This comparison illustrated that the permeability predicted by using the Swanson based model was much reasonable. The Swanson based permeability model was much valuable than those of the SDR and Timur-Coates models. Hence, in our target heavy oil-bearing formation, the Swanson based permeability model was given priority to recommend, it can be well used.

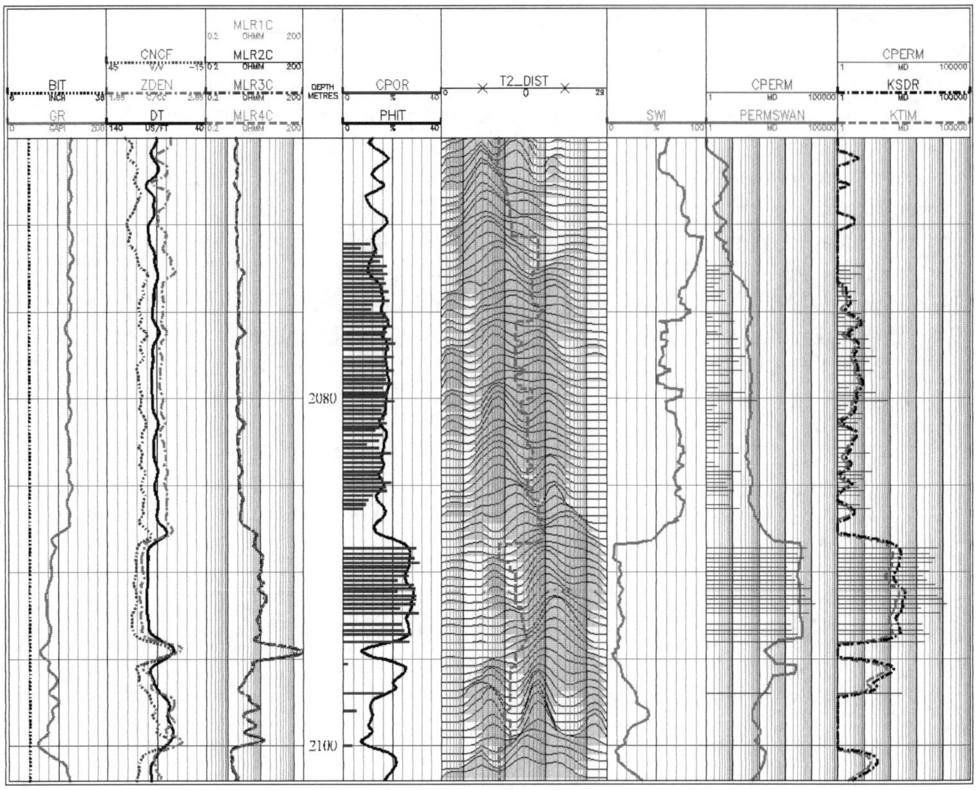

Figure 5. The Reliability Analysis and Comparisons of the Permeability Estimated Models

5. Conclusions

Permeability is an important input parameter in formation estimation and deliverability prediction, not any conventional well logging data can be directly used to calculate permeability. The NMR logging data was advantage in permeability prediction. The SDR and Timur-Coates models were two classical NMR based permeability models. The Swanson based model can be well used to predict permeability after formation pore structure was first characterized.

Based on the laboratory NMR and MICP experiments of 24 core samples, the classical SDR and Timur-Coates models were calibrated, and an improved Swanson parameter based model was also established. These models were applied in field applications in the target and adjacent regions to predict permeabilities. Comparisons of predicted permeabilities with core derived results illustrated that the Swanson parameter based model was optimal. It can be well used to predict permeability in our target heavy oil-bearing formation.

6. Reference

[1] Yuan X T, Peng S M and Lin C Y 2015 An interpretation method for permeability based on flow units and its applicability *Acta Petrol Sin* **26(6)** pp 78-82

[2] Zhang J C and Song K P 2007 Eigen curve of relative permeability and its application *Acta Petrol Sin* **28(4)** pp 104-107

[3] Zhao J, Hui Y A and Wang P 2007 Permeability synthesis method for heterogeneous reservoirs *Acta Petrol Sin* **28(2)** pp 102-104

[4] Manika P 2003 Velocity-permeability relations within hydraulic unit *Geophysics* **68(1)** pp 108-117
[5] Kenyon W E 1997 Petrophysical principles of applications of NMR logging *The Log Analyst* **38(2)** pp 21-43
[6] Coates G R, Xiao L Z and Primmer M G 2000 NMR Logging Principles and Applications. Houston: Gulf Publishing Company, Houston, pp 42-78
[7] Dunn K J, Bergman D J and Latorraca G A 2002 Nuclear magnetic resonance: petrophysical and logging applications. Handbook of Geophysical Exploration. Pergamon, New York, pp 1-176
[8] Liu Z H, Zhou C C and Liu G Q 2007 An innovative method to evaluate formation pore structure using NMR logging data *48th SPWLA Logging Sym Tran* paper S
[9] Timur 1972 Nuclear magnetic resonance study of carbonate rocks *The Log Analyst* **13(5)** pp 3-11
[10] Mao Z Q, Xiao L, Wang Z N, Jin Y, Liu X G and Xie B 2013 Estimation of permeability by integrating nuclear magnetic resonance (NMR) logs with mercury injection capillary pressure (MICP) data in tight gas sands *Appl Magn Reson* **44(4)** pp 449-468
[11] Xiao L, Liu X P, Zou C C, Hu X X, Mao Z Q, Shi Y J, Guo H P and Li G R 2014 Comparative Study of Models for Predicting Permeability from Nuclear Magnetic Resonance (NMR) Logs in Two Chinese Tight Sandstone Reservoirs *Acta Geophys* **62(1)** pp 116-141
[12] Xiao L, Mao Z Q and Jin Y 2012 Calculation of Irreducible Water Saturation (Swirr) from NMR Logs in Tight Gas Sands *Appl Magn Reson* **42(1)** pp 113-125
[13] Xiao L, Mao Z Q, Li J R and Yu H Y 2008 Effect of hydrocarbon on evaluating formation pore structure using nuclear magnetic resonance (NMR) logging *Fuel* **216** pp 199-207
[14] Xiao L, Li J R, Mao Z Q, Lu J, Yu H Y, Guo H P and Li G R 2018 A method to determine nuclear magnetic resonance (NMR) T2cutoff based on normal distribution simulation in tight sandstone reservoirs *Fuel* **225** pp 472-482
[15] Huet C H, Rushing J A, Newsham K E and Blasingame T A 2007 Estimating klinkenberg-corrected permeability from mercury-injection capillary pressure: A new semianalytical model for tight gas sands, *SPE102890*
[16] Swanson B F 1981 A simple correlation between permeabilities and mercury capillary pressure *J Petrol Tech* **6** pp 2498-2503
[17] Xiao L, Mao Z Q, Zou C C, Jin Y and Zhu J C 2016 A new methodology of constructing pseudo capillary pressure (Pc) curves from nuclear magnetic resonance (NMR)logs *J Petrol Sci Eng* **147** pp154-167.

IOP Conf. Series: Materials Science and Engineering **735** (2020) 012058 doi:10.1088/1757-899X/735/1/012058

Uncertainty Analysis for the Estimated Hydraulic Conductivity from the Reduced Duration Pumping Test

Ying Zhang[1], Chengpeng Lu[2,*], Rui Zeng[3], Liyong Jiang[2] and Lei Wei[2]

1 Water Resources Planning Bureau of Jiangsu Province, Nanjing 210029, China
2 College of Hydrology and Water Resources, Hohai University, Nanjing 210098, China
3 Lower Changjiang River Bureau of Hydrological and Resources Survey, Changjiang Water Resources Comission, Nanjing 210011, China
Email: chplu@foxmail.com

Abstract. Use of a homogeneous aquifer method to represent a heterogeneous aquifer for parameter estimation will inevitably result in uncertainties in the estimated parameter values. The objective of this paper is to evaluate the impacts of the reduced duration of the pumping test on the estimated hydraulic conductivity using the observations from pumping tests. Firstly, a heterogeneous hydraulic conductivity *K* field was generated using the Monte Carlo method. A MODFLOW groundwater flow model was then constructed to perform numerical pumping tests. The *K* value was estimated by the inverse method based on the Theis solution for an unconfined aquifer. The *K* fields were generated based on the hydraulic conductivity data sets collected for a karst aquifer in southwest of China. The effective hydraulic conductivity (*Ke*) was calculated to represent the hydraulic conductivity of a heterogeneous aquifer. For a single test in a heterogeneous aquifer, while the duration of the test is less than a critical value, the reduced duration will lead to the big uncertainty of the estimated hydraulic conductivity. However, *Ke* can achieve a relative stable value while the number of reduced duration tests is big enough to eliminate the impacts of the uncertainty of the separate *K* values.

1. Introduction

The study on the heterogeneity of an aquifer is important to the analysis of groundwater flow and contaminant transport. The heterogeneity of an aquifer can include the spatial variation of hydraulic properties, such as hydraulic conductivity (*K*), specific yield (S_y), and coefficient of dispersity. Moench et al. (2001) noted that the differences (errors) between simulated and measured drawdowns in groundwater models are caused by local variations of hydraulic properties, primarily in *K*. Sudicky (1986) also reported that *K* is a key factor in controlling groundwater flow and solute transport, and it can vary significantly over short distances.

In the last century, the methods to determine *K* for a homogenous aquifer were developed from the point scale to regional scale, such as the slug test, permeability test, pumping test and inverse modeling. Therefore, numerous approaches were proposed to determine the *K* values of a homogenous aquifer. However, for a heterogeneous aquifer, the spatial heterogeneity results in difficulty in the determination the *K* values accurately under different scales. In the field studies on the estimation of *K* values of a heterogeneous aquifer, many researchers focused on the scale effects of *K* (Sudicky 1986; Illman 2006) and the methods of upscaling *K* values from a small scale (Felletti et al. 2006). In most of these studies, the actual heterogeneous aquifers were assumed to be homogenous aquifers, therefore, when the approaches derived from a homogenous aquifer are applied to a heterogeneous aquifer, they might lead to large uncertainties in the *K* values.

Devlin and McElwee (2007) used numerical modeling to study the effects of measurement error on the estimations of horizontal hydraulic gradient, which led to more attention to the uncertainty analysis of field tests and measurements. In practice, some pumping tests may be forced to stop due to the short depth of pumping well, the energy disruption, and other unpredictable factors, especially in a karst aquifer with strong heterogeneity, which might result in fewer observations with reduced duration of pumping test. The reduced duration tests (RDTs) induce a concern whether the fewer observations may significantly affect the estimations of hydraulic conductivity of a heterogeneous aquifer. The impacts of the reduced duration of test on the estimates of hydraulic properties were studied by Moench et al. (2001), and they reported that results for an 8-hour test were as valid as the results for a 72-hour test, which was conducted in Cape Cod, Massachusetts. In our study, the impact of the reduced duration of pumping tests on the estimated hydraulic conductivity will be further studied in several hypothetic aquifers with different heterogeneities.

In general, the understanding of heterogeneity is generally based on numerous field investigations, such as pumping test, slug test, ground tomography and other geophysical works; however, it is impossible to perform field investigations at each test site in a small scale to analyze the hydraulic properties of an aquifer. Due to the lack of data and sampling biases of field tests, numerical simulations and laboratory tests are implemented to achieve some insights of nature, which easily satisfy the specific statistical assumptions (Zhang et al. 2007). In comparison with the lab tests and numerical simulations, for the known physics the numerical simulation is more convenient to conduct than the lab test (Zhang et al. 2007; Beckwith et al. 2003). On the other hand, pumping test was always used to yield reasonable estimates of the average hydraulic conductivity of an aquifer (Butler and Healey 1998; Moench et al. 2001). This study focused on parameter estimation from pumping tests, and the drawdown datasets were primarily produced by the numerical pumping tests.

2. Materials and Methods

2.1. Aquifer Tests and the Solutions

Three pumping-recovery tests, nine slug tests and nine permeability tests (three constant-head tests and six falling-head tests) were conducted in an unconfined karst aquifer within the Houzhai basin from August 13th to 21st, 2008. The Houzhai basin is near the town of Puding, southwest of China. Drawdowns from the pumping and recovery periods were recorded in every ten minutes at the beginning of the tests. At the end of the pumping-recovery tests, the changes of drawdowns were minimal so that the groundwater flow in the aquifer was regarded as steady-state. The Dupuit solution for an unconfined aquifer was used to calculate the *K* value under the steady-state conditions. The drawdowns collected at each observation well in the pumping-recovery tests were used to estimate the hydraulic parameters using the Moench solution (Lu et al. 2011). Totally 13 *K* estimates were obtained using the Dupuit solution and the Moench solution based on the pumping-recovery tests. Both of the constant-head and falling-head tests were analyzed by Darcy's Law, and the slug tests were evaluated by the solutions of Hvorslev (1951) and Bouwer and Rice (1976). In all, 9 estimates and 18 estimates of *K* were obtained by the permeability tests and slug tests, respectively.

2.2. Numerical Simulations for the Pumping Test in a Heterogeneity Aquifer

2.2.1. Generation of the Random K Fields. Zhang et al. (2007) noted that the natural deposits often exhibit long range correlation. Webb and Anderson (1996) demonstrated that the pure random field with assumptions of spatial correlation and spatial trends might result in unrealistically smooth distribution of *K*. However, in a karst area, the spatial correlation exists slightly because of the strong heterogeneity of aquifer. Therefore, the Monte Carlo simulation is more feasible than the Gaussian sequential simulation in generating the random *K* fields in the study.

In order to study the impacts of the variability of an aquifer on the estimated *K*, the *K* field was given randomly according to the specific *EK* and *Cv*. The generation of random *K* value in this study followed the steps of the Monte Carlo method. The pseudorandom uniform numbers were produced by the algorithm described by Lehmer (1951), and the pseudorandom normal deviates were transformed

IOP Conf. Series: Materials Science and Engineering **735** (2020) 012058 doi:10.1088/1757-899X/735/1/012058

from the uniform numbers using Box–Muller (1958) method. This random number generator has passed all the standard tests for random number generators.

2.2.2. Forward Modelling. The hypothetical pumping tests were conducted in heterogeneous aquifers generated based on the mean and deviation of *K* values from actual aquifer tests. MODFLOW was used to simulate the hydraulic heads in response to groundwater pumping. The simulated aquifer was unconfined, and the dimension of the model was 2500 m in length, 2500 m in width and 50 m in depth. The model area was discretized into 100 columns and 100 rows in the x and y directions, respectively. Since the determination of the vertical *K* was not the focus of this study, the model had only one layer of a depth of 50 m, thereby inducing a two-dimensional groundwater flow system. A pumping well was placed at the center of the model. The pumping test lasted for 24 hours with a constant pumping rate of 1728 m^3/d, which is the actual rate of a pumping test conducted near the town of Laoheitan. In all, the pumping duration was divided into 10 stress periods and 15 time steps in each stress period with a 1.1 multiplier used within each stress period. Therefore, 150 drawdowns were recorded at each of the two observation wells. In the study of the heterogeneous aquifer, two observation wells, labeled as Wmin and Wmax, were placed at different directions of the pumping well. Well Wmin was located in the short axis of the influence since pumping, and well Wmax was located in the long axis of the influence.

2.2.3. Estimation of the Hydraulic Conductivity. In this study, the Theis solution for the unconfined aquifer, labeled as Quasi-Theis, thereafter (Chen 2001), was used to calculate the hydraulic heads of the simulated aquifer, which is expressed as $h^2 = h_0^2 - \frac{Q}{2\pi K}W(u)$. Here, h_0 is the hydraulic head prior to pumping, *h* is the hydraulic head since pumping begins, and *W*(*u*) is the well function. Using of the Theis solution for this study is to document the difference between the derived *K* value from the simulated drawdown datasets and the actual *K* values built into the test heterogeneous aquifers. The gradient method reported by Chen and Ayers (1997) was regarded as the inverse method to estimate the *K* of the hypothetic aquifers, which was to minimize the square errors between the observed hydraulic heads from the forward modeling and the calculated hydraulic heads using the Quasi-Theis solution.

The effective hydraulic conductivity, labeled as *Ke*, was generally regarded as the hydraulic conductivity of the heterogeneous aquifer (Renard and de Marsily 1997), and *Ke* can be calculated from the discrete *K* values obtained from the previous parameter estimations. The equations proposed by Gutjahr et al. (1978) were adopted in this study because they are simple to conduct without considering the boundary conditions. *Ke* was calculated as follow, respectively,

$$K_e = K_g[1 + (\frac{1}{2} - \frac{1}{n})\sigma_Y^2],\ K_g = \left(\prod_{i=1}^{N} K_i\right)^{1/N} \tag{1}$$

Where σ_Y^2 the variance of LnK, Kg is is the geometric mean of *K* values, *n* is the dimensionality of media, *N* is the number of the *K* values. Since the numerical modeling was two-dimensional in the study, *Ke* is equal to *Kg*.

Ke represent the estimated hydraulic conductivity of the heterogeneous aquifer. *EK* and *Cv* for the *K* of model grids represent the average *K* and the variability of the hypothetical aquifer, respectively. In the numerical simulations, the *EK* values were set to be 10, 50 and 100 m/d, respectively, and the *Cv* values of given *K* at each mean value were set to be 0.1 (small variance), 0.5 (medium variance) and 1 (large variance). The *K* values of the girds in the model were assigned from the ten thousand *K* values. Under the condition of a set of the assigned *EK* and *Cv* values, one hundred realizations were run to obtain an estimate of *Ke* for an observation well. The impacts of the aquifer heterogeneity were assessed by generation of the heterogeneous *K* fields, running the numerical pumping tests in these heterogeneous aquifers and performing inverse modeling for *K* estimation.

3. Results and Discussion

3.1. The Estimated K Values in the Karst Aquifer

Figure 3 shows the wide range of the estimated *K* values from 10^{-7} to 10^{3} m/d in the karst aquifer located in Houzhai basin. The estimated *K* values range from 26.74 to 469 m/d, 6.39×10^{-5} to 20.7 m/d, and 6.67×10^{-7} to 32.5 m/d in terms of the pumping tests, permeability tests, and slug tests, respectively. The results show that the estimated *K* values from the pumping tests are greater than those from the permeability tests and slug tests, which may be attributed to the fact that the scales of pumping tests are larger than those of other aquifer tests (Rovey 1998). In the meanwhile, the *Cv* of the estimated *K* values from the pumping tests are lower than those from other tests, which indicates that pumping tests may yield a higher *K* value with less variability.

Scale effect of *K* means that *K* values vary with the data support (Zhang et al. 2007). The estimated *K* values from the pumping-recovery tests, permeability tests and slug tests were plotted versus the support volume (Schulze-Makuch et al. 1999) and the nominal length (Illman 2006) to evaluate the scale effect and characterize the variability of the *K* values. For the pumping tests, the pumpage in the pumping test and the radius of influence were regarded as the support volume and the nominal length, respectively. For the permeability tests and slug tests, the volume of the injected water and the depth of the test hole were considered as the support volume and the nominal length, respectively. Butler and Healey (1998) suggested that the fact of different *K* values with different measurement scales cannot be attributed to the scale effect without the examination of hydraulic tests. Here, when the support volume is regarded as the scale of measurement, the observed *K* values have an increasing trend with the support volume (Figure 1a). On the other hand, due to the lack of field investigations in the karst aquifer, when the nominal length is regarded as the scale of measurement, there is no increasing trend of *K* values with the nominal length (Figure 1b).

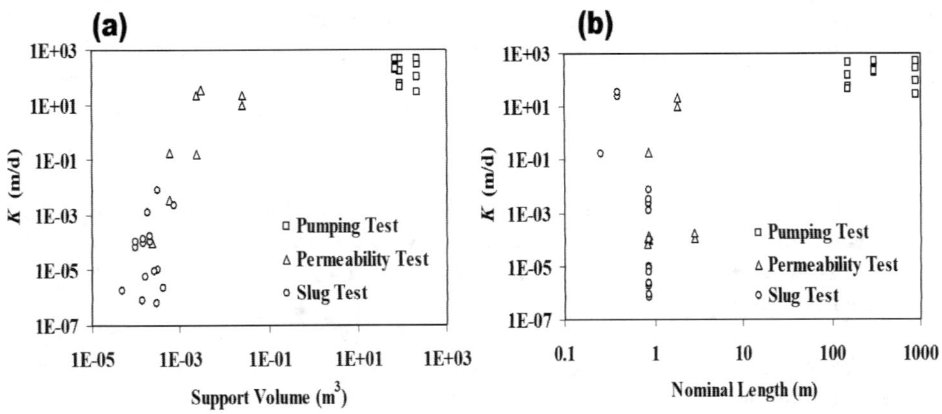

Figure 1. The Hydraulic Conductivities through the Field Tests with the Support Volume (a) and the Nominal Length (b). The Support Volume is calculated by the Involved Water Volume in Each Test. The nominal length represents the depth of test hole in permeability test and slug test, and represents the radius of influence in pumping test, respectively.

All the estimated *K* values show a log-normal distribution, which can be tested through the Kolmogorov-Smirnov (K-S) test (Figure 2a). The mean of log-K is -1.20 and the deviation of log-K is 3.22. The mean and *Cv* of the estimated *K* are 80.81 m/d and 1.73, respectively. In terms of the pumping tests, 13 different *K* values were estimated, and they are in a normal distribution from the K-S test (Figure 2b). The mean and *Cv* of the 13 estimated *K* values are 223.35 m/d and 0.71, respectively. Sanchez-Vila et al. (1996) suggested a log-normal distribution for the *K* values in the aquifers. However, due to the non-stationarity of the log-normal distribution, extreme variations of the *K* field may lead to the nonconvergence of the computation in numerical models. In addition, since the

K value of a model grid represents the hydraulic conductivity of the entire grid, including the matrix and fissure within the area, the normal distribution for the 13 estimated *K* values from pumping tests may be more suitable for numerical simulation in this study. Therefore, the normal distribution was selected as the probability distribution function in the following numerical simulations.

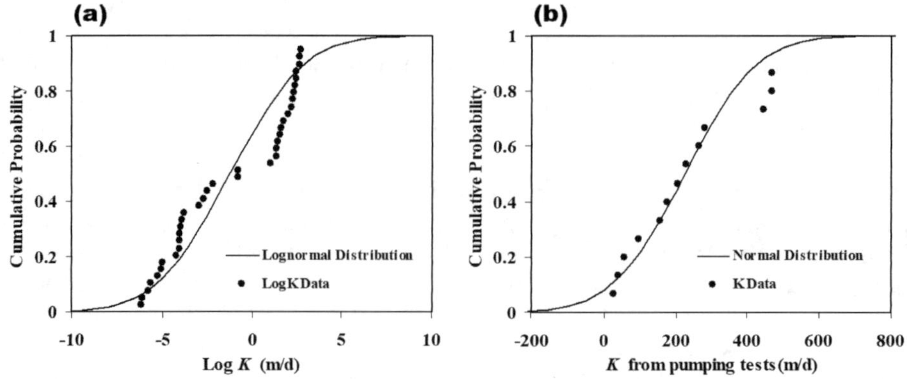

Figure 2. The hydraulic conductivities through the field tests with the support volume (a) and the nominal length (b). The support volume is calculated by the involved water volume in each test. The nominal length represents the depth of test hole in permeability test and slug test, and represents the radius of influence in pumping test, respectively.

3.2. Occurrences of Reduced Duration Test by Aquifer Heterogeneity

Given the *EK* and *Cv* values from the previous pumping-recovery tests in the karst aquifer, thousands of different *K* values representing the heterogeneous *K* field were generated randomly based on the Monte-Carlo method. As *Cv* increases, the variability of the given *K* distribution becomes stronger, and the probability of that the pumping well is located in a grid with extremely small *K* become greater. Because the pumping rate was constant in the numerical models, the grid cell with a small *K* value placed the pumping well will occasionally change into a dry cell in the process of the simulation. Since the durations of these tests were not equal to one day, here, these reduced duration tests (RDTs) were not taken into account to estimate *K*. However, in the next section, numerous artificial reduced duration tests were conducted to study the impacts of the reduced duration tests on the estimated hydraulic conductivity. Table 1 gives the occurrences of RDTs at different *EK* and *Cv* values. The occurrences increase as it has a small *EK* and a large *Cv*.

Table 1. Occurrences of Reduced Duration Test (RDT) Owing to the Variability of the Random K Field. One hundred realizations were performed with each set of EK and Cv.

Cv	0.1			0.5			1		
EK (m/d)	10	50	100	10	50	100	10	50	100
Occurrences	0	0	0	6	3	1	19	16	14

3.3. Impacts of RDTs on the Estimated K

As the description of the numerical simulation in the previous section, due to the strong spatial variability of *K*, the model cell will become dry during pumping, resulting in the RDTs. In practice, various reasons can cause RDTs. With respect to the estimate of *K*, the direct impact of RDTs is expressed by the shorter length of the observations from the reduced duration of tests. In this section, the impacts of reduced duration tests on the estimated *K* of a single pumping test and on the *Ke* value through one hundred realizations are discussed.

3.3.1. *Impact on the Estimated K Using the Datasets from a RDT.* According to the results of the numerical tests, there are two identified types of pumping tests (see Figure 3). The calculated heads and observed ones from the first type of pumping test easily agree well; however, the calculated head curve is difficult to fit the observed heads in the other type of pumping test. In terms of the two types of pumping tests, a series of artificially reduced observations were used to estimate *K*, such as 10% off, 20% off, 30% off and up to 90% off, respectively. The comparison of the observed time-head with the calculated time-head and the goodness of fit for the two curves are shown in Figure 3.

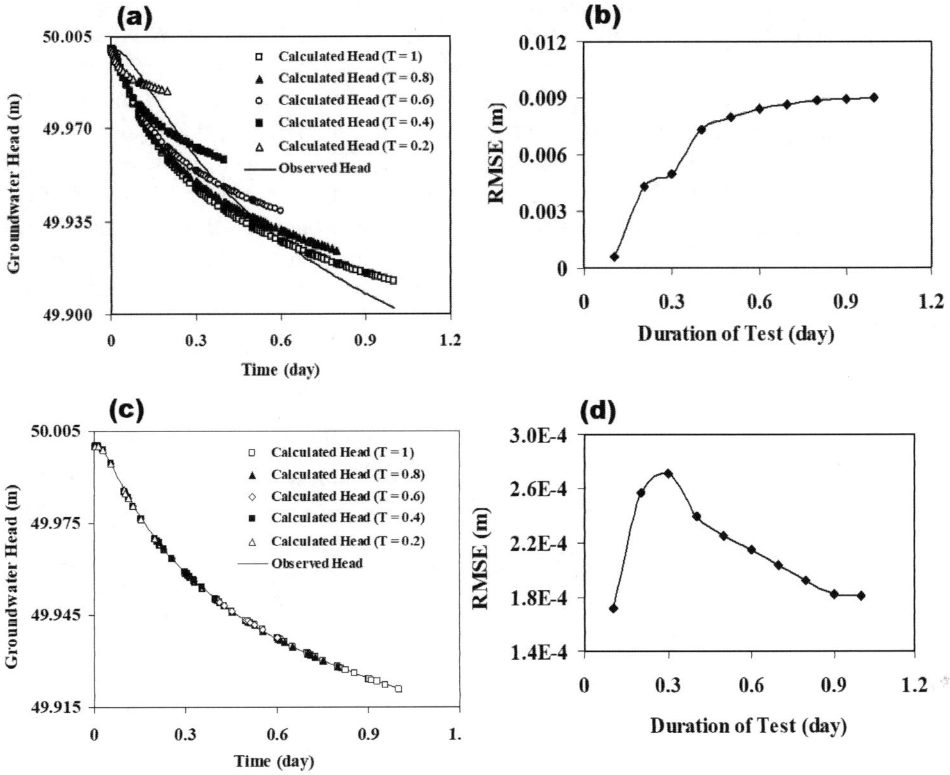

Figure 3. Comparison of the Observed Head with the Calculated Heads in Two Types of Pumping Test for a Heterogeneous Aquifer. (a) shows the observed time-head and calculated time-head using the estimated K with different observed duration, and the groundwater flow is significantly affected by the heterogeneity of aquifer in this type of pumping test. (b) shows the root mean square error (RMSE) between the calculated values and observed ones in the case of (a). (c) shows the observed time-head and calculated time-head, and the groundwater flow is not affected by the heterogeneity of aquifer obviously in this type of pumping test. (d) shows the RMSE in the case of (c).

Figure 3a shows the observed time-head and the calculated time-head of the first type of pumping test. The groundwater flow is significantly affected by the heterogeneity of aquifer (see Figure 3a) so that the curve of time-head is different with the curve in theory. The dissimilarity between the two curves results in the difficulty in fitting the curve of observed time-head using analytical solution. Though the root mean square error (RMSE) between the calculated heads and observed heads drops significantly (see Figure 3b) when the duration of pumping test decreases, it cannot be concluded that the more accurate estimation is achieved with the fewer observations. Because the reduced duration with the same pumping rate will lead to a smaller cone of depression, the differences between observed heads and calculated heads are less than the cases with longer pumping duration. Therefore, the smaller RMSE value resulted from the smaller drawdowns. With respect to the long-duration prediction, the estimates of *K* from the reduced duration test in terms of this type of pumping test will produce greater errors of the drawdowns.

The drawdown curves of the second type of pumping test are shown in Figure 3c. Since the groundwater flow is not affected obviously by the heterogeneity, the drawdown curves are generally fitted well by the analytical solutions. The absolute values of RMSE (Figure 3d) are one order of magnitude lower than the RMSE values of the first type of pumping test (see Figure 3b). In Figure 3d, when the duration is less than 0.3 day, the RMSE has lower value, which may result from the smaller drawdowns. While the duration is larger than 0.3 day, the RMSE value decreases with the increasing duration. The results indicate the accuracy of the estimated *K* increases with the increasing duration of test while the duration is greater than a critical value.

In particular, the different length of observations were used to estimate the *K* value while *EK* was equal to 100 m/d, and *Cv* was equal to 0.5 and 1, respectively. Figure 4 shows the estimated *K* varied by the duration of test at well Wmin and well Wmax. The estimates of *K* at each well can reach a relative stable value when the duration is greater than 0.6 day. Meanwhile, the influences of estimates are shown in Figure 4 while the duration is less than 0.4 day. The results indicate that a long duration test will reduce the uncertainty of the estimate of *K* in a heterogeneous aquifer. Moreover, the accuracy of the estimate cannot be improved significantly while the duration of test is larger than a critical value (specifically 0.6 day in this case). Real-time estimation of the K value can be regarded as an efficient correction method, and it increases the efficient of pumping tests and reduces the cost of field work.

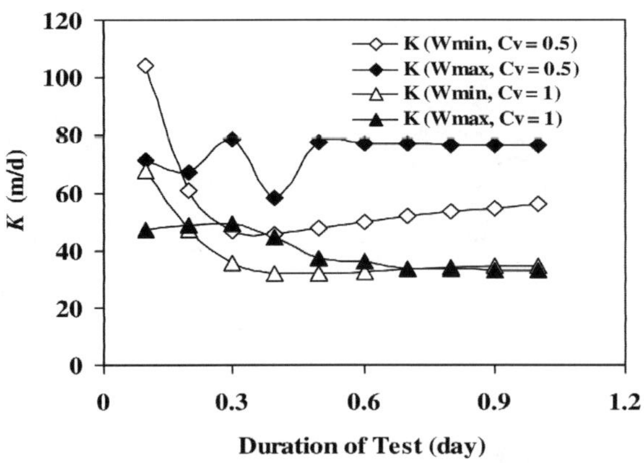

Figure 4. The Estimated K Using the Datasets of Different Durations at Observation Well Wmin and Wmax while the EK value is equal to 100 m/d, and the Cv values are equal to 0.5 and 1, respectively.

3.3.2. Impact of RDTs on the Estimated Ke. Numerous numerical tests were conducted to observe the effect of reduced durations on the estimated *Ke*. The *EK* value was set to be 100 m/d, and the *Cv* values were set to be 1 and 0.5, respectively. One hundred realizations were conducted to calculate a *Ke* value. In order to evaluate the relation between *Ke* and duration of test more accurately, anther one hundred realizations were run to obtain more estimates for each observation well with the same *EK* and *Cv* value.

Figure 5 gives the *Ke* values varied by the different durations of the tests. The different durations represent the length of observations 0% off, 10% off, 30% off, 60% off and 90% off, respectively. When *Cv* is equal to 1, the estimated *Ke* values are smaller than the corresponding estimates when *Cv* is 0.5. Because two *Ke* values were estimated in terms of an observation well in this section and there are two observation wells in the model, each point in Figure 5 represents the average value of the four *Ke* values from well Wmin and well Wmax. We cannot derive any significant trend from the results; therefore, the *Ke* values calculated from large numbers of tests are slightly influenced by the duration. That may result from that the averages of many tests (greater than or equal to one hundred realizations in this study) eliminate the random fluctuations of single estimated *K* of RDT.

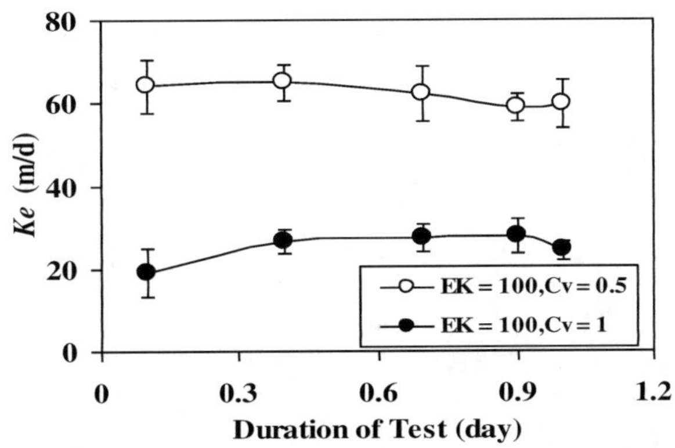

Figure 5. The estimated Ke varied by duration of test while the EK value is equal to 100 m/d, and the Cv values are equal to 0.5 and 1, respectively. The average value and error bar of each point are calculated from four separate values.

4. Summary and Conclusions

The estimated hydraulic conductivity is strongly affected by the heterogeneity of aquifers and the choice of determination methods. Especially for a pumping test, the duration of the test leads to the significant variation to the estimated hydraulic conductivity. In this study, the heterogeneity of the natural aquifer was observed using the field data collected by the field tests in Puding, southwest of China. The numerical model was used to study the impacts of the heterogeneity of aquifer on the estimated hydraulic conductivity.

The results of the field tests, including pumping tests, permeability tests and slug tests, indicate that the estimated hydraulic conductivities range widely from 10^{-7} to 10^{3} m/d. The estimates of *K* exhibit the strong heterogeneity of the karst aquifer. The results of pumping tests are generally larger than the results of permeability tests and slug tests. With respect to the support volume as the measurement scale, the scale effect is shown in our datasets. Due to the paucity of the nominal length as the measurement scale, the structure trend of hydraulic conductivity versus the nominal length does not display explicitly. All the estimated *K* values were log-normally distributed. However, the K estimations from pumping tests show the normal distribution. It should be noted that the estimated *K* from pumping tests are much smoother compared with all the estimates of *K*.

For a heterogeneous aquifer, while the drawdown curve is difficult to fit using a theory curve in the whole pumping duration, the reduced duration will bring high uncertainty of estimated hydraulic conductivity even the goodness of fit between the observed heads and simulated ones becomes better with the reduction of the duration. In terms of a pumping test conducted in a heterogeneous aquifer, the estimated hydraulic conductivity is influenced by the length of observation. A critical duration of test can be obtained to achieve a stable estimate of *K* and to reduce the uncertainty of single estimation. At the same time, the reduced duration of test can make a great economic benefit through the reducing cost of field work. However, the duration of test is not of importance to evaluating an effective hydraulic conductivity under the condition of large numbers of tests, because the random influences may be eliminated between each other. In other words, with respect to the effective hydraulic conductivity of a heterogeneous aquifer, reduced duration tests cannot change the estimates significantly.

5. Acknowledgement

This study was supported by the National Key R&D Program of China (2018YFC0407701), National Natural Science Foundation of China (41971027), Natural Science Foundation of Jiangsu (BK20181035), Water Conservancy Science and Technology Project of Jiangsu Province (2017012). This study does not necessarily reflect the views of the funding agencies.

6. References

[1] Beckwith, C.W., Baird, A.J., Heathwaite, A.L., 2003. Anisotropy and depth-related heterogeneity of hydraulic conductivity in a bog peat. II: modelling the effects on groundwater flow. Hydrol. Process.17: 103-113.
[2] Bouwer, H., Rice, R.C., 1976. A slug test method for determining hydraulic conductivity of unconfined aquifers with completely or partially penetrating wells. Water Resources Research, 12(3): 423-428.
[3] Box, G.E.P., Muller, M.E., 1958. A Note on the Generation of Random Normal Deviates. Ann. Math. Statist. 29(2): 610-611.
[4] Butler, J.J., Healey, J.M., 1998. Relationship between pumping test and slug-test parameters; scale effect or artifact? Ground Water 36 (2): 305–313.
[5] Chen, X., 2001. Migration of induced-infiltrated stream water into nearby aquifers due to seasonal ground water withdrawal. Ground Water 39(5): 721-728.
[6] Chen X., Ayers J.F., 1997. Utilization of the Hantush solution for the simultaneous determination of Aquifer Parameters. Ground Water 35(5): 751-756.
[7] Devlin, J.F., McElwee, C.D., 2007. Effects of measurement error on horizontal hydraulic gradient estimates. Ground Water, 45(1): 62-73.
[8] Felletti, F., Bersezio, R., Giudici, M., 2006. Geostatistical simulation and numerical upscaling model ground-water flow in a sandy-gravel, braided river, aquifer analogue. Journal of Sedimentary Research, 76: 1215–1229.
[9] Gutjahr, A.L., Felhar, L.W., Bakr, A.A, MacMillan, J.R., 1978. Stochastic analysis of spatial variability in subsurface flow. 2: Evaluation and application. Water Resources Research, 14: 953-960.
[10] Hvorslev, M.J., 1951. Time Lag and Soil Permeability in Ground-Water Observations. bul. no. 26, Waterways Experiment Station, Corps of Engineers, U.S. Army, Vicksburg, Mississippi.
[11] Illman, W.A., 2006. Strong field evidence of directional permeability scale effect in fractured rock. Journal of Hydrology 319: 227–236.
[12] Lehmer, D.H., 1951. Mathematical methods in large-scale computing units. Annu. Comput. Lab. Harvard Univ. 26: 141-146.
[13] Lu, C., Shu, L., Chen, X., Cheng, C., 2010. Parameter estimation for a karst aquifer with unknown thickness using the genetic algorithm method, Environmental Earth Sciences, 63(4), 797-807.
[14] Moench, A.F., Garabedian, S.P., LeBlanc, D.R., 2001. Estimation of Hydraulic Parameters from an Unconfined Aquifer Test Conducted in a Glacial Outwash Deposit, Cape Cod, Massachuesetts USGS. Professional Paper 1629
[15] Renard, P., de Marsily, G., 1997. Calculating equivalent permeability: a review. Advances in Water Resources 20: 253–278.
[16] Rovey, C.W. II., 1998. Digital simulation of the scale effect in hydraulic conductivity. Hydrogeology Journal 6: 216–225.
[17] Sanchez-Vila X., Carrera J., Dirardi J.P., 1996. Scale effects in transmissivity. Journal of Hydrology 183: 1-22.
[18] Schulze-Makuch, D., Carlson, D.A., Cherkauer, D.S., Malik, P., 1999. Scale dependency of hydraulic conductivity in heterogeneous media. Ground Water, 37(6): 904-919.
[19] Sudicky E.A., 1986. A natural gradient experiment on solute transport in a sand aquifer: spatial variability of hydraulic conductivity and its role in the dispersion process. Water Resources Research, 20(19): 2069-2082.
[20] Webb, E.K., Anderson, M.P., 1996. Simulation of preferential flow in three-dimensional heterogeneous conductivity fields with realistic internal architecture. Water Resources Research, 32(3): 533 545.
[21] Zhang Y., Person M., and Gable C.W. (2007) Representative hydraulic conductivity of hydrogeologic units: Insights from an experimental stratigraphy. Journal of Hydrology, 339: 65–78.

A Novel BP Neural Network with Wavelet Transform Inputs for Air Quality Index Prediction

Zheng Fang, Lei Zhang and Yakun Huang

School of Mechanical & Electrical Engineering, Xuzhou University of Technology, No. 2, Lishui Road, Yunlong District, Xuzhou City, Jiangsu Province, China.
Email: triple-stone@foxmail.com

Abstract. Air quality index prediction research is of great significance to people's life. Various prediction methods are used to pursue its accurate prediction. A novel BP neural network with wavelet transform inputs for air quality index prediction is proposed in this paper. The proposed method introduces wavelet transform in the front of the neural network. It decomposes the original time series signal into a linear superposition of the main signal and some noise signals. After establishing the predictive BP neural network for each decomposed signal, the air quality index is restored based on the linear additivity of wavelet decomposition. The simulation experiments show that the prediction accuracy of proposed method is high. The relative error of long-term prediction for 50 consecutive days is only 4.3994% for Xuzhou City's air quality index. The research work provides a useful reference for air quality index prediction.

1. Introduction

Air Quality Index (AQI) is an important indicator for evaluating air quality. Research on its prediction has been a hotspot. Various methods are used to pursue AQI's accurate prediction. Some of them relies on mining AQI historical data itself to predict, such as time series [1], gray model [2], Markov Model [3], and so on. The others need more data support besides AQI for data training and learning, such as neural network [4], random forest [5], principal component analysis [6] support vector machine [7] machine learning [8], etc. If there is no other data support except AQI, these methods' forecasting accuracy will be greatly reduced. For example, using BP neural network to predict AQI, if there is no supplementary data on major pollutant indicators, generally speaking, its prediction accuracy is unsatisfied. However, this is not necessarily the case. Previously, the authors of this paper only used AQI time series as the input of neural network, and used GA-BP neural network to predict, which received some good results [9]. And this paper will continue the previous study, discussing the AQI prediction problem caused by using time series directly as the input of BP neural network. In order to improve the prediction accuracy, wavelet transform will be introduced. The proposed method of this paper is shown in figure 1.

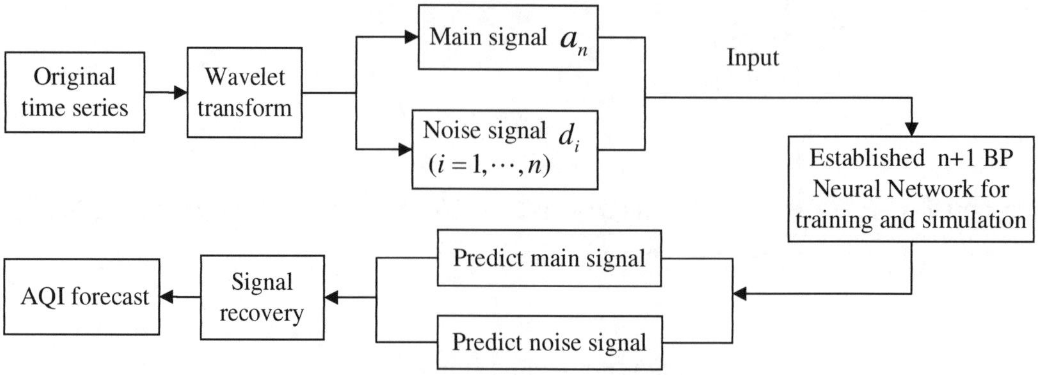

Figure 1. Research Route

As shown in figure 1, the proposed method can be described as follows: Firstly, the original time series which is regarded as unstable input signal is decomposed into a linear combination of the main signal a_n and some noise signals d_i (i =1, ..., n) by wavelet transform. Then, using a_n and d_i as sample training database, n+1 BP neural networks are established to predict main signal and noise signals respectively. Finally, the AQI time series is restored according to the linear combination, and its value is the predicted value. The proposed method does rely on AQI data itself without any auxiliary data, and it is different from the existing method of AQI prediction using wavelet neural network [10-13].

In figure 1, several issues need to be discussed deeply. The remainder of this paper will clarify them in four sections. Section 2 shows the selection of wavelet function and the results of wavelet decomposition. Section 3 introduces the establishment of BP network, and a simulation case study is discussed in section 4. Finally, a summary is provided in section 5. The AQI data shown in figure 2 in this paper is from December 2013 to November 2018 in Xuzhou City, Jiangsu Province, China. A total of 1814 AQI data are obtained from the website of Jiangsu Meteorological Bureau.

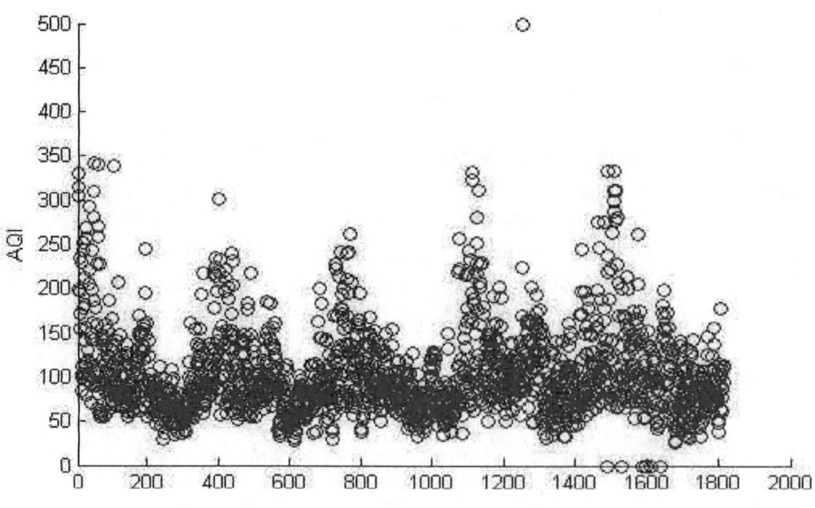

Figure 2. AQI Data of Xuzhou from December 2013 to November 2018

2. Wavelet Transform

The original AQI time series shown in figure 2 could be regarded as a group of unstable signals. Following the wavelet transform theory, it could be decomposed as a linear superposition of the main signal and some noise signals. The decomposition process could be illustrated in figure 3.

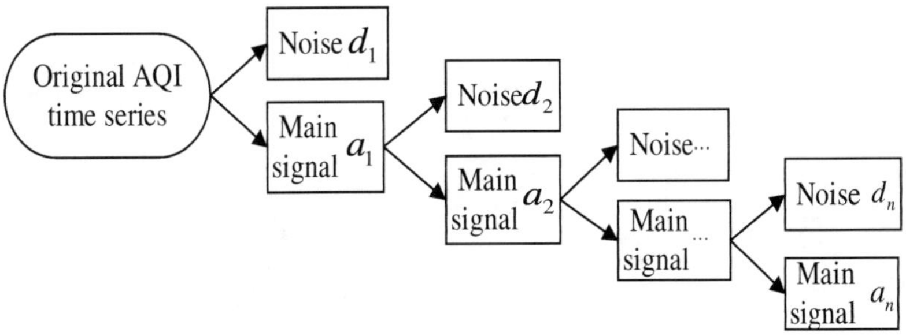

Figure 3. Wavelet Decomposition Schematic

From figure 3, the original AQI time series by wavelet decomposition can be expressed as follows,

$$\mathrm{AQI} = a_n + \sum_{i=1}^{n} d_i \,, \tag{1}$$

where a_n is the main signal and $d_i\,(i=1, \ldots, n)$ are noise signals, i.e., AQI is a linear superposition of the a_n and $d_i\,(i=1, \ldots, n)$.

In order to decompose AQI into linear combinations shown in equation (1), the selection of wavelet functions is a very important work. Different wavelet functions will result in different decomposition results which will have a great impact on BP neural network prediction. However, for the selection of wavelet functions, there are still no unified selection principles today. We used the common wavelet functions to make lots of tests in Matlab 2014a based on data from figure 2. The following results are obtained.

For *Haar* wavelet function, it is discontinuous and has no smoothness. When it is used to decompose the signal of figure 2, the decomposition signal is seriously distorted. The *dbN* wavelet function is asymmetric. When it is used to decompose the signal of figure2, the signal distortion is also very obvious, especially in the edge. Therefore, *Haar* and *dbN* cannot be used as wavelet function in this situation. Although *coifN* wavelet function has good symmetry, its performance decreases with the increase of its support length, and the computational cost decreases with the increase of its computational efficiency. However, it is inconsistent with the noise signal integrity in our proposed model. Hence, it is not suitable either. *Morl* function and *Mexh* function are not taken into consideration due to their insufficiency in compact support, scale function and orthogonality. Compared with *dbN*, *symN* function has the advantages of approximate symmetry and compact orthogonality. It is consistent with our application. Therefore, we use *symN* wavelet function to perform wavelet decomposition.

On the choice of N value, the symmetry of *symN* is not good as $N<4$. When $N>4$, the function distribution is too dense to decompose. Hence, $N=4$ is proposed. For the layer of wavelet decomposition, the tests are carried out in Matlab 2014a. When layer s < 7, the main signal's decomposition is insufficient. When layer s > 7, the main signal is over-decomposed and becomes a square wave. Therefore, the decomposition layer s = 7 is determined. After above analyses and tests, the original signal is decomposed by *sym*4 wavelet function. The decomposed results are shown in figure 4.

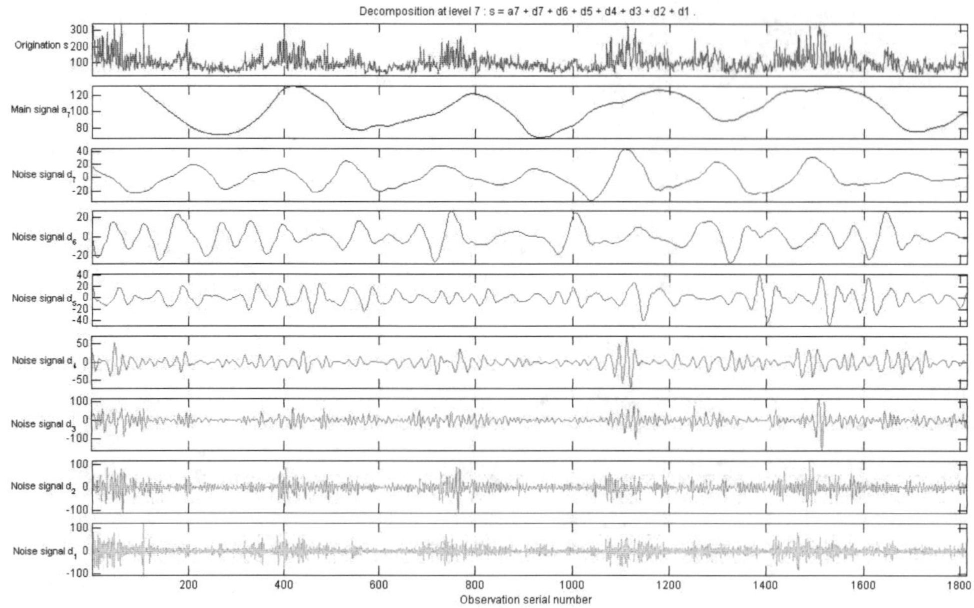

Figure 4. AQI Wavelet Decomposition Signal Diagram

After decomposition, the modeling work could be started. It will be demonstrated in the next section.

3. Establishment of BP Neural Network

3.1 Determination of the Number of Input Nodes in BP Neural Network

From figure 4, one group of main signals and seven groups of noise signals are obtained. Hence, eight BP neural networks are needed to be established according to figure 1. Since the data itself is used as input, it is crucial to determine the number of input nodes for each BP neural sub-network.

In this paper, the optimal number of input nodes is confirmed by calculating Mean Square Error (MSE) value and the Correlation Test (CT) value of test samples. Taking the noise signal d_1 in figure 4 as an example, calculating MSE and CT values of 13 input nodes, respectively in Matlab 2014a (Here, 13 is an approximate number, and more input nodes could be taken into account). The results are shown in the following table.

Table 1. Test Results of Different Input Nodes of Noise Signals d_1

The number of input node	MSE	CT	The number of input node	MSE	CT
2	0.104	0.81366	8	0.000998	0.98355
3	0.0036	0.93959	9	0.00096	0.98436
4	0.0016	0.97362	10	0.00092	0.98497
5	0.0013	0.97926	11	0.000994	0.98362
6	0.000992	0.98364	12	0.000998	0.98347
7	0.000985	0.98389	13	0.000994	0.98355

From table 1, When the number of input node is 10, MSE is the smallest and CT is the largest among 13 input nodes. This shows that at this time the performance of the BP neural network is better. Hence, 10 is determined to be optimal for d_1. That is to say, 10 consecutive AQI values are input into the BP neural network of d_1, and the 11st AQI value is used as label data. By analogy, the BP neural network of d_1 has 1804 teams of data. Following this method, the number of input nodes for $d_i(i=2,\ldots,7)$ and a_7 is determined as shown in table 2.

Table 2. The Number of Input Nodes for each Signal BP Neural Network

Signal Sequence	Number of input nodes	Data teams	Signal Sequence	Number of input nodes	Data teams
a_7	3	1811	d_4	5	1809
d_1	10	1804	d_5	5	1809
d_2	10	1804	d_6	5	1809
d_3	5	1809	d_7	5	1809

3.2 Determination of the Number of Hidden Layer Nodes

At present, for BP neural network, the determination of the number of hidden layer nodes usually depends on empirical formula. Based on empirical formula and MSE of network loss function which is provided by computer under setting different hidden layer nodes, the hidden layer nodes for each BP neutral sub-network are determined as follows.

Table 3. Hidden Layer Node Settings

Signal Sequence	Number of Hidden Layer Nodes	Signal Sequence	Number of Hidden Layer Nodes
a_7	35	d_4	50
d_1	30	d_5	50
d_2	30	d_6	50
d_3	35	d_7	50

After determining the number of input nodes and hidden layer nodes, BP neural network could be constructed by the neural network toolbox of Matlab 2014a. In next section, the simulation results of figure 2 using the proposed method will be given and discussed.

4. Simulation Case Study

4.1 Simulation Tests of Network Performance

Eight BP neural sub-networks were established respectively by computer. The default training accuracy is 0 and the default training function is *trainlm* in the neural network toolbox of Matlab 2014a. The first 1764 data of 1814 data were used as training set and the last 50 data were used as test set. When the simulation results of eight sub-networks are obtained, the AQI simulation results of the whole system can be calculated according to formula (1). The root mean square error (RMSE) between the predicted value and the real value of the test set and 50-day prediction accuracy were calculated for eight sub-networks and whole system, respectively. They are shown as table 4.

Table 4. Prediction Error of each Group of Signals by Simulation

Signal sequence	RMSE	50-day prediction accuracy	System MSE
a_7	0.1302	0.1301%	0.0007784
d_1	3.3265	4.8766%	0.0009819
d_2	3.1169	5.2634%	0.0009860
d_3	3.1013	9.8701%	0.0009815
d_4	1.2647	6.1459%	0.0006421
d_5	0.2556	4.5628%	0.0001189
d_6	0.0993	3.5435%	0.0001106
d_7	0.0406	2.3040%	0.0003459
Predictive value of AQI simulation	5.1623	4.3994%	0.0006186

The first eight rows of table 4 are the simulation test performance of eight sub-networks. What the last row of table 4 shows is the whole system's performance. From table 4, the RMSE values of eight sub-networks are about 3 or less, whereas the whole system's is 5.2, which indicates that the predicted values deviate very small from the real values. The whole system's 50-day prediction accuracy is about 4.4% and MSE value is almost 0, which illustrate that the restored prediction system has a good accuracy.

4.2 Comparison of Simulation Prediction

In order to demonstrate the superiority of our proposed method, the traditional BP neural network and GA-BP neural network proposed in reference [9] are input into the time series shown in figure 2. The predictive results are shown in figure 5 and the predictive performances are shown in table 5.

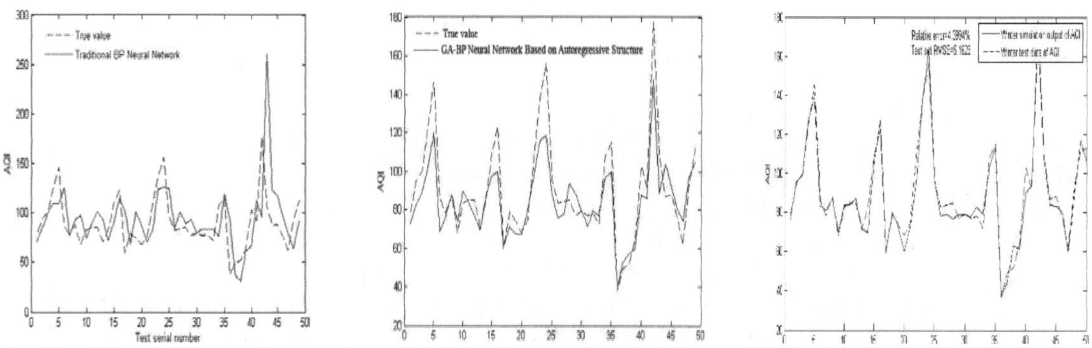

(a) Traditional BP neural network (b) GA-BP proposed in [9] (c) Method proposed in this paper

Figure 5. Predictive Contrast Diagram

Table 5. Performance Comparison of three Prediction Methods

Prediction System	System MSE	Prediction Relative Error	Test Set RMSE
Traditional BP neural network	0.0621	36.6628%	42.3130
GA-BP neural network proposed in reference [9]	0.0462	9.1375%	11.4722
BP neural network established in this paper	0.0006186	4.3994%	5.1623

From figure 5 and table 5, it is obvious that the method proposed in this paper has best accuracy for AQI prediction.

5. Conclusion

The BP neural network with wavelet transform inputs is an improvement to the traditional BP neural network. It is suitable for the situation where time autoregressive data are input into BP neural network. In future, we will conduct a comparative study between this network and time series model to increase the understanding of this network.

6. Acknowlegement

The research work was supported by the key research and development program in Xuzhou city (No. KC17007, KC17225) and college students' practical innovation training program (xcx2018107).

7. Reference

[1] Li Weidong, Li Li, Xu Yan. The Concentration Research of PM2.5 in Beijing with Time Series Analysis. Operational Research Transactions, 2018, 22 (2):115-126.
[2] Zhao Meiling, Xue Rui. Application of Grey Theory in Ambient Air Quality Trend Analysis. Environment and Development, 2013, 29 (2): 76-79.
[3] Yang Jinwei, Sun Baolei. The Prediction of Air Pollutant Concentration in Pingdingshan Based on Grey Markov Model. Practice and Recognition of Mathematics, 2014, 44(2): 64-70.
[4] Du Xiaoliang. Application of Artificial Neural Network in Air Quality Prediction. Shandong University, 2009.
[5] Lin Kaichun, Shao Fengjing. Research on Air Quality Prediction Based on Random Forest and Neural Network. Journal of Qingdao University (Engineering Edition), 2018, 33 (02): 32-36.
[6] Wei Zhen, Lin Fang. Air Quality Evaluation of Cities Based on Cluster Analysis and Principal Component Analysis. Journal of Huaiyin Institute of Technology, 2018, 27 (03):86-96.
[7] Guo Fei, Xie Liyong. Air Quality Index Prediction based on Meteorological Factors and Improved Support Vector Machine. Environmental Engineering, 2017, 35 (10): 151-155.
[8] Yin Qi. Air Quality Classification and Prediction based on Machine Learning. Zhongbei University, 2018.
[9] Fang Zheng, Zhang Lei, Wang Yuqin, et al. AQI prediction based on Autoregressive GA-BP neural network. Environmental Monitoring and Warning, 2019, 11 (02): 22-25.
[10] Wang Xiaohan, Zhang Yanping, Zhao Shu, et al. Air quality prediction based on dynamic granularity wavelet neural network. Computer Engineering and Application, 2013, 49 (6): 221-224.
[11] Dunea D, Pohoata A, Iordache S. Using wavelet–feedforward neural networks to improve air pollution forecasting in urban environments. Environmental Monitoring and Assessment, 2015, 187(7): 477-493.
[12] Bai Y, Li Y, Wang X, et al. Air pollutants concentrations forecasting using back propagation neural network based on wavelet decomposition with meteorological conditions. Atmospheric Pollution Research, 2016,7 (6): 557-566.
[13] Morabito F C, Versaci M. Wavelet neural network processing of urban air pollution. International Joint Conference on Neural Networks. IEEE, 2002.

IOP Conf. Series: Materials Science and Engineering **735** (2020) 012060 doi:10.1088/1757-899X/735/1/012060

FAHP on the Development of Exhibition Ecological Education

Lin Yang

Jinhua Polytechnic, Jinhua, ZheJiang,321000, China
Email:860814235@qq.com

Abstract. Based on the research findings of some experts, this article tries to construct the index system of the sustainable development of exhibition ecological education and the assessment model with the utilization of FAHP. In the meantime, it conducts the fuzzy Comprehensive assessment with the case study of Jinhua Polytechnic.

1. Preface

Fuzzy analytic hierarchy process (FAHP) is the method of the union of the analytic hierarchy process and fuzzy comprehensive assessment. Its main idea can be used to guide the quatity assessment of exhibition ecological education.

The theory and practice of exhibition ecological education are still in the exploratory stage in China.Theoretically, in the VIP using "exhibition ecological education" as the keyword search can only find 9 relevant literature, and research, focused on ecological rather than education; in practice, the author can only collect 20 items of related exhibition projects of ecological education, including some still in the enactment of laws and regulations related to the implementation stage yet, or still in the pilot phase.

The status of ecological education constitutes two difficulties in practice, but also two major difficulties in the completion of this paper: (1) how to establish the assessment indicators of the exhibition ecological education projects of sustainable development; (2) how to evaluate construction completed indicators by quantitative and qualitative analysis.

2. Construct Sustainable Development Index System Of exhibition Ecological Education

China's exhibition ecological education mechanism is reflected in the subject of education, education objects, methods and means, in order to filter out the most closely, the most representative indicator of the three areas. In the research process, the article chose the current being implemented or planned 10 exhibition projects as ecological education study sample, first of all, collected relevant data online, found out the problem, then conducted a telephone survey to separated interests body in the implementation of various projects to collect the figures, to grasp the main factors affecting project implementation, determine the relevant assessment system.

2.1. Sample Analysis

As ecological education of exhibition pilot project is still in the exploratory stage, data accuracy is so difficult to be assured that the relevant assessment index system is not established. Therefore, this paper adopts the combination of qualitative and quantitative survey methods, the establishment of ecological education of exhibition sustainable development assessment system.

Table 1. Overview Analysis of Samples of Exhibition Ecological Education

number	Project name	Education Performance
1	Jinhua Polytechnic	Drawing lessons from the concept of "ecological" education and taking "ecological integration" and "ecological sustainable development" as the guiding principles of education, a comprehensive training plan for convention and exhibition education is formulated
2	Tourism College of Zhejiang	Reasonably construct the framework of in-class education, form a three-dimensional intersecting spatial relationship, and form a healthy ecosystem
3	Fujian Institute of Economics and Trade	Through the intergrowth and complementarity of theory, practice and students, school teaching and Exhibition industry, and through the combination of ecological education among various disciplines, elements and professionals.
4	Shenzhen polytechnic	Guided by the "ecological education concept" and following the professional rules, the teaching structure system should be constructed.

2.2. Indicators Establishment

First of all, based on the related research, determing exhibition ecological education level indicators: the education subject, education object, education standard, education style and education financial security. On this basis, using the Delphi method, invite experts to filter the primary evalution indicators and keep the balance with the changeof its real impact on the community.

Secondly, the use of survey questionnaires, using Likert five scale implementation of the project of tourist ecological education to measure the impact of assessment indicators. Statistical analysis of survey results follows the table.

Table 2. The Main Factors of Exhibition Ecological Education Effect (Note: R values refer to Table 3)

factor / number	R11	R12	R13	R21	R22	R23	R24	R31	R32	R33	R41	R42	R43	R51	R52	R53
1			4			4.2	3	5	5	4			5		4	2
2		4	4.8	4.4	4		4	5	5		4		5	4.2		4.5
3			3.6			4	4.5	5	5	4		4	5		4.5	4.7
4			4		4.8			5	5	4.6			5		4	
5		4.8	4	4.2	3.8	4.3		5	5				5		4	
6			4.2		4			5	5	4		4.1	5		4	
7	4.2	4	3.8	3	4	3.8		5	5	4.5	4.5		5	4.5	4	
8	4.5	4	4.6	4.5			4.5	5	5	3.9	4		5	4	4	4
9		5	4	3.8		3.7		5	5	4	3.7	4	5	4.2	4.2	4.2
10	3.6		4.2					5	5				5	2.8	4.2	

2.3. The Establishment of Index System

In the above survey and assessment framework, under the basic idea of FAHP and the practical needs of sustainable development of exhibition, this article will divide ecological education assessment index system of sustainable development into three levels, the target layer (highest level), criteria layer (middle layer) and the index layer (bottom).

Table 3. Exhibition Indicators of Sustainable Development of Ecological Education System

Target layer G	Criteria layer C	Index layer P
The indicators system of sustainable development of ecological education	education Subject R_1	education audience R_{11}
		education Implementer R_{12}
		education controller R_{13}
	education object R_2	Regional ecological function R_{21}
		exhibition resource R_{22}
		Ecological Environment R_{23}
		education resources R_{24}
	education Standard R_3	Equity of education Standard R_{31}
		Education amount R_{32}
		Affordability of education R_{33}
	education Style R_4	Policy education R_{41}
		Project education R_{42}
		ecostype education R_{43}
	education Financial securityR_5	Government education R_{51}
		Market education R_{52}
		Legal penalty R_{53}

3. The Implementation Steps of the Sustainable Development Assessment on exhibition Ecological Education Project

3.1. Solve the Index Weight Of Assessment With FAHP

In this paper, based on preliminary investigation, issued 60 questionnaires, 50 valid questionnaires. According to survey statistics order to obtain the average value of each index, and establish into the following six comparison matrix.

Matrix of G-R: $$\left\{\begin{matrix} 2 & 1 & 1/4 & 1/2 & 1/3 \\ 1/3 & 1 & 1/2 & 1/2 & 1/3 \\ 3 & 2 & 2 & 1/3 & 1/2 \\ 3 & 2 & 1 & 2 & 3 \\ 2 & 3 & 1/3 & 1/2 & 1 \end{matrix}\right\}$$

Matrix of R1: judgement matrix $$\left\{\begin{matrix} 1 & 2 & 1/4 \\ 2 & 1/3 & 1/2 \\ 1 & 2 & 1/3 \end{matrix}\right\}$$

Matrix of R2: $$\left\{\begin{matrix} 1 & 2 & 1/3 & 1/4 \\ 1/2 & 1/3 & 1/3 & 2 \\ 2 & 2 & 1/3 & 1/2 \\ 2 & 1 & 1 & 1 \end{matrix}\right\}$$

Matrix of R3: -P $$\left\{\begin{matrix} 1 & 2 & 2 \\ 1 & 1 & 2 \\ 1/2 & 1/3 & 1/4 \end{matrix}\right\}$$

Matrix of R4-P: $\begin{Bmatrix} 1 & 2 & 1 \\ 1/3 & 1/4 & 1/3 \\ 1 & 2 & 2 \end{Bmatrix}$ Matrix of R5: $\begin{Bmatrix} 1 & 2 & 1 \\ 2 & 1/2 & 1/2 \\ 1/2 & 1/2 & 1/3 \end{Bmatrix}$

Make use of FAHP, according to statistics data, expert opinions and analysis personnel recognition, and with reference to pairs of qualitative analysis of various factors, various indicators compare the relative importance of two, tectonic model of fuzzy consistent matrix, and then to single-level sorting, solving the weight of each index (accordiing to Table 5).

Table 5. Weights of Exhibition Ecological Education Projects

Criteria layer	R_1		R_2		R_3		R_4		R_5	
weight	0.0977		0.0763		0.3458		0.3879		0.1924	
index layer& weight	R_{11}	0.310	R_{21}	0.1182	R_{31}	0.4922	R_{41}	0.3404	R_{51}	0.4558
	R_{12}	0.3402	R_{22}	0.2885	R_{32}	0.3533	R_{42}	0.2254	R_{52}	0.3886
	R_{13}	0.3498	R_{23}	0.2412	R_{33}	0.1545	R_{43}	0.4342	R_{53}	0.1556
			R_{24}	0.3521						

The Judgement matrit of the largest eigenvalue, respectively 5.1765,3.0845,4.2541,3.0340,3.0701,3.0464, CR values were 0.0394,0.0728,0.0941,0.0371,0.0604,0.040, are less than 0.10, so all through the consistency test, the above weight of each index level are valid.

3.2. Comprehensive Assessment of Exhibition Ecological Education Projects with Fuzzy Comprehensive Assessment

Using FAHP to calculate ecological education of exhibition relative importance of each indicator quantitatively, you can use fuzzy comprehensive assessment method to calculate for sustainable development level of a exhibition ecological education projects. For example, Jinhua Polytechnic scenic ecological education programs, fuzzy comprehensive assessment are as follows:

3.2.1. Single-level Assessment of Ecological Education of the Jinhua Polytechnic

The establishment of a scenic area staff, Jinhua Polytechnic residents and experts as panel of judges, according to the index contains specific assessment requirements, the assessment team reviews the given set V = (very good, good, ordinary, poor, very bad). And the establishment of this assessment based on fuzzy assessment matrix (Accoring to Table 6).

IOP Conf. Series: Materials Science and Engineering **735** (2020) 012060 doi:10.1088/1757-899X/735/1/012060

Table 6. Fuzzy Assessment Matrix of Sustainable Development of Ecological Education on Jinhua Polytechnic

assessment Factors		assessment grade				
		Well	good	normal	bad	worse
education Subject R_1	education audience R_{11}	0.3	0.2	0.2	0.2	0.1
	education Implementer R_{12}	0.2	0.3	0.2	0.2	0.1
	education controller R_{13}	0.1	0.4	0.1	0.2	0.2
education object R_2	Regional ecological function R_{21}	0.2	0.2	0.3	0.2	0.1
	exhibition resource R_{22}	0.3	0.1	0.3	0.2	0.1
	Ecological Environment R_{23}	0.1	0.2	0.3	0.2	0.2
	education resources R_{24}	0.1	0.2	0.3	0.2	0.2
education Standard R_3	Equity of education Standard R_{31}	0.1	0.2	0.2	0.3	0.2
	education amount R_{32}	0.1	0.2	0.2	0.2	0.3
	Afford ability of education R_{33}	0.1	0.3	0.2	0.2	0.2
education Style R_4	Policy education R_{41}	0.2	0.1	0.2	0.4	0.1
	Project education R_{42}	0.2	0.3	0.3	0.1	0.1
	ecostypes education R_{43}	0.1	0.2	0.2	0.3	0.2
education Financial security R_5	Government education R_{51}	0.2	0.1	0.2	0.2	0.3
	Market education R_{52}	0.2	0.2	0.2	0.3	0.1
	Legal penalty R_{53}	0.2	0.2	0.2	0.2	0.2

Specific method is to invite 10 experts to assess every indicator, finally, take the probability in judgment set. For example, the education audience indicator in education subject is 0.3; means in 10 experts 3 have said that indicator is better. Among them, the education subject indicators layer is the fuzzy matrix B_{1j}, the weight coefficient is A_1, calculate the index level of comprehensive assessment matrix.

$$R_1 = A_1 \times B_{1j} = (0.340\ 0.3017\ 0.3493) \times$$

$$= (0.1371\ 0.3274\ 0.3681\ 0.1342\ 0.1342$$

$$\begin{Bmatrix} 0.3 & 0.3 & 0.3 & 0.1 & 0.1 \\ 0.1 & 0.3 & 0.3 & 0.3 & 0.3 \\ 0.1 & 0.1 & 0.3 & 0.3 & 0.3 \end{Bmatrix}$$

The above results were normalized as

$$R_1 = (0.1386\quad 0.3295\quad 0.3705\quad 0.1357\quad 0.1357)$$

Using the same method to get other indicators' comprehensive assessment matrix.

$$R_2 = (0.1187 \quad 0.3000 \quad 0.3220 \quad 0.3000 \quad 0.1593)$$

$$R_3 = (0.1000 \quad 0.3495 \quad 0.3505 \quad 0.3000 \quad 0.3000)$$

$$R_4 = (0.1572 \quad 0.3360 \quad 0.3000 \quad 0.1534 \quad 0.1534)$$

$$R_5 = (0.1844 \quad 0.3000 \quad 0.3000 \quad 0.3000 \quad 0.1156)$$

3.2.2. Ecological Education Project of the Jinhua Polytechnic Scenic Multi-level Assessment

The assessment results will be integrated, form index layer fuzzy assessment matrix B_i. According to AHP to determine the combination factor A, get the final result, and for normalization.

$$R = A \times B_i = (0.1428 \quad 0.3928 \quad 0.3116 \quad 0.1774 \quad 0.1335)$$

3.2.3. Assessment Results and Analysis

According to the principle: maximum degree of membership to determine the assessment rating, which in the final membership matrix, in composite indicator which indicator assessed higher levels of membership, will have to determine the assessment target as the assessment grade. The comprehensive assessment index for the maximum degree is 0.3908, the corresponding assessment rating of “good”. This shows that current Jinhua Polytechnic’ ecological education is in a healthy state of sustainable development. Through the establishment of score ratings vector, specific score can be calculated, setting the corresponding score assessment for each grade vector.

$$E = (93\ 85\ 75\ 60\ 42)^T$$

Calculated the final assessment score

$$Z = R \times E = 71.35474$$

4. Conclusion

According to the calculation results, through analysis of sustainable development level of exhibition ecological education projects in China. Find gaps and shortcomings and propose improvement measures. Jinhua Polytechnic’s ecological education programs generally are in the forefront of similar projects in our country, but the final score was low, indicating the project of exhibition ecological education overall level in China is low. And the Jinhua Polytechnic has significant gaps in the education subject, education object recognition; need to take appropriate measures for improvement. In addition, the randomness of the assessment process, evaluator’s subjective uncertainty and ambiguity in understanding cause a certain degree of subjective assumptions, so there are some errors in assessment results.

4.1. Constructing Ecotype Teaching System Based on Exhibition Specialty Characteristics

Exhibition is a multi-disciplinary comprehensive cross-disciplinary specialty. It is an information dissemination activity aiming at information exchange and visual communication. It is also a highly integrated platform for information dissemination and exchange in art space, involving a wide range of knowledge and strong technical operability. Due to the different scale and themes of "convention and exhibition activities", there are different conferences and exhibition forms with different sizes and themes, such as politics, economy, culture, sports and religion. They involve economics, management, informatics, tourism, architecture, linguistics, transportation, port science, art, environmental science, statistics and beauty. Learning, body etiquette, etc.

4.2. Subdividing the Direction of Specialty and Cultivating it in Accordance with Each Other to Embody the "Specialty and Precision"

In the training of exhibition planning, management and design, undergraduate education in Colleges and universities can draw lessons from the concept of "ecological" education, take "ecological integration" and "ecological sustainable development" as the guiding principles of education, formulate the training plan of exhibition education in an all-round way, reasonably construct the framework of in-class education, and form a three-dimensional education. Cross spatial relationships form a healthy ecosystem. Eco-education concept aims at improving students' comprehensive quality and creativity. It aims at establishing teachers and students through the multi-level and mutually reinforcing relationship among exhibition theory, professional practice and students, school teaching and exhibition industry, and through the combination of multi-disciplines, multi-elements and multi-professionals in eco-education. Students learn and practice together in a learning environment to cultivate high-quality professionals for exhibition majors.

4.3. Setting up "Wide and Deep" Degree of Basic Teaching of Exhibition

The undergraduate education of exhibition is one of the important ways to cultivate core and auxiliary talents of exhibition. Good teaching environment, basic teaching system and sound educational system provide students with opportunities to learn basic knowledge of exhibition specialty. Especially in comprehensive universities, the advantages of multi-disciplines are embodied in the blending of knowledge and professional exchanges, which provide students with a wide range of knowledge sources, can meet the requirements of exhibition talents for "wide" knowledge and "deep" professional knowledge and theory, and lay a good foundation for professional learning. "Subdividing professional orientation and counterpart training" has created conditions for students' professional training. For Convention and exhibition majors with strong creativity, the superior hardware teaching conditions and abundant educational technology of undergraduate education provide guarantee for students to open up their thinking and cultivate their creative consciousness, which is conducive to the cultivation of students' creative ability.

5. References

[1] ZHANG Jijun. Fuzzy Analytical Hierarchy Process. Fuzzy Systems and Mathematics [J].2012 (2): 80-82

[2] LAI QIfu, CHEN Guisong, HUANG Xiujuan. Benefit assessment of Forest exhibition Lodging Facilities Based on AHP [J]. Journal of Xiangfan University, 2019 (1): 84-88

[3] Chang Dayong, Zhang Dayong, Zhang Lili. Economic management in fuzzy mathematics [M].Beijing: Beijing Economic College Press, 2015.

IOP Conf. Series: Materials Science and Engineering **735** (2020) 012061 doi:10.1088/1757-899X/735/1/012061

Development of an Anti-corrosion and Anti-pollution Stone Material

Yuning Lu, Xiang Lin, Pei Zhao , Yanhao Zhang, Enyao Li and Su Zhao *

Institute of Materials Science and Engineering, Shenyang Jianzhu University, Shenyang 110168, Liaoning.

Email:zpljh123@163.com

Abstract. Stone is a non-renewable resource. In order to implement sustainable development in the field of building materials, it is necessary to carry out feasible protection for the existing stone. At present, the performance of the protection materials is relatively single. How to give consideration to both waterproof and anti-corrosion and anti-pollution properties is the focus of this study. Using tetraethyl orthosilicate (TEOS) as silicon source and adding appropriate modifier (silicone oil), the modified silicon stone protective material was prepared by sol-gel method. The ratio ranges of water, ethanol, hydrochloric acid and TEOS were determined by experiments, which were 1:1, 1: 1 ~ 1: 2, 1: 10 ~ 1: 20 respectively, the reaction temperature was 60℃, and the reaction time was 3.5h .Under the experimental conditions, adding modifier hydroxyl silicone oil (0.2%) can improve the anti-corrosion and stain resistance of stone protection materials.

1. Introduction

Entering the 21st century, China's stone industry ranks first in the world for many years. Stone resources belong to non-renewable resources. From the perspective of circular economy, no matter whether the resources are abundant or not and whether the utilization of resources is efficient or not, the final result will be the exhaustion of resources. How to protect the existing stone and slow down the loss of natural resources caused by environmental pollution and acid rain corrosion are the biggest problems currently facing the stone field [1].

At present, organic silicon protective materials have been tried to be applied in the field of material protection [2-3]. Its mechanism of action is through dehydration condensation reaction with silanol groups (Si-OH) on the stone surface, and a hydrophobic film is formed on the stone surface. Organic silicon protective materials can be prepared by sol-gel method. However, the biggest defect of stone protective materials prepared by sol-gel method is that they are easy to crack in the drying process of stone pores [4-5], so modification is needed. In this paper, silicone oil modifier is introduced to increase the pore diameter of the protective layer to make it evenly distributed, and to increase the toughness of the protective layer to prevent cracking, so that it has the effects of waterproof, anticorrosion and stain resistance.

2. Experiment and Test Methods

2.1. Stone Processing

Two commercially available white and yellow sandstones were cut into cuboids with a size of 5x5x2cm^3 ,soaked in distilled water for 12 hours, cleaned with an ultrasonic cleaner for 30min, and then coated with stone protection materials after naturally drying for 24 hours.

2.2. Preparation Method of Organic Silicone Stone Protective Material
A proper amount of tetraethyl orthosilicate, a certain amount of modifier and absolute ethyl alcohol were added into a three-necked flask, the temperature was increased after full stirring, a small amount of hydrochloric acid solution was added dropwise after reaction for a certain time, then distilled water was added dropwise, and a constant temperature hydrolysis condensation reaction was carried out for a certain time to obtain transparent hydrolysate. Finally, the solid content was determined to prepare a modified silicon protective agent with a certain concentration.

2.3. Performance Test Methods

2.3.1. Contact Angle Test. Contact angle can quantitatively reflect the hydrophobicity of stone surface. Static contact angles of stone samples before and after sealing were measured by surface contact angle measuring instrument (GW360A), and compared with hydrophobic performance grade, with hydrophobic grade ranging from grade I to grade v [6], as shown in table 1.

Table 1. Relationship between Surface Hydrophobic Effect and Contact Angle

Hydrophobic capacity grade	Contact angle θ/°	Hydrophobic effect
I	θ≤30°	Fully moist
II	30°<θ≤90°	Obvious wetness
III	90°<θ≤110°	Slightly moist
Ⅳ	110<θ≤130°	Good hydrophobicity
V	θ>130°	Very good hydrophobicity

2.3.2. Acid Resistance Test. Weighed the treated stone sample to M_0, soaked it in sulfuric acid solution with PH=5 for 48h, the liquid level should be 50mm higher than the surface of the sample, then took out the sample and washed it with deionized water for three times, and dried it, weighed it to M_1, dried it at 60℃, and calculated the mass change rate *W* of the stone sample:

$$W=\frac{M_1-M_0}{M_0}\times 100\%$$

The acid corrosion is reflected by the mass change rate. The larger the mass change, the more serious the corrosion and the smaller the mass change rate, the better the acid resistance [7].

2.3.3. Pollution Resistance Testa. The physical standard for making stain-resistant experiments with white marble [2]: Prepared solutions of different proportions with ink and distilled water and dropped them onto the white marble, then let it stand for 5 minutes, rinsed with distilled water, air dry, and graded them after comparing with the evaluation standard. Table 2 for stain resistance rating.

Table 2. Pollution Resistance Grade Table

Pollution resistance grade	1	2	3	4	5	6
V (ink) / V(distilled water)	1:0	1:2	1:4	1:6	1:8	1:10

3. Results and Discussion

3.1. Determination of Material Proportion

On the premise of fixed catalyst hydrochloric acid dosage, solvent ethanol dosage and other reaction conditions (constant temperature 65℃, stirring reaction time 2.5h, constant TEOS dosage, modifier hydroxyl silicone oil dosage 0.1%), the ratio of reaction materials was determined by single factor experiment.

3.1.1. Effect of Water on Hydrophobicity. The mass ratio of water to ethyl orthosilicate was 1:2, 1:1, 2:1, and 3:1. The influence of water content on hydrophobicity was discussed with reference to contact angles of different proportions. The contact angle is shown in Figure 1.

As shown in Figure1, the effect of water content on hydrophobicity increases first and then decreases, reaching the best effect when the mass and TEOS are 1:1.This is because water is an important reactant in hydrolytic condensation, too little will affect the progress of the reaction, and too much water will affect the solubility of the final product, thus reducing the waterproof effect of the product.

3.1.2. Effect of Ethanol on Hydrophobicity. The mass ratio of ethanol to TEOS was 1:2, 1:1, 2:1, and 3:1. The influence of ethanol content on hydrophobicity was discussed with reference to contact angles of different ratios. The contact angle is shown in Figure 2.

As shown in Figure 2, the amount of ethanol has an effect on the hydrophobicity of both sandstones, reaching the best effect approximately around 2:1 of ethanol mass and TEOS. This is because when ethanol is used as a blending agent, the reactants cannot be completely dispersed when the amount is small, thus affecting the full progress of the reaction, and too much ethanol will evaporate with the increase of temperature, taking away some reaction products and reducing the reaction performance.

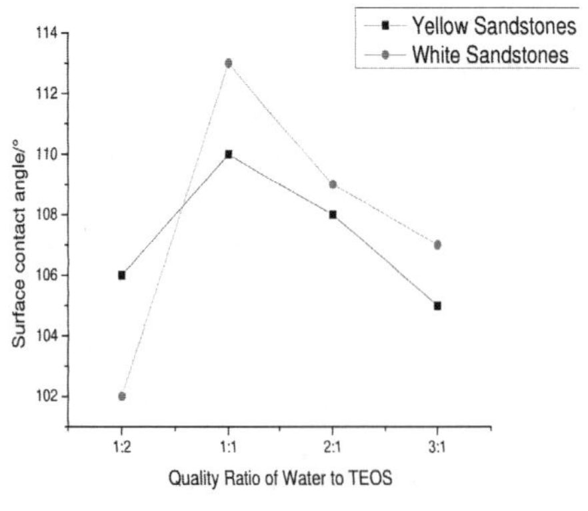

Figure 1. Relationship between Water Content and Contact Angle

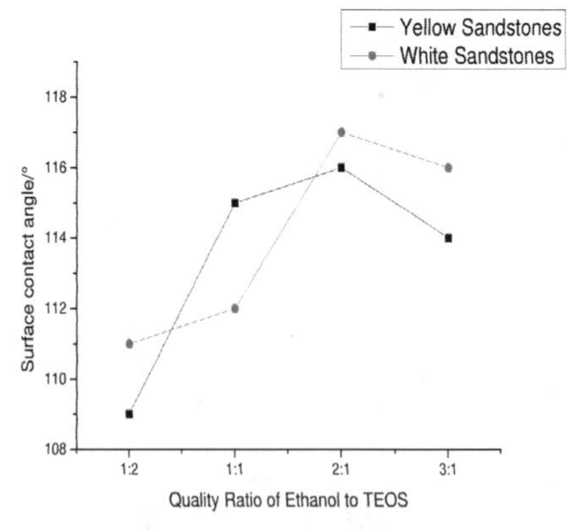

Figure 2. Relationship between Ethanol Content and Contact Angle

3.1.3. Effect of Hydrochloric Acid on Hydrophobicity. The mass ratio of hydrochloric acid to TEOS was 1:4, 1:10, 1:20 and 1:30. The influence of hydrochloric acid content on hydrophobicity was discussed with reference to contact angles of different ratios. The contact angle is shown in Figure 3.

As shown in Figure 3, the influence of hydrochloric acid as a catalyst on the hydrophobic property fluctuates slightly, and the contact angle changes of the two stones are not different, but the amount of hydrochloric acid has an influence on the reaction time, so the amount of hydrochloric acid in this experiment is generally between 1:10 and 1:20.

To sum up, under the experimental conditions, the ratio range of water, ethanol, hydrochloric acid and TEOS is: 1:1, 1: 1 ~ 1: 2, 1: 10 ~ 1: 20.

3.2. Determination of Reaction Conditions

3.2.1. Effect of Temperature on Hydrophobicity. The reaction temperature was chosen to be 50°C, 55°C, 60°C, 65°C, 70°C to discuss the effect of temperature on hydrophobicity. The contact angle is shown in Figure 4.

As shown in Figure 4, the influence of temperature on the hydrophobic property of both sandstones increases slightly first and then decreases, too low a temperature will lead to incomplete reaction, while when the temperature is higher than 60°C, the contact angle drops sharply, indicating that too high a temperature will lead to the condensation reaction to start earlier without complete hydrolysis, thus greatly reducing the hydrophobicity of the final product. Therefore, the experimental reaction temperature is chosen to be about 60°C.

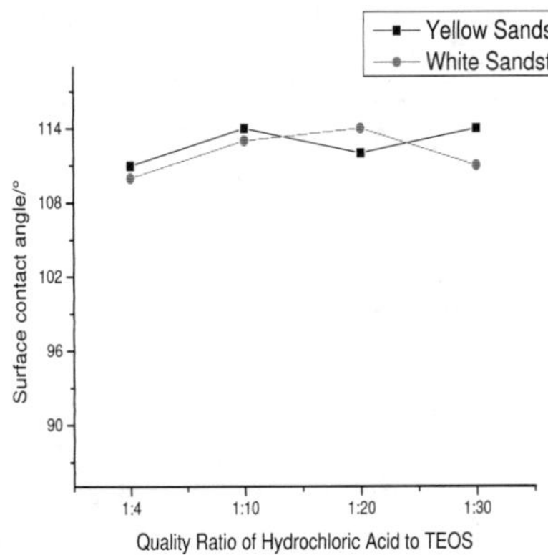

Figure 3. Relationship between Hydrochloric Acid Dosage and Contact Angle

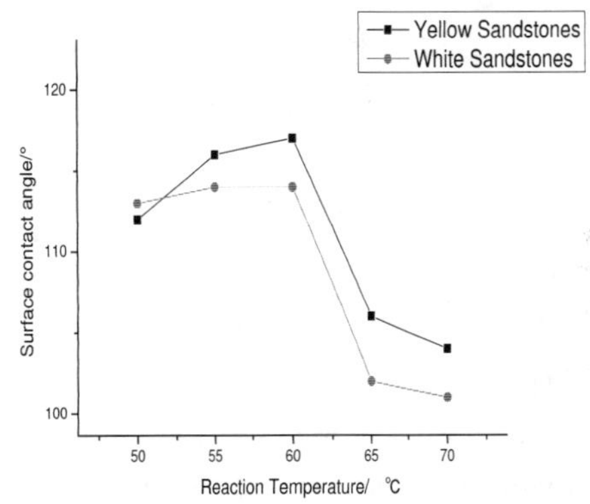

Figure 4. Relationship between Reaction Temperature and Contact Angle

3.2.2. Effect of Reaction Time on Hydrophobicity. The reaction time was chosen to be 2.5h, 3h, 3.5h, 4h, 4.5h, and the effect of reaction time on hydrophobicity was discussed.The contact angle is shown in Figure 5.

As shown in Figure 5, the contact angle continuously increases before the reaction time reaches 3.5h, and tends to be stable after reaching 3.5h, indicating that the reaction is complete at this time, so the reaction time is selected as 3.5h.

3.3. Determination of Modifiers

Hydroxy silicone oil and dimethyl silicone oil were selected as modifiers, and the dosages are 0, 0.1%, 0.2%, 0.3%, 0.4% and 0.5% respectively. The influence of modifier on hydrophobicity was discussed with reference to contact angles of different dosages. The results are shown in Figure 6.

As can be seen from Figure 6, the hydrophobicity of hydroxyl silicone oil is significantly better than that of dimethyl silicone oil. Hydroxy silicone oil can undergo hydrolytic condensation reaction due to the presence of hydroxyl, and $-CH_3$, the outermost hydrophobic group of hydroxyl silicone oil, is successfully linked to the hydrophobic membrane and acts on the outermost side of the whole system, thus improving the hydrophobic capability. However, dimethyl silicone oil has no hydroxyl group that can undergo dehydration condensation and can only achieve common effect, so its hydrophobic ability is not strong.

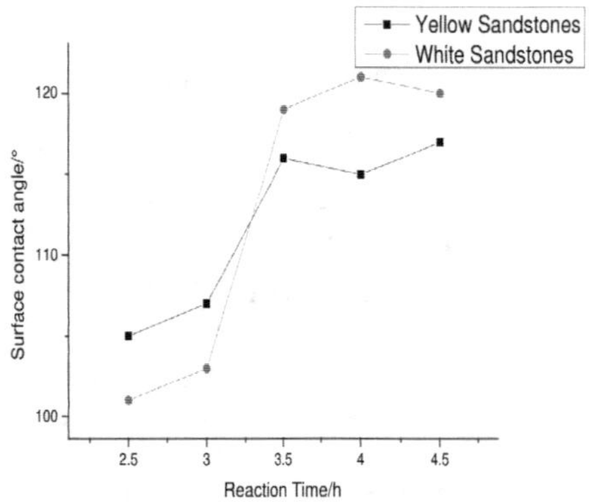

Figure 5. Relationship between Reaction Time and Contact Angle

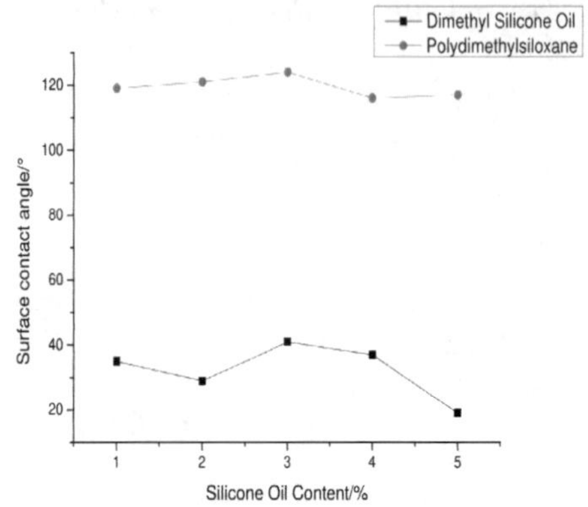

Figure 6. Relationship between Silicone Oil Content and Contact Angle

3.4. Relevant Performance Test

3.4.1 Acid and Corrosion Resistance Test. Stone protective materials before and after modification with hydroxyl silicone oil (content: 0.2%) were coated on the surfaces of two kinds of sandstones, and acid and corrosion resistance tests were carried out according to the method of 1.3.2. The results are shown in Table 3.

From Table 3, it can be seen that the mass change rate of the modified protective material sample is obviously reduced, indicating that the acid resistance is enhanced. The addition of hydroxyl silicone oil can increase more Si-O-Si bonds, and the increase of Si-O-Si bonds can weaken the oxidation ability of acid, thus improving the ability of stone sample surface to resist acid corrosion. Through comparison, it can be seen that the corrosion resistance of the protective material to the yellow stone is improved more obviously, because compared with the white stone, the main mineral composition of the yellow stone is CaCO3, and the existence of CaCO3 causes the stone to be more easily corroded by acidic substances and damages the integrity. The corrosion resistance of the yellow stone sealed by the organic silicon protective material will be improved more obviously.

Table 3. Acid Corrosion Resistance Test Results

Stone type	Hydroxy silicone oil	Poor quality /g	Integrity
Yellow sandstone	No	-0.66	Incomplete
Yellow sandstone	Yes	-0.23	More complete
White sandstone	No	-0.22	Incomplete
White sandstone	Yes	-0.12	More complete

3.4.2 Pollution Resistance Test. Stone protective materials before and after modification with hydroxyl silicone oil (content: 0.2%) were coated on the surfaces of two kinds of sandstones, and the stain resistance was tested according to the method of 1.3.3. The results are shown in Table 4.

From Table 4, it can be seen that the stain resistance level of the modified protective material sample is increased from no stain resistance effect to Grade 5, and the stain resistance ability is greatly improved, because the surface of the stone sample sealed by the modified silicon protective agent forms a compact and smooth hydrophobic film, thus reducing the roughness of the surface of the stone sample, reducing the attachment points of water molecules and contaminated particles, and improving its stain resistance ability.

IOP Conf. Series: Materials Science and Engineering **735** (2020) 012061 doi:10.1088/1757-899X/735/1/012061

Table 4. Pollution Resistance Test Results

Stone type	Hydroxy silicone oil	Pollution resistance rating
Yellow sandstone	No	______
Yellow sandstone	Yes	Five
White sandstone	No	______
White sandstone	Yes	Five

4. Conclusion

(1)The ratio range of water, ethanol, hydrochloric acid and TEOS is determined by experiments, which are 1:1, 1: 1 ~ 1: 2 and 1: 10 ~ 1: 20 respectively.

(2)Under the conditions of 60℃ temperature and 3.5h reaction time, the prepared modified stone protective material has better hydrophobic effect.

(3) Adding 0.2% modifier hydroxyl silicone oil can not only improve the hydrophobicity of stone protection materials, but also greatly improve their corrosion resistance and stain resistance.

5. References

[1] Chen Weixu, Zhang Jiyu. Cherish the Stone Resources in China and Protect the Sustainable Development of Mines [J]. Journal of Fuzhou University (Philosophy and Social Science Edition), 2009, 23(02):35-37+48.
[2] Su Jingliang, Wu Ge. Research and Application Progress of Stone Protective Agents [J]. Liaoning Chemical Industry, 2013, 42(04):349-351.
[3] Liu Jia. Preparation of Modified Silicone Protective Agent and Its Application on Baodingshan Sand Rock [D]. Chongqing University of Technology, 2015.
[4] SUPPLIT R, SCHUBERT U. Corrosion protection of aluminum pigments by sol-gel coatings [J]. Corrosion Science, 2007, 49: 3325-3332.
[5] Zhao Qiang, Qiu Jianhui, Peng Cheng. Study on Application of Silicate Oligomer in Stone Protection [J]. Cultural Relics Protection and Archaeology, 2007(03):26-31.
[6] Maria J. Mosquera, Desiree´ M. de los Santos, Antonio Montes. New Nanomaterials for Consolidating Stone [J]. Langmuir 2008, 24(6), 2772-2778.
[7] Sun Shunjie, Hong Yongshun, Zhang Lin. Study on Properties of Silicone Stone Protective Agent [J]. Stone, 2008(03):8-13.

Ecological Toxicity of Triclosan (TCS): A Review

Xiaolu Xu, Deyong Zhang* and Wuchun Zhang

College of Biology and Environmental Engineering, Zhejiang Shuren University, 8 Shuren Road, Hangzhou, China.
Email: deyonghz@163.com

Abstract. TCS has been detected in water systems and organisms all over the world. TCS in water body mostly exists at ng/L or g/L level. The highest content of TCS is mostly from sewage plant water samples, some of which reach 26.80 g/L. TCS concentration was 0.4-38.0 ng/g in maternal plasma and 0.022-0.95 ng/g in breast milk. TCS has been proved to be toxic to algae, plants, lower animals, amphibians, fish, mammals and other organisms at all levels, mainly in growth inhibition, lethal effect, endocrine interference, reproductive toxicity, DNA toxicity and so on.

1. Introduction

As an important antimicrobial agent, triclosan (TCS) has been widely used in personal care products (such as toothpaste, antimicrobial soap, skin cream and decolorizer), daily consumer products (such as fibre fabrics, Plastic Kitchenware and sports shoes and socks), medical supplies (such as dental consumables, medical preservatives and bactericides) and household cleaning products (such as household detergents and disinfectants) for more than 30 years. The annual use of it in the world is estimated to be over 750×10^3 tons [1]. Due to its strong lipophilicity and accumulation ability in organisms, TCS are posing a great ecological threat.

2. TCS in the Environment and in the Organisms

Environmental TCS has been reported worldwide. The Geological Exploration Association of United States (USGS) has studied 139 rivers in 30 states of the United States and found that TCS is the most frequently detected organic pollutant [2]. Van der Ohe et al. detected TCS in 802 sites(representing 572 water systems in Saxony, northern Germany) in the Elbe River Basin, indicating that TCS should be considered as a pollutant of priority concern. Ying et al. monitored TCS in water and sediments of the Pearl River, Yangtze River, Yellow River, Haihe River and Liaohe River in China. TCS showed a high occurring rate in water and sediments, and the high concentration is generally located in the polluted tributaries. The maximum detectable concentrations in water and sediment reached 478ng/L and 1329 ng/g respectively [3]. The detected concentration of TCS in sewage treatment plant is higher, reaching 1.86-26.8 g/L. Through cluster analysis and principal component analysis, it was found that urban domestic sewage discharge was the main source of TCS [4].

TCS has strong lipophilicity and can accumulate in various organisms [5]. According to some reports, TCS in various organisms was 50-400 g/kg in Chaetophyta algae, 50-300g/kg in muscle tissue of freshwater snail, 14000-80 000 g/kg in bile of Oriental blackbream (male), 0.75-10 g/kg in ocean fish plasma, 0.12-0.27 g/kg in South Carolina Atlantic bottlenose dolphin plasma, 0.025-0.11g/kg in Atlantic bottlenose dolphin plasma in Florida Indian River, and 9.0g/kg in killer whale plasma in Vancouver Aquarium. Also, TCS has been detected in human urine, breast milk and plasma by American and Swedish scientists. In 2008, American researchers published 2517 urine samples of volunteers, of which the concentration of TCS was about 2.4-3790ng/mL. The plasma TCS concentration of lactating mothers using TCS products was 0.4-38.0ng/g, and that of breast milk was 0.022-0.95ng/g. The plasma TCS concentration of mothers without TCS products was 0.01-19 ng/g, and that of breast milk was 0.018-0.35 ng/g [6, 7].

3. Toxic Effects of TCS on Various Organisms

3.1. Algae

Among aquatic species, algae are considered to be the most susceptible to TCS, mainly showing acute toxicity. For example, both green algae (freshwater algae) and Dunaliella (an oceanic algae) are very sensitive to TCS. TCS could significantly inhibit the growth of freshwater algae, with a 72h-EC_{50} of about 2.8 g/L [8]. The toxicity of TCS to Scenedesmus obliquus fluctuates due to the influence of phosphorus concentration. The effects of different phosphorus concentrations (0-50mg/L) and TCS (50, 200 g/L) on the growth, chlorophyll fluorescence characteristics and physiological and biochemical indices of Scenedesmus obliquus were analyzed. Under three phosphorus concentration conditions (0, 2.5, 50.0 mg/L), TCS showed a dose-effect relationship on the growth inhibition of Scenedesmus obliquus, and its chlorophyll fluorescence characteristics indicated a dose-effect relationship. Standard Yield-F/Fo and NPQ were affected to varying degrees. The activities of antioxidant enzymes (SOD, CAT) and phosphorus metabolism related enzymes (ACP, AKP) also changed accordingly [9].

3.2. Plants

TCS also affects the growth and development of plants. Wu et al. found that TCS could accumulate in the root of legumes and transfer to the upper part of legumes in the soil through which organic solid waste and wastewater flowed [10]. TCS in soil can inhibit the growth of many kinds of plants. EC50 of rice seeds and cucumber seeds germination are 57 mg/kg and 108 mg/kg [11]. Amorim et al. confirmed that TCS has an effect on the growth of wheat and Brassica plants [12].

3.3. Lower Animals

TCS is toxic to various lower aquatic and terrestrial animals, including daphnets, snails, tetrahedrons, earthworms and so on. The toxic effects include growth inhibition, reproductive inhibition and lethal effects. TCS was recognized as RED insecticide by the U.S. Environmental Agency and is hazardous to lower organisms [13]. TCS showed acute toxicity to Daphnia magna, with a 48h-EC_{50} of 390 g/L [8]. TCS at μg/L level caused growth inhibition on *Tetrahymena*, with 24h-EC_{50} of 141μg/L, minimum ineffective response concentration of 2μg/L and minimum effective concentration of 4μg/L. In addition, when the concentration of TCS reached 1000 μg/L, the cell membrane of *Tetrahymena* was damaged obviously. TCS also affected the lysosomal activity in *Tetrahymena*. Treatment with TCS of 1 μg/L for 2 h resulted in a 88.62% lysosomal activity [14]. TCS of 1.0 μM/L significantly decreased the sperm activity of sea urchin, and 2.0μM/L leaded to sperm inactivation. TCS of 0.5μM/L significantly reduced the fertilization rate and 1.5 μM/L TCS resulted in totally failed fertilization [15]. Four weeks exposure TCC (TCS analogues) to a freshwater snail confirmed that the lowest effective concentration (LOEC) and the lowest ineffective response concentration (NOEC) were 0.2μg/L and 0.05μg/L respectively, and EC_{50} were 2.5μg/L [16]. Chronic exposure to 50-300 mg/kg TCS significantly reduced the reproductive capacity in earthworms, and the EC_{50} was 142.11 mg/kg. Comet assay showed that TCS caused DNA damage with EC_{50} value of 8.85 mg/kg. The expression level of Hsp70 gene in 50 mg/kg group was up-regulated, which was 2.28 times higher than that in control group. EC_{50} value based on Hsp70 biomarker was 1.79 mg/kg [17].

3.4. Fishes

When exposed to TCS solution, especially at high concentration, the incubation time of fish embryos was significantly delayed and teratogenic effects were produced on larvae and adults. TCS showed acute toxicity to Pimephales promelas, with a 24h-LC_{50} of 360 g/L [8]. TCS and its derivatives have an effect on the secretion of male hormones in fish, and belong to newly recognized endocrine interferons. Also, It may cause many problems including cancer, reproductive dysfunction and developmental abnormalities [18]. The 96h-EC_{50} to Brassica brasiliensis was 10-13 g/L, and 96 h-LC_{50} to blue mumps and rainbow trout was 97μg/L and 180μg/L respectively [19]. The effects of TCS on the reproduction of blue gill fish were also confirmed. Fourteen days of exposure to 313 g/L TCS resulted in the decrease of hatching ability of fertilized eggs and the prolongation of hatching time [20].

3.5. Mammals

Thirty days exposure to TCS (40 to 632 mg/kg) in rats caused hepatotoxicity, nephrotoxicity, damaged immune system. The observed effects included decreased weight gain, spleen-body ratio, and increased liver-body ratio, male-kidney-body ratio, serum alanine aminotransferase, urea nitrogen and creatinine [21]. TCS exposure showed effects on the pregnancy outcome of pregnant rats. The pregnant rats had abnormal feeding and weight gain ability, blood sugar, serum malondialdehyde, serum protein and other indicators. The composition of live fetus and absorbed fetus of pregnant rats also changed [22]. TCS caused DNA damage in human hepatocytes. DNA breakage in normal human hepatocytes cultured in vitro was observed, with a significant dose-effect relationship and time-effect relationship [23]. After short-term intake of TCS in male mice, TCS was detected in testis, epididymis and prostate. It was found that TCS tended to accumulate in epididymis, showing a significantly longer half-life and accumulation time. High dose of TCS (200 mg/kg) was also shown to reduce sperm production and cause sperm malformation in rats [24]. A short time exposure to TCS in male rats indicated that the expression levels of several important proteins such as testosterone emergency monitoring protein, androgen receptor and so on were significantly decreased. The levels of luteinizing hormone, follicle stimulating hormone, cholesterol, progesterone and testosterone in serum were significantly decreased, showing a significant impact on sex hormones [25].

4. Discussion

As mentioned above, TCS has been widely detected in the environment and organisms worldwide, and its toxic effects on organisms have been confirmed. While, most studies are conducted in recent years, and the research of TCS is not deep or systematic enough. For example, the following questions are worth discussing: (1) Most studies focused on acute and subacute experiment, while chronic toxicity are poorly studied. In fact, people and animals are generally exposed to TCS substances via long-term low-dose exposure. The negative effects of this contact mode are quite different from that of acute mode. Therefore, chronic toxicity studies over six months should be strengthened to reflect its toxic effects better. (2) There have been many studies on TCS itself, while less attention is paid to its derivative or degradation products. TCS has been confirmed to undergo photolysis. When exposed to sunlight or ultraviolet radiation, about 12% of dissolved TCS can be converted to dioxins. In addition, TCS can react rapidly with manganese oxides such as δ-MnO2 and MnOOH. TCS partially degrades or derives many toxic by-products, such as phenoxyphenol, chlorinated phenol, trihalomethane and dioxins. The toxicity, persistence and enrichment of degradation products of TCS may be higher than that of TCS. Therefore, attention should be paid to the mechanism, conditions and influencing factors of the transformation and degradation of TCS. (3) There have been more studies on lower organisms than on higher animals and human beings. Because TCS is used as a product of direct contact with human body, its threat to human health should be taken as an important concern. Such problem as the potential carcinogenic effect of TCS, which is widely concerned, has still no certain conclusion. (4) There are many observational studies on toxic effects of TCS but too few mechanism studies. The toxic mechanisms observed in many studied are still poorly understood. (5) There has been much detection in water but too less in other environmental medium like soil and air.

5. Acknowledgement

This study was financially supported by Zhejiang Provincial Natural Science Foundation of China (LY18C030003), Zhejiang provincial philosophy and social sciences plan (19NDJC262YB) and College Students Innovation and Entrepreneurship Training Program of China *Removal of Cu2+ and Al3+ by Biocharcoal Adsorption.*

6. References

[1] Bradley OC, Stephen RS, 2011, *Environ Int.*, 37(1): 226-247.
[2] Kolpin DW, Furlong ET, Meyer MT. 2002, *Environ Sci Technol*, 36(6): 1202-1211.
[3] Zhao JL, Zhang QQ, Chen F, Wang L, Ying GG, Liu YS, Yang B, Zhou LJ, Liu S, Su HC, Zhang RQ. 2013, Water Res., 47(1): 395-405.
[4] Chalew TE, Halder RU. 2009, *J Am Water Resour As*, 45(1): 4-13.
[5] Snyder EH, O'Connor GA, McAvoy DC. 2010, *Sci Total Environ*, 48 (13):2667-2673
[6] Gilles B, Benoit R, Olivier T, Virginie D, Barbara LB. 2012, *Environ Sci Pollut R*, 19(4): 1044-1065.
[7] Von der Ohe PC, Jaroslav Slobodnik MSJ, Wemer Brack. 2012, *Environ Sci Pollut Res*, 19: 585-591.
[8] Orvos DR, Versteeg DJ, Inauen J, Capdevielle M, Rothenstein A, Cunningham V. 2002, *Environ Toxicol Chem*, 21(7): 1338-1349.
[9] Wu YY, Jiang LJ, Xiao L. 2013, *Journal of anhui agricultural sciences*, 41(14): 6412-6415.(in Chinese)
[10] Wu CX, Spongberg AL, Witter JD, Fang M, Czajkowski K. 2010, *Environ Sci Technol.* 44:6157–6161
[11] Liu F, Ying GG, Yang LH, Zhou QX. 2009, *Ecotoxicol Environ Saf*, 72:86–92
[12] Amorim MJB, Oliveira E, Soares AMVM, Scott-Fordsmand JJ. 2010, *Environ Int*, 36:338–343
[13] Zhao JL, Ying GG, Liu YS, Chen F, Yang JF, Wang L. 2010, *J Hazard Mater*, 179(3):215-222.
[14] Bai QF, Gao L, Yuan T. *Environmental Chemistry*, 2012, 31(5):720-725(in Chinese)
[15] Hwang J, Suh SS, Chang M, Yun Park S, Ryu TK, Lee S, Lee TK. 2014, *Ecotoxicol Environ Saf.* 100:148-52
[16] Ben DG, Thomas MY. 2010, *Environmental Toxicology and Chemistry*, 29(4):966-970.
[17] Lin DS, Li Y, Zhou QX, Xu YM, Wang D. 2014, *Ecotoxicology*. 23(10):1826-1832.
[18] Foran CM, Bennett ER, Benson WH. 2000, *Mar Environ Res.*, 5(1): 153-156.
[19] Consortium T. 2002, *Iuclid Data Set TCC Consortium.* (New York)pp 44.
[20] Ishibashi H, Matsumura N, Hirano M, Matsuoka M, Shiratsuchi H, Ishibashi Y. 2004, *Aquat Toxicol.* 67(2):167-79.
[21] Jiang SS, Zhou L, Li YL, Qiao SS, Dong J. *China Occupational Medicine*. 2006, 33(6):432-434(in Chinese).
[22] Zhang GX, Wang LL, Dong JY, Wei XY, Ding JC. *Journal of Chongqing Medical University*. 2013, 38(10): 1163-1168(in Chinese).
[23] Li LP, Ma HM, Hu JJ, Li YH, Wang YP, Sheng GY. *Ecology and Environmental Science*, 2010, 19(12): 2897-2901(in Chinese).
[24] Zhou L, Tae HK, Kai SB, Xiao HC, Hyung SK. 2015, *Environ Toxicol*. 30(1):83-91
[25] Kumar V, Chakraborty A, Kural MR, Roy P. 2009, *Reprod Toxicol*. 27(2):177-85.

Design of A New High Efficiency Heat Storage System for Small and Medium-sized Photothermal Power Stations

Yipu Wang and Bingzheng Wang

School of Power and Mechanical Engineering, Wuhan University, Wuhan, Hubei, 430072, China
Email: yipuwang_whu@163.com

Abstract. In order to solve the problems of low temperature and high cost of traditional photothermal power generation heat storage system, a new high-temperature solid heat storage system consisting of three subsystems of heat extraction, heat storage and heat utilization is designed. Taking air as heat medium and safe heat-resisting regenerator as heat storage material, the heat preservation characteristics, overall thermal efficiency and flow resistance of the regenerator were analyzed and calculated. The results show that the regenerative system runs steadily, the heat loss of the regenerator is less than 5%, and the overall thermal efficiency of the system is more than 85%, which is better than the general regenerative system.

1. Introduction

Intermittence and instability of solar energy directly affect the continuity and stability of the system in the process of photothermal utilization, and high-temperature heat storage technology is the key to solve this problem. In recent years, it has received great attention from domestic research institutions and related enterprises [1]. In a regenerative system, a complete energy storage cycle consists of three stages: heat storage, storage and heat release [2].

Scholars at home and abroad have done a lot of research on numerical calculation and experiment of photothermal regenerative system. Mu Zhijun [3] et al. carried out an experimental study on the solar photovoltaic integrated photothermal system. The analysis results show that the electrical efficiency of the system is about 7% higher than that of the conventional photovoltaic system. Zheng Jiantao [4] et al. simulated the non-uniform heating of the solar column heat absorber, and N.G. Barton [5] analyzed and compared the effects of two different operation modes on the overall heat storage performance of the system through numerical simulation.

Based on the above research, a high-temperature solid heat storage system is proposed for small and medium-sized solar photothermal power generation projects. High temperature resistant solid particles are used as heat storage and the air is used as a heat medium. Compared with other types of heat storage systems, this system has obvious advantages: (1) The working temperature range of hot medium air is wide, non-toxic and harmless. (2) The regenerator has a wide range of sources, low cost, high-temperature resistance, and the regenerative temperature can reach more than 1000 °C. (3) There is no need to use corrosion resistant materials for the shell and pipeline of the regenerator. (4) The system has good security, low investment and low operating cost. It cannot only generate photothermal power, but also meet the actual industrial production and daily heat demand, and improve the utilization efficiency and scope of solar energy.

2. Overall Design Schemes of the Regenerative System

The design of the regenerative system should meet the following principles:

- The thermal efficiency of the regenerative system is over 85%.
- The system is running continuously and steadily, and the solar energy provides the energy needed by the system.
- The system can meet the requirements of multi-function.
- The flow resistance of the regenerative system is small and the operating cost is low.
- The system should be easy to adjust and operate flexibly to meet other operations such as maintenance.

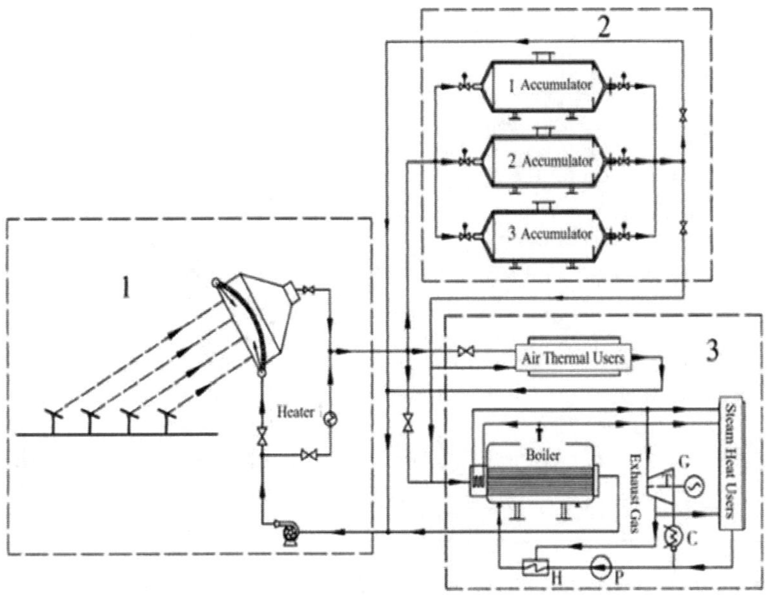

Figure 1. Schematic Diagram of High-Temperature Solid Heat Storage System for Photothermal Power Generation

As shown in Figure 1, the high-temperature solid heat storage system of photothermal power generation consists of three parts: heat extraction subsystem, heat storage subsystem and heat utilization subsystem. A closed loop can be formed between the two subsystems. When the tower solar concentrator arranged in the heat extraction subsystem 1 works [6], the heliostat reflects the solar beam to the collector, and the heat medium at the collector absorbs the solar radiation energy and converts it into high-temperature heat energy. Electric heaters are installed in the end bypass of the heating subsystem to prevent the influence of continuous rainy weather on the operation of the whole system. The regenerator subsystem 2 is composed of several regenerators. The regenerator is filled with regenerative materials, and the high-temperature medium exchanges heat with the regenerative medium to realize the effective storage of heat energy, which solves the problems of intermittent and unstable solar energy. The heat subsystem 3 includes air heating user and steam boiler cogeneration system. Under good lighting conditions, heat is obtained from the collector. One part is supplied to the heating subsystem and the other part is stored in the heat storage subsystem. When the illumination condition is not good, the collector of the collector cannot meet the need of the heat system or in the absence of illumination, the regenerator of the regenerative system carries out heat release to supply the heat subsystem.

3. Optimum Design of Regenerator

3.1. Determination of Structure and Capacity of Regenerator

The structure of the regenerator is shown in Figure 2. The regenerator is arranged horizontally and consists of two parts: the main body of the regenerator and the thermal insulation layer. The filling material of the inner filling layer of the regenerator is ceramic ball-high Al ball ($Al_2O_3 \cdot SiO_2$). The design maximum temperature is 900°C, the external surface temperature of the regenerator is 50°C, and the ambient temperature is 20°C.

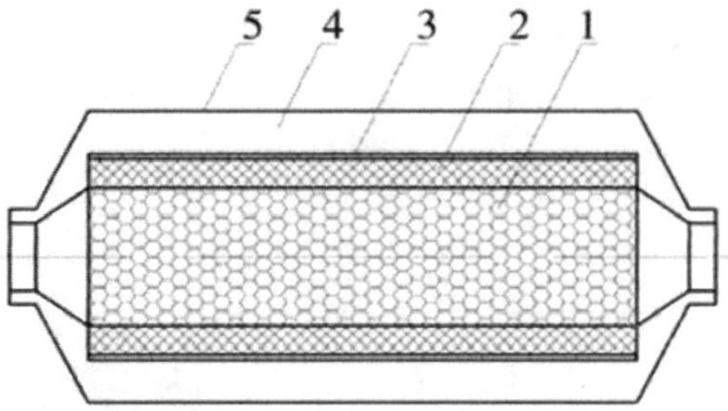

Figure 2. Structural Chart of High-Temperature Solid Regenerator.

In order to minimize the environmental heat loss and take into account the cost and load-bearing factors, the inner wall of the thermal insulation layer of the regenerator is built with refractory brick 2, the tank made of Superalloy steel 3 on the outer wall, and the expanded perlite 4 on the outer wall is used as the thermal insulation layer. In order to reduce the radiation loss of the regenerator, the outer layer of the regenerator is coated with aluminum foil 5 with high reflectivity. The size calculation of each structural material of the regenerator is shown in Table 1.

Table 1. Internal Diameter of Regenerator and Thickness of Insulation Material

Regenerator diameter / mm	Refractory bricks / mm	Steel plate / mm	Expanded perlite / mm
2000	289	6	300

In order to simplify the analysis and calculation, the following assumptions are made for the regenerator:

- The thermal insulation layer of the regenerator has better thermal insulation effect, and the starting temperature of the regenerative material is 300°C.
- The average daily heat storage time of the regenerator is 8 hours.
- The heat storage period is 3 days and 2 days sunshine is better.
- The maximum regenerative temperature of the regenerator is 900°C.

The total heat power of the regenerative system is 1.0 MW. 80% of the heat collected by the collector is used for heat storage and 20% for heat release. There are three regenerators in the regenerative system. In a working cycle, the total heat collected by the system is 57600 MJ, and the heat stored by the regenerative system is 46080 MJ. The relevant calculation of the regenerator is shown in Tables 2 to 4.

Table 2. Relevant Calculation of Regenerative Spheres in a Single Regenerator

Total heat storage of single regenerator / MJ	Diameter of regenerative sphere / m	Volume of single regenerative sphere / m^3	Specific heat capacity / $kJ \cdot (kg \cdot ℃)^{-1}$	Regenerative sphere density / $kg \cdot m^{-3}$
1.5360	0.03	1.41×10^{-5}	1.0	2000
Heat transfer temperature difference / ℃	Total volume of regenerative sphere / m^3	Number of regenerative balls/ Pieces	Total mass of regenerative ball / kg	Total surface area of regenerative sphere / m^2
600	12.8	9.078×10^5	25600	2569

Table 3. Relevant Calculation of Hot Air in a Single Regenerator

Thermal Storage	1.293	2.52	1.95	1.066	300	805.8	15360	5.3
Exothermic	1.293	0.71	0.55	1.066	257.7	195.08	13132.8	18.7

Table 4. Relevant Calculations of Heat Accumulators

Diameter of regenerative sphere / mm	Regenerator volume / m^3	Volume voidage of regenerator	Regenerator area voidage	Regenerator Section Area / m^2	Regenerator length / m
0.03	18.055	0.29	0.164	3.14	5.75
Minimum cross-sectional area of flow passage/ m^2	Volume Flow Rate of Gas under Standard Conditions / $m^3 \cdot s^{-1}$	600℃ Volume Flow of Gas / $m^3 \cdot s^{-1}$	Flow Velocity at Minimum Section / $m \cdot s^{-1}$	Flow Velocity at Maximum Cross Section / $m \cdot s^{-1}$	Average velocity / $m \cdot s^{-1}$
0.514	1.95	6.24	12.14	1.99	7.065

3.2. Analysis of Thermal Insulation Characteristics of the Regenerator

Ignoring the thermal resistance between the filling layer and the inner wall of the regenerator, the heat transfer coefficient of the regenerator is as follows:

$$k = \frac{1}{\frac{\sigma_1}{\lambda_1}+\frac{\sigma_2}{\lambda_2}+\frac{1}{h}} \quad (1)$$

In the equation

k - heat transfer coefficient of regenerator / $W \cdot (m^2 \cdot \ )^{-1}$;

δ-thickness/m;

λ - thermal conductivity / $W \cdot (m \cdot \ )^{-1}$;

h-Convection Heat Transfer Coefficient/ $W \cdot (m^2 \cdot \ )^{-1}$;

Subscript 1 - Refractory bricks;

Subscript 2 - Expanded perlite.

In this paper, the thermal insulation characteristics of solid heat accumulator are analyzed by the lumped parameter method. The differential equation of thermal conductivity is as follows:

$$\frac{\partial t}{\partial \tau} = \frac{\dot{\Phi}}{\rho_s c_s} \quad (2)$$

In the equation

t - temperature / ;

τ- heat storage time / h;

ρ-density/$kg \cdot m^{-3}$;

c -constant pressure specific heat capacity / $kJ \cdot (kg \cdot \ )^{-1}$;

$\dot{\Phi}$-volume heat source, $\dot{\Phi} = \frac{-A_0 k(t-t_f)}{V_0}$ (volume of V_0 filling layer, heat transfer area of A_0 filling layer)/$W \cdot m^{-3}$

Subscript s - heat storage material.

It can be sorted out:

$$\rho s c s V_0 \frac{dt}{d\tau} = -kA_0 (t-t_f)$$

$$\frac{\theta}{\theta_0} = \frac{t-t_f}{t_0-t_f} = exp(-\frac{A_0 k}{\rho_s c_s V_0}) \quad (3)$$

In the formula, θ-excess temperature / ℃.
Filling layer temperature t can be obtained by substituting relevant data.

$$t = 10 + exp(- 1.59 \times 10^{-3}\tau) \quad (4)$$

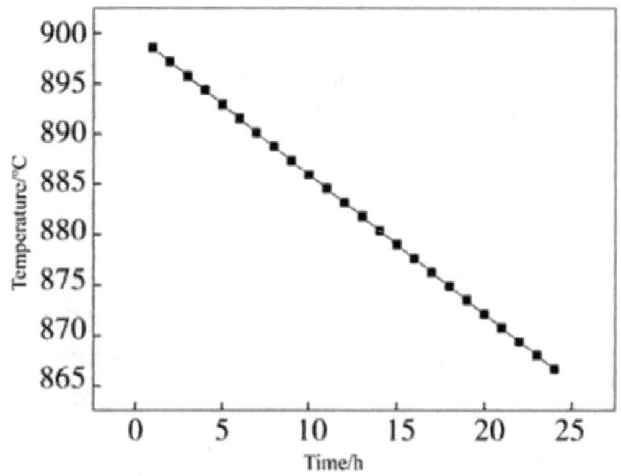

Figure 3. Temperature Variation of Regenerator Filling Layer with Time in One Day

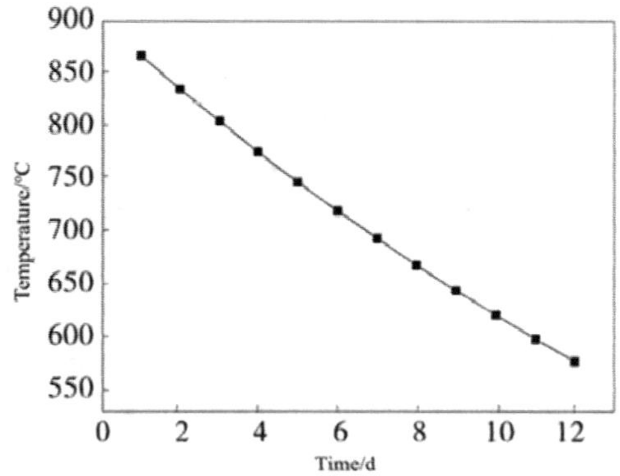

Figure 4. Temperature Variation of Regenerator Filling Layer with Days

Figure 3 and Figure 4 are the relationship between the temperature of the regenerator filling layer and the time in a day and the temperature of the regenerator filling layer with the number of days. From figs. 3 and 4, it can be seen that the temperature of the regenerator filling layer varies linearly with time and the temperature of the regenerator filling layer varies linearly with days. The regenerator keeps the heat for one day, and the temperature drop is about 4.8%. In the practical application process, multiple regenerators are prepared to store heat simultaneously, which basically meets the requirements of industrialization and ensures that the regenerator system can keep heating and generating electricity without long rainy and rainy weather.

4. Thermal Balance and Thermal Efficiency Computation of Systems

4.1. System Heat Balance

The heat loss of the whole system includes the heat loss of the collector, the heat loss of the accumulator, the heat loss of pipes and accessories, and the heat loss of valves and measuring instruments. According to the principle of heat balance, the total power input to the heat storage system should be equal to the algebraic sum of heat storage and heat loss of each part in the heat storage system. That is:

$$Q_0 = Q_1 + Q_2 + Q_3 + Q_4 + Q_5 \quad (5)$$

In the equation:
Q_0 - Total power input to regenerative system / MW;
Q_1 - Heat stored in the regenerator / MW;
Q_2 - heat loss of collector / MW;
Q_3 - heat loss of regenerator / MW;
Q_4 - Heat loss of pipes and accessories / MW;
Q_5 - Heat loss of valves and measuring parts / MW.

4.2. Calculating the Thermal Efficiency of the System

Thermal efficiency of regenerative system:

$$\eta_{xr} = \frac{Q_1}{Q_0} = q_1 \times 100\% = \frac{Q_0-(Q_2+Q_3+Q_4+Q_5)}{Q_0} = 1-(q_2+q_3+q_4+q_5) \times 100\% \tag{6}$$

In the equation:
Q_1 - Total thermal efficiency of regenerative system / [%];
Q_2 - heat loss rate of collector / [%];
Q_3 -heat loss rate of regenerator / [%];
Q_4 -heat loss of pipes and accessories / [%];
Q_5 - Heat dissipation loss rate of valves and measuring parts / [%].
The thermal efficiency of the system is calculated as shown in Table 5 below. The total efficiency of the system can reach 85.4%.

Table 5. Heat Dissipation and Thermal Efficiency of Various Parts of High Temperature Solid Heat Storage System

Total power /MW	Collector heating / [%]	Regenerator heating / [%]	Pipeline heat dissipation / [%]	Valve isothermal compress / [%]	Total heat loss / [%]	efficiency / [%]
0.8	4	3.2	5.6	1.8	14.6	85.4

5. Calculation of Runner Resistance of the Regenerative System

5.1. Calculation of Flow Resistance

The total resistance Δp_z of regenerator system is composed of pipeline resistanceΔp_1, local resistance Δp_2 and resistance loss Δp_3 of solid particle filling layer in the regenerator.

$$\Delta p_z = \Delta p_1 + \Delta p_2 + \Delta p_3$$

The size of the pipes used in the regenerative system is φ630 × 10, the total length of the pipes is 150 m, the mass flow in the pipes is q_m = 2.52 kg / s, and the air calculating temperature is 300 ℃. The velocity u = 14.04 m/s and ρ = 615 kg/m^3 of air in a pipe can be calculated as turbulent motion. When 103 < Re < 108, the resistance in pipe can be calculated by Kypnakob formula 7.

$$\Delta p_1 = (1.8\ln Re_1 - 1.5)^{-2} \cdot \frac{L_1}{d_e} \cdot \frac{\rho u^2}{2} \tag{7}$$

In the equation:
u -the velocity of the fluid in the channel / m·s^{-1};
d_e - equivalent diameter / m of flow passage;
L_1- the length of the runner / m.
(1) The local resistance in the system is due to the energy loss caused by the change of flow direction or cross-section area (e.g. valve, elbow, tee and variable diameter section) when air flows. Its local resistance formula is

$$\Delta p_2 = \sum \xi_i \frac{\rho u^2}{2} \tag{8}$$

In the system, the elbow is 90 degrees $\xi_1 = 0.75$ and the valve is ball-core valve and if is fully open $\xi_2 = 6.4$.

(2) General Formula for Resistance Loss of Fluid Flow through Solid Particle Filling Layer 8

$$\Delta p_3 = \left(150 + 1.75 \frac{Re_2}{1-\varepsilon}\right) \cdot \frac{\mu G_2}{\rho d_b^2} \cdot \frac{(1-\varepsilon)^2}{\varepsilon^3} \cdot L_2 \tag{9}$$

In the equation:

$$\mathrm{Re}_2 = \frac{G_2 d_b}{\mu}$$

G_2 - mass flow rate at the entrance of filling layer in regenerator / kg · $(m^2 \cdot s)^{-1}$;
L_2 - length of regenerator / m;
d_b - diameter of regenerative ball / m.

5.2. Fan Selection

Because the total resistance of the system is relatively large, the selection of the fan in the regenerative system has certain requirements. The selected fan should ensure that the regenerative system meets the requirements of hot air, airflow rate and pressure head when the regenerative system operates at full load under the given working conditions.

The actual power of the fan is:

$$P = \frac{\beta Q \Delta P_c}{\xi_1 \xi_2} \tag{10}$$

In the equation
Q - fan airflow / $m^3 \cdot h^{-1}$;
β-The capacity reserve coefficient of the motor is 1.3;
ξ_1 - Internal efficiency, 0.8;
ξ_2 - Mechanical efficiency, 0.97.

After calculation, the actual power required by the fan is 46.84 kW, so the final fan type is 6.3A centrifugal blower, and the flow rate is 14763.8 $m^3 \cdot h^{-1}$.

6. Conclusion

(1) The designed high-temperature solid heat storage system can reach 900°C, and can use solar energy for continuous, stable and efficient operation throughout the day.

(2) According to the basic requirements of solid heat storage materials and heat media, air is chosen as heat medium and high Al ball is chosen as heat storage material through comparative optimization analysis; the size and storage time of heat storage device are analyzed and calculated according to the total amount of heat storage and the physical parameters of heat storage materials.

(3) According to the requirements of structural safety, thermal insulation and economy, the structure of regenerator is as follows: inner firebrick + heat-resistant supporting steel plate + expanded perlite + outer Al foil. The lumped parameter method is used to analyze the heat preservation performance of the heat accumulator, and the heat loss of the high-temperature solid heat accumulator is less than 5% per day.

(4) The overall efficiency of the regenerative system is more than 85%. The resistance of the system is analyzed, and the formulas for calculating the resistance of each part are given. The total resistance of the system is calculated based on the average temperature of circulating air, and the type of fan is selected accordingly.

7. References

[1] Wu Yuting, Zhang Lina, Ma Chongfang. High temperature thermal storage technology for solar hermal power generation [J]. Solar energy, 2007 (3): 23-26.

[2] Hanjingxiao, Yang Yongping, Hou Hongjuan. Progress in sensible heat storage technology for solar thermal power generation [J]. Renewable energy, 2014 (7): 901-905.

[3] Mu Zhijun, Guan Xin, Liu Peng. Experimental study on operation of solar photovoltaic integrated photothermal system [J]. Energy-saving technology, 2009, 27 (5): 445-447.

[4] Zheng Jiantao, Han Linwu, Xu Haiwei, et al. Numerical simulation of non-uniform heating of solar column heat recoverer [j]. Energy saving technology, 2016, 34 (1): 7-11.

[5] N.G. Barton. Simulations of air-blown thermal storage in a rock bed [J]. Applied Thermal Engineering, 2013 (3): 1798-1806.

[6] Cao Chuanzhao, Zheng Jiantao, Liu Mingyi, etc. Development of Tower Solar Thermal Power Generation Technology [J]. Renewable Energy, 2013, 31 (12): 21-25.

[7] Wen Liangying, Yong Haiquan, Zhang Zhengrong, et al. Resistance characteristics of regenerative chamber of regenerative burner [J]. Industrial heating, 2001 (4): 8-10.

[8] Li Zhongyou. Theoretical calculation and analysis of honeycomb regenerative heat exchanger [J]. Journal of Kunming University, 2007, 18 (4): 14-16

The Co-Pilot Position of Vehicle Noise Quality Prediction Based on SVM

Xiaojuan Zhang

School of mechanical engineering, Dalian Institute of Science and Technology Dalian, China
dalianxiaojuan@163.com

Abstract. In three different types of domestic cars for example, the paper studied interior noise sound quality when engine idle at a constant speed in different speed operation. Noise test using HEAD Acoustics multi-channel noise test system, sound quality subjective evaluation experiment was carried out in a semi-anechoic. Vehicle sound quality was studied using support vector machine (SVM) model, the results showed that car interior sound quality mainly by four psycho-acoustical parameters: A-weighted sound pressure level, loudness, roughness and jitter degree. The support vector machine forecasting method can guarantee the accuracy of prediction results for small sample data, which has higher accuracy in predicting, error lower.

1. Introduction

As people's requirements increasing to acoustic environment quality about vehicle interior, engine noise problem is particularly prominent. Vehicle interior sound quality has become an important index of vehicle performance. The engine noise is main noise source on the interior of vehicle, which accounts for 50% to 70% [1]. In order to evaluate the level of vehicle interior noise, it is needed to systematic study on sound quality of engine noise, and given enough attention. Based on psychoacoustics theory, establishing sound quality prediction model of engine idle at a constant speed in different speed operation, which is important for comprehensive evaluation of vehicle performance.

2. Test Equipment and Points

In this paper, noise samples come from three different models domestic vehicle, which tested in the semi-anechoic chamber. Noise test points were set in vehicle cab near firewall and microphone pointing to engine. Studying its effect on sound quality of vehicle cab by collecting acoustic signal of engine idle at different constant speeds. Test equipment used HEAD Acoustics multi-channel noise test system, which consists of SQLab Ⅱ data collected recorder, data collected front, HMSⅢ (Head Manual System) binaural information collector, Artemis(Advanced Research Technology for Measurement and Investigation of Sound and Vibration) measurement and analysis software and acoustic sensors. The system has higher accuracy than similar products, which can guarantee noise test reliability and accuracy.

3. Test Conditions

Engine noise signals were collected by test equipment under different idle speeds of engine, each test condition tests 3 times, and test time sustains 10 seconds.

Further processing of noise samples through samples collected, intercept, equal-loudness processing, it can obtain 15 noise samples. The noise sample conditions were1000r/min, 2000r/min,

3000r/min, 4000r/min, 5000r/min, their time length was 5s, which for subjective evaluation experiments. Noise samples contents as shown in Table 1.

Table 1. Test Condition of Noise Samples

Sample Number	1	2	3	4	5
Test Condition (r/min)	1car-2000	1car-3000	1car-4000	1car-5000	2car-3000
Sample Number	6	7	8	9	10
Test Condition (r/min)	2car-4000	2car-5000	3car-1000	3car-2000	3car-3000
Sample Number	11	12	13	14	15
Test Condition (r/min)	4car-4000	4car-5000	2car-2000	1car-1000	2car-1000

4. The Basic Theory of SVM

Support Vector Machines (SVM) is a new method for data mining, which a new tool by means of optimization method to solve machine learning problems, it is putted forward by Vapnik [2]. SVM method is based on seeking structural risk minimization to improve learning machine generalization ability, and achieve empirical risk and confident range minimization. It can obtain good statistical law under a small amount of statistical samples.

Assume that training samples are $\{(x_i \quad y_i), i=1,2...N\}$, N is number of samples, x_i is input value, y_i is expected value.

Support vector machine (SVM) regression model is:

$$f(x) = \omega\Phi(x_i) + b \tag{1}$$

In formula:
ω is weighted vector;
b is threshold value, $b \in R$;
$\omega\Phi(x_i)$ is two vectors' inner-product;
Define ε is insensitive loss function,

$$|y - f(x_i)|_\varepsilon = \begin{cases} 0 & ? \; |y - f(x_i)| \le \varepsilon \\ |y - f(x_i)| - \varepsilon & |y - f(x_i)| > \varepsilon \end{cases} \tag{2}$$

ε is crucial to the numbers of support vector, ε increasing, the regression curve become flat, calculation accuracy will become lower, ε value for 0.01.

In order to solve problem introducing penalty factor c and slack variable $\xi, \xi^*(i=1,2...N)$,so regression model (1) changing to:

$$\min \frac{1}{2}\|\omega\| + c\sum_{i=1}^{N}(\xi_i + \xi_i^*)$$

$$s.t?y_i - \omega\Phi(x_i) - b \le \varepsilon + \xi_i$$

$$\omega\Phi(x_i) + b - y_i \le \varepsilon + \xi_i^*$$

$$\xi_i, \xi_i^* \ge 0(i = 1, 2 \cdots N)$$

Using Lagrange multiplier method and introducing GAUSS RBF kernel function $K(x_i \quad x_j) = \exp(-g\|x_i - x_j\|^2)$(g is radial basis function, which can transform high-dimensional space into low dimensional space function calculation. It can obtain support vector regression machine function follow as:

$$f(x) = \omega\Phi(x_i) + b = \sum_{i=1}^{N}(a_i - a_i^*)K(x_i, \quad x) + b \tag{3}$$

In the formula, penalty factor c and RBF kernel function g have a great influence on model precision. If support vector machine (SVM) model is unable to correctly choose parameters c and g, prediction result's error may larger. This paper selects the optimal kernel parameter using web search method, to evaluating prediction result basis on multi–terms error analysis.

5. Subjective Evaluation of Sound Sample

5.1. Subjective Evaluation Experiment

There are twenty university students for this subjective evaluation, evaluators hearing is good. Using grouped pair-wise comparisons in this experiment, voice samples including 105 pairs i-j comparison and 8 pairs i-i comparison, a total of 113 samples. Evaluation vocabulary is preferences, evaluators given result according to listening. Sample continuous 5s, the interval of 2s between samples in the same sample. It takes 23min to finish s set of evaluation, that meets abroad sound quality evaluation time requirement.

The experiment is carried in semi-anechoic room, whose background noise is 34dB, which can ensure experiment without disrupt.

5.2. The Effective Analysis of Subjective Evaluation

During experiment, judgment of evaluator's is taken subjective sense. If evaluators don't pay enough attention, using different standards lead to result unstable. Thus, subjective data need to be arranged, computed, statistics, and eliminating high false rate evaluators ensure correctness of data post-analysis. In this experiment use A-A same event test and triangular cyclic error check test. By triangular cyclic error check test theory compile matlab program; compute times of every evaluator false judgment which is data effectiveness validity check foundation.

Using Kendall proposed weighting consistency coefficient to inspect Misjudgment result's right, inspect ing subjective evaluate data's effectiveness and repeatability.

Weighting false judgment rate is:

$$C_W = \frac{C_1E_1 + C_2E_2}{(E_1 + E_2)^2} \tag{4}$$

Weighting consistency of coefficient is:

$$\xi_w = 1 - C_w \tag{5}$$

In formula:

C_w is weighting misjudgment rate, E_i is possible times of misjudgment, C_i is actual misjudgment rate, ξ_w is weighting consistency coefficient.

The same sound sample's misjudgment times is 8, triangle circular misjudgment times is 91, calculating the consistency coefficient is presented in table2.

Table 2. The Weighting Coefficient of Consistency

evaluators	consistency coefficient	evaluators	consistency coefficient	evaluators	consistency coefficient	evaluators	consistency coefficient
Tp1	0.808	Tp6	0.795	Tp11	0.898	Tp16	0.981
Tp2	0.846	Tp7	0.635	Tp12	0.794	Tp17	0.898
Tp3	0.821	Tp8	0.961	Tp13	0.955	Tp18	0.936
Tp4	0.572	Tp9	0.770	Tp14	0.834	Tp19	0.821
Tp5	0.821	Tp10	0.757	Tp15	0.891	Tp20	0.930

The other evaluators' consistency coefficient is above 0.75 except TP4 and TP7 who own low consistency coefficient. To further testing effectiveness of evaluation result, using Pearson correlation coefficient in SPSS. By calculation, the Pearson correlation coefficient is above 0.85. The higher evaluators' correlation coefficient is meaning evaluation data is effective and reliable.

6. Prediction of Sound Quality Model Based on SVM

The paper collects 15 sound samples at engine idling for 5 uniform operations; in operation cab of different types of domestic vehicle. The sound samples objective parameter analysis using Head Artemis10.0, as shown in table 3.

Table 3. Objective Parameters and Subjective Evaluation Results of Sound Samples

Sample number	SPL dB	SPL(A) dB(A)	Loudness Sone	Roughness Asper	Sharpness Acum	Fluctuation strength Acum	Preference
1	76.2	60	12.2	0.655	1	0.0406	10.13
2	77.6	66	17	1.14	1.16	0.0579	6.75
3	84	71.5	27.5	1.69	1.3	0.0617	5.38
4	94.1	81.5	45.9	2	1.36	0.0881	1.75
5	84.4	66.9	21	1.33	1.35	0.0531	5.81
6	84.6	71.5	28.5	1.96	1.52	0.0621	3.44
7	87.1	76.3	38.5	2.6	1.79	0.0719	1.50
8	82.5	54.7	7.71	0.354	0.587	0.0429	10.19
9	82.1	61.7	15.1	0.701	1.06	0.0452	8.25
10	82.2	70	21.8	1.13	1.27	0.0587	5.44
11	82.8	70	21.8	1.13	1.27	0.0587	1.94
12	87.2	78	33.7	2.04	1.45	0.0705	0.38
13	82.1	59.4	13.6	0.665	1.13	0.0446	9.19
14	77.1	53	7.24	0.313	0.848	0.0379	11.75
15	74.7	49.3	6.16	0.321	1.01	0.0317	12.55

In order to obtain accurate results of correlation analysis, to determine what objective parameter has a significant effect on sound quality index of automobile engine speed under uniform speed running state, then setting up sound quality evaluation model with vehicle interior, which described through objective parameters has high correlation with subjective evaluation results.

According to the statistics related theory, using software SPSS to calculate the correlation coefficient between preference of subjective evaluations and objective parameters, Calculation results are shown intable 4.

It can be seen from table 4, there is negatively correlated between objective parameters and preference. Among them, A weighted sound pressure level, loudness, roughness and fluctuation strength and preference have high correlation, its absolute value is above 0.9; there is little correlation between sound pressure level and preference; as the engine is at uniform speed in the idle state, the correlation between sharpness and preference is not high.

Table 4. Pearson Correlation between Subjective Evaluation and Objective Parameters

		SPL	SPL(A)	Loudness	Roughness	Sharpness	Fluctuation strength
Preferen ce	correlation coefficient	-0.788	-0.961	-0.906	-0.905	-0.833	-0.911
n=15	two-tailed test	0.0005	0.0000	0.0000	0.0000	0.0001	0.0000

Therefore, this paper takes four objective parameters including A weighted sound pressure levels, loudness, roughness, fluctuation strength as input, takes subjective evaluation results preference as output, using support vector machine to establish the vehicle interior engine sound quality prediction model. According to the theory of support vector machine algorithm, related procedures are compiled to achieve prediction of sound quality model in Matlab environment.

Using the samples 1-12 in table 3 as training samples, for the establishment of vehicle interior sound quality SVM prediction model; and using the sample 13, 14, 15 as test samples, to test the correctness of the prediction model. The paper selects to optimize the SVM parameters, uses cross validation method to find out the optimal parameters combination (c, g).When penalty factor c and radial basis kernel function g were optimized, this paper choose the optimal c =5.6569, g = 0.044194.Using the trained the vehicle interior sound quality SVM prediction model to predict the test sample, the results shown in figure 3. Thus it can be seen, the relative error of the predicted value is in the range of 0.5%, the prediction result is comparatively ideal.

7. Conclusions

7.1 By analyzing the correlation of subjective evaluation results and objective parameter, it showed that sound quality and A sound pressure level loudness, roughness, fluctuation strength has a high negative Correlation relationship, so using these parameters to description driving indoor sound quality is feasible.

7.2 Under the condition of automobile engine idle state uniform operation, collect the car sound samples and subjective evaluation experiment.

7.3 On the basis of the small sample training data, by using the SVM model to predict the vehicle sound quality, Prediction results are accurate and the prediction error is low.

8. References

[1] Sang-Kwon L. Objective evaluation of interior sound quality in passenger cars during acceleration [J].Journal of Sound and Vibration, 2008, 310(5):149-168

[2] Vapnik N.The nature of statistical learning theory [M].New York.Spring' 1995

[3] Genuit k.The sound quality of vehicle interior noise: a challenge for the NVH engineers [J].Vehicel noise and Vibrationg, 2004,1 (1):158-168.

[4] DUAN Min, WANG Yansong, SHI Jing. Prediction of the noise in engine based on neural network [J]. Automotive Engineering, 2002, 24(6): 508-510.

[5] Hao peiyi, Chiang J H.Fuzzy regression analysis by support vector learning approach [J].IEEE Transactions vector learning approach [J].IEEE Transactions on mobile Computing,2007,6(3):311-321

[6] Jayadeva, Khemchandan IR, Chandra S.Twin support vectormachines for pattern classification [J].IEEE Transactions on Pattern Analysis and Machine Intelligence, 2007, 29(5):905-91

IOP Conf. Series: Materials Science and Engineering **735** (2020) 012065 doi:10.1088/1757-899X/735/1/012065

Development Model and Analysis of Single Well Circulating Shallow Geothermal Energy Based on Mathematical Modeling

Liwu Yu

Hefei No.Eight Senior High School, Hefei 230071, China.
Email:francisyu24@hotmail

Abstract. Shallow geothermal energy has been used as an important energy because of its wide distribution, large reserves, rapid regeneration and high utilization value. As a new exploitation method of shallow geothermal energy, single well circulation has been widely concerned. In this paper, the mathematical model of single well circulation mining mode is analyzed from the point of view of mathematical modeling, and the practical problems encountered in the process of single well circulation mining are considered by using concepts of conceptual model, mathematical equation and numerical calculation method, and hydrogeology and hydrodynamic process are considered by assigning specific constants and variables. The influence of heat and mass transfer on geothermal energy exploitation. Based on the above model, the problems encountered in shallow geothermal exploitation are analyzed and discussed in depth, and the corresponding conclusions are drawn. To solve the current problems encountered in shallow geothermal energy exploitation, and provide a theoretical basis for the future development and utilization of new energy.

1. Introduction

Shallow geothermal energy is a new type of renewable energy sourced from the heat and solar radiation inside the Earth. Shallow geothermal energy is widely found in many media such as rocks, soils, and water sources, and is located in a constant temperature zone within 200m of the surface. Its mining is an energy and environmental protection technology that has gradually developed with the global energy crisis and environmental problems. Its research is particularly important for China, which relies heavily on the rapid development of petrochemical energy economy.

Many scholars and organizations at home and abroad have made unremitting efforts in the research of shallow geothermal energy and achieved fruitful results [1-3]. At present, the main utilization methods of shallow geothermal energy are water source heat pump with surface water source as heat source and soil source heat pump with surface rock and soil medium as heat source. The research object of this paper is the single well circulation heat transfer ground energy collection well [4], which is an original new shallow geothermal energy development technology suitable for various geological conditions. This single well circulation shallow geothermal energy development method combines the advantages of the above two traditional geothermal energy collection methods, and realizes the complete recharge of circulating water in the same aquifer, thereby successfully realizing the single geothermal circulation shallow geothermal energy. The equipment has low production cost, no water pollution, convenient maintenance, and is easy to promote in areas with shallow groundwater, realizing real green energy utilization. Because China's shallow geothermal energy is widely distributed and rich in resources, it is of great significance to carry out this research and vigorously develop this single well circulation shallow geothermal energy to ensure energy security and reduce

IOP Conf. Series: Materials Science and Engineering **735** (2020) 012065 doi:10.1088/1757-899X/735/1/012065

environmental pollution. In this paper, the complete conceptual model and mathematical model of the single well circulation heat transfer energy collection well are established, which is of great significance for the in-depth understanding of its mathematical principles, hydrogeology and physical processes, and the efficient development and utilization of this single well circulation shallow geothermal energy.

2. Physical Concept Model and Porous Media Characteristics of Collection Wells in Single Well Circulating Heat Exchange Wells

2.1. Physical Concept Model

The single well circulation heat transfer energy collection well uses circulating water as a medium to collect thermal energy with a subsurface temperature lower than 25 °C to achieve complete recharge of the same layer of groundwater. This type of well does not consume or contaminate groundwater and is therefore safe for groundwater quality.

This type of ground energy collection well is divided into two types: energy storage particle type and non-storage particle type.

The physical process of collecting wells using single well circulation heat transfer and energy storage particles is that the circulating water is pumped out through the submersible pump in the pumping area of the insulated well wall and sent to the heat pump unit after the heat release. Or endothermic. Return to the ground to collect the inside of the pressurized backwater area above the well. Water flows in the annular space between the insulating wall and the separator, with energy storage particles, most of which enter the lower pumping zone, and a small portion of the wall of the through hole is separated from the pressurized return water zone by the permeate flow. Enter the soil medium. Finally, return from the pump area. All backwaters from the surface collection well is pumped from the heat pump to the heat pump unit via a submersible pump.

Single well cyclic heat transfer without energy storage particles can collect the space between the adiabatic well wall and the isolation membrane in the well, without energy storage particles, filled with medium water, the rest of the structure and mass transfer heat transfer physical process and energy storage particle single well cycle The heat exchanger can collect the same well as shown in Figure 1.

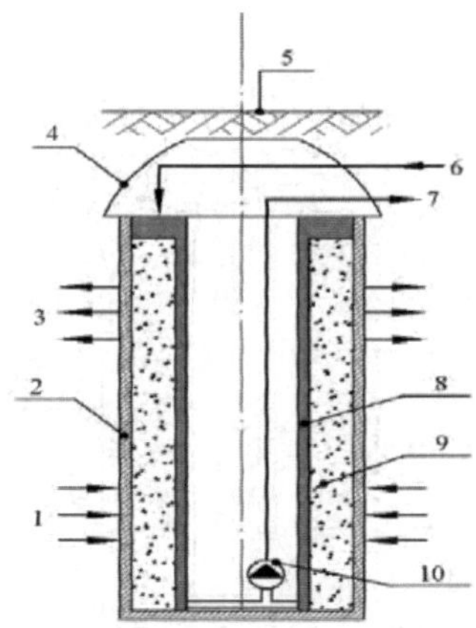

Figure 1. Chematic Diagram of Energy Collection Wells in Single Well Circulation Heat Transfer

2.2. Characteristics of Porous Media

There are two types of porous media that can collect wells in a single well cycle heat exchange. One is to fill the heat storage particles between the insulating wall and the partition, which is characterized by good heat storage properties. The heat storage coefficient of the heat storage particles is large, that is, the ratio of the amplitude of the heat flow fluctuation to the temperature amplitude is large. The other is a saturated soil porous medium outside the separator for exchanging heat and water, ie heat and mass transfer, between the water and the energy storage particles. It can be seen that the single well circulating heat transfer energy collecting well is a hydrothermal coupling body containing convective heat transfer and mass transfer of two porous media, wherein water plays the most important role.

3. Mathematical Model of the Collection Well Of Single Well Circulation Heat Transfer

The mathematical model of the energy storage particle collection well in the single well circulation collection well has only one calculation area of porous medium, and the non-storage particle collection well includes two calculation areas of porous medium and water filling. The selection of the mathematical model is more complicated. The former can be considered by considering the hydrodynamic model of the solid skeleton. In addition, the latter needs to increase the turbulence model to calculate the calculation area only filled with water. When the filled water is laminar. The turbulence model degenerates into a laminar flow model calculation. It can be seen that a typical single well cyclic heat transfer can collect three flow regimes in the well, namely turbulent flow and laminar flow in the aqueous medium and laminar flow in the soil (energy storage particles), see table 1.

Table 1. Two Kinds of Single Well Circulation Heat Transfer Ground Energy Collection Well Mathematical Model List

<table>
<tr><td>Kind / Index</td><td>Single-cycle heat exchange with energy storage particles</td><td colspan="2">Single well circulation heat transfer energy collection well without energy storage particles</td></tr>
<tr><td>Full media type</td><td>Soil (porous medium)
Energy storage particles (porous media)</td><td colspan="2">Soil (porous medium)
water</td></tr>
<tr><td rowspan="2">Computational domain flow</td><td>Laminar flow in soil</td><td rowspan="2">Turbulence when the Re number is greater than the critical value</td><td rowspan="2">Laminar flow when the Re number of water is less than the critical value</td></tr>
<tr><td>Energy storage particles are laminar</td></tr>
<tr><td>Mathematical model governing equation</td><td>Laminar flow equations for soil and energy storage particles (three equations of mass, momentum, and energy conservation)</td><td colspan="2">1. Laminar flow equations for soil (three conservation equations of mass, momentum, and energy)
2. The turbulent flow model k-s equation for water (water degenerates into a laminar flow equation for water at low Re numbers, where there is no soil solid skeleton)</td></tr>
<tr><td>Determining conditions for mathematical models</td><td>Initial condition (no turbulent flow energy)
Boundary conditions (no turbulent flow energy)</td><td colspan="2">Initial conditions (with turbulent flow energy)
Boundary conditions (full flow energy)</td></tr>
</table>

3.1. Mathematical Model of Collecting Wells with Energy Storage Particles

3.1.1. Control Equation

Because of the single-well circulation of the energy storage particles, all parts of the collection well are porous media, and the filled fluid is water. The flow rate of water in the porous medium is very slow and is calculated by the laminar flow of the viscous incompressible layer. The flow of water in a porous medium takes into account the solid skeleton, following the mass conservation equation, the momentum conservation equation, and the energy conservation equation [5].

(1) Mass conservation equation

$$\frac{\partial(\gamma\rho)}{\partial t} + \nabla\cdot(\gamma\rho V_i) = 0 \quad (1)$$

Where γ is the porosity of the porous medium, ρ is the fluid density, t is the time variable, and V_i is the velocity tensor of the fluid, i = 1, 2, 3.

(2) Momentum conservation equation

$$\frac{\partial(\gamma\rho V_i)}{\partial t} + \nabla\cdot(\gamma\rho V_i V_i) = -\gamma\nabla\text{p} + \nabla\cdot(\gamma J_k) + \gamma\rho g + S_m \quad (2)$$

Where p is hydrostatic pressure, $J_k$is the stress tensor, j, k = 1, 2, 3, g is the gravitational acceleration, S_m is the momentum source term of the porous medium, γ, ρ, t, V_i have the same meaning as (1).

(3) Energy conservation equation

$$\frac{\partial(\gamma\rho_f E_f + (1-\gamma)\rho_s E_s)}{\partial t} + \nabla\cdot\left(V_i\left(\rho_f E_f + p\right)\right) = \nabla\cdot\left(k_e\nabla T - \left(\textstyle\sum_l h_i J_i\right) + \left(J_k V_i\right) + S_f^h\right. \quad (3)$$

Where E_f is the total energy of the fluid in the porous medium, E_s is the total energy of the solid skeleton in the porous medium, h_i is the baking of the fluid and various solid skeletons in the porous medium, and J_i is the fluid and various solid skeleton components in the porous medium. i is the mass flow rate per unit area relative to the mass average velocity, i = 1, 2, ..., S_f^h is the source term of the fluid, p γ, ρ, t, V_i, J_khave the same meaning as (1) , k_e is porous The effective thermal conductivity of the medium is weighted according to the thermal conductivity of the fluid in the calculation zone and the thermal conductivity of the solid framework. It is apparent that the above equation degenerates into a laminar flow control equation for an incompressible fluid when the porosity is 100%.

3.1.2. Determining Conditions

There are many types of settlement conditions, which can be listed according to the specific conditions in the actual project. The common settlement conditions are as follows.

(1) Initial conditions

Calculate the flow field in a static state, that is, the initial time velocity V(0) is zero, the initial time pressure P(0) is the external atmospheric pressure P0, and the initial time temperature T(0) is a constant T0, the expression is as follows

$$V\ (0) = 0 \quad (4)$$

$$P\ (0) = P_0 \quad (5)$$

$$T\ (0) = T_0 \quad (6)$$

(2) Boundary conditions

1. Calculate the import boundary of the domain

Import flow Qin is the constant Q1, i.e.

$$Q_{in} = Q_1 \quad (7)$$

Inlet temperature Tin is constant T1, which is

$$T_{in} = T1 \quad (8)$$

Import pressure Pin is a constant P1, i.e.

$$P_{in} = P_1 \quad (9)$$

2 Calculate the exit boundary of the domain

The outlet flow rate Qout is constant Q2, i.e.

$$Q_{out} = Q_2 \quad (10)$$

The outlet temperature Pout is unknown, is not set, and is calculated from known conditions.

The outlet pressure P. Is a constant P, i.e.

$$P_{out} = P_2 \quad (11)$$

3.1.3. Calculate Other Boundaries of the Domain

The soil isothermal boundary Γ_1 and the adiabatic outer boundary Γ_2 are expressed as follows.

$$T|_{\Gamma 1} = t_0 \quad (12)$$

$$q|_{\Gamma 2} = 0 \tag{13}$$

The fixed heat flux boundary conditions are as follows

$$q|_{\Gamma 2} = c \tag{14}$$

The flow velocity V on the boundary is zero, ie.

$$\vec{V}\big|_{\Gamma 1 \cup \Gamma 2} = 0 \tag{15}$$

In the formulas (12) to (15), T represents the temperature of the boundary Γ_1, and q represents the heat flux density of the boundaryΓ_2.

3.2. Mathematical Model of Collecting Wells without Energy Storage Particles

3.2.1. Control Equation

The calculation area of the non-storage particle single well circulation energy collection well is divided into two parts: a porous medium calculation area and a non-porous medium calculation area. The two parts use different control equations.

(1) There is a porous medium calculation area.

The flow of porous soil in a single well with no energy storage particles can be laminar. The governing equation is exactly the same as the governing equations (1) to (4) for collecting wells with energy storage particles in a single well.

(2) No porous medium calculation area.

The non-storage granules can collect the inlet water pressurization pipe and the pumping return pipe part of the well without porous media, all filled with water, and adopt the standard K-ε model country of viscous incompressible turbulence.

The general form of the K-ε differential equation is as follows

$$\frac{\partial}{\partial t}(\rho\varphi) + div(\rho V_i \varphi) = div(\Gamma_\varphi \text{grad}\varphi) + S_\varphi \tag{16}$$

Where t is a time variable, ρ is the fluid density, φ is any dependent variable studied, V_i is the fluid velocity tensor, i=1, 2, 3, Γ_φ is the diffusion coefficient, S_φ Is the source item. The four items in equation (16) are the unsteady term, the convection term, the diffusion term and the source term, respectively. For details, please refer to the relevant literature.

3.2.2. Determining Conditions

The conditions of the enthalpy flow energy k and the enthalpy flow energy dissipation rate ε are added to the initial conditions and the inlets and boundaries, and the other contents are the same as in 3.1.2.

4. Calculation Model and Parameter Setting of the Collection Well of Single Well Circulation Heat Exchange

4.1. The Calculation Conditions of the Calculation Model

The single well cycle energy collection well simulation calculation needs to make the following reasonable assumptions:

(1) The heat transfer process is instantaneously completed between the solid skeleton of the soil and the fluid:

(2) The solid skeleton and fluid of the soil are homogeneous, isotropic or anisotropic continuous media:

(3) The filled fluid medium is an incompressible water body, and the water is set according to turbulent conditions. When the Reynolds number of water reaches laminar flow conditions, the flow pattern degenerates into laminar flow:

(4) The water is laminar at the interface where the unconsumed particle calculation area intersects with the animal energy particle calculation area:

(5) The inertial force is only the downward gravitational acceleration;

(6) In the momentum equation, the porous medium is calculated as a source term of the computational domain to simplify the programming calculation of fluid and heat transfer:

(7) The apparent velocity of the fluid is used in the simulation calculation, not the actual velocity of the fluid in the solid skeleton tunnel, which is convenient for calculation:

(8) In the simulation calculation, the thermal process such as the upper bound solar radiation in the calculation domain and the heat exchange with the atmosphere can be easily calculated which is more conducive to reflecting the physical nature of the problem.

The assumptions of the calculation model need to be adjusted in combination with the specific situation in the actual project, and can be determined after repeated argumentation by scientific assumptions.

4.2. Numerical Calculation Method

The research methods of shallow geothermal energy utilization mainly include field test methods, heat storage estimation methods, analytical methods, numerical simulation methods, and combinations of the above methods. The numerical simulation methods include the finite element method, the finite volume method, the boundary element method, the mixed element method, etc., and the related content can be referred to related materials, and will not be described herein.

In the field of computational fluid dynamics, the finite volume method is widely used, which divides the calculation area into many control volumes, and then integrates the partial differential equations into each control volume to obtain discrete equations. The key problem of the finite volume method is discrete. The process is assumed by the distribution of the function and its derivatives. The coefficients of the obtained discrete equations are clear and conservative. The method is computationally efficient due to the small amount of computation.

In recent years, with the rapid development of numerical heat transfer and computational fluid dynamics, many related professional software have been developed and applied, and CFX, FLUEVT, STAR2CD, and ANSYS software can be used. When the interior of the calculation area is homogeneous and homogenous, the MATLAB software can also be used to solve the above partial differential equations, and the graph corresponding to the calculation result can be drawn.

4.3. Physical Model

The generalized physical model is illustrated by taking a single well circulation heat transfer energy collection well with energy storage particles as an example. A three-dimensional cylindrical radial symmetric region with a diameter of 100m and a depth of 50m is set up. The single well circulation energy collection well is located at the upper part of the geometric center with a depth of 20m, as shown in Figure 2.

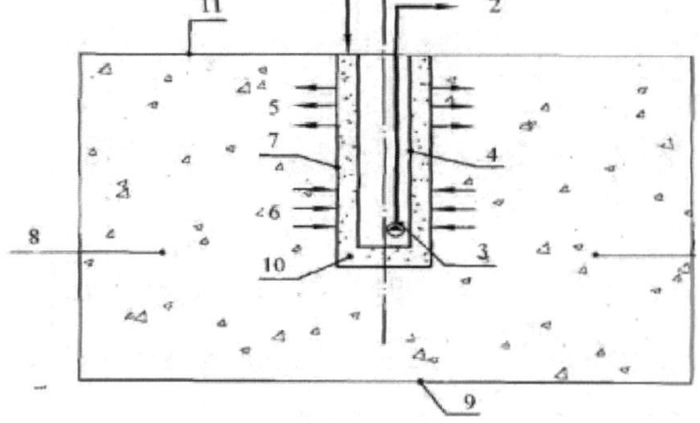

Figure 2. Schematic Diagram of the Energy Collection Well and Study Area of Single Well Circulation Heat Exchange

The inlet flow rate of the ground energy collection well is 50L/s and 80L/s, respectively, and the inlet water temperature is 30K and 290K respectively. The soil temperature and the outer boundary of the soil are calculated according to the constant temperature. The annular cavity between the separator and the insulated inner wall is filled with heat storage particles and water; the outside of the separator is the soil medium, and the water inlet and the submersible pump suction nozzle. The ground can collect the inner wall of the well as an adiabatic boundary condition.

4.4. Geometry Model

Taking the area of Figure 2 as an example to illustrate the establishment of the geometric model, this example uses the appropriate software to establish the space model of the study area, and then uses the meshing software to generate a certain number of rectangular cell grids for each area of the outer boundary. For a face, each intersection is a node of the unit. The area where the well can be collected near the ground is densely divided to meet the accuracy requirements of the calculation results. See Figure 3 and Figure 4 for details. When the calculation area is three-dimensional, its geometric model is usually divided into three-dimensional grids.

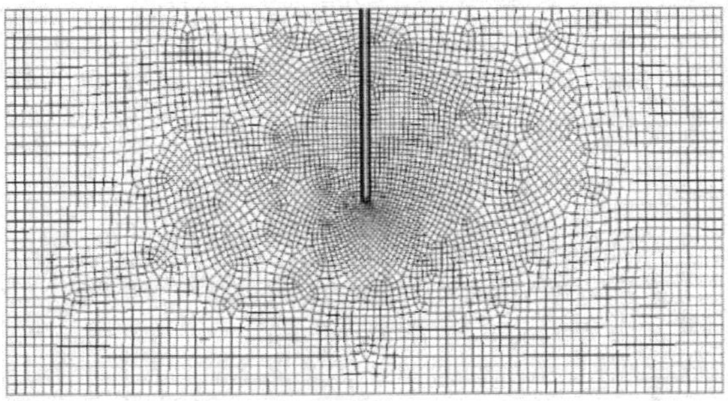

Figure 3. Mesh Section of the Study Area of Single Well Circulation Heat Transfer Ground Energy Collection Well

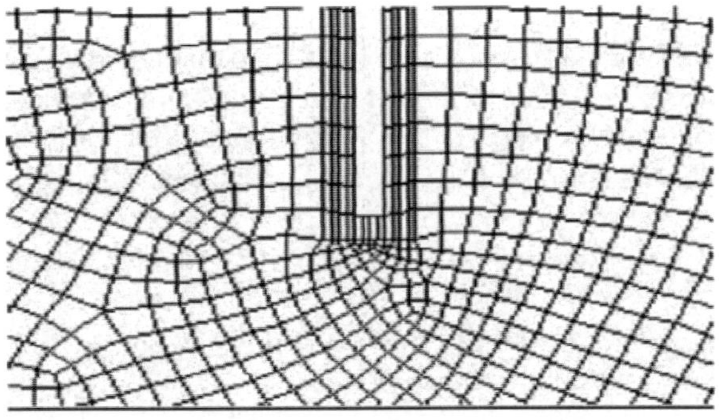

Figure 4. A Schematic Diagram of Local Grid Encryption near the Well Pumping Well

4.5. Parameter Settings

The type of calculation model is mainly including flow state (laminar flow or turbulent flow), whether it is a steady flow, the magnitude and direction of gravity acceleration, and whether temperature is considered.

The medium in the calculation area that needs to set parameters includes water, soil, and energy storage particles.

(1) Water

It is necessary to set the temperature, density, specific heat capacity, viscosity coefficient, standard state enthalpy, and standard state entropy.

(2) Energy storage particles

Need to set density, specific heat capacity, heat transfer coefficient, porosity, viscous drag coefficient, inertial drag coefficient.

(3) Soil

Need to set density, specific heat capacity, heat transfer coefficient, porosity, viscous drag coefficient, inertial drag coefficient.

The setting conditions of the solution include: initial time velocity, initial time pressure, initial soil temperature, inlet water flow, outlet pressure, soil boundary temperature, adiabatic boundary heat flux density, flow field velocity on the boundary, and rough constant of the soil outer boundary.

Since the flow rate of water in the porous medium is very slow, it belongs to the laminar flow state, and the viscous drag coefficient of the energy storage particles and the soil solid skeleton is 1/a.

$$1/a = \rho g/(K\mu) \tag{17}$$

Where 1/a is the viscous drag coefficient; g is the gravitational acceleration; K is the permeability coefficient, and μ is the kinematic viscosity coefficient of water.

The calculation of the inertia drag coefficient C2 can be borrowed from the Eugen formula commonly used in the chemical industry. It is simplified to the following formula only after considering the laminar flow term:

$$C_2 = \frac{3.5}{D_p} \cdot \frac{1-\varepsilon}{\varepsilon^3} \tag{18}$$

Where C_2 is the inertia drag coefficient; D_p is the equivalent diameter of the particles; ε is the porosity of the porous medium. Since the inertial drag coefficient of the laminar flow regime in the porous medium is small, mainly because the viscous drag coefficient acts, a smaller value or a straight zero can be taken. After the above steps are completed, a simulation calculation is performed to obtain a desired series of results.

5. Conclusions and Recommendations

In this paper, the physical and hydrogeological models and mathematical models of single well circulation heat transfer energy collection wells with energy storage particles and non-storage energy particles are given, which provides a theoretical basis for quantitative calculation.

Due to the complexity of the heat and mass transfer process of a single well cyclic heat transfer, the numerical simulation method is needed for temperature calculation. Accurately simulating and predicting the temperature values of soil and effluent medium has important guiding significance and reference value for theoretical research and production practice of shallow geothermal energy development. Compared with model test and field test, it can save a lot of money and short construction period.

In this paper, the complete process of mathematical equations, numerical calculation methods, computational models, geometric models, and various parameter assignment methods including constants and variables for the ideal single well circulation heat transfer energy collection well is given, and the hydrogeological process, the hydrodynamic process, heat and mass transfer process are unified in a mathematical model to facilitate research. In the actual research and production process, combined with the observation data and the inversion of the differential equation parameters, the mathematical model of the actual situation of the reaction is obtained through model identification. According to the above method, the simulation calculation and solution can be used for production research.

By assigning different values to the parameters in the numerical model to obtain different results, the influence of different hydrogeological parameters, thermodynamic parameters and hydrodynamic parameters on the collection well of single well circulation heat transfer can be analyzed. The choice of parameters in the design provides the main basis.

Through the numerical simulation and analysis based on the development of shallow geothermal energy in single well circulation, it is beneficial to deepen its scientific understanding.

The scientific implementation of numerical simulation of shallow geothermal energy development in single well circulation also depends on the further development of forced convection, natural convection, heat conduction and heat dispersion theory in the flow and heat transfer process. The development of various turbulence modes in fluid mechanics and heat transfer promotes the development of the thermal energy discipline is a powerful proof.

The technology of collecting wells in single well circulation heat transfer has been applied in many important projects at home and abroad including China National Grand Theater, and has obtained good economic and social benefits. Through numerical simulation analysis, this kind of green energy can be used more effectively to benefit human society and provide strong support for the sustainable use and development of population-resource-environment.

6. References

[1] Li Wei, Wu Qiang. Discussion on the genesis and influencing factors of the Nanding geothermal field J]. Journal of China University of Mining and Technology, 2002, 31(1): 53~57.

[2] Li Heng, Zhang Xiwen, He Feng. On the Numerical Simulation Method of Fluid Flow in Porous Media[J].Journal of the University of Petroleum, Natural Science Edition,2000,24 (5): 111~116.

[3] Lu Yubei. Research on shallow geothermal energy recharge in the development of water source heat pump [J]. Hydropower Energy Science, 2011, 29(7): 126~128.

[4] Beijing local standard. Technical specification for the process of collecting wells in single well circulation heat transfer (DB11/T93502012) [S]. Beijing: Beijing Municipal Bureau of Quality and Technical Supervision, 2011.

[5] Xu Jiangrong, Zhai Zheyong. Mathematical model and mathematical simulation of heat flow process [M]. Beijing; National Defense Industry Press, 2012.

[6] Tao Wen. Numerical Heat Transfer (Second Edition [M]. Xi'an: Xi'an Jiaotong University Press, 2001.

Review of GIS Technology and Its Applications in Different Areas

Ziye Liu[1] and Liusan Cheng[2]

1 Department of Information Management and Information system, College of Information Science and Technology, Shihezi University, 221 Beisi Road, Shihezi city, Xinjiang Uygur Autonomous Region.
2 Department of Geology and Surveying Engineering, Chongqing Vocational Institute of Engineering, Jiangjin, Chongqing, China.
Email: 643865104@qq.com

Abstract. The 21st century, with the rapid development of information technology (IT), digital life gradually penetrated into every aspect of production and life. To investigate the use of digital life, each country throughout the world, including of China, started planning for digital nation and digital city project. Effective information tools, such as geographic information system (GIS), help to deal with spatial data and complex interactions in city and nation construction. In order to systematically realize the goal of urban planning and constantly promote the level of urbanization, this literature is reviewed for the applications of GIS in the past, of land management, urban and rural planning, traffic management and environmental management. Besides, it puts forward the digital nation expect of GIS applications in the future.

1. Introduction

At present, our society has stepped into the new development situation, in such a background, the city development has made considerable achievements. It is the geological information system connecting the computer science and practical improvement of digital city construction that works no matter in the city management or the city design. Therefore, one of what matters to Urban and rural planning, environmental management etc., is reviewing the applications of geographic information system in the past and proposing the expect of GIS applications in the world in the future.

According to a lot of studies before, we can conclude that GIS has taken an indispensable participation everywhere in our world. Notably, GIS and spatial analysis can be powerful tools to understand, prevent, and help organize cities development strategic. For example, from the view of land management, as a visualization aid, GIS software can be used with multivariate analyses (principal component analysis (PCA), redundancy analysis (RDA) combined with forward selection of variables and cluster analysis) to investigate links between land-use pattern and water quality in the KRB (Kharaa River Basin) [1]. So, we exactly can have a spectacular blueprint on digital city in the world to use this for reference.

For this article, mainly from the introduction of geographic information system technology content, such as the concept, characteristics and functions, firstly set out for discussion of geographic information system technology which is applied to digital nation and digital city theory support points. Moreover, review current global geographic information system on the digitalization construction of typical applications, especially around land management, urban and rural planning, traffic management and environmental management. Besides, personally location of the geographic information system technology in the construction of digital expectation in the future is put forward.

IOP Conf. Series: Materials Science and Engineering **735** (2020) 012066 doi:10.1088/1757-899X/735/1/012066

2. GIS Technology

This section is discussed the concept, characteristics and functions of geographical information system, to strengthen the extent of the combination of theoretical knowledge and practical application, which is conducive to a better understanding of the relation between construction of digital and geographic information system.

2.1. Geographic Information Systems (GIS)

Geographic Information Systems (GIS) are systems designed to store, manage and display spatial data and aid in analysis and interpretation of spatial data [2] [3]. It is a frontier technology discipline based on information theory, cybernetics, computer science, system engineering, and artificial intelligence [4]. The system can enable geoscientific researchers, management, and production departments of various industries to obtain timely and accurate geographic (spatial) information with high accuracy related to regional analysis, scheme optimization, strategic decision-making, etc. [5]. In the process of practical application, GIS, under the circumstances of a cooperation of every discipline, can provide a scientific and systematic service to construction of digital nation and digital city.

2.2. Characteristics and Functions of Geographical Information System Technology

Geographic information system (GIS) technology, namely spatial geographic information system, can output geospatial information through its own ability of spatial data processing and analysis, and achieve the effect of data spatial visualization [6][7][8]. With the development of computer, GIS, as a common technique, has provided a new way to realize, organize and use geographic information.
To conclude the basic features of GIS technology, it should be illustrated to 5 points including cross-disciplines, variety of data type, complex data structure, widely-use application and mainly based on spatial analysis [9]. Take the second point for instance, GIS technology uses huge amounts of data, mainly comes from two aspects, one is the geographic data, which is the object managed by GIS; the other is derived data, mainly from spatial data analysis. These various types of data for GIS bring a series of conveniences in the system operation, data organization and network transmission. Meanwhile, they also come with many problems in these fields above.

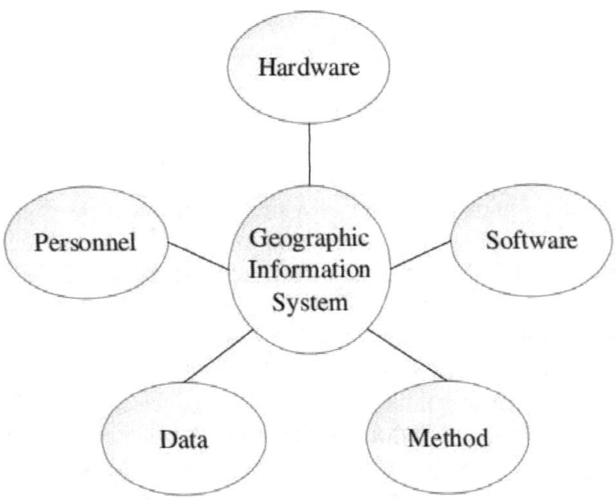

Figure 1. The Figure that Shows the Basic Composition of GIS

The basic composition of GIS generally includes five main parts, which are hardware, software, data, personnel, and methods. The specific relationship is shown in Fig. 1. Therefore, within the components of GIS, the functions of GIS technology include data acquisition and editing, mapping, spatial database management, and spatial analysis [10]. For data acquisition, there are many facilities helping to collect data conducive to digital city construction, such as digitizer, scanners, photography measurement equipment, etc.

3. GIS Application

This section is discussed the applications of GIS in the past, of land management, urban and rural planning, traffic management and environmental management. Besides, personally location of the geographic information system technology in the construction of digital expectation in the future is put forward in this section.

3.1. Application of Land Management

GIS spatial database management function is of great significance to realizing the high integration and comprehensive application of land resources and management. Obviously, as we all know, managing land properly is the first and vital step to realize the digital city and digital nation.

One of applications of land management on the realization of digital nation is that we need to manage cadastral reasonably, such as number, location, purpose, ownership, quality, etc [11]. And when it comes to the ownership of land, it seems harder because it refers to not only record in first time but also change of registration. However, within the use of GIS, collecting and analysing the data and material about cadastral management has been easier than before. Meanwhile, inserting the function of searching information in GIS does help to count the statistics and analyse the nowadays information to more useful information and resource, referring to derived data, which enable GIS to help people to correctly orient this land use promoting to its digitization. Step by step, each part of land in a city even in a whole country can be translating to the data on GIS then be organized appropriately.

Considering in the future way using GIS to realize digital city, with the development of mobile GIS technology, land resource information system can be set on mobile phone, flat computer and some other mobile devices that the related manager will have a more convenient way to glance over and search the data about land use no matter when and where. The staff can also apply this data to work more efficiently, such as carrying out thematic layer statistics and selecting site for any land use [12].

3.2. Application of Urban and Rural Planning

Urban and rural planning reflects one city's development and changes, so a system which can make everything from not only a space level but also a time level to truth is in need. Be sure, GIS is an essential tool to be used on urban and rural planning. Otherwise, it can be applied with other intelligence systems in order to construct a digital city.

Building information modelling (BIM) and geographical information systems (GIS) provide digital representation of architectural and environmental entities. BIM focuses on micro-level representation of buildings themselves, and GIS provide macro-level representation of the external environments of buildings [13]. Studies have shown that the Industry Foundation Classes (IFC) substantial model used in BIM can be converted into a GIS superficial model by transferring high-level geometric and semantic information obtained from BIM to a geospatial environment [14][15]. Specifically, it can be effectively integrated into a software architecture based on a GIS-based facilities management system [16]. The above means that Geographic information system has promoted the trend of the management of the urban and rural planning, broken through the planning business of each phase in the relatively "independence", promoting the business process data sharing, so as to realize the overall lifecycle management of digital city.

Because the use of big data is versatile and adaptable and may generate new information, we should make full use of BIM and GIS data to optimise solutions and planning decisions [17]. So as for the expect of GIS use in the future, we hope geographic information system can be finer and more quantified. For example, when it comes to handle problems about different types of data, such as conflict between power lines planning, housing location planning and spatial, GIS can refine these to overlay analysis, find out the conflict area, and automatic adjustment of the region.

3.3. Application of Traffic Management

In recent years, the number of studies about the tools for analysing accidents and road design has increased considerably. Among these tools, Geographical Information Systems (GIS) stand out for their ability to perform complex spatial analyses [18].

Some Chinese department of traffic construction management geographic information system made some beneficial development research attempt, such as western Hubei transportation construction management system, which uses the Mapx as the GIS development platform. And based on the consideration of the network environment, C/ SJ/ S three layers architecture is adopted, which can meet the requirement in the application of traffic construction management. Meanwhile, it used Web Service distributed technology as a means of the realization of the remote data access [19].

With the development of mobile cloud technology, cloud GIS is a new direction to develop. As we all know, cloud computing is a process in which computers at the nodes of the network work together to realize powerful computing power at a low cost, thus providing shared resources, software and information to terminal devices as needed. Therefore, CGIS (Cloud Geographic Information System) can highly integrate richer spatial data with more complex computer functions, which can greatly improve GIS application. And CGIS truly has a much more visible foreground in traffic management.

3.4. Application of Environmental Management

In China, we often say that green mountains are mountains of gold and silver. And the application of geographic information system can overall grasp the characteristics of ecology and environment, so as to scientifically protect and improve the ecological environment of the related areas and realize the target of building intelligent and digital city at the same time.

For the aspect about GIS application of environment management, spatial data management and analysis functions of GIS contribute to the scientific management of auxiliary environmental function zoning, for instance, many GIS techniques were used to integrate social and ecological data including: descriptive maps, site prioritisation techniques, and predictive modelling [20]. For example, in the forest quota cutting management, with the help of GIS, we can make detailed cutting plan, determine the cutting purpose, location, species, area, storage, and cutting method and update measures [21].

And these years, with the surge of big data, relevant environmental management data will be effective and reasonable. So, we can analyse unlimited data from different latitudes to achieve the goal of complete digitalization of environmental management.

4. Summary

In the process of effectively promoting the application of actual geographic information system and the planning and design of digital city and digital country, it is very important to have a detailed understanding of the application of previous geographic information system. In this process, make full use of geographic information system, keep up with the pace of the new technology, and constantly innovate the traditional management mode and management method of each department and work of the city and the country, so as to better avoid unnecessary waste of resources no matter of land management, urban and rural planning, traffic management or environmental management.

In conclusion, we ought to cherish resources, apply geographic information system, rationalize management process, and constantly seek a harmonious ecological development path.

5. Acknowledgements

The research was financially supported by the Scientific and Technological Research Program of Chongqing Education Committee (No. KJ1732436) and the Chongqing Natural Science Foundation Program (No. cstc2019jcyj-msxmX0291), the supports was gratefully acknowledged

6. References

[1] Gunsmaa B, Martin P, Martin K and Daniel K 2019 Development and Application of GIS-based Assessment of Land-use Impacts on Water Quality: A Case Study of the Kharaa River Basin *Ambio* p 1155

[2] Cromley E and McLafferty S 2011 *GIS and Public Health 2nd* (New York: The Guilford Press)

[3] Maguire D, Goodchild M and Rhind D 1991 *An Overview and Definition of GIS* (Geographical Information Systems: Principles and Applications) pp 9-20.
[4] Chen G 2019 GIS Method and its Application for Harmonious Evaluation of Urban Construction Land and Geological Environment *Arab. J. Geosci.* https://doi.org/10.1007/s12517-019-4761-x
[5] Kim Y, Hyun S, Cheong J, Woo N and Lee S 2018 Hydrogeochemistry in the Coastal Area during Construction of Geological Repository *J. Hydrol.* **562** pp 40-49
[6] Wu Y, Lyu H, Shen J and Arulrajah A 2018 Geological and Hydrogeological Environment in Tianjin with Potential Geohazards and Groundwater Control during Excavation *Environ. Earth. Sci.* p 392
[7] Suluguru A, Surana S and Guharay A 2018 Experimental Investigations on Building Derived Materials in Chemically Aggressive Environment as a Partial Replacement of Soil in Geotechnical Applications *Geotech. Geol. Eng.* pp 1-17
[8] Wang G, Wu Y and Lu L 2017 Investigation of the Geological and Hydrogeological *Environment* with Relation to Metro System Construction in Jinan, China *Bull. Eng. Geol. Environ.* pp 1-20
[9] Hu T 2018 *Research on Geographic Information System Technology and Application* (Beijing: China Waterpub Press) pp 3-5
[10] Chen G 2019 GIS Method and its Application for Harmonious Evaluation of Urban Construction Land and Geological Environment *Arab. J. Geosci.* https://doi.org/10.1007/s12517-019-4761-x
[11] Li C 2018 Explore the Application of GIS Technology in Land Management *Intelligent City* p 59
[12] Hu T 2018 *Research on Geographic Information System Technology and Application* (Beijing: China Waterpub Press) p 28
[13] Wang H, Pan Y and Luo X 2019 Integration of BIM and GIS in Sustainable Built Environment: A review and Bibliometric Analysis *Automation in Construction* **103** pp 41-52
[14] Isikdag U, Underwood J and Aouad G 2008 An Investigation into the Applicability of Building Information Models in Geospatial Environment in Support of Site Selection and Fire Response Management Processes *Advanced Engineering Informatics* pp 504-519
[15] Liu H, Shi R, Zhu L and Jing C 2014 Conversion of Model File Information from IFC to GML *IEEE Joint International Geoscience and Remote Sensing Symposium / 35TH Canadian Symposium on Remote Sensing* doi: https://doi.org/10.1109/igarss.2014.6947141
[16] Kang T and Hong C 2015 A study on Software Architecture for Effective BIM/GIS-based Facility Management Data Integration *Automation in Construction* **54** pp 25-38
[17] Kim H 2016 Implementing a Sustainable Decision-making Environment-cases for GIS, BIM and Big Data Utilization *Journal of KIBIM* pp 24-33
[18] Romi S and María C 2016 GIS Tools for Analyzing Accidents and Road Design: A Review *Transportation Research Procedia* **18** pp 242-247
[19] Wu X, Li R, Wang M and Dong G 2013 *Geographic Information System Practice and Industry Application* (Beijing: Tsinghua University Press) p 363
[20] Mae M, David H, Jamie P and Bruce D 2019 Linking the Social to the Ecological Using GIS Methods in Marine Spatial Planning and Management to Support Resilience: A review *Marine Policy* 103657
[21] Wu X, Li R, Wang M and Dong G 2013 *Geographic Information System Practice and Industry Application* (Beijing: Tsinghua University Press) p 270

Experimental Study on the Damping Effects of Various Vibration Reduction Measures

Tanghui Liu [1], Rui Rao [2 *], Jianwu Liu [2], Mao Ye [2] and Yue Huang[2,]**

1. Guangzhou Metro Design and Research Institute Co., Ltd, Guangzhou, 510000
2. Tamkang University- Guangzhou University Joint Research Center for Engineering Structure1, Guangzhou 510006

Abstract. Field measurements are the most effective and direct way to evaluate the effects of vibration reduction measures. Through field measurements, this study examined the actual damping effects and vibration transfer characteristics of five vibration reduction measures for subway systems, namely, steel spring floating slabs, resilient short crossties, double-layer nonlinear damping fasteners, Vanguard fasteners and rubber floating slabs. The vibration level and 1/3 octave frequency division vibration level was used as vibration evaluation indices for analysis. This study aimed to provide a reference for the actual damping effects of vibration reduction measures. The results show the following: steel spring floating slabs outperformed resilient short crossties by 23.2 dB. Double-layer nonlinear damping fasteners outperformed Vanguard fasteners and rubber floating slabs by 1.6 and 2 dB, respectively. Resilient short crossties, Vanguard fasteners and double-layer nonlinear damping fasteners were effective in reducing vibration from steel rails to track beds (floating slabs). Steel spring floating slabs and rubber floating slabs were effective in reducing vibration from track beds (floating slabs) to tunnel walls.

1. Introduction

China has been developing subway transit systems for over 50 years. By the end of December 2016, 193 subway lines (including extensions), with a combined length of over 4,700 km, were under construction in full swing in 41 cities in China. With strong support from the government, more Chinese cities will have subway transit systems in future [1]. Unlike automobiles, subway systems neither take up ground space nor emit exhaust that pollutes the air. Additionally, subway systems can effectively relieve the pressure caused by shortages of ground space and are more environmentally friendly. By transporting passengers, subway systems can efficaciously mitigate urban ground traffic congestion.

Field measurement is the most effective and direct way to evaluate the effects of vibration reduction measures. Researchers in China and elsewhere have extensively measured subway train-induced vibrations in the field. By regression analysis of field-measured data, Japanese researchers Okwnura and Kuno [2] found that the types of train, train length, distance from the centerline, train speed, track structure and background vibration are the primary factors affecting vibrations. Through field measurements of the sources of train-induced vibrations on the ground and inside the surrounding buildings, Danish researcher Jakobsen [3] determined the patterns of vibration attenuation with distance in soils of various textures. Through field measurements, Saurenman and Philips [4] found that a resilient short crosstie track structure was less effective in reducing vibrations than expected. These researchers recommended testing of the performance of softer subgrades. Additionally, researchers in other countries, including Spain, the Czech Republic and the Netherlands, have conducted extensive field measurements, surveys and studies to examine environmental vibrations induced by rail traffic. Through field measurements and an analysis of Line 5 of the Beijing Subway, Li et al. [5] found that type III rail vibration absorbers and steel spring floating slab tracks outperformed ordinary fasteners in reducing vibrations by 10–15 and 15–25 dB, respectively. Through field measurements of Line 1 of the Guangzhou Metro, Yu [4] discovered a vibration amplification zone in the ground vibration propagation process. By field measurements of Line 3 of the Guangzhou Metro, Huangfu [7] found that acceleration first increased and then decreased as the distance from the centerline of the train increased.

In summary, subway train-induced vibration remains a focal topic of recent research. However, most studies examined only two or three vibration reduction measures and were therefore unable to provide comprehensive understanding of the damping effects of various vibration reduction measures. Track vibration reduction is a complex systematic problem that involves a multitude of factors, including geological conditions, track structure, vehicle system and building structure. Therefore, there are limitations in the available results obtained from field measurements. This study examined subway train-induced environmental vibrations. Five vibration reduction measures, namely, resilient short crossties, steel spring floating slabs, Vanguard fasteners, rubber floating slabs and double-layer nonlinear fasteners, were investigated through field measurements. The relative damping effects and vibration transfer characteristics of the five vibration reduction measures when used on the same line were compared using three indices, namely, vibration level, 1/3 octave frequency division vibration level and frequency spectrum. By increasing the number of vibration reduction measures investigated, a more comprehensive understanding of the damping effects of vibration reduction measures can be obtained.

2. Test Overview

Line 1 of a certain urban subway system has a total length of 18.497 km, of which 2.048 km are aboveground and 16.499 km are belowground. The maximum train speed on this line is 80 km/h. Line 3 of this subway system has a total length of 64.41 km. The maximum train speed on this line is 120 km/h. Indeed, there is a relatively large difference between the train speeds of Lines 1 and 3 during operation. To reduce errors, the vibration reduction measures implemented on the same line were jointly analyzed. Two vibration reduction measures, namely, resilient short crossties and steel spring floating slabs, were implemented on the underground section of Line 1. Three vibration reduction measures, namely, Vanguard fasteners, rubber floating slabs and double-layer nonlinear fasteners, were implemented on Line 3. The vibration responses of the steel rails, track beds (floating slabs) and tunnel walls on both Lines 1 and 3 were measured.

3. Test Equipment

A LAN-XI 3050-A-040 data acquisition system (manufactured by Brüel & Kjær Sound & Vibration Measurement A/S, Denmark) and a DH8303 data acquisition system (manufactured by Jiangsu Donghua Testing Technology Co., Ltd, China) were used as the acquisition instrument, shown as Figure 1. The PULSE professional data acquisition and analysis software package was used to acquire and analyse the data. Integrated electronic piezoelectric (IEPE) accelerometers (manufactured by YMC Piezotronics Inc., Yangzhou, China) and piezoelectric accelerometers (manufactured by Jiangsu Donghua Testing Technology Co., Ltd, China) were used as sensors. Table 1 summarizes the relevant sensor parameters.

4. Test Arrangement

The vibration evaluation indices and test and data processing methods were selected according to Measuring Methods for Environmental Vibration in Urban Areas (China GB 10071-88) [8] and Standards for Environmental Vibration in Urban Areas (China GB 10070-88) [9]. Regarding the arrangement of the measuring points, the above two specifications stipulate that the measuring points should be placed at locations sensitive to vibrations. Thus, measuring points were set on the steel rails, track beds (floating slabs) and tunnel walls. One vibration pickup was placed at each measuring point. Because subway train-induced vibration is primarily vertical [10, 11], only vertical vibration signals were picked up in this study. One IEPE accelerometer was placed at each of the three measuring points, as shown in Figure2. Based on the actual field conditions, the data were continuously collected 20 times from each measuring point.

Table 1. Sensor Parameters

Type	Parameter range	Application
121A100 IEPE accelerometer	Measuring range: -50–50 g Frequency range: 1 Hz–10 kHz Sensitivity: 9.9 mV/ms-2	Acceleration sensors on track beds and tunnel walls
121A05 IEPE accelerometer	Measuring range: -1,000–1,000 g Frequency range: 1 Hz–10 kHz Sensitivity: 5.28 mV/ms-2	Acceleration sensors on steel rails
1A102E piezoelectric accelerometer	Measuring range: -500–500 g Frequency range: 1 Hz–10 kHz Sensitivity: 1mV/ms-2	Acceleration sensors on steel rails

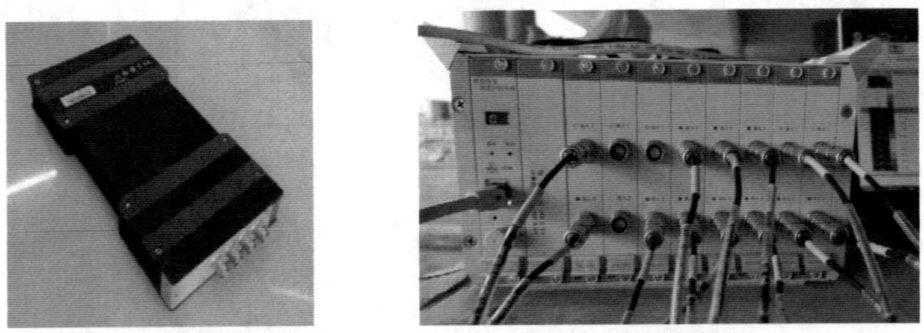

（a）LAN-XI 3050-A-040 (b) DH8303

Figure 1. Data acquisition system

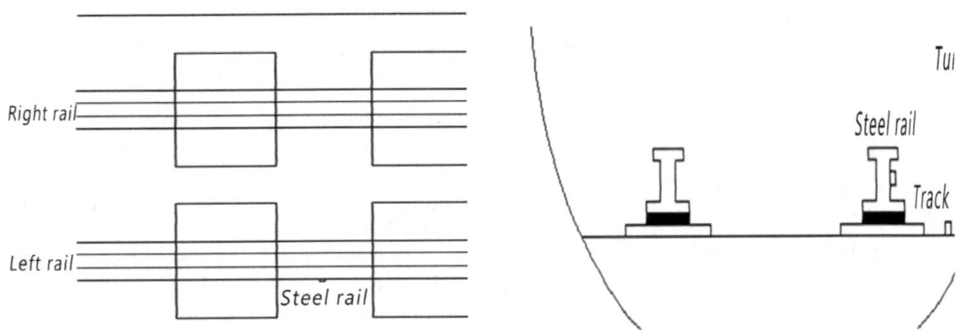

(a) Planar view (b) Sectional view

Figure 2. Schematic Diagram of the Arrangement of Measuring Points

5. Test Results

The peak acceleration and vibration level are important parameters in environmental vibration research [12]. Tests were performed during day-time rush hours. In the field tests, the vertical acceleration time histories were recorded for at least 20 trains passing through each test section to ensure that the data were reliable. Due to the limited length of this article, a relatively good typical acceleration history for one pass is presented. Figure 3–7 show the time histories at the steel rails, track beds and tunnel walls recorded when a train passed the sections where resilient short crossties, steel spring floating slabs, Vanguard fasteners, rubber floating slabs and double-layer nonlinear fasteners were implemented, respectively. As demonstrated in Figure 3-7, the peak acceleration at the steel rails was the highest, followed by that at the track beds and that at the tunnel walls.

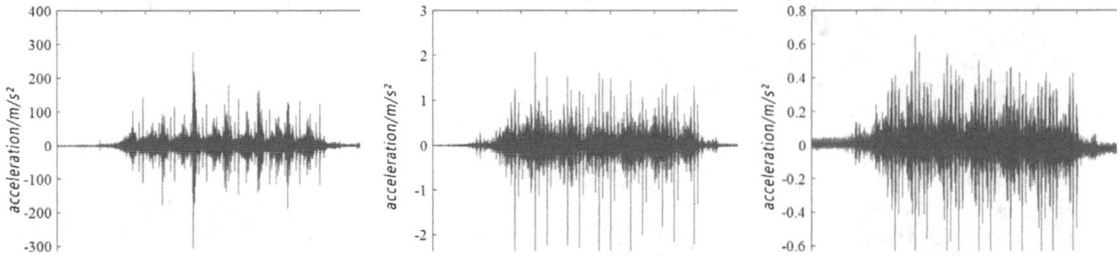

(a) At the steel rail (b) At the track bed (c) At the tunnel wall

Figure 3. Acceleration Time Histories Recorded at the Section where Resilient Short Crossties were Implemented (speed: 120 km/h)

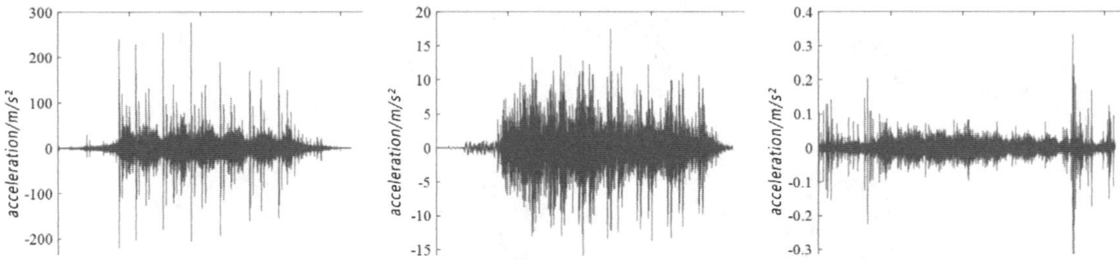

(a) At the steel rail (b) At the track bed (c) At the tunnel wall

Figure 4. Acceleration Time Histories Recorded at the Section where Steel Spring Floating Slabs were Implemented (speed: 120 km/h)

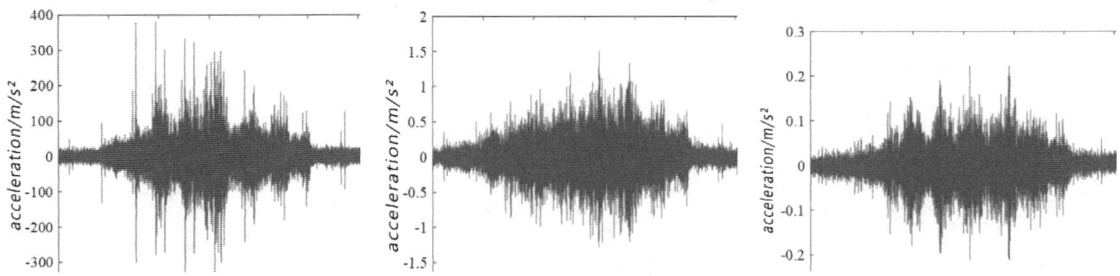

(a) At the steel rail (b) At the track bed (c) At the tunnel wall

Figure 5. Acceleration Time Histories Recorded at the Section where Vanguard Fasteners were Implemented (speed: 80 km/h)

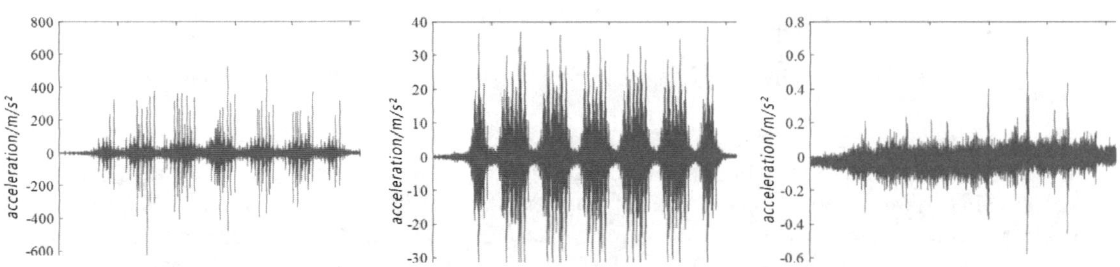

(a) At the steel rail (b) At the track bed (c) At the tunnel wall

Figure 6. Acceleration Time Histories Recorded at the Section where Rubber Floating Slabs were Implemented (speed: 80 km/h)

IOP Conf. Series: Materials Science and Engineering **735** (2020) 012067 doi:10.1088/1757-899X/735/1/012067

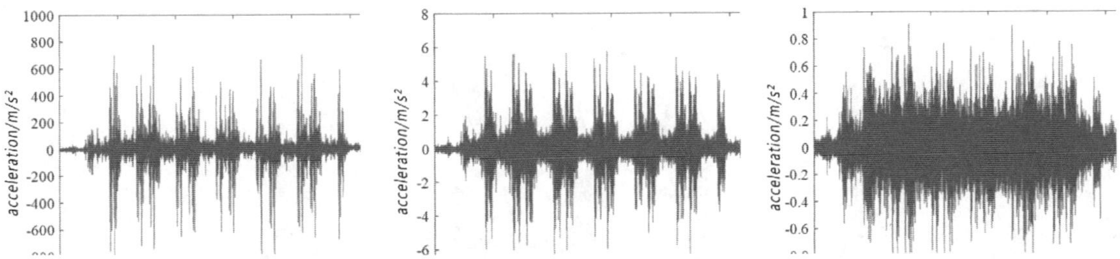

(a) At the steel rail (b) At the track bed (c) At the tunnel wall

Figure 7. Acceleration Time Histories Recorded at the Section where Double-layer Nonlinear Fasteners were Implemented (speed: 80 km/h)

6. Analysis of Damping Effects and Vibration Transfer Characteristics

6.1 Vibration Level Analysis

Environmental vibrations have complex effects on the human body. The vibration acceleration level is the most common evaluation index for environmental vibrations. Chinese Standard GB 10070-88 stipulates the following. The evaluation range is 1–80 Hz. For the effects of environmental vibration on the human body, effective acceleration must be corrected based on frequency. By vertically weighting the vibration level, an evaluation quantity—the vibration level VLz—is obtained. This quantity is the human body vibration evaluation index, which is referred to as the Z vibration level for short (unit: dB). VLz is calculated using the following equation.

$$VLz = 20\log\frac{a_{\omega}}{a_0} \tag{1}$$

Where αo is the reference acceleration (αo=10-6 m/s2) and α is the vibration acceleration (unit: m/s2) obtained after correction based on the full-body-vibration Z-weighting factor stipulated in the International Organization for Standardization 2631/1-1985 Standard.

Tables 2 and 3 summarize the peak acceleration and VLz values for the steel rails, track beds (floating slabs) and tunnel walls when 20 trains passed sections with various vibration reduction measures. Fig. 7 shows the decrease in the VLz value from the steel rails to track beds (floating slabs) and from the track beds (floating slabs) to tunnel walls.

The peak acceleration and VLz values in Tables 2 and 3 and the decrease in VLz value in Figure 8 demonstrate the following.

(1) On Line 3 of the subway system (speed: 120 km/h), the VLz value at the tunnel walls where steel spring floating slabs were implemented was 23.2 dB lower than that at the tunnel walls where resilient short crossties were implemented. Evidently, steel spring floating slabs were more effective in reducing vibrations.

(2) On Line 1 of the subway system (speed: 80 km/h), the average VLz values at the tunnel walls where double-layer nonlinear fasteners were implemented were 1.6 and 2 dB lower than that at the tunnel walls where Vanguard fasteners were implemented and that at the tunnel walls where rubber floating slabs were implemented, respectively. The double-layer nonlinear fasteners performed the best in reducing vibrations, followed by the Vanguard fasteners and rubber floating slabs. The less-than-ideal damping effects of the Vanguard fasteners and rubber floating slabs might be a result of the reduction in the damping effects of these components due to the relatively significant wear on this line.

(3) Resilient short crossties, Vanguard fasteners and double-layer nonlinear damping fasteners were more effective in reducing the vibrations from the steel rails to track beds than the vibrations from the track beds to tunnel walls. The steel spring floating slabs and rubber floating slabs were more effective in reducing vibrations from the floating slabs to tunnel walls than vibrations from the steel rails to floating slabs.

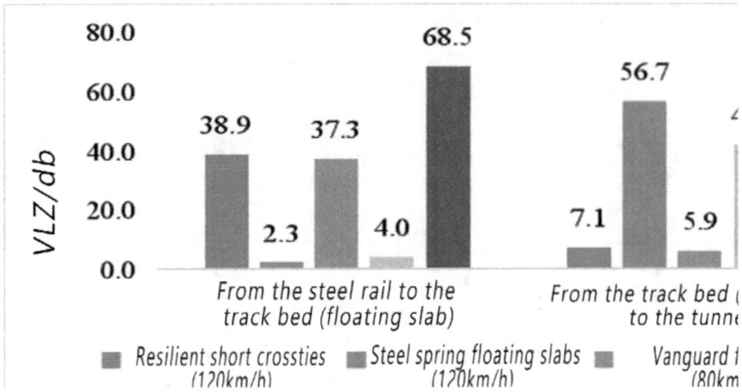

Figure 8. Decreases in *VLz* Values

Table 2. Peak Acceleration Values under Various Vibration Reduction Measures

Vibrati on reduction measure	Steel rail		Track bed (floating slab)		Tunnel wall	
	Acceleration (m/s2)		Acceleration (m/s2)		Acceleration (m/s2)	
	Average	Range	Average	Range	Average	Range
Resilient short crossties (120 km/h)	245.4	177.6–340.8	2.8	2.2–3.5	0.9	0.7–1.1
Steel spring floating slabs (120 km/h)	313.6	168.8–455.9	17.2	13.9–23	0.5	0.4–1.0
Vanguard fasteners (80 km/h)	324.1	163.8–417.1	1.9	0.9–3.9	0.3	0.2–0.5
Rubber floating slabs (80 km/h)	632.9	487.7–781.1	44.9	34.9–59.0	0.6	0.3–1.2
Double-layer nonlinear fasteners (80 km/h)	831.0	657.9–1142.3	7.5	6.4–9.6	1.1	0.8–1.9

Table 3. VLz Values under Various Vibration Reduction Measures

Vibration reduction measure	Steel rail		Track bed (floating slab)		Tunnel wall	
	VLz (dB)		VLz (dB)		VLz (dB)	
	Average	Range	Average	Range	Average	Range
Resilient short crossties (120 km/h)	122.4	109.9–143.7	83.5	80.8–89.9	76.4	74.0–83.4
Steel spring floating slabs (120 km/h)	112.2	108.2–114.4	109.9	105.9–111.2	53.2	52.7–54.0
Vanguard fasteners (80 km/h)	109.9	107.6–116.4	77.3	73.2–81.0	71.1	68.8–74.7
Rubber floating slabs (80 km/h)	113.2	110.7–116.0	109.7	107.6–114.1	71.5	68.9–75.2
Double-layer nonlinear fasteners (80 km/h)	142.3	138.5–144.9	73.8	72.3–76.4	69.5	68.4–70.9

6.2 One-third Octave Frequency Division Vibration Level Analysis

The vibration acceleration time history data measured were processed using the 1/3 octave frequency division vibration level [13]. The distribution figures of the 1/3-octave central frequency vibration levels at the measuring points were compared. Based on the energy attenuation in each frequency band at the steel rails, track beds and tunnel walls, the damping effects of various vibration reduction

IOP Conf. Series: Materials Science and Engineering **735** (2020) 012067 doi:10.1088/1757-899X/735/1/012067

measures were analyzed. The Standard for Limits and Vibration Test Methods for Building Vibration and Secondary Radiation Noise Caused by Urban Rail Transit (JGJ/T 170-2009) [13] stipulates that 1/3-octave central frequencies should be corrected based on various Z-weighting factors for frequencies within the range of 4–200 Hz.

Figure 12–16 show the 1/3-octave curves of the vibration acceleration responses under each vibration reduction measure.

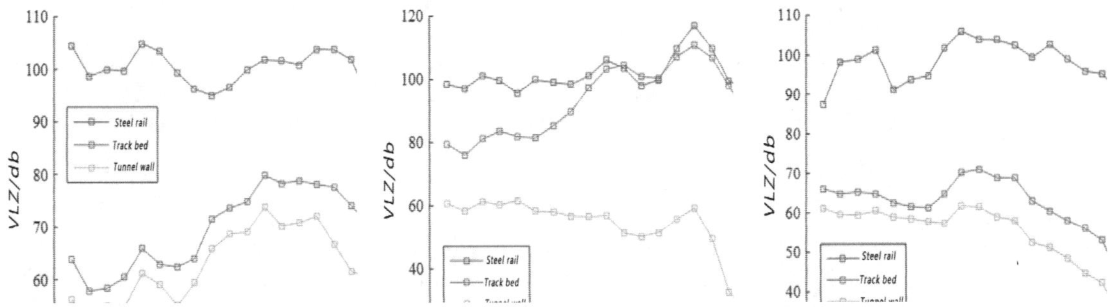

(a) under resilient short crossties (b) under rubber floating slabs (c) under Vanguard fasteners

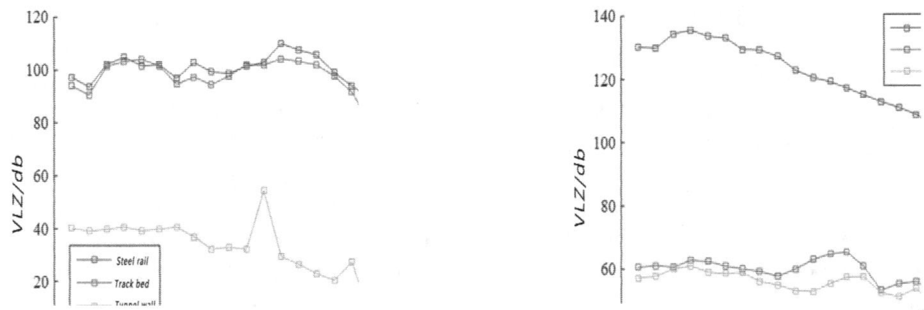

(d) under steel spring floating slabs (e) under double-layer nonlinear fasteners

Figure 9. Frequency Division Vibration Levels

The following can be found from Figure 9.

(1) Within the frequency range of 4–200 Hz, on Line 3 of the subway system (speed: 120 km/h), the vibration level value decreased by approximately 20–40 and 3–12 dB from steel rails to track beds and from track beds to tunnel walls, respectively, in the sections where resilient short crossties were implemented. In the sections where steel spring floating slabs were implemented, the vibration level values at the steel rails and floating slabs were basically the same. Additionally, in these sections, the vibration level values decreased by approximately 50–80 dB from the floating slabs to tunnel walls. The resilient short crossties were more effective in reducing the vibrations from the steel rails to track beds (floating slabs) than the steel spring floating slabs. The steel spring floating slabs were more effective in reducing the vibrations from the track beds (floating slabs) to tunnel walls.

(2) Within the frequency range of 4–200 Hz, on Line 1 of the subway system (speed: 80 km/h), the vibration levels decreased by 20–45 and 3–11 dB from the steel rails to track beds and from the track beds to tunnel walls, respectively, in the sections where the Vanguard fasteners were implemented. In the sections where the rubber floating slabs were implemented, the vibration levels decreased by approximately 8–20 dB from the steel rails to floating slabs within the frequency range of 4–20 Hz. Additionally, in these sections, the vibration levels at the steel rails and floating slabs were basically the same within the frequency range of 20–200 Hz. Moreover, the vibration levels decreased by approximately 20–65 dB from the floating slabs to tunnel walls in the sections where the Vanguard fasteners were implemented. In the sections where double-layer nonlinear fasteners were implemented, the vibration levels decreased by approximately 50–70 dB from the steel rails to track beds, and the

vibration levels at the track beds and tunnel walls were basically the same. The double-layer nonlinear fasteners were the most effective in reducing the vibrations from the steel rails to track beds (floating slabs), followed by the Vanguard fasteners and rubber floating slabs. The rubber floating slabs were the most effective in reducing the vibrations from the track beds (floating slabs) to tunnel walls, followed by the Vanguard and double-layer nonlinear fasteners.

(3) Within the frequency range of 4–200 Hz, the vibration levels decreased to varying extents from the steel rails to track beds (floating slabs) and from the track beds (floating slabs) to tunnel walls under various vibration reduction measures.

(4) The frequency at which the peak acceleration value in each structure appeared varied between the five vibration reduction measures examined in this study. This result is related to the properties of each structure. These five vibration reduction measures were effective in reducing the vibrations in different frequency bands.

7. Conclusions

In this study, through field measurements of track structures within a subway system under various vibration reduction measures, the vibration responses of steel rails, track beds and tunnel walls on the same line under various vibration reduction measures were determined. Additionally, the vibration level and 1/3 octave frequency division vibration level was analyzed. The test results show the following:

(1) On Line 3 of the subway system (speed: 120 km/h), the steel spring floating slabs outperformed the resilient short crossties in reducing vibrations by 23.2 dB. The steel spring floating slabs were more effective in reducing vibrations. The resilient short crossties were more effective in reducing the vibrations from the steel rails to track beds (floating slabs). The steel spring floating slabs were more effective in reducing the vibrations from the track beds (floating slabs) to tunnel walls.

(2) On Line 1 of the subway system (speed: 80 km/h), the double-layer nonlinear vibration reduction fasteners were the most effective in reducing vibrations, followed by the Vanguard fasteners and rubber floating slabs. The double-layer nonlinear vibration reduction fasteners outperformed the Vanguard fasteners and rubber floating slabs by 1.6 and 2 dB, respectively. The double-layer nonlinear fasteners were the most effective in reducing the vibrations from the steel rails to track beds (floating slabs), followed by the Vanguard fasteners and rubber floating slabs. The rubber floating slabs were the most effective in reducing the vibrations from the track beds (floating slabs) to tunnel walls, followed by the Vanguard fasteners and double-layer nonlinear fasteners.

To reduce errors, only the vibration reduction measures on the same line were compared. This limitation ensured consistent test conditions (e.g., tunnel structure, sectional form and burial depth). However, limited by the conditions, it was impossible to ensure a completely uniform train speed at the measuring points under each vibration reduction measure. Additionally, unpredictable external factors interfered with the test process. Conclusions were derived from this study while ignoring the aforementioned factors. While there are certain errors in the conclusions, the conclusions can, to a certain extent, demonstrate the relative damping effects of various vibration reduction measures and provide a certain reference and basis for implementation of vibration reduction measures in urban subway systems in future.

8. Acknowledgments

This work is supported by Guangzhou Education Bureau project (Project No. 1201610505) and the National Natural Science Foundation of China (No. 51608135)

9. References

[1] Research team for annual reports on urban rail transit traffic in China. Annual reports on urban rail traffic in China [M]. Beijing Jiaotong University Press (Beijing), 2017: 2-2.

[2] Okumura Y, Kuno K. Statistical analysis of railway noise and vibration field data collected in an urban area [J]. Applied Acoustics, 1991, 33(4):263-280.

[3] Jakobsen J. Ground Vibration from Rail Traffic [J]. Journal of Low Frequency Noise Vibration & Active Control, 1987:96-103.

[4] Saurenman H. In-service tests of the effectiveness of vibration control measures on the BART rail transit system [J]. Journal of Sound & Vibration, 2006, 293(3–5):888-900.
[5] Li K, Liu W, Xun X, et al. In situ test of vibration attenuation of underground line of Beijing Metro Line 5 [J]. Journal of the China Railway Society, 2011, 33(4): 112-118.
[6] Yu Z. Test on vibration caused by subway and response research of structure and pile-soil [D]. Guangdong University of Technology, 2015.
[7] Huangfu Chanyuan. Study on subway-introduced vibration response of ground surface and building structure [D]. South China University of Technology, 2014.
[8] Measuring methods for environmental vibration in urban areas (Chinese GB Standard 10071-88) [s]. Standards Press of China (Beijing), 1988.
[9] Standards for environmental vibration in urban areas (Chinese GB 10070-88) [s]. Standards Press of China (Beijing), 1988.
[10] Lou M, Jia X, Yu J. Field measurement and analysis of ground vibration induced by subway trains [J]. Journal of Disaster Prevention and Mitigation Engineering, 2009, 29(3): 282-288.
[11] Lou M, Li S. Evaluation research of building vibrations induced by subway running train [J]. Journal of Vibration and Shock, 2007, 26(8): 68-71.
[12] Zhang X. A study on viaduct traffic-induced vibration, building vibration reduction, and control methods [D]. Beijing University of Technology, Beijing, 2009.
[13] Wang J. Application of MATLAB in vibration signal processing [M]. China Water & Power Press (Beijing) and Intellectual Property Publishing House Co., Ltd. (Beijing), 2006.
[14] Standard for limits and vibration test methods for building vibration and secondary radiation noise caused by urban rail transit (JGJ/T 170-2009) [s]. China Architecture & Building Press (Beijing), 2009.

IOP Conf. Series: Materials Science and Engineering **735** (2020) 012068 doi:10.1088/1757-899X/735/1/012068

Wet Oxidation Treatment of Waste Activated Sludge of a Pharmaceutical Factory

Pengfei Guo, Xu Zeng*, Guodong Yao, Yangyuan Zhou and Jianfu Zhao

College of Environmental Science and Engineering, State Key Laboratory of Pollution Control and Resources Reuse, Tongji University, Shanghai, China
Email: zengxu@tongji.edu.cn

Abstract. As the development of pharmaceutical industry, large amounts of sludge was produced, which is a hazardous waste with high environmental risks and high disposal fees. In this study, the wet oxidation treatment of pharmaceutical sludge was investigated. The effects of the reaction parameters on the process performance were discussed, such as reaction time, temperature, initial oxygen supply and catalysts. The results illustrated that the highest removal rate of volatile suspended solids (VSS) and the total chemical oxygen demand (COD) removal rates could reach up to 60.7% and 91.3% separately under 260°C with initial oxygen pressure 1.0 MPa and the reaction time 60 min. Thus, the wet oxidation process provides an economical and feasible alternative for the industrial activated sludge treatment, especially for the pharmaceutical sludge.

1. Introduction

Due to the rapid development and increasing demand of pharmaceutical industry, large amounts of activated sludge were generated in recent years. The sludge from the pharmaceutical industries is especially problematic as it contains recalcitrant antibiotics. Conventional technologies including biological, thermal, and physicochemical treatments are always used for the treatment of these sludges. Generally, biological treatment is the most common and economical treatment method. Solidification/stabilization processes were very popular for the final treatment of the sludge [1]. However, these methods are insufficient to remove all potentially hazardous constituents. This problem has attracted much attention for the environmentally friendly treatment methods.

Recently, advanced oxidation processes (AOPs) have attracted much attention in the area of the treatment of hazardous pollutants [2] and are effective in the decomposition of a wide range of organics [3-4]. Wet oxidation (WO) technology is very promising for the treatment of effluents containing high concentrations of organic or toxic contaminants. In the WO process, the generation of hydroxyl radicals occurs under high temperatures and pressures. Then, the pollutants are degraded to carbon dioxide, water, and small molecule carboxylic acid, etc. Therefore, WO is an interesting alternative for the degradation of activated sludge [5]. Recently, the WO of activated sludge was investigated [6] and the technic-economic and environmental sustainability was assessed [7]. However, in recent years, only a few studies were performed to the examination of real industrial sludge. Therefore, the wet oxidation of real industrial pharmaceutical and hazardous sludge is well worth examining.

In this paper, some experiments were conducted by using a batch reactor. The effects of WO conditions, namely, reaction time, temperature, oxygen supply, and the catalyst, on the process performance were investigated.

2. Material and Methods

2.1. Materials
The chemical activated sludge was collected in a pharmaceutical factory, which is located in the east of China. The characteristics of the raw activated sludge are as follows: the mixed liquor COD is 30 000~40 000 mg/L, VSS is 27.8 ~29.6 g/L and pH 7.5~8.5.

2.2. WO Reaction System
Experiments were conducted in a SUS316 reactor, which was bought from Anhui Kemi Machinery Technology Co. Ltd, China, as shown in Figure 1. The experimental procedure is: pharmaceutical sludge solution with different water ration and catalyst were put into the reactor, and the initial oxygen pressure was 1.0 MPa. Then the reactor was heated. The reactor was removed to cool water to terminate the reaction, after the desired reaction temperature was achieved.

Figure 1. Picture of Experimental Reactor

2.3. Analysis
COD was measured by using Hach Heating System, Hach Corporation, USA, with the potassium dichromate oxidation method. VSS was measured by the Ignition loss method. The COD and VSS removal rates were calculated to assess the treatment efficiency. pH was measured using pH-201, Hanna Corporation, Italy.

3. Results and Discussion

3.1. Effect of Reaction Temperature on the Treatment Efficiency
The effect of reaction temperatureon the COD and VSS removal rates was investigated in the range of 180-260 °C with reaction time 60 min and oxygen pressure of 1.0 MPa. As shown in Figure 2, the removal rates of COD and VSS were very high under high reaction temperature, which indicates highly treatment efficiency. COD and VSS removal rate showed a rising trend, and the effect of reaction temperature was also very significant. The results illustrated that the highest removal rates of VSS and COD could reach up to 60.7% and 91.3% separately under 260 °C with reaction time 60 min and the initial oxygen pressure 1.0 MPa. It was reported that the intermediates from wet oxidation of organics were some carboxylic acids with small molecular weights, such as acetic acid and propionic acid, which were resistant to oxidation. Therefore, it could be concluded that the reaction parameters, such as higher temperatures, were favourable for the removal of COD and VSS. On the other hand, higher temperatures could lead to higher operating costs. Therefore, the reaction temperature of 260 °C was used in the later experiments.

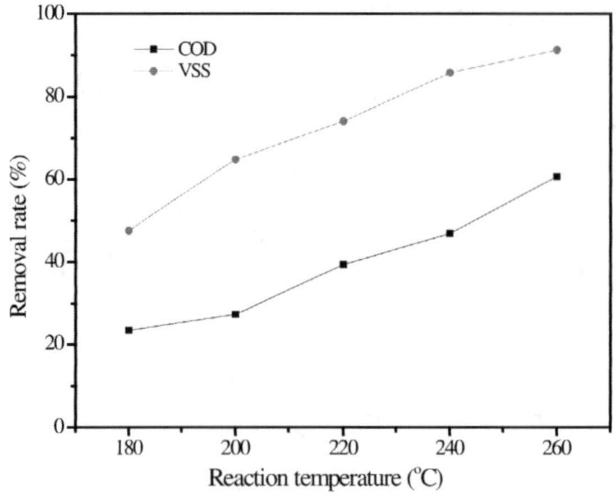

Figure 2. Effect of Reaction Temperature (60 min, O_2 initial pressure 1.0 MPa)

3.2. Effect of Reaction Time on the Treatment Efficiency
Figure 3 shows the effect of reaction time on the treatment efficiency at 260 °C with an initial oxygen pressure 1.0 MPa. In these experiments, the reaction time was changed in the range of 15 to 60 min. The results indicate that reaction time has a significant effect, as shown in Figure 3.The removal rates of COD and VSS gradually increased with the increase of reaction time. At the same time, the degradable organics in liquid accumulated. Therefore, with the increase of the reaction time, the removal rates of COD and VSS increased, because of the degradation of the organic compounds.

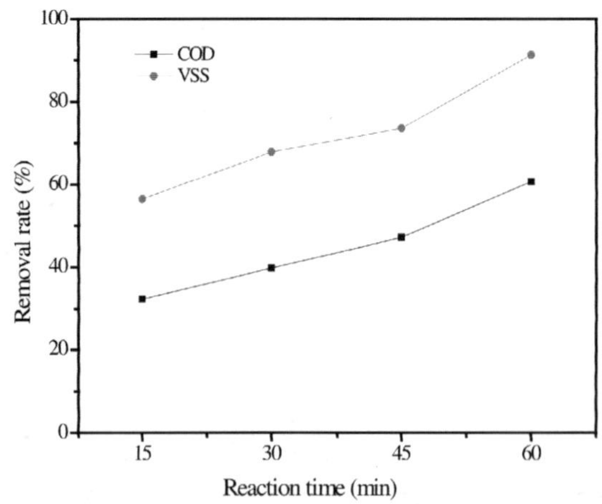

Figure 3. Effect of Reaction Time (260 °C, O_2 initial pressure 1.0 MPa)

3.3. Effect of Oxygen Pressure on the Treatment Efficiency
Experiments were performed by changing the initial oxygen pressures in the range of 0.6 to 1.0 MPa to investigate the effect of oxygen pressure on the treatment efficiency. The results could be seen in the Figure 4. As shown in the Figure 4, the initial oxygen pressure has a great impact on the removal rates of COD and VSS. The probably reason is that the reaction took place under the hydrothermal conditions is an oxidation process, which means that the oxygen was sued as an oxidation agent. Therefore, the increase of oxygen pressure means the increase of the reaction materials. On the other hand, the increase of the oxygen pressure also increased the concentration of the dissolved oxygen in the solution. It would be helpful for the oxidation of the oxidation species of the organic pollutants.

Therefore, with the increase of oxygen pressure, the removal rates of COD and VSS increased. On the other hand, once the oxygen was surplus, the increase of the removal rates stopped.

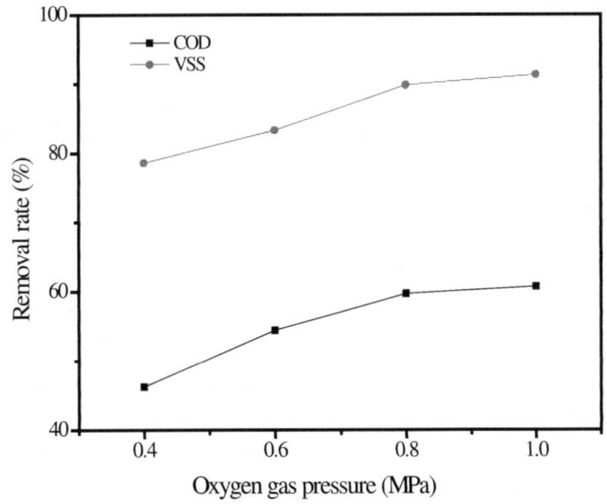

Figure 4. Effect of Initial Oxygen Pressure (60 min, 260 °C)

3.4. Effect of Catalysts on the Treatment Efficiency

It was reported that homogeneous catalysts can enhance the wet oxidation efficiency significantly. Therefore, in this section, some homogeneous catalysts were collected to study the effects of homogeneous catalysts, including $FeSO_4$, $MnSO_4$, $Ni(NO_3)_2$, $Cu(NO_3)_2$. As shown in Figure 5, the COD degradation was accelerated significantly when the homogeneous catalyst was added. The effect of $Cu(NO_3)_2$ was the highest. After the reaction, small molecule carboxylic acids, such as formic acid or acetic acid, were produced. These small molecule carboxylic acids are not easily to be oxidized. The VSS removal rates showed that there was almost no difference. In addition, it should be noted that the obtained data indicate that some of them are non-oxidizable under the wet oxidation conditions.

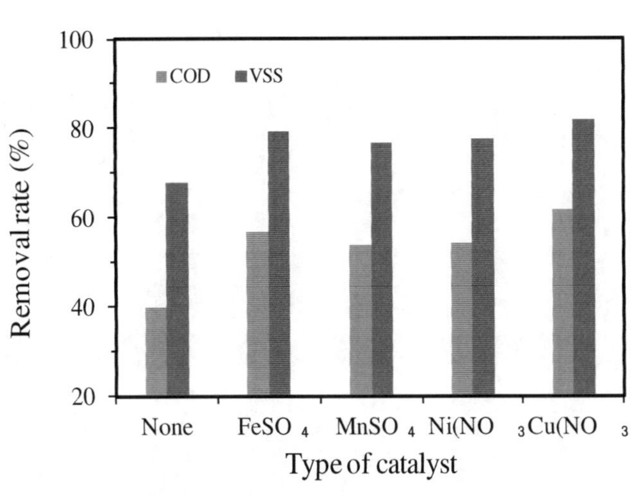

Figure 5. Effect of Catalysts (240 °C, 60 min, initial oxygen pressure 1.0 MPa, catalyst 0.1 g).

4. Conclusions

In this study, the wet oxidation of industrial pharmaceutical sludge was investigated by using a batch reactor. The effects of reaction parameters on the process performance were discussed, such as reaction time, temperature, initial oxygen pressure and catalyst. The results illustrated that the highest removal rates of VSS and COD could reach up to 60.7% and 91.3% separately under 260 °C with initial oxygen pressure 1.0 MPa and the reaction time 60 min. Thus, the wet oxidation process provides a favourable and feasible alternative for the industrial activated sludge treatment, especially for the pharmaceutical sludge.

5. Acknowledgments

This work was financially supported by the National Key R&D Program of China, Key Technologies and Engineering Demonstration for the Treatment and Resource Utilization of Pharmaceutical Sludge by Hydrolysis-Wet Oxidation Process (2018YFC1902103).

6. References

[1] Cioffi R., Lavorgna M., Santoro L. Environmental and technological effectiveness of a process for the stabilization of a galvanic sludge. *J Hazard Mater,* 2002, 89, 165–175.
[2] Malato, S., Blanco, J., Vidal A., Richter, C. Photocatalysis with solar energy at a pilot-plant scale: an overview. *Appl. Catal. B*, 2002, 37, 1-15.
[3] Arana, J., Rend´on, E. T., Rodnguez, J. D., Meli´an, J. H., Diaz O. G., Pena, J. P. Highly concentrated phenolic wastewater treatment by the Photo-Fenton reaction, mechanism study by FTIR-ATR. *Chemosphere*, 2001, 44, 1017-1023.
[4] Luck, F. Wet air oxidation: Past, present and future. Catal. Today, 1999, 53, 81-91.
[5] Tu, Y. T., Xiong Y., Tian S. H. Catalytic wet air oxidation of 2-chlorophenol over sewage sludge-derived carbon-based catalysts. *J. Hazar. Mater,* 2014, 276, 88-96.
[6] Bertanza, G., Canato, M., Heimersson, S., Laera, G., Salvetti, R., Slavik, E., Svanström, M. Techno-economic and environmental assessment ofsewage sludge wet oxidation. *Environ. Sci. Pollut. Res.*, 2015, 22, 7327-7338.
[7] Bertanza, G., Galessi, R., Menoni, L., Zanaboni, S. Wet oxidation of sewage sludge from municipal and industrial WWTPs. *Desalin. Water Treat,* 2015, 57, 2422-2427.

Design and Optimization of the Collector of Electric Sugarcane Leaves Returning Field Machine

Yisen Lin[1], Meng Yu[1], Fujun Wen[1], Zengxue Zhang[1] , Yanqin Tang[1] and Yuxing Wang[1*]
1Collage of Engineering, South China Agricultural University, 483 Wushan road, Tianhe district , Guangzhou, 510642, China.
Email: scauwyx@scau.edu.cn

Abstract. In this paper, the collector is designed for electric sugarcane leaves returning field machine. According to the working principles of the collector and the moving trace rules of collecting components, we confirm the important structural parameters of the collector by using the ADAMS software for motion simulation optimization, including the rotor radius, crank length, the locus curve in the center of cam disc slide and et al. The ADAMS software is used for simulating the movement of the collector with different combinations of two factors (the forward speed of machine and rotate speed of roller plate), and we can obtain the results about the angular acceleration at the end of collecting components and the leakage area per meter. The data obtained from the simulation test is analyzed by Central Composite Design (CCD) with the aim of assessing the best working parameter combination on low leakage rate. The best parameter combination of two variables are determine, i.e., the forward speed is v=0.4 m/s, and the rotate speed is ω=49.41 rpm. The sum area of leakage area is 26.95cm2/m, and the largest angular acceleration at the end of collecting components is 200 rad/s2. The collector based on the theoretical design results is produced, and it would be tested in the forward speed of v=0.4 m/s, and rotate speed of ω=50 rpm for analyzing its performance. The test results show that the collecting effect of the collector is well, and the leakage rate is 2.93 %, which can be up to the requirement of agricultural technical index. The collector can complete the collecting work of the sugarcane leaves returning field machine.

1. Introduction

Sugarcane is the main raw material of sugar in China, its planting yield is just less than Brazil and India, and it is mainly distributed in Guangdong, Guangxi, Yunnan, Hainan and other South places of China. The main by - product of it is its leaves, which have a large quantities of inorganic substances, so that they can be used to increase soil nutrients when returning them to the field. At present, there are many foreign developed countries adopt sugarcane combine harvesters, which harvest sugarcane while cut off the sugarcane leaves and returning them to field. The development of sugarcane harvesting machinery is still lagging in China. When the machine harvests sugarcane, at the same time, it can't cut sugarcane leaves, so that most of leaves are left in the field [1-6]. The treatment of sugarcane leaves is still dominated by incineration, which brings environmental pollution and fire safety. Our research on the sugarcane leaves returning field machines are still in the developing stage. Yueqian Jin designed the 3S-Y140 type sugarcane leaves returning field machine, which collected and cut the sugarcane leaves by means of high speed rotating knives, but the knives were easy to damage themselves and the perennial root; Ming Liang added spring - finger cylinder pickup for the sugarcane leaves returning field machine, which could collected the sugarcane leaves to a height and then the leaves would be brought to the cutting area by the flail knives; Ming Li designed a sugarcane leaves returning field machine with profiling pick-up device, which could lift the leaves out the earth, so that the leaves could be brought into the cutting area by the flail knives; Yuxing Wang designed the flexible knives with nylon, which could cut the leaves in the field by the high - speed revolution of double knives roll, and the cutting effect was good but the flexible knives were easy to abrasion. The foreign collecting machines have been mature basically, and they mainly engage in the development of the machine performance and increasing the technological content [7-10]. While in China, most of our returning field machines are large - scale machines, mainly based on the tractors, which consume more energies. The machines have higher requirements on the height and width of field ridge, and will compact the soil and damage the perennial root. Besides, the high - speed rotating knives will cut the stones in the soil and the perennial root easily. In addition, there are mainly hilly lands in the south China, where the adapTableility of machine is poor. Thus, it is not suitable to push the development of large - scale returning machines. This returning field mode is not conducive to the sustainable development of sugarcane cultivation [11].

Therefore, it is necessary to design such a small - scale electric sugarcane leaves returning field machine. Its advantage lies in light volume and weight, low power consumption, and it can be controlled remotely and has low damage to perennial root. The workflow of the whole machine is to pick up the leaves by means of collector, and then the leaves will be delivered to the working area by brush roller for cutting and returning field. Sugarcane leaves are picked up by pickers, and then is cut by knives, so that knives will not cut the stones in the soil and the perennial root. The collector is a critical part of the sugarcane leaves returning field machine, which has connection with the effect of collecting and the framework of the whole machine. In order to study the principles of structure and motion, in this research, we use the ADAMS software to make motion simulation on the collector, so that it can confirm the key structural parameters of the collector. The Central Composite Design (CCD) is used to analyze the influence of motion parameters with the aim of assessing the best motion parameters combination on low leakage rate.

2. Structure and Working Principles of the Collector

The collector is composed by cam disc, roller, crank, roller plate, power shaft, collecting components, clamping rings, carbon fiber bar, and fender apron, which can be seen in the Figure1. The motor drives the power shaft to rotate through the chain drive and the roller plates connected with the power shaft will also rotate together. The two ends of the four carbon fiber rods are connected to the two roller plates by means of bearings. The clamping rings are isometric distributed in the carbon fiber rods, and the collecting components are fixed at the clamping rings. Besides, the one side of the carbon fiber rods are fixed with cranks and the other end of the cranks are fitted with rollers which can move along the cam disc slot. The fender aprons cover the whole machine; so that the collecting components push sugarcane leaves along the surface of them with the leaves can't be involved inside.

The collecting components can pick up the sugarcane leaves leaved spreading on the field and achieve the complex motion, which requires three movements to be compounded including the forward moving of the machine, rotation of the roller plates and the swing of the collecting components. (1) The forward moving: the forward power of the whole machine is provided by the two back driving wheels; (2) Rotation of the roller plates: the power shaft drives the roller plates to rotate, thereby the four carbon fiber rods mounted on the roller plate can rotate around the power shaft. The collecting components distributed on the carbon fiber rods evenly also can rotate around the power shaft; (3) Swing of the collecting components: the carbon fiber rods and cranks are connected fixedly, and the cranks drive the rollers to move along the slot of the cam disc. The irregular motion of the rollers is fed back to the carbon fiber rods through the cranks. Because the carbon fiber rods are in articulated connection with the roller plates, they can be repeatedly rotated by the influence of the rollers. Thus, the collecting components can swing repeatedly. By means of the complicated move, the cycle of the move of collecting components can have four different stations, which are picking up, lifting, pushing and returning. In order to keep ordered operation of the cycle of the four stations, it should be controlled by the critical component cam disc, so that it is necessary to have parameter design and simulation assistance analysis on the slide curve of cam disc [12-15]. Cam disc can be regarded as a "reverse swinging follower disc cam mechanism" [16]. The designed cam disc slide central trajectory is composed by three sections of arcs and one straight line, which can be seen in Figure 2.

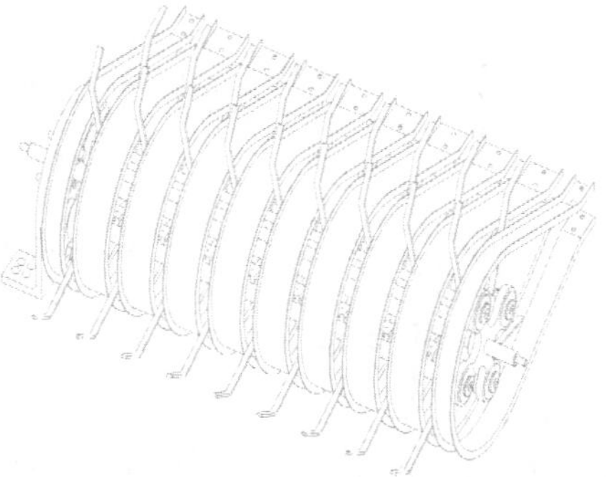

Figure1. Schematic Diagram of the Collector

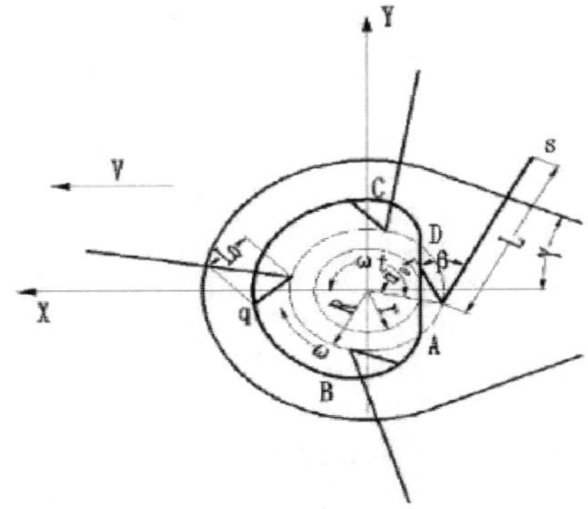

Figure 2. Motion Analysis of the Collector

In one work cycle, when the rollers comes to the AB segmental arc of cam disc slide central trajectory, the collecting components stretch out from the fender apron quickly, and pick up the leaves on the ground. Because of the flexible features of the sugarcane leaves, they are easy to be attached on the collecting components. The radius of AB segmental arc should not be too small, in case that there a big impact force to the rollers, which will reduce the service life of the rollers. BC segmental arc is the lifting stage of the sugarcane leaves. In order to keep the smooth movement of the sugarcane leaves, the collecting components should not swing, so the BC segmental arc should share the same center with the roller plate. When the rollers move to the point C, the collecting components should keep a vertical mode with the fender apron. The movement of rollers from point C to point D is the pushing stage of the sugarcane leaves; the collecting components push the leaves to the brush roller area along the surface of the fender apron. The reason for this design is the leaves can avoid to be brought into the fender aprons by the collecting components. DA line segment is the returning stage, and the collecting components can hide in the fender apron for the next work of collecting [19-20].

3. Mathematical Modelling

The motion of collecting components is composed with three single movements, and the mathematical modelling will be composed by these three movements. The moving direction of the collector is X axial, the vertical direction is Y axial, and the coordinate system is as the Figure 2. Thus, we can get the displacement equation of Q in the roller center, and get the velocity equation and acceleration equation by derivation (Eq. 1).

$$\begin{cases} x_q = Vt + R\cos\omega t - L_0\cos(\omega t + \alpha_0 + \alpha) \\ y_q = R\sin\omega t - L_0\sin(\omega t + \alpha_0 + \alpha) \end{cases}$$

$$\begin{cases} V_{x_q} = V - R\omega\sin\omega t + L_0\omega\sin(\omega t + \alpha_0 + \alpha) \\ V_{y_q} = R\omega\cos\omega t - L_0\omega\cos(\omega t + \alpha_0 + \alpha) \end{cases} \tag{1}$$

$$\begin{cases} a_{x_q} = -R\omega^2\cos\omega t + L_0\omega^2\cos(\omega t + \alpha_0 + \alpha) \\ a_{y_q} = -R\omega^2\sin\omega t + L_0\omega^2\sin(\omega t + \alpha_0 + \alpha) \end{cases}$$

The rationality of the designed collector can be analyzed and modified by the trace of the end of the collecting components. In the coordinate system of Figure2, the displacement equation, velocity equation and acceleration equation are at the end of the collecting components S (Eq. 2).

$$\begin{cases} x_s = Vt + R\cos\omega t - L\cos(\omega t + \alpha_0 + \alpha + \beta) \\ y_s = R\sin\omega t - L\sin(\omega t + \alpha_0 + \alpha + \beta) \end{cases} \tag{2}$$

$$\begin{cases} V_{x_s} = V - R\omega\sin\omega t + L\omega\sin(\omega t + \alpha_0 + \alpha + \beta) \\ V_{y_s} = R\omega\cos\omega t - L\omega\cos(\omega t + \alpha_0 + \alpha + \beta) \end{cases}$$

$$\begin{cases} a_{x_s} = -R\omega^2\cos\omega t + L\omega^2\cos(\omega t + \alpha_0 + \alpha + \beta) \\ a_{y_s} = -R\omega^2\sin\omega t + L\omega^2\sin(\omega t + \alpha_0 + \alpha + \beta) \end{cases}$$

Where the V is forward velocity of the machine, (m/s); t is operating time, (s); R is roller radius, (m); ω is rotating speed of the roller, (rpm); L_0 is length of the crank, (m); L is length of the collecting components, (m); α_0 is the initial angular of the collecting components, (°); α is tilt angle of the collecting components, (°); β is the intersection angle of the collecting components and the crank, (°).

4. Preliminary Determination of Critical Parameters in the Collector

At present, the drum - type collectors are mostly used for the collecting of straw, herbage, and oilseed rape, which have the features of large capacity in collecting and large size. The size of the collector studied in this paper is small, and collectors sold on the market do not meet the design requirements of this article, so we need to redesign it. Analyzing the working principles of current collector, and considering the total size of the machine, we determine the radius of roller plates is R=60 mm, the length of cranks is L0=40 mm, and the length of collecting components is L=150 mm. The following design is the central orbit curve in the slide of cam disc; it is also the most critical part, which decides that if the collector can meet the requirements of the design. It can be optimized and modified by the ADAMS software motion simulation, without complicated mathematical calculation to get the reasonable trajectory curve.

As to the modelling of the collector, in order to reduce the simulation time, model conveniently, and be added constraints easily el at., it can shorten the length of carbon fiber bar and save some features of the power crank and fasteners. We use the three - dimensional software SolidWorks for modelling and saved it into the form of Parasolid, and then led it to the ADAMS software; the total modelling can be seen in the Figure 3. In the ADAMS software, the parts of the modelling are endowed with material performance, and also the relationships for restraining them. When adding the drive, the moving vice between the cam disc and the earth simulates the forward movement of the machine, and the rotary vice between the roller plate relative and cam disc simulates the rotating of roller plate [21]. Such an example of the collector, it is required in the program that the initial selection of the forward speed is v=0.5 m/s, and the rotating speed of the roller plates is ω=40 rpm. The simulated results could be checked and analyzed in Post Processor of the ADAMS software.

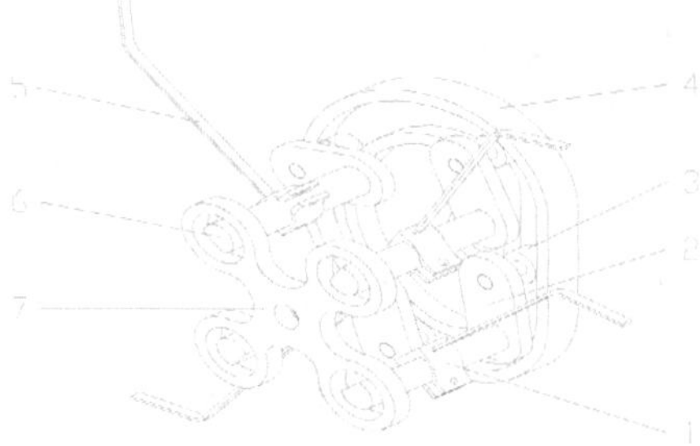

1. Clamping ring 2. Crank 3. Roller 4. Cam disc

5. Collecting component 6. Carbon fiber bar 7. Roller plate

Figure 3. Three-dimensional Simplified Model of the Collector

The Figure 4 is the trajectory picture of the end of the collecting components, and the simulation time is two cycles. We can see the buckles in the trajectory line on the figure, which means that when the machine is working, the collecting components have opposite level speed with the forward direction of the machine, so that the collecting components can push the materials backward. That is the essential condition of collector to realize the collecting and pushing of materials [22].

The Figure 5 is the curve graph of line speed, line acceleration speed, and angular acceleration speed of the end of the collecting components with the changes of the time, and the simulation time is a cycle. It can be seen in the figure that at the time of t=1.13 ~1.2 s and 0~0.53 s, the line acceleration speed is decreasing, while the two kinds of acceleration speed nearly have no change. During these times, the collecting components are in the lifting, which lift the sugarcane leaves up smoothly. During the time of t=0.53~0.71 s, the line acceleration speed is beating down, and the collecting components push the sugarcane leaves to the cutting area along the fender apron. During the time of t=0.71~1 s, the line speed is rising, and two acceleration speeds slightly floating, that is the returning stage of the collecting components. During the time of t=1~1.13 s, there is big change on the three kinds of speeds, which will reach to the summit at the time of t=1.05 s. The collecting components stretch out from the fender apron quickly, which have big impact force on the sugarcane leaves, do bad to the collecting of the sugarcane leaves. In addition, the roller has a large impact force on the cam disc, which reduces the service life of roller. It is possible to increase the radius of the trajectory of the cam plate at this period, as to increase the route of the cam disc at this period, so that rollers can have a better transition, reduce the acceleration speed, and reduce the impact force to cam disc.

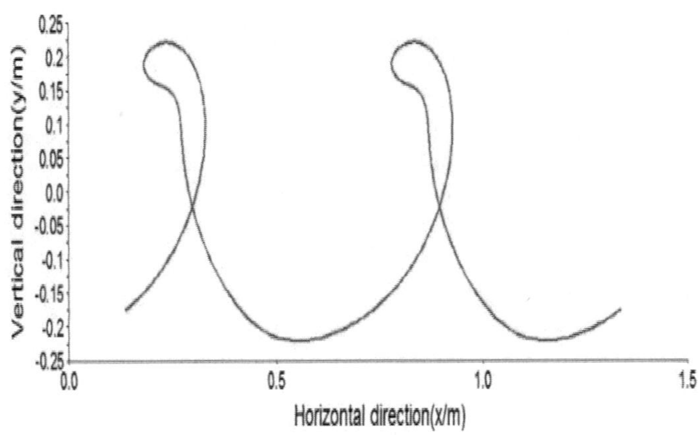

Figure 4. The Trajectory Picture of the End of the Collecting Components

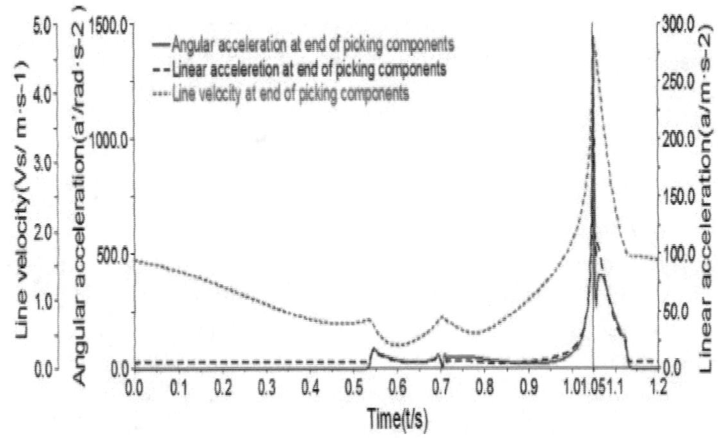

Figure 5. Curve of Line Velocity, Linear Acceleration and Angular Acceleration at End of Collecting Component

Because the cam disc slide central trajectory is a closed curve which is composed by three sections of arcs and one straight line, the reasonable parameters of the collector are finally confirmed after

IOP Conf. Series: Materials Science and Engineering **735** (2020) 012069 doi:10.1088/1757-899X/735/1/012069

several times of parameter design and simulation optimization of the ADAMS software, which can be seen in Table 1.

5. Optimization of Movement Parameter of the Collector

From the mathematical modelling, we can know that after the confirmation of critical parameters of the collector, the trajectory line at the end of the collecting components just has connection with the forward speed and the rotating speed of the roller plate. In order to get the reasonable trajectory line, it is necessary to optimize the best group of the forward speed and the rotating speed, so that the collector can pick up the sugarcane leaves and has a lower leakage rate. In this article, we use the ADAMS software for simulation experimental analysis on the model, and obtain the related data in different matching combinations of the forward speed and the rotating speed. And then, the data are processed by Central Composite Design (CCD) to optimize the best group of the forward speed and the rotating speed with minimum leakage rate [23-24].

5.1. Experimental Factor

In the simulated test, the factors are the forward speed of the whole machine and the rotating speed of the roller plate. According to the project requirements and the operating conditions of the sugarcane leaves returning field machine, the test adopts two factors and three levels, which can be seen in the following Table 2.

Table 1. Structure Parameters of the Collector

Parameters	Numerical value
Base circle radius(r/mm)	41
Roller plate radius(R/mm)	60
Brank(L_0/mm)	40
Collecting component length(L/mm)	150
Angle between the collecting components and the fender apron (β/°)	52
Three-section arc radius(mm)	41, 53, 89
Upper guard fender apron angle(γ/°)	24
Number of carbon fiber bars (z/pcs)	4

Table 2. Test Factors and Levels

Levels	Factors	
	Forward speed (m/s)	Rotating speed of the roller plate (rpm)
-1	0.4	35
0	0.5	50
1	0.6	65

5.2. Optimizing Index and Method

The ADAMS software simulation can't directly get the leakage rate, so it is indirectly analyzed by two optimization indicators, which are leakage area per meter and the largest angular acceleration at the end of the collecting components. (1) Leakage rate is an important standard to check the performance of the collector. When the collector is working, the uncovered areas at the end of the four trajectory lines are the leakage area, which can be seen in the shadow areas of the Figure6. As to different group for two factors, the leakage area by simulated analysis is different. In order to make the analysis results comparable, it is necessary to calculate the sum area of leakage area per meter. The calculation of leakage area per meter is gotten by the MATLAB software, and makes the results as one of the indicators of the performance of the collector; (2) If the angular acceleration speed is larger at the end of the collecting components, which mean that the rollers have a large impact force on the cam disc. Thus, the rollers are easy to be damage and, and also it will influence the working of other parts, so that the collector is unsTable to work. Therefore, the angular acceleration at the end of the collecting components should be small, and it is used as one of the indicators for simulation analysis. The angular acceleration speed at the end of the collecting components in this test is no more than 200 rad/s^2.

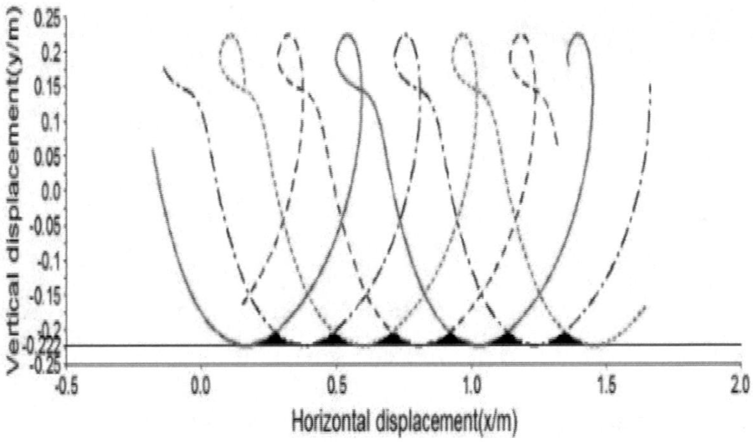

Figure 6. The Leakage Area of the Collector

5.3. Analysis on the Experimental Result

In this test, a 2-factor 3-level full factorial design method are adopted, which contained 13 sets of test data. The ADAMS software obtains date from different group of two factors, the data were as the Table.3. The data in the Table are put into the Design - Expert software for processing, so that we can obtain the response surface of the two factors, which are like the Figure7 ~ 8.

Table 3. Test Plan and Analysis Results

Forward speed(m/s)	Roller plate rotation speed(rpm)	Leakage Height(cm)	Area of leakage area (cm^2)	Number per meter	Area of leakage area per meter (cm^2)	Maximum angular acceleration in end of collecting components(rad/s^2)
0.4	35	1.26	7.22	9	64.98	105.57
0.6	35	2.40	20.54	4	82.16	105.57
0.4	65	0.47	1.34	11	14.74	308.02
0.6	65	0.87	4.09	7	28.63	308.02
0.4	50	0.72	2.77	8	22.16	203.98
0.6	50	1.34	8.21	5.5	45.16	203.98
0.5	35	1.82	12.93	5	64.65	105.57
0.5	65	0.66	2.50	8.5	21.25	308.02
0.5	50	1.03	5.06	6.5	32.89	203.98
0.5	50	1.03	5.06	6.5	32.89	203.98
0.5	50	1.03	5.06	6.5	32.89	203.98
0.5	50	1.03	5.06	6.5	32.89	203.98
0.5	50	1.03	5.06	6.5	32.89	203.98

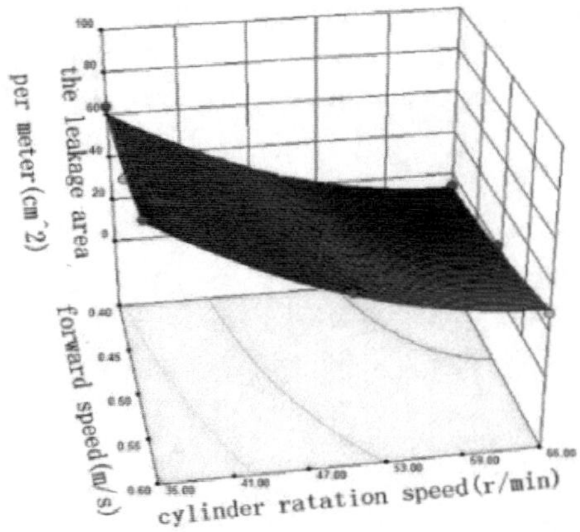

Figure 7. The Response Surface of the Leakage Area Per Meter

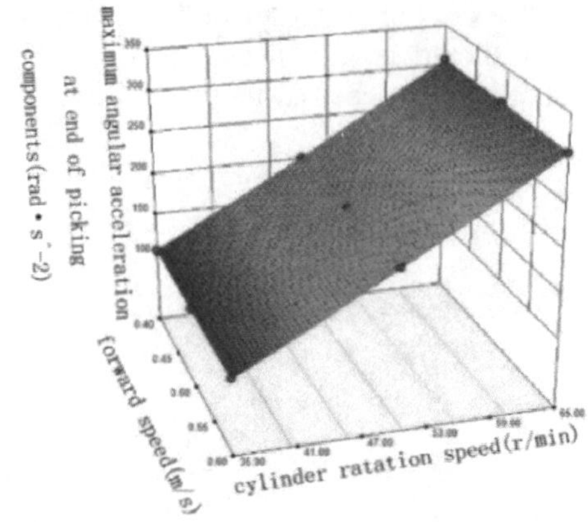

Figure 8. The Response Surface of Maximum Angular Acceleration in End of Collecting Components

(1) From the aspect of single factor, when the rotating speed of the roller plate is constant, the sum area of leakage area per meter is larger with the increasing of the forward speed; when the forward speed is constant, sum area of leakage is smaller with the increasing rotating speed. Thus, the response surface will lean in the direction of slower forward speed and faster rotating speed, and the sum area of leakage is smaller and smaller. (2) It can be seen in Figure8 and Table.3, the largest angular acceleration at the end of the collecting components has no connection with the forward speed, and it will be faster with the increasing rotating speed of the roller plate, but it is bad to the sTableility of the collector. With slower forward speed and faster rotating speed, the leakage area will be smaller, but it does not meet the design requirements. It is necessary to find a suitable combination of forward speed and rotating speed to meet the low leakage rate and the job requirements of the total machine. The boundary conditions are input in the software of Design - Expert, which include the angular acceleration speed at the end of the collecting components that is no more than 200 rad/s^2 and forward speed that is more than 0.4 m/s. The result shows that the most optimized parameter assembly for the two factors are that: the forward speed is v=0.4 m/s, the rotating speed of the roller plate is ω=49.41 rpm. The sum area of leakage area per meter is 26.95 cm^2, and the largest angular acceleration at the end of the collecting components is 200 rad/s^2. Based on the above motion parameters, the trajectory chart, pendulum angle, linear acceleration, and angular acceleration at the end of the collecting components are as Figure9 ~ 10.

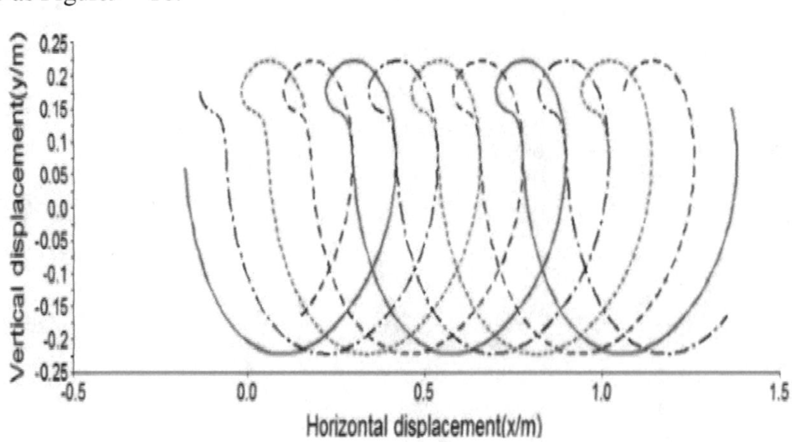

Figure 9. The Trajectory Chart in End of Collecting Components

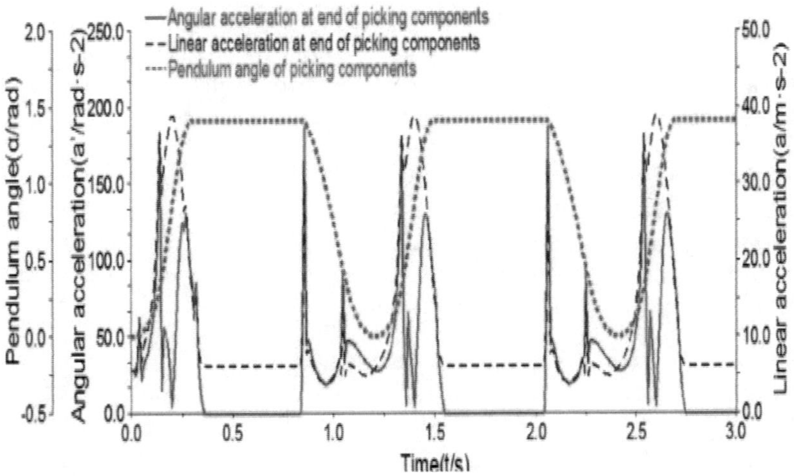

Figure 10. The Pendulum Angle, Linear Acceleration and Angular Acceleration in End of Collecting Components

6. Experimental Verification

6.1. Experimental Equipment and Site

Based on the above theoretical analysis and optimized result, the cam disc and other components can be assembled to the electric sugarcane leaves returning field machine. In order to verify whether the machine has a low leakage rate under theoretical motion parameters, it is necessary to carry out field test on the machine, which can be seen in Figure11. The test instruments include electronic scale, meter ruler, rotating speed measuring instrument, and computer.

Figure 11. Experimental Site Diagram

6.2. Experimental Performance Index and Method

The leakage rate of the test will be measured by weighting method. The leakage rate of the collector means that the proportion of the weight of non-pick up materials in the total weight of all materials (Eq. 3).

$$\eta = \frac{m_2}{m_1} \times 100\% \quad (3)$$

Where η is leakage rate of the collector (%), m_1 is total weight of the materials during the collecting course (kg), m_2 is the weight of non-picking up materials during the collecting course (kg). The yield of sugarcane leaves is 7.5 t/hm^2, which mean that the yield of sugarcane leaves of per square meter is 0.75 kg [25-26], so when the test, the sugarcane leaves are laid on an area of 2 m^2, where the length is 2 m and the width is 1 m, and the total weight of the sugarcane leaves is 1.5 kg. When testing, the forward speed and the rotating speed will be adjusted into theoretical parameter, which means that the forward speed is v=0.4 m/s, the rotating speed is ω=50 rpm, and we use the weighing method to calculate the leakage rate.

6.3. Analysis on the Experimental Result
The experimental materials are selected from the ROC 22 sugarcane leaves. The test conducting five sets of experiments, the results are showed in the Table 4. The average leakage rate is 2.93 %, which can meet the job requirements. This shows that the structural and working parameters obtained by theoretical analysis can meet the design requirements of the returning field machine.

Table 4. Experimental Data

Experimental number	Total weight of materials (kg)	Weight of non - pick up materials (kg)	Leakage rate (%)
1	3	0.0832	2.77
2	3	0.0905	3.17
3	3	0.1056	3.52
4	3	0.0745	2.48
5	3	0.0809	2.70

7. Conclusion
(1) This paper presents the design of a collector for sugarcane leaves returning field machine. The mathematical model of the collector is esTablelished by analysis its working principle. The important parameters of the collector are confirmed, including the roller plate radius, crank length, and the locus curve in the center of cam disc slide through the motion simulation of the ADAMS software.

(2) The combination of the minimum leakage rate of the collector have obtained under the two factors of the forward speed of 0.4~0.6 m/s and the roll plate speed of 35~65 m/s. Firstly, the method of the simulation analysis by the ADAMS software is used for analysis data of different combinations of factors. And then we use the Central Composite Design (CCD) to screen the best combination. The conclusions are as follows: the forward speed is V=0.4 m/s, the rotating speed is ω=49.41 rpm, the sum area of leakage per meter is 26.95 cm^2 , and the most angular acceleration at the end of collecting components is 200 rad/s^2.

(3) The test results show that the actual leakage rate of the collector meets the design requirements under the forward speed of v=0.4 m/s and the rotational speed of the roller plate of ω=50 rpm. The optimal working parameters combination obtained by the simulation meet the requirements of the agricultural technique index and the sugarcane leaves returning field machine well.

8. Acknowledgement
Subject was supported by the Fund for the Science and Technology Program of Guangzhou (№. 201704020022).

*Corresponding author: Yuxing Wang

9. References
[1] Chang Ge, Ming Li, Lijiao Wei, et al. 2016. Discussion and research on mechanized returning technology of sugarcane leaves in hot areas. *Chinese Tropical Agriculture*, No1, pp32-34.

[2] Graham MH, Haynes RJ and Meyer JH. 2002. Soil organic matter content and quality: effects of fertilizer applications, burning and trash retention on a long-term sugarcane experiment in South Africa. *Soil Biology and Biochemistry*, vol 34, No1, pp93-102.

[3] Galdos M V, Cerri C C and Cerri C E P. 2009. Soil carbon stocks under burned and unburned sugarcane in Brazil. *Geoderma,* vol 153, No3-4, pp347-352.

[4] Guanbao Li, Dong Liang, Ming Li, et al. 2017. Design of Deep Buried Single Flat Ditching Equipment in Sugarcane Leaves Shattering and Returning field machine Design. *Journal of Agricultural Mechanization Research*, No5, pp95-99.

[5] Lijiao Wei, Ming Li, Jingming Lu, et al. 2011. A Literature Review on the Technology of Mechanically Crushing Sugarcane Leaves and Tilling into Soil and Its Effects on Soil. *Chinese Agricultural Mechanization*, No1, pp88-91.

[6] Haishui Yang, Jinxia Feng, Silong Zhai, et al. 2016. Long-term ditch-buried straw return alters soil water potential, temperature and microbial communities in a rice-wheat rotation system. *Soil and Tillage Research*, vol 163, pp21-31.

[7] Yuejin Jin, Jian Yang, Zhaoxin Liang, et al. 2004. Test Research on Sugarcane Broken-leaves Machine. *Journal of Agricultural Mechanization Research*, vol 120, No4, pp137-138.

[8] Ming Li, Jinli Wang, Yiguo Deng, et al. 2008. Structural design and experiments on sugarcane leaves shattering and returning field machine. *Transactions of the Chinese Society of Agricultural Engineering,* vol 24, No2, pp121-126.

[9] Mathanker S K, Maughan J D, Hansen A. C., Grift T E and Ting K C. 2014. Sensing miscanthus swath volume for maximizing baler throughput rate. *Transactions of the Asabe*, vol 57, No2, pp355-362.

[10] Bortolini Marco, Cascini Alessandro and Gamberi Mauro. 2014. Sustainable design and life cycle assessment of an innovative multi-functional haymaking agricultural machinery. *Journal of Cleaner Production*, vol 82, pp23-36.

[11] Jingming Lu, Ming Li, Lijiao Wei, et al. 2012. Problems in Technical Extension of Crushing Sugarcane Leaves and Returning to Fields and Suggestions. *Tropical Agricultural Engineering*, vol 36, No6, pp43-45.

[12] Haiquan Ding, Zhihong Yu, Weifeng Liu, et al. 2015. Theory analysis on kinematics characteristics of spring-finger cylinder pickup machinery device. *Journal of Agricultural Mechanization Research*, vol 37, No10, pp76-82.

[13] Tao Xu, Yongzhe Shen, Lianxing Gao, et al. 2016. Spring-finger Peanut Pickup machinery Mechanism Based on Two-stage Harvest. *Transactions of the Chinese Society for Agricultural Machinery*, vol 47, No3, pp90-97+111.

[14] Caiyun Yuan, Peisong Diao and Daolin Zhang. 2011. Design and motion simulation of spring-finger cylinder pickup machinerys. *Journal of Agricultural Mechanization Research*, vol 33, No5, pp73-76.

[15] Zhihong Yu, Shoucheng Huai and Wenming Wang. 2018. Leakage rate and optimization of working parameters for cylinder pickup machinery collector based on spring-finger trajectory. *Transactions of the Chinese Society of Agricultural Engineering*, vol34, No4, pp37-43.

[16] Kai Sheng and Nanhong Zeng. 1990. An Analytical Method of Mechanism Designing of Spring-Finger Cylinder Pickup machinerys. *Journal of Jilin Institute of Technology*, No2, pp50-57.

[17] Kai Sheng and Nanhong Zeng. 1991. The mechanical feature and motional math model of spring-finger cylinder pick-ups. *Transactions of the Chinese Society for Agricultural Machinery,* vol 22, No1, pp51-57.

[18] Kai Sheng. 1986. General program of CAD of plate cam mechanism with swing driven member. *Journal of Jilin Institute of Technology*, No, pp121-129.

[19] Zhenhua Wang, Decheng Wang, Guilin Liu, et al. 2010. Pickup machinery parameters design of square baler. *Transactions of the Chinese Society for Agricultural Machinery*, vol 41, pp107-109.

[20] jisiguleng Wu. 2010. Simulation and Experimental Study on Performance Parameters of Spring-tooth Grass Pickup machinery Roller (Hohhot: Inner Mongolia Agricultural University Press).

[21] Changzhi Jia, Junhui Yin, Wenxing Xue, et al. 2010. MD ADAMS virtual prototype from entry to proficient (Beijing: Mechanical Industry Press).

[22] Wenming Wang and Chunguang Wang. 2012. Parameter analysis and simulation of spring-finger cylinder pickup machinery collector. *Transactions of the Chinese Society for Agricultural Machinery*, vol 43, No10, pp 82-89.

[23] Jian Zhao, Yun Chen, Yalei Wang, et al. 2019. Experimental Research on Parameter Optimization of PorTable Vibrating and Harvesting Device of Chinese Wolfberry. *Journal of Agricultural Mechanization Research*, vol 41, No3, pp176-182.
[24] Hongyi Li, Mingliang Wu, Haifeng Luo, et al. 2016. Experimental Study on Rape Picking Up of Spring-finger Cylinder Pickup machinery Collector. *Chinese Agricultural Science Bulletin*, vol 32, No18, pp176-182.
[25] Qing Liao, Guangpo Wei, Guifen Chen, et al. 2011. Effect of Trash Returning on Microbial Communities, Physical and Chemical Properties of Soil and Plant Growth of Sugarcane. *Southwest China Journal of Agricultural Sciences*, No2, pp 658-662.
[26] Xiongwei Cui, Yuebin Zhang, Jiawen Guo, et al. 2010. Effects of Different Patterns of Sugarcane Leaves Returning Field on Soil Moisture and Sugarcane Yield. *Sugar Crops of China*, No4, pp21-23.

Analysis of Removal Function Results of Permanent Magnet Polishing Device Test

Xiaoqin Zhou[1], Baoquan Sun[2], Haocheng Wang[2] and Hang Yu[3] and Xiaoyang Wu[3]

School of Mechanical and Aerospace Engineering, Jilin University, Jilin Province, China
Email: ss15643249600@163.com

Abstract. Based on the test platform of the polishing device, the material removal function model of the workpiece is obtained by using the Preston empirical equation and the motion state of the polishing tool. The test platform is built and the material removal model is verified by polishing the workpiece. Using Matlab software, draw the mathematical model material removal function curve and the test material removal function curve, analyze and compare the material removal law of the workpiece, and analyze the surface quality before and after the workpiece polishing.

1. Introduction

Ultra-precision machining technology is a key technology for a country to manufacture high-end precision equipment and become a powerful country in equipment manufacturing. It is of great significance to people's quality of life and national development and construction [1]. In many fields, the shape of parts of mechanical equipment is more and more complicated, and even many parts have internal cavity structure, which greatly increases the difficulty of ultra-precision machining technology [2]. Because the surface roughness is universal and random on the surface of the workpiece, the performance of many devices is severely and complicatedly affected by the limited surface quality of the workpiece cavity [3].

Aiming at this situation, this paper designs a magnetic field teleoperation polishing tool, which is a test platform for directional polishing of the inner cavity surface of the workpiece. N52 NdFeB permanent magnet is used as the source of space magnetic field. The KR60-3 robot arm of KUKA and the DCX26L DC motor of Maxon are selected as the power drive system of the magnetic field generating device to control the six degrees of freedom of the permanent magnet generated by the magnetic field. . The smaller permanent magnet polishing tool is brought into contact with the inner cavity of the workpiece under a certain pressure by the external permanent magnet, and rotates on the surface of the inner cavity of the workpiece at a certain rotation speed, thereby polishing the inner cavity surface of the workpiece.

Based on the test platform of the polishing apparatus, a material removal function model was established to analyze the material removal amount and material removal error of the workpiece cavity surface. Select the appropriate polishing parameters, polish the surface of the workpiece cavity and measure the accuracy, and analyze the polishing results.

2. Permanent Magnet Device Polishing

The permanent magnet magnetic field generating device generally uses the property of the magnetic field around the permanent magnet to control the movement of the permanent magnet, and generates a permanent magnet of the spatial magnetic field, which is called a magnetic field generating permanent magnet, and the remotely operated permanent magnet is called a target permanent magnet[4].When the magnetic field occurs, the permanent magnet is stationary in a stable space environment, and the distribution of the spatial magnetic field around it is fixed. Therefore, when a changing magnetic field is required, it is generally achieved by changing the position of the permanent magnet to cause the magnetic field to generate permanent magnet motion.

Based on the permanent magnet magnetic field generating device, a device capable of polishing the surface of the inner cavity of the workpiece by using a magnetic field remote manipulation polishing tool is designed. The permanent magnet is selected as the magnetic field generating source of the magnetic field teleoperation system; the DC motor is used to drive the permanent magnet to rotate around its own axis to generate a degree of freedom of the permanent magnet; the six-axis mechanical arm is used to drive the permanent magnet to generate the remaining five degrees of freedom of the permanent magnet. Design the polishing tool structure to meet the requirements of use; and configure the polishing fluid. Finally, a polishing test platform was set up, and a suitable workpiece was selected to polish the inner cavity surface to explore the polishing effect.

3. Preston Empirical Equation

The material removal function refers to the functional relationship between the amount of material removed during processing and the individual polishing parameters. The construction of the polishing material removal function makes the whole process deterministic and is an important basis for the analysis of the workpiece polishing process [5]. The material removal function is used to set the polishing parameters, and the surface accuracy of the workpiece is corrected by planning the polishing tool motion path, eliminating the edge effect, calculating the processing residence time of the polishing tool, etc. [6-8]. At present, the widely used material removal function model is the Preston empirical equation proposed by Princeton in 1927 [9-10].
Preston empirical equation:

$$\frac{dh}{dt}=kPV \tag{1}$$

Where h is the amount of material removed; V is the speed at which the polishing tool moves on the surface of the workpiece; k is the Preston coefficient, which represents all other factors except P, V, and t, including the composition of the polishing fluid, the material of the workpiece, and the polishing environment. Such as the comprehensive factors, there is no mature theory can be analyzed and calculated, which can be obtained through experimental measurements (test measured: $k=2.79\times10^{-9}$). P is the pressure applied by the polishing tool to the workpiece. Assuming that the surface of the inner cavity of the polished workpiece is in perfect contact with the polishing tool and the pressure of the workpiece is uniform, then:

$$P=\frac{F_N}{\pi R^2} \tag{2}$$

For the material removal function, the following conditions are required:

(1) The material removal efficiency of each point on the surface of the workpiece to be polished is the same;

(2) In the case where the polishing parameters are determined to be constant, the material removal amount of the workpiece surface is proportional to the processing time thereof;

(3) The material removal function remains stable throughout the polishing process and does not change due to an increase in the amount of removal and a change in the polishing position.

The polishing tool is mainly divided into fixed-point polishing, linear motion polishing and curved motion polishing on the surface of the workpiece cavity. In this paper, the polishing tool is used to polish the surface of the workpiece cavity along the straight line, as shown in Figure 3.1.

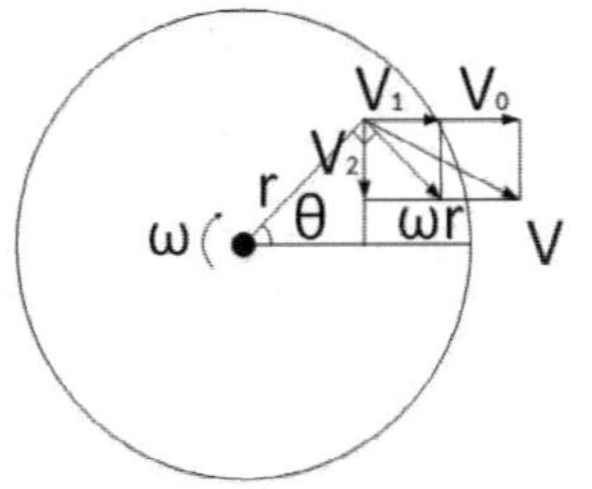

Figure 3. 1. The Speed at which the Polishing Tool Moves on the Surface of the Workpiece

According to Figure 3.1, the rotation speed of any point on the polishing tool is decomposed in the horizontal direction and the vertical direction:

$$\begin{cases} V_1 = \omega r \sin\theta \\ V_2 = \omega r \cos\theta \end{cases} \tag{3}$$

Where ω is the rotation angular velocity of the polishing tool; r is the radius length of the polishing tool at any point on the polishing tool from the center point of the polishing tool; θ is the angle between the arbitrary point on the polishing tool and the center point of the polishing tool relative to the horizontal line.

Find the speed at which any point on the polishing tool moves on the surface of the workpiece:

$$V = \sqrt{(V_1 + V_0)^2 + V_2^2} \tag{4}$$

In the formula, for the linear speed of the polishing tool, the finishing speed is V0:

$$V = \sqrt{(\omega r \sin\theta + V_0)^2 + (\omega r \cos\theta)^2} \tag{5}$$

Substituting V0=0, the speed at which the polishing tool moves on the surface of the workpiece during fixed-point polishing is:

$$V = \omega r \tag{6}$$

Union Equation1, Equation 2 and Equation 5:

$$\frac{dh}{dt} = -\frac{3k\mu_0 |\vec{M}||\vec{m}| \cos\alpha}{5.6\pi^2 x^4 R^2} \sqrt{(\omega r \sin\theta + V_0)^2 + (\omega r \cos\theta)^2} \tag{7}$$

Substituting V_0=0,the material removal function of the polishing tool during spot polishing can be obtained:

$$\frac{dh}{dt} = -\frac{3k\mu_0 |\vec{M}||\vec{m}| \omega r \cos\left(\tan^{-1}\frac{12\lambda R}{7x}\right)}{5.6\pi^2 x^4 R^2} \tag{8}$$

In the actual polishing process, the linear speed of the polishing tool is small and much smaller than the rotation speed of any point on the polishing tool, namely:

$$V_0 \leq V_1 \tag{9}$$

Therefore, when analyzing the material removal function model, we think that $V_0 \approx 0$ and V_0 are ignored. Using Matlab, the relationship between the material removal rate $\frac{dh}{dt}$ and the distance from the position of the polishing tool to the center of the polishing tool at different distances x is plotted.

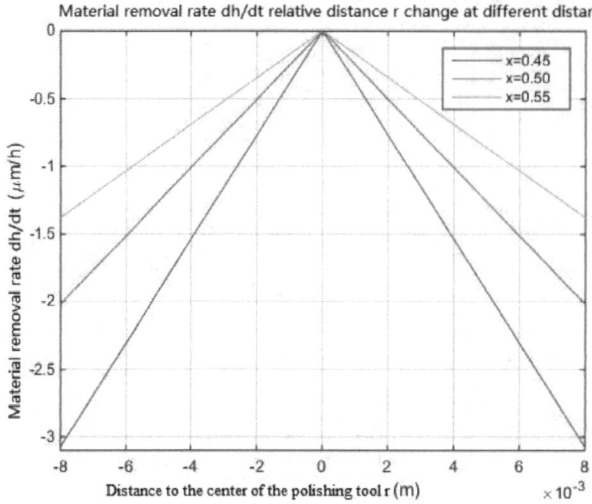

Figure 3. 2. Relationship between Material Removal Rate $\frac{dh}{dt}$ and Distance r at Different Distances x

As can be seen from Fig. 3.2, when the distance between the polishing tool and the permanent magnet x in the magnetic field remote control device is constant, the greater the distance r from the polishing point to the center of the polishing tool, the smaller the material removal rate $\frac{dh}{dt}$, and the larger the absolute value, that is, the same time. The greater the amount of material removed, and the linear relationship. When the distance from the polishing point to the center of the polishing tool is constant, the larger the distance between the polishing tool and the permanent magnet in the magnetic field remote control device, the larger the material removal rate $\frac{dh}{dt}$, and the smaller the absolute value, that is, the material removal in the same time. The smaller the amount.

4. Polishing Test Removal Function Analysis

4.1. Material Removal

When the polishing tool is polished in a linear motion on the surface of the workpiece cavity, the amount of material removed by the polishing tool through a fixed point is the sum of the material removal points of the points on the polishing tool passing through the fixed point, as shown in Figure.

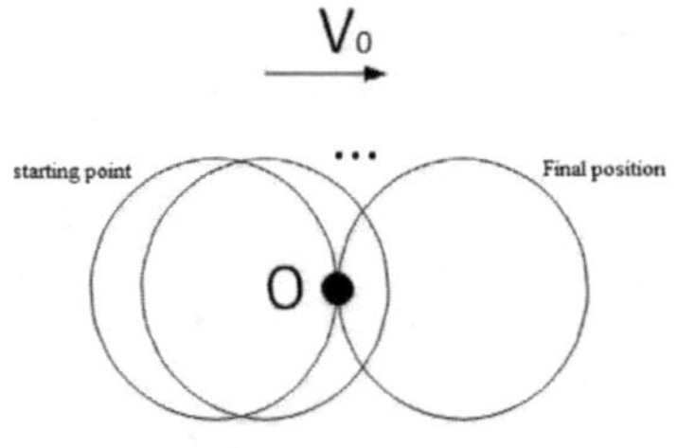

Figure 4.1.a Material Removal at a Fixed Point on the Workpiece

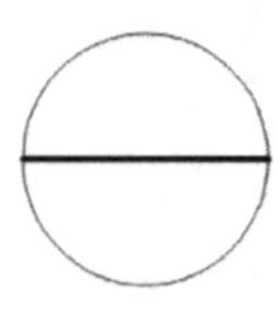

Figure 4.1.b Schematic Diagram of the Simplified Analysis of Removal

Figure 4.1. Analysis of Material Removal

In Figure 4.1a, the polishing tool polishes the O point and moves linearly from the starting position to the final position at a speed of V0. The material removal amount at point O is the sum of the material removal amount of each point on the polishing tool p$\frac{dh}{dt}$ assing through the fixed point, that is,

IOP Conf. Series: Materials Science and Engineering **735** (2020) 012070 doi:10.1088/1757-899X/735/1/012070

the integral of the polishing material removal efficiencyat each point of the horizontal diameter on the polishing tool with respect to time t in Fig. 4.1b, namely:

$$h=\int_0^{\frac{2R}{V_0}}\int_0^{R}-\frac{6k\mu_0\left|\vec{M}\right|\left|\vec{m}\right|\omega r\cos\left(\tan^{-1}\frac{12\lambda R}{7x}\right)}{5.6\pi^2x^4R^2}drdt \tag{10}$$

Solved by Matlab, the material removal of the O point is:

$$h=\int_0^{R}-\frac{6k\frac{2R}{V_0}\mu_0\left|\vec{M}\right|\left|\vec{m}\right|\omega r\cos\left(\tan^{-1}\frac{12\lambda R}{7x}\right)}{5.6\pi^2x^4R^2}dr \tag{11}$$

4.2. Polishing Test Verification

A polishing test platform is built to perform fixed-point polishing on the surface of the workpiece cavity. In this paper, the 2207 surface profilometer is used to measure the surface removal of the workpiece. Model 2207 surface profile measuring instrument, including contour measurement, roughness measurement, straightness measurement, waviness measurement and other systems, can be used to measure surface roughness, material removal, straightness and other parameters.

When the polishing parameter is $\omega=10(s^{-1})$, $x=0.05(m)$, $t=1(h)$, the relationship between the material removal amount h and the distance r is obtained, as shown in Fig. 4.2.

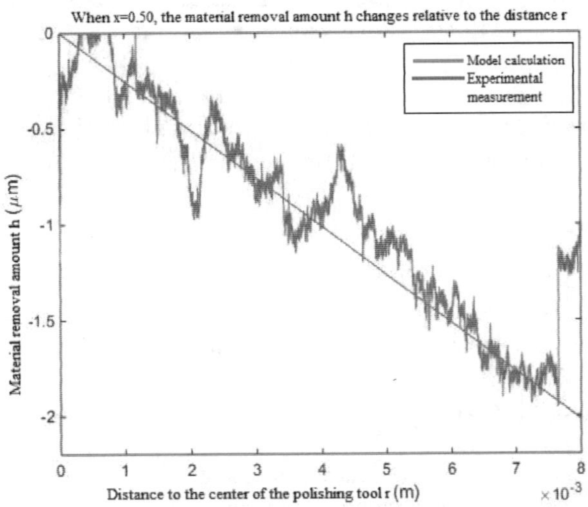

Figure 4.2. Comparison of the Relationship between Material Removal Amount h and Distance r

In this paper, the Zygo NewView 8300 white light interferometer is used to detect the surface quality of the workpiece after polishing. The white light interferometer measures the surface profile of the workpiece based on white light interference technology.

When the polishing parameters are $\omega=10(s^{-1})$, $x=0.05(m)$, $t=1(h)$, the surface roughness of the inner cavity surface after polishing of the workpiece is shown in Fig. 4.3, Fig. 4.3a is the surface roughness of the inner cavity before polishing, and Fig. 4.3b is the surface roughness of the inner cavity after polishing. The surface roughness of the workpiece is processed from 0.102 μm (Sa) to 0.029 μm (Sa).

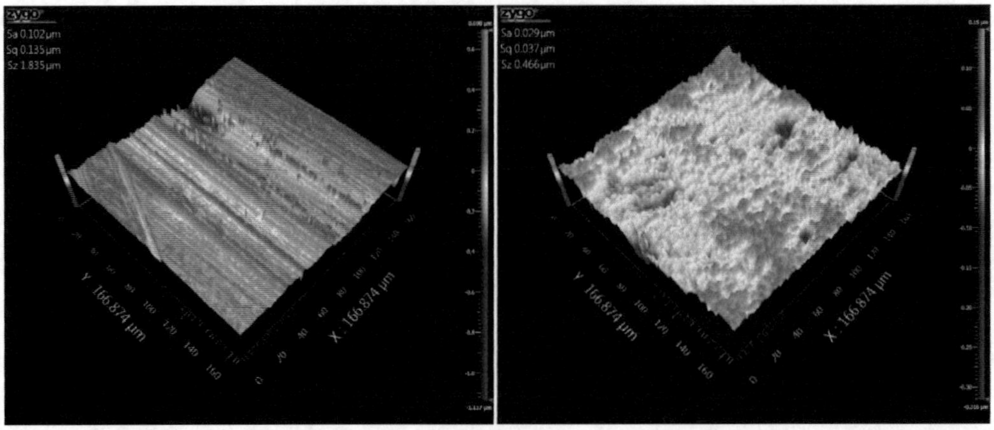

Figure 4.3a Before the Workpiece is Polished Figure 4.3b After the Workpiece is Polished

Figure 4.3. Surface Quality before and after Workpiece Polishing

5. Conclusion

In this paper, we use the Preston empirical equation to obtain the material removal function model of the workpiece. Using Matlab software, draw the material removal function curve, and obtain the material removal rule of the workpiece: when the distance between the polishing point and the permanent magnet x in the magnetic field teleoperation device is constant, the distance r from the polishing point to the center of the polishing tool is larger, the same The greater the amount of material removal during the time; when the distance from the polishing point to the center of the polishing tool is constant, the distance between the polishing tool and the permanent magnet x in the magnetic field remote control device is larger, and the amount of material removal in the same time is smaller. A polishing test platform was built to perform fixed-point polishing on the surface of the workpiece cavity, and the material removal function of the workpiece was verified.

6. Acknowlegment

This work was funded by International Cooperation Project of the Ministry of Science and Technology, China (No. 2016YFE0105100).

7. References

[1] Li Fuli. Ultra-precision machining technology and its development trend [J]. Shandong Industrial Technology, 2016(24):34-34.

[2] Zhou Zhixiong, Zhou Qinyuan, Ren Yinghui. Research Status and Development Trend of Complex Surface Machining Technology [J]. Journal of Mechanical Engineering, 2010, 46(17): 105-113.

[3] Li Na. Research on the influence of surface roughness on the electrical properties of high precision microwave electronic equipment [D]. Xidian University, 2012.

[4] Mahoney A W, Abbott J J. Managing magnetic force applied to a magnetic device by a rotating dipole field [J]. Applied Physics Letters, 2011, 99(13): 134103.

[5] Yang Hang, He Jianguo, Huang Wen, et al. Micro-decomposition coupling method for magnetorheological polishing removal function acquisition [J]. Intense Laser and Particle Beams, 2015, 27(8).

[6] Zhang Wenzhao. Theoretical study on removal function based on computer controlled small grinding head polishing [D]. Tianjin University, 2014.

[7] Zhang Yanfu, Wang Achun, Wang Bo, et al. Removal Function Model of Atmospheric Pressure Plasma Jet Processing Material[J]. Nanotechnology and Precision Engineering, 2013, 11(4): 354-359.

[8] He Ting, Shao Jianda, Wei Chaoyang. Research on Femtosecond Laser Polishing Removal

Function Characteristics [C]// Strong Laser Materials and Components Symposium and Laser Destruction Symposium.
[9] Zhang Feng. Numerical Control Chemical Mechanical Polishing of Silicon Modified Layer on Aspherical Silicon Carbide Surface [J]. Editorial Office of Optics and Precision Engineering, 2013, 21(12): 3015-3020.
[10] Yuan Bo. Polishing of viscoelastic materials polishing tools [D]. Jilin University, 2015.

IOP Conf. Series: Materials Science and Engineering **735** (2020) 012072 doi:10.1088/1757-899X/735/1/012072

Abandonment Research and Environmental Impact Analysis for Retired Dong-Huang Pipelines

Xuesong Zhang, Rangang Yu and Jinping Chen

College of Pipeline and Civil Engineering, China University of Petroleum(East China), Qingdao, China.

Email: snowpiner@upc.edu.cn

Abstract. Nowadays, China is not experienced in the research and practice of the abandonment approaches to oil and gas pipelines retirement. Using the experience of the North America for reference in abandonment of onshore pipelines, including retired pipeline safety environmental disposal scheme choice, waste disposal technology and retired environmental impact assessment of the key problems were discussed. The comparatively complete construction technology was put forward to dismantle pipelines. Besides, the construction technology of filling the controlled low strength material to the decommissioning oil pipelines was put up. Based on the case of Dong-Huang pipeline abandonment, the selection disposal scheme was optimized, key construction practices such as the cleaning of pipeline residues and low strength controlled material filling pipeline were expounded, potential environmental influence of different disposal scheme was analyzed, besides, some measures were represented to prevent and relieve the pollution corresponding to the construction technology. Research results have guiding significance for perfecting safe abandonment research of domestic retired oil and gas pipelines.

1. Introduction

By the end of 2015, the accumulative length of in-service pipelines in Mainland China has been 120,000 kilometers [1]. Among them, there are over 10,000 kilometers land pipelines constructed in the 1970s and 1980s, and partial ones have been kept in service for more than 40 years, which brings noticeable potential safety hazard [2]: the aging and corrosion phenomena are critical, besides, there are more and more pipelines being destroyed and occupied. In 2013, Qingdao Economic and Technological Development Zone witnessed an explosion hazard, which is a wake-up call to society that it's time to pay attention to the safety problems of pipelines. When pipelines are aging severely or have no durable years left, which means they can't continue to operate safely, it's obliged to decommission them. But if decommissioning pipelines can't be dealt with properly, they may pollute surrounding air, soil and water bodies, even lead to safety problems like fires and explosions [3]. Now there is not much research and practice about how to deal with decommissioning pipelines in China, corresponding decommissioning technique research, systematic pipeline decommissioning technique and regulations are also in lack [4, 5]. Pipeline environment impacts evaluation mainly focuses on the construction and operation period, but not the decommissioning period. And environmental management rarely makes imperative requirements of decommissioning environment impacts [6]. Reasonable decommissioning plan and systematic and well-established decommissioning technique (including decommissioning technique and analysis of environment impacts) are the basis of decommissioning pipelines safely. The article is based on the case of Dongying-Huangdao decommissioning pipelines, choosing reasonable decommissioning plan and technique, to analyze environment impacts during decommissioning period and offer corresponding measures to prevent pollution and reduce negative impacts.

2. Decommissioning Pipeline Disposal Scheme and Selection Basis

The typical ways to decommission pipelines are dismantling and local decommissioning. Decommissioning pipelines and surface facilities can remove safety and environment hazards drastically. In China, the common way of decommissioning pipelines is totally dismantling, which needs large quantity of construction works, high costs, which exists potential safety hazard and environmental pollution. Local decommissioning concludes the process of pigging, segmenting, plugging and filling pipelines. And if necessary, corresponding technical measures are also needed. This way of decommissioning costs relatively low. Decommissioning plans should be decided on the actual conditions of pipelines. North America and other developed countries usually combine the two methods together to deal with practical projects [6]. Selection basis of pipelines decommissioning plans include:

The conditions of land utilization is a vital factor. When choosing pipelines decommissioning plans, it's vital to investigate present and future utilization planning of land to avoid negative effects caused by local decommissioning on land use (like excavation, pile driving and construction of underground engineering structure). According to utilization, land can be divided into agricultural land, non-agricultural land, etc. (like environmental sensitive region, road, railway, river and other facilities). Summing up other countries' experience of decommissioning pipelines, these types of land need to adopt the measure of dismantling: agricultural land that needs deep tillage in the future, non-agricultural land that plans to construct and land with pipelines which are buried in shallow embedded depth (less than 1 meter). The land, which is suitable to adopt the measure of local decommissioning, mainly concludes urban land, road, railway, river and other pipeline crossing parts, most of agricultural land and land with pipelines which are buried in deep embedded depth. Choose pipelines decommissioning plans on the basis of environmental impact assessment in environmental sensitive region, which is in high risk of land subsidence, soil pollution and groundwater pollution. Pipeline crossing parts (crossing river, road, railway and public facilities) are also suitable for these plans.

Actual condition of pipelines is also an important factor. When choosing pipelines decommissioning plans, issues like type of media, transportation technology anti-corrosion and insulation measures should be taken into consideration. For example, it's obliged to dismantle pipelines when residue contains toxic substance which is difficult to clean or conventional cleaning method can't meet environmental protection requirements. Pipelines, which can't be dismantled, should be sealed by nitrogen or through filling slurry [4,7,8].

Technical feasibility and economic factors should also be taken into consideration. Technical feasibility mainly focuses on evaluation of technology and mechanical equipment, including security, reliability, complexity, responsive capability and convenience. Economic factors are mainly affected by engineering budget evaluation, including resource consumption, valuation of wastes, remedial measures and security measures.

3. Construction Technology of Decommissioned Pipelines

3.1. The Demolition of the Decommissioned Pipelines

The technology of the demolition of the decommissioned pipelines include: the termination of the pipelines' transportation, the recycle and the disposal of the residues in pipelines, the recycling of decommission of pipelines. The aim of the termination of pipelines' transportation is to unload the pressure inside the pipelines, as well as construct a safe environment of working.

3.1.1. Recycling and Cleaning of the Residues in the Pipelines. Not only can the residues in pipelines make it difficult for the construction of wastes, but also it can cause the pollution, even lead to the safety accidents. Thus, it is an important process of the disposal of the decommissioned pipelines to recycle and dispose the residues in pipelines. The main methods of the recycling of the oil gas: the first method is to send the oil gas into the pipelines which without replacement through the special pipeline pump for plugging equipment, the second method is to put the residual oil gas into tanker to recycle. The different methods of cleaning residues on the basis of different medium of transportation: it is difficult for the pipelines of crude oil to be cleaned, so the method of combining physical cleaning with chemical cleaning can be used; as for other pipelines, for example, it is easier to clean the residues of product oil, so the measure of nitrogen purging or using foam cleaner can be taken. In addition, an inspection is necessary after finishing the disposal, mainly to monitor the petroleum content in washing water.

3.1.2. The Recycle to the Dismantled Pipelines. The machine of excavation mainly include excavator, leveling machine and backhoe. Cutting pipelines and welding pipelines use the hydraulic plate shear and high pressure welding machine respectively. Caution is needed when excavating, in order to preventing the breakdown of pipelines and avoiding the residual pollution, that pollutes the environment. To make the transportation convenient, the pipelines can be cut before lifting demolition, and be removed from the scene timely. When the removed pipelines need to be stored, being directly exposed to the soil and water should be avoided. Post-settlement should be considered in the process of backfill, and the residues should be cleared thoroughly, in order to make sure that the ground has been recovered.

3.2. Construction Technology of Controlled Low-strength Material for Decommissioning Pipeline Filling

Construction technology of disposal on site mainly includes: the termination of the pipelines' transportation, the recycle, disposal, cut, sealing and the plugging after filling treatment of the residues in pipelines. According to the related criteria of oil and gas of pipeline abandonment of Canada, America, England and Australia, the common methods to abandon pipelines all have their requests and suggestions of disposal on site. Adopting the method of water injection, mud injection and Nitrogen injection sealing can guarantee the safety of disabled pipelines temporarily, but there are still some hidden troubles that the residues of pipelines may cause environment pollution. The pipelines should be abandoned after being solidified, filled and sealed when they still don't meet the request of environment protection after disposal, position cross makes it easier for settlement, or the diversion effect threaten the environment potentially.

Decommissioning pipeline filling can use the Controlled low-strength Material, which has the characteristics of high fluidity, low strength and long setting time. There is no need or just a little need of vibration under the deadweight. And it can flow automatically or pump to fill the abandoned pipeline and get to a high density. The relevant index is shown on table 1 [9]. Controlled Low-Strength Material can eliminate the hidden trouble thoroughly, filling material cannot cause the secondary pollution to the surroundings on the later stage, it can also prevent the ground setting.

Table 1. The Performance Indicators of CLSM Used for Filling Retired Pipeline

Filling material	0 min fluidity	30 min fluidity	0 min setting time	8 h bleeding rate	7 d Compressive strength	28 d Compressive strength
	/ mm	/ mm	/ h	/ %	/ MPa	/ MPa
CLSM	≥ 300	≥ 280	≥ 8	≤ 5.0	≥ 0.5	≥ 2.0

4. Disposal of Decommissioned Pipelines and Analysis of Environmental Impact during the Decommissioning Period of East Yellow Line

4.1. Case Outline

Dongying-Huangdao Crude Oil Pipeline (East Yellow Line for short), were completed and put into production in 1974 and 1986 respectively. In recent years, some pipelines have been heavily occupied and adjacent to buildings, which makes the pipeline difficult to overhaul and repair. There are more security risks. In order to ensure the safe operation, the construction unit intends to carry out safety hazard management for pipelines and oil transportation stations. Renovation project includes change lines locally and then build new pipelines and disposal of decommissioned pipelines.

4.2. The Disposal Plan of the Decommissioned Pipelines

The decommissioned pipeline adopts a scheme for dismantling and filling the disposed combination. The pipelines with demolition conditions are all dismantled; the pipelines that cannot be dismantled are filled with controlled low-strength material (CLSM). The disposition plan and the pipeline length condition are shown in table 2.

Table 2. Disposal Plan for Decommissioned Pipelines

Pipeline name	Removal Length / km	Filling treatment length / km
East Yellow Line	5.01	27.99
East Yellow Double Line	10.38	18.62

4.3. The Construction Technology of Filling Controlled Low-strength Material In-Situ Disposal

4.3.1. Nitrogen Replacement Oil Pumping. The two ends of the decommissioned pipe section are excavated for oil extraction operation pits (bottom 4 m * 6 m). Cleaning the anticorrosion layer of the valve position, and nitrogen replacement is carried out with the original equipment of the pipeline.; welding open holes, install DN 50 ball valve (pump valve) at bottom of pipe (from 1.5 m); after sealing open the hole with the special opening machine of the pipe, closed the pump valve. The other end of the pumping valve is connected to the pump, and opening the pump valve and pump, then Pumping crude oil into tank trucks for recycling.

4.3.2. The Nitrogen Pushes the Ball for Cleaning Oil. First, install the tee at the lower end of the pipe, and the ball tube is installed on the other side. Next, gasify the liquid nitrogen at the side of the tee and push the ball after pressure and remove the crude oil and gas in pipeline. Meanwhile, open the pumping device next to the ball tube. Then pull it in special tank truck for recycling. And then the receiver emits oil and gas. Finally, the number of promotion shall not be less than 3 times and ensure that oil and gas in the pipeline are drained, and fill the whole pipe with nitrogen.

4.3.3. Sectional Cutting and Sealing Pipeline. Combined with the layout of the site pipeline and use the filling performance of controlled low-strength materials to divide the filling construction segment, Each length L can be controlled at 600~1000 m.

Excavate the pit at the truncation point of the pipe, and the size and requirements are the same as the pumping pit. According to the local highs and lows of the survey pipeline, cutting and segmenting with mechanical cutting machine. Finally, install the grouting valve and exhaust valve at each high point. Mason butter mud wall at 300 mm from the pipe mouth at each pipe section, and install blind sealing tube. Meanwhile, the installed receiver can be reserved for sealing. Install DN 50 exhaust valve and observation valve respectively at each end of the pipe segment from the blind plate (or the receiver) inside 1-1.5 meters. The high point valve is used for exhaust, and the low valve is used to observe (and exhaust) whether the filler is full of ends. If the depth of the pipeline between the grouting port and the distal blind plate is undulating, and increase the installation of exhaust valve at a local high point exceeding the pipe diameter section.

4.3.4. Filling of Controllable Low-Strength Materials. Put the foam ball in the tee, and use the slurry pump truck fill in the controlled low-strength filling materials from the middle of the pipe continuously. Do well in on-site grouting quantity measurement, and ensure that grouting volume is matched with pipe capacity. Open exhaust valve and observation valve when filling. Next, confirm whether the filler arrives and fills the end position, the middle fluctuation segment and the high point. After each confirmation, turn off the end position observation valve, middle fluctuation section exhaust valve and high vent valve. Then stop grouting when filling is filled with pipe. Finally, through the check valve confirms that the material is solidified, and sealing and plugging. The construction section can be continuously operated, and different construction sections can be operated simultaneously. Backfill the job pits after completing, and recover the site.

4.4. Environmental Impact Analysis of Pipeline Decommissioning Period

4.4.1. Environmental and Air Impact Analysis. In this case, the construction of the decommissioning pipeline is carried out little by little, there are a little sensitive points in the excavation section, and date of the work is generally 1-2 months. The impact on ambient air is mainly from the dust of trench excavation, dumping temporary piling up, backfill, vehicle transportation and so on. In addition, during the process trench excavation, a small amount fuel combustion flue gas will be generated during construction machinery operation, all of which are discharged at low attitude with a small amount of exhaust gas and a short impact time. The preparation of controllable low-strength materials preferred to select areas that far away from the residences and avoid personnel activities, which has short preparation time and the impact on the surrounding environment is small.

4.4.2. Environmental Impact Analysis of Surface Water. The wastewater generated during the construction period mainly comes from the domestic sewage of construction workers. The site does not set camp, and the accommodation of the construction staff is rented local hotels or houses. The construction section has the large disposability and the local discharge is very small. Domestic sewage relies mainly on the local domestic sewage treatment system, and has a slight impact on the surface water to environment. Construction should be strengthened management, prohibit the discharge of domestic sewage into nearby water bodies. Filing and curing pipeline construction section only at both ends of the construction activity, the middle of no construction activities and the construction of the two ends of the river should be avoided to ensure that the surface water has little effect.

4.4.3. Analysis of the Environmental Impact of Groundwater. The depth of buried pipeline is about 1.5m, and shallow's excavation does not involve underground aquifers. Have no influence on the groundwater environment. During the construction, the residual oil in the pipeline should be ensured not to drip and leak, so as to reduce the influence on the groundwater in the construction area.

4.4.4. Analysis of Sound Environment Impact. The source of noise in construction is mainly from construction machinery, such as excavators, hydraulic pipe-breaking machines, oil pumps, grouting machines and so on. The machinery, equipment and vehicles used in different construction's processes. Each construction's machine is approximated as a point source, and only the attenuation calculation of the distance is taken into consideration to obtain the noise contribution's values at different distances. The results are shown in table 3.

The standard of construction site boundary noise (GB 12523-2011) stipulates that the daytime noise limit of the construction site boundary is 70 dB (A) and night time of the noise limit is 5dB(A). As can be seen from Table 3, the noise of main construction machinery does not exceed the noise limit off daytime beyond 10m, and the noise of night time does not exceed 50 m. The impact of noise on the surrounding acoustic environment during construction is acceptable, but also short-term and reversible. Once the construction is completed, which will stop immediately.

Table 3. Estimated Value of Noise in Main Construction Equipment at Different Distances Unite: dB(A)

No.	The name of the machine	Noise values at different distances from the construction site				
		10 m	50 m	100 m	150 m	200 m
1	Bulldozer	52	38	32	28.5	26
2	Excavator	55	41	35	31.5	29
3	Pipe layer (crane)	52	38	32	28.5	26
4	Transport vehicles	60	46	40	36.5	34
5	Oil bump grouting bump	55	41	35	31.5	29
6	Diesel generators	60	46	40	36.5	34
7	Hydraulic pipe breaking machine	55	41	35	31.5	29

4.4.5. Environmental Impact Analysis. Take demolition measures to deal with decommissioning pipelines, and the process of digging will destroy the vegetation on the surface and will affect the construction of the soil within the scope of the filling controllable low-strength materials for curing treatment at both ends of the pipeline, which will have a small amount of ecology affected, but the middle area basically will not have ecological impact.

The soil and negation will be destroyed and disturbed, which in the excavation and backfilling construction zone. Especially in the excavation ditch about 2-3m in the range. The vegetation is severely damaged. The disturbance of soil makes for the structure, composition physicochemical properties of the soil. Furthermore, the properties of soil and the restoration and growth of natural vegetation were affected.

The plants affected by the destruction of construction sites are widely distributed and common species along the road, and distributed relatively evenly. The original vegetation is damaged locally, but it is not to change the species and composition in the area. Most of the pipe removed in urban areas, which will not cause damage to crops.

The destruction of natural vegetation in the construction area will cause some wildlife to lose some foraging places, habitats and active areas. Due to the very narrow area of destroyed vegetation, the adverse impact on the living environment of the animals is slight.

4.4.6. Impact Analysis of Solid Waste. The length removal of the pipe is about 15.39 km, according to the original welding position, 1283 scrap steel in total, the external environmental impact would be less, if the disposal of the qualified recycling unit was given to the qualified recycling department.

When installing a receiver device, welding flanges, attached to the pipe need to dismantle the original asphalt anticorrosive coating, quantity about 1 m^3, oil-free by qualified recycling disposal recycling units, and contains oil leakage is to have the corresponding qualification of environmental technology companies collect disposal, after dispose of external environment.

The plugging operation produce no oil pollution when adopt airtight opening way. Place oil groove below the pipe cutting position, collect the dirty oil produced by cutting. In construction, there is no oil pollution, which has little impact on the external environment. A small amount of crude oil drips on the land surface in the construction of pipe cutting, the yield is about 3 m^3, as hazardous waste, to have the corresponding qualification of environmental technology companies collect disposal, which do less damage to the external environment if being handled properly.

Domestic garbage produced by construction workers amount about 1.8 tons during 60 days, during the construction of decommissioning pipeline section. It will do less damage to the external environment if being collected and disposal by the local sanitation department.

The pipeline service a long time, punch stolen oil and pipeline leakage occurs frequently, which will pollute near soil. After excavation, the soil is repaired according to the soil condition in the trench, and the oily soil is taken out, which is collected and disposed by the environmental protection technology company with the corresponding qualification. After proper treatment, the external environment is less affected.

In the process of pigging, there is a part of oil pollution on the pipe cleaner, which belongs to hazardous waste and should be collected and disposal by the local sanitation department, thus do less damage to the external environment.

4.4.7. Analysis of Environmental Influence Factors of Filling Materials. The controllable low strength material is non-toxic and harmless, and the strength after filling and solidification is close to the soil strength, and the property is stable, thus it will not affect the environment. There is no oil and gas in the pipeline, after filling treatment, which has less impact on the environment.

5. Conclusion

The disposal scheme of decommissioning pipeline mainly includes demolition and disposal. The choice of disposal scheme mainly depends on the land utilization along the pipeline, the condition of the pipeline as well as the environmental impact assessment, while considering the technical feasibility and economic factors at the same time. The combined disposal scheme is not only capable to eliminate the safety and environmental risks, but also reduce the cost.

In view of the important link of the pipeline abandonment process, combined with the actual case of pipeline waste disposal in foreign countries, the pipeline shutdown, pipeline residue recovery and cleaning should be completed before or after the operation, thus it is suggested to adopt a combined cleaning method to clean up the whole line. Decommissioning pipeline demolition work carefully, to avoid leakage of residual pollutants, destruction of the environment, restoration of the ground. The construction technology of controllable low strength material for decommissioning pipeline is put forward. The controllable low strength material CLSM has the characteristics of high fluidity, low strength and long setting time, and can be compacted by self-flow or pumping to fill the waste pipeline.

The actual case in the east of the yellow line retired pipe is disposed, select the disposal scheme of combination, and discusses the key technology of filling pipeline cleaning, residue, analyzed the potential environmental impact of the pipeline retirement period, put forward the corresponding measures to prevent pollution and reduce the effect, to provide reference for the scientific disposal of waste oil and gas pipeline in China.

For the oil and gas pipeline should be abandoned, abandoned by plan reasonable, make the construction of a scientific and complete engineering, environmental protection design and decommissioning of environmental management, so as to guarantee the implementation of safety, environmental protection, economy of onshore oil and gas pipeline decommissioning.

6. Acknowledgements

This research work was supported by the National Natural Science Foundation of China (NSFC) (Grant No. 51309238) and Sinopec pipeline storage and transportation branch fund (Grant No. 2014QWX003). This financial support is gratefully acknowledged. In addition, the authors wish to thank the editors and the anonymous reviewers of this manuscript for their elaborate work.

7. References

[1] Peng Gao, Peihong Wang, Yaohui Yang, et al. 2016. China's oil and gas pipeline construction in 2015. *International Petroleum Economics*, Volume **24**, Issue 3, pages 60-65.

[2] Yan Di, Jian Shuai, Xiaolin Wang, et al. 2013. Study on methods for classifying oil & gas pipeline incidents. *China Safety Science Journal*, Volume **23**, Issue 7, pages 109-115.

[3] Guofu Niu, Yanmin Wang, Honglei Pang, et al. 2015. The accrual of abandonment cost for oil and gas pipelines in China and abroad. *International Petroleum Economics*, Volume **23,** Issue 11, pages 67-75.

[4] Yewei Kang, Li Zuo, Zhenghong Guo, et al. 2015. Abandonment of onshore oil and gas pipelines. *Oil & Gas Storage and Transportation*, Volume **34,** Issue 2, pages 122-127.

[5] Shichao Liu, Yichi Liu, Xiaoxiao Zhang, et al. 2015. The summary of foreign oil and gas pipeline abandonment engineering instance. *Total Corrosion Control*, Volume **25,** Issue 11, pages29-31.

[6] Changlin Li, Yunshi Xiong, Bao Geng, et al. 2012. Characteristics of environmental impact assessment of oil and gas transmission pipeline projects. *Environment Protection of Oil & Gas Fields*, Volume **22,** Issue 1, pages 40-45.

[7] Swanson J M, Kunicky T, Poohkay P. 2010. A case study from abandonment of a southern Alberta pipeline. *The 8th International Pipeline Conference*, Calgary, Alberta, Canada, September 27 - October 1, pages 179-185.

[8] Mingyang Xie, Bin Li, Lei Cui, et al. 2014. How to abandon onshore oil and gas pipelines Oil & Gas Storage and Transportation, Volume **33,** Issue 8, pages 825-828.

[9] Xuesong Zhang, Rangang Yu, Jinping Chen, et al. 2017. Properties of Controlled Low-Strength Materials Using in Waste Oil Pipeline. Bulletin of the Chinese Ceramic Society, Volume **36,** Issue 6, pages 1824-1829.

IOP Conf. Series: Materials Science and Engineering **735** (2020) 012073 doi:10.1088/1757-899X/735/1/012073

Effect of Peg on Polysulfone Support and Reverse Osmosis Membrane Property

Jin Wang[1], Zhijie Zhang[1], Jiarui Li[1], Jiani Zhu[1], Xiaorong Meng[2], Xiaojun Li[3] and Xudong Wang[1]*

1 School of Environmental and Municipal Engineering, Xi'an University of Architecture and Technology, Xian 710000, China
2 School of Chemistry & Chemical Engineering, Xi'an University of Architecture and Technology, Xian 710000, China
3 Xi'an Thermal Power Research Institute CO., LTD, Xian 710000, China
Corresponding author e-mail:18629407889@126.com

Abstract. A study on the thin film composite on polysulfone support membranes prepared by different casting solution was conducted. The following conclusions can be drawn from this work, the membrane flux increased with increasing PEG concentration and the pore size and the porosity on the membrane surface became bigger by the increase PEG concentration; the permeability of thin film composite membrane was effected by the membrane continuity and the amount of polyamide formed within PSf membrane pores.

1. Introduction

Water resources become to a crucial problem directly related to the sustainable development of the society and economy after human civilization entering a new century [1]. Recently, reverse osmosis (RO) has become a promising water technology, utilizing the selective permeability of semi permeable membrane to eliminate impurity and the supply the clean water [2-3].

The typical RO TFC membranes consist of two parts [4-5]: (1) support substrate(50 m), the polysulfone (PSF) ultrafiltration membrane is the most common use for supporting, (2) thin barrier layer (200 nm), the desalination layer is generated during the interfaci- -al polymerization (IP) interaction formed on the substrate. Researchers focus on the properties of thin barrier layer and neglect the effect of PSF substrate on the water permeability and salt rejection of RO membrane. Until recently, the importance of supporting membrane received more attention [6-8]. Ghosh et al reported that the pore structure and chemistry have a significant influence on interfacial polymerization reaction [7]. However, the manually preparation of PSF supporting membrane could no guarantee the identical casting speed and residence time nascent membrane both in air and coagulation bath [9-10].

In the paper, the preparation of PSF supporting membrane is precisely controlled by a semi-automatic film casting machine, the effect of concentration of PEG-600 on the structure and chemistry property of both supporting membrane and ultra-thin barrier layer additive was investigated, the optimal preparation parameters was confirmed which could provide valuable data and theory foundation on processes optimization of RO membrane.

IOP Conf. Series: Materials Science and Engineering **735** (2020) 012073 doi:10.1088/1757-899X/735/1/012073

2. Experimental

2.1 Materials and Chemicals

Polysulfone (PSf) P-3500 was supplied by Solvay. DMF (liquid, analytical reagent grade), poolyethlene glycol (PEG, 600Da) m-phenylenediamine (MPD, powder), triethylamine (TEA, liquid, 99.5%), D(+)-10-Camphorsulfonic acid (CSA, powder, 99.0%) and trimesoyl chloride (TMC, powder) was purchased from Sigma-Aldrich. Isopar G used as organic solvent was obtained from Exxon Corp.

2.2 Experimental

2.2.1. Polysulfone (PSf) Membrane Formation

The PSf was measured and separated into 2 pieces and then dried at 125 °C for 3 hours to get rid of the moisture in the PSf particle. A piece of PSf was added into an airtight bottle in which a specified amount of DMF was in a stirring state at 25 °C. The solution was heated to 80 °C after PSf completely dissolved and the rest of the PSf was added and agitated for several hours. PEG was finally added into the above solution when the effects of the PEG additive were studied. The specific composition of casting solutions was provided in Table 1. The casting solution was kept still overnight for degassing. The membrane was formed by spreading the solution over the non-woven support fabric using semi-automatic film casting machine with a gap of 120 μm and then, the non-woven support was immediately immersed into a DI water bath. The PSf membrane was washed thoroughly and stored in the DI water.

Table1. Specific Composition of Casting Solutions and Membrane Number

Membrane	Casting solution (wt.%)			Membrane	Casting solution (wt.%)		
	Polysulfone	DMF	Additive		Polysulfone	DMF	Additive
PSf-1	15	85	-	PSf-4	18	82	-
PSf-2	16	84	-	PSf-5	15	80	5
PSf-3	17	83	-	PSf-6	15	70	15

2.2.2. TFC membrane formation

The PSf membrane was immersed into an aqueous solution contained MPD and other additives for 15 s. The excess aqueous solution was then eliminated from the skin surface via an air knife. After then, the membrane was immersed into the TMC-Isopar G solution for 15 s to initiate the polymerization reaction. The resulting TFC membranes were subsequently heat cured for 6 min at 90 °C. The TFC membranes were finally washed thoroughly to remove any residual solution from the surface by DI water.

2.2.3. Characterization of polysulfone and polyamide membrane

Separation performance of PSf membrane and TFC membrane was evaluated using a cross-flow filtration system with an effective membrane area of 70 cm2 at 25 °C. For the PSf membrane, the pure water flux was performed under the pressures of 0 and 0.3 MPa, respectively. For the TFC membrane, the flux and the selectivity were characterized by a 2000 ppm NaCl feed solution under 1.55 MPa. Salt rejections were determined from the conductivity of the feed and permeate water.

The surface and cross-sectional morphology of membranes is visualized by scanning electron microscopy, SEM (Nova Nano SEM 430, FEI Company, Hitachi, Japan). The chemical structure of PA-TFC membranes was characterized using attenuated total reflectance Fourier transform infrared spectroscopy (ATR-FTIR).

3. Result and Discussion

3.1. Effect of PSf Concentration on PSf Membrane

The water flux changes of PSf membranes prepared with different PSf concentration are presented in Figure 2. It indicated that the flux of the PSf membrane decreased significantly with the increasing PSf concentration under both 0 MPa and 0.3 Mpa, the PSf-1 membrane was the most permeable membrane with the flux of 960 LMH (0 Mpa) and 2400 LMH (0.3 Mpa), whereas the PSf-4 membrane has the lowest flux values of 415 LMH (0 Mpa) and 1048 LMH (0.3 Mpa), respectively. As shown in SEM micrographs, a typical asymmetric membrane structure was observed in all the cross-sectional images, the top-layer of the membrane was dense and the sub-layer where macrovoids existed was sparse. Furthermore, the membrane structure became denser and the quantity of the macrovoids was suppressed with the increasing PSf concentration, the membrane changed to sponge-like structure with few macrovoids when PSf concentration reached 18%.The changes of the membrane structure and permeability could be attributed to the transport rates, the inflow rate of non-solvent (water) and the outflow rate of the solvent (DMF).

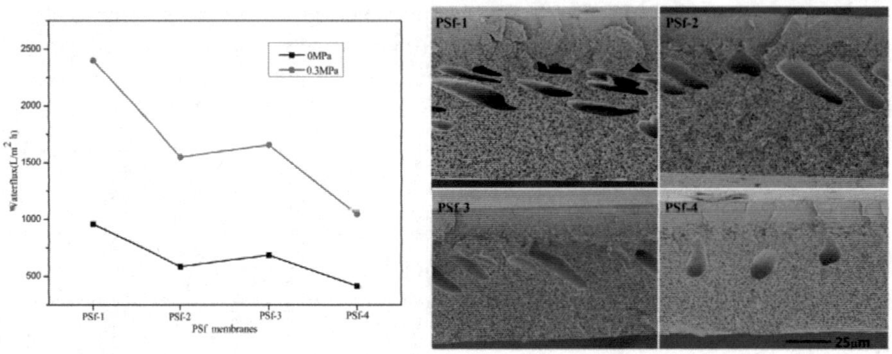

Figure 1. Effect of Polysulfone Content on PSf Membrane Permeability (left), Cross-section Images of PSf Membrane with Different Casting Solution (right)

3.2. Effect of Additive Concentration on PSf Membrane

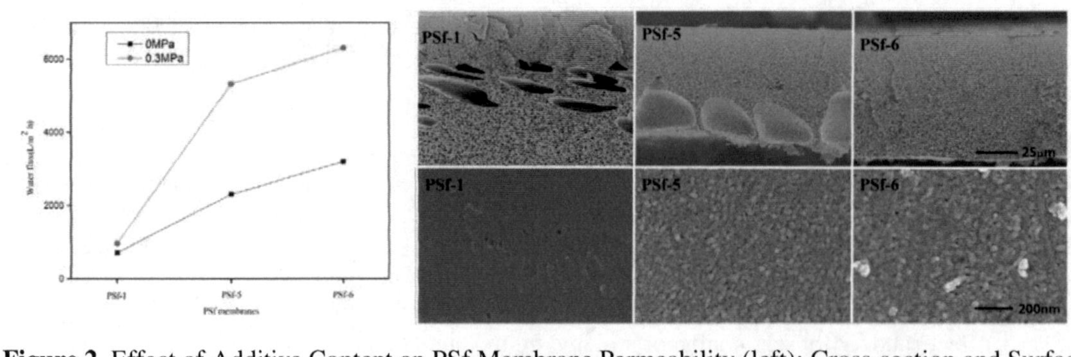

Figure 2. Effect of Additive Content on PSf Membrane Permeability (left); Cross-section and Surface Images of PSf Membrane with Different Additive Concentration (right)

The flux changes of the PSf membrane after adding PEG-600 were presented in Figure 3. From the plot we could clearly observed that the membrane flux increased with increasing the concentration of PEG. The structure of membrane still showed a dense top-layer and sparse sub-layer feature. Moreover, the pore size and the porosity on the membrane surface became bigger by the increase PEG concentration. The structure changes of the membranes could also be considered as the result of the different diffusion rates. It could be observed from Table 1. That the viscosity of the casting solution increases with increasing the ratio of PEG to DMF. According to the literature, PEG possesses the nonsolvent characteristic for PSf comparing with NMP, the addition of PEG result in the inter-molecular aggregations and entanglements of the polymer chains.

3.3. Effect of PSf Membrane on TFC Membrane

The surface morphology and chemical composition of the TFC membrane on the PSf-1 support membrane are analyzed by SEM and ATR-FTIR as shown in Figure 3.(a) and (b). We could observe a nodular membrane surface under the experimental condition which is different from the classic ridge-and-valley surface features in the literatures. In order to further characterize TFC thin-film layer, ATR-FTIR spectroscopy was used to identify the main functional groups, indicated that a successful polymerization process had occurred.

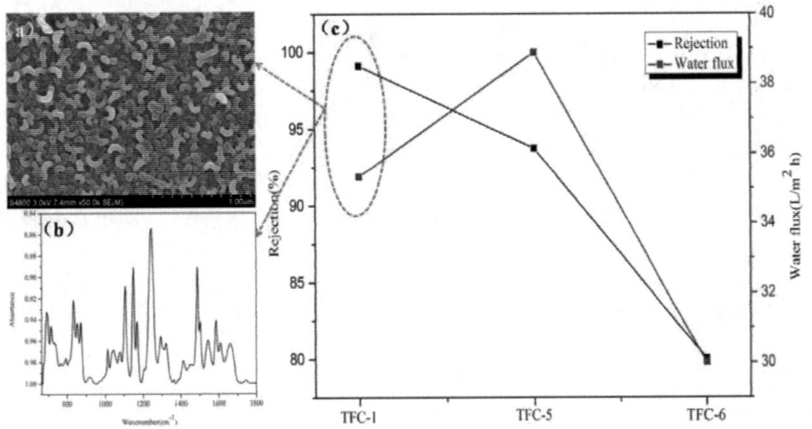

Figure 3. Influence of PSf Membranes on TFC Membranes. (a) The Surface Image (TFC-1); (b) The ATR-FTIR Spectra (TFC-1); (c) The Permeability Performance of TFC Membranes

Figure 3.(c) resents the changes of TFC membranes permeability synthesized on different support layer surface. It clearly appeared that the PSf membrane has significant effect on the TFC membrane performance. According to Figure 6.(c), the TFC-5 membrane possessed the highest water flux 10.7 LMH with 94.8% salt rejection, TFC-1 membrane had the highest salt rejection 99.1% and the flux was just little lower than that of TFC-5. For the PSf membrane prepared with PEG, the aqueous solution diffuse into the surface pores more easily because of the lager pore size and hydrophilic. However, when the polymerization reaction began, MPD diffused slowly out of the pores due to favorable interactions with polar functional groups of PEG. As a result, amount of polyamide formed deeper within PSf membrane pores and no continuous and complete polyamide formed on the surface especially when the PEG was added to 15%. This factor increased the overall path length for water transport and caused the less salt rejection and water flux.

4. Conclusion

The concentration of PEG additive lead to a structural change from dense top-layer and sparse sub-layer feature to a symmetric, sponge type with no macro-void structure, which could be considered as the result of the different diffusion rates.

The polyamide formed deeper within PSf membrane pores result in the increased the overall path length for water transport and the less salt rejection and water flux.

5. Acknowledgments

This work was financially supported by National Natural Science Foundation of China (Grant No. 51804240), Key Industrial Chain Project of Shaanxi Province (Grant No. 2017ZDCXL-GY-07-03).

6. References

[1] Global risks 2015, Insight Report, 10th Edition, World Economic Forum, (2015)

[2] Li D, Wang H. Recent developments in reverse osmosis desalination membranes [J], Journal of Material Chemistry, 2010, (20):4551-4557.

[3] Lee K, Arnot T, Mattia D, A review of reverse osmosis membrane materials for desalination -Development to date and future potential [J], Journal of Membrane Science, 2011, 370:1–22.

[4] Prakash R A, Joshi S V, Trivedi J J. Structure-performance correlation of polyamide thin film composite membranes: effect of coating conditions on film formation [J]. Journal of Membrane Science, 2003, (211):13-24.

[5] Ng L, Mohammad A, Leo C. Polymeric membranes incorporated with metal/metal oxide nanoparticles: A comprehensive review [J], Desalination, 2013, (308): 15–33.

[6] Kang G D, Gao C J, Cao Y M. Study on hypochlorite degradation of aromatic polyamide reverse osmosis membrane [J], Journal of Membrane Science, 2007, (300): 165-171.

[7] Asim G, Eric H. Impacts of support membrane structure and chemistry onpolyamide–polysulfone interfacial composite membranes [J], Journal of Membrane Science, 2009, (336):140-148.

[8] Asim G, Jeong K, Huang B. Impacts of reaction and curing conditions on polyamide composite reverse osmosis membrane properties[J], Journal of Membrane Science, 2008, (311): 34-45.

[9] Puyam S, Joshi S, Trivedi J. Probing the structural variations of thin film composite RO membranes obtained by coating polyamide over polysulfone membranes of different pore dimensions [J], Journal of Membrane Science, 2006, (278):19-25.

[10] Kenneth M, Vassilis G, Gun T. Study of membrane compaction and its influence on ultra-filtration water permeability [J], Journal of Membrane Science, 1995, (100):155-162.

Preparation of NaOH Modified Spent Grain Adsorbent and Adsorptive Properties for Dyes

Yawei Wang [1]*, Yaxin Lv [1]*, Haiqi Shen [1], Shenyang Xu [1], Yu Guo [1], Zhijie Bao [1], and Yan Feng [1]**

School of Textile Science and Engineering, Tiangong University, 399 Binshui West Road, Xiqing District, Tianjin
Email: fengyan@tjpu.edu.com

Abstract. The spent grain which is the by-product of beer production contains cellulose, lignin and other substances. It can be used as an adsorbent material not only for waste utilization, but also to reduce environmental contamination. The structure and morphology of spent grain (SG) were discussed. The NaOH modified spent grain (ASG (alkaline spent grain)) was prepared. And the effects of the parameters such as NaOH concentration, temperature and time on the absorption performance of Methylene blue wastewater were analyzed. The results showed that, when the modifier concentration reached 5 g/L, the temperature and time was 60°C and 120 min respectively, the maximum decolorization could reach 95.87%. The decolorization rate of 55.82% of the original grain increased significantly. The adsorption process of themodified materials was chemical adsorption, and the adsorption behavior could be described by Freundlich and pseudo-second-order kinetics equation.

1. Introduction

With the rapid development of printing and dyeing industry, the discharge of dye wastewater has increased dramatically, which has resulted in serious environmental pollution. Dye wastewater is characterized by complex components, high chromaticity, more suspended substances, unstable water quality and more refractory substances. Wastewater often contains a large number of organic compounds with stable properties and certain absorption ability to external light, such as hydrocarbons, aromatic compounds, and polycyclic aromatic hydrocarbons. Further, several organic compounds found in wastewaters are non-biodegradable, thus reducing the transparency and self-purification ability of the water body, resulting in serious ecological and environmental problems. At present, several techniques have been tested for treating dye wastewaters including physical treatment, chemical treatment and biological treatment. Among them, physical and chemical treatment methods are more efficient, but the cost is also higher and easy to cause secondary pollution [1-3]. Biological treatment methods are environmentally friendly and cost-effective, but they are affected by treatment conditions and treatment environment. Adsorption is one of the physical methods to treat dye wastewater. It mainly purifies dye wastewater through porous and high specific surface area substances, such as activated carbon and resin. Among all the techniques, adsorption processes using solid media prove to be very efficient for the removal of dye from wastewaters. Adsorption processes offer several benefits in such as low capital, operational costs and widely used. Therefore, it is of great significance to develop new adsorbents with excellent adsorption function for dye wastewater.

Spent grains (SG), the major by-products in beer industry, most of them are discarded directly and a small amount is used in feed industry, resulting in waste of resources and environmental pollution. SG that basically consists of cellulose and lignin, is an attractive alternative for removing dye from wastewater as it has a large specific surface area and biodegradability, with many functional groups such as carboxyl, hydroxyl and amino that are responsible for the binding of organics. Therefore, the adsorption of organics by SG has potential for industrial use. At present, the research on adsorbing materials of spent grain is focused on the chemical modification of SG by alkalization, acidification, esterification and mercaptoylation, which can be used for the adsorption of dye wastewater and heavy metal ion wastewater such as cadmium and lead. In this article, SG was chosen as a starting material for an adsorbent and NaOH as modifier [3-10]. The effects of process parameters and concentration of methylene blue solution on the adsorptive performance of alkaline spent grain (ASG) and its adsorption mechanism were studied. Finally, the application of SG in the treatment of dye wastewater was investigated.

2. Experimental

2.1. Materials

Spent grain was obtained from Junling brewery located in Tianjin, China. The waste was washed with deionized water to remove soil particles or dusts, then dried at 60 °C for 5h and ground to pass through a 28 mesh sieve in order to be used in further chemical modification. Sodium hydroxide (NaOH) and Methylene Blue (MB) were purchased from Tianjin Fengchuan Chemical Reagent Technology Co., Ltd., Tianjin, China and Tianjin Zhiyuan Chemical Reagent Co., Ltd., Tianjin, China, respectively.

2.2. Preparation of ASG and Simulated MB Dye Solutions

Measure a certain concentration of NaOH, followed by blending with 0.5 g of spent grain prepared as above and stirring at room temperature for a certain time. Thereafter, quaternization of SG with NaOH was carried out at room temperature for a certain time. After the reaction, the unreacted NaOH was removed by washing with deionized water. Finally, the residues were dried at 60 °C to use.
A stock solution of MB dye was prepared (20 mg L^{-1}) by dissolving a required quantity of MB dye powder in deionized water.

2.3. Structural Characterization of SG, ASG

The surface and cross-sectional morphologies of SG and ASG were examined by field emission scanning electron microscopy (FESEM, TM-3030, Hitachi, Japan).
The surface functional groups of SG and ASG affecting the adsorption were identified by Fourier transforms infrared spectrometer (FTIR, Nicolet iS50, Thermo Fisher Scientific). The spectra were recorded from 4000 to 400 cm^{-1}.
The specific surface area of SG and ASG were determined by N2 adsorption method at 70 °C and the diameter of the tube was 6 mm (BET, ASiQM000000-6, Quantachrome).

2.4. Adsorption Experiments

Dried SG and ASG added to MB dye solutions, respectively. The test solutions were obtained by shaking, adsorbing and filtering. The clarified supernatants were collected from the flasks at predetermined time intervals. MB dye concentration in the solution before and after treatment was calculated from the calibration curve, which was prepared by measuring the absorbance of different concentrations of MB dye solutions at λ_{max} 664 nm using a UV Spectrophotometer (UV-1800, Shanghai Metropolitan Instruments Co). Then, the decolorization rate and adsorption of MB dye by SG and ASG were calculated. The percent of dye removal was calculated by the following equation:

$$\%\ \text{removal of MB dye} = \frac{c_0 - c}{c_0} \times 100\% \tag{1}$$

Where C_0 and C are the initial and final MB dye concentration (mg L^{-1}) in the solution, respectively.

The amount of dye adsorbed at equilibrium onto BSG, q_e (mg g^{-1}), was calculated as follows:

$$q_e = \frac{v(c_0 - c)}{m} \quad (2)$$

Where V is the dye solution volume (L) and m is the mass of adsorbent used (g).

2.5. Calibration Curve

The calibration curve was drawn in Figure 1 by measuring the absorbance of different known concentrations of MB dye solutions. The relationship between solution concentration C and absorbance A is: A=k·C. And a linear correlation was found between A and C. From Figure 1, it can be seen that the plots are linear, the fitting equation of the curve is Y=62.91057 X+0.07759, the correlation coefficient is $R^2 = 0.98226$, and the intercept is 0.07759.

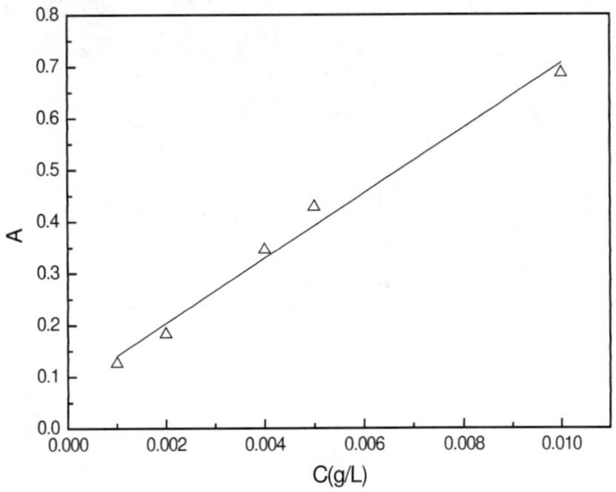

Figure 1. MB Calibration Curve

3. Results and Discussion

3.1. Characterization of SG, ASG

3.1.1. SEM of SG, ASG

Figure 2 shows the SEM micrographs of SG, ASG ((a) and (b) SEM micrographs of SG, (c) and (d) SEM micrographs of ASG). There is some ridges and gullies on the surface of the SG sample, and most of the surface is rough and porous. Compared with SG, great changes are observed on the surface of the modified SG. The number of ridges and gullies on the surface of ASG increased greatly, which further increased the specific surface area, facilitated the exposure of surface functional groups, provided sufficient adsorption space for ASG as adsorbent material, and promoted its adsorption of dye.

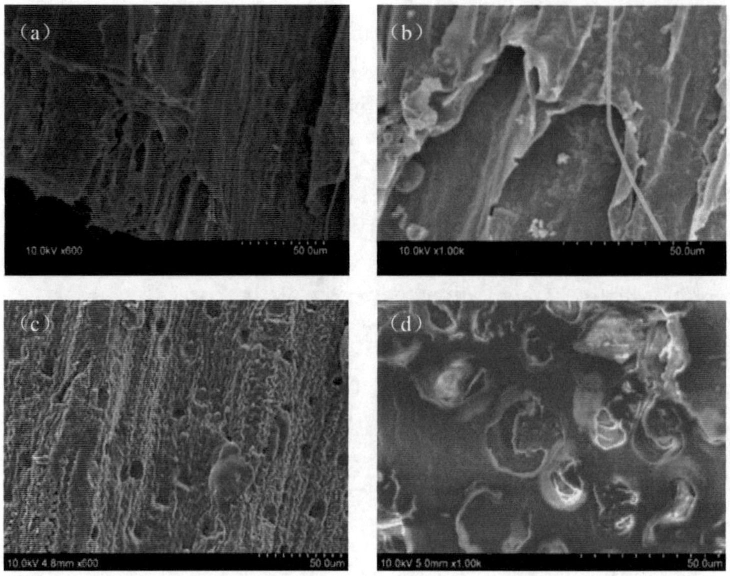

Figure 2. SEM Images of Spent Grain (a, c. 600 Times Larger, b, d. 1000 Times Larger， a, b. SG, c, d. ASG)

3.1.2. Infrared spectra of SG, ASG

The spent grains are predominantly composed of cellulose and lignin. The lignocellulosic polymers lignin and cellulose comprise over 45% of SG's dry weight, while proteins account for another 21%.The infrared spectra of SG and ASG are shown in Figure 3, respectively. The broad peaks at 3000-3700 cm^{-1} are attributed to the hydroxyl groups which formed hydrogen bonds between carbohydrate molecules in spent grain; the absorption peaks at 2920 cm^{-1} and 2850 cm^{-1} are caused by asymmetric-CH_2; the absorption peaks at 1650 cm^{-1} are attributed to carbonyl C=O; the absorption peaks at 1030 cm^{-1} are mainly caused by C-O bonds; and the absorption peaks at 529 cm^{-1} and 443 cm^{-1} are mainly caused by C-Cl bonds. Therefore, spent grain contains hydroxyl, carboxyl and other active functional groups, which can participate in active adsorption.

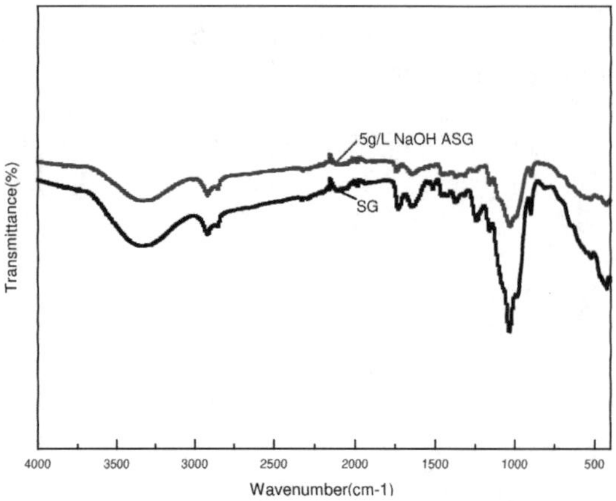

Figure 3. FTIR of Spent Grain

3.1.3. BET of SG, ASG

The results of specific surface area, pore volume and pore sizes of spent grains are shown in Table 1. It can be seen that the grain grains have smaller pore size and higher specific surface area. They are porous biomaterials and can be used to prepare adsorption materials.

Table 1. Structural Characterization Parameters of SG

Surface area (m^2/g)	Pore volume (mL/g)	Pore size (nm)
67.231	0.123	1.751

3.2. Effect of Technological Parameters on MB Dye Adsorption

The modification technology includes NaOH solution concentration, modification temperature and time, which have important influence on the adsorption performance of SG.

3.2.1. Effect of Modification Temperature and Time on MB Dye Adsorption

The modification temperature is a key parameter affecting the adsorption performance of ASG. Figure 4 and Figure 5 show the effects of modification temperature and time on the adsorption properties of SG and ASG. It can be seen from Fig. 4 that with the increase of temperature, the decolorization rate and adsorption of the SG and ASG on the MB solution increase, and the equilibrium was attained, after which it remained constant. The optimum modification temperatures are 50°C and 60°C, respectively. This is due to the activation of partially inactivated adsorption sites with the increase of temperature, which generates more adsorption sites and increases its adsorption effect on MB. This effect is stronger for ASG. As the temperature continues to rise, the adsorption sites of the two tend to be saturated, and the number does not continue to increase. As a result, the curves tend to be stable.
Another important parameter in the adsorption process is modification time because it gives an idea of the equilibrium time for the adsorption process. It can be clearly observed from Figure 5 that the decolorization rate and adsorption increase with an increase in contact time and the equilibrium is attained during the time interval of 120 min for ASG, after which it remained constant. For SG, the decolorization rate and adsorption only slightly fluctuated with time, and the maximum is attained during the time interval of 90 min. With the increase of modification time, the effect of NaOH on spent grain is obvious, and the surface activity of spent grain is continuously improved. But when modification time reaches a certain value, the activity of spent grain tends to be stable, so its adsorption effect is no longer obvious change.

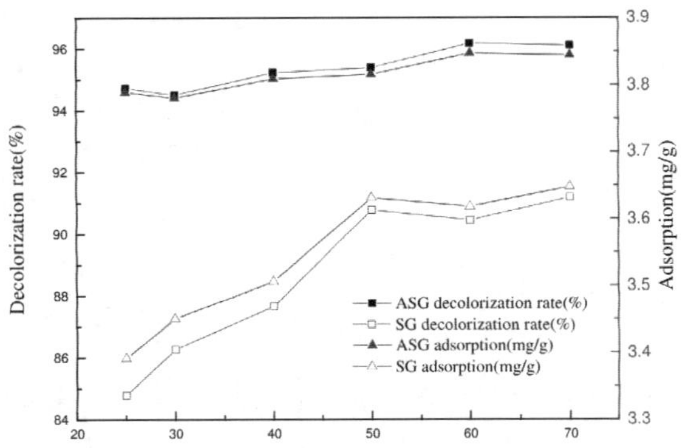

Figure 4. Effect of Modified Temperature on Adsorption Effect

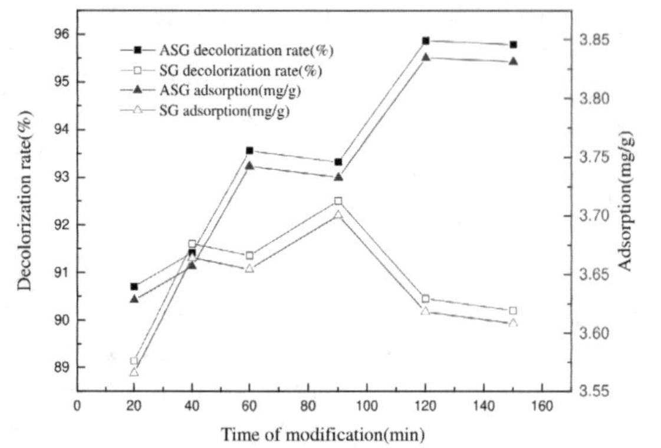

Figure 5. Effect of Modification Time on Adsorption Effect

3.2.2. Effect of NaOH Solution Concentration on MB Dye Adsorption

When modification temperature is 60℃ and time is 120 min, the effect of NaOH solution concentration on the adsorption performance of ASG is shown in Figure 6. In the initial stage, with the increase of NaOH solution concentration, the decolorization rate and adsorption of ASG to MB dye increased. When the concentration reached 5 g/L, the decolorization rate and adsorption of ASG to MB dye reached the maximum. When the concentration of NaOH solution continued to increase, the decolorization rate and adsorption decreased, and when the concentration reached 8 g/L, the decolorization rate and adsorption of ASG to MB dye were lower than SG. The reason is that when the concentration of modifier is low, the alkalinity of the solution increases with the increase of the concentration of modifier, which is conducive to hydrogen bonding, improving the surface activity of wort and improving its adsorption effect. But when the concentration of NaOH solution is too high, the surface structure of spent grain will be destroyed and the adsorption effect will be affected. Therefore, the optimum concentration of NaOH solution was taken as 5 g/L.

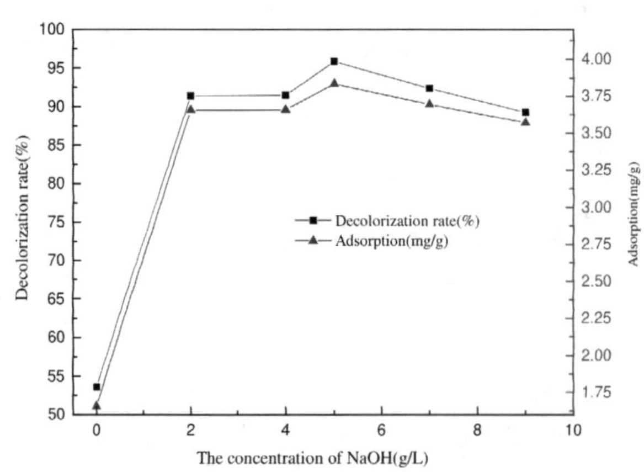

Figure 6. Effect of NaOH Concentration on Adsorption Effect

3.3. Effect of Adsorbent Dose on MB Dye Adsorption

The effect of adsorbent dose on the decolorization rate was investigated under the optimal modification temperature and time conditions and the results are presented in Figure 7. The experimental results show that the decolorization rate of SG and ASG to MB increases with increase in adsorbent dose loading, up to 0.5 g/L and 1.5 g/L respectively. However, the adsorption of SG and ASG to MB decreased above the adsorbent dosage value. This is because when the amount of spent grain reaches a certain value, the reaction between them and MB reaches equilibrium. When the spent grain is added, the adsorption amount per unit mass of spent grains is reduced, that is, the adsorption amount is decreased.

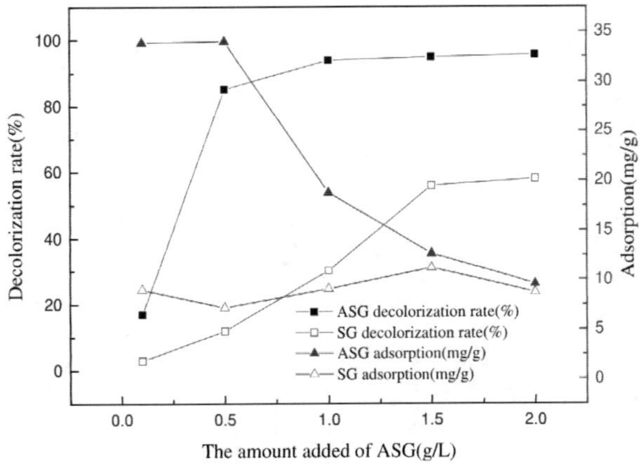

Figure 7. Effects of Adsorbent Dose on MB Dye Adsorption

3.4. Effect of Initial Concentration on MB Dye Adsorption

The effect of initial concentration of MB dye solution by SG and ASG on adsorption efficiency was investigated and the experimental results are shown in Figure 8. With the increase of the initial concentration of MB dye solution, the adsorption of SG and ASG increased, while the decolorization rate of ASG increased first and then decreased, while the decolorization rate of SG to MB dye solution remained stable and slightly decreased. This is because when the concentration of MB solution is within a certain range, the active groups on the surface of ASG can fully interact with MB molecule and have a good adsorption effect. However, when the MB dye solution is too high, the adsorption of MB dye on spent grain reaches saturation, and its adsorption changes slightly, while the decolorization rate decreases.

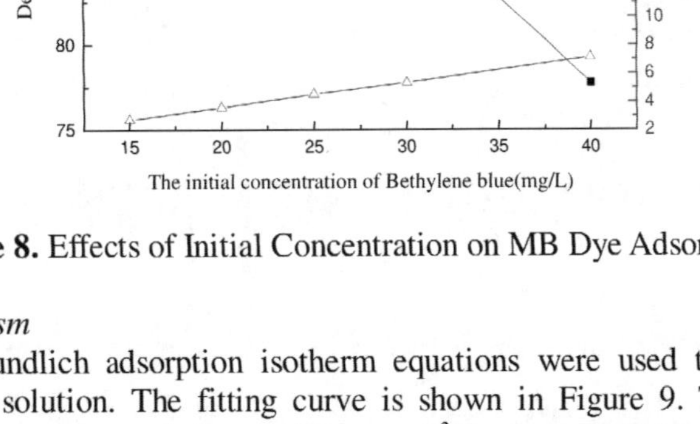

Figure 8. Effects of Initial Concentration on MB Dye Adsorption

3.5. Adsorption Mechanism

The Langmuir and Freundlich adsorption isotherm equations were used to fit the data of ASG adsorbed with MB dye solution. The fitting curve is shown in Figure 9. The relevant adsorption constants (q_{max}、b、K_f、n) and correlation coefficients (R^2) is shown in Table 2.

IOP Conf. Series: Materials Science and Engineering **735** (2020) 012074 doi:10.1088/1757-899X/735/1/012074

Table 2. ASG Adsorption Isotherm Model Parameters

Langmuir			Freundlich		
q_{max} (mg/g)	B (L/mg)	R^2	K_f (mg/g)	n	R^2
571.428	0.338	0.861	2.101	1.057	0.999

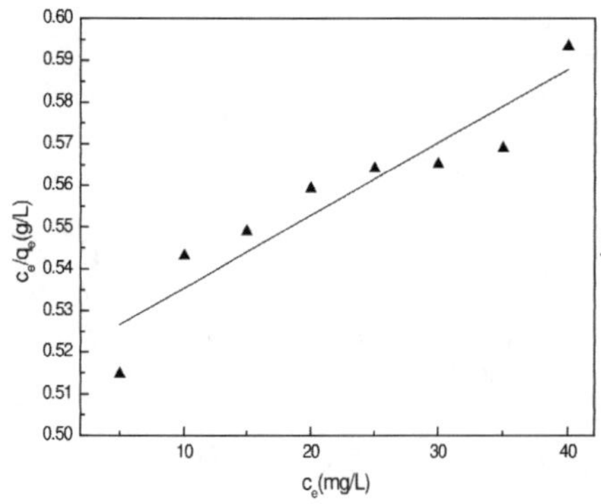

(a)

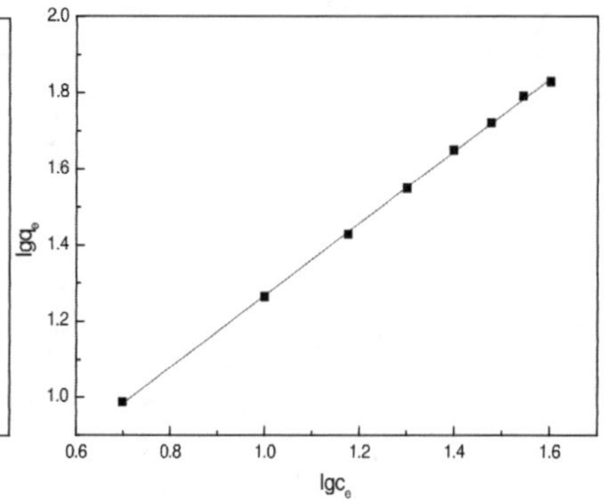

(b)

Figure 9. ASG Langmuir (a) and Freundlich (b) Adsorption Isotherm Model

It can be seen from the correlation coefficient of the fitted curve in Table 2 that the adsorption isotherm of ASG to MB dye solution has a better fit to the Freundlich equation, R^2>0.99. The K value can indirectly reflect the adsorption surface area of the adsorbent, reflecting the total capacity of the adsorbent activity or decolorization, and n is the adsorption characteristic of the adsorbent itself, representing the coefficient of adsorption strength. When $1<n<10$, it is favorable for the adsorption reaction happened. The value of n in the table is 1.057, and the reaction of ASG to adsorb methylene blue solution is more likely to occur.

The adsorption kinetics of adsorption of MB dye by ASG was studied by using different kinetic equations. The fitting curve is shown in Figure 10. The relevant kinetic model parameters and correlation coefficient (R^2) are shown in Table 3.

Table 3. ASG Adsorption Kinetics Model Parameters

Serial number	Model name	Model parameter		R^2
a	pseudo-first-order kinetic equation	K_{ad} : 0.036	q_e : 40.629	0.836
b	pseudo-second-order kinetic equation	K_2 : 15.625	q_e : 70.126	0.973
c	Intraparticle diffusion equation	K_t : 17.357	A : -28.298	0.864
d	Elovich equation	K_p : 5.114	C : 0.743	0.726

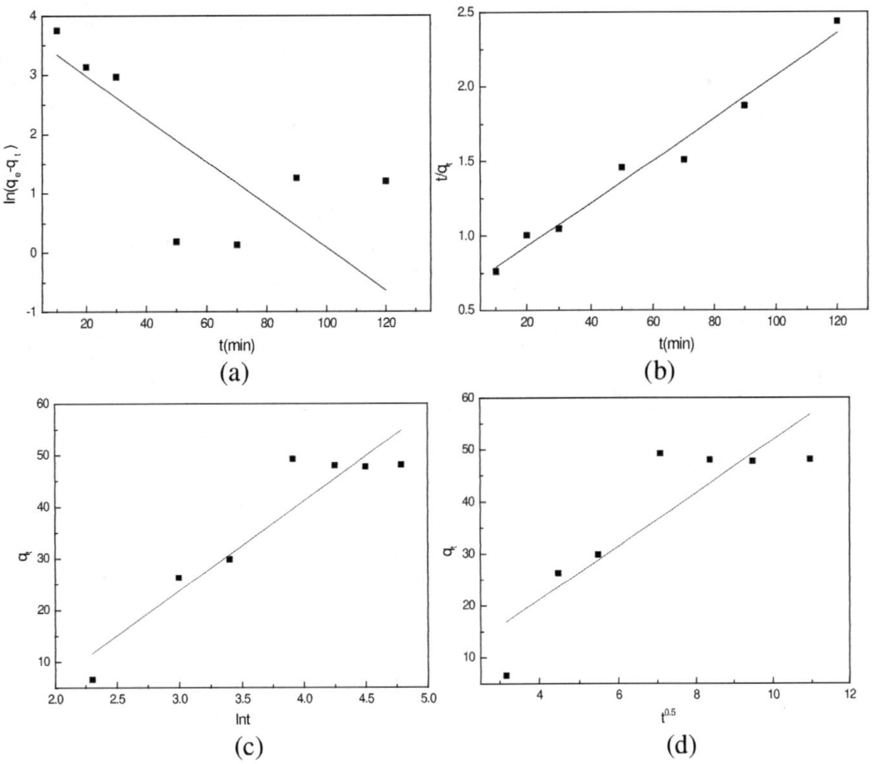

Figure 10. Adsorption Kinetics Curve of ASG

It can be seen from Figure 10 and Table 3 that the pseudo-second-order kinetic equation can better describe the adsorption test kinetics of ASG. The correlation coefficient is 0.973. The pseudo-second-order kinetic equation can be considered as chemical adsorption.

4. Conclusion

(1)Using NaOH as modifier, the modified spent grain adsorption material was prepared. The maximum decolorization rate of MB dye solution (95.87%) was attained under the conditions of 5g/L modified concentration, 60℃ modified temperature and 120 min modified time. The ASG showed better adsorption performance to MB solution.

(2)The adsorbent dose and the initial concentration of dye solution affected the adsorption performance of ASG to MB. When the adsorbent dose was 0.5 g/L and the concentration of MB dye solution was 30 mg/L, the decolorization rate and adsorption were higher.

(3)The Freundlich equation can better fit the adsorption behavior of ASG with R^2>0.99. The kinetic adsorption process can be characterized by pseudo-second-order kinetic equation with a correlation coefficient of 0.973.

5. References

[1] Pospiskova K, Safarik I. Magnetically modified spent grain as a low -cost, biocompatible and smart carrier for enzyme immobilization [J]. Journal of the Science of Food & Agriculture, 2013, 93(7):1598-1602.

[2] Xuexia Guo, Huiyuan Zhang, Chuangye Lai, et al. Application of beer by-products in feed industry [J]. Feed Industry, 2007 (21): 59-62.

[3] Pedro Silva J, Sousa S, Rodrigues J, et al. Adsorption of acid orange 7 dye in aqueous solutions by spent brewery grains [J]. Separation and Purification Technology, 2004, 40 (3):309-315.

[4] Yunnen Chen. Study on deep purification of arsenic and cadmium in water by waste wheat grain biosorbent [D]. Central South University, 2009.

[5] Yunnen Chen, Changshi Xiong, Jian Wu, et al. Mechanism and conditions optimization of adsorption of trivalent arsenic ions in modified wheat grains [J]. Journal of Environmental Engineering, 2016(09): 4673-4678.
[6] Qingzhu Li, Wen-qing Yan, Liyuan Chai, et al. Adsorption Characteristics of Zn(II) by Thiol-modified Wheat Grains[J]. Chinese Journal of Nonferrous Metals, 2014(04):1076-1083.
[7] Li Q, Chai L, Yang Z, et al. Kinetics and thermodynamics of Pb(II) adsorption onto modified spent grain from aqueous solutions[J]. Applied Surface Science, 2009, 255(7): 4298-4303.
[8] Qingzhu Li, Wenqing Yan, Li-yuan Chai, et al. Adsorption Characteristics of Pb(II) by Esterified Modified Wheat Grain[J]. Chinese Journal of Nonferrous Metals, 2013(04):1152-1159.
[9] Changshi Xiong, Yunnen Chen, Zhenpeng Dai. Experimental Study on Adsorption of Acidic Lake Blue A and Alkaline Lake Blue BB in Modified Wheat Grains [J]. Environmental Science and Technology, 2015(12):41-44.
[10] Zhenpeng Dai, Changshi Xiong, Dachao Zhang, et al. Experimental study on adsorption of methylene blue simulated wastewater by modified wheat grains[J]. Journal of Jiangxi University of Science and Technology, 2015(05): 6-10.

IOP Conf. Series: Materials Science and Engineering **735** (2020) 012075 doi:10.1088/1757-899X/735/1/012075

Quenching Process Simulation of Spur Gear Based on ANSYS

JingDong Zhang, Kuan Yang, Shenglian Zhao and Bin Zheng*

School of Transportation and Automobile Engineering, Panzhihua University No. 4, New Hong international community, Airport Road, East District, Panzhihua, Sichuan.
*Corresponding author: Bin Zheng (22198334@qq.com)

Abstract. Gear transmission is one of the most common mechanical transmissions in contemporary machinery. It is the main form of transmission of mechanical power and movement, and it is an important part of all kinds of machinery. For gear transmission, better hardness and wear resistance are required, which can be achieved by quenching process. But the spur gear in quenching process, its size and shape will appear different degree of deformation, this paper uses the Pro/Engineering software to establish solid parameterized model of the spur gears. Then, the simplified single tooth solid model of the spur gear imported to ANSYS software to obtain the finite element model. The temperature field of the single tooth spur gear during quenching were carried out. Thus, it is concluded that the relationship between the factors influencing the quenching. The results provide a certain reference basis for the establishment of the hot working processes of the spur gear.

1. Introduction

Spur gear is one of the most common mechanical parts in modern machines. Spur gear hardening is used to make gear with hard tooth surface in order to improve the hardness and bearing capacity of the spur gear. In the heat treatment process of harden tooth surface, the same torque of hardened gear is transferred. It makes smaller in size, lighter in weight and lower in cost, so it becomes the main processing method [1, 2].

At present, many scholars have studied the quenching process of the spur gear. Song used ANSYS software to simulate and analyze the surface quenching of a spur gear, and obtained the results of temperature field, thermal stress and thermal deformation in the quenching process of the spur gear [3]. Zhang conducted simulation analysis on the quenching process of a cylindrical gear. The conclusions are that the tooth center temperature of the cylindrical gear is the highest, the tooth top temperature is the lowest, and the stress is mainly concentrated in the tooth root [4]. Wang summarized the shortcomings of traditional quenching process; combined with the computer simulation technology, put forward the numerical simulation of gear carburizing and quenching process as the development direction of the current industry [5]. The paper takes a spur gear as the research object, establishes the three dimensional model of the spur gear by using Pro/Engineering and imports it into ANSYS for quenching numerical simulation analysis. The quenching process of spur gear was studied and analyzed from the aspects of temperature field. It provides a certain reference for the formulation of gear hot working technology.

2. Spur Gear Three Dimensional Modelling

The software Pro/Engineering is used to establish three dimensional solid model of a spur gear. By measuring the dimension of gear, the main dimension parameters are obtained. Module of the spur gear is 3 mm, teeth is 25, tooth profile angle is 20 °, thickness of the spur gear is 20 mm, high coefficient of the spur gear is 1.0, tip clearance coefficient of the spur gear is 0.25. The three dimensional geometric model of the spur gear is established by using Pro/Engineering software. The three dimensional model of the spur gear is shown in figure1.

Figure 1. Three Dimensional Model of the Spur Gear

In order to save space and improve the simulation speed, a simplified single tooth three dimensional model of the spur gear was established by using Pro/Engineering. The single tooth model of the spur gear is shown as figure 2.

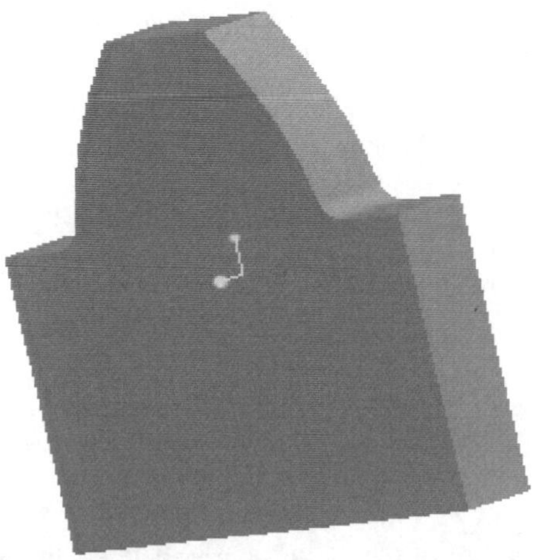

Figure 2. Single Tooth Model of Spur Gear

3. Thermodynamic Parameters of Materials

The material of spur gear is 20CrNi2MoA. The density, specific heat capacity, heat conductivity and other parameters of the material is given as table 1.

Table 1. 20CrNi2MoA Thermal Parameters

Density ρ (kg/m3)	20°C			7820			
Specific heat capacity Cρ (J(Kg.k))				460			
Temperature t(°C)	0	100	200	400	600	800	1000
Thermal conductivity Γ(w/(m.k))	14.7	16.6	18	20.8	23.5	26.3	26.8

The analysis of the spur gear quenching process by ANSYS also needs to provide mechanical performance parameters at different temperatures, such as elastic modulus, yield strength, poisson's ratio, and coefficient of linear expansion and shear modulus. Tab.2 gives the mechanical parameters of 20CrNi2MoA [6].

Table 2. 20CrNi2MoA Mechanical Parameters

temperature T /°C	Elastic modulus E(Pa)	Shear modulus G(Pa)	Yield strength Δ(Pa)	Coefficient of linear expansion 1/°C	Poisson ratio
20	1.93e11	1.93e10	0.2e9		
500	1.5e11	1.5e10	0.933e9		
1000	0.7e11	0.7e10	0.435e9	1.78e5	0.29
1500	0.1e11	0.1e10	0.07e9		
2000	0.01e11	0.01e10	0.007e9		

4. Mathematical Model of Quenching Process

During quenching, the heat transfer mode includes heat conduction and heat convection. In the three dimensional rectangular coordinate systems, the heat conduction equation is shown in equation (1). Heat conduction is the exchange of heat between two bodies in complete contact or between different parts of an object due to a temperature gradient. Heat transfer follows Fourier theorem.

$$\frac{\partial}{\partial x}\left(\lambda\frac{\partial T}{\partial x}\right)+\frac{\partial}{\partial y}\left(\lambda\frac{\partial T}{\partial y}\right)+\frac{\partial}{\partial z}\left(\lambda\frac{\partial T}{\partial z}\right)+q_v=\rho c\frac{\partial T}{\partial t} \quad (1)$$

Here:
ρ is the density of the material;
λ is the heat conduction coefficient;
c is specific heat capacity;
q_v is latent heat of phase change;
T is the temperature.

For the spur gear quenching process, heat conduction coefficient, specific heat capacity and material density are related to the temperature in quenching process. The heat flux, temperature, thermal boundary conditions and internal energy in the transient heat transfer system all change with time [7]. According to Fourier law, the transient heat transfer formula is shown in equation (2).

$$[C]\{t^k\}+[k]\{t\}=\{q\} \quad (2)$$

Here:
$[k]$ is heat transfer matrix;
$\{t^k\}$ is the time derivative of temperature;
$[C]$ is specific heat capacity matrix;
$\{q\}$ is the node heat flux vector, including heat generation.

5. Temperature Field Analysis and Discussion

For the general heat conduction problem, the content of single value condition includes physical condition, geometric condition, initial condition and boundary condition. Geometric conditions refer to the geometric dimensions and shapes of the spur gears involved in the heat conduction process. Physical conditions are the main physical parameters and physical characteristics of the spur gear guiding heat. Questions such as whether the physical parameters vary with temperature, whether there is an internal heat source, whether they are evenly distributed and so on. The initial condition refers to the temperature field in the spur gear at the beginning of heat conduction process. The boundary condition is to guide the heat exchange between the hot spur gear and the external environment.

The model was imported into ANSYS for pre-processing, including material property setting, grid division and material parameter setting. Finite element mesh generation is the transformation of geometric model into a finite element model composed of nodes and elements. The density of grid can be adjusted by controlling the size of unit and dividing precision. Automatic mesh generation method is adopted for mesh generation of the spur gear model. The final single tooth finite element model is shown in figure 3.

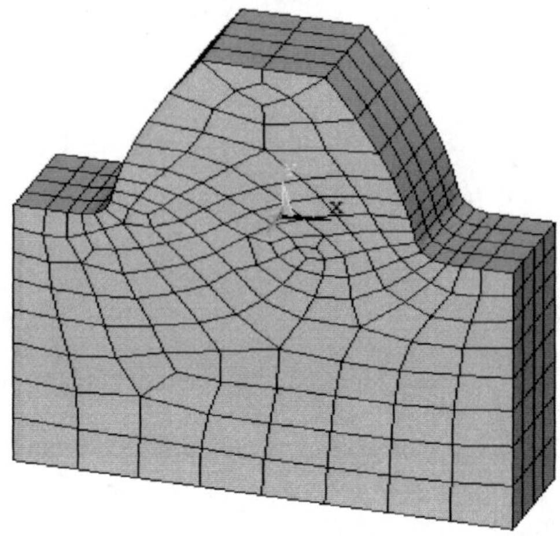

Figure 3. Single Tooth Finite Element Model

The single tooth spur gear with the initial temperature of 850℃ is quenched in the oil tank with the ambient temperature of 25℃. The analysis of temperature field set the time as 3 minute and the time interval as 10 second. The cloud graph of temperature field distribution at quenching time intervals between 30 second and 120 second is given, as shown from figure 4 to figure 7.

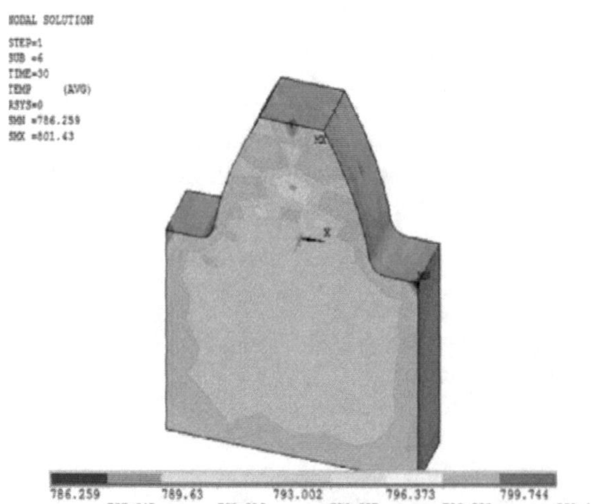

Figure 4. Cloud Graph of Temperature Distribution at 30 Second

IOP Conf. Series: Materials Science and Engineering **735** (2020) 012075 doi:10.1088/1757-899X/735/1/012075

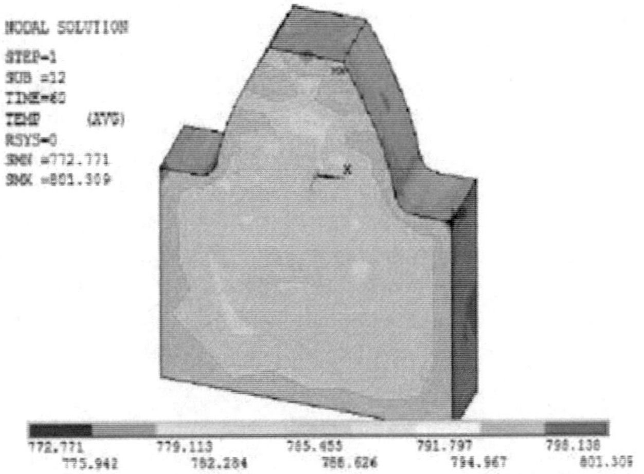

Figure 5. Cloud Graph of Temperature Distribution at 60 Second

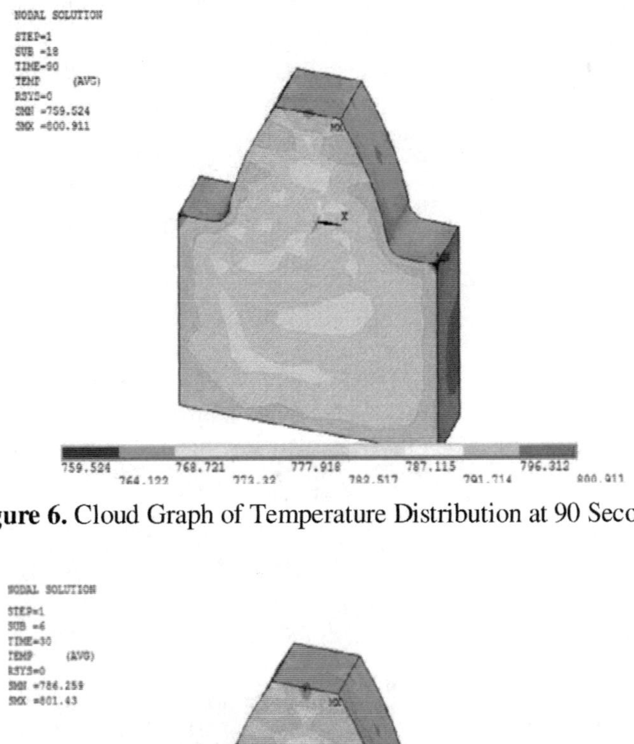

Figure 6. Cloud Graph of Temperature Distribution at 90 Second

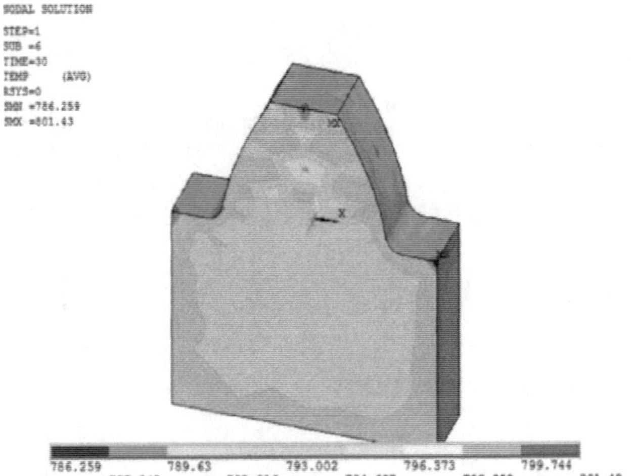

Figure 7. Cloud Graph of Temperature Distribution at 120 Second

It can be seen from figure 4 to figure 7 that the temperature difference between the tooth centre and the tooth surface is large. The difference between the maximum value and minimum value of the quenching temperature field of the gear at 10s is about delta 1=5℃. At the time 20s, the maximum value of the gear temperature field is about delta 2=10℃. At the time 60s, delta 3 is about 29 ℃. Therefore, it can be concluded that the temperature difference between tooth surface and tooth centre gradually increases with time, and the corresponding thermal stress is also gradually large, so it is the main process of accumulating thermal stress at the beginning of quenching.

In the quenching process of single tooth spur gear, the temperature difference between tooth surface and tooth centre is large, for the whole single tooth spur gear, the smaller size of the gear is more likely to cause the deformation of the gear, so in the single tooth spur gear processing time for small size gear, to ensure that there is enough machining allowance to ensure the function of the gear processing requirements.

6. Conclusions

According to heat transfer and finite element theory, a three dimensional heat transfer model of the spur gear is established to simulate real working conditions. By adding specific boundary conditions, the finite element analysis of gear quenching process is carried out based on the actual working conditions. The quenching temperature values of different time are obtained, which provides theoretical basis for further determination of gear quenching process. By analysis the temperature distribution of single tooth spur gear after quenching, the change trend of the temperature inside the gear can be obtained. The research results provide a certain reference for the design of the spur gear and the formulation of thermal processing technology.

7. Acknowledgment

This work was supported by national training program of innovation and entrepreneurship for undergraduates (201811360017) and scientific research cultivation project of Panzhihua University (2017ZD007).

8. References

[1] Li Qiang, Wang Ge, Cheng Nailu. Computer simulation and experiment of work piece temperature field during cooling process [J]. Yanshan university press, 2002, 26 (4); 298.

[2] Wang Yuanlong, Wang Liangmo. Finite element analysis and experimental investigations of the back case of a transmission with a central brake [J]. Proceedings of the Institution of Mechanical Engineers. 2015, 25(12):13-14.

[3] Song Ke, Yang Bangcheng, Yin Chunsong. Simulation and Analysis of Surface Quenching Process of Spur Gear Based on ANSYS [J]. Hot working technology, 2015, 44(24):197-203.

[4] Zhang Jinzhou, Ye Lirong, Li Xiaoming, et al. Finite Element Simulation of Straight Tooth Cylindrical Gear Quenching Based on ANSYS [J]. Hot working technology, 2016, 45(20):245-247.

[5] Wang Xin, Li Baokui, Gu Min, et al. Distortion control of large carburized gear shaft based on finite element analysis[J]. Heat Treatment of Metals, 2019, 44(4): 239-242.

[6] Ji Yangrong, Huang Chaoqun. Computer Simulation and Verification of 20crmnti Gear Carburizing and Quenching [J]. Foundry Technology, 2017, 38(5):1047-1050.

[7] Zheng Bin, Xiong Jie. FE Numerical Simulation of Heavy Rail During Quenching Process Based on ANSYS Software [J]. Hot working technology, 2016, 45(2): 238-240.

IOP Conf. Series: Materials Science and Engineering **735** (2020) 012076 doi:10.1088/1757-899X/735/1/012076

Reform of University Physics Experiment Course under the Combination of Big Data and Industrial Internet

Chunjuan Han

Northeast Petroleum University, School of Electronic Information Engineering, 550 Hebei Street, Qinhuangdao, Hebei, China.
Email: 64768170@qq.com

Absrtact. This paper puts forward the combination of big data and university physics experiment teaching based on the combination of big data and industrial internet, and tries to get rid of the traditional teaching mode of university physics experiment and develop innovative experiment teaching, focusing on the application of computer data acquisition and data analysis in the process of university physics experiment teaching.Let our students, like scientists and engineers, open up the relevant concepts in reality in their study, reduce the chance of gross errors in the experiment process, use cloud computing to achieve real-time data feedback, improve the understanding of concepts, increase students'interaction, and promote students' thinking, thus becoming active participants, allowing students to spend more time in analysis and development. High-level thinking skills.

1. Introduction

Physics plays an important role in the cultivation of human scientific quality. Experiments are the basis of physics. It reflects the commonness and universality of science and engineering experiments, and plays an irreplaceable important role in the cultivation of talents' scientific quality. In recent years, the new scientific and technological revolution represented by computer information science and technology, life science, space science and material science has greatly accelerated the development of science and technology and the intersection and infiltration of various disciplines. The new trend of integration has become the mainstream of scientific development. Therefore, physical experiment curriculum system, teaching content, teaching methods and means must be changed from closed to open.

This paper mainly studies the methods and significance of College Physics Experiment Teaching Reform under the combination of big data and industrial internet. The practical significance of the Reformed College Physics Experiment to the experimental classroom is as follows:

- Large amount of data, you can see more details;
- Obtain multi-node experimental data in university physics experiment through remote internet.
- The gross error in the experiment is reduced, and the relative uncertainty of the experiment is lower.
- The focus of students shifts from observation and recording data to thinking about the differences between experimental phenomena and theories.
- We can really try to obtain data with different possibilities in time and conditions.
- There is no way for students to copy experimental data from each other in class.

2. Analysis of Research Status at Home and Abroad

2.1. Overseas Research Status

As we all know, outstanding physicists in Europe and the United States emerge in large numbers. On the one hand, physicists' endless thirst for knowledge is indispensable, on the other hand, they also benefit from the clear educational principles of European and American countries: cultivating innovative talents as the basic starting point. Based on this open teaching environment, some outstanding physicists, such as Newton and Bohr, and Nobel laureates, have emerged. In the compilation of textbooks, they will give students a space for independent design and thinking, and let them design feasible experimental schemes and operate them independently according to the experimental principles. At the same time, they will pay close attention to the advanced scientific research achievements and constantly update the experimental projects, so as to provide more possibilities for the discovery of new physical phenomena. In terms of instrument configuration, with the continuous progress of science and technology every year, Instruments are also updated, with the characteristics of short replacement cycle and high utilization rate. On the teaching object, strictly abide by the status of student subject and teacher object, give the initiative to students, through active exploration, grasp the essence of physical laws.

The teaching of experimental data processing in physics laboratories of famous universities in the United States and Canada has the following characteristics:

- Computer technology is widely used in data processing. In physical experiments, computer has been widely used to collect and process data, and various data analysis software has also been widely used in the processing of experimental results.
- Data processing theory is closely integrated with experimental courses. The theory and
- Method of data processing that students have learned before the experiment class should also be fully used in the experiment.
- Teachers pay attention to the training of students' data processing skills. The experimental
- Technicians of these schools attach great importance to improving the data processing ability of students in the experimental operation. Teachers encourage students to carry out experimental evaluation in the experiment.

It can be seen that the data processing of physical experiments abroad is much more advanced than that of our country. Therefore, we need to study the experimental data processing methods, study various kinds of computing software, and find suitable software and teaching methods for our college physics experiment data processing, in order to improve the teaching level of experimental data processing in our country.

2.2. Domestic Research Status

In our country, university physics experiment courses are offered in Universities of science and technology. Traditional teaching methods are generally adopted, which can be roughly divided into several steps: telling experiment principles, introducing experiment instruments, explaining experiment contents, data recording and processing, etc. There are few inspiration and guidance teaching methods. At present, some university physics teachers in China have made preliminary progress in using computer-aided processing of physical experiment data.

We are in the information age, and more and more information needs to be acquired and processed in our work, study and daily life. For the students who are in the learning stage in school, how to use computer technology to process a large number of useful information, so as to make more efficient use of these resources, will become one of the necessary abilities of students. Modern educational technology must be adopted in teaching, which is one of the indispensable means and methods to promote quality education. Therefore, it is necessary to integrate big data assisted experiment with university curriculum, which is conducive to promoting teaching reform.

3. Rresearch Contents

Mr. John Dewey, a 20th century philosopher of education, said, "If we teach today's children the way we taught them yesterday, we will rob them of their tomorrow!" Yesterday was the industrial age. Physics experiment teaching was mainly manual. The teaching mainly took the form of blackboard writing, single-machine experiment of industrial analog instrument, table recording and physical archiving. Today is the PC era, digital, teaching mainly Flash, PPT projection, lesson preparation and learning efficiency has been greatly improved, some single computer data acquisition, data image processing. Tomorrow is the era of industrial interconnection. Big data. The main teaching methods are interactive homework, framework system, remote control, multi-point sensing, data acquisition and transmission, processing, judging and application of big data.

As the first scientific experiment course for college students, college physics experiment should not only train students with strict and systematic experimental skills, master the basic knowledge, methods and skills of scientific experiment, but also cultivate students' rigorous scientific thinking ability and innovative spirit, and train students' ability to integrate theory with practice, analyze and solve practical problems. They are the comprehensive ability to adapt to the development of science and technology, the innovation ability to adapt to the development of the times and the progress of science and technology.

The main contents of this paper are as follows:

(1) The combination of big data and industrial internet. Each industrial equipment will be labeled as a node, installed sensors, informationized the operation of the equipment, and even the key components of the equipment can be tracked, and then the production and manufacturing environment can be monitored and collected. Through the industrial internet, the information from tens of thousands or even tens of millions of sensors can be aggregated, and then based on the large data platform, according to the corresponding indicators and rules to be filtered and analyzed, a lot of useful information will be found.

(2) From the location and video surveillance of the concept of the Internet of Things, it really evolved to add value to university physics experiments.

(3) Cloud computing is oriented to IT infrastructure; big data is more closely related to the curriculum of College Physics experiments.

The experimental method based on big data computer greatly saves the time of recording, observing, calculating and analyzing in the process of traditional scientific experiment, and makes the focus of experiment fall on the assumption, analysis and thinking of the problem. And these are the basic qualities that we need to cultivate students in scientific research and engineering. Using large data computer to collect and analyze data can realize real-time data acquisition, and improve the accuracy, repeatability and stability of data acquisition. It can achieve high sampling rate, real-time capture of instantaneous changes. Large data cloud computing software is easy to understand, powerful, and can save, analyze and export data in real time.

The comparison results between several experiments combined with big data and college physics experiments and traditional physics experiments are as follows:

Figure 1. Frank Hertz Experiment

	Conventional method	Big data acquisition
Experimental installation	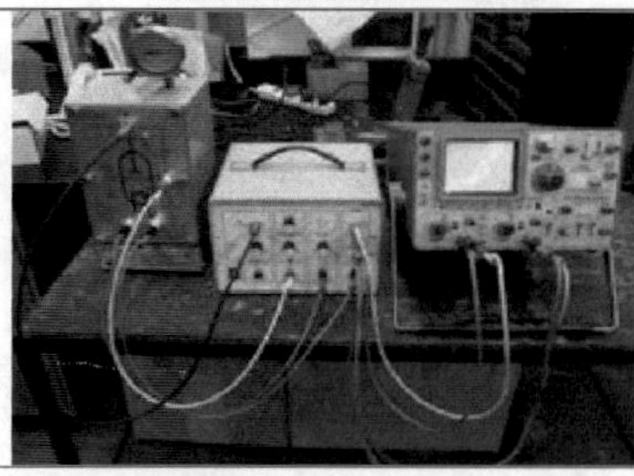	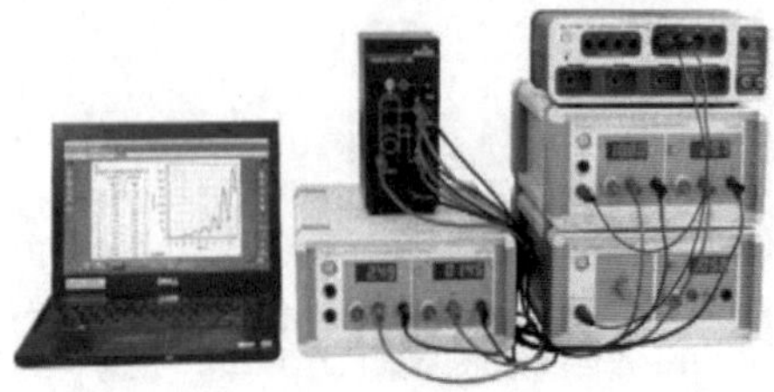
Data logging	It is necessary to design the step size of the collected data before the data is collected, and the point to be taken at the point where the curvature changes. However, when the student is in the shape of an unknown curve, this is difficult to achieve.	Just set the appropriate sampling frequency (up to 10MHz), such as 100MHz
Experimental period	It takes a school hour to collect about 80 sets of 100 sets of data, and it takes a picture to draw at the end of the experiment.	It takes tens of seconds to collect 600 point data, and the data is synchronized with the image.
Experimental result	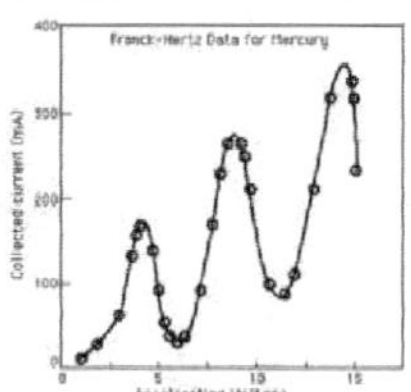 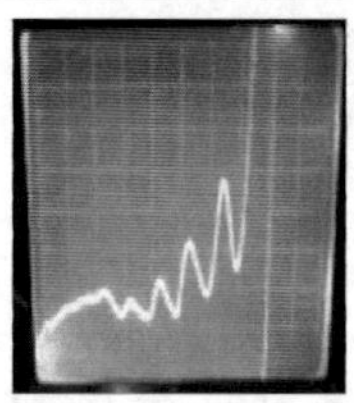	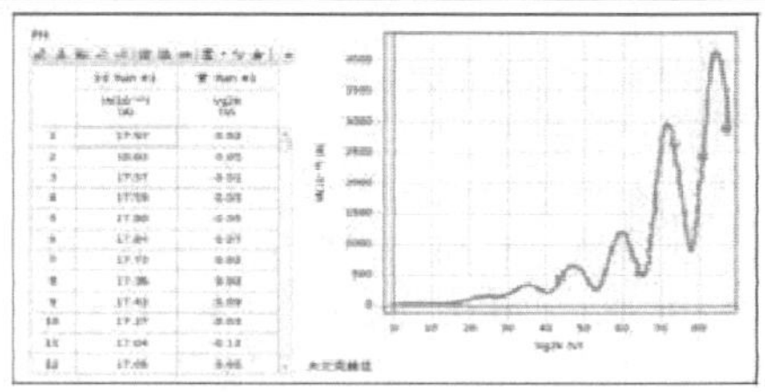

IOP Conf. Series: Materials Science and Engineering **735** (2020) 012076 doi:10.1088/1757-899X/735/1/012076

Figure 2. Photoelectric Effect Experiment

<table>
<tr><th></th><th>Conventional method</th><th>Big data acquisition</th></tr>
<tr><td>Experimental installation</td><td>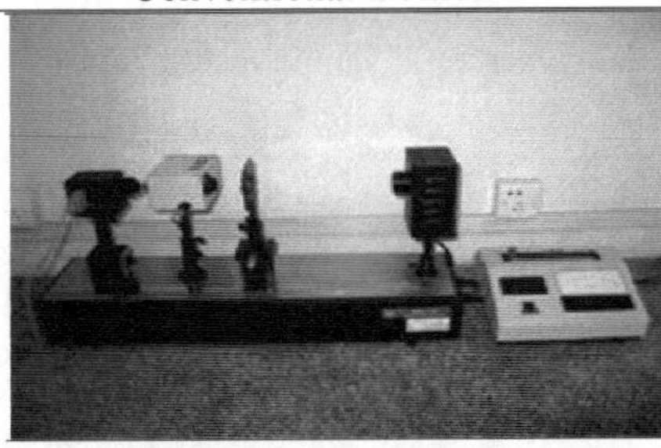</td><td>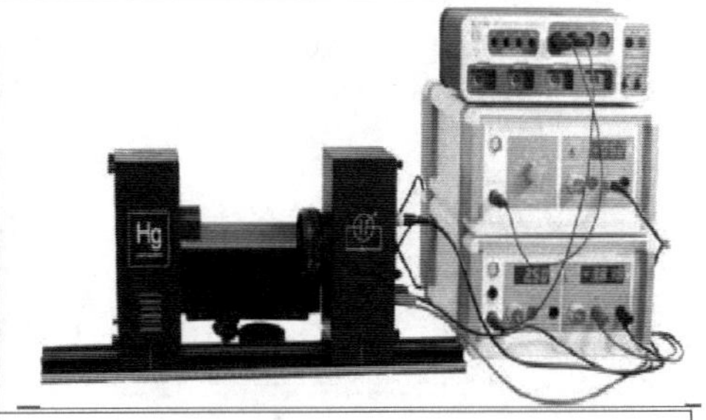</td></tr>
<tr><td>Data logging</td><td>The step size of the collected data needs to be designed before the experimental data is collected, and a little more should be taken at the point where the curvature changes. However, when students do not know the shape of the curve, this is relatively difficult to do.</td><td>Just set the appropriate sampling frequency (up to 10MHz), such as 100MHz</td></tr>
<tr><td>Experimental period</td><td>Point by point recording, about 3 data per wavelength, 4 wavelengths, about 120 data, about 1 lesson</td><td>It takes 5-10 minutes for 600-800 point data to be collected for each wavelength, and total of 3000 sets of data is required. Data synchronization with image</td></tr>
<tr><td>Experimental result</td><td><table>
<tr><td>V</td><td>-3.00</td><td>-1.80</td><td>-1.45</td><td>-1.40</td><td>-1.20</td><td>-1.00</td><td>-0.80</td><td>-0.60</td><td>-0.40</td><td>-0.20</td></tr>
<tr><td>μA</td><td>-0.3</td><td>-0.2</td><td>-0.1</td><td>0.0</td><td>0.2</td><td>0.7</td><td>1.3</td><td>1.9</td><td>2.8</td><td>3.7</td></tr>
<tr><td>V</td><td>0.00</td><td>0.20</td><td>0.40</td><td>0.60</td><td>0.80</td><td>1.00</td><td>1.20</td><td>1.40</td><td>1.60</td><td>1.80</td></tr>
<tr><td>μA</td><td>4.5</td><td>5.4</td><td>6.3</td><td>6.8</td><td>7.5</td><td>7.9</td><td>8.2</td><td>8.6</td><td>9.1</td><td>9.3</td></tr>
<tr><td>V</td><td>2.00</td><td>2.50</td><td>3.00</td><td>5.00</td><td>10.00</td><td>15.00</td><td>20.00</td><td>25.00</td><td></td><td></td></tr>
<tr><td>μA</td><td>9.5</td><td>10.2</td><td>10.5</td><td>12.0</td><td>13.0</td><td>13.9</td><td>14.2</td><td>14.5</td><td></td><td></td></tr>
</table></td><td>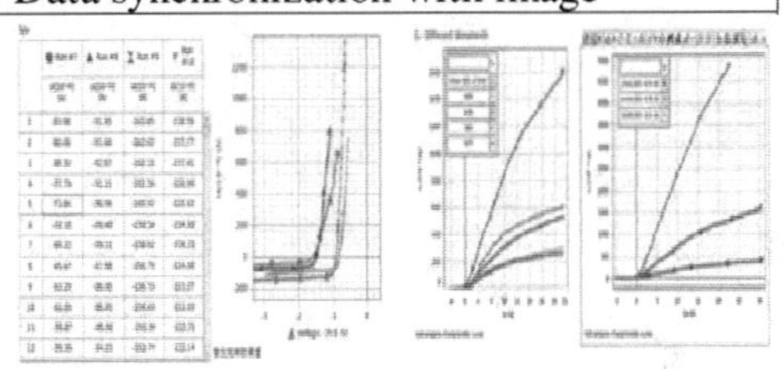</td></tr>
</table>

4. Summary

In a word, in the process of physics experiment teaching, using big data and traditional physics experiment instruments to reproduce and implement knowledge points can simplify physical research, strengthen experimental conditions, accelerate or delay physical process, visually and vividly display details, scientifically and accurately transmit a large amount of information, turn abstraction into image, stimulate students' interest in learning physics, and improve teaching. Learning quality. In physics experiment teaching, the combination of computer and physics experiment correctly and skillfully can break through the limitations of conventional experimental instruments and equipment, operate those difficult to observe, complex and difficult experiments, and bring unprecedented changes to our teaching and learning.

5. References

[1] Industrial Big Data Technology and Application Practice [M]. Electronic Industry Publishing House, Industrial Internet Industry Alliance Industrial Big Data Ad Hoc Group, 2017

[2] Accelerate the development of Jiangsu Industrial Internet Platform [J]. Zhang Rui, Shi Xiaopeng, Chen Yingwu. Realism. 2019 (02)

[3] Development of Industrial Internet Platform in China [J]. Lin Xuan. China's Industry and Information Technology. 2019 (01)

[4] "Industry 4.0": A New Generation of Intelligent Production Mode [J]. Rui Mingjie. World Science. 2014 (05)

[5] Reform and Innovation of College Physics Course in Applied Universities [J]. Xu Wei,

Changjiang, Wanxiaodong. Industry and Science Forum. 2018 (14) Computer Aid and Promotion for College Physics Teaching [J]. Lan Yi. Reference for Middle School Physics Teaching. 2018 (18)

[6] How Smart, Connected Products Are Transforming Competition. Porter M, Heppelmann J. Harvard Business. 2014

Characteristics of the Violent Typhoon Events Seriously Affecting Hainan Island and Its Relationship with PDO

Jingjing Zhu[1,2], Xiaoping Zhao[3], Shengan Wu[1,2], Caiying Xing[1,2] and Jing Yang[1,2]

1 Hainan Climate Center, Haikou 570203;
2 Key Laboratory of South China Sea Meteorological Disaster Prevention and Mitigation of Hainan Province, Haikou 570203;
3 Xichang Satellite Launch Center, Xichang 615000
Email: jingjingzjpq@163.com

Abstract. With the typhoon cyclone data, NCEP/NCAR reanalysis datasets and PDO index data from 1949 to 2018, the characteristics of violent typhoon events seriously affecting Hainan Island (VTESAHs for short) in the past 70 years are analyzed. On this basis, the characteristics of VTESAHs in the different phases of PDO and their relationship with SST are analyzed. The results show that there is obvious inter-annual variation in the frequency of VTESAHs. The VTESAHs have good continuity in both calm and active periods, and have obvious clustering. VTESAHs are more likely to occur in the cold phase years of PDO, almost generated in the western Pacific. The source of VTESAHs in the cold phase years of PDO are mainly concentrated in the sea area of 5 °N -15°N and 125 °E -150°E,while the source of VTESAHs in the warm phase years of PDO are mainly around 10°N. When the PDO phase is cold (warm), there is a significant negative (positive) correlation between the frequency of VTESAHs and the sea surface temperature in the western Pacific Ocean east of the Philippines at low latitudes.

1. Introduction

Typhoon is one of the main disastrous meteorological events, so the inter-annual, seasonal and monthly prediction of typhoon is great significance to disaster prevention. The relationship between typhoon activities and atmospheric and oceanic conditions has been studied extensively. It is pointed out that large-scale circulation and Pacific warm pool can affect the frequency and path of typhoon [1-3]. As a climatic variability signal, the Pacific decadal oscillation (PDO) has been used widely in the study of Pacific typhoons in recent years. According to the sea surface temperature anomaly in the northeast and tropical Pacific Ocean, Steven [4] divided PDO into cold phase and warm phase. Yang [5] summarized three kinds of views on the formation of PDO: tropical origin theory, mid-latitude origin theory and common origin theory. Wang [6], Li [7] and Gao [8] found that PDO is closely related to typhoon. The frequency of typhoons affecting China is more in the cold phase of PDO and less in the warm phase of PDO, as are the frequencies of typhoons north-going in midsummer and those affecting Shandong Province.

Violent typhoons affect Hainan Island every year from 2014 to 2016, bringing significant impacts and disasters to Hainan Island. The study on the activity characteristics and influencing factors of violent typhoons affecting Hainan Island is conducive to providing services for disaster prevention and mitigation in Hainan. Some studies [9-12] have been carried out by Hainan meteorologists. The climatic characteristics, formation mechanism and influencing factors of typhoons have been understood, which provides theoretical support for typhoon prediction. However, there is a lack of systematic understanding and application of violent typhoons due to relatively few studies. Therefore, it is necessary to study such typhoons. The characteristics of violent typhoon events seriously affecting Hainan Island (VTESAHs for short) from 1949 to 2018 are analyzed in this study. On this basis, the relationship between VTESAHs and PDO is carried out in order to provide scientific basis for violent typhoon prediction and decision support for disaster prevention and mitigation in Hainan.

2. Datasets and Methods

The typhoon cyclone data in 1949-2018 come from Typhoon Almanac compiled by China Meteorological Administration and Tropical Cyclone Best Track Dataset compiled by Shanghai Typhoon Institute. The Pacific Decadal Oscillation (PDO) Index is derived from NOAA. Sea Surface Temperature (SST) is derived from the National Centers for Environmental Prediction-National Center for Atmospheric Research (NCEP-NCAR) reanalysis data, with a horizontal resolution of 2° latitude × 2° longitude.

After entering the seriously affected area, the tropical cyclone with intensity greater than or equal to 14 grades (violent typhoon) is defined as VTESAH [13].

3. Characteristics of VTESAHs

The variation characteristics of VTESAHs from 1949 to 2018 are displayed in Figure1a. VTESAH occurred 26 times in recent 70 years with an annual average of 0.4. It occurred more frequently in the 1950s-1990s, less in the 1990s-2010s and increased significantly after 2010s. There is obvious inter-annual variation in the frequency of VTESAHs. The VTESAHs have good continuity in both calm and active periods.

It occurred four times in 1953-1956, four times in 1963-1966, five times in1970-1974, four times in 1981 1987, and none at othcr timcs bcforc 1990s. VTESAIIs occurred in 1991 and 2005 respectively in 1990s and 2000s, and after six consecutive years of calm, it occurred five times in 2011-2016. It indicates that VTESAHs have obvious clustering.

Statistics on the seasonal distribution of VTESAHs (Figure 1b), 26 VTESAHs in the past 70 years occurred in July-November, mainly concentrated in July-October. VTESAH occurred 10 times in summer, evenly distributed in July and August and none occurred in June.

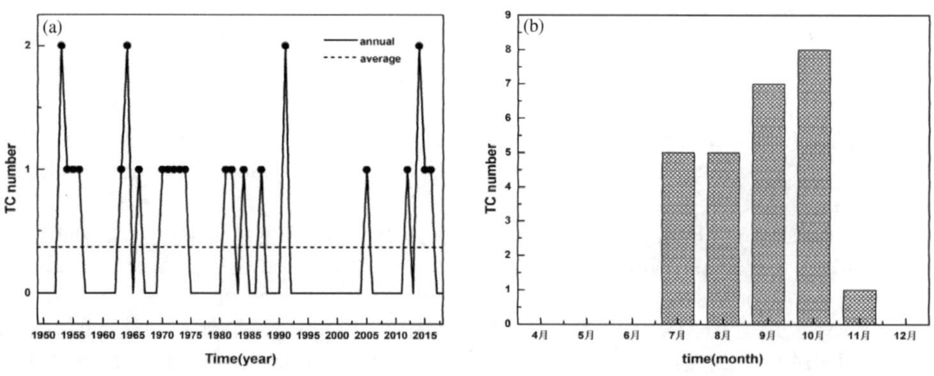

Figure 1. Inter-annual and Seasonal Distribution of VTESAHs from 1949 to 2018

4. Relationship between VTESAHs and PDO

4.1. Influence of PDO on VTESAH Frequency

The Pacific Decadal Oscillation (PDO) is the variability of the North Pacific SST on an inter-decadal time scale, which has an important impact on the inter-decadal characteristics of climate change in the Northern Hemisphere [14].The PDO index is defined as the projections of monthly mean SST anomalies onto their first EOF vectors in the North Pacific (north of 20ºN), which can be divided into cold phase and warm phase. When the index is positive (negative), it is the warm phase (cold phase) of PDO, the Central and Eastern Pacific is unusually warm (cold) at low latitudes, the North Pacific is unusually cold (warm) at mid-latitudes, and the West Bank of North America is unusually warm(cold) [5].

It can be seen from Figure2 that the cold phase year of PDO is 1949-1976 and 1999-2013, and the warm phase year of PDO is 1977-1998 and 2014-2018. 17 VTESAHs occurred in the PDO cold phase, accounting for 65.4%, and 9 occurred in the PDO warm phase. Therefore, the phases of PDO have a certain relationship with VTESAHs. VTESAHs have a good correspondence with the cold phase of PDO. It indicates that VTESAHs are more likely to occur in the cold phase years of PDO.

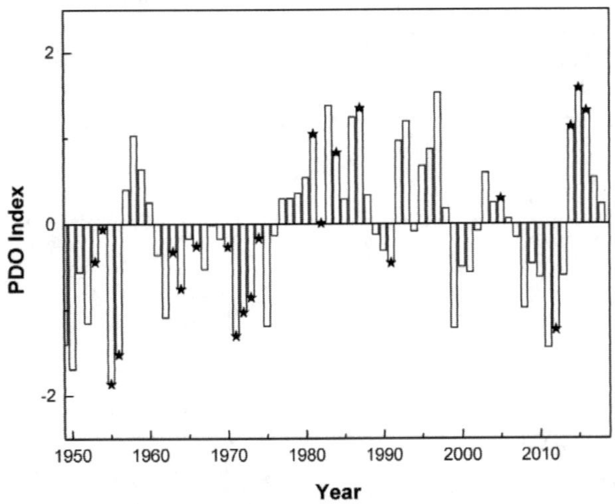

Figure 2. The Series of PDO Index from 1949 to 2018 (★Indicates that Violent Typhoon Event Occurred That Year)

4.2. *Influence of PDO on the Source of VTESAHs*

In the past 70 years, 25 VTESAHs were generated in the western Pacific Ocean east of the Philippines, and 1 in the South China Sea. There is a certain difference between the cold and warm phase of PDO on the source of VTESAHs (Figure3). The source of VTESAHs in the cold phase years of PDO are mainly concentrated at 5 °N -15°N and 125 °E -150°E, and the north side can be up to 16.7°N. However, the source of VTESAHs in the warm phase years of PDO are mainly around 10°N, which is more south than the cold phase.

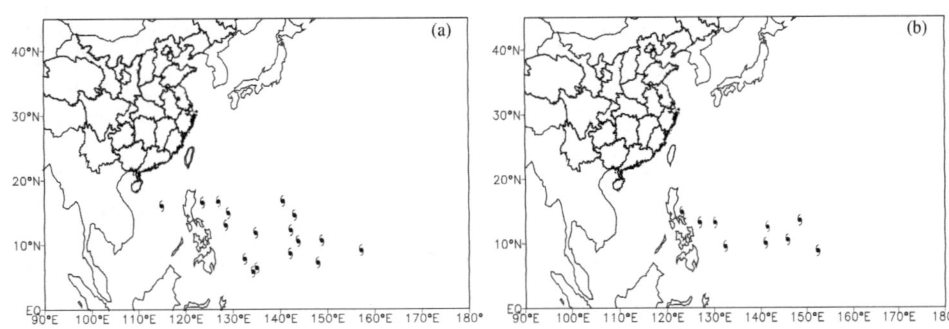

Figure 3. The Influence of PDO on the Source of VTESAHs (a: The Cold Phase of PDO, b: The Warm Phase of PDO)

4.3. *Correlation between SST and the Frequency of VTESAHs*

The correlation between the Pacific SST and the frequency of VTESAHs in different phases of PDO is analyzed in order to reveal the effect of PDO on VTESAHs (Figure4, Figure5). Figure4 shows the correlation between SST and the frequency of VTESAHs in the cold phase years of PDO. The SST in the western Pacific east of the Philippines (near 140ºE-160ºE) and the frequency of VTESAHs are significantly negatively correlated at low latitudes in spring, summer, autumn and winter. It indicates that there is a negative correlation between SST in the area and the frequency of VTESAHs in the cold phase years of PDO. The cold phase of PDO usually corresponds to the positive anomaly of sea surface temperature in the western Pacific, which is beneficial to the formation and development of typhoon in the area. The SST anomaly may affect the ocean-atmosphere interaction through thermal exchange. It will have a certain impact on the frequency and intensity of the typhoon. However, more frequency of strong typhoons further strengthens the ocean mixing of the ocean surface and the subsurface, resulting in a decrease in surface sea temperature. The combined effect of sea surface temperature and strong typhoon frequency makes the sea temperature in the western Pacific Ocean and the frequency of VTESAHs negatively correlated in the cold phase of PDO. Combined with the source of VTESAHs in the cold phase years of PDO (Figure3), it can be found that the significant negative correlation area is the main source of VTESAHs.

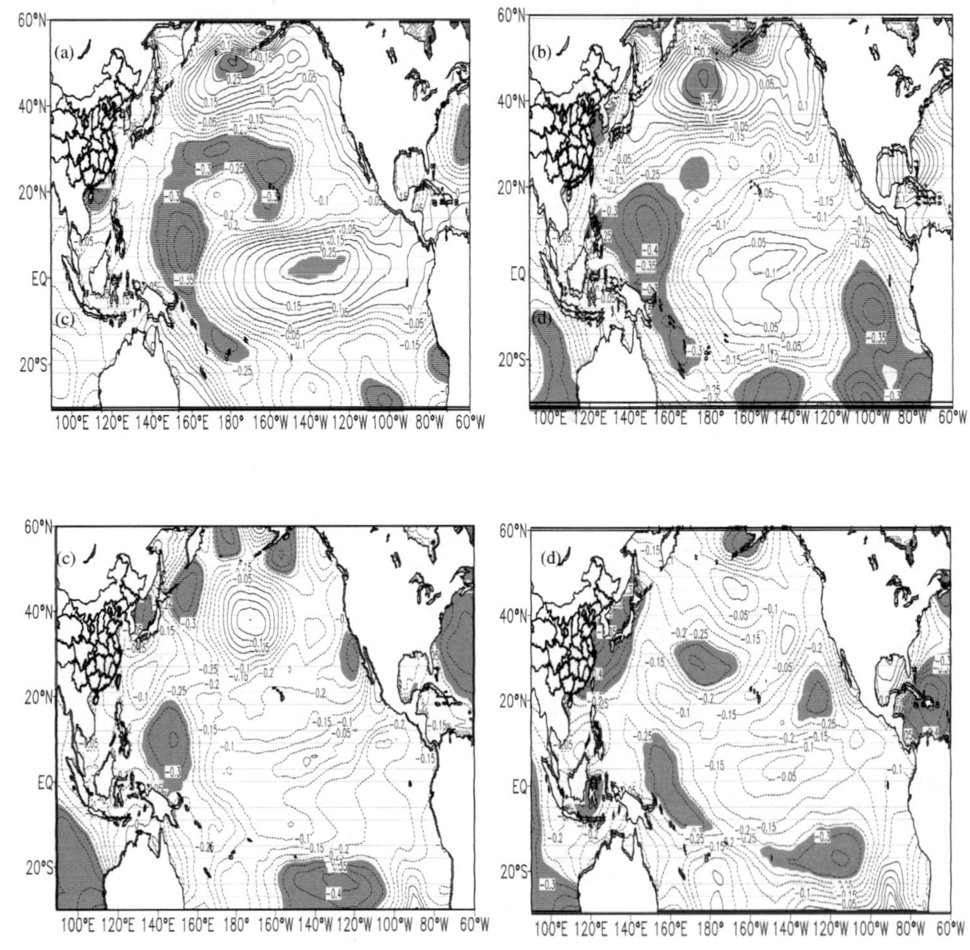

Figure 4. The Correlation between SST and the Frequency of VTESAHs in the Cold Phase Years of PDO. a: Winter, b: Spring, c: Summer, d: Autumn (Shadow Area Passed the Significance Test of 95% Level)

Figure 5 shows the correlation between SST and the frequency of VTESAHs in the warm phase years of PDO, which is quite different from the cold phase years. The Pacific SST and the frequency of VTESAHs are mainly positively correlated at low latitudes in spring, summer, autumn and winter.

IOP Conf. Series: Materials Science and Engineering **735** (2020) 012077 doi:10.1088/1757-899X/735/1/012077

The significant regions at low latitudes are mainly located in the western Pacific Ocean east of the Philippines and the Central Pacific Ocean from 160 °E to 170 °E. It indicates that there is a positive correlation between SST in the area and the frequency of VTESAHs in the warm phase years of PDO. The warm phase of PDO usually corresponds to the negative anomaly of the sea surface temperature in the western Pacific, which is not conducive to the occurrence of VTESAHs. It is consistent with the conclusion that the frequency of VTESAHs in the warm phase is relatively less. However, the SST anomaly affects the frequency and intensity of typhoon through ocean-atmosphere interaction. The frequency of VTESAHs is positively correlated with SST in the warm phase years of PDO. High sea temperature is also conducive to the formation and development of strong typhoon.

In summary, When the PDO phase is cold (warm), there is a significant negative (positive) correlation between the frequency of violent typhoons and the sea surface temperature in the western Pacific Ocean east of the Philippines at low latitudes.

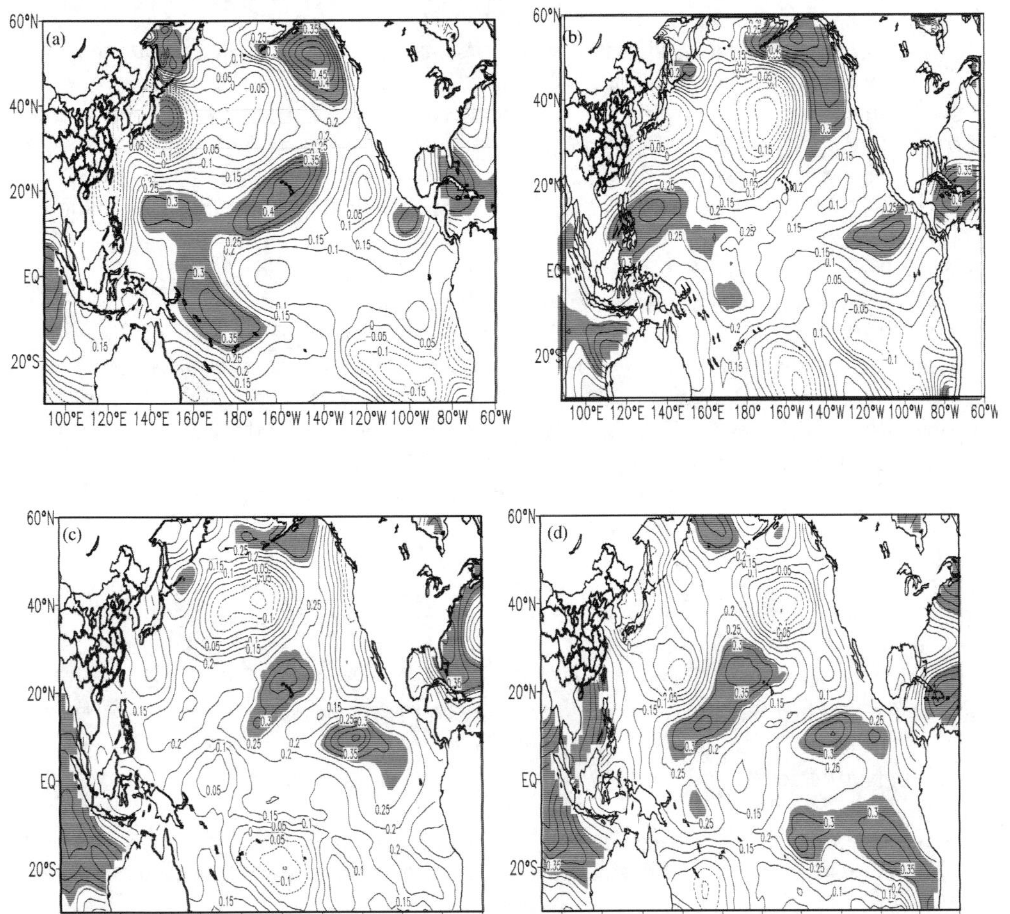

Figure 5. The Correlation between SST and the Frequency of VTESAHs in the Warm Phase Years of PDO. a: Winter, b: Spring, c: Summer, d: Autumn (Shadow Area Passed the Significance Test of 95% Level)

5. Summary

There is obvious inter-annual variation in the frequency of VTESAHs. The VTESAHs have good continuity in both calm and active periods, and have obvious clustering mainly occurring from July to October.

VTESAHs are more likely to occur in the cold phase years of PDO, almost generated in the western Pacific. The source of VTESAHs in the cold phase years of PDO are mainly concentrated in the sea area of 5 ºN -15ºN and 125 ºE -150ºE,while the source of VTESAHs in the warm phase years of PDO are mainly around 10ºN.

When the PDO phase is cold (warm), there is a significant negative (positive) correlation between the frequency of VTESAHs and the sea surface temperature in the western Pacific Ocean east of the Philippines at low latitudes.

6. References

[1] TAO Shiyan, XU Shuying, GUO Qiyun. The characteristics of the zonal and meridional circulation over tropical and subtropical regions in eastern Asia in Summer [J]. Acta Meteorologica Sinica (in Chinese), 1962, 32(1): 91-103.

[2] DING Yihui, Late E R. Large-scale circulation conditions of affecting typhoon over Western Pacific [J]. Acta Oceanologica Sinica (in Chinese), 1983, 5(5): 561-574.

[3] CHEN Guanghua, HUANG Ronghui. The effect of warm pool thermal states on tropical cyclone in West Northwest Pacific [J]. Journal of Tropical Meteorology (in Chinese), 2006, 22(6): 527-532.

[4] Steven R H, Mantua N J. Empirical evidence for North Pacific regime shifts in 1977 and 1989[J]. Progress in Oceanography, 2000, 47(1): 103-145.

[5] YANG Xiuqun, ZHU Yimin, XIE Qian, et al. Advances in Studies of Pacific Decadal Oscillation [J]. Chinese Journal of Atmospheric Sciences (in Chinese), 2004, 28(6): 978-992.

[6] WANG Yongmei, LI Weijing, REN Fumin, et al. Study on climatic characteristic of China-influencing typhoons and the interrelations between them and their environmental factors [J]. Journal of Tropical Meteorology (in Chinese), 2007, 23(6): 538-544.

[7] LI Ji, LI Fei, LIN Rong, et al. Interdecadal variations of midsummer northward-going typhoons over East Asia and the relationships with Pacific Decadal Oscillation[J]. Journal of Tropical Meteorology (in Chinese), 2011, 27(5): 731-737.

[8] GAO Xiaomei, JIANG Jing, LIU Chang, et al. Frequency variation characteristics of typhoons affecting Shandong in recent 67 years and their relationship with several climate factors [J]. Journal of the Meteorological Sciences (in Chinese), 2018, 38(6): 749-758.

[9] WU Hui. Climatic features of tropical cyclone in Hainan Island and its relationship with ENSO [J]. Meteorological Monthly (in Chinese), 2005, 31(12): 61-64.

[10] WU Shengan, LI Tao, KONG Haijiang. Sensitivity factors for typhoon genesis over the Western North Pacific during July-September [J]. Journal of Tropical Meteorology (in Chinese), 2011, 27(6): 797-804.

[11] ZHAO Xiaoping, ZHU Jingjing. Impacts of atmospheric quasi-biweekly oscillation of tropical and subtropical on typhoon tracks in WNP [J]. Natural Science Journal of Hainan University (in Chinese), 2014, 32(4): 368-377.

[12] ZHU Jingjing, ZHAO Xiaoping, WU Hui, et al. Abnormal abundant causes of tropical cyclone in Hainan Island in 2013[J]. Journal of the Meteorological Sciences (in Chinese), 2016, 36(2): 262-268.

[13] WU Shengan, XING Caiying, ZHU Jingjing, et al. Climate characters of disaster climate events of Hainan Island[C]. Haikou: Second Symposium on Disaster Risk Analysis and Management in Chinese Littoral Regions (in Chinese), 2014, 86-90.

[14] DING Yihui, WANG Huijun. Newly acquired knowledge on the scientific issues related to climate change over the recent 100 years in China [J]. Chinese Science Bulletin (in Chinese), 2016, 61(10): 1029-1041.

Research on A New Type Topology of 4-Quadrant Converters

Dawei Pan

School of Mechanical and Control Engineering, Shengli College, China University of Petroleum, Shandong Dongying, 257061, China
Email: pandawei@slcupc.edu.cn

Abstract. In consideration of the demerits of the traditional four-quadrant converter whose operating voltage of the dc side is very high, a new type topology converter is proposed whose topology belongs to quasi-buck topology and the voltage of the dc side is reduced by about 50%. Therefore this kind of four-quadrant can be used in the lower voltage filed. The modelling of the converter is described in this paper. The validity of the converter topology and the model along with the theory is proved by MATLAB simulation and the experiment result.

1. Introduction

Four-quadrant converters (4QC) are widely used in many fields, such as wind power technology, reversible PWM rectifier and active power filter. For the moment, most of the four-quadrant topologies belong to bridge voltage circuits, and the dc-side voltage of this kind of the topology is very high, which depends on the boost work mechanism. The demerit mentioned above limits the range application to a large extent. In this paper a new type topology converter is proposed whose topology belongs to quasi-buck topology and the voltage of the dc side is reduced by about 50%. Therefore this kind of four-quadrant can be used in the lower voltage filed. The modelling of the converter is described in this paper. The feasibility of the converter topology, the model and the theoretical analysis is proved by MATLAB numerical simulation and the hardware experiment result.

2. Topology Conversion and Principle Analysis

Both of the traditional one-phase and three-phase bridge four-quadrant converter can be considered to be composed of the half-bridge units, which are shown in figure 1(a). The kind of the half-bridge topology is extensively used, but it has the demit that the voltage U_d of the dc-side must be higher than the ac-side peak-peak voltage, which is determined by the boost work mechanism. This kind of topology can't meet the need of dc-side low voltage load if the ac-side voltage isn't reduced. To meet the need of the low voltage, the load can be transferred from the positive and negative bus u_d of the dc-side to its upper half terminal u_1, and the mode is called dc-side unsymmetrical output. But in this mode, shapes of u1 and u_2 are of distortion seriously, and the circuit can't work normally. This is because the direct component of the load at u_1 flows through the source e_S inevitably to form one single dc path, which makes the average value of i_S not to be zero, and it doesn't conform to the fundamental role of the hypostasis of four-quadrant converter. It won't work well unless the identical load is connected to the side of u_2 to form another dc circuit to balance the direct component of i_S. In order to solve the problem, the load R_1 is considered to connected to the capacitor C_1 in parallel, and at the same time the capacitor C_2 and the power switch S_2 exchange places according to the original polarity to keep the mean value of the ac-side current zero, and the topology is shown is figure 1(b). Compared with figure 1(a), the work mechanism of figure 1(a) doesn't change essentially, but the half-bridge circuit has passed out of existence, and the previous u_d has been split in half, set at each side, whereas u_1 and u_2 still exist. If U_1 and U_2, the dc mean value of u_1 and u_2, keep equal to each

IOP Conf. Series: Materials Science and Engineering **735** (2020) 012078 doi:10.1088/1757-899X/735/1/012078

other and S_1 and S_2 are applied with the PWM control similar to the half-bridge circuit, the PWM waveform of the point K of the two topologies will be of the same. Comparing with the half-bridge circuit, this topology has the advantage to avoid and protect the positive and negative short circuit easier, and point P and point N' are of the same potential to dc side. In order to keep point P and point N' equipotential to dc side to reach the dc balance, one inductance $L1$ is set between point K and point N' according to the Principle of steady-state voltage-second balance of inductance, which is shown in figure 1(c). Accordingly, in the unsymmetrical case that only the capacitor $C1$ is connected to the dc side load, on one hand that the dc side voltage of U_C and U_O are of balance, on the other hand that a new dc circuit is formed along *K-L-G-N'-L1-P-K*, which makes it become possible to remove the dc component of is in the circuit of *K-P-R*L*-G-L* and remove the dc component and that i_S will not contain the dc component because of the unbalance of the load. The dc side output voltage of the four-quadrant converter can be reduced about 50% by the topology conversion, and the topology can be called Quasi-Buck 4-Quadrant Converter (QBC).

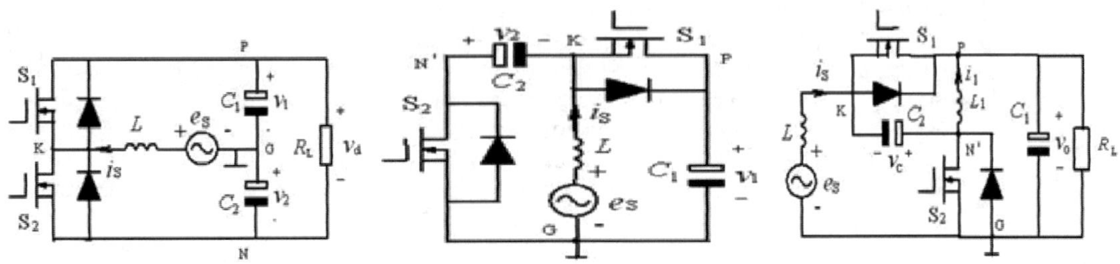

Figure 1. Topology Conversion

3. The State Space Average Modelling

To the Quasi-Buck 4-Quadrant Converter shown in figure 1(c), it is assumed that the conducting time of switches of S_1 and S_2 are dT and $d'T=(1-d)T$, where T is PWM switching period and d is the duty cycle. When S_1 is on and S_2 is off, the sate space equations of the linear circuit can be obtained, which is in the form of matrix:

$$\begin{bmatrix} L\dot{i}_S \\ L_1\dot{i}_1 \\ C_2\dot{u}_C \\ C_1\dot{u}_O \end{bmatrix} = \begin{bmatrix} 0 & 0 & 0 & -1 \\ 0 & 0 & 1 & 0 \\ 0 & -1 & 0 & 0 \\ 1 & 0 & 0 & -1/R_L \end{bmatrix} \begin{bmatrix} i_S \\ i_1 \\ u_C \\ u_O \end{bmatrix} + \begin{bmatrix} 1 \\ 0 \\ 0 \\ 0 \end{bmatrix} e_S \tag{1}$$

With S_1 off and S_2 on, we have

$$\begin{bmatrix} L\dot{i}_S \\ L_1\dot{i}_1 \\ C_2\dot{u}_C \\ C_1\dot{u}_O \end{bmatrix} = \begin{bmatrix} 0 & 0 & 1 & 0 \\ 0 & 0 & 0 & -1 \\ -1 & 0 & 0 & 0 \\ 0 & 1 & 0 & -1/R_L \end{bmatrix} \begin{bmatrix} i_S \\ i_1 \\ u_C \\ u_O \end{bmatrix} + \begin{bmatrix} 1 \\ 0 \\ 0 \\ 0 \end{bmatrix} e_S \tag{2}$$

It is assumed that the system matrix and input matrix of equation (1) and equation (2) are A_1, B_1 and A_2, B_2 respectively. On the basis of reference [1] and reference [2], equation(1) and equation (2) can be synthetically transformed into the state space average (SSA) model of the converter with $A = dA_1 + (1-d)A_2$ and $B = dB_1 + (1-d)B_2$.

$$\begin{bmatrix} L\dot{i}_S \\ L_1\dot{i}_1 \\ C_2\dot{u}_C \\ C_1\dot{u}_O \end{bmatrix} = \begin{bmatrix} 0 & 0 & 1-d & -d \\ 0 & 0 & d & -(1-d) \\ -(1-d) & -d & 0 & 0 \\ d & 1-d & 0 & -1/R_L \end{bmatrix} \begin{bmatrix} i_S \\ i_1 \\ u_C \\ u_O \end{bmatrix} + \begin{bmatrix} 1 \\ 0 \\ 0 \\ 0 \end{bmatrix} e_S \tag{3}$$

Left Multiplying equation (3) with $[i_S \quad i_1 \quad u_C \quad u_O]$, we can get

$$Li_S\dot{i}_S + L_1 i_1\dot{i}_1 + C_2 u_C\dot{u}_C + C_1 u_O\dot{u}_O =$$

$$= -d'u_C i_S + du_O i_S - du_C i_1 + d'u_O i_1 + d'u_C i_S + du_C i_1 - du_O i_S - d'u_O i_1 - \frac{u_O^2}{R_L} + e_S i_S = e_S i_S - \frac{u_O^2}{R_L}$$

Since $Li_S\dot{i}_S = \frac{d}{dt}(\frac{1}{2}Li_S^2)$ means the instantaneous power of the inductance L and other items can be analogized, so the equation above proves that the power balance can be satisfied at any instance and the power from the source equals to the sum of output power and the power absorbed by all the reactive power components.

4. The Simulation and Experiment Results

The system control structure designed for the QBC simulation and experiment is shown in figure 2. Parameters of the converter are shown below: L=20mH, L_1=100mH, C_1=C_2=2200uF, R_L=50Ω, E_S= $\sqrt{2}$ *110V, U_O=180V. In this system equation (3) is regarded as the QBC simulation model, and the calculation of the model control variable-duty cycle d and the current closed-loop control are realized with a small inertia current tracking control algorithm [3, 4]. The MTALAB numerical simulation waveform is shown in figure 3. In the simulation result it can be found that u_O and u_C both contain the fundamental wave and second harmonic and the current i_1 barely contains the second harmonic just as the steady state analysis. The above is more obvious in the FFT spectral analysis of MATLAB shown in figure 4.

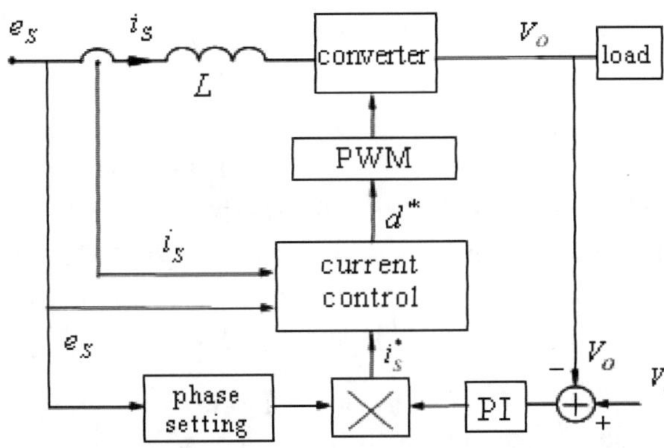

Figure 2. 4QC System Control Configuration

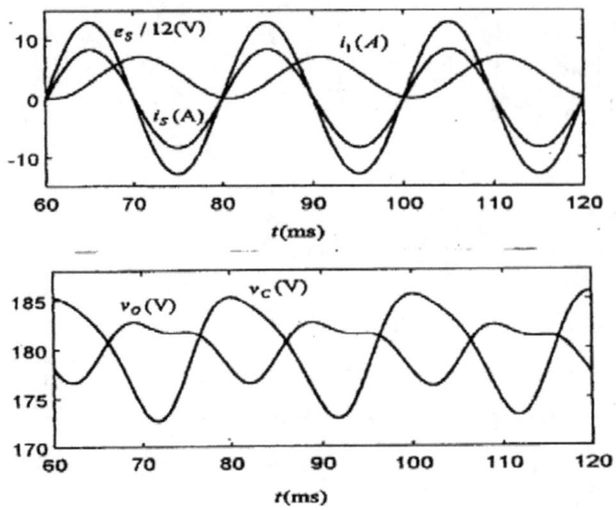

Figure 3. QBC Simulation Waveforms

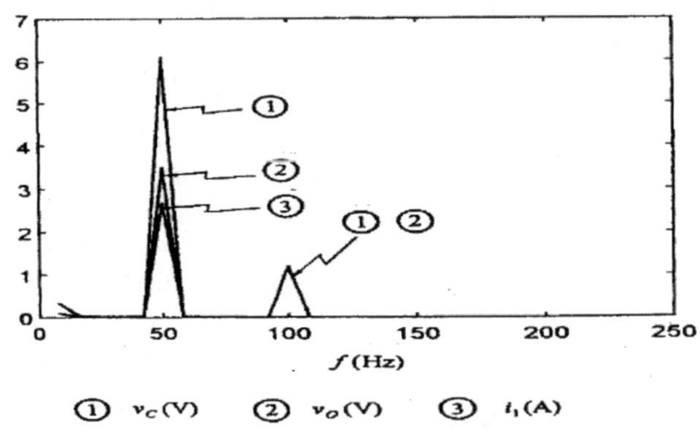

Figure 4. QBC FFT Spectrum

The current-hysteresis-loop tracking control technique is applied to compare the feedback current and the given current i_S^* to form the switch signals in the experiment system. The ac-side input current is forced to track i_S^* and the deviation is limited in a desired hysteresis band by on-off action of the power switch. The measured experiment waveform is shown in figure 5 according to the converter parameter same as the simulation system. It can be found that the current is almost a sinusoidal waveform and the power factor pf≈1, which agrees with the simulation and theoretical analysis basically. The only difference is that the simulation waveform doesn't reflect the switching frequency ripples, which is decided by the "averaging" character of the SSA model.

5. The Formation of the Single Phase and 3-Phase Converter

As the half-bridge topology shown in figure 1(a) can be the component unit of one-phase and tree-phase four-quadrant converter, the QBC unit shown in figure 1(c) can also form one-phase and three-phase four-quadrant converter, which is shown in figure 5. The dc-side output port of each QBC unit is connected in parallel, and one filter capacitor is used by the units together. In figure 5(a), the ac-side input voltage e_1 and e_2 is required to have the same amplitude and the reverse phase. In practice, an ac power $e=2e_1$ is connected between two input terminals to form a one-phase converter, and the two filter inductances of the input circuits can also be combined into one. A symmetrical three-phase source at the ac-side is required in the three-phase structure shown in figure 5(b).

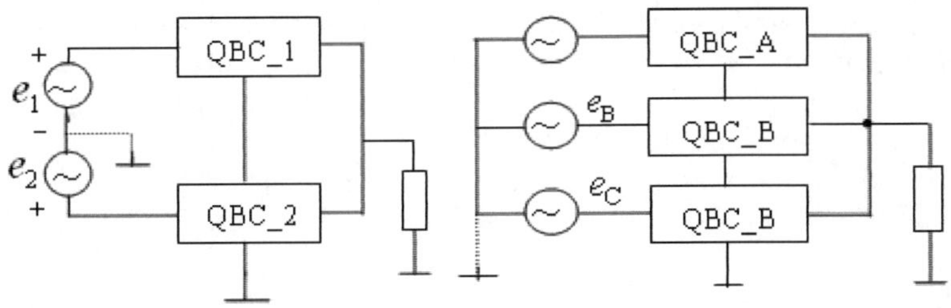

Figure 5. Configurations of Single-phase and 3-pahse Buck 4QC

As the lower limit value of the QBC dc-side voltage is reduced about 50% compared with the half-bridge unit, the dc-side voltage lower limit value of one-phase and three-phase combined by QBC is reduced greatly. It is assumed that the peak value of each unit dc-side input voltage is E_S, and the compare of the dc-side voltage lower limit value between QBC and bridge structure is shown in table 1, which shows that the dc-side voltage lower limit value of QBC is reduced about 45% compared with the bridge structure. The dc-side output voltage of three-phase QBC doesn't contain fundamental wave and the second harmonic because of the three-phase symmetry of the three units output ripple. The one-phase QBC doesn't contain the fundamental wave because the fundamental waveforms of the two units are of reversal phase. However, the second harmonic still exits like the one-phase bridge structure.

Table 1. Comparison Of the Dc-side Voltage Lower Limit

Configurations	Unit Circuit	Single-phase Circuit	3-phase Circuit
Bridge Configuration	$2E_S$	$2E_S$	$\sqrt{3}\,E_S$
QBC Configuration	E_S	E_S	E_S

6. Conclusion

In this paper a new type quasi-buck four-quadrant topology is described in detail on the basis of the analysis to the traditional half-bridge four-quadrant, and the dc-side voltage of the topology can be reduced by about 50%. The MATLAB simulation and experiment results show:

(1) The quasi-buck four-quadrant topology is correct, reasonable and feasible. Though one more inductor L1 is employed in this topology, the lower limit value of the dc-side output voltage can be reduced by about 50%, and the lower limit of one-phase and three-phase converters composed of QBC unit can be reduced by 45% compared with the traditional bridge structure. Thus this topology is especially appropriate for the application of lower dc-side voltage.

(2)The SSA model of QBC unit is correct, and it stands for the practical operating characteristic of the high switching frequency basically.

(3) The analysis results of QBC unit model match up with the MATLAB numerical simulation and experiment waves, and the results describe the relationship between each state variable and circuit parameters quantitatively to supply the important theoretical basis for the QBC system parameter design.

7. References

[1] Jiasheng Zhang, Rongtai Hao. A New Type PWM Reversible Rectifier [J]. *Transactions of China Electrotechnical Society*, 1998, 13(5):37-41

[2] Jiasheng Zhang, Rongtai Hao. State Space Averaging Model and Simulation of the 4-Quadrant Converter[J]. *Joural of Northern Jiaotong University*,1998, 22(1):67-72

[3] Jiasheng Zhang, Rongtai Hao. A New Type 4-quadrant Converter-Topology, Modeling and Simulation[J]. *Electric Drive for Locomotive*, 1997(6):15-17

[4] Jiasheng Zhang, Rongtai Hao. A New Type Topology of 4-Quadrant Converters-Modeling. Analysis and Simulation [J]. *Acta Electronica Sinica*, 1998, 26(11):73-76+72

[5] Wei Li. Improvement of Single-phase Four-quadrant Converter Dual-loop Control [J]. *Electrical Drive for Locomotives*, 2017(5):22-25+34

[6] Yijie Shi, Xiao Li and Jianxin Guo. Research on a New Type of SVPWM Control Strategy in Four-Quadrant Converter [J]. *Journal of North University of China (Natural Science Edition)*, 2016, 37(6):601-605

[7] Minyao Huang, Wenhuan Yang and Tiantian Hou. Three-Phase Double Pwm Four Quadrant AC Power Transmission System [J]. *Electronic Sci. & Tech.*, 2016, 29(5):183-186

IOP Conf. Series: Materials Science and Engineering **735** (2020) 012079 doi:10.1088/1757-899X/735/1/012079

Analysis and Experiment of Flow Field in Sugarcane Leaf Cutting and Returning Machine

Haiyan Dai[1], Yuxing Wang[1]*, Fujun Wen[1] and Yanqin Tang[1]

1 Collage of Engineering, South China Agricultural University, 483 Wushan road, Tianhe district , Guangzhou, 510642, China.

Email: scauwyx@scau.edu.cn

Abstract. In order to improve the sugarcane leaves cutting effect of electric sugarcane leaves returning machine, the sugarcane cutting process is modeled and simulated by ANSYS/LS-DYNA software. It shown that in the double-support cutting mode, the cutter roller ensures the multi-layer sugarcane leaves were cut at the rotate speed of 800r/min. During the cutting process of sugarcane leaves, the stress is mainly concentrated at the incision, and the sugarcane leaves on the side of the blade are subjected to greater stress. The smaller the distance between the rotary blade and the fixed blade of the same combination, the easier the sugar cane leaves are cut off. In the cutting test of sugarcane leaves, the cutting rate was 91.47%, which echoed the simulation results, indicating that modeling and simulation have reference value.

1. Introduction

Sugarcane is a kind of perennial plant. In the harvest season, a large number of sugarcane leaves are left in the field. They must be processed for their large length and high toughness; otherwise, they will affect the subsequent mechanized farming seriously. Burning on the spot is the simplest and cheapest way to process the sugarcane leaves, but the burning is likely to cause serious problems such as air pollution, nutrient loss, fire and traffic accidents, and so on [1]. The most effective way to solve the problems is mechanized sugarcane leaves returning to the field. Since the 80s of last century, research on mechanized sugarcane leaves returning to field technology have been conducted in China. 3SY-120, 3SY-140 and 3SY-180 types of sugarcane leaf crushing machines are designed and manufactured by Guangxi Agricultural Machinery Research Institute [2]. 1GYFH, 1GYF-120, 1GYF-150, 1GYF-200 and GYF-250 types of sugarcane leaf crushing and returning machines are developed by the Institute of Agricultural Machinery Research Institute of Chinese Academy of Tropical Agricultural Sciences [3]. The above research mainly focuses on the design and test of the whole structure of sugarcane leaf machine. Although the types of machine have a good effect on the pulverization of sugarcane leaves, the structure is complex, and needs a large amount of power which would cause large weight. It has a serious compaction effect on the soil and affects the growth of sugarcane perennial roots. To solve this problem, research on remote-controlled self-propelled sugarcane leaf cutting and returning machine are conducted in South China Agricultural University. Flexible knives to cut the sugarcane leaves into pieces are used on the machine. Low power, light weight and having nice operating environment are the advantages of the machine, but it is found that the cutting effect of sugarcane leaves is not ideal for the winding caused by the self-excited airflow through experiments. In order to reduce the winding of sugarcane leaves and seek the best way to work without changing the structure, the internal flow field under the different working modes of the machine was analyzed by using the LS-DYNA software and the arbitrary Lagrange-Euler method (ALE) [4-9]. Finally, the wind speed test and the high speed photography [10-15] were carried out to verify the accuracy of the simulation results.

1.1. The Structure of The Whole Machine and Its Working Principle and the Structure of Sugarcane Leaf Returning Machine and Its Working Principle

The sugarcane leaf cutting and returning machine adopts crawler type chassis, and its structure is shown in Figure 1. It is mainly composed of battery, walking mechanism, cutting device, frame, dome and so on. The walking mechanism mainly consists of walking motor, driving wheel, caterpillar track, tensioning wheel and driven wheel. There are two walking motors, which are installed on the side of the left and right wheels to drive the wheels. The cutting system includes working motor, belt 1, belt 2, belt wheel 1, belt wheel 2, main knife roller, driven knife roller, knife plates and flexible knives. The outer end of the working motor is equipped with a belt wheel, each of which is installed with two knife plates, and the flexible knife is consolidated on the knife roller through the knife plate.

In the process of the returning machine, the electric energy of the battery can be distributed to supply the walking motor and the working motor. The size and direction of the walking motor's speed can be adjusted by the hand-held remote controller, and the forward, backward, left turn and right turn motions of the returning machine can be realized. When the two motor is moving clockwise at the same speed, the returning machine is straight forward; when the two walking motor is moving anti-clockwise at the same speed, it is straight backward; when the rotational speed of the left motor is less than the right speed, it is turned left movement; conversely, right turn movement.

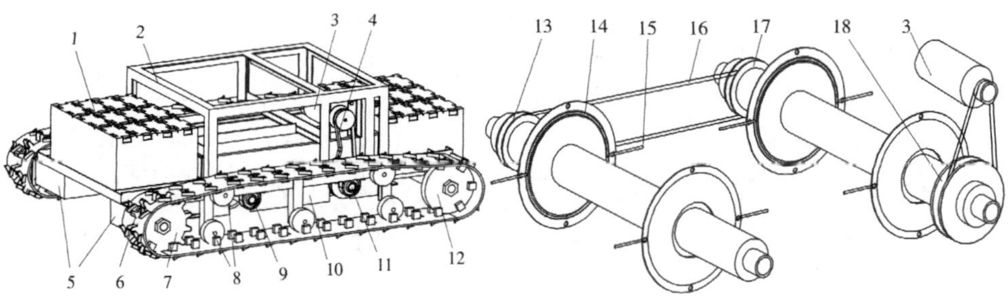

a. The structure of the returning machine b. The structure of the cutting system

1. Battery 2. Frame 3.Working motor 4. Belt 1 5. Walking motor 6. caterpillar track
7. Driving wheel 8. Tensioner pulley 9. The driven knife roller 10.Dome
11. The main knife roller 12. Driven wheel 13. Pulley 1 14. Knife plate
15. Flexible knife 16. Belt 2 17. Pulley 2 18. Pulley 3

Figure1. Engineering Drawing of Sugarcane Leaf Cutting and Returning Machine

When the working motor is started, the motor passes the power to the main knife roller through the belt wheel and the belt, and the main knife roller passes the power to the driven knife roller through the belt driving. The knife plates mounted on the knife rollers rotate at high speed. It is used to drive the flexible knife to beat sugarcane leaves, thus cutting off them. In order not to damage the ratoon, the cutting knife is made of flexible PA66 material. During the working process, the cutting device and the dome form a common working space.

1.1.1. Whole Machine Parameter

The length of sugarcane leaf is generally 80~150cm, and the width is about 2~8cm [16]. According to agronomic requirements, if the length of the leaves is less than 25cm, it can get good effect of sugarcane leaf returning to field, and achieve reutilization of resources. Based on the design requirements, the main structural and technical parameters of the machine are shown in Table 1.

Table 1. Structure and Technical Parameters of Sugarcane Leaf Cutting and Returning Machine

structure and technical parameters	design value
shape size (long × width×height)	1186×800×404
rated power of walking motor/kw	350
rated power of working motor/kw	400
curb weight/kg	≤150
the rotating speed of the main roller / $r{\cdot}min^{-1}$	3000
the rotating speed of the driven roller/$r{\cdot}min^{-1}$	2800
working speed/ $m{\cdot}s^{-1}$	0.4
operation width/mm	800
number of knife rollers	2
number of flexible knifes	4
number of batteries	24

2. Simulation Model and Related Settings

2.1. Basic Assumptions of Simulation

1) Assuming that the machine walks without wind, the speed of the returning machine is 0.4m/s, then the wind caused by travel is neglected in initial state.

2) The self-excited wind field generated by the rotary motion of the flexible knife has little influence on the ground, the knife roller and the dome, so the three parts are regarded as a rigid body.

3) In reality, the volume of the ground and air is endless, but the surface and air outside the dome have little influence. Therefore, only the finite volume that surrounds the whole dome, the knife rollers and the flexible knives can be considered for the air model.

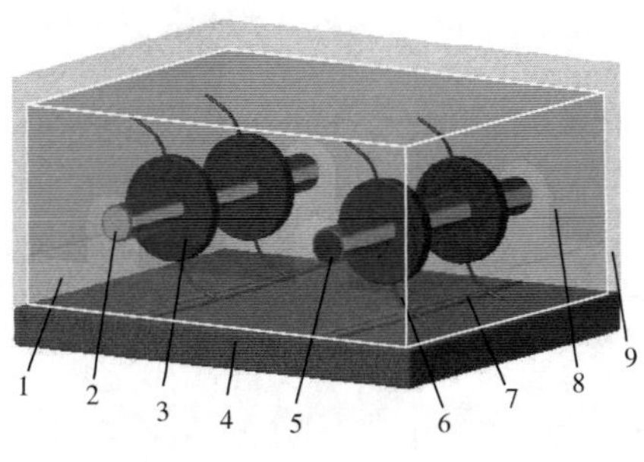

1. Air outlet 2. The main knife roller 3.Knife dish 4. Ground
5. Driven Knife roller 6. Flexible knife 7. Sugarcane leaf 8. Dome 9.Air

Figure 2. Simulation Model

The simplified structure is shown in Figure 2. The dome is located outside of the two knife rollers, wrapping the cutting device, the distance is 50mm from the ground to the bottom of the dome, and the gap is the air outlet of the self-excited wind field. The finite volume which is enclosed by the dome, the floor, the knife rollers and the flexible knives, is the volume corresponding to the self-excited wind field.

2.2. The Establishment of Finite Element Model

2.2.1. The Knife Roller, Dome and Ground Material Model

According to the assumptions of the model, the structural components and fluid inside the dome are mainly considered, the material model of knife roller and knife plate adopts rigid material model

MAT_RIGID, the type is set to solid164 unit with a density of 7820 kg/m^3, a modulus of elasticity of 207GPa, and a Poisson's ratio of 0.28. Flexible knives use nylon PA66 material, select solid164 unit, set material density is 1400kg/m^3, shear modulus is 3.01GPa. The dome model is set to shell163 element with a thickness of 2mm and 304 stainless steel with a density of 8000 kg/m^3, a modulus of elasticity of 210GPa, and a Poisson ratio of 0.3.

2.2.2. Air Material Model

The air is regarded as an ideal gas with a density of 1.25 kg/m^3 and a dynamic viscosity coefficient of 1.7456× 10^5 N·s/m^2. The equation of state is defined as a linear polynomial; the air state equation is that:

$$p = C_0 + C_1\mu + C_2\mu^2 + C_3\mu^3 + \left(C_4 + C_5\mu + C_6\mu^2\right)e_{ipv0} \tag{1}$$

The C0~C6 are the constant coefficients, and eipv0 is the initial internal energy. Because it is an ideal gas, there is an initial pressure, but no initial pressure input in the parameters of the equation of state. It should be converted into the initial internal energy input, and the initial internal energy is input as follows:

$$P = (\gamma - 1)\frac{1}{v_{rel}}e_{ipv0} \tag{2}$$

The γ is the specific heat of the gas, the value is 1.4 kJ/(m^3·℃), the relative volume is 1, and the P is the gas pressure. The initial internal energy e_{ipv0} can be calculated to be 2.5331× 10^5 J. To ensure the calculation progress, and to reduce the calculation time and divide the air grid, the number of grid units is controlled by line size, and the size of the air mesh in contact with the flexible knife is smaller and the grid is much denser. In order to prevent leakage, the mesh density of the dome is larger than the air.

2.2.3. Definition of Fluid-Solid Coupling and Setting of Initial Conditions

After establishing the simulation model, we set up the related constraints and contact conditions, add the fluid-solid coupling keywords, set up the air as the main component; combine the rigid body such as the ground, the dome, the knife rollers, the knife plate and the flexible knives, set them into the subordinate parts, and use the penalty function method to get the fluid-solid coupling.

In order to simulate the infinite volume of air and ground, the boundary between the air and the ground model is set to the non-reflecting boundary. When the flexible knives are moving, it produces large deformation which influences the flow field. So the fluid-solid coupling is considered, and the air is given as the ALE model, so that Arbitrarily Lagrange-Euler algorithm is used [8, 17-18].

Before setting the speed of the knife rollers, the speed of the knife rollers is test. The result shows that, the motor's rated speed is 5000 r/min, when fully loaded; the maximum speed is 3000 r/min, so the main knife roller's speed is set to be 3000 r/min. The transmission efficiency of the drive belt is 93.3%, so the driven knife roller's speed is 2800 r/min. The speed curves of the two rollers are defined respectively, and the rotation directions of the cutter rollers are respectively set in three situations as follow:

1) Both of the knife roller clockwise motion

2) The main knife roller rotates anticlockwise with the clockwise movement of the driven knife roller

3) The main knife roller rotates clockwise with the anticlockwise movement of the driven knife roller

3. Calculation Results and Analysis to Them

3.1. Velocity Vector Analysis under Different Working Modes

The flexible knife rotates in the middle plane of the knife plate, so the flow field in this plane is analyzed at different modes. As shown in Figure 3, the velocity vector of the flow field is analysed under three different working modes, when the flexible knife moves to 0.10 s and 0.118 s. When the flexible knife rotates under the third working mode, 4 vortices are formed at the two different times respectively. Eddy current will disturb the flow field, so the maximum flow field disturbance formed by the third working modes.

The number of sugarcane leaves left on the ground is different, and the thickness of the accumulated leaves is about 0~40 mm. The sugarcane leaves which are closer to the left and right outlet are often taken out of the body and not easily to cause entanglement. Therefore, the velocity vector is analyzed at 0~40 mm height from the ground. According to Figure 3, the area of the red box which is close to the ground between the two knife rollers is amplified as that in Figure 4.

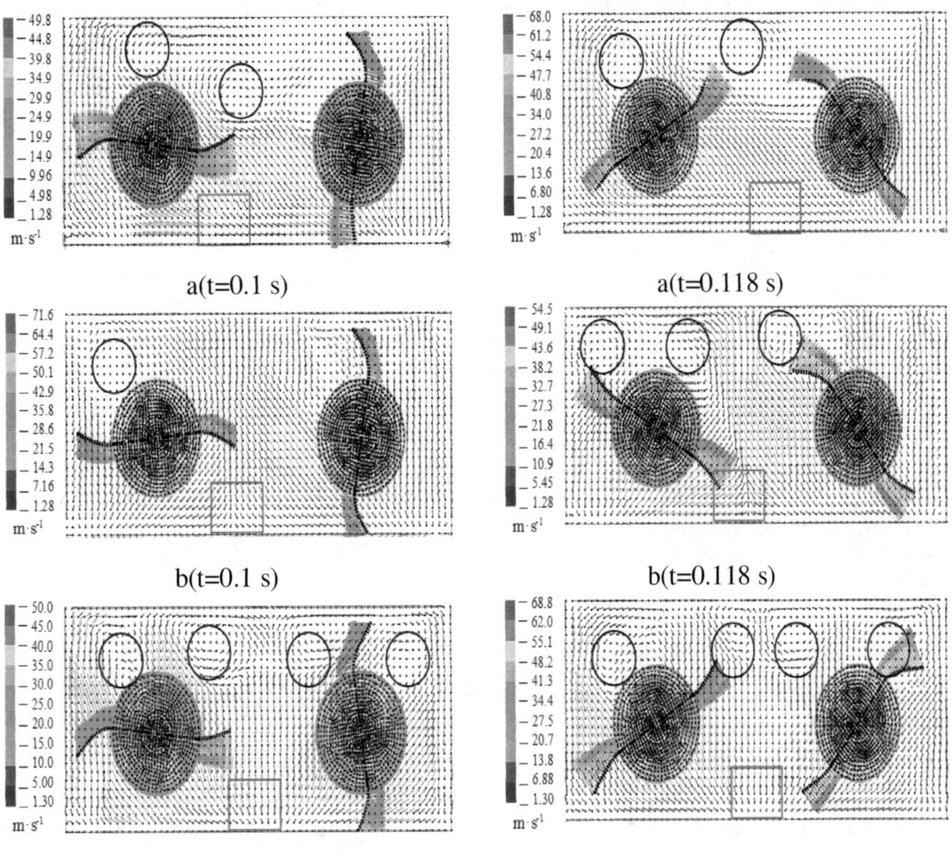

c(t=0.1 s) c(t=0.118 s)

(a)Both of the Knife Roller Clockwise Motion

(b) The main knife roller rotates anticlockwise with the clockwise movement of the driven knife roller

(c) The main knife roller rotates clockwise with the anticlockwise movement of the driven knife roller

Figure 3. Velocity Vector

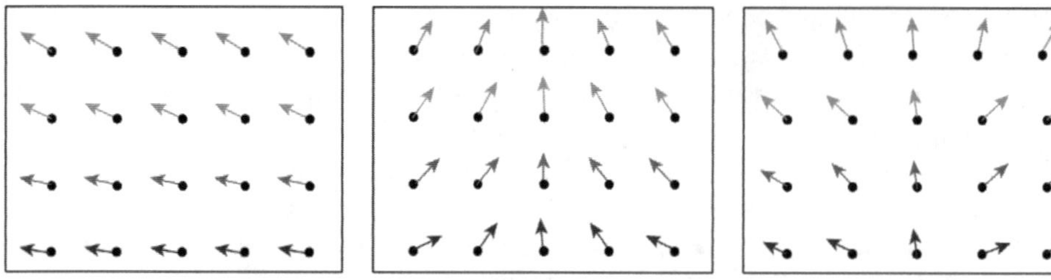

(a) Both of the knife roller clockwise motion

(b) The main knife roller rotates anticlockwise with the clockwise movement of the driven knife roller
(c) The main knife roller rotates clockwise with the anticlockwise movement of the driven knife roller

Figure 4. Speed Vector Local Large Map

The result shows that: With the same direction clockwise rotation, the velocity direction of the region is above the left, and the angle is changed in the range of 10 ~30 degrees. The angle and the height have an increasing function relationship, the greater the height, the larger the angle. When the main knife roller rotates anti-clockwise with the driven knife roller clockwise rotation, the angle change is in the range of about 20 ~90 degrees, the velocity vector direction near the side of the main knife roller is inclined to the right, and the other side velocity vector is inclined to the left. Meanwhile, the velocity vector is basically vertical between of the two knife rollers, equivalent to the Y direction. When the main knife roller rotates clockwise with the driven knife roller anti-clockwise rotation, the speed vector is changed in the range of about 30 ~90 degrees. The velocity vector direction near the side of the main knife roller is inclined to the left, and the other side velocity vector direction is inclined to the right. The direction of velocity vector is exactly opposite to the second working methods, and the middle position is basically the same, which is vertical upward. According to the above analysis, the third working mode is more likely to cause disturbance in the flow field, and the direction of velocity vector in the middle region is more likely to cause sugarcane leaves to be swirled.

3.2. Analysis to the Velocity of Different Height

In order to discuss the problem further, 4 different height's velocities were analyzed respectively. The distance from the ground is 10mm, 20mm, 30mm and 40mm, every height in the horizontal direction is taken at the 10mm distance respectively, 30 points are taken on every line, so 120 nodes are obtained on 4 heights. As shown in Figure 5, the speed of every node is analyzed. Because the vertical direction speed will greatly influence the blowing of the sugarcane leaves, the Y direction speed of every node is extracted in the whole motion process of the system respectively. And then, the maximum speed of every node in the direction of Y in the movement process is found.

According to the previous analysis, the third working mode is easily caused by the flow field disturbance and the blowing of the sugarcane leaves. Therefore, only the two modes are analyzed: both of the knife rollers rotate at the same clockwise direction, and the main roller anti-clockwise rotates with the driven knife clockwise rotation.

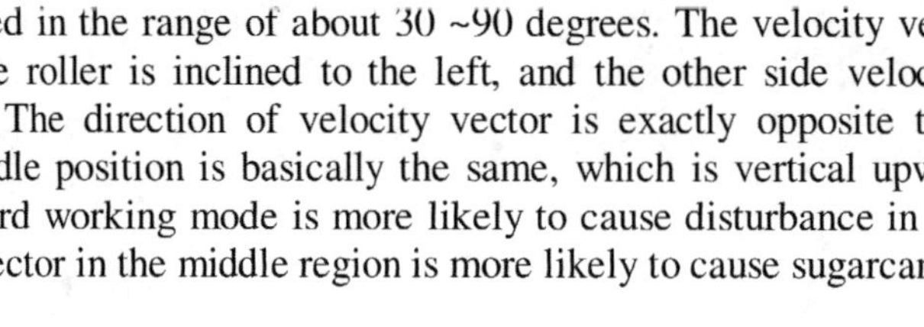

Figure 5. Key-point Position

In order to describe the maximum velocity change at the same height and position more clearly, Gauss function is applied to fitting the data points. The Gauss function fitting equation is as follows:

$$v_y = y_0 + k\, e^{-2\left[(x-a)/b\right]^2} \tag{3}$$

Two different working modes and four different height conditions correspond to eight sets of data. Every group of data is fitted by Gauss. The coefficients in the fitting equation are shown in Table 2:

Table 2. The Coefficient Value

Height	Working mode	y_0	k	a	b
10mm	1	0.744	26.676	18.510	5.680
	2	4.213	43.623	15.856	6.866
20mm	1	1.201	27.684	15.384	10.394
	2	3.144	44.655	15.578	6.487
30mm	1	3.137	44.734	15.567	6.495
	2	0.106	27.932	14.988	10.304
40mm	1	12.649	29.056	2.142	1.072
	2	0.004	51.198	15.646	7.615

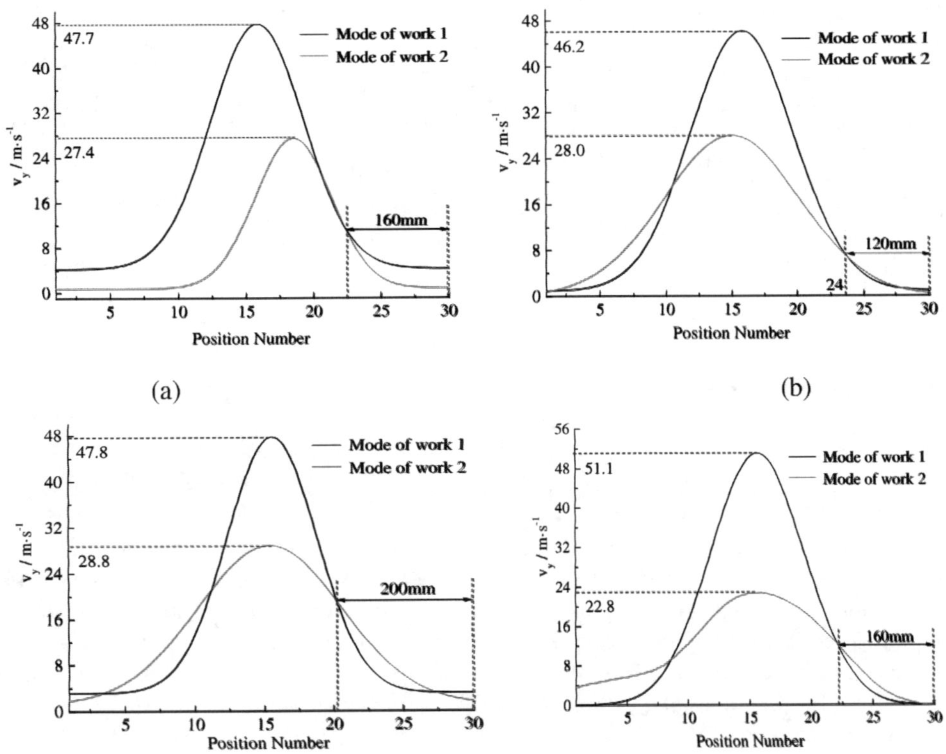

(a) (b)

(c) (d)

(a)10mm (b)20mm (c) 30mm (d) 40mm

Figure 6. Maximum Velocity Variation in Y Direction at Different Positions

As shown in Figure 6, the working mode 1 is the anti-clockwise rotation of the main knife roller with the clockwise rotation of the driven knife roller, and the working mode 2 is the clockwise rotation of the two rollers at the same time. At first, the fitting curve shows that the maximum velocity of the same height is different from left to right in the whole period. The speed near the outlet of the left and right sides is smaller, and the speed of the middle position between the two knife rollers is the largest, the reason is that the rotation of the flexible knife causes the superposition of the velocity in the middle of the part.

Secondly, in the right side which is near the outlet of the air outlet, the maximum speed of the Y direction is close in the two working modes at the same height, but there is a large difference in the middle area. The superposition of the self-excited wind field caused by the anti-clockwise rotation of the main knife roller with the driven knife roller clockwise rotation is obvious, and the difference is large compared with the clockwise rotation of the two knife rollers.

The maximum speed is found at every height, as shown in Table 3. It is obvious that the maximum speed of the Y direction increases with the height increasing in the 10~30mm under the working mode 1; and the velocity increases with the height increasing in the 20~40mm under the working mode 2. In addition, at the same height, the maximum speed of the Y direction produced under the working mode 1 is greater, the maximum velocity is more than 50m/s when the height is in 40mm, and the difference between the two is up to 55.4%.

Combined with the above analysis, it is found that the working mode 2 is more beneficial than the working mode 1 to reduce the flow field in the body and reduce the effect of blowing and winding of the sugarcane leaves.

Table3. Maximum Velocity of Y Direction m/s

	Different heights (mm)			
	10mm	20mm	30mm	40mm
The working mode 1	47.7	46.2	47.8	51.1
The working mode 2	27.4	28.0	28.8	22.8
The ifference value	42.5%	39.4%	39.7%	55.4%

Table4. Results of Wind Speed m/s

	Different heights (mm)			
	10mm	20mm	30mm	40mm
Left outlet	1.974	3.137	3.432	3.928
Middle position	0.766	0.790	0.958	1.024
Left outlet	-1.235	-1.289	-1.35	-1.367

4. Test Verification and Analysis

4.1. Wind Speed Test

To verify the Modeling and the simulation result, the wind speed measurement of the returning machine is carried out. The test principle is shown in Figure 7.

The velocity is measured under both of the rollers clockwise rotation. The rotational speed is in accordance with the simulation parameters, and the GM8903 hot wire anemometer is selected to measure the wind speed of each key point. Due to the limitation of the installation position of the dome, the wind speed can only be measured under the dome and near the ground. 3 measurement positions are selected; the positions are the left, the middle and the right. Four points are taken at the height of 10mm, 20mm, 30mm and 40mm respectively, in the vertical direction of every position. A total of 12 points are measured. In order to improve the accuracy of measurement, the test was repeated 3 times at every position, and the average value was obtained. The test device gets a set of data every 0.5s, it is equivalent to the average speed of a certain position in every 0.5s. The results of the test are shown in Table 4:

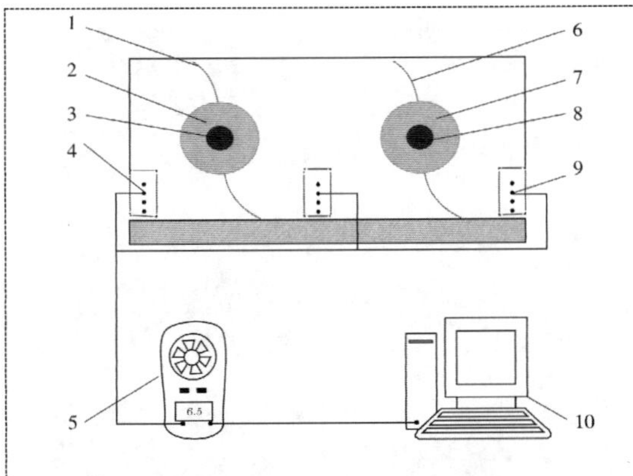

1.№1 flexible knife 2.The main knife roller 3.№1 knife dish 4.№1 test point 5.Anemometer 6.№2 flexible knife 7.The driven knife roller 8.№2 knife dish 9. №2 test point 10. Analysis software

Figure 7. Experimental Schematic Diagram

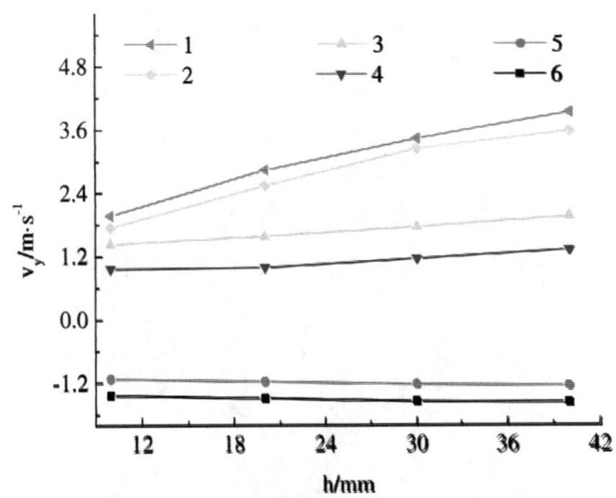

1.Left simulation value 2.Left test value 3.Middle simulation value 4.Middle test value 5.Right simulation value 6.Right test value

Figure 8. Test Value and Simulation Value

Compared with the simulation value, the experimental values of the Y direction of the average wind speed are shown in Figure 8:

The results show that, when the two knife rollers rotate clockwise at the same time, the change trend of the test wind speed is the same as the simulation analysis's result, and it increases with the increase of the height. The test value of the wind speed in each position is close to the simulation value. The maximum gap appears at the middle distance of the ground 20mm positions. The difference between the velocity simulation value and the test value is 17.7%, which may cause certain disturbance to the internal flow field of the anemometer, the instrument itself measure also produces certain deviation, and the deviation is less than 20%, it is in the acceptable range. So the Modeling and the simulation are all reliable.

4.2. High Speed-photography Test

In order to further prove that the both of the knife rollers clockwise movement can reduce the winding of sugarcane leaves, the high-speed photography is carried out to analyze the motion law of sugarcane leaves in the internal flow field.

4.2.1. Experimental Material

The sugarcane leaves were gotten from HEIPI sugarcane variety at Panyu, Guangzhou, China. The harvest time was February in 2018. The leaves of the sugarcane leaves were complete and no breakage, the total length of the leaf (including the leaf sheath) was 80~150 mm.

4.2.2. Experimental Equipment

The main equipment for the high-speed photography test is the returning machine, high-speed camera, flash lamp and computer. The field test is shown in Figure 9. The high-speed photography equipment is M/R/LC110 high-speed camera; the shooting rate is set to 500 frames / sec in the test. In order to avoid the problem of unclear shooting caused by insufficient internal light, two flash lamps are placed near the returning machine. It is equipped with a notebook computer with PCC 2.6.749.0 program. In the experiment, the yellow wire is connected to the computer to realize data transmission. The related parameters of the high-speed camera can be adjusted by the computer, and the results are post-processed.

1. Dome 2. Flash lamp 3. Power cord 4. Network line 5. Returning mechine 6. High-speed camera 7. Computer

Figure 9. High Speed Photography Field Test Chart

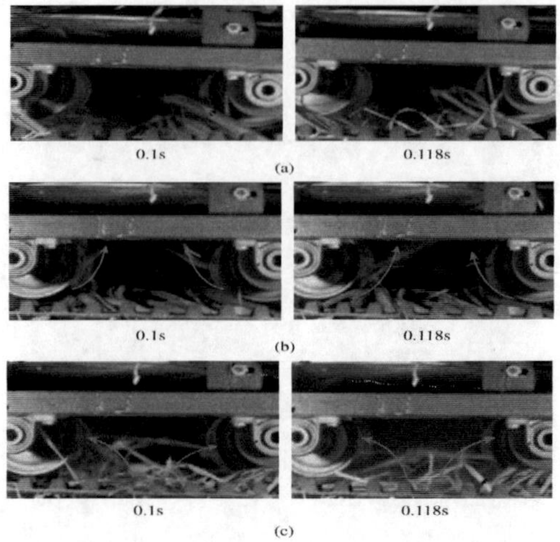

(a) Both of the knife roller clockwise motion (b) The main knife roller rotates anticlockwise with the clockwise movement of the driven knife roller (c) The main knife roller rotates clockwise with the anticlockwise movement of the driven knife roller

Figure10. The Movement State of Sugarcane Leaves under Different Working Conditions

4.2.3. Test Method

On the clean and flat ground, there is a regular flat layer of tested sugarcane leaves in parallel direction with the knife roller, start the sugarcane leaf returning machine and drive it to the ground covered with sugarcane leaves. The flash lamp and the high-speed camera are placed on the same side of the caterpillar track and have a certain sharp angle with it. It ensures that the flow field is light enough and layout of the high-speed camera to meet the requirements of the test area. In accordance with the simulation test, high speed photography tests are carried out under the three different working modes. The speed of the main knife roller is 3000 r/min.

4.2.4. Test Results and Analysis to Them

The results at 0.1s and 0.118s were selected for analysis under different working modes. The motions of the sugarcane leaves are shown in Figure 10.

When the two knife rollers rotate clockwise at the same direction, the sugarcane leaves on the left of the main knife roller can be discharged along the left side; the sugarcane leaves are flying in the middle area of the two knife rollers, Flexible knives further attack and cut the flying sugarcane leaves during the course of rotation, it doesn't the lead sugarcane leaves to twine the knife roller. When the main knife roller rotates clockwise with the driven knife roller anti-clockwise rotation, the sugarcane leaves gather in the middle area. After a period of time, the sugarcane leaves twist the knife rollers. When the main knife roller rotates anti-clockwise with the driven knife roller clockwise rotation, the movement of the sugarcane leaves in the middle of the two knife rollers is more and more disorderly, and there is a tendency to move to the two knife rollers respectively. At the end of the test, there are more sugarcane leaves on the knife rollers.

It shows that the results of high-speed photography coincide with the simulation movement of sugarcane leaves in the sugarcane leaf returning machine, and further proves that the same clockwise movement is beneficial to reduce the problem of winding caused by the blowing of sugarcane leaves.

5. Conclusion

1) Due to the influence of the self-excited wind field caused by the flexible knife on both sides, the internal flow field will produce certain eddy current in the three different working modes; it will cause the flow field to have a turbulent motion. When the main knife roller rotates clockwise with the driven knife roller anticlockwise rotation, there are more turbulent motions.

2) On the basis of the maximum velocity analysis in Y direction, the maximum velocity distribution at the same height from left to right shows a trend of first increase and then decrease. The speed of the two sides near the outlet is small, the speed of the middle area is large, and the peak appears in the middle position. When the main knife roller is revolving anti-clockwise with the driven knife roller clockwise rotation, the Y direction maximum speed in the middle is higher than both of the two rollers movement as the same clockwise rotation. The difference of them is up to 55.4%.

3) The Y direction velocity of three different positions is tested under the two knife rollers in the same clockwise movement. The test results are close to the simulation results. The maximum difference is 17.7%; it is in the acceptable range. The test verifies the accuracy and the reliability of the Modeling and the simulation results.

4) The high-speed photography test is consistent with the simulation results, it verifies the accuracy of the simulation results, and proves that the working mode of the same direction is beneficial to reduce the problem of blowing and winding of sugarcane leaves.

In summary, when the machine works, it is better to choose the working mode of the two knife rollers clockwise rotation, it is beneficial to reduce the blowing of the sugarcane leaves and reduce the winding of the sugarcane leaves as much as possible. This paper provides a reference for further optimization of the flow field in future.

6. Acknowledgement

Subject was supported by the Fund for the Science and Technology Program of Guangzhou (№. 201704020022).

*Corresponding author: Yuxing Wang

7. Reference

[1] Patcharaporn Pongpat, Shabbir H. Gheewala and Thapat Silalertruksa. 2017. An assessment of harvesting practices of sugarcane in the central region of Thailand, *Journal of Cleaner Production*, vol 142, pp 1138-1147.

[2] Jian Yang, Zhaoxin Liang, Jianlin Mo, Yueqian Jin and Meiying Gu. 2005. Vibration simulation of 3SY-140 sugarcane leaf-breaker. *Transactions of the Chinese Society for Agricultural Machinery*, vol 36, No11, pp74-77.

[3] Ming Li, Jinli Wang, Yiguo Deng, Hui Huang and Jin Zhang. 2008. Structural design and experiments on sugarcane leaf shattering and returning machine. *Transactions of the Chinese Society of Agricultural Engineering,* vol 24, No 2, pp121-126.

[4] Hirth A, Haufe A and Oluvsson L. 2007. Airbag simulation with LS-DYNA past - present - future. *Cheminform*, vol 24, No 9, pp394-395.

[5] Escalona J L. 2017. An arbitrary Lagrangian–Eulerian discretization method for modeling and simulation of reeving systems in multibody dynamics. *Mechanism & Machine Theory,* vol 112, pp1-21.

[6] Nassiri A and Kinsey B. 2016. Numerical studies on high-velocity impact welding: smoothed particle hydrodynamics (SPH) and arbitrary Lagrangian-Eulerian (ALE). *Journal of Manufacturing Processes*. Vol 24, part 2, pp376-381.

[7] He Jia, Wei Rong and Guoliang Chen. 2010. The simulation of parachute inflation process based on LS-DYNA software. *Spacecraft Environment Engineering*, vol 27, No 3, pp367-373.

[8] Helenbrook B T and Hrdina J. 2018. High-order adaptive arbitrary-Lagrangian-Eulerian (ALE) simulations of solidification. *Computers & Fluids*, vol167, pp40-50.

[9] Sanbin Lu, Liping Dong, Guang Chen, Junyuan Zhang and Zhichao Yang. 2009. Numerical simulation of curtain airbag deployment based on arbitrary lagrangian - eulerian algorithm. *Automotive Engineering*, vol 31, No 12, pp1158-1161.

[10] Hongxin Liu, Xiaomeng Xu, Junxiao Liu and Chen Wang. 2016. Working characteristics of vertical shallow-basin type seed-metering device based on high-speed photography and virtual simulation. *Transactions of the Chinese Society of Agricultural Engineering*, vol 32, No 2, pp13-19.

[11] He Xing, Ying Zang, Xiaoman Cao, Zaiman Wang and Xiwen Luo. 2015. Experiment and analysis of dropping trajectory on rice pneumatic metering device. *Transactions of the Chinese Society of Agricultural Engineering*, vol 31, No 12, pp 23-30.

[12] Zhichao Hu, Bing Wang, Zhaoyang Yu, Baoliang Peng, Yanhua Zhang and Lvke Tan. 2017. Design and test of semi-feeding test-bed for peanut pod picking. *Transactions of the Chinese Society of Agricultural Engineering*, vol 33, No 17, pp42-50.

[13] Liqing Chen, Dong Zhang and Wuwei Chen. 2014. Numerical simulation and test on two-phase flow inside shell of transfer case based on fluid-structure interaction. *Transactions of the Chinese Society of Agricultural Engineerin*, vol 30, No 4, pp 54-61.

[14] Xiangwei Mou,Yinggang Ou and Hao Wu.2012. Damage of sugarcane leaf sheath under action of elastic leaf-stripping elements based on high-speed photography. *Transactions of the Chinese Society for Agricultural Machinery*, vol 43, No 2, pp85-89.

[15] Xinping Li, Yuli Ma, Lianxing Gao. 2009. High-speed photograph analysis on threshing process of corn seed. *Transactions of the Chinese Society for Agricultural Machinery*, vol 40, No 11, pp46-49.

[16] Shenchang Huang, Wang Yang, Jian Yang and lei Liang. 2018. Study on finite element modeling method of sugarcane stalk-sugarcane leaf system. *Journal of Agricultural Mechanization Research*, vol 40, No 6, pp19-23.

[17] Waltz J and Bakosi J. 2018. A coupled ALE-AMR method for shock hydrodynamics. *Computers & Fluids*, vol 167, pp357-371.

[18] Sean Mauch, Dan Meirona, Raul Radovitzky and Ravi Samtaney. 2003. Coupled Eulerian-Lagrangian simulations using a level set method. *Computational Fluid and Solid Mechanics 2003*. Proceedings Second MIT Conference on Compurational Fluid and Solid Mechanics, June 17–20, 2003, pp1448-1453.

IOP Conf. Series: Materials Science and Engineering **735** (2020) 012080 doi:10.1088/1757-899X/735/1/012080

The Study of Thermal Environment during the Movement of Air-conditioned Train

Hongmin Li

School of Energy Science and Engineering, Central South University, Changsha, Hunan province, P.R, China
Email: lhongmin@csu.edu.cn

Abstract. During the movement of air-conditioned train, outdoor temperature is constantly changing with the shift of time and space, and the thermal environment of compartment is affected by periodic temperature variation. In this paper, we simulated outdoors thermal environment in the period of the movement of air-conditioned train, and obtained outdoor temperature wave, sol-air temperature waves and temperature waves of interior wall for all orientations of compartment. As the effect of periodic temperature variation, it leads to the variety of quantity of heat transmitted from enclosure to compartment. By calculated the attenuation and delay for temperature wave, the fluctuation of quantity of heat transmitted from enclosure is obtained. It will establish theoretical basis for the study of thermal comfort about compartment and real time regulation about air-conditioned unit.

1. Introduction

It is a hot topic about the study of thermal comfort in the field of air conditioning technique. 1980s, Fanger P O. et al studied the perception if draught in ventilated spaces [1,2], 1990s, there were many investigations about the influence that indoor temperature made on personal thermal comfort [3-6], and there were many scholars who had put forward some measures to improve indoor thermal environment [7,8]. However, up to now, there is little study about the influence that outdoors thermal environment makes on indoor thermal environment and comfort. Though in references [9], authors pointed out this question, but didn't do deep into it.

In China, air-conditioned train is the main means for passenger transport. The thermal environment of compartment has the direct influence to thermal comfort. Vast in territory, complicated climate and far travelling route for air-conditioned train, these all lead to outdoors environment parameters varying with time and space. The variety will affect on thermal environment and thermal comfort of compartment. In order to study the influence of outdoors thermal environment, above all, we should grasp the law of the variety of outdoors thermal environment during the movement of air-conditioner train. Only by grasping the law, can we obtain the quantity of heat transmitted from enclosure and take measures to improve the thermal comfort of compartment.

2. Imitation of Outdoor Air Temperature

During the movement of air-conditioned train, the variety of outdoor air temperature has a certain law. We take train T97 and T98 as the illustration to find the law. Table 1 and 2 show the main stations that these two trains pass, and figure 1 shows the specific position of these stations on map.

Table 1. Schedule for Train T97

Station	Beijing	Shi Jiazhuang	Zhengzhou	Wuchang	Changsha
Time	10.06	12.45	16.42	22.22	1.50
Station	Chenzhou	Shaoguan	Guangzhou	East of Guangzhou	Shenzhen
Time	5.20	7.01	9.15	9.31	13.10

Table 2. Schedule for Train T98

Station	Shenzhen	East of Guangzhou	Guangzhou	Shaoguan	Chenzhou
Time	15.00	17.47	18.49	21.02	22.41
Station	Changsha	Wuchang	Zhengzhou	Shi Jiazhuang	Beijing
Time	2.10	5.41	11.23	15.21	18.05

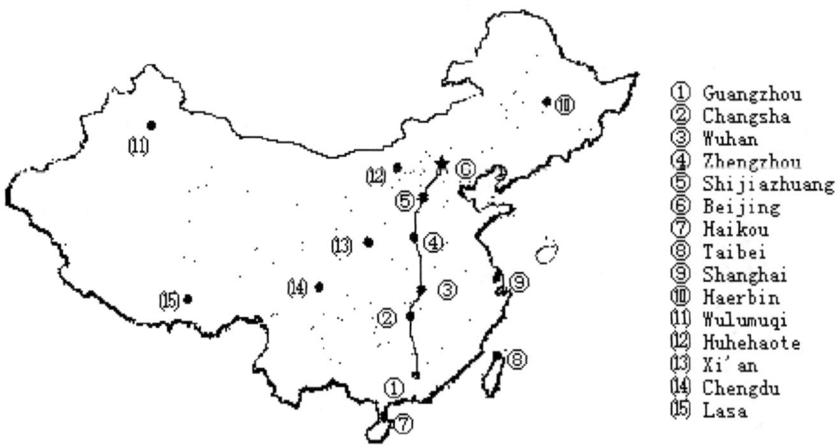

Figure 1. The Main Stations for Train T97 and T98

In order to grasp the law of periodic temperature variation, above all, we should know the average temperature and the instantaneous temperature during a running period of train (out and home once from jumping-off point to terminus as a period, to train T97 and T98, a period is 72 hours). We can obtain instantaneous temperature by formula (1) (during calculation, referenced design conditions come from design standards manual) [10].

$$t_{w,\tau} = t_{w,\max} - \alpha_\tau \Delta t_w \tag{1}$$

In this formula: $t_{w,\tau}$—— outdoor temperature at τ hour, °C; $t_{w,\max}$—— highest outdoor temperature, °C; α_τ—— modulus ratio; Δt_w —— difference in temperature between highest temperature and lowest one, °C.

Outdoor temperature has the characteristic of periodic variation. Its fluctuation can be expressed by harmonic wave. We use cosine function to express harmonic wave, formula (2) shows it [11].

$$t(\tau) = H\cos(\frac{360\tau}{Z} + \varphi) \tag{2}$$

IOP Conf. Series: Materials Science and Engineering **735** (2020) 012080 doi:10.1088/1757-899X/735/1/012080

In this formula: H —— Amplitude for harmonic wave, °C; τ —— starting time, hour; Z—— fluctuant periods, hour; φ —— original phase for harmonic wave, degree.

Actually, the variety of temperature involves disharmonic variables, and these variables also have their periodicity. We expand these varieties for Fourier progression, and use the sum of cosine functions to express it, and formula (3) shows it.

$$t(\tau)=H_0+H_1\cos(\frac{360\tau}{Z}-\varphi_1)+H_2\cos(\frac{2\times 360\tau}{Z}-\varphi_2)+H_3\cos(\frac{3\times 360\tau}{Z}-\varphi_3)+\cdots\cdots \quad (3)$$

In this formula: H_0 is the average value for temperature wave, and the second item of Fourier progression is the first step harmonic wave. Relative the first step, periods for all the other steps reduces 2, 3, 4…times.

Curve imitation is a method for data processing that uses a proper functional relation to express the inherent law of several discrete values. Fourier progression that is expressed by trigonometric function is an infinitude progression, and it is feasible to apply it to express seasonal function. In order to obtain the best outcome (simulative error smallest), we use 36 steps Fourier progression to simulate the fluctuation of temperature. Formula (4) is the mathematic model for curve imitation.

$$t=\frac{1}{72}\sum_{\tau=0}^{71}t_\tau+\sum_{n=1}^{N}\left\{\left[\frac{1}{36}\sum_{\tau=0}^{71}t_\tau\cos(5n\tau)\right]\cos(5n\tau)+\left[\frac{1}{36}\sum_{\tau=0}^{71}t_\tau\sin(5n\tau)\right]\sin(5n\tau)\right\} \quad (4)$$

We predigest formula (4) and obtain formula (5), it will benefit program [12].

$$t=\frac{1}{72}\sum_{\tau=0}^{71}t_\tau+\sum_{n=1}^{N}\left[\sqrt{M_n^2+L_n^2}\cdot\cos\left(5n\tau-arctg\frac{L_n}{M_n}\right)\right] \quad (5)$$

In this formula: N —— the sum of steps for harmonic wave; n -- step; M_n —— amplitude of cosine wave, $M_n=\frac{1}{36}\sum_{\tau=0}^{71}t_\tau\cos(5n\tau)$; L_n —— samplitude of sine wave, $L_n=\frac{1}{36}\sum_{\tau=0}^{71}t_\tau\sin(5n\tau)$.

Instantaneous temperature can be obtained by formula (1), and use formula (5) for curve imitation. We obtained outdoor temperature wave for train T97and T98. Fig 2 and 3 show the results.

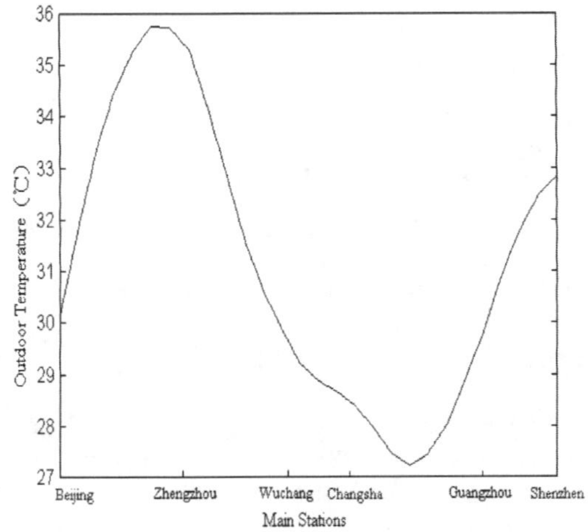

Figure 2. Outdoor Temperature Wave for Train T97

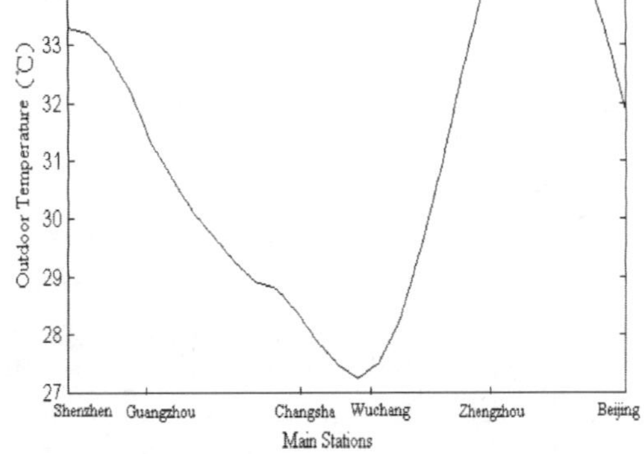

Figure 3. Outdoor Temperature Wave for Train T98

Fig 2 and Fig 3 show: to train T97 and T98, when they enter Zhengzhou, their outdoor temperature reach the maximum, and when they enter Changsha, outdoor temperature reach the minimum, the fluctuation of temperature is over 3 degrees.

3. Imitation of Sol-air Temperature

Sol-air temperature is a weather parameter that is synthesized by air temperature and solar radiation, and it can be obtained by formula (6) [10].

$$T_{zh,\ \tau} = T_{w,\ \tau} + \rho J(t) / a_w \tag{6}$$

In this formula: $T_{zh,\tau}$ — outdoor sol-air temperature in summer, °C; $T_{w,\tau}$ — outdoor instantaneous air temperature in summer, °C; ρ -- absorption coefficient of compartment for solar radiation; J(t) — solar radiation intensity, W/(m^2·°C); α_w — surface heat transfer coefficient for compartment, W/(m^2·°C), αw=9+3.5V0.66 [13], V is train velocity, km/h.

Sol-air temperature function can be expressed by formula (7).

$$T_{zh,\tau} = A_0 + \sum_{n=1}^{m} A_n \cos(\omega_n \tau - \varphi_n) \tag{7}$$

In this formula: A0 is the average value for sol-air temperature, the second item is the fluctuant value for instantaneous sol-air temperature.

Sol-air temperature is varies with different orientation and absorption coefficient for solar radiation. Solar radiation values that we used for calculation are each station's solar radiation value in design day [14]. Absorption coefficient of compartment is 0.7 (assume exterior is deep paint surface) [13]. Train velocity is 120km/h. After calculation, use formula (5) for curve imitation. We obtained outdoor sol-air temperature wave for train T97 and T98 when their movement. Fig 4 and 5 show the results.

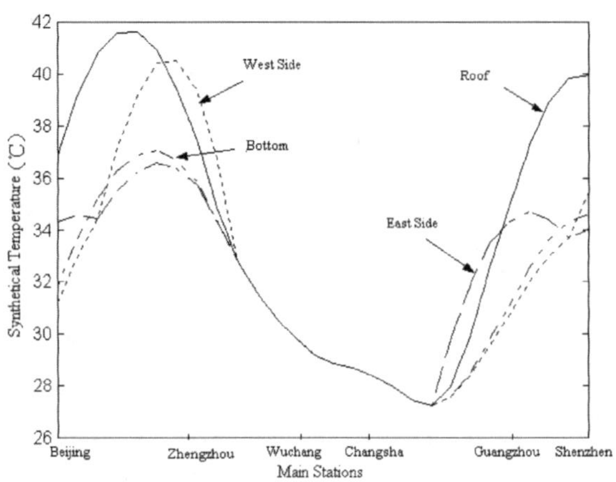

Figure 4. Sol-air Temperature Wave for Train T97

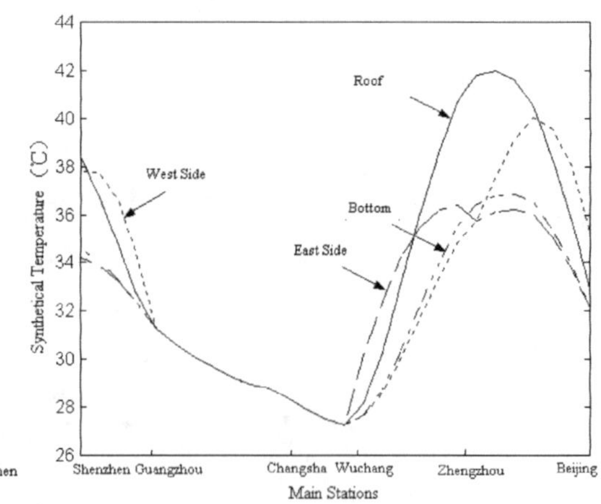

Figure 5. Sol-air Temperature Wave for Train T98

4. Imitation Temperature Wave for Compartment Surface

As compartment's wall has thermal stability, when outdoor sol-air temperature wave passing through wall to the inside surface, the wave will be delayed and attenuated. When temperature wave passing through a certain material, every step of Fourier progression will be attenuated and delayed, it is showed by formula (8), Fig 6 is temperature variation figure by unsteady heat conduction.

IOP Conf. Series: Materials Science and Engineering **735** (2020) 012080 doi:10.1088/1757-899X/735/1/012080

$$t(\tau)=t_{av}+\frac{H_1}{v_1}\cos(\frac{360\tau}{Z}-\varphi_1-\varepsilon_1)+\frac{H_2}{v_2}\cos(\frac{2\times360\tau}{Z}-\varphi_2-\varepsilon_2)+\frac{H_3}{v_3}\cos(\frac{3\times360\tau}{Z}-\varphi_3-\varepsilon_3)+\cdots\cdots \quad (8)$$

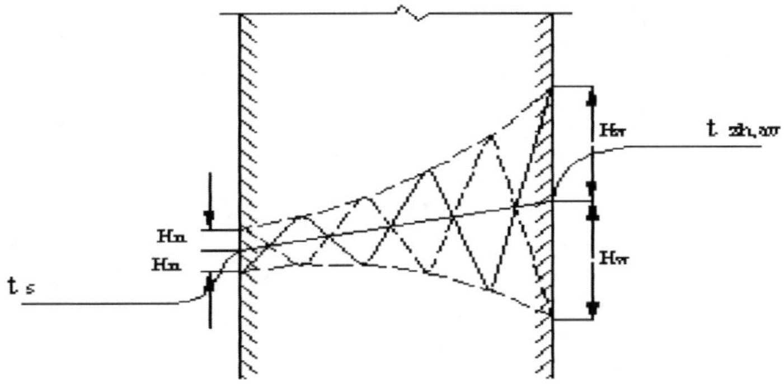

Figure 6. Temperature Variation at the Condition of Unsteady Heat Conduction

In this figure: $t_{zh,av}$ —— the average value for outdoor sol-air temperature; A_w — amplitude of temperature wave for exterior surface; A_n — amplitude of temperature wave for interior surface; t_s — average temperature for indoor air.

From reference [13] we know the whole constitution of wall for compartment. The main constitution of a compartment is following: roof is composed of 2mm steel plate, 70mm thermal insulation material and 5mm plywood; east and west of compartment are composed of 3mm steel plate, 65mm thermal insulation material and 10mm plywood; bottom is composed of 2mm steel plate, 85mm thermal insulation material and 20mm board. The main material for thermal insulation material is polystyrene. Table 3 shows the properties for all these materials.

Table 3. Properties for Materials

Material	Density (kg/m^3)	Specific Heat (kJ / kg.K)	Thermal Conductivity [W / (m·K)]
Plywood	600	2.513	0.151
Board	500	2.51	0.14
Polystyrene	25	0.837	0.035
Steel plate	7850	0.481	50

We should calculate delay and attenuation firstly so that we can obtain temperature wave for compartment surface. If periodic temperature variation influences on one side of a material, its surface temperature wave will has the same period. Storage heat coefficient of a material is the ratio of amplitude of rate of heat flow passing through surface to the amplitude of temperature wave for its surface, and formula (9) is the formula. Storage heat coefficient can show sensitive degree of a material to the fluctuation of heat flow. Material with big Storage heat coefficient has better thermal stability than that with small storage heat coefficient.

$$S=\sqrt{\frac{2\pi}{Z}\lambda C\rho} \quad (9)$$

In this formula: S — exterior storage heat coefficient of material, W / (m^2 · K); Z — the fluctuant periods of rate of heat flow, hour; λ, C, ρ — thermal conductivity of material [W / (m · K)], specific heat [(W · h) / (kg · K)], density (kg/m^3).

Interior storage heat coefficient is an index that can show surface's sensitivity to the fluctuation of heat flow. At the same condition, temperature of a material is more stable if its storage heat coefficient bigger. If inner material layer doesn't thick, the amplitude of inner surface temperature wave is not only affected by its physical property, but also affected by that of the following material. Formula (10) is the formula for the interior storage heat coefficient.

$$Y_n = \frac{R_n S_n^2 + Y_{n-1}}{1 + R_n Y_{n-1}} \tag{10}$$

In this formula: Y_n — interior storage heat coefficient for n layer material; R_n — thermal resistance for n layer material, m^2 · K/W; S_n — exterior storage heat coefficient for n layer materials.

$$v = 0.9e^{\frac{\sum D}{\sqrt{2}}} \cdot \frac{S_1 + \alpha_i}{S_1 + Y_1} \cdot \frac{S_2 + Y_1}{S_2 + Y_2} \cdot \frac{S_3 + Y_2}{S_3 + Y_3} \cdot \frac{Y_3 + \alpha_w}{\alpha_w} \tag{11}$$

In this formula: V — thermal amplitude decrement for wall; ∑ D — thermal inertia index, ∑ D = R1S1+R2S2+…+ RnSn.

In formula (11), the last multiplier $\frac{Y_3 + \alpha_w}{\alpha_w}$ is the attenuation for temperature wave from outdoor to exterior surface of wall.

$$\varepsilon = 40.5\sum D - arctg\frac{\alpha_i}{\alpha_i + Y_i\sqrt{2}} + arctg\frac{Y_e}{Y_e + \alpha_w\sqrt{2}} \tag{12}$$

In this formula: ε— delay for wall, degree; arctg -- inverse trigonometric function; Yi — interior storage heat coefficient for wall; Ye -- exterior storage heat coefficient for wall; α_i—coefficient of convective heat transfer for interior surface of wall, in this paper, it equals to 8W/ m^2 · K [13].

In formula (12), the last item is the delay of temperature wave for exterior wall to outdoor sol-air temperature wave. The following process can calculate exterior storage heat coefficient for each layer.

Exterior storage heat coefficient for the first layer:

(1) If D_1 >=1 then $Y_1 = S_1$; if $D_1 < 1$ then $Y_1 = \frac{R_1 S_1^2 + \alpha_{in}}{1 + R_1 \alpha_{in}}$;

(2) To the following layers (m>1):

If D_m >=1 then $Y_m = S_m$, if D_m <1 then $Y_m = \frac{R_m S_m^2 + Y_{m-1}}{1 + R_m Y_{m-1}}$.

The delay and attenuation for each step harmonic wave of each orientation can be obtained by formula (9), (10), (11) and (12).

The average temperature of exterior wall and interior wall can be obtained according to the process of steady heat conduction.

$$\begin{cases} (t_{zh,av} - t_{w,av}) \times \alpha_w = (t_{w,av} - t_{n,av}) \times \dfrac{1}{\sum R} \\ (t_{zh,av} - t_{w,av}) \times \alpha_w = (t_{n,av} - t_s) \times \alpha_i \end{cases} \tag{13}$$

In this formula: $t_{zh,av}$ — average of outdoor sol-air temperature, °C; $t_{w,av}$ — average temperature for exterior wall, °C; $t_{n,av}$ — average temperature for interior wall, °C; t_s — indoor temperature, °C, in this paper, it equals to 26 °C; $\sum R$ — the sum of thermal resistance for materials $(m^2 \cdot K)/W$.

Table 4. The Temperature and Thermal Resistance

Orientation	average of outdoor sol-air temperature(°C)	the sum of thermal resistance for materials from each layer $(m^2 \cdot K)/W$	average temperature of exterior wall(°C)	average temperature of interior wall(°C)
Roof	33.9275	2.0344	33.8877	26.4568
East side	32.6591	1.9234	32.6237	26.4042
West side	31.7929	1.9234	31.7617	26.3516
Bottom	31.1462	2.5715	31.1255	26.2376

By calculation, we obtained temperature wave of interior wall for train T97 and T98 during their movement. Fig 7and figs 8 show the results.

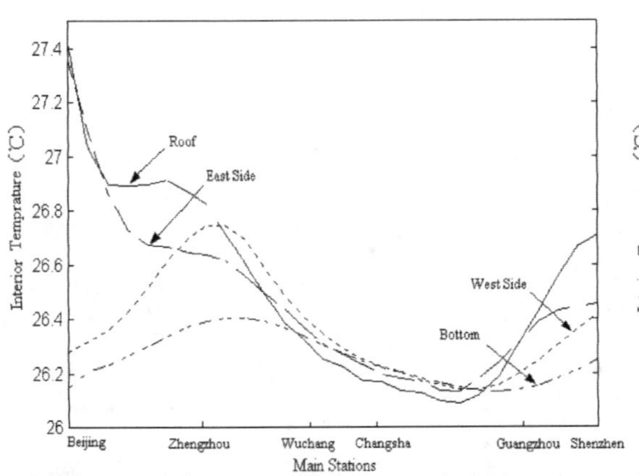

Figure 7. Temperature Wave of Interior Wall for T97

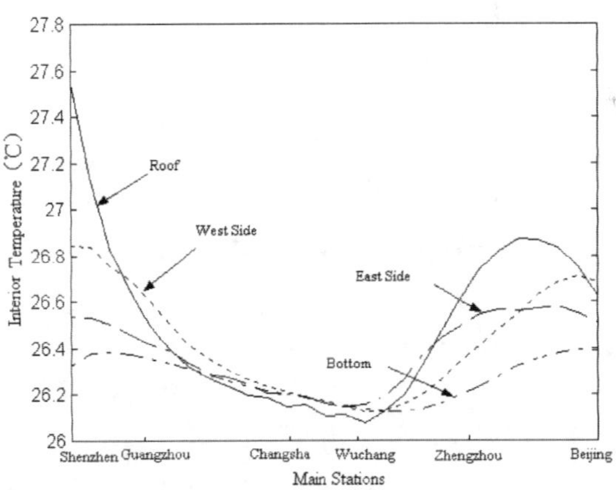

Figure 8. Temperature Wave for Train T98

Temperature wave of exterior wall can be obtained by calculated the delay and attenuation. Fig 9 and Fig 10 show the results.

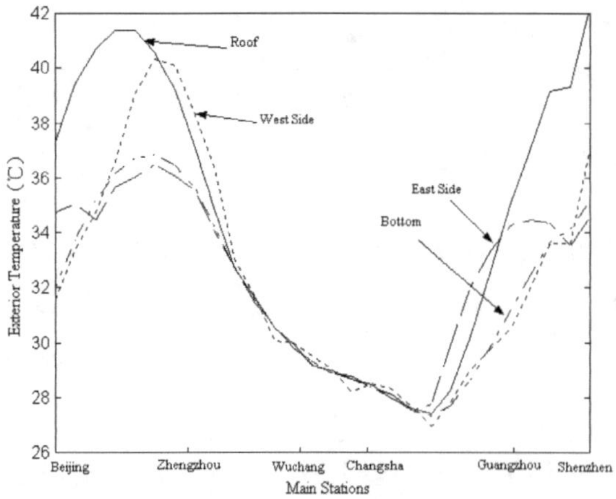

Figure 9. Exterior Temperature for Train T97

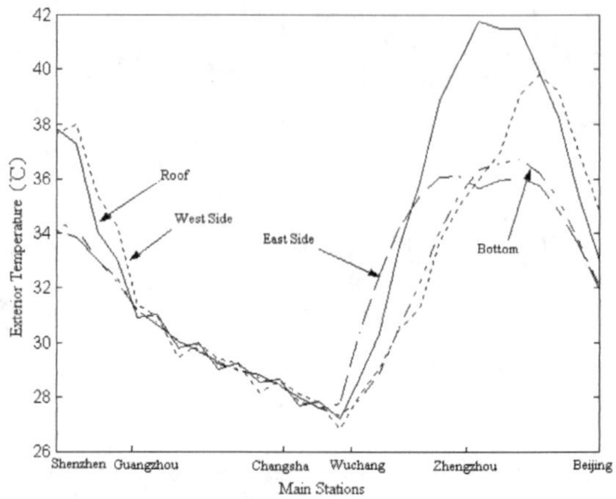

Figure 10. Exterior Temperature for Train T98

5. Quantity of Heat Transmitted from Enclosure

Quantity of heat transmitted from enclosure includes two parts, one is the quantity of heat transmitted from wall, and the other is from windows. Quantity of heat transmitted from wall can be obtained by formula (14).

$$Q = K(t_{zh,av} - t_s) + \alpha_{in}\frac{H_1}{v_1}\cos(\frac{360\tau}{Z} - \varphi_1 - \varepsilon_1) + \alpha_{in}\frac{H_2}{v_2}\cos(\frac{2\times 360\tau}{Z} - \varphi_2 - \varepsilon_2) + \cdots\cdots \quad (14)$$

In this formula: K -- overall coefficient of heat transfer for wall, W/ (m^2· K).

According to the delay and attenuation for each step harmonic wave and the formula for sol-air temperature, quantity of heat transmitted from wall to compartment can be obtained. Fig11 and 12 show the results.

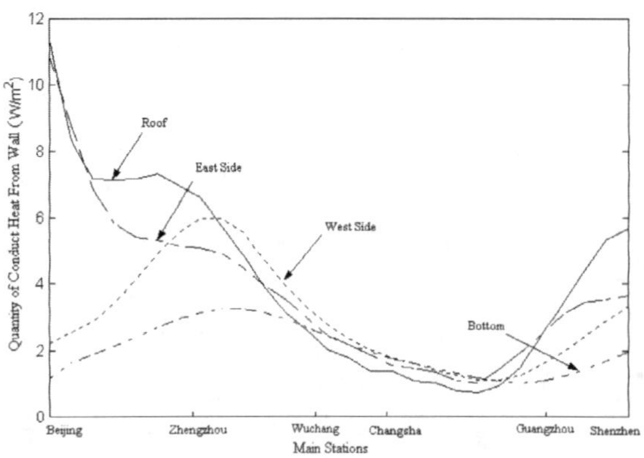

Figure 11. Quantity of Conduct Heat for T97

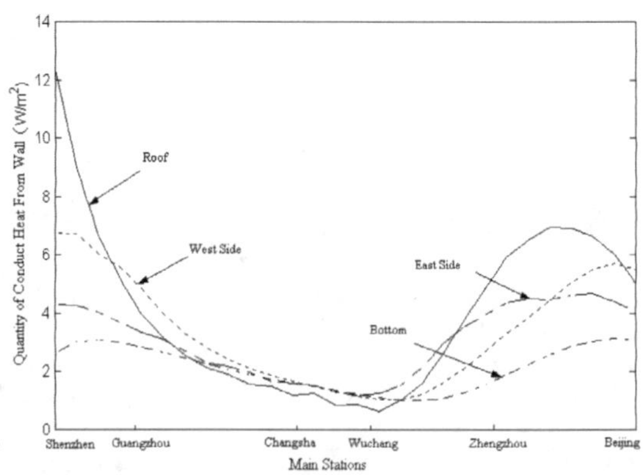

Figure 12. Quantity of Conduct Heat for T98

Quantity of heat transmitted from window is affected by solar radiation and the fluctuation of temperature. Solar radiation entering into compartment has two parts: one is that a portion of solar radiation gets straight into compartment with the form of short-wave radiation; the other is that windows improve temperature for itself by absorbed additional solar radiation, and then release the quantity of heat with the form of convection and long wave radiation. Quantity of heat transmitted from windows can be calculated by formula (15). Fig13 and 14 show the results.

$$Q = C_g C_n D_s + K_c(t_{zh} - t_s) \quad (15)$$

In this formula: Cg — shading coefficient for windows; Cn — shading coefficient for curtain; Ds — daily obtained heat factor.

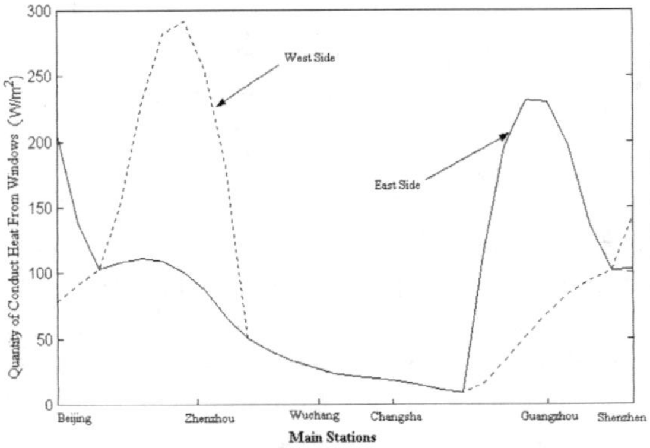

Figure 13. Quantity of Heat from Windows for T97

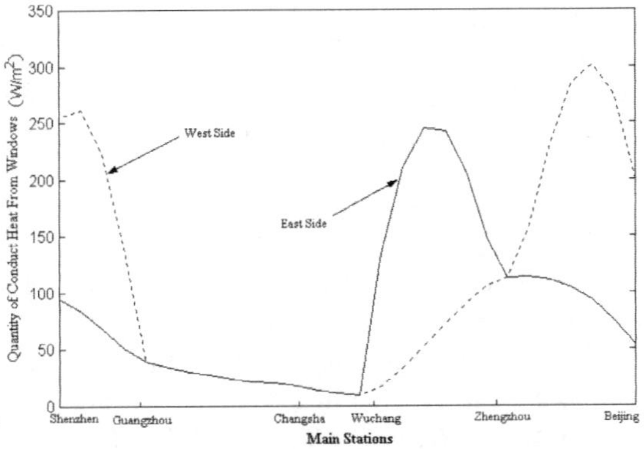

Figure 14. Quantity of Heat from Windows for T98

Fig (11), (12), (13), (14) show: at the same condition, quantity of heat transmitted from windows is bigger than that from wall, so different positions in compartment obtain different quantity of heat. It will lead to a condition that quantity of heat is not evenly distributing in compartment, and it would form superheat region (near window) and super-cooling region (apart from windows).

As the fluctuation of quantity of heat transmitted from window is big, even at the same region, the gained quantity of heat varies with time and space. In order to understand the influence that outdoors thermal environment make on the whole thermal load of compartment, the following task is obtaining the quantity of heat transmitted from the whole enclosure. To a certain compartment, it is 25.5meters length, 3.1 meters width, 2.42 meters height; the area of windows is 14m2 or so. On the basis of these conditions, we obtained the instantaneous quantity of heat transmitted from enclosure. Fig 17 and 18 show the results.

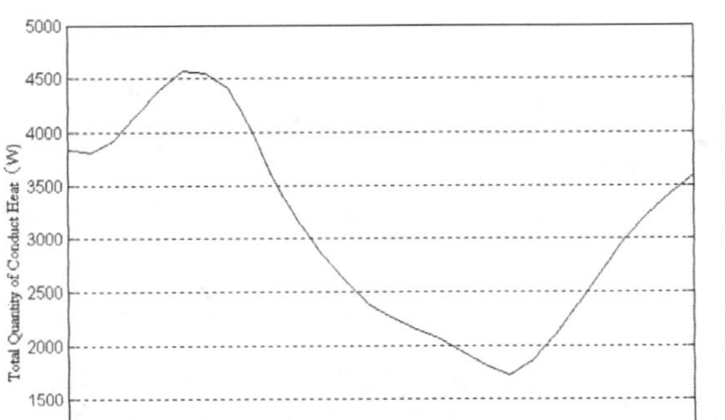

Figure 15. Quantity of Heat from Enclosure for T97

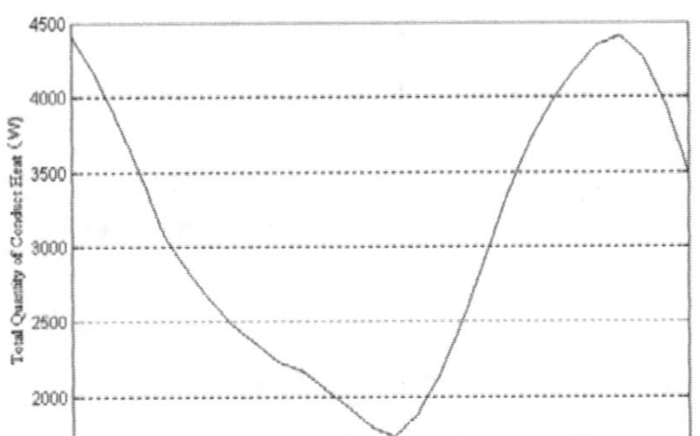

Figure 16. Quantity of Heat from Enclosure for T98

Figure (15) and (16) show: as solar radiation varying with time and orientation, and the fluctuation of outdoor temperature wave, these all lead to the quantity of heat transmitted from enclosure varied with time. To train T97, when it enter Zhengzhou, the quantity of heat reach the climax, at that time, we should supply much more quantity of cooling to restrain the increase of thermal load in compartment, when train enter Changsha, the quantity of heat decreased, this time, we should increase the quantity of cooling. To train 98, when it runs between Shenzhen and Wuchang, the quantity of heat gradually descend, at that time, we should relevant drop quantity of cooling, when train enter Wuchang, the quantity of heat increasing, this time, we should increase the quantity of cooling.

6. Conclusion

This paper simulated outdoors thermal environment of train, and obtained outdoor temperature wave, outdoor sol-air temperature wave for all orientations, temperature wave of exterior surface and interior surface, then at last, grasp the law of variation of quantity of heat transmitted from enclosure. According to the above study, we come to the following conclusion:

(1)Air-conditioned train is a movable building in a periods, it will pass through many different climatological districts, we shouldn't only consider a certain district when determining outdoor parameters, on the contrary, we should consider these climatological parameters synthetically, and make them corresponding to reality.

(2)The calculation shows: different train has different outdoors thermal environment, and the influence on indoor thermal environment also different, so it is different for the regulative scheme of refrigerating capacity.

(3)This paper offered a way to simulated the fluctuation of temperature and quantity of heat, by this way, we can grip the law for the change of air conditioning system load, and it also established theoretical basis for the study of thermal comfort for compartment and air conditioning system real time control.

7. References

[1] Fanger P O, NK. Christensen. Perception if draught in ventilated spaces. Ergonomics, 1986, v29 (2): 215~235.
[2] Fanger P O. et al. Air turbulence and sensation of dranght. Energy and Buildings, 1988, V12:21~29;
[3] deDear R J, A Auliciems. Validation of the predicted mean vote model of thermal comfort in six Austrilian field studies. ASHRAE Trans, 1985, V91 (2): 452~468.
[4] Schiller G E, et al. A field study of thermal environment and comfort in office building ASHRAE Trans, 1988, V94 (2): 280~308.
[5] Schiller G E, A comparison of measured and predicted comfort in office buildings. ASHRAE Trans, 1990, V96 (1): 609~622.
[6] Busch J F. A tale of two-populations: thermal comfort in air-conditioned and naturally ventilated office in Thailand. Energy and Buildings. 1992, 18: 235~249.
[7] D. J. Croome, G. Gan and H. B. Awbi. Thermal comfort and air quality in offices. Indoor Air 3, Proceedings of the 6th International Conference on Indoor Air Quality and Climate, Col 6, 37~42, 1993.
[8] P. Taffe. A qualitative response model of thermal comfort. Building and Environment, Vol 32, No 2, 115~121, 1997.
[9] Haghighat F, Allard F, Megri AC, Blondeau P, Shimotakahara R. Measurement of thermal comfort and indoor air quality aboard 43 flights on commercial airlines. INDOOR AND BUILT ENVIRONMENT 8 (1): 58~66 JAN-FEB 1999.
[10] Chen Peiling, Chao Suwei and Guo Jianxiong compile. Theory and Method for the Calculation of Air-conditioning Load. Publishing Company of Tongji University, 1987
[11] Chen Zaikang and Cai Zukang compile, Transference of heat at the condition of periodic temperature variation. Publishing Company of Chinese industry, 1964
[12] Chen Zaikang, etc. compile. Computer Method for heating and ventilation. Publishing Company of Chinese Architecture Industry, 1985
[13] Zhang Ying, etc. compile. Reference Manual for the Design of Vehicle. Publishing Company of Chinese railway, 1993
[14] Dan Jiping.Applied Method for the calculation of air-conditioning load. Publishing Company of Chinese Architecture Industry, 1989

[15] Edit group for architecture physics teaching material compile. Architecture physics. Publishing Company of Chinese industry, 1961
[16] Ye compile. Thermal Environment of architecture. Publishing Company of, 1996.

IOP Conf. Series: Materials Science and Engineering **735** (2020) 012081 doi:10.1088/1757-899X/735/1/012081

Drilling Fluid Leakage Model Based on the Dual-Medium Theory

Jinjiang Liu, Wenlin Wu, Peng Qian and Shuo Wang

Institute of Drill and Production, Liao He Oilfield Limited Company of CNPC, Panjin, 124100, China

Email: liujinj@126.com

Abstract. In view of the lost circulation in carbonate formation, based on the dual-medium model, a two-dimensional transient model for fracture-porous formation is established. All the factors that influence the leakoff rate in the fracture-porous formation are analyzed. The result shows that the leakoff rate increases as the normal stiffness of the crack decreases; the leakoff rate increases as the initial fracture aperture increases; The larger the pressure difference is, the larger the drilling fluid leakage is; The larger the porosity is, the more drilling fluid flows from fracture to matrix, the slower the leakoff descent is; The larger the permeability of matrix is, the larger of leakoff rate is at the initial stage of leakage.

1. Introduction

Lost circulation is one of the most common downhole problems during the drilling operation, especially in the fractured reservoir. According to statistics, in the development process of Fuling shale gas field, lost circulation occurs in 67.2% wells. Drilling fluid leakage increases the cost of drilling and even causes downhole accidents [1-2].By analyzing the influence of different factors on the leakage rate of drilling fluid, we can judge and evaluate the development of fractures in the formation, and provide reference for subsequent sealing operations.

Several examples and mud loss modele has been made in the literature. Lietard O et al. considered incompressible drilling fluid having Bingham rheology. They developed a model of radial mud flow into a smooth fracture with constant aperture and infinite length [3]. Shahri M P et al. considered influence of the angel of the facture on the drilling fluid loss rate [4]. Based on the model of Lietard O, Majidi.R et al. developed a model which regards drilling fluid as Herschel-Buckley fluid [5-7]. Li daqi et al. developed a model considering the roughness of the fracture [8].

All the models above don't consider the influence of matrix. If the formation with higher permeability, it will make larger error. Based on the dual-medium theory, a new mud leakage model is presented in the paper.

2. Model Formation

A square fracture, which height is equal to its width, is considered. The well is in the middle of the fracture. The fracture is horizontal so the gravity effect does not need to be taken into account. At the beginning of mud leakage, the fracture aperture is same in all points of fracture and the surface of fracture is smooth, which means the influence of roughness of fracture surface does not need to be considered.

According to the law of conservation of mass and Reynolds equation, the mass balance equation for an incompressible fluid is given by [9]:

$$-\nabla(\rho w v_f) = 2\rho q_t + \frac{\partial \rho w}{\partial t} \tag{1}$$

Where w is the fracture aperture, m; v_f is the fluid flow velocity in the fracture, m/s; q_t is the flow rate between fracture and rock matrix, m^3/s; ρ is the drilling fluid density, kg/m^3.

At the present, most of drilling fluid is water-based mud, which can be regarded as a kind of incompressible fluid. For most of drilling fluid, power-law rheology is acceptable, which can be described as follow:

$$\tau = K\gamma^n \tag{2}$$

where τ is the shear stress, Pa; K is the consistency index, $Pa \cdot s^n$; γ is the shear rate, s^{-1}; n is the power law exponent.

For the power-law fluid, the drilling fluid velocity in the fracture can be expressed as follows:

$$v_{fx} = \frac{n}{1+n}\left(\frac{1}{2^{n+1}K}\right)^{\frac{1}{n}} w^{\frac{n+1}{n}}\left(\frac{\partial p_f}{\partial x}\right)^{\frac{1}{n}} \tag{3}$$

$$v_{fy} = \frac{n}{1+n}\left(\frac{1}{2^{n+1}K}\right)^{\frac{1}{n}} w^{\frac{n+1}{n}}\left(\frac{\partial p_f}{\partial x}\right)^{\frac{1}{n}} \tag{4}$$

Where n is the power exponent; K is the consistency index, $Pa \cdot s^n$; v_{fx} is the drilling fluid velocity in the fracture in the x-direction, m/s; v_{fy} is the drilling fluid velocity in the fracture in the y-direction, m/s; p_f is the fracture pressure, MPa.

Compared with the fracture, the permeability of rock matrix is small, so the flow of drilling fluid in the rock matrix can be ignored. The only function of rock matrix is storing drilling fluid. The control equation can be expressed as follow:

$$q_m = \emptyset \frac{\partial p_m}{\partial t} \tag{5}$$

Where ∅ is the porosity of rock matrix; p_m is the formation fluid pressure, MPa; q_m is the flow rate from fracture to the rock matrix, m^3/s.

The fracture aperture has a great influence on the loss rate of drilling fluid. Fracture aperture is not invariable during the leakage. Fracture aperture will response to the change of pressure in the fracture. It will cause error if the width of fracture is assumed to be constant. When the drilling fluid flows into the fracture, the pressure in the fracture will increase, and then the fracture aperture will increase as well. At the present, there are two kinds of fracture deformation models, one is exponential model, and the other one is linear model. In this paper, linear deformation model is adopted. The formula is as follow [10]:

$$w = w_o + \frac{p_f - p_o}{K_n} \tag{6}$$

Where w_o is the fracture aperture when t=0, m; p_f is the pressure in the fracture, MPa; p_o is the formation fluid pressure when t=0, MPa; K_n is the normal fracture stiffness, Pa/m.

Substituting (3)(4)(6) into (1)(5), the following equations are obtained:

$$\begin{cases} \nabla \cdot \left[w \frac{n}{2n+1}\left(\frac{1}{2^{n+1}K}\right)^{\frac{1}{n}} w^{\frac{n+1}{n}} (\nabla p_f)^{\frac{1}{n}} \right] - q_f = \frac{1}{K_n}\frac{\partial p_f}{\partial t} \\ q_m = \emptyset \frac{\partial p_m}{\partial t} \end{cases} \tag{7}$$

Where q_f and q_m are the flow rate between fracture and rock matrix, their value are equal to each other, but signs are opposite.

Flow between fracture and matrix is slow, so it can be regarded as steady seepage. Based on the dual-medium theory, the formula of power-law fluid flow between fracture and matrix can be expressed as follows [11]:

$$\begin{cases} q_f = a\left[\frac{k_m}{\mu_{eff}}\frac{(p_f-p_m)}{dz/2}\right]^{\frac{1}{n}} \\ q_m = -a\left[\frac{k_m}{\mu_{eff}}\frac{(p_f-p_m)}{dz/2}\right]^{\frac{1}{n}} \end{cases} \tag{8}$$

$$\mu_{eff} = K\left[\frac{K}{12}\left(9+\frac{3}{n}\right)^n (150K_m\emptyset)^{\frac{1-n}{2}}\right]^{n-1} \tag{9}$$

Where a is the shape factor; dz is the height of cap rock, m; n is the power exponent; K is the consistency index, Pa·s^n; k_m is the rock matrix permeability, um.

3. Initial and Boundary Conditions

Compared with the permeability of fracture, the permeability of rock matrix is very small. To reduce count quantity, it's assumed that drilling fluid can only flow from wellbore to the fracture but not flow from wellbore to the rock matrix [12-14].

$$\nabla\cdot\left[w\frac{n}{2n+1}\left(\frac{1}{2^{n+1}K}\right)^{\frac{1}{n}} w^{\frac{n+1}{n}}(\nabla p_f)^{\frac{1}{n}}\right] - q_f + \frac{Q_f}{dx\cdot dy} = \frac{1}{K_n}\frac{\partial p_f}{\partial t} \tag{10}$$

$$Q_f = 2\pi w\left(\frac{n}{2n+1}\right)\left(\frac{w}{2}\right)^{\frac{n+1}{n}}\left[\frac{1-n}{K}\frac{1}{(r_w^{1-n}-r_e^{1-n})}\right]^{\frac{1}{n}}(p_w-p_f)^{\frac{1}{n}} \tag{11}$$

Where Q_f is the drilling fluid loss rate from wellbore to the fracture, m^3/s; r_w is the supply radius, m; re is the wellbore radius, m; p_w is the wellbore pressure, MPa.

At the time t=0, it's assumed that the pressure all over the fracture is same, and the fracture pressure is equal to the formation fluid pressure, which can be expressed as follows:

$$p_m=p_f \tag{12}$$

Reservoir boundaries are considered impermeable and thus act like no-flow boundary:

$$\begin{cases} \frac{\partial p}{\partial x} = 0 \quad x = 0, x = L_x \\ \frac{\partial p}{\partial y} = 0 \quad y = 0, y = L_x \end{cases} \tag{13}$$

4. Mesh Generation and Stimulation

In this paper, rectangle central grid is adopted. At the beginning of the leakage, the leakage rate is large and the pressure near the wellbore changes fast. In order to guarantee the calculation speed and accuracy, the variable density grid is used in the paper. The grid near the wellbore should be smaller and the grid far from the wellbore should be bigger. In the initial stage of leakage, the time step should be as small as possible to ensure the accuracy of calculation.

The model developed in the previous section and comprising the nonlinear partial differential equation (7), describes radial mud flow in a fracture and the associated ballooning phenomenon. The equation was integrated numerically by using an explicit finite difference scheme [15, 16]. Newton Raphson method is used to solve the problem [17].

5. Case Study

The effect of various factors on the drilling fluid leakage rate are analysed in this section. The inputs for the calculation are given in Table 1.

Table 1. Basic Date for the Calculation

Parameter Name	Value	Parameter Name	Value
Fracture Length in x Direction	100m	Permeability	1×10^{-3}um^2
Fracture Length in y Direction	100m	Drilling Fluid Density	1.2×10^3kg/m
Formation Height	0.5m	Consistency Index	0.8Pa·s^n
Normal Fracture Stiffness	10^6Mpa/m	Power Law Exponent	0.5
Initial fracture aperture	0.5mm	Formation pressure	50Mpa
Formation porous	0.1	Borehole Pressure	45Mpa

The effect of normal fracture stiffness on mud leakage is illustrated in figure 1. According to equation (6), normal fracture stiffness has a direct influence on the width of fracture. Higher normal fracture stiffness means that the rock is harder and is difficult to be compressed. In the case of the same value of pressure difference between fracture and formation, the bigger normal fracture stiffness is, the narrower the fracture is.

The effect of initial fracture aperture on mud leakage is illustrated in figure 2. Initial fracture aperture has a great influence on the initial drilling fluid leakage rate. The larger the fracture aperture is the faster drilling fluid loss at the beginning. But after a few seconds, once the fracture is filled with drilling fluid, the drilling fluid leakage rate will be almost same. That is because once the fracture is filled with drilling fluid, the initial fracture aperture will have limited influence on the mud leakage rate. On the other hand, pressure difference between fracture and formation, permeability of formation and other influence factors will play a leading role.

The effect of pressure difference between borehole and formation on mud leakage rate is illustrated in figure 3. As figure 3 shows, larger pressure difference between borehole and formation will increase the peak in the mud leakage curve. From figure 3, we can see that the larger the pressure difference between borehole and formation is, the higher the mud leakage rate is at the beginning. Compared with figure 2, influence of pressure difference on the leakage rate will last longer than fracture aperture. This is because pore pressure increases more slowly; it will take longer time to get a balance between fracture pressure and pore pressure.

The effect of the rock matrix porosity on mud loss rate is illustrated in figure 4. As figure 4 shows, the matrix porosity does not influence the peak in the mud loss curve, which means the value of the rock matrix porosity has no effect on the seepage rate between fracture and rock matrix. On the other hand, the larger the matrix permeability is, the slower the drilling fluid flow rate decrease. That is because that the fluid pressure in the matrix will increase slower if the matrix porosity is higher.

The effect of the rock matrix permeability on mud leakage rate is illustrated in figure 5. As figure 5 shows, at the beginning, larger permeability promotes the fluid flow in the rock matrix, and the pore fluid pressure increases slower, which makes the pressure difference between fracture and rock matrix keep large. After a while, rock with large permeability will first be filled with drilling fluid and pore fluid pressure will increase, then the mud leakage rate will decrease.

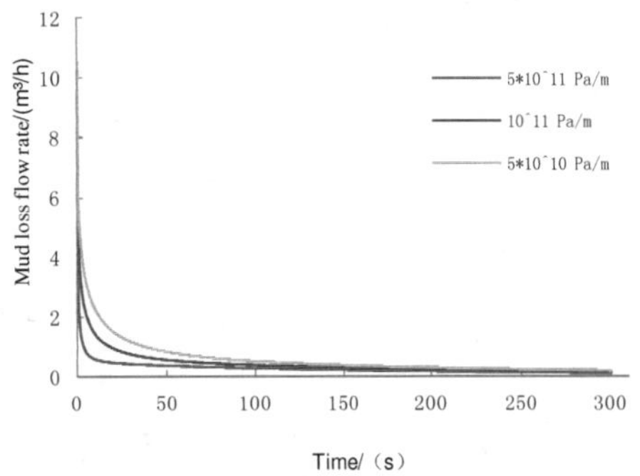

Figure 1. Influence of the Normal Stiffness of the Fracture on the Fluid Loss Rate

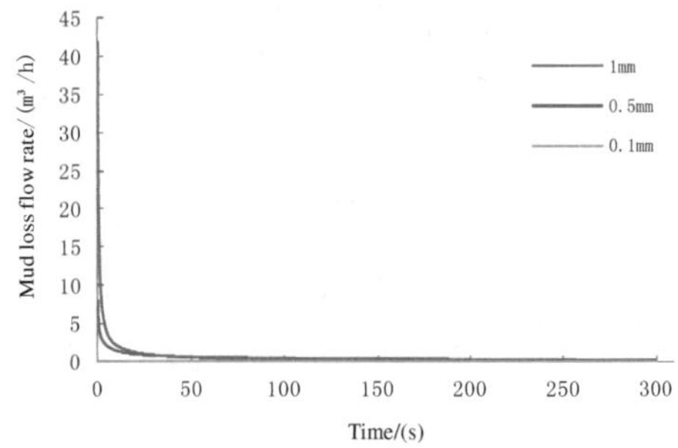

Figure 2. Influence of the Initial Fracture Aperture on the Fluid Loss Rate

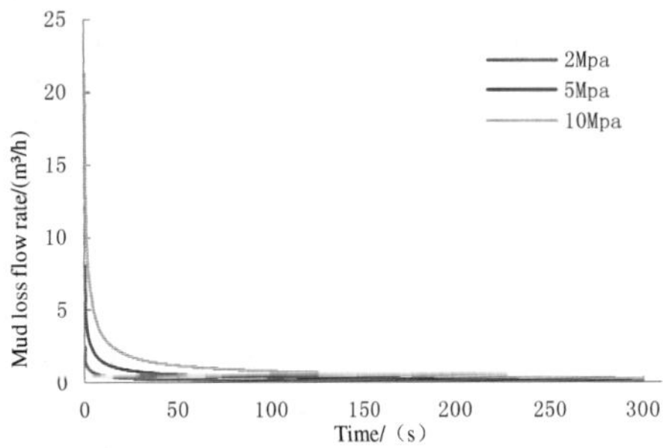

Figure 3. Influence of the Pressure Difference on the Fluid Loss Rate

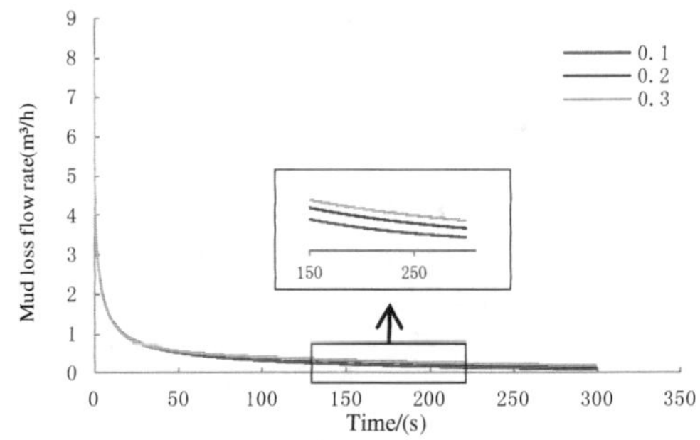

Figure 4. Influence of the Rock Matrix Porosity on the Fluid Loss Rate

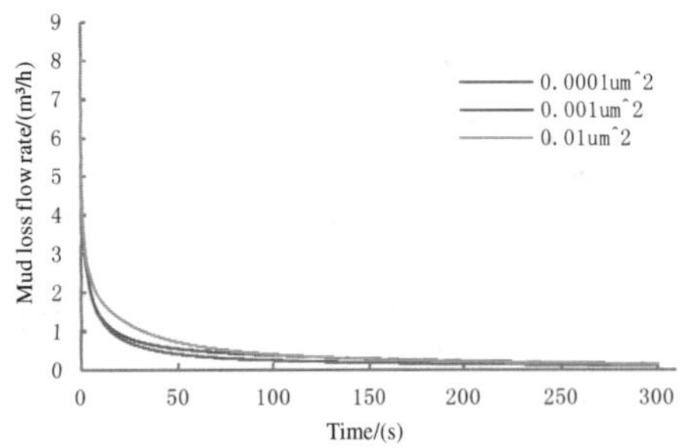

Figure 5. Influence of the Rock Matrix Permeability on the Fluid Loss Rate

6. Conclusions

Based on the dual-medium theory, a model of mud leakage into a single isolated deformable horizontal fracture was developed for a power-law drilling fluid. Numerical simulation with the model has shown that:

(1) Increasing normal stiffness of fracture decreases the mud leakage rate.

(2) Increasing initial fracture aperture increases the height of the peak of the mud loss curve, but after the beginning, the impact of initial fracture aperture is small.

(3) Increasing pressure difference increases the height of the peak of the mud loss curve.

(4) The larger the porosity of matrix is, the larger the mud leakage at the beginning is.

(5) Permeability of matrix has little influence on the leakage rate at the beginning, but larger permeability will make the leakage last longer in a larger value.

7. Reference

[1] Li Daqi, Kang Yili, Liu Xiushan. The lost circulation Pressure of carbonate formations on the basis of leakage Mechanisms [M]. Acta Petrolei Sinica, 2011,32(5);900- 904

[2] Jin Yan, Chen Mian, Liu Xiaoming et a1.Statistic analysis of leakage pressure of Ordovician carbonate formation in middle Tarim Basin [J]. Oil Drilling & Production Technology, 2007, 29(5): 82—84.

[3] Lietard O, Guillot D, Hodder M. Fracture width LWD and drilling mud/LCM selection guidelines in naturally fractured reservoirs [R]. SPE-36832, 1996.

[4] Shahri M P, Mehrabi M. A new approach in modeling of fracture ballooning in naturally fractured reservoirs[R]. SPE-163382, 2012.

[5] Lavtov A, Tronvoll J. Modeling mud loss in fractured formations [R].SPE-88700, 2004.

[6] Majidi R, Miska S Z, Yu M, et al. Quantitative analysis of mud losses in naturally fractured reservoirs: The effect of rheology [J]. SPE Drilling & Completion, 2010, 25(4): 509-517.

[7] Majidi R, Miska S Z, Yu M, et al. Fracture ballooning in naturally fractured formations: Mechanism and controlling factors[R]. SPE-115526, 2008.

[8] Li Daqi. Numerical and experimental investigations of drilling fluid losses in fractured formations [D]. Chengdu: Southwest Petroleum University, 2012.

[9] Jia Lichun, Chen Mian, Hou Bing, et al. Drilling fluid loss model and loss dynamic behavior in fractured formations [J]. Petroleum Exploration and Development, 2014, 41(1); 95-101.

[10] Murata S, Saito T. Estimation of tortuosity of fluid flow through a single fracture [J]. Journal of Canadian Petroleum Technology, 2003, 42(12): 39-45.

[11] Liu Huiqing. The subject of petroleum reservoir simulation methods [M]. Dongying: China University of Petroleum Press.

[12] Warren J E, Root R J. The behavior of naturally fractured reservoirs [J]. SPE-4261963.

[13] Zhou Yuliang. The recovery evaluation of underbalanced flow back of overbalance water influx damage in dual media [M].Chengdu: Southwest Petroleum University, 2012.

[14] Murata S, Saito T. Estimation of tortuosity of fluid flow through a single fracture [J]. Journal of Canadian Petroleum Technology, 2003, 42(12): 39-45.

[15] Chen Linsong. Mechanics of porous media flow [M]. Beijing: Petroleum Industry Press.

[16] Hemphill T, Campos T, Pilehvari A. Yield-Power Law Model More Accurately Predicts Mud Rheology [J]. Oil &Gas Journal, 1993, 91(34): 45-50.

[17] Peng Xiaolong, Liu Xueli, Du Zhiming. Numerical Model and the Characteristics of Flowing Through Porous Media of Fracture and Vug Dual-Media [J].Journal of Southwest Petroleum University(Science & Technology Edition), 1995, 10(3): 160-166.

IOP Conf. Series: Materials Science and Engineering **735** (2020) 012082 doi:10.1088/1757-899X/735/1/012082

The Influence of Particle Sizes and Inlet-Outlet Configuration on The Hydraulic Characteristics of Horizontal Subsurface Constructed Wetlands

Hua Zhong, Peng Li, Xuan Guo and Chengjun Zhang

Institute of Plant Nutrition and Resources, Beijing Academy of Agriculture and Forestry Sciences, Beijing 100097, China.
Email: Remyzhong@163.com

Abstract. The objective of this study was to evaluate the influence of inlet and outlet location and matrix particle size on the hydraulic efficiency of horizontal subsurface constructed wetlands. A quasi-two-dimensional model was set up to evaluate its hydraulic efficiency by salt tracer test, and the change of flow state in the system was observed by dye tracer test. Three inlet and outlet settings: top inlet-bottom outlet, middle inlet-middle outlet and bottom inlet-top outlet; three particle size distributions are 3 mm, 5 mm and 3/5 mm (one half part 3 mm, the other 5 mm). According to the tracer test, the position of inlet and outlet and the size distribution of matrix have an effect on the hydraulic characteristics of constructed wetlands. Larger substrate size with bottom inlet–top outlet configuration could improve hydraulic behaviour, improve water treatment efficiency.

1. Introduction

In this study, effects of particle sizes, inlet -outlet configuration Horizontal subsurface flow constructed wetland, a water treatment technology, uses the substrate, microorganisms and plants to remove pollutants by physical, chemical and biological treatment [1-3]. It not only has the advantage of low treatment cost, but also has high removal rate of nitrogen, phosphorus, organic matter and pathogen [4-6]. In the subsurface flow constructed wetland, hydrodynamic conditions can change the physical and chemical properties, as well as determine the interaction time among pollutants in the water flow and matrix, plant roots, microbial population and hydraulic efficiency [7-9].

In constructed wetlands, different configurations can change the hydraulic performance. The inlet and outlet configuration influence the flow path to form dead zone. The influence of filter size on flow distribution is mainly shown in priority path and dead zone; a larger flow distribution can be obtained by increasing the number of packing sections. In general, the hydraulics of constructed wetlands can be evaluated by numerical simulation. A quasi-two-dimensional (2d) plexiglass boxes, filled with colourless glass beads, as miniature artificial wetlands [10-12].

In this study, effects of inlet-outlet configuration, filter size may lead to change hydraulic efficiency. A quasi-two-dimensional box was established for NaCl tracer and dye tracer experiments. The NaCl tracer test was used to calculate the hydraulic efficiency, and the dye tracer test was used to visualize the propagation of tracer in porous media reactor, the law of the internal hydraulic performance of constructed wetlands can be revealed.

2. Test Set

The horizontal subsurface flow constructed wetland, made of plexiglass, 400 mm×200mm×10mm (length × width × height). One side of the device there were three inlets, and the other side three outlets (Figure 1). Water distribution area and catchment area set the sponges (size 200mm×10mm×10mm, simulated porous media. The inside of device was full of glass beads which were simulated the substrates. The glass beads had two size, 3 mm and 5 mm. There are three filling modes: 3 mm (porosity 25%), 5 mm (porosity 36%), another is 5 mm glass beads at the bottom, 3 mm glass beads at the top (porosity 30%). The device is 10 mm thick, approximately two-dimensional.

3. Tracer Experiments

The tracers used were NaCl and Rhodamine B. Dissolved NaCl 0.5 g in 10 ml of water. The constant flow pump (Baoding Lange BT300-2J) was used to inject NaCl solution at a constant speed of 25r/min to the device. Water conductivity was measured by Conductivity meter (HQ14d) every minute. 0.2 g Rhodamine B was soluble in 20ml water, after the solution entered the installation, take pictures every 10 seconds.

4. Hydraulic Parameters

According to the fluid reactor theory, in the pulse tracer experiment the concentration measured corresponds to the distribution density of hydraulic residence time. The measured conductivity is standardized by formula (1) [13-15].

$$N(t) = (C(t) - C_w)\frac{1}{\lambda_{Na^+}+\lambda_{Cl^-}} M Q_s \frac{1}{MI} \quad (1)$$

$C(t)$ conductivity at the outlet; C_w conductivity of wastewater composition; $\lambda_{Na^+}\lambda_{Cl^-}$ the molar conductivity limit, M the sodium chloride molar weight; Q_s the outflow rate; *MI* the total mass injected.

The theoretical residence time T_n, the ratio of the effective volume of constructed wetland to the flow.

$$T_n=V/Q \quad (2)$$

The actual residence time T_m is the centre of mass of the *N(t)* curve:

$$T_m = \int_0^\infty tN(t)\,dt / \int_0^\infty N(t)\,dt \quad (3)$$

The efficiency of hydraulic λ_t

$$\lambda_t=T_m/T_n \quad (4)$$

σ^2 is the variance of RTD, describes the proportions of hybrid system.

$$\sigma^2 = \int_0^\infty (t - T_m)^2 N(t)\,dt \quad (5)$$

σ_θ^2 is the standard variance of RTD variance normalized, a measure of the RTD distribution.

$$\sigma_\theta^2 = \sigma^2/\lambda_t^2 \quad (6)$$

5. Results and Discussion

Under different particle size distribution, the order of actual hydraulic retention time is bottom-top > middle-middle > top-bottom (table 1). The same inlet and outlet, the actual hydraulic retention time sorting is 3/5 mm < 3 mm <5 mm, but the bottom-top 3 mm <3/5 mm <5 mm, because the water resistance from the bottom to the top in the process, from the big pores into small pores, , increased the head loss, smaller pore velocity, substrate in contact with the water flow more fully, the average hydraulic retention time is longer than the other. Bottom-top has the highest hydraulic efficiency.

The distribution of RTD density curve is consistent, all of them tend to 0 after reaching the peak value, and are bell-shaped with a long trailing tail (figure 1). The peak value of homogeneous matrix with 5 mm particle size was the smallest. In the middle-middle, the peak value of RTD density curve with

IOP Conf. Series: Materials Science and Engineering **735** (2020) 012082 doi:10.1088/1757-899X/735/1/012082

different particle sizes was not significantly different, while the peak value of 3 mm and 3/5 mm was much higher than 5 mm in the other two inlet-outlet.

Table 1. The Hydraulic Parameters of the Different Substrate Sizes and Inlet-outlet Configurations

particle size	Inlet-outlet	Tm/h	Tn/h	λ_t	σ^2	σ_θ^2
	top to bottom	0.124		2.588	0.729	0.109
3 mm	middle to middle	0.143	0.0481	2.982	0.568	0.064
	bottom to top	0.206		4.276	1.565	0.086
	top to bottom	0.131		2.720	0.886	0.132
5mm	middle to middle	0.175	0.0693	3.644	1.039	0.078
	bottom to top	0.431		8.966	2.736	0.034
	top to bottom	0.087		1.808	0.308	0.046
3/5 mm	middle to middle	0.137	0.0587	2.847	0.565	0.070
	bottom to top	0.231		4.811	1.752	0.076

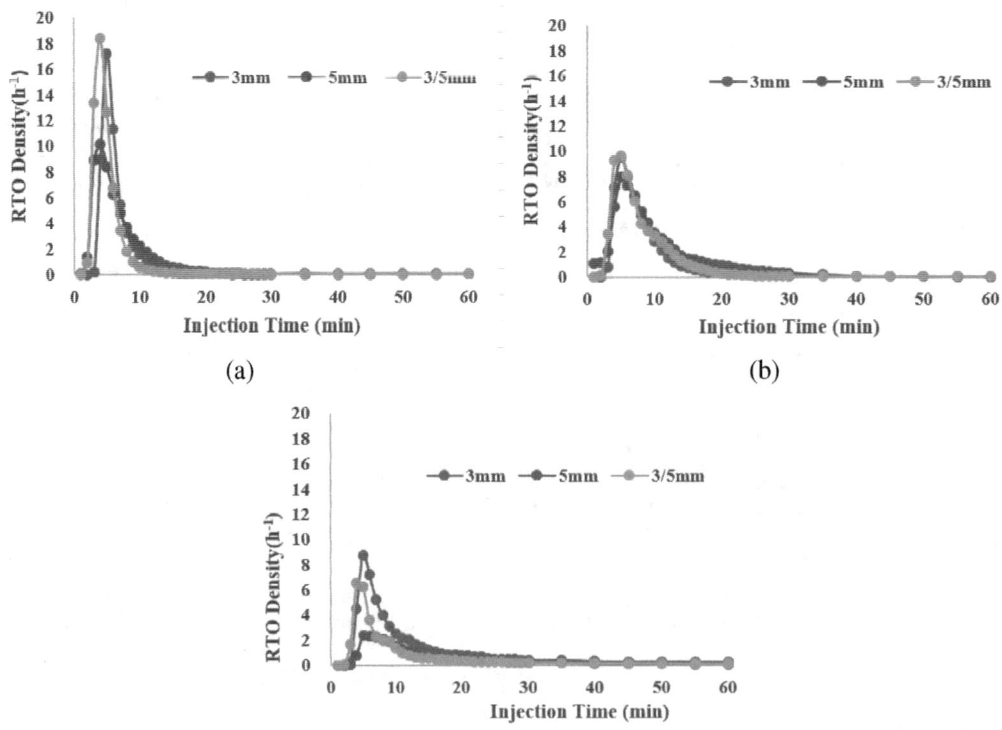

(a) (b) (c)

(a) Top inlet-bottom outlet (b) Middle inlet-middle outlet (c) Bottom inlet-top outlet

Figure 1. Breakthrough Curves of NaCl Tracer for Different Inflow Rates

The table 2 shows, for the same inlet-outlet, large size particles have large porosity, so dye diffusion speed fast; small size particles have small porosity, the flow diffusion has to overcome more resistance. Obviously 3/5 mm, middle to middle.

Table 2. Dye Tracer Propagation of the Different Substrate Sizes and Inlet-outlet Configurations

particle size	time	top to bottom	middle to middle	bottom to top
	2 min			
3 mm	4 min			
	6 min			
	2 min		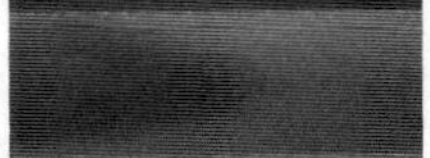	
5 mm	4 min			
	6 min			
	2 min			
3/5 mm	4 min			
	6 min			

6. Conclusion

Among all kinds of inlet-outlet layouts, bottom-top set has longer flow, lower pore velocity, maximum average hydraulic retention time and hydraulic efficiency, while top-bottom set has shorter flow, minimum average hydraulic retention time and hydraulic efficiency. Larger size of matrix has larger porosity, long hydraulic retention time and high hydraulic efficiency.

7. Acknowledgments

This paper thanks to the following supports: Youth Foundation of Beijing Academy of Agriculture and Forestry Sciences (Fund key. QNJJ201708).

8. References

[1] Yuhui W, Xinshan S, Weihong L, Ruihua N, Wei W, Yi D, Yuan W and Denghua Y 2014 *Ecol. Eng* **69** 177
[2] Cristina A, Jaime N, Linda O, Kinfe K, Tom H, Roland A. M, Josep M. B and Joan G 2014 *Sci Total Environ* **494** 211
[3] Natalia P, Nikolaos Th. S and Vassilios A. T 2017 *Chem. Eng.J.* **308** 1120
[4] Emily E. R, Jhonatan B, Paula L. P and Johannes G. J 2016 *J.Hydro. Res.* **12** 91
[5] Roger S, Joan G, Pascal M and Nicolas F 2016 *J.Environ. Manag.* **165** 271
[6] Haiwen X, Jun Z, Rongsen D and Guanbin L 2012 *Acta. Sci. Circum.* **32** 2998
[7] Yixuan S, Yuhui W, Xin C and Xinshan S 2019 *J. contam. Hydro.* **226** 103537
[8] Garcia Zumalacarregui, J. A. and von Sperling, M 2018 *Water Sci, Technol.* **78** 848
[9] Rengers, E. E., da Silva, J. B., Paulo, P. L.and Janzen, J. G. 2016 *J. Hydro-Environ. Res.* **12** 91
[10] Junfeng W, Xinshan S, Dengming Y, Yuhui W, Gang W and Pengchao Z 2015 *Environ. Sci. & Tech.* **38** 75
[11] Ruihua N, Yuhui W and Xinshan S 2014 *Chinese J. Environ. Eng.* **8** 4163
[12] Shaoyuan B, Zhixin S, Yanli D, Shaohong Y and Shan H 2014 *Environ. Sci,* **35** 593
[13] Florent C, Gerard M and Yves G 2003 *Ecol. Eng* **21** 165
[14] Qin L, Xinshan S, Huajie L and Denghua Y 2012 *Environ. Pol.& Con.* **34** 23

[15] Ping T, Bohai Y, Yongchao Z, Yiping Z and Jin L 2017 *Environ. Sci. & Pol. Res.* **24** 9210

Study on the Theory of High Pressure Water Jet Coal Breaking and Its Main Controlling Factors

Yongjiang Zhang[1,2]

1 State Key Laboratory of the Gas Disaster Detecting, Preventing and Emergency Controlling, Chongqing 400037, China
2 China Coal Technology and Engineering Group Chongqing Research Institute, Chongqing 400039, China
E-mail: zhangyj_1026@163.com

Abstract. In order to solve the problems of gas extraction and pressure relief and permeability enhancement under low permeability and high surrounding rock stress conditions in deep coal seams, the ultra-high pressure hydraulic cutting technology is proposed. The mechanical mechanism of coal breaking by high pressure water jet is analyzed. The process of coal breaking by jet is divided into stress wave effect and secondary crack propagation effect. The effects of physical and mechanical properties of coal and rock and jet parameters on jet cutting depth are analyzed theoretically. The results show that the shear strength, tensile strength, fracture and pore development degree of coal and rock correspond to different types of coal and rock failure. With the increase of jet pressure and flow rate (nozzle diameter), the jet cutting depth increases gradually. The influence of jet rotating speed on cutting depth is analyzed. It is concluded that reducing rotating speed should be adopted to improve cutting depth in field practice. The research results can effectively improve the stress state of coal body, play the role of pressure relief and permeability enhancement, and improve coal seam permeability and gas release capacity, which can well meet the control needs of coal mine gas complex dynamic disasters.

1. Introduction

The mining depth of coal mines in China extends to the depth at the rate of 10m~30m per year. The mining coal seams show remarkable characteristics of "low coal seam permeability, high surrounding rock stress and high coal seam gas pressure". Faced with the complex dynamic disasters of coal and rock in deep mines, it is difficult to achieve ideal extraction effect under the conditions of high surrounding rock stress and low permeability by adopting single measures such as surface well or borehole pre-pumping. it is unable to eliminate the elastic strain energy accumulated in coal and rock and reduce the stress threat in the inland of mining area. The effective pressure relief and permeability enhancement measures are needed to meet the needs of the treatment of composite power disasters. It is the direction of joint efforts of coal mine scientific researchers to study a technical means which can not only meet the requirements of rapid gas extraction, but also reduce the stress of coal seam [1]. Water jet cutting technology is to cut coal seam with high pressure water as technical means, and man-made fissures as gas flow space to improve the gas flow state in coal seam and create favorable conditions for extraction. It is an important development direction of prevention and control of coal and rock composite dynamic disasters in coal mine [2].

As a new technology developed rapidly in recent years, ultra-high pressure hydraulic slotting technology is suitable for cleaning, drilling, cutting and rock breaking, which reduces costs and improves work efficiency. The technology of pressure relief and permeability enhancement by high-pressure hydraulic drilling is to use water as power to scour and strip the coal around the borehole and increase the fissures in the coal [3-5], which can greatly improve the gas flow state in the coal seam, create favorable conditions for gas extraction, change the original stress and fissure state of the coal body and reduce the stress in the coal body and surrounding rock. Force can not only weaken or eliminate the power of outburst, but also greatly change the physical and mechanical properties of outburst coal seam, play the role of pressure relief and outburst prevention, and improve the permeability and gas release capacity. It is one of the effective measures to prevent and control gas dynamic disasters in coal mines.

2. Mechanics Mechanism of Coal Breaking by High Pressure Water Jet

The velocity and direction of jet impacting on coal wall change dramatically and the distribution of impact load is complex. With the increase of cutting depth, the coal wall in the impacting area will be destroyed. On the one hand, the restriction conditions of flow field change, which affects the velocity distribution inside the jet, on the other hand, changing the force of jet on coal and rock will cause the jet edge to be destroyed. The boundary conditions have changed further, so the mechanism of coal seam cutting by water jet has not yet formed a mature and perfect theoretical system. At present, the accepted hypothesis divides the process of coal breaking by water jet into two stages: water hammer pressure and stagnation pressure, which correspond to the stress wave effect and crack propagation effect of water jet respectively. According to the characteristics of soft coal seam through drilling, the best effect of strengthening pumping is the high pressure water discharge and increase penetration technology. The technique of water jet pressure reduction is mainly used to cut and flush the coal around the borehole to form a large area of pressure relief and penetration in the borehole to achieve the purpose of increasing the permeability. The impact process of high pressure water jet shown in Figure 1.

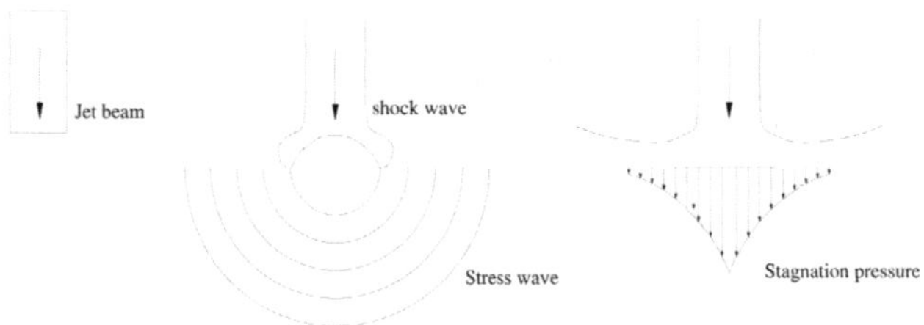

Figure 1. Impact Process of High Pressure Water Jet

2.1. Stress Wave Effect

The main body of coal and rock damage caused by water jet is the water hammer pressure stage in the first few microseconds. When the jet first contacts the surface of coal and rock, the reaction force of coal and rock to the jet transfers deformation in the center of the jet, and the deformation in the compressed area forms in the jet. The propagation velocity in the coal is faster than that in the jet, which produces water hammer pressure. When the jet liquid column contacts coal and rock sufficiently and the jet flow field is rebalanced under the external restraint, the impact of the jet decreases from the water hammer pressure to about one tenth of the stagnation pressure. In this process, the impact effect in a very short time makes the coal and rock in a compressive state. Because of the sharp decrease of the compressive stress, the coal and rock in a highly compressive state are rapidly released to form a tensile load. When the tensile stress produced by the unloading exceeds the strength limit of the coal and rock itself, the failure occurs. Because of the entrainment effect of the jet boundary layer, the axial velocity center of the jet is larger and the edge is lower, so the elastic strain energy at the impact edge of the jet is released most sufficiently, and the damage first occurs in the region. With the destruction of coal and rock at the impact edge, the coal body with strong compression at the jet center expands along the radial direction, and the coal body damage area expands from outside to inside.

2.2. *Secondary Crack Propagation Effect*

Coal and rock are typical porous media materials. Before water jet impingement, there are micro-structural defects such as pore, fracture and bedding. Under the impact of water jet, cracks in coal and rock mass develop and propagate, and eventually penetrate with the free surface to separate the coal block from the coal and rock surface as a whole. Crack propagation under water jet impingement is shown in Figure 2.

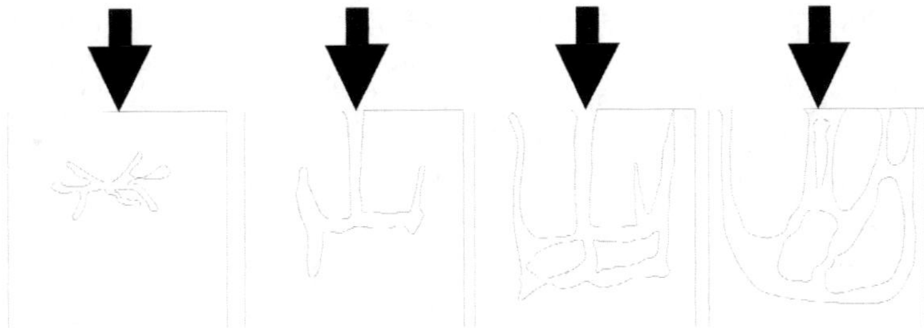

Figure 2. Crack Propagation under Wateejet Impingement

According to the different mechanism of crack development, there are two different types of failure: one is that the damage effect is caused by the direct impact force of the jet, the jet impact produces the tensile stress zone in the coal and rock, the tensile action forms the radial crack, the crack propagates in the radial direction, and with the crack expansion, the impact pit is formed. Another kind of damage effect is caused by water wedge. In the process of water jet impact, high-pressure water propagates along the crack and produces stress concentration at the crack tip. The crack destroys coal and rock mass under hydrostatic pressure.

2.3. *Jet Erosion Effect*

Coal and rock mass are heterogeneous materials. The structural characteristics of plants and geological processes in coal-forming process will destroy the integrity of coal and rock, reflecting the random distribution characteristics of coal and rock strength. Under the impact of high-pressure water jet, some coal-rock units with higher strength can resist the erosion and destruction of jet. The lower strength part falls off from the surface of coal-rock under the action of water erosion, and then the overall strength of coal-rock is weakened. With the increase of jet pressure, the number of separated coal-rock units increases gradually, and the stability of the higher strength part is affected, which is taken away by jet under impact.

3. Effect of Physical and Mechanical Properties of Coal on Jet Cutting Depth

3.1. *Effect of Coal Mechanics Strength of Coal on Jet Cutting Depth*

In the process of high pressure water jet impacting on coal body, under the action of the axial compression of the jet, the coal body displaces downward, and the core area of impacting and the edge area produce relative dislocation. From the impact edge area to the unaffected area, the radial displacement of coal and rock near the axis caused by the axial settlement results in tensile deformation far away from the impact position of the jet. As a kind of rock material, the compressive strength of coal and rock is far greater than the shear strength and tensile strength. When the axial impact pressure of jet is still lower than the compressive strength of coal, the shear stress in coal and rock exceeds the limit of shear strength due to the mechanical characteristics of brittle material of coal and rock, which leads to the relative occurrence of coal and rock at the impact axis position. Distortion or the tensile stress exceeds the ultimate tensile strength, which results in tensile fracture of coal and rock far away from the impact zone.

3.2. Effect of Coal Cracks on Cutting Depth

The influence of coal-rock fracture development on cutting depth includes two aspects. The stress wave generated during the initial impact process of jet reflects when it encounters cracks. Therefore, it is advantageous for jet to break coal body by limiting the propagation range of stress wave and increasing the unloading intensity of stress wave. On the other hand, when the jet impinges on the coal body stably, the high-pressure water is transmitted to the tip of the coal body through the cracks to produce the water wedge effect, which makes the cracks continue to expand and finally cause the macro damage in the coal body. Therefore, when the slotting pressure is low, the jet flow rate is large and the slotting time is long, the stress wave effect produced by the lower jet impact velocity is not significant. The long-term impact of the jet makes the high-pressure water have enough time to make the coal body continue to crack and expand. The coal breaking by jet mainly depends on the secondary crack propagation effect. When the cutting pressure is high, the jet flow is small and the cutting time is short, the stress wave propagation is limited by the development of the fracture, and the stress wave relief strength is enhanced. The jet coal breaking mainly depends on the stress wave effect.

4. Effect of Jet Parameters of Coal on Cutting Depth

4.1 Effect of Pressure and Flow Rate of Coal on Cutting Depth

4.1.1. Effect of Jet Pressure

The water jet pressure determines the velocity of water jet and is the most important parameter affecting cutting ability. Practice has proved that there is a threshold pressure when water jet impacts coal and rock. Coal and rock under the threshold pressure only produces plastic deformation and hardly destroys. Coal and rock will produce leap-forward failure when it is higher than the threshold pressure. For plunger pumps, the greater the static pressure in the high-pressure cylinder liner, the greater the initial jet velocity at the nozzle outlet. The force acting on coal and rock under stable impact is expressed in the form of jet flux. The increase of jet flux makes the load on the surface of coal and rock increase when the direction of jet velocity is changed, and then the jet cutting depth increases.

4.1.2. Effect of Jet Flow Rate of Coal on Cutting Depth

The water jet flow rate is determined not only by the jet velocity, but also by the jet volume flow rate. According to Bernoulli equation, when the pump output power is enough, the jet velocity is only related to the pump pressure. The larger the nozzle diameter is, the larger the jet volume flow is at the same water jet velocity. Whether in the first stage or the second stage of slotting process, the main factors, such as water hammer pressure and stagnation pressure, are affected by jet flux. Therefore, the larger the nozzle diameter, the stronger the slotting ability of the jet.

4.2 Effect of Rotating Speed on Cutting Depth

The rotational velocity of high-pressure water jet has three main effects on the cutting depth: changing the stress state on the surface of coal and rock mass, reducing the water cushion effect in the cutting process, and reducing the impact time of the jet on the unit area.

The stress state of coal and rock mass surface under the impact of rotating jet is changed. The stress state of coal and rock mass surface under the impact of rotating jet is changed. The rotation of jet produces tangential force along the surface of coal and rock mass besides the vertical force perpendicular to the surface of coal and rock mass. Compared with the simple axial impact on coal mass, the axial shear force along the slot greatly improves the rock breaking efficiency of jet. Secondly, the jet rotation makes the direction of inflow and reflux different, which reduces the water cushion effect in the slotting process. Because the dynamic viscosity of water is much larger than that of air, the attenuation of jet velocity in water is more severe, so the weakening of water cushion effect improves the cutting ability of jet. Finally, the increase of jet rotation speed reduces the cutting time of coal in unit area. Because the continuous propagation of internal cracks in coal is a dynamic process, the reduction of cutting time will stop the development of the expanded cracks due to the disappearance of external loads. In summary, coal with higher hardness coefficient and worse fracture development, higher rotation speed is conducive to changing the stress state on the surface of coal body and increasing the depth of coal breaking; coal with lower hardness coefficient and fully developed cracks, higher rotation speed is not conducive to the re-expansion of internal cracks in coal body and thus the overall separation and decline. Low coal breaking depth.

In practice field, coal and rock are affected by geology and excavation stress. Usually, the strength of coal and rock is low and there are many internal micro-structures. For a jet with cutting pressure up to 100MPa, the exit velocity of nozzle is 447m/s, and the rotation speed of drilling rig is about 40r/min~180r/min. The radial velocity is about 0.6m/s. Therefore, in practice field, it should be adopted. Reduce the rotation speed to increase the cutting depth.

5. Mainly Conclusion

(1) In terms of physical and mechanical properties of coal and rock, the main factors affecting jet cutting depth include shear strength, tensile strength, fracture and pore development, which correspond to different failure types of coal and rock respectively. The corresponding failure mechanism of shear strength is shear slip failure of coal and rock surface, the corresponding failure mechanism of tensile strength is transgranular fracture of coal and rock interior, and the corresponding failure mechanism of pore and fracture is secondary propagation of cracks in coal and rock interior.

(2) In terms of jet properties, the main factors affecting jet cutting depth include: jet parameters and jet rotation speed, jet parameters affecting high-pressure water jet flux, thereby affecting the impact of jet on coal and rock, the rotational velocity of jet mainly eliminates the water cushion effect in rock breaking by jet, affects the time when coal and rock are impacted by jet in unit area, and changes the size and direction of jet impact force.

6. Acknowledgments

This work was financially supported by the national key research and development program of China (2017YFC0804206); Special fund for science and technology innovation and venture capital of China Coal Science and Technology Group Co., Ltd. (2018MS011); Tiandi Science and Technology Co., Ltd. Science and Technology Innovation Venture Capital Special Project (2018-TD-MS076); Project of Chongqing Research Institute Co, Ltd. of China Coal Science and Technology Group (2019YBXM16).

7. Reference

[1] Yuan Liang, Lin Baiquan and Yang Wei (2015) "Research Progress and Development Direction of Gas Control with Mine Hydraulic Technology in China Coal Mine," Coal Science and Technology. J., (01): 45-49.

[2] Zhang Yongjiang, Meng Xianzheng (2018) "Study and application of antireflection technology for ultra-high pressure hydraulic cutting of long borehole in bedding," Mining Safety and Environmental Protection. J., v.45; No.241 (05):1-5+11.

[3] Zhang Yongjiang, Huang Zhenfei (2018) "Investigation and Application of High Pressure Water Jet Annularity Slotting Self Pressure Release Mechanism," International Journal of Coal Science & Technology. J., 43(11):3016-3022.

[4] Meng Xianzheng, Cao Jianjun and He Qin, et al (2009) "Study and Application on Hydraulic Highly Efficient Outburst Prevention Integration Technology," Mining Safety and Environmental Protection. J., (S1): 49-53.

[5] Zhang Yongjiang, Meng Xianzheng (2012) "Investigation and Study on Drainage Radius of Hydraulic Reaming with High Pressure Water Jet," Mining Safety and Environmental Protection. J., (S1): 45-46+188.

AUTHOR INDEX